W9-AYF-693

MOONS & PLANETS

O vast Rondure, swimming in space
Covered all over with visible power and beauty,
Alternate light and day and the teeming spiritual darkness,
Unspeakable high processions of sun and moon and countless stars above,
Below, the manifold grass and waters,
With inscrutable purpose, some hidden prophetic intention,
Now first it seems my thought begins to span thee.

Down from the gardens of Asia descending
Adam and Eve appear, then their myriad progeny after them,
Wandering, yearning, with restless explorations,
with questionings, baffled, formless, feverish,
with never-happy hearts, with that sad incessant refrain,
—"Wherefore unsatisfied soul? Whither O mocking life?"

Ah who shall soothe these feverish children?
Who justify these restless explorations?
Who speak the secret of the impassive earth?

Walt Whitman

Opposite: Sample collection on Europa. (Painting by Pamela Lee)

Wynken, Blynken, and Nod one night,
Sailed off on a silvery shoe.
Sailed on a river of misty light
Into a sea of dew.
"Where are you going and what do you wish?"
The old moon asked the three.
"We have come to fish for the herring fish
That live in this beautiful sea.
Nets of silver and gold have we,"
Said Wynken, Blynken, and Nod.

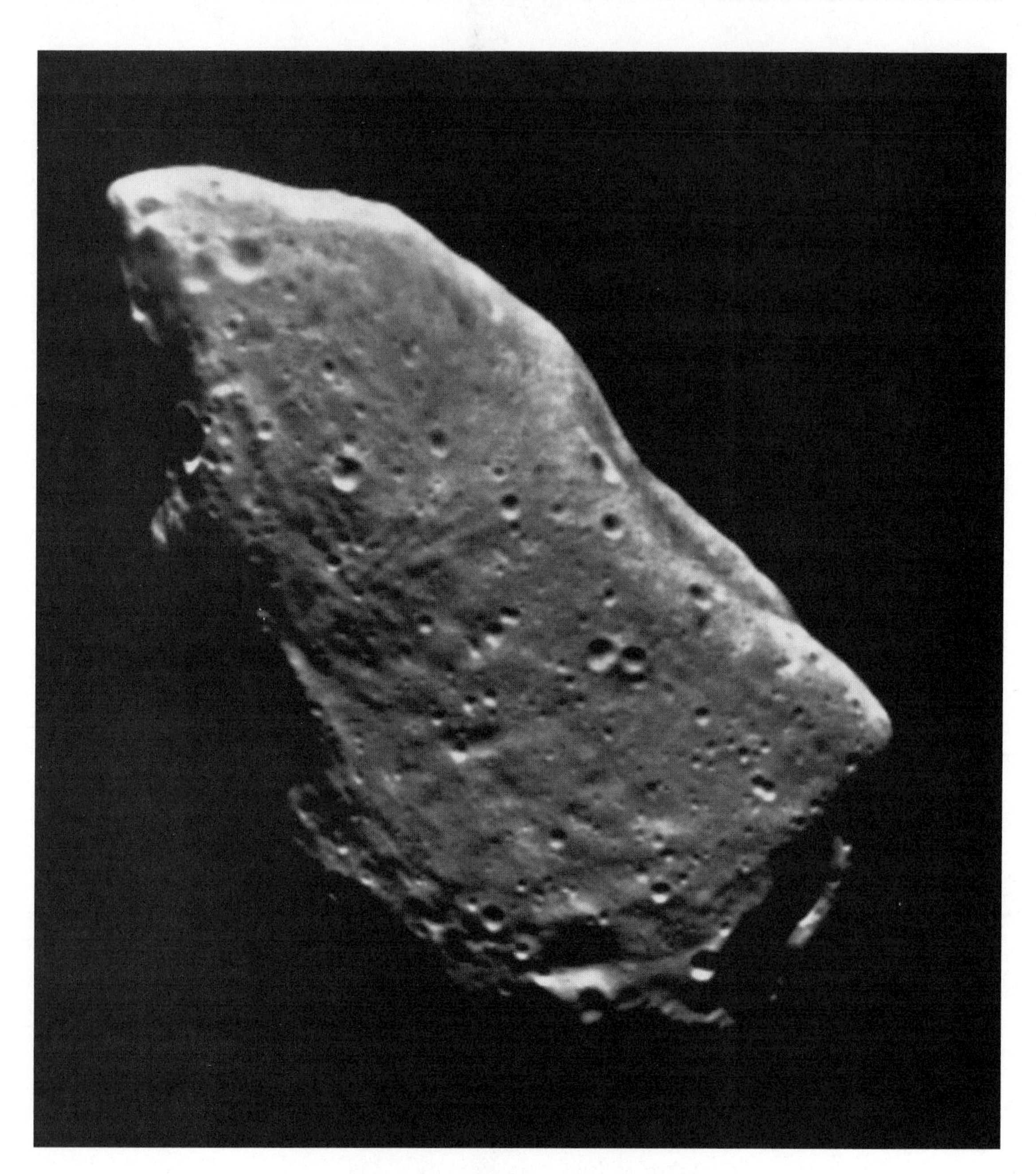

The first close-up photos of an asteroid were returned by the Galileo spacecraft after its close approach in 1991. The first-returned photo of this 12 × 16 km asteroid, 951 Gaspra, had somewhat lower resolution and is compared with two small moons on page 204. This asteroid is believed to be a fragment of a larger parent asteroid, smashed apart during interasteroid collisions hundreds or thousands of millions of years ago. (NASA)

MOONS & PLANETS

THIRD EDITION

WILLIAM K. HARTMANN

Wadsworth Publishing Company
A Division of Wadsworth, Inc.
Belmont, California

Astronomy Editor: Anne Scanlan-Rohrer

Editorial Assistant: Leslie With

Production Editor: Gary Mcdonald

Managing Designer: Ann Butler

Print Buyer: Barbara Britton

Art Editor: Kelly Murphy

Permissions Editor: Robert Kauser

Designer: Detta Penna

Copy Editor: John Bell

Technical Illustrators: Catherine Brandel, Victor Royer, Carole Lawson, Jill Turner

Cover Designer: Ann Butler

Cover Illustrator: William K. Hartmann

Compositor: GTS Graphics

Printer: Fairfield Graphics

This book is printed on acid-free recycled paper.

2 3 4 5 6 7 8 9 10—97 96 95 94 93

Library of Congress Cataloging in Publication Data

Hartmann, William K.
Moons and planets/William K. Hartmann. — 3rd ed.
p. cm.

Includes bibliographical references and index.
ISBN 0-534-18894-X (alk. paper)
1. Planetology. 2. Solar system. I. Title.
QB601.H34 1993
523.2—dc20 92-8018
ISBN 0-534-18894-6

FRONT COVER

An asteroidal impact makes a momentary flash on the night side of Rhea, one of Saturn's icy satellites. Saturn is setting behind the limb of Rhea, and subtends 13° from this distance. See Chapter 10 for discussion of impact processes and Chapter 11 for discussion of Rhea surface. (Painting by the author)

SPINE

Voyager spacecraft photo of Io emphasizes its orangish color, caused by sulfur compounds, and the strange pattern of its dark volcanic vents. See Chapter 11 for discussion of Io volcanism and surface features. (NASA)

BACK COVER

The swirling clouds of Jupiter form a surrealistic abstract pattern. Colors are believed to be caused by complex chemical compounds, possibly including organics. The Great Red Spot is a giant storm system larger than planet Earth. Sworls are caused by interaction of cloud masses caused by wind motions in Jupiter's atmosphere. See Chapter 12 for discussion of atmospheric dynamics. (NASA Voyager photo)

PREFACE

The preface to the 1972 first edition and the 1983 second edition of this book stated that "the last decade of space exploration has revealed a new solar system. This book is for any reader—scientist, amateur astronomer, or student—who wants a contemporary view of that new solar system." Clearly the solar system has expanded even more dramatically in the years since the early editions, in terms of both scientific detail and sheer natural wonder. Now we know something of the soil on Mars, the texture of slabby rocks on Venus, the active volcanoes on Io, the smoggy atmosphere of Titan, the geysers of Triton, and the coma of Chiron.

The first edition preface also noted that ". . . probably no one has a wide enough background to do full justice to such a broad subject." Today that is even more true. Perhaps it is crazy to try. Yet the need exists. Since planetary science is an interdisciplinary field, it may be first encountered by students at any level between freshman and first-year graduate student. It may be encountered in an astronomy department, a geology department, or a newly created planetary science department. A teacher of such a course is likely to have his or her own specialty and emphasis. Thus, while it is easy for our educational system to produce specialists, it remains hard for individual students and readers to get a broad overview of the discoveries and terminology of our explorations into space. In order for all of us to share the excitement of these explorations, we need to share a *first-order understanding* of the basic scientific facts. I have been reminded of this while participating in international conferences where meteorologists, astronomers, geologists, spectroscopists, chemists, and physicists struggled with each others' terminology and world views while trying to share hypotheses and conclusions about the volcanoes, landscapes, internal evolution, volatiles, and magnetic fields of strange and distinctly different worlds, and again while such scientists argued with paleontologists about the radical consequences of the apparent asteroid impact 65 My ago, which wiped out three-fourths of then-existing species on Earth.

For such reasons, my publishers and I have retained one of the best-liked features of the earlier editions, the basic framework for presenting planetary science. We do not work our way through the solar system, planet by planet. Other books do that. One good example, which would make excellent supplemental reading for this course is *The New Solar System* (ed. Beatty and Chaikin, Cambridge, Mass., Sky Publishing Corp.), now in its third edition. But the basic principles of planetary science do not work planet by planet.

They are universal. The equation for escape velocity, the principles of mantle and atmosphere convection, or the identification of basic mineral types are not unique to Mars, Jupiter, or Enceladus. When you master these basic ideas, you can apply them across the solar system, and understand phenomena on many worlds.

This third edition is not a total revision, but a necessary update. We have retained the original pages in many sections of the book—for example, in the celestial mechanics section where updates are minimal. The updates are concentrated in about a third of the book where new information from telescopes, Voyager, Giotto, Magellan, PHOBOS-2, etc. has expanded and altered our perceptions.

The basic text material is a virtually nonmathematical presentation, starting with background on basic facts and planetary motions, and moving into topical areas such as planetary origins, interiors, surfaces, and atmospheres. Numerous equations are developed in isolated, boxed, grey-shaded mathematical notes. Using the book *without* the math notes should give a good descriptive overview. Using the book *with* the math notes should allow a moderate level of physical sophistication.

Study aids at the ends of chapters include "problems" (descriptive questions) and "advanced problems" (more mathematical problems, usually based on the boxed mathematical notes in the preceding chapter or chapters). A hurdle in entering any new field is learning the relevant terminology, contributed in this case from astronomical, geological, meteorological, and physical traditions. We approached this problem by presenting important terms in **boldface** with adjacent definitions, and by repeating these terms in a concept list at the end of each chapter. A good studying technique for each chapter is to *review these concepts*, and *define or use each one in a sentence.* Most of these concepts are also listed in the index, which thus serves as a glossary to guide the reader to each term's definition.

I extend thanks to the staff of Wadsworth Publishing Company for cooperation, hospitality, and friendship (even during tight schedules and MS cutting), with special thanks to Anne Scanlan-Rohrer. I also thank many scientific colleagues in research and teaching who offered suggestions, criticisms, and/or encouragements since the first edition, including Fran Baganal, Sasha Basilevsky, Jeff Bell, Richard Binzel, Dominique Bockelée-Morvan, Leigh Broadhurst, Nicole Borderies, both Robert Browns, Joe Burns, Dale Cruikshank, Don Davis, Mike Drake, Jim Elliot, Steffi Engel, Ron Greeley, Richard Greenberg, Dave Grinspoon, Alan Harris, Jim Head, Floyd Herbert, Alan Hildebrand, Don Hunten, Bill Hubbard, John Kerridge, Baerbel Luchitta, Lisa McFarlane, Bill McKinnon, Carlé Pieters, Carl Sagan, Bob Singer, Alan Stern, Dave Tholen, Joe Veverka, Stu Weidenschilling, George Wetherill, Charles Wood, and many other colleagues for their helpful conversations. Thanks also to my students, especially Michelle Gates, who helped locate typos and errors in the 1983 edition. Thanks to Catherine Date and Elaine Owens of the Planetary Science Institute of Science Applications International Corporation for assisting me with the part-time status that allowed me to work on this project in the midst of other demanding projects.

Special thanks to Al McEwen (USGS, Flagstaff), and at the Jet Propulsion Lab and NASA, thanks to the Magellan team and PIO staff for grace under pressure in supplying up-to-date images from various spacecraft.

I also appreciate the knowledgeable craftsmanship and interest in space exploration of artists such as Michael Carrol, Don Davis, Jim Hervat, Pamela Lee, Ron Miller, Hiroki Morinoue, James Nichols, Andrei Sokolov, and other members of the International Association for the Astronomical Arts, for helpful conversations and for allowing me to represent their work in my Wadsworth textbooks.

Finally, special thanks to Gayle and Amy Hartmann for putting up cheerfully (more or less) with the cosmic misfortune of having a scientist and writer around the house.

CONTENTS

CHAPTER THREE
CELESTIAL MECHANICS 57

CHAPTER FOUR
THE FORMATION OF STARS AND PLANETARY MATERIAL 89

CHAPTER EIGHT
COMETS 215

CHAPTER NINE
PLANETARY INTERIORS 243

CHAPTER ELEVEN

PLANETARY SURFACES 2: VOLCANISM AND OTHER PROCESSES OF SURFACE EVOLUTION 343

CHAPTER TWELVE

PLANETARY ATMOSPHERES 403

CHAPTER THIRTEEN
LIFE: ITS HISTORY AND OCCURRENCE 449

APPENDIX: PLANETARY DATA TABLE 471

For Gayle and Amy

CHAPTER ONE

INTRODUCTION: PLANETARY SCIENCE AND PLANETARY PERSPECTIVE

Fittingly enough, the first edition of this book was started on July 16, 1969. That morning in Florida, Armstrong, Aldrin, and Collins sat in their command module atop the 98-meter Saturn V. Walter Cronkite announced that there was one minute to lift-off. Lift-off came with a burst of smoke and brilliant flame. The first flight to the moon had begun. At that moment we didn't know if they would make it.

Apollo 11 achieved lunar orbit on July 20, 1969. The ship came around the limb of the moon and reestablished contact with Houston. The astronauts were "go" for the landing. Armstrong and Aldrin left Collins for their long voyage down. Newsmen Huntley and Brinkley, who had been keeping up a commentary for hours, announced that they simply had nothing more to say during these historic minutes, so they let us listen to the dialogue between Houston and the lunar module: "Picking up some dust." Contact. "Tranquillity Base here. The Eagle has landed." 1:18 P.M. M.S.T.

Departure of Apollo 16 for the moon in 1972 symbolizes the direct exploration of the solar system by space vehicles, which has revolutionized our understanding of planets in the last two decades. (NASA)

At 7:30 P.M. M.S.T. Armstrong and Aldrin opened the hatch. Armstrong came down the ladder, moving faster and surer than we had expected. He stepped down onto the landing pad and as quickly off, with his left foot, onto the moon, as if there was nothing to it. "That's one small step for a man; one giant leap for mankind." 7:56 P.M. M.S.T., July 20, 1969.

WHY STUDY THE PLANETS?

That week won't be forgotten. The president called it the greatest event since Creation. We witnessed an evolutionary quantum jump: humankind's first step across the void from the world where we first appeared to another world.

What can we hope to find there? We hope to find knowledge of how planets evolve, knowledge about what governs their crustal structure and their climates, knowledge that can help us live more successfully in our own environment. We expect to find abundant solar energy sources and new supplies of raw material such as nickel-iron. We may even find new worlds to inhabit or other life forms with whom we will have to coexist. We hope to leave for our children better answers to the questions that people have asked for 10 000 years; questions about where we and our world came from—questions dealt with some 3000 years ago in a book called Genesis, which three men read to us as they made the first human flight around the moon on Christmas Eve, 1968.

Our generation faces the problem of whether to turn inward on itself in response to what Congressman Morris Udall (1969) described as "a serious crisis of the spirit" in which "there is a real questioning about whether life is really going to be better." Our generation also has begun—for the first time in history—operations at a scale that can affect the whole planetary environment. Such effects could be very bad, whether they result from industrial activity (nuclear wastes and carbon dioxide emissions), accidents (nuclear meltdowns or oil spills), unforeseen byproducts of consumer goods (fluorocarbon "spray-can" emissions that damaged the ozonospheres), warfare (nuclear fallout and effects of biological weapons), or purposeful environmental changes (projects to alter the climate or retard desertification).

The more we limit our population and attention to the finite globe of Earth, the bleaker the future seems, since we lose our sense of frontier. We decrease our chances to learn how worlds work and our chances to escape our own mistakes. If we study Earth and other planets together, in both their present states and their (remarkably different) earlier states, we gain a better idea of the future possibilities for human life, on Earth and elsewhere. The activities of humanity off Earth replace the threat of an earthbound civilization with the promise and excitement of a civilization in which outward exploration and growth can resume productively.

The exploration of space has already given us a new perspective, as shown by two passages quoted by NASA Associate Administrator Oran Nicks (1970). The first passage was written in 1948 by British astrophysicist Fred Hoyle:

> Once a photograph of the earth, taken from the outside, is available—once the sheer isolation of the earth becomes plain—a new idea as powerful as any in history will be let loose.

The second passage, written after the exploration of space had become reality, is quoted from a letter by John Caffrey:

> I date my own reawakening of interest in man's environment to the Apollo 8 mission and to the first clear photographs of the Earth from that mission. My theory is that the views of the Earth from that expedition and from the subsequent Apollo flights have made many of us see the Earth as a whole, in a curious way—as a single environment in which hundreds of millions of human beings have a stake. I suspect that the greatest lasting benefit of the Apollo missions may be, if my hunch is correct, this sudden rush of inspiration to try to save this fragile environment . . . if we still can.

The exploration of space may affect humanity's outlook in another way. Catherine Drinker Bowen (1963) writes of the sense of optimism engendered in Europe by the exploration of the New World in the late 1500s:

> Were God's heavens then no longer immutable? And not only heaven but earth was shifting its geography. Francis Drake sailed home, having seen the limitless Pacific. . . . People spoke of America as today we speak of the moon, yet far more fruitfully. [Francis Bacon spoke of] "that great wind blowing from the west . . . the breath of life which blows on us from that New Continent." Columbus, he said, had made hope reasonable.

Contemporary scientists have sounded a similar theme while reflecting upon the lunar landings. Physicist Freeman Dyson (1969), of Princeton's Institute for Advanced Study, has written:

> We are historically attuned to living in small exclusive groups, and we carry in us a stubborn disinclination to treat all men as brothers. On the other hand, we live on a shrinking and vulnerable planet which our lack of foresight is rapidly turning into a slum. Never again on this planet will there be unoccupied land, cultural isolation, freedom from bureaucracy, freedom

for people to get lost and be on their own. Never again on this planet. But how about somewhere else?

. . . Many of you may consider it ridiculous to think of space as a way out of our difficulties, when the existing space program . . . is being rapidly cut down, precisely because it appears to have nothing to offer to the solution of social problems. . . . If one believes in space as a major factor in human affairs, one must take a very long view.

Our **goals in exploring space** include not just the collection of new facts and third-order theories about the universe, but also the gaining of a basic understanding of how our own world works, and how other worlds work, as symbolized in Figures 1–1 and 1–2. Perhaps just as important to us is an awareness that there really is a frontier in space—scientifically, psychologically, and physically.

LIVING ON A SPHERICAL FRONTIER

There is a planetary perspective that goes a step beyond the conventional environmental perspective that we adopted (either by foresight or by necessity) during the 1970s. The planetary perspective recognizes that, while we have evolved in the finite ecosystem of Earth, we really live in a much larger environment that we have just begun to probe. While our old geographical frontiers have swept around the globe and closed in on themselves, we have just realized that we have been living all the time on the edge of a new frontier—Earth's surface, with its very thin atmosphere, is the edge of interplanetary space.

Living on that spherical frontier has forced us to face new issues of how to interact with our planetary environment as a whole as well as with the environment around us. An example of such an issue comes from a project on which I recently consulted. At the direction of the Department of Energy, a national research institute in 1977 initiated a study of the long-term hazards associated with burying nuclear wastes. The specific issue was: If one or more societies on the planet pursue nuclear-power production and bury the radioactive wastes, what are the probabilities that geological or human activities might breach the burial chamber and release the radioactive materials into the open environment in the next million years?* The point here is not to discuss possible nuclear-power dangers, but to point out an example of a serious technological issue that requires information about geological history. Processes considered in this study included meteorite impact, glacial erosion during ice ages, sea-level changes, earthquakes, and volcanic eruptions. All of these processes ultimately stem from larger processes that are planetary in scope. For example, the meteorite threat results from an interaction with our planetary environment (see Chapter 6), and ice ages are believed to be governed partly by our planet's orbital variations (see Chapters 12 and 13). The project provided a remarkable example of engineers asking planetary scientists for what had once been "purely academic" data. We are really beginning to live on a planetary scale.

The natural tendency for a successful species to expand its numbers will intensify planetary-scale problems if we remain confined to Earth. Earth's population is now roughly 4.0×10^9 (4 billion) and is expected to double in 40 years or less, in spite of low birth rates in technically advanced countries. The mean population density for Earth's land area is about 27 persons/km^2, including uninhabited regions. This compares to 1976 figures for the United States (23 persons/km^2), Japan (301), Canada (2), India (154), and England (225). Since the days of economist Thomas Robert Malthus (1766–1834), we have been warned that such population numbers can't keep increasing in a finite world without outstripping the raw materials needed for their support.

*The million-year figure was arbitrarily chosen. In some nuclear wastes, many of the radioactive atoms would decay to safe by-products in less than a million years.

Figure 1-1. Symbolic of continuing discoveries in the solar system is this Voyager 2 image of Uranus' 484-km-diameter satellite, Miranda. Scientists were completely surprised to find deep canyons, faulted cliffs, and arrays of linear ridges with sharp turns, such as the vaguely L-shaped swath in the center, on a world that had been thought too small and too cold to have much geologic resurfacing activity. This image has been reprojected by computer to place the linear array at center; faint latitude and longitude lines have been added. (NASA)

But the sky is no longer the limit for human operations. Carrying human operations into the interplanetary environment not only is exciting but also provides a safeguard against the dangers of a finite homeland with a closed frontier.

ENVIRONMENTAL ISSUES ON THE SPACE FRONTIER

Space exploration raises new environmental issues. For instance, old science fiction and new engineering studies suggest the opportunity of utilizing interplanetary materials to further space exploration, establish space colonies, and alleviate raw material and energy shortages on Earth. Soviet and American experiments show that very large space structures may be built in the next few decades as research stations, power-generating stations, and manufacturing facilities. In 1981, the Soviet Union demonstrated the first modular space station by docking the 15-ton Cosmos satellite with the 21-ton Salyut station. Soviet scientists placed the much larger *Mir* space station in orbit in the 1980s, to which they planned enough additions to make the original "look like a mere Christmas ornament" (Cosmonaut Alexie Leonov, private communication, 1989). Modules and docking ports soon expanded *Mir* into the largest space station in history. International crews occupied *Mir* for ever-longer stays. By 1988, individuals had lived in *Mir* for as long as a year—time enough to fly to Mars or Venus. By 1988 Soviets had also successfully launched the world's largest rocket booster, *Energia*, and a space shuttle. However, by the early 1990s, economic problems caused by the Soviet collapse forced cutbacks in Russian space science. Even the *Mir* space station's future is uncertain.

In 1984, President Reagan responded to the Soviet program by proposing a larger American station, and in 1989, President Bush proposed an American permanent base on the moon and eventual human missions to Mars. However, Congress cut the budget for space science and the U.S. space station in the early 1990s remained underfunded and unbuilt.

Figure 1-2. Two astronauts on the moon during the last lunar voyage in 1972, which was televised live on Earth, symbolize possible planetary explorations in the future. Lunar astronauts began the transfer of traditional geological field techniques to the surfaces of other planets. (NASA)

A more visionary idea has yet to be seriously discussed by politicians: demonstrating collaborative exploration as a role model for solving 21st-century problems. By refurbishing the *Mir* station in a joint program, international teams could move toward lunar and Martian exploration, develop solar energy resources in space, and explore asteroids as a source of metals and other raw materials to relieve environmental pressures on Earth. As later chapters will show, asteroids contain metals and other resources.

It is believed possible to acquire such asteroid resources instead of digging ever deeper into Earth for lower-grade ores, and dumping the waste products into our ecosphere.

A dreamy idea? Perhaps. But who in 1776 would have predicted railroads surging across the American West within a century, or jets spanning intercontinental skies a century after that?

A scheme for modifying small asteroids' orbits to make them more accessible is attributed to science fiction writer Arthur C. Clarke, and has been built in lab models. This so-called **mass driver** magnetically accelerates scoops of asteroid soil, using the soil itself as a rocket propellant (Criswell, 1977).

Various researchers (Gaffey and McCord, 1977; Hartmann and others, 1984) have emphasized the enormous economic value of even a small asteroid composed of nickel-iron. Such objects might be moved by a mass driver to orbits around Earth, and the material mined outside the ecosphere, to replace dwindling mineral reserves on Earth. Freewheeling Princeton physicist Freeman Dyson has gone a step further, noting that mined, hollowed-out asteroids or comets could provide radiation-free living spaces, and that comets have all the ingredients necessary for human settlement, including water (see Chapter 8).

Of course, speculations about utilizing space materials have provoked an opposing viewpoint. Asteroid researcher George Wetherill (1979) concluded an article on the nearby asteroids as follows: "Perhaps it would be better if the region of space occupied by [these asteroids] were declared an inviolable 'wilderness area' " so that they are not mined.

Thus, while progress toward colonizing space increases our chances of surviving natural, accidental, or purposeful destruction of Earth's environment, it also underscores the age-old question of how we should best handle our natural resources. Recent environmental awareness taught us three planetary lessons that can be applied here: (1) Major social changes are beginning to occur on Earth because we are running out of the finite earthbound supplies of cheaply available materials; (2) large-scale tampering with Earth's ecosystems is dangerous because complex interactions occur between both biological and nonbiological components of these systems; (3) gluttonous consumption destroys beautiful, unique, and scientifically valuable resources. The first point favors a vigorous space exploration program to search out new resources beyond Earth. The second point implies that great restraint and moderation should be exercised in altering environments, but that processing materials such as asteroidal metal in space would be less damaging to our ecosphere than mining and smelting metal ores on Earth. The third point suggests that we should minimize our consumption of planetary resources, to an extent consistent with the necessity of improving living standards on Earth and in space. Mindful of these three points, I would argue for the most vigorous possible program of exploring and learning about the planets, perhaps using our own exploitation of America in the last century as a guide to possible pitfalls. Perhaps the Yosemites of space will be set aside as monuments, while regions of economic value can be first studied and then made available for use by Earth's people. But this book is not the place to imagine future history. I merely invite you to consider, as you read on, the challenging issues that are just beginning to emerge as our penetration into the solar system begins in earnest.

A VIEW OF SCIENCE

Many people erroneously envision science as a tree branching outward; in time the divisions of science branch finer and finer so that science becomes more and more complex, more and more specialized. Science is like this only in the sense that data proliferate.

Science should really be viewed in the opposite way, as a process that synthesizes many observations into theories that enable a person to understand a wide range of natural phenomena. By "understand" I mean predict the right answers to problems. Take mechanics, for example. During the Renaissance, scientists made endless observations of levers, falling objects, inclined planes, rotation, orbital motion, and so on. Yet even the most learned people were not able to predict very many experimental results until Newton and his successors saw the relationships among the observations. Today, a college sophomore studying physics can understand (that is, quantitatively predict) most of the phenomena that Newton studied, and a senior's understanding surpasses that of Newton.

Thus, science is like working *down* a tree instead of up. First, we can understand only individual events—the twigs. Then we see how they all join together, and we can grasp a branch of knowledge. Eventually we can understand whole portions of the tree. (Nobody has yet reached the trunk!)

In this book I hope to emphasize that, by mastering a few basic concepts from several different fields, the reader can understand a remarkable variety of processes that affect Earth and other planets.

PLANETARY SCIENCE

Planetary science is the study of the origin, evolution, and current environments of planetary bodies, whether they orbit around the sun, as in our solar system, or (hypothetically) around other stars. **Stars** can be defined as masses of material (generally gas) so large that the temperature and pressure in their centers are great enough to push the nuclei of atoms together, thus causing nuclear fusion reactions. These reactions release the energy that the stars radiate. **Planetary bodies,** on the other hand, can be defined as bodies generally composed of silicates and ices that orbit the sun (and probably other stars) and are not large enough to generate nuclear fusion reactions in their interiors. Planetary science, sometimes called planetology, can be defined as the physics of planetary bodies.

Originally, **physics** was defined as "the branch of knowledge treating the material world and its phenomena." In that sense, all science is physics. The discipline now called physics concerns the behavior of matter and energy in close-to-ideal circumstances. Physicists may study atoms, or friction, or electromagnetic fields, but they usually try to isolate a specific phenomenon as much as possible in order to determine the basic laws involved.

Other sciences deal with the physics of more complicated systems, where many phenomena may operate at once and specific subjects can't be isolated. Chemistry, for example, is the physics of matter interacting at the electron level (that is, without atomic nuclei interacting). Astronomy is the physics of cosmic matter, mostly in the form of stars and systems of stars. Meteorology is the physics of the atmosphere, geophysics that of Earth as a body, geology that of Earth's (and nowadays all planets') outer layers. Biology is the complicated physics of living material.

Planetary science is more integrative than geology, meteorology, and biology, for it combines them all. An analyst trying to understand the peculiar riverbed-like channels on Mars may deal with geology (could water or some other fluid have eroded these channels?), geochemistry (could water have been produced on Mars in the past?), ordinary chemistry (under what conditions would the water freeze instead of flow?), or meteorology (what was the Martian air pressure in the past as gases erupted from volcanoes and dissipated?).

Similarly, a planetary scientist may have to deal with a wide variety of data sources. An analyst trying to understand the satellites of Jupiter will deal with theoretical data (the dynamics of satellite orbits), conventional astronomical data (spectrophotometric data that indicate composition), and data sent back from space vehicles (digitized and computer-enhanced photographs and magnetic-field measurements).

Implementing Planetary Science

Planetary research in America is supported primarily by grants from the National Aeronautics and Space Administration (NASA) and, to a lesser extent, the National Science Foundation (NSF). NASA supports research that fulfills its mission of exploring the planets for the benefit of humanity. This mission was set by Congress and is funded according to the degree of public interest as perceived by Congress. In the post-Apollo 1970s this funding dropped somewhat. Shuttle flights are now beginning to provide opportunities for launching new space probes, although the disastrous explosion of the shuttle Challenger, in 1986, delayed the launch of many spacecraft, because America relied on the shuttle for all major space missions. Shuttles started flying again in 1988, but Air Force missions took a growing portion of the payloads. (Indeed, in terms of funding, NASA became a smaller space program than that of the Defense Department.) Many experts believe that a future mixed fleet of shuttles and unmanned boosters will allow better flexibility in launches.

During the 1980s and 1990s, NASA tried to maintain the shuttle program, begin a new program to build a large, human-occupied space station, and respond to President Bush's 1989 mandate to return to the moon and go on to explore Mars. At the same time, Congress declined or delayed funding for these programs. As a result, funding for basic planetary science research and probes to other planets has been severely squeezed, and many planetary scientists had their research grants cut or terminated around 1990.

To apply for such funding, researchers write proposals (descriptions of planned studies). The proposals are sent to funding agencies such as NASA, and are ranked for quality by panels of scientists appointed by the agency. (Membership of the review panels rotates year by year.) Available research money is apportioned among the best proposals.

Planetary scientists must often make a creative compromise between being a specialist and being a generalist. Too specialized a background can leave one too narrowly trained to see the "big picture" in data from different fields; the scientist becomes too much the narrow expert of the type described by a variant of Newton's third law: "For every Ph.D. there is an equal and opposite Ph.D." Too broad a background can leave one with too superficial an understanding of physical detail.

Thomas Kuhn (1962) points out in his essay on scientific revolutions that a new field of scientific inquiry is not established until the participants agree on a **paradigm,** a body of basic facts, theories, and approaches to their work. Planetary science is a new field, and its agreed-upon body of knowledge is only now emerging. Journals, such as *Icarus, Earth and Planetary Science Letters*, and *Meteoritics*, consolidate planetary studies; in the meantime, planetary science papers are still being published in such varied journals as *Science, Nature*, and *The Astronomical Journal*. Certain areas of vigorous study have developed names of their own. **Space science** has been used to designate the study of the interplanetary medium, uppermost atmosphere, solar–planetary interactions, and cosmic rays; **astrogeology** (something of a misnomer) designates geological and stratigraphic studies of lunar and planetary surfaces and is a popular term within the U.S. Geological Survey. **Exobiology** refers to hypothetical life not native to Earth and the attempt to detect such life.

EVOLUTIONARY VIEWPOINT OF THIS BOOK

Because planetary scientists must understand *processes* as they apply to a variety of worlds they cannot merely study descriptions of the present-day planets one at a time. Therefore, we will not describe the planets one by one. Instead, we will describe processes such as the origin of planets and the evolution of their surfaces. We will treat the planets as four-dimensional, extending in three dimensions of space and a fourth dimension of time.

The four-dimensional approach is being applied in modern studies of Earth as well as of other planets. Until recent decades, geological investigations were often limited to traverses over two-dimensional landscapes, with results plotted on two-dimensional maps. Little evidence was gathered by drilling into Earth's crust (the third dimension) or from radioisotope dating techniques (the fourth). An excellent example of the new approach is the contemporary geologic study of the volcano Kilauea in Hawaii. The whole mountain has been fitted with seismographs,

Figure 1-3. Ringed planets. Exemplifying unexpected planetary discoveries is the difference between the Saturn and Neptune ring systems. (**a**) *Saturn is surrounded by an intricate system of broad rings composed of billions of ice particles. Narrow gaps are believed to be caused by gravitational effects of Saturn's moons. This image, unlike some reproductions, shows realistic contrast between the overexposed crescent-lit globe and the fainter rings.* (**b**) *Neptune has narrow rings separated by wide gaps. These rings are so faint that two long exposures were required to photograph them; the highly overexposed crescent of Neptune in the center was not recorded, to avoid glare. Puzzling local concentrations, or "arcs," were also found in Neptune's rings, but were not in position to be recorded when these images were made. (NASA photos by Voyager spacecraft 1 and 2, respectively;* (**a**) *courtesy of Al McEwen, U.S. Geological Survey)*

tiltmeters, and other geophysical instruments whose data output is transmitted to a central observatory on the rim of the main crater. You can stand in one room and watch an array of graphs that are recording the workings of the volcano's interior as well as of its surface. Motions of molten lava, for example, can be traced from deep below the mountain's summit as the lava works its way toward the surface.

Our study of planetary processes will be subject oriented, not method oriented. That is, we will arrange chapters by major subjects and introduce methodology only as needed to clarify the subjects.

The chapters are arranged from a chronological or evolutionary viewpoint. Early chapters describe basic observations, definitions, and matters of celestial mechanics. We then begin with a view of the origin of the solar system (Chapters 4 and 5). Small bodies of the solar system are then discussed with a reminder that they provide evidence of the earliest state of the planets (Chapters 6 through 8). Planet interiors and surfaces are then discussed (Chapters 9 through 11), followed by their atmospheres (Chapter 12). Finally, we consider the development of life (Chapter 13), whose flowering (modestly referring to us) was extremely recent.

As a final introductory note, consider this sticky philosophical problem. Most of our contemporaries would not quarrel with the assertion that we live in an evolving system. Yet this belief is not universal, as the British philosopher and cosmologist H. Dingle (1960) reminds us:

> Nearly 100 years ago Philip Gosse, in order to reconcile the facts of geology with the Hebrew scriptures, advanced the theory that . . . "there had been no gradual modification of the surface of the earth, or slow development of organic forms, but that when the catastrophic act of creation took place, the world presented, instantly, the structural appearance of a planet on which life has long existed." The beginning of the universe on this theory occurred some 6000 years ago. There is no question that the theory is free from self-contradiction and is consistent with all the facts of experience we have to explain; it certainly does not multiply hypotheses beyond necessity since it invokes only one; and it is evidently beyond refutation by future experience. If, then, we are to ask of our concepts nothing more than that they shall correlate our present experience economically, we must accept it in preference to any other. Nevertheless, it is doubtful if a single person does so.* It would be a good discipline for those who reject it to express clearly their reasons for such a judgment.

Can we, as "enlightened" planetary scientists, express clearly our reasons for judging that the evolution of the cosmos is a meaningful concept? We find in nature landscapes showing ordered strata, in which fossils appear in an orderly sequence terminating in the present, and rocks contain patterns of radioactivity that correspond to the fossil sequence in age. We also find in physical theory a scheme that enables us in our minds—or rather in our computers—to construct models that begin with hypothesized ancient conditions, evolve into the present conditions, and successfully predict future conditions.

This internal consistency of our evolutionary viewpoint does not prove that it is the ultimate truth. However, it does assure us that our philosophy is pragmatic and that future discoveries will continue to refine present beliefs. This is the strength of the scientific method: It continues to converge on a more and more detailed, consistent, satisfying, and practical view of the universe. And the scientific method has, on the whole, been much more successful than have occult

*Here the philosopher is optimistic. Although this theory is hardly popular in our society, fundamentalist evangelical radio programs dominating the evening radio waves in vast areas of America insist on precisely this view—that all the seemingly evolved strata, fossils, and radioisotopes were put into their complex pattern just to tempt scientists into error. In 1981 two states have passed legislation to require that some sort of "creationist" theory be presented alongside any evolutionary theory of biology, geology, or astronomy. These laws are being tested in court; in other states, creationists have brought still more lawsuits to require that their beliefs be taught in science courses. We begin our planetary studies with our feet planted in an interesting sociological soil and with roots extending far back in history.

methods in predicting observable phenomena. With a profound interest and a rather religious sense of approaching an ultimate reality, we pursue the evolutionary thread that leads backward to what seems—according to recent findings—to have been the creation of the universe and leads forward into the future.

SUMMARY

Planetary science is a highly integrative field that combines knowledge from other fields of science. Planetary scientists try to understand the origin and evolution of planets and their surface environments. Planetary scientists are, to borrow Barbara Novak's (1980) description of naturalist painters on the Western frontier in the last century, "archaeologists of the Creation." At the same time, planetary scientists try to apply their results to a better understanding of humanity's present and future environment.

The last decade saw the first data radioed back by American, Russian, European, and Japanese space probes from Uranus, Neptune, Halley's comet, and an asteroid, not to mention detailed mapping of Venus and new data from ground-based and orbiting telescopes about possible planetary systems around other stars. Milestones during the "operational lifetime" of this edition of this book may include (if all goes well) orbiters, balloons, and landers operating at Mars, more missions to asteroids, and the first probes parachuted into the atmosphere of Jupiter. Over the longer term, planetary scientists anticipate manned missions to nearby asteroids and Mars, and may begin to deal with human utilization of resources in space.

CONCEPTS

goals in exploring space
mass driver
science
planetary science
stars
planetary bodies
physics
paradigm
space science
astrogeology
exobiology

PROBLEMS

1. How can data obtained on other planets help illuminate environmental processes and evolutionary processes on Earth?

2. Can data on the history and conditions of our own planetary system shed light on whether planet systems exist around other stars? If so, how?

PROJECTS

1. See if your school library receives any journals devoted to planetary science, such as *Icarus* or *The Moon and Planets*, or journals in which planetary science results are published, such as the *Journal of Geophysical Research*. If so, report on some areas of current planetary science research.

2. In recent issues of general scientific journals, such as *Science*, *Nature*, and *Scientific American*, determine the percentage of articles related to planetary science. Does planetary science make a contribution to the modern scientific outlook?

CHAPTER TWO

THE SOLAR SYSTEM: AN OVERVIEW

The solar system is commonly said to have nine **planets,** shown in Figure 2–1. In order from the sun, they are Mercury, Venus, Earth, Mars, Jupiter, Saturn, Uranus, Neptune, and Pluto. A traditional mnemonic device for remembering this sequence is "*M*en *v*ery *e*arly *m*ade *j*ars *s*tand *u*pright *n*icely (*p*eriod)." (Surely today's students can do better than this!) Because confusion often sets in at *S*aturn, *U*ranus, and *N*eptune, remember that the *SUN* is in the system too.

The first four planets are sometimes called the **terrestrial planets** because of their nearness to Earth and the similarity of their rocky, metallic compositions. The four planets from Jupiter through Neptune are sometimes called the **giant,** or *Jovian*, planets because of their size and similarity to Jupiter.

A view of Jupiter's satellite Io from Voyager 1 spacecraft in 1979 provides an example of the variety of planetary bodies. Mottled patches on Io result from active volcanism that constantly modifies the sulfur-rich surface. Doughnut-shaped ring in center is a cloud of debris being erupted from an active volcano. (NASA)

WHAT IS A PLANET?

One might think the term *planet* could be easily defined; after all, there are nine bodies that seem uniquely distinguished by size, position, orbit, and so on. But the situation is not really so clear. Among the ancients, a planet was any of seven bodies that changed position from day to day with respect to the stars. But these seven included the sun and the moon, as well as the first five planets (excluding Earth) out to Saturn. The rest had not yet been discovered.

A recent definition is that a planet is any body except a **comet, asteroid, meteoroid,** or **satellite** orbiting the sun. But this definition is ambiguous unless comets, asteroids, meteoroids, and satellites are defined. We beg the question by calling those objects "interplanetary" bodies. And scientifically, how do we justify excluding the asteroid **Ceres,** which orbits between Mars and Jupiter and has the sizable diameter 1000 km, nearly twice as big as those of other asteroids? To confuse matters more, in 1977 Hale Observatories researcher Charles Kowal discovered a new object estimated at about 350 km in diameter orbiting mostly between Saturn and Uranus (though occasionally crossing just inside Saturn's orbit), a region hitherto not known to

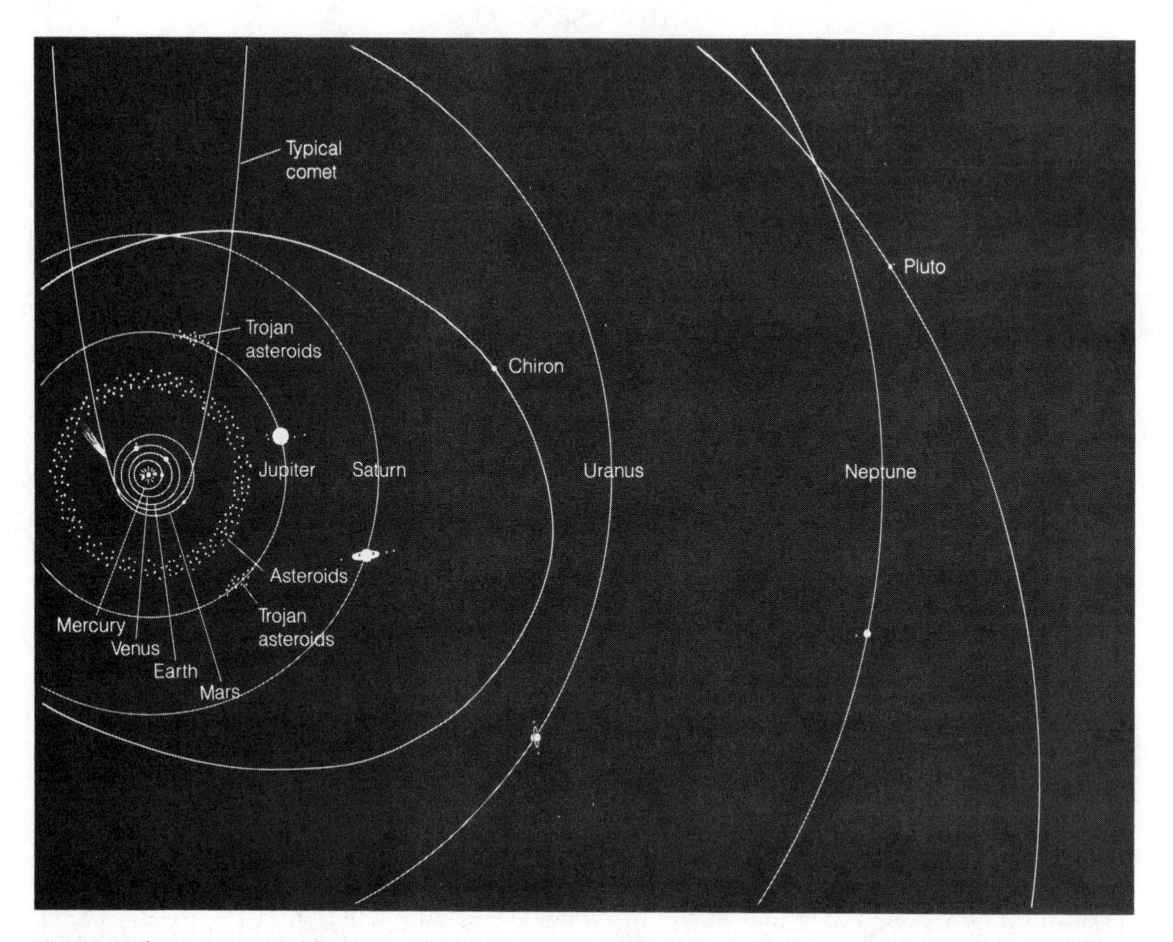

Figure 2-1. The arrangement of the nine principal planets and other lesser bodies of the solar system with orbits drawn to scale. This figure illustrates the relative closeness of the four terrestrial planets, the relatively vast distances in the outer solar system, the noncircular orbits of the small bodies Pluto and Charon, and the presence of comets and different classes of asteroids.

contain any such large body. He named it **Chiron,** and it was cataloged as asteroid number 2060. But was it an asteroid, a small planet, or what? In 1978, Kowal said, "For now, Chiron is just—Chiron." But in 1988, as its remote orbit brought it slightly closer to the sun, it suddenly brightened dramatically, and observers discovered that it was giving off the diffuse gas and dust characteristic of a comet!

If we try to define planets by the regular spacing of their nearly circular orbits, we face the fact that not only Chiron crosses a giant planet's orbit; Pluto has a markedly noncircular orbit that crosses inside Neptune's. In fact, Pluto is now, in the 1990s, inside Neptune's orbit!

If we try to define planets by size alone, we are confounded by the fact that some satellites are larger than the planets Mercury and Pluto, as shown in Figure 2-2, and that Pluto, about the

Figure 2-2. (right) Relative sizes of the terrestrial planets and smaller bodies, drawn to scale. Diameters are given in km. Giant planets' satellites are indicated by letters J, S, U, and N and a number code, as well as names. (The number code is a traditional usage, indicating either the order outward from the planet, or, usually for larger numbers, the order of discovery.) Asteroids have a number and a name.

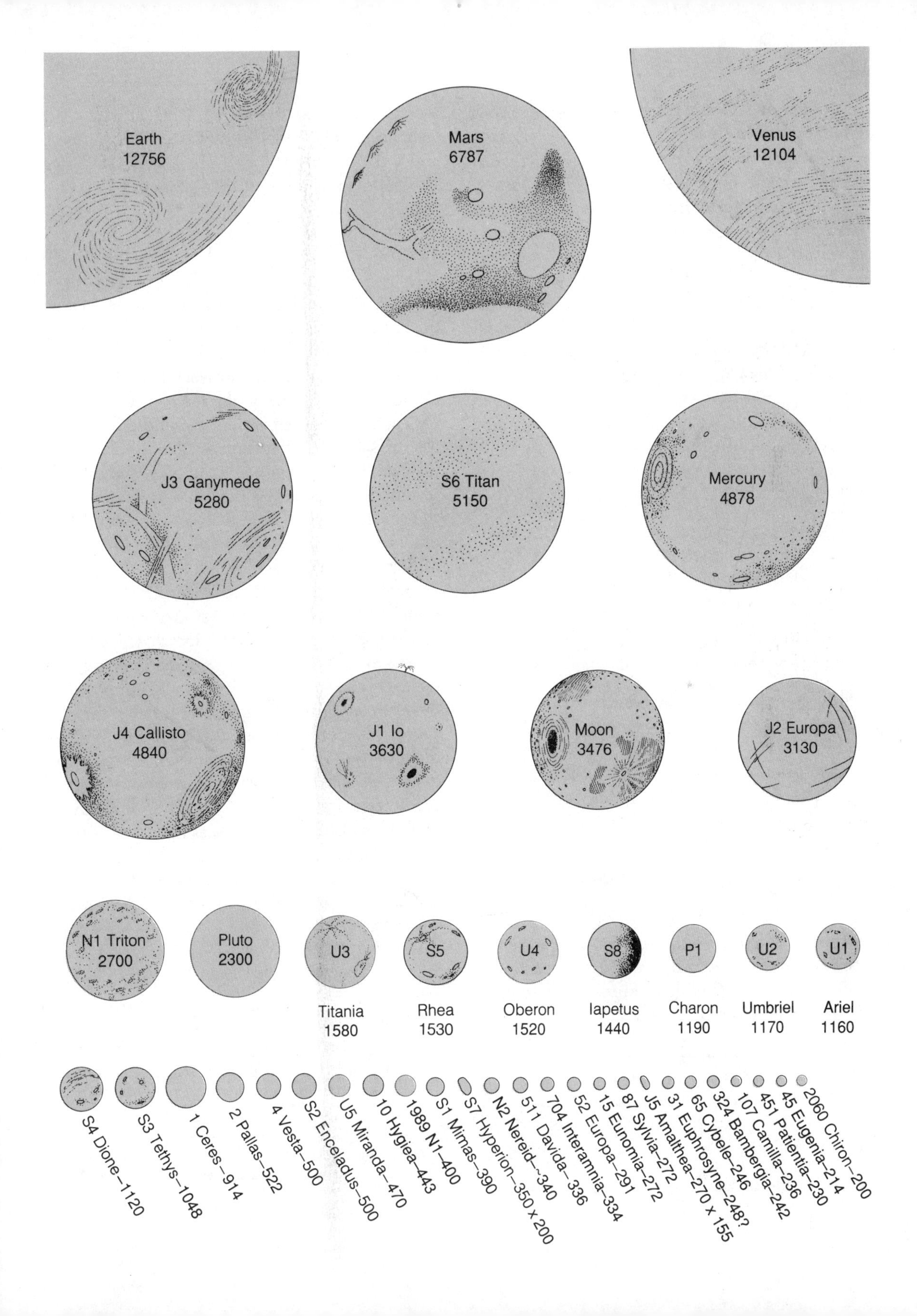

Earth
12756
Mars
6787
Venus
12104
J3 Ganymede
5280
S6 Titan
5150
Mercury
4878
J4 Callisto
4840
J1 Io
3630
Moon
3476
J2 Europa
3130
N1 Triton
2700
Pluto
2300
U3
S5
U4
S8
P1
U2
U1
Titania
1580
Rhea
1530
Oberon
1520
Iapetus
1440
Charon
1190
Umbriel
1170
Ariel
1160
S4 Dione–1120
S3 Tethys–1048
1 Ceres–914
2 Pallas–522
4 Vesta–500
S2 Enceladus–500
U5 Miranda–470
10 Hygiea–443
1989 N1–400
S1 Mimas–390
S7 Hyperion–350 x 200
N2 Nereid–340
511 Davida–336
704 Interamnia–334
52 Europa–291
15 Eunomia–272
87 Sylvia–272
J5 Amalthea–270 x 155
31 Euphrosyne–248?
65 Cybele–246
324 Bambergia–242
107 Camilla–236
451 Patientia–230
45 Eugenia–214
2060 Chiron–200

size of our own moon, is accompanied by a companion that is a sizable fraction of its own size.

Finally, if we try to define planets generically, we face the problem that the origins of planets, asteroids, comets, and satellites are not precisely understood. For these reasons we need to clarify our terminology as follows:

Planetary body: As defined in Chapter 1, this term refers to any star-orbiting body too small to initiate energy-releasing nuclear reactions in its interior. Planetary bodies are usually solid (with some liquid or gas components possible) and mostly composed of silicates or ices. Planetary bodies include the dust observed around some stars, the bodies in our own solar system, and the hypothetical planets orbiting other stars.

Principal planet: Mercury, Venus, Earth, Mars, Jupiter, Saturn, Uranus, Neptune, or Pluto—the bodies called planets by tradition—or an analogous body near other stars.

Planet: (1) One of the principal planets. (2) Loosely, any relatively large planetary body.

Planetesimal: One of the relatively small planetary bodies. Particularly, one of the small planetary bodies in the early solar system. The term usually implies a size between that of a microscopic dust grain (less than 1 mm in diameter) and a large asteroid (about 1000 km in diameter).

Satellite: Any planetary body in orbit around a larger planetary body.

Asteroid: A planetesimal composed mostly of rock or metals; most orbit the sun between Mars and Jupiter, and all of those known to date have diameters of less than 1000 km.

Comet: A planetesimal that can emit an observable gaseous halo and is to some degree (probably mostly) composed of ices. Most comets are located in the outer solar system beyond the asteroid belt.

World: One of the larger planetary bodies (at least 1000 km in diameter). This term gained increased use in 1979 when Voyager researchers used it to group the large Jovian satellites ("worlds") together with the planets.

BODE'S RULE

In 1772 the German astronomer Johann Bode* popularized a simple empirical rule, **Bode's rule,** which is a helpful tool for remembering the distances of the planets from the sun. This rule should be memorized. Write down a sequence of 4s, one for each planet except Neptune and one for the asteroids, and add the sequence 0, 3, 6, 12, 24, . . . (see Table 2-1). Dividing by 10 gives the relative distances of the planets from the sun in terms of the mean distance from Earth to the sun. This distance is called **astronomical unit** (AU). Thus, Jupiter is 5.2 AU from the sun.

Bode's rule lacked any theoretical justification, but it passed its first test in 1781 when the English astronomer William Herschel discovered Uranus at 19.2 AU. Afterwards, a search was made for the "missing planet," which was supposed to lie between Mars and Jupiter. This led to the discovery of the first asteroid, Ceres, on the first night of the new century, January 1, 1801, by the Italian astronomer Giuseppe Piazzi, exactly at the predicted solar distance of 2.8 AU. Discoveries of more asteroids followed in subsequent years, most having solar distances around 2.8 AU.

To this day, Bode's rule lacks any theoretical justification, although it has proved unsettlingly accurate. The formation of the solar system must have involved some dynamical effect that created regularly spaced zones, each dominated by a single planet.

A set of symbols for designating the planets is the one useful contribution of astrology. The symbols, listed in Table 2-1, form convenient

*The rule was apparently first discovered by another German astronomer, Johann Titius; thus it is often called the Titius-Bode rule.

Table 2-1
Bode's Rule and Planetary Symbols

	Mercury	**Venus**	**Earth**	**Mars**	**Asteroids**	**Jupiter**	**Saturn**	**Uranus**	**Neptune**	**Pluto**
	4	**4**	**4**	**4**	**4**	**4**	**4**	**4**		**4**
	0	**3**	**6**	**12**	**24**	**48**	**96**	**192**		**384**
Predicted distance (AU)	0.4	0.7	1.0	1.6	2.8	5.2	10.0	19.6		38.8
Actual distance (AU)	0.4	0.7	1.0	1.5	2.8	5.2	9.5	19.2	30.0	39.4
Symbol	☿	♀	⊕	♂	ⓝ[a]	♃	♄	♅	♆	♇
No. known satellites	0	0	1	2	?	16[b]	>17[b]	5	2[c]	1

[a]The well-observed asteroids (roughly 2000 known) are numbered; the symbol is the encircled number.
[b]Voyagers 1 and 2 discovered several small satellites on the outskirts of the ring systems of Jupiter and Saturn. More may exist and the smallest may grade into the largest ring particles. At least four more probable moons were seen by Voyagers near Saturn.
[c]Search for a Neptune ring during an occultation turned up no ring but did reveal evidence for a third satellite, whose orbit is still unknown.

subscripts and datum-point symbols in theoretical discussions and graphs.

A SURVEY OF THE PLANETS

Before proceeding to the chapters describing the dynamics, origin, and nature of the planets, we will give a thumbnail sketch of each planet in order from the sun. These sketches include some basic facts, a description of each planet as studied from Earth, and a background leading up to our present explorations by spacecraft.

Mercury

Mercury is the smallest principal planet and the closest to the sun. It has no detectable atmosphere. Its surface has a slightly pinkish cast, and through the telescope one can see faint mottlings reminiscent of the appearance of the moon to the naked eye (Figure 2-3). Some dark regions may be lava-covered areas similar to the dark lava plains of the moon. Like the dark regions on the moon and Mars, these dark regions are called *maria* (singular **mare**) from the Latin word for sea, because early astronomers erroneously thought these regions were oceans.

Because Mercury's orbit is so close to the sun, Mercury is usually hidden by the sun's glare, as shown in Figure 2-4. As a result, a number of astronomers never saw the planet. It can be seen only at twilight or in the daytime sky, and even with a telescope it is hard to observe. Only a few of its markings can be reliably detected.

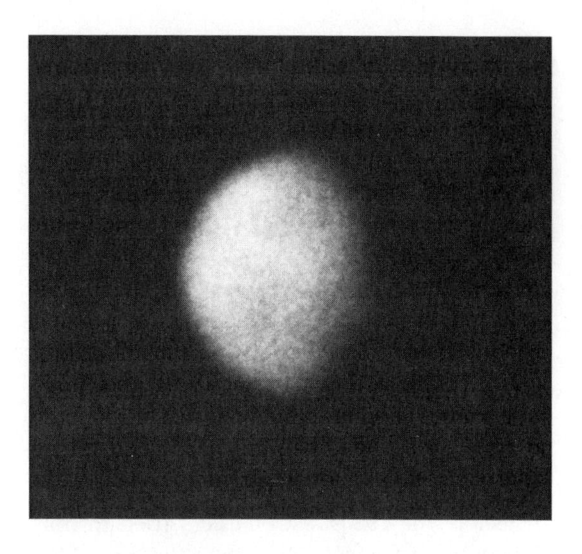

Figure 2-3. Mercury. This 1969 photo is one of the few in existence to show the planet's surface markings. (New Mexico State University Observatory)

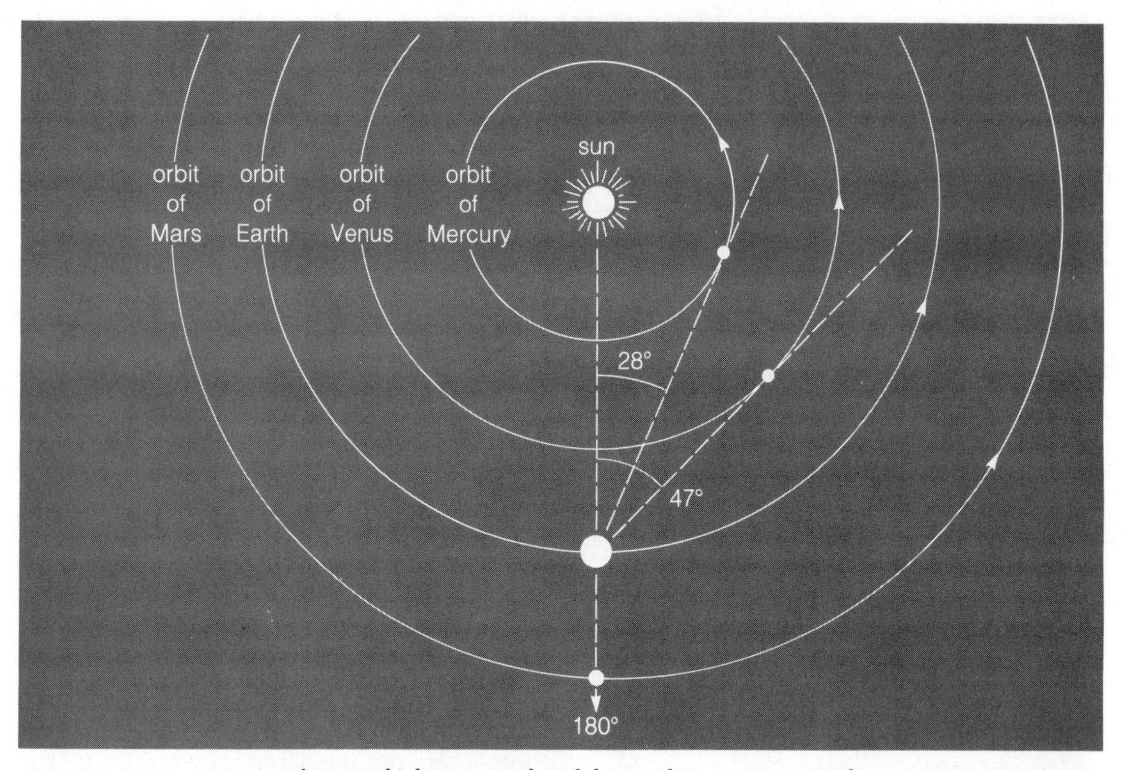

Figure 2-4. Because Mercury and Venus orbit between Earth and the sun, their maximum angular separations from the sun are 28° and 47°, respectively. Because the 28° angle is so small, solar glare makes Mercury difficult to see. Exterior planets, however, can appear at opposition, that is, at 180° from the sun.

On three occasions in 1974, the American spacecraft *Mariner 10* sailed within 6000 km of Mercury and made the first close-up photos of the planet, showing roughly half the planet in detail. The pictures, such as Figure 2-5, show the planet to be moonlike and crowded with **craters,** or circular depressions, like those on the moon. Some plains may be lava-flooded areas, but they are not as clearly defined as are those on the moon. Rugged cratered areas on planets like Mercury are called **uplands,** or sometimes *terrae* (Latin for lands). As seen in Figure 2-6, they dominate the planet. They also contain many areas of crater-damaged, dusky plains that may be very old lava flows.

Before the 1960s, astronomers tried to map the dusky markings to clock the rotation of Mercury. They had concluded that Mercury kept one side toward the sun during its 88-d orbital period, just as the moon keeps one face toward Earth. To their embarrassment, radar work in the 1960s showed that Mercury rotates not in 88 d but in 59 d, and it does not keep one side toward the sun (Pettengill and Dyce, 1965)! The saving grace for the visual observers was that the 59-d rotation (exactly two-thirds of the 88-d revolution) causes a peculiar pattern of recurring configurations between Earth, Mercury, and the sun called a **resonance** (see Chapter 3), so that on certain occasions when Mercury is well placed for observation from Earth, it tends to have the same side toward the sun (Cruikshank and Chapman, 1967).

Venus

Venus is sometimes called Earth's sister planet. It most nearly matches Earth in size and its orbit

Figure 2-5. A mosaic of Mariner 10 photos of Mercury shows a heavily cratered planet that resembles the moon. (NASA)

is closest to Earth's orbit. Its very dense atmosphere is composed mostly of carbon dioxide (CO_2). Opaque white or yellowish clouds hide its surface.

Because of the highly reflecting clouds and its close approaches to Earth, Venus appears very bright. Like Mercury, it moves only a small angular distance from the sun as seen from Earth (see Figure 2-4). When it moves into a position to be our **evening star** or **morning star** (setting or rising a few hours after or before the sun), it is some 15 times brighter than the brightest star (Sirius) and can cast shadows.

Through the telescope, Venus appears as a nearly blank disk, occasionally mottled by barely visible dusky bands or patches. As seen in Figure 2-7, these cloud patterns are much more prominent in ultraviolet light because some of the cloud or atmospheric constituents absorb the ultraviolet. Spacecraft photographs taken with ultravi-

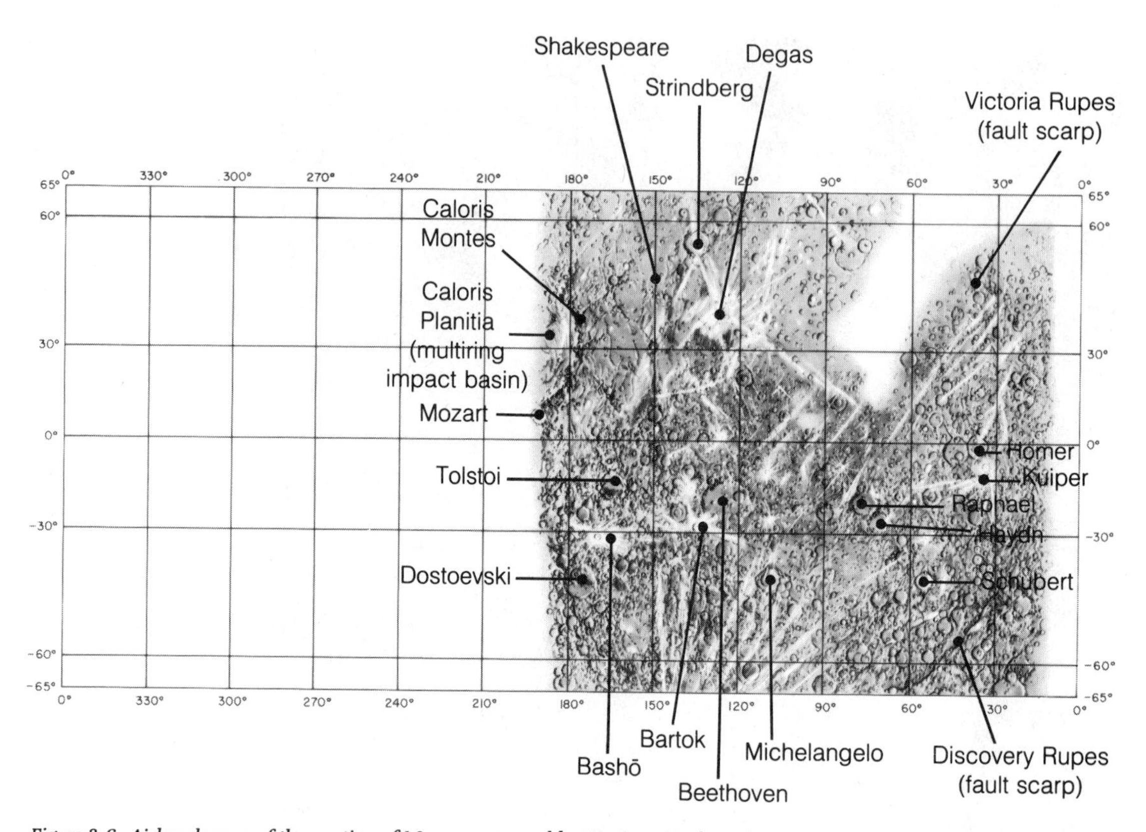

Figure 2-6. Airbrush map of the portion of Mercury mapped by Mariner 10, the only portion presently known in detail. Most features are named after historical figures in the arts. (Base map courtesy R. M. Batson, U.S. Geological Survey)

olet filters more clearly reveal the roughly banded cloud patterns, which lack the great cyclonic whorls characteristic of Earth's low-level clouds.

Because of Venus' cloud cover, we cannot watch the rotation of the solid surface of the planet, and until recently the period of rotation was a mystery. Repeated attempts to determine the rotation by watching motions of the cloud markings on Earth-based ultraviolet photographs were frustrated by rapid changes in the clouds and their ill-defined patterns. A rotation period of about 20 to 30 d gained popularity among one group of observers, while another favored a 24-h period, and another a rotation that kept one side always facing the sun.

In the early 1960s radar techniques yielded the totally unexpected finding that Venus rotates in 243.1 d, not in the same direction as Earth and most other bodies rotate, but backward, from east to west (Dyce, Pettengill, and Shapiro, 1967)! This peculiar situation may be abetted by dynamical resonance between Earth and Venus (Goldreich and Peale, 1967), although the resonance is apparently not exact. This east-to-west motion, whether in a planetary body's orbit or spin, is called **retrograde motion,** as contrasted to the usual west-to-east or **prograde motion.**

Ultraviolet photos show that the high-atmosphere clouds circulate east to west in roughly 4 d (Smith, 1967), a result confirmed by spectroscopic studies from Earth (Betz and others, 1977) and by spacecraft photography (Murray and

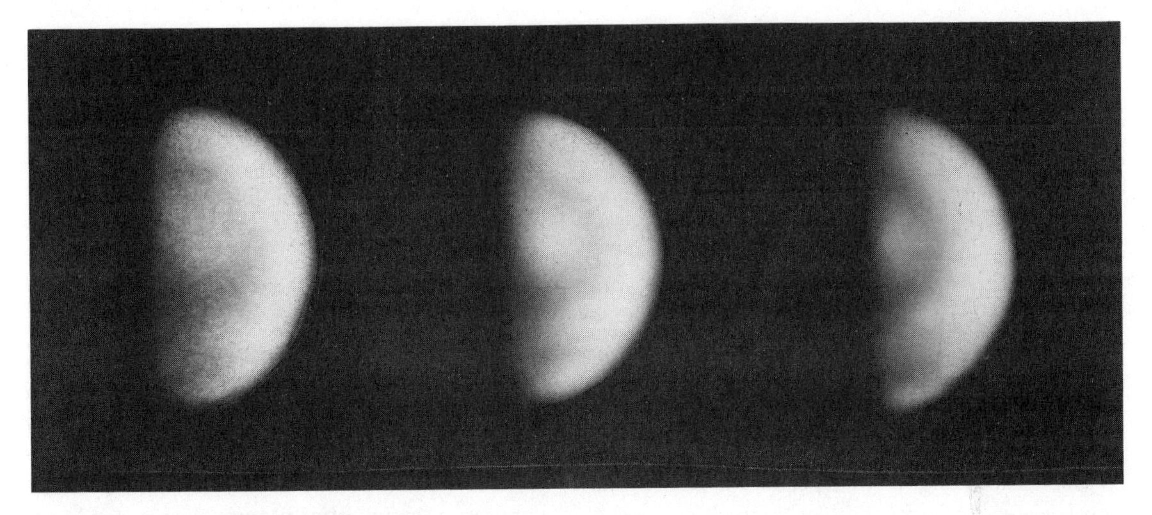

Figure 2-7. Sequence of ultraviolet photographs of Venus made on different days with Earth-based telescopes. These pictures, which are among the best made with ground-based equipment, show faint cloud patterns. (Lunar and Planetary Laboratory, University of Arizona)

others, 1974). This corresponds to a circulating wind near the cloud tops (40 to 60 km above the surface) of about 100 m/s. Other winds have been measured by spacecraft and spectroscopy, ranging from very gentle surface breezes to a 125-m/s wind from daytime to nighttime hemispheres at an altitude of around 115 km (Benz and others, 1977).

What lies hidden beneath the clouds of Venus? Decades of speculation produced many theories. Some writers thought that the clouds were composed of water droplets and that the surface must be a steamy jungle of vegetation swept by torrential rains. Others suggested a windy, dry desert or an ocean of oily liquid. The decades of speculation were ended by a series of spacecraft encounters with the planet and new techniques of intensive Earth-based observation in the 1960s. Venus was first observed at close range on December 14, 1962, by the U.S. spacecraft Mariner 2, which passed 38 854 km (21 645 mi) from the surface of the planet. The first contact with the planet was achieved October 18, 1967, when the Soviet spacecraft Venera 4 parachuted into the atmosphere and radioed back data. On December 15, 1970, **Venera 7** made the first successful unmanned landing on Venus, sending back more atmospheric data. Finally, on October 22 and October 25, 1975, Veneras 9 and 10 sent back the first photos of the surface (Chapter 11), which reveal a desolate, dry, rock-strewn landscape (Florensky, Basilevsky, and Pronin, 1977; Keldysh, 1977). The atmosphere at ground level is surprisingly clear. Gravel and rounded bedrock outcrops give evidence of erosive and weathering processes that are absent on the moon.

Earth-based instruments and the Soviet vehicles on Venus' surface show that the surface temperature and pressure are extremely high, about 750 K (890°F) and 90 atmospheres (90 times the sea level air pressure on Earth), respectively. The mystery of the cloud composition was solved when Sill (1973) and Young (1973) independently found that the cloud properties could be explained by droplets of concentrated sulfuric acid (H_2SO_4 dissolved in water). The droplets are typically 2 μm across with an acid concentration of 78% to 90% (Pollack and others, 1978). Venus is thus a fearsome place—hot and dry with a dense CO_2 atmosphere and sulphuric acid clouds.

Space probes and Earth-based radio telescopes have bounced radar waves through the clouds and off the surface, yielding topographic maps of surface features. The mapping was accomplished at improving levels of detail by the American Pioneer Venus probe in 1978, Soviet Venera 15 and 16 probes starting in 1983, and the American Magellan probe in 1990–92. The results reveal a volcanically active sister to planet Earth. Venus' volcanic plains are broken by a few Australia-sized uplands 2 to 5 km high, with a few volcanic peaks as high as 11 km above the plains—exceeding Mt. Everest's 8 km above sea level. Meteorite impact craters, sparsely scattered on the plains, suggest a surface nearly as young as Earth's, about 800 My in average age. Tectonic fractures testify to seismic unrest, and some of the volcanoes may be active today. Circular features called coronas, a few hundred km across, may be due to local upwelling currents in Venus' mantle (Phillips and others, 1991; Saunders and Pettengill, 1991).

Earth

The next planet from the sun is characterized by shifting white clouds and a bluish color resulting from sunlight scattered by molecules in its atmosphere. Contrary to the CO_2-dominated atmospheres of its neighbor planets, Venus and Mars,

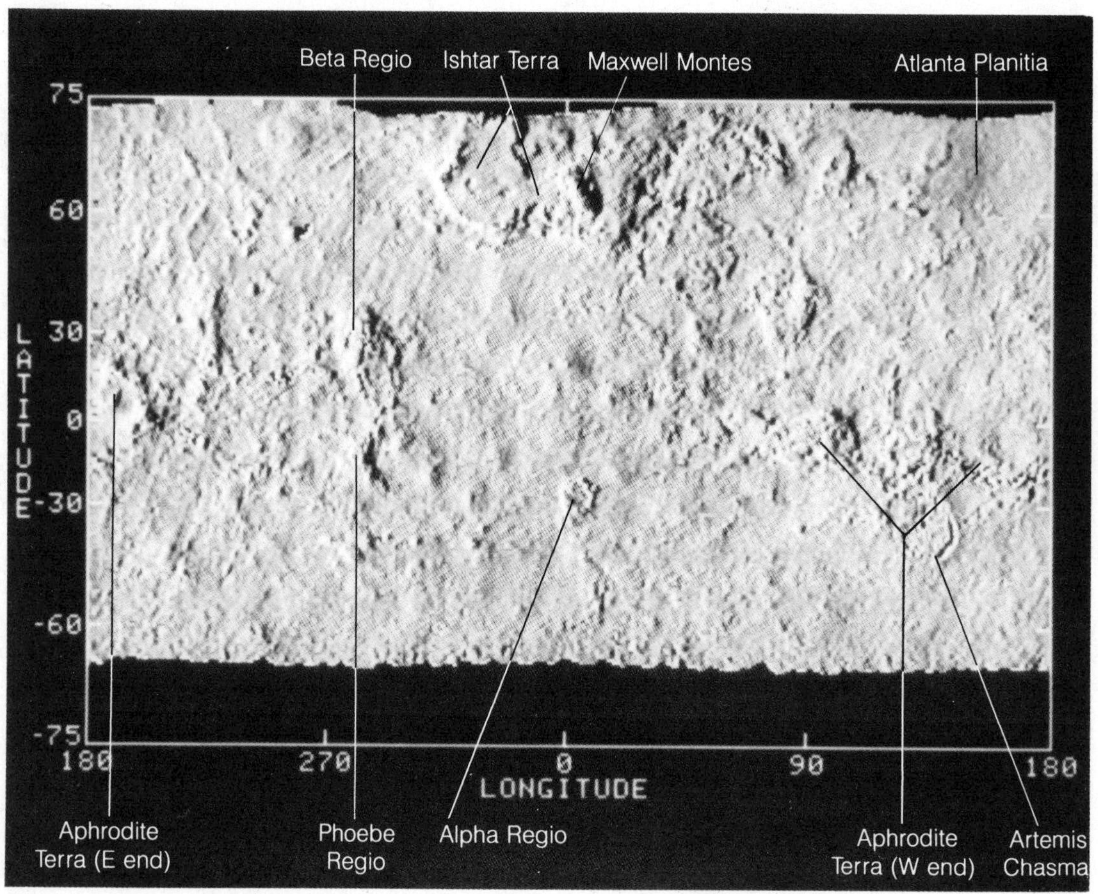

a

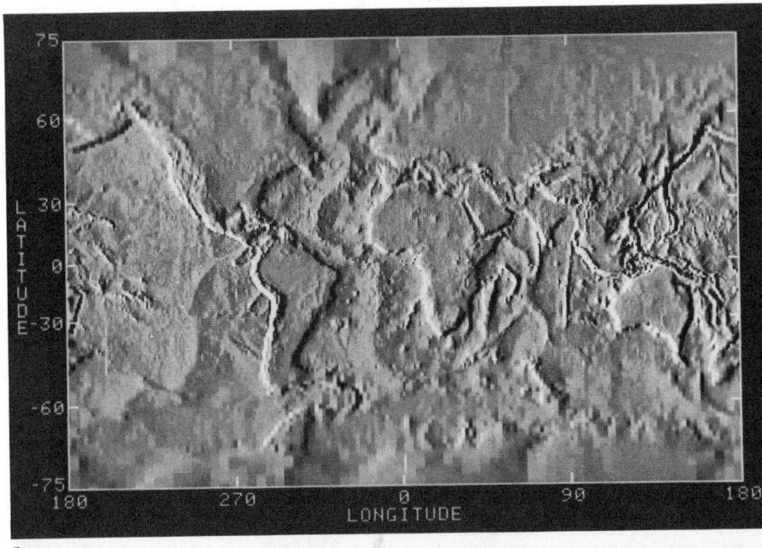

*Figure 2-8. Comparative relief maps of (**a**) Venus, (**b**) Earth, and (**c**) Mars. Altitude data have been digitized and presented as shaded relief with light from the left. Venus has broader expanses of rolling plains than Earth, but also has some elevated, continent-like masses (such as Aphrodite Terra) and features resembling sea-floor trenches (such as Artemis Chasma). Mars retains a more primitive topography with traces of old impact basins (such as Hellas). Martian relief is also dominated by large domes of volcanic lavas surmounted by high volcanic peaks (such as Olympus Mons atop the Tharsis dome). (Images courtesy Michael Kobrick, Jet Propulsion Laboratory)*

b

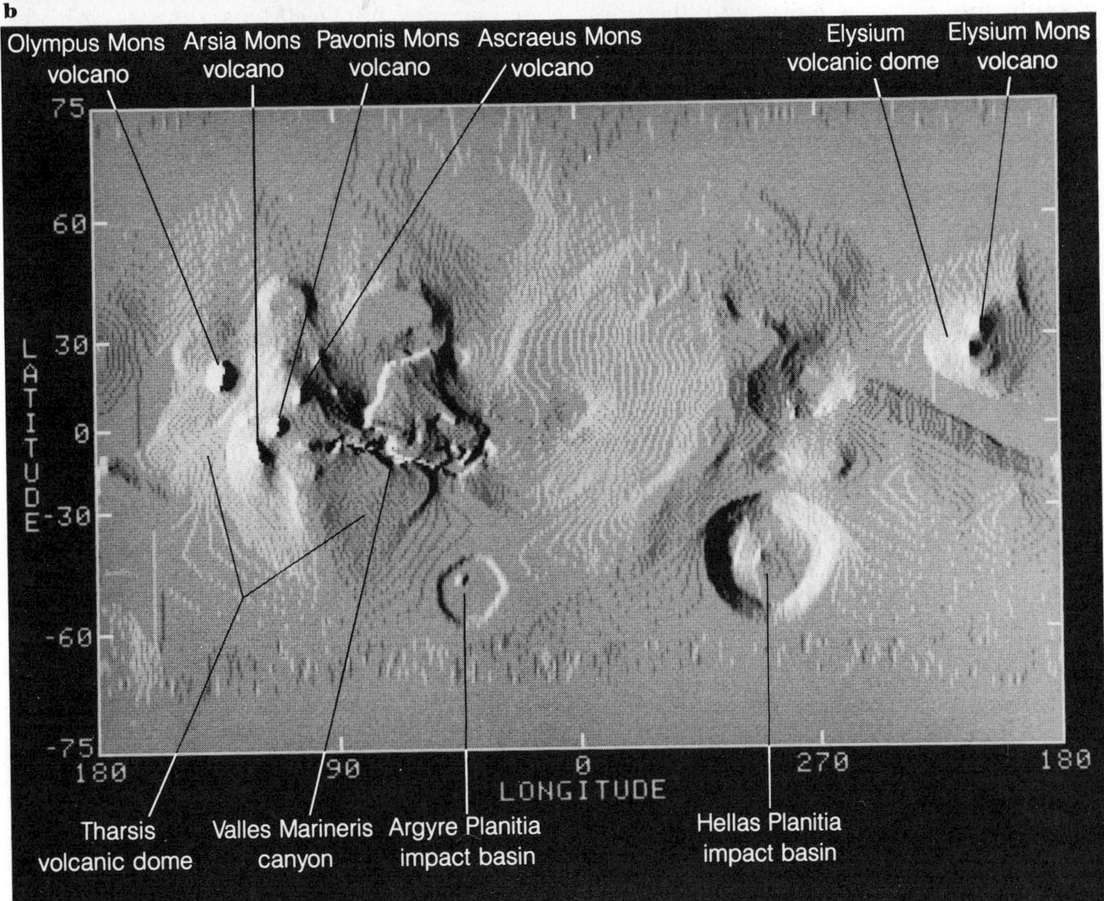

c

its atmosphere is mostly molecular nitrogen (N_2) and molecular oxygen (O_2), with variable traces of water vapor (H_2O). Its surface is very unusual, being temperate and 71% covered with liquid water.

The most important and unique characteristic of this planet—**Earth**—is its widespread life, both on land and in the oceans.

Evidence of this life is not easy to detect optically from other planets, but it can be detected by radio signals. Protecting Earth's life appears to be a major challenge to Earth's current civilization.

Moon

Earth has one natural satellite (Figure 2-9), the **moon,** whose character remained rather mysterious during centuries of telescopic study. The broad dark patches composing the "man in the moon" were first thought to be seas but were finally recognized as plains formed by lava flows. Early observers gave these plains Latin names such as Mare Tranquillitatis (Sea of Tranquillity) and Mare Imbrium (Sea of Rains). Rugged cratered regions of lighter color were called uplands or terrae. After some debate about a volcanic origin, researchers concluded that most of the larger craters were caused by meteorite impact. As shown in Figure 2-10, craters and lava plains have been mapped in detail on both sides of the moon. Especially noteworthy was detailed telescopic mapping by the U.S. Geological Survey in the 1960s in preparation for Apollo.

Unmanned landers in the 1960s revealed that the lunar surface is covered by a dusty layer called a **regolith**—a layer of powdery soil and scattered rock fragments created by eons of bombardment by meteorites.

The moon's history remained enigmatic until 1969, when astronauts explored the ghostly lunar landscape and brought back the first rock and soil samples. As of 1982 lunar samples have been returned from six Apollo landing expeditions and three sites visited by Soviet unmanned vehicles. These samples indicate that the moon has a bulk chemical composition roughly similar to that of Earth's outer mantle layers, with many rock types known to terrestial geologists. However, the moon lacks water and certain other compounds, while

Figure 2-9. Earth and the moon portrayed in scale as seen from a distance. This view emphasizes that the Earth-moon system forms a double planet. The moon is larger relative to its planet than any other satellite except probably Charon relative to Pluto. Its surface area is 7% that of Earth's, and about half of Earth's land area. (Painting by author)

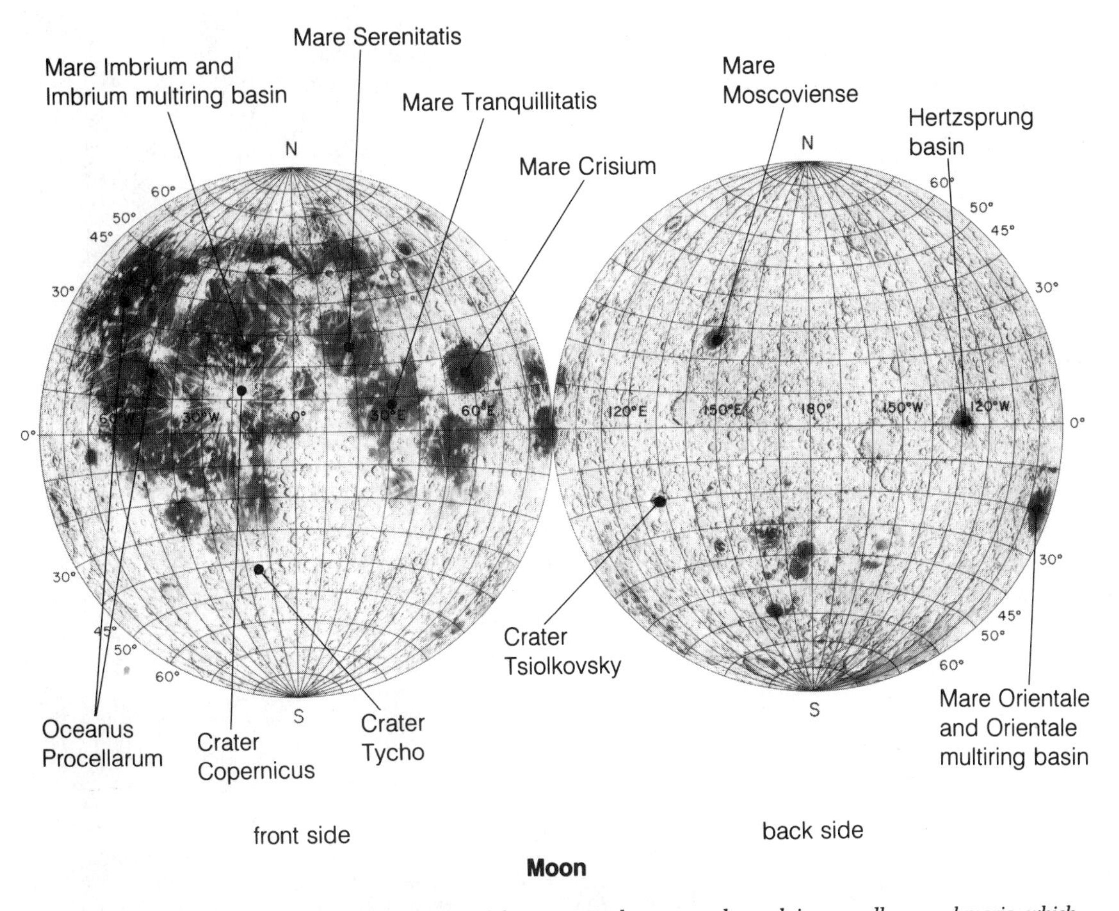

Figure 2-10. Airbrush map of front and rear faces of the moon. Dark areas are lava plains, usually named maria, *which occupy floors of large impact basins. (Base map courtesy R. M. Batson, U.S. Geological Survey)*

being enriched in others compared with Earth. These chemical differences are clues to the moon's origin and history and are the subject of much vigorous research. The rocks show that the moon, like Earth, formed about 4.5 billion years (Gy)* ago. The mode of formation is uncertain. The early moon underwent intense bombardment by meteorites, forming most of the many craters shown in Figure 2-11 by 4.0 Gy ago. The early moon was volcanically active, and about 3.5 Gy ago massive lava flows created the dark plains or maria. Most lunar volcanism ceased about 2 Gy ago. Most of the Apollo astronaut landings occurred on the maria, whose dusty, rock-strewn landscapes became familiar when the lunar exploration was broadcast live from the moon's surface into our living rooms.

*In the SI (International System) system of metric units, the prefix "giga," abbreviated G, stands for 10^9.

Mars

If Venus is Earth's sister planet, **Mars** is a smaller brother, half the size of Earth. It has an atmosphere with clouds and has ice deposits at its

Figure 2-11. The moon is dominated by rugged regions covered by craters mostly formed during intense meteorite bombardment about 4.5 to 3.5 Gy ago. In the upper left are dark, smooth lava plains called maria, *formed about 3.5 Gy ago. (NASA, Apollo 16)*

poles. It rotates in 24 h, just as Earth does. Unfortunately for the proposed family resemblance, as well as for future manned exploration, the atmosphere is very thin, cold, and dry. The famous red **surface coloration** is due to oxidized iron minerals. This color is prominent even when viewed with the naked eye from Earth. Compared to Earth's sea level surface pressure of 1000 mbar, Mars' surface pressure in most regions is about 5 to 10 mbar and would require a spacesuit.

As shown in Figure 2-12, intriguing dusky markings have been mapped on Mars since 1659. They are all the more intriguing because they change shape and intensity from season to season and year to year and have led to suggestions of life on Mars. Observers in the 1800s interpreted the seasonal changes as strong evidence of vegetation that darkened in summer and faded in winter.

Around 1800 observers such as William Herschel discovered clouds and white polar ice fields, called **polar caps.** He thought these features bolstered the philosophical idea called the **plurality of worlds**—the idea that other planets were rather like Earth, profusely inhabited with others of God's creatures. By the late 1800s, astronomers began to realize that the Martian air was colder, drier, and thinner than had been thought, but Darwin's 1859 theory of evolution and adaptation made it reasonable to suppose that the Martians might have evolved and adapted to their planet's conditions.

During the 1877 approach of Mars, the Italian observer Giovanni Schiaparelli called attention to the streaky Martian markings, which he called *canali.** Contrary to common belief, Schiaparelli did not discover these streaks; they had been drawn earlier as shown in Figure 2-12. However, Schiaparelli drew them differently, as straight or curved narrow lines.

Percival Lowell, a flamboyant American astronomer, pulled these data into a radically new

*This followed the tradition of naming dark lunar and planetary markings after bodies of water because of the mistaken assumption that they actually were bodies of water.

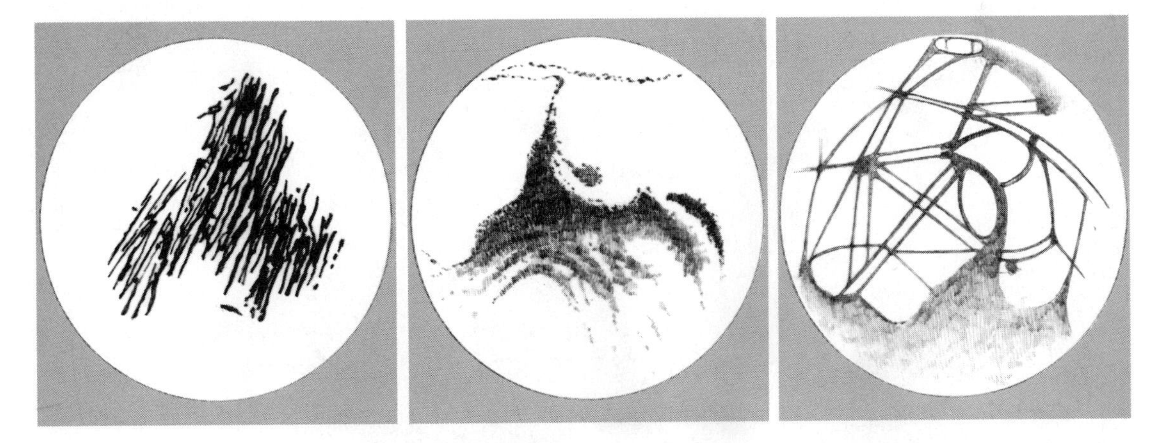

Figure 2-12. **(a)** *One of the earliest known sketches of Mars was drawn by Christian Huygens on November 28, 1659. The northward-extending triangle is believed to be a dark region still visible today and known as Syrtis Major. (After Huygens)* **(b)** *In a drawing by English observer W. R. Dawes during the 1864–1865 opposition, Syrtis Major extends northward but with a streaky extension of a type later called a canal. Clouds covering the north polar ice cap are outlined by a dotted line at the top. (After Dawes)* **(c)** *Mars as drawn by Italian observer Giovanni Schiaparelli on June 4, 1888, with the north polar ice cap at top. The dark triangle of Syrtis Major again extends north with a streaky, dark tail. Schiaparelli first popularized the canals, shown by him here as nearly straight lines and line pairs covering Mars. However, spacecraft show that the canals do not exist in this form. (After Schiaparelli)*

conception of Mars. Lowell saw the lines, which he called **canals,** as even more sharply defined than Schiaparelli had. "Each line," Lowell wrote in 1895, "not only goes with wonderful directness from one point to another, but at the latter spot it contrives to meet, exactly, another line." The Martians, said Lowell, were having trouble surviving because the weak gravity of their planet was allowing its water to leak off slowly into space.* So the Martians had constructed a vast network of canals to carry the spring runoff from the melting polar ice down to the warmer equatorial regions, where they could cultivate the vegetation, seen as the dark, variable markings.

Lowell's was a wonderful theory, but it was all wrong. Its underpinnings, the canals, essentially do not exist. As shown in Figure 2-13, other astronomers saw the canals not as geometric lines but as splotchy patterns and boundaries of different-toned regions. And as seen in Figure 2-14, later photos confirmed only splotchy markings and a few broad streaks, but no fine lines. The Martian canal affair occurred because the eye tends to interpret rough alignments as linear streaks. Lowell was especially inclined toward this and even drew lines on Venus.

Lowell's theory, though it collapsed, had the virtue of forcing scientists and intellectuals to recognize that theories of planetary and biological evolution raise the probability of life on other worlds. The ramifications of this idea shook intellectuals and the public at large and remain exciting today.* And the theory of primitive plant life on Mars remained viable up to at least 1976.

Observations of Mars continued, and maps were drawn showing many of the major dark and light features that were named by Schiaparelli for mythical lands and historic regions around

*This idea of gases escaping into space more easily on weak-gravity planets is scientifically correct, as we will see in Chapter 12.

*A more detailed, nontechnical history of the changing theories is given by Hartmann and Raper (1974). An excellent biography of the colorful Lowell is given by Hoyt (1976).

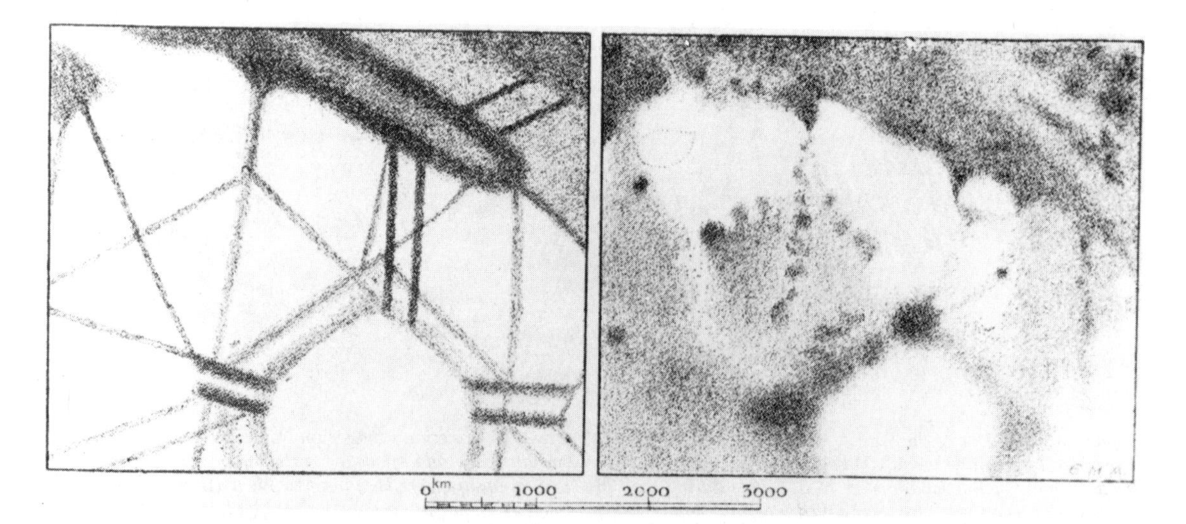

Figure 2-13. The explanation of the canals proposed by the French observer E. M. Antoniadi around 1930. Under mediocre observing conditions, some observers perceive canallike streaks as Lowell described them (left). Under ideal observing conditions, observers with large telescopes find that these streaks break apart into a complex pattern of mottling and dark patches. The canals result from poor observing conditions combined with the eye's tendency to connect dots into lines.

the Mediterranean (Figure 2-15, compare with relief map in Figure 2-8c). By the early 1960s, observations had established that the Martian ground temperature exceeds freezing only on the warmest summer days (though the air temperature usually stays below freezing), that the surface is dry, and that the air has only minuscule traces of water vapor.

On July 15, 1965, the first close-up spacecraft photos of Mars were taken by the American spacecraft **Mariner 4.** There were no signs of dying Martian cities or canals—only a moonlike cratered surface. In 1969, Mariners 6 and 7 flew by, showing more craters and a peculiar depressed, hummocky formation that was named **chaotic terrain**. By now the spacecraft data had confirmed Mars' thin CO_2 atmosphere as having a surface pressure of only around 6 mbar. Although the nature of the changing dark markings was still unknown, many astronomers now concluded that Mars was dead and moonlike—similar to the moon but with a little wind to blow the dust around.

On November 14, 1971, Mariner 9 became the first spacecraft to go into orbit around Mars, beginning a highly successful mapping program that charted the whole planet in detail for the first time. The **resolution**, or scale, of the smallest visible details was as low as a few hundred meters. On the arid plains of Mars, Mariner 9 photographed many examples of landforms that looked astonishingly like dry riverbeds! Examples appear in Figure 2-16. Some of these landforms emanated from the previously discovered chaotic terrain. After much controversy, most planetary geologists concluded that these features, called **channels** (not to be confused with the illusory canals), were carved by running water. Their evidence, which will be discussed in more detail in Chapters 11 and 12, suggests that the climate of Mars may have been more clement in the ancient past.

Mariner 9 also indicated that substantial amounts of H_2O still exist, frozen at the poles under a transient winter cap of CO_2 snow and accompanied by stratified polar sedimentary deposits. Did this mean that water once flowed on Mars? Did it mean that life might have evolved on Mars in the past when liquid water was available? Since no one had yet seen Martian details smaller than a couple hundred meters, astronomers wondered once again if the surface supported plants or even animals after all!

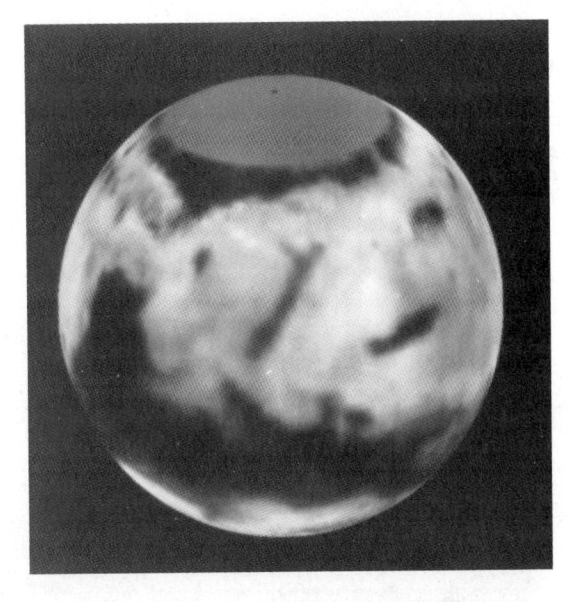

Figure 2-14. Viking orbiter instruments measured albedos (reflectivities) of Martian surface regions; these were computer-plotted onto globes to show Martian markings at higher resolution than can be seen from Earth. A few broad, streaky markings (center) are prominent in some seasons of some years, and may have contributed to the myth of the canals. The dark feature at left center is Syrtis Major, the most prominent marking on Mars. North polar cap has been arbitrarily shaded gray and the contrast is enhanced. (Jet Propulsion Laboratory, courtesy Jerome Apt)

The first human-made devices to reach the surface of Mars might have answered these questions but failed before or after touchdown. These were three Soviet spacecraft: Mars 2 crashed in November 1971; Mars 3 landed but failed 20 s after touchdown in December 1971; Mars 6 failed moments before touchdown in February 1974.

July 20, 1976, the seventh anniversary of the first lunar landing, was a better day: The Viking 1 lander parachuted into the atmosphere of Mars

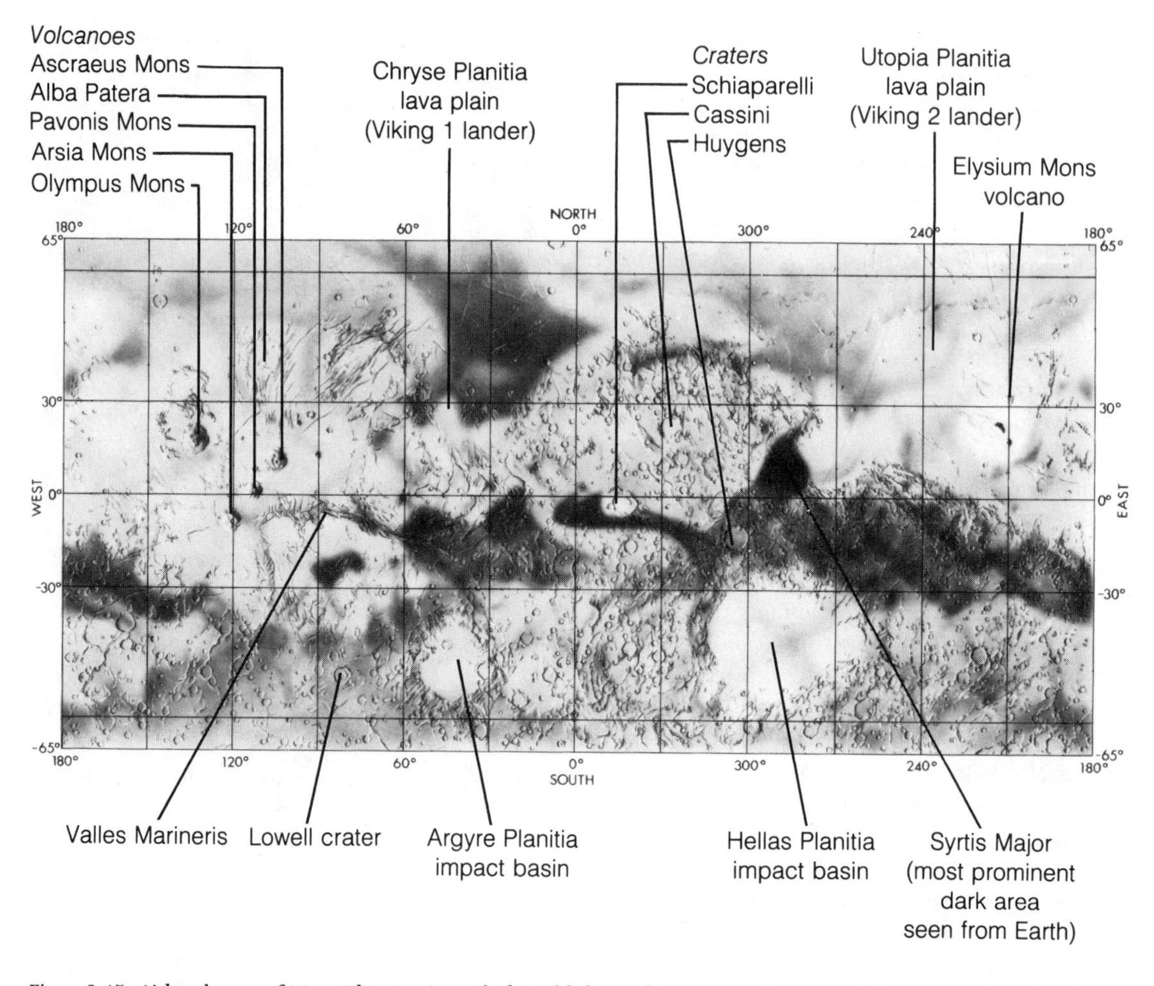

Figure 2-15. Airbrush map of Mars. The prominent dark and light markings are semipermanent features seen from Earth, created mostly by dust deposits. The underlying topographic features, such as craters and mountains, were mapped only in the 1970s by spacecraft. (Base map courtesy R. M. Batson, U.S. Geological Survey)

and made the first successful touchdown on the planet (Figure 2-17). In the following months, Viking 1 and its sister ship, Viking 2 (which landed September 3, 1976) found a striking desert landscape devoid of any obvious life. There was not even any organic material in the Martian soil, measured to a precision of a few parts per billion. Although three experiments deisgned to seek life found peculiar reactions in the soil at both landing sites, most scientists relate those reactions to unusual soil chemistry. Nonetheless, the Viking landers and their two mother ships in orbit around Mars did make fascinating discoveries that we will discuss in later chapters.

Mars has two small satellites, Phobos and Deimos, discovered in 1877 by the American astronomer Asaph Hall. Curiously, literary works such as Swift's *Gulliver's Travels* (1720) and Voltaire's *Micromegas* (1750) referred to two moons of Mars 150 y before they were discovered! This has led some pseudo-science writers to suggest occult knowledge of the two moons centuries ago, but the explanation is simpler. Johann Kepler, who discovered the laws of planetary motion around 1610, believed in numerology and suggested that if Earth had one moon and Jupiter four (the four discovered by Galileo at that time), then Mars ought to have two to maintain the progression. Later spacecraft have searched for smaller satellites and found none down to diameters of 1.6 km (Pollack and others, 1973).

Close-up photos of Phobos by Mariner 9 and later, more detailed close-ups of Phobos and Deimos by Viking orbiters show irregularly shaped satellites about 20 × 28 km and 10 × 16 km, respectively, with heavily cratered surfaces, as shown in Figure 2-18.

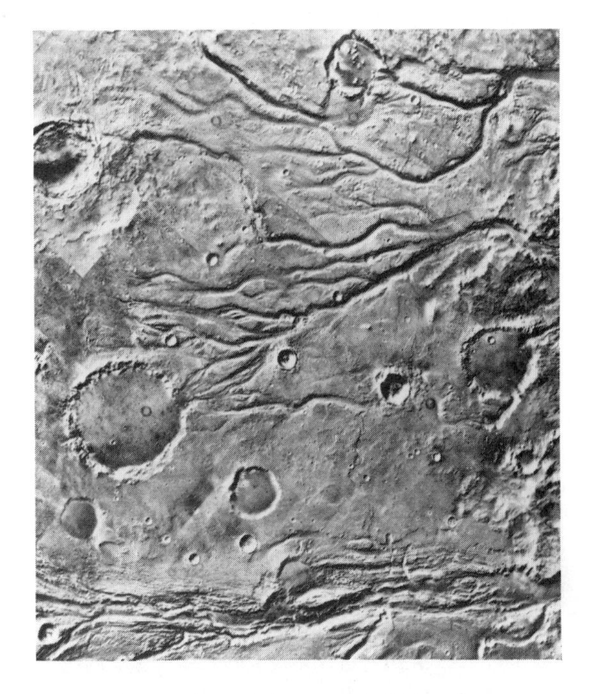

Figure 2-16. The channels (not canals) of Mars: flow features that may have been eroded by running water. Like arroyos in Earth's terrestrial regions, these channels have tributary systems and get wider and deeper in the downhill directions. (NASA, Viking orbiter)

Jupiter

Jupiter is by far the largest planet, having more mass than all the other planets put together. It has more than three times the mass of the next largest planet, Saturn, and about a 20% larger diameter. Dynamically the solar system can be thought of as composed of two main bodies, the sun and Jupiter; Jupiter has 0.001 the mass of the sun, and the other bodies are negligible.

Figure 2-17. The first photograph transmitted from the surface of Mars showed soil near the footpad of the Viking 1 lander. During the touchdown, some soil was blown into the footpad (lower right). The camera scanned vertically, beginning at the left and proceeding to the right, within minutes after touchdown. Light and dark streaks at left, therefore, may be light haze and shadows caused by dust clouds kicked up during the landing. The largest rock (center) is about 10 cm (4 in.) across. (NASA)

Figure 2-18. A crescent view of Mars' satellite, Phobos, with the night side faintly illuminated by Mars. The sun is out of the picture to the left, and Mars is out of the picture to the right. The irregular shape of Phobos can be traced around the entire outline. (NASA, Viking Project; courtesy T.C. Duxbury)

Jupiter has a dense atmosphere of molecular hydrogen (H_2, 79% by mass), helium (19%), water vapor (H_2O), methane (CH_4), and ammonia (NH_3), characterized by bands of clouds parallel to the equator, shown in Figures 2-19 and 2-20. Dark cloud bands are called **belts** and brighter ones **zones.** The belts and zones are semipermanent and have been named, as shown in Figure 2-20. The clouds show a variety of colors—browns, tans, yellows, and reds. Various kinds of features typify these cloud belts. Large irregularities, such as dark, oval-shaped clouds that may last for years, are called **disturbances.** Small dark areas (at the limits of telescopic resolution from Earth—around 2000 km across) are called **knots;** delicate, wispy streaks are called **festoons.**

Another long-lived feature is the **Great Red Spot**, prominent in Figures 2-19 and 2-21. Probably first seen by Robert Hooke in 1664 or Giovanni D. Cassini in 1665, it was named in 1878 when it became very prominent and was rediscovered (Peek, 1958; Chapman, 1968). Observations by Reese and Smith (1968) and time-lapse movies by Voyagers 1 and 2 in 1979 show that the Great Red Spot is characterized by circulating currents—small clouds caught in the Red Spot spiral around counterclockwise like a leaf caught in a whirlpool. It is a giant whirlpool indeed; somewhat variable in size, it can reach 4 Earth diameters!

Jupiter and Saturn present an interesting problem with respect to rotation. Different parts of their atmospheres can move independently (with accompanying shear and turbulence). The equatorial zones rotate faster than the higher latitudes.

In the case of Jupiter, two different rotation rates are used to keep track of the cloud markings: $9^h50^m30^s$ for the equator (System I) and $9^h55^m41^s$ for the high latitudes (System II). The famous Great Red Spot drifts along with a variable period, drifting sometimes ahead of and sometimes behind other features in System II. Peek (1958) gives values $9^h\ 55^m\ 31^s$ to $9^h\ 55^m\ 44^s$ over a 76y period. Which rate represents the true rotation of the solid planet? Probably none. Clearly, the Great Red Spot cannot be related to a fixed surface feature, and radio signals originating from electromagnetic phenomena in the intense magnetic field around Jupiter give a third rotation rate used to suggest an internal (System III) rotation period of $9^h\ 55^m\ 30^s$.

The first spacecraft to Jupiter, **Pioneer 10,** left Florida at the fastest speed of a vehicle until that time, sailing past the moon's orbit in only 11 h (compared to the 3 d taken by Apollo spaceships). After 21 mo, on December 4, 1973, Pioneer 10 sailed "under" Jupiter's south pole and then "up," out of the ecliptic plane at a velocity high enough to escape the solar system permanently and sail on toward the stars. Pioneer 11 (December 1974), Voyager 1 (March 1979), and Voyager 2 (July 1979) were the next vehicles to fly through the Jupiter system and then on toward the outer

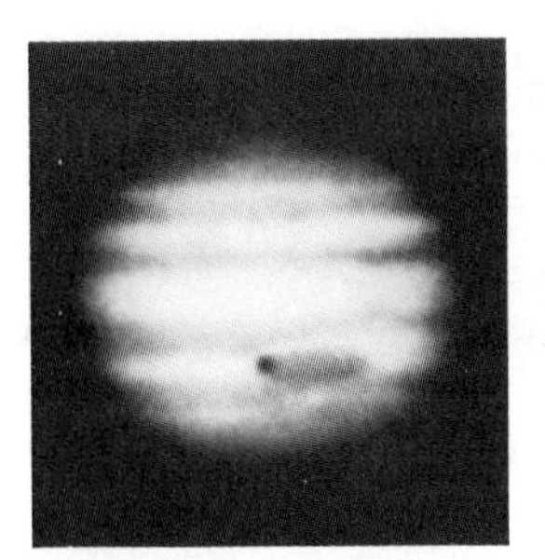

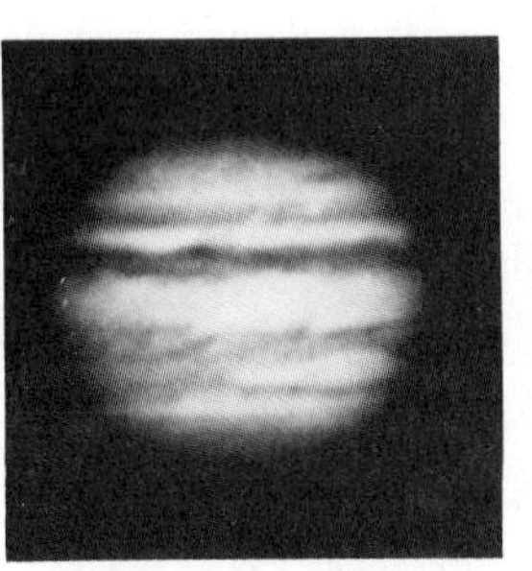

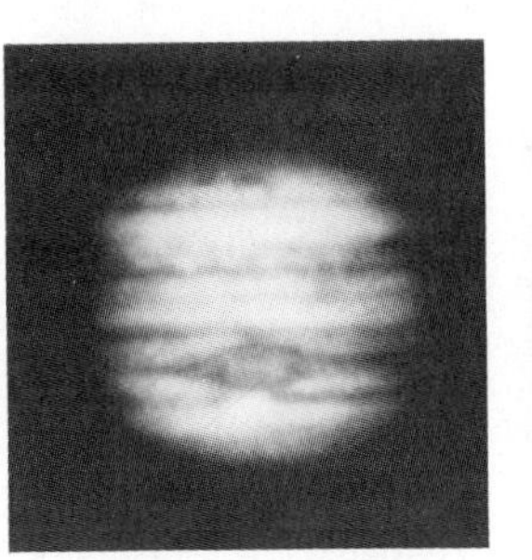

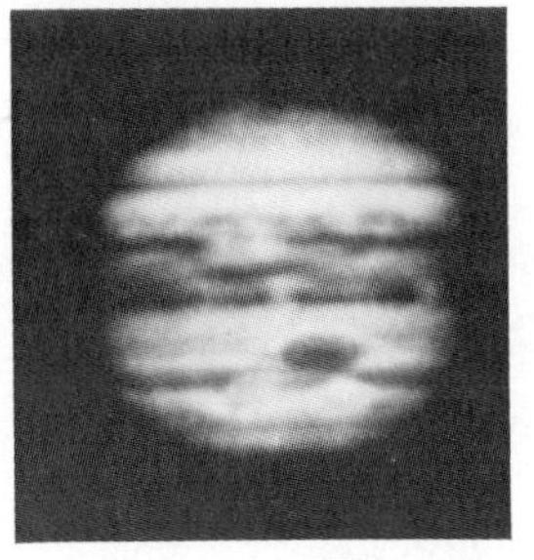

Figure 2-19. Photographs of Jupiter spanning 81 years, showing the changing array of Jupiter's belts and zones. The dark north temperate belt is relatively permanent, but the equatorial zone changes from bright (1891) to dark (1972). Many of these images show the Great Red Spot. (Lowell Observatory)

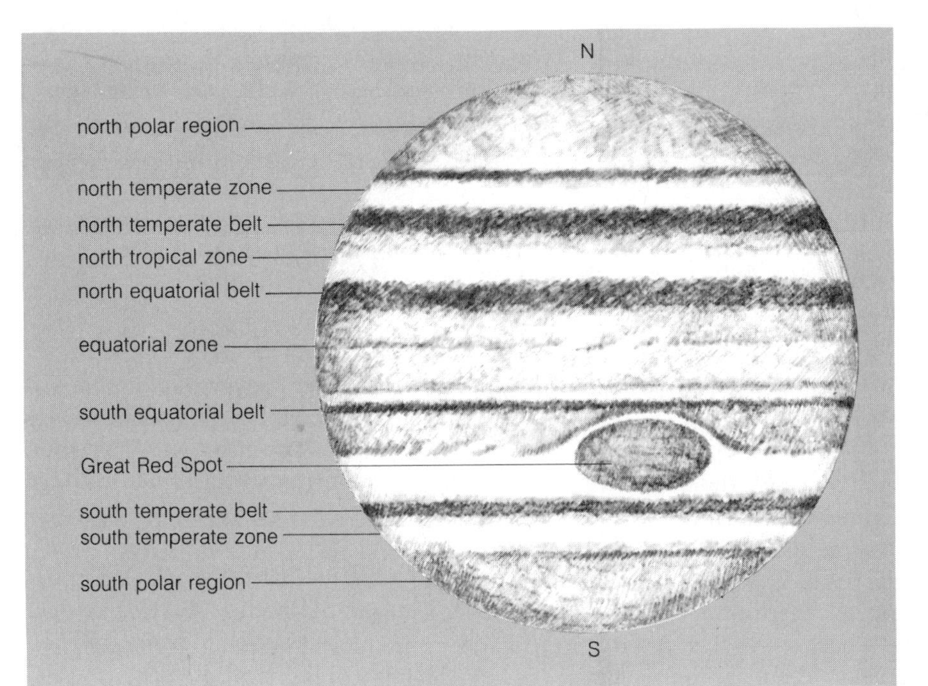

Figure 2-20. Semipermanent cloud formations of Jupiter, visible from Earth through small telescopes.

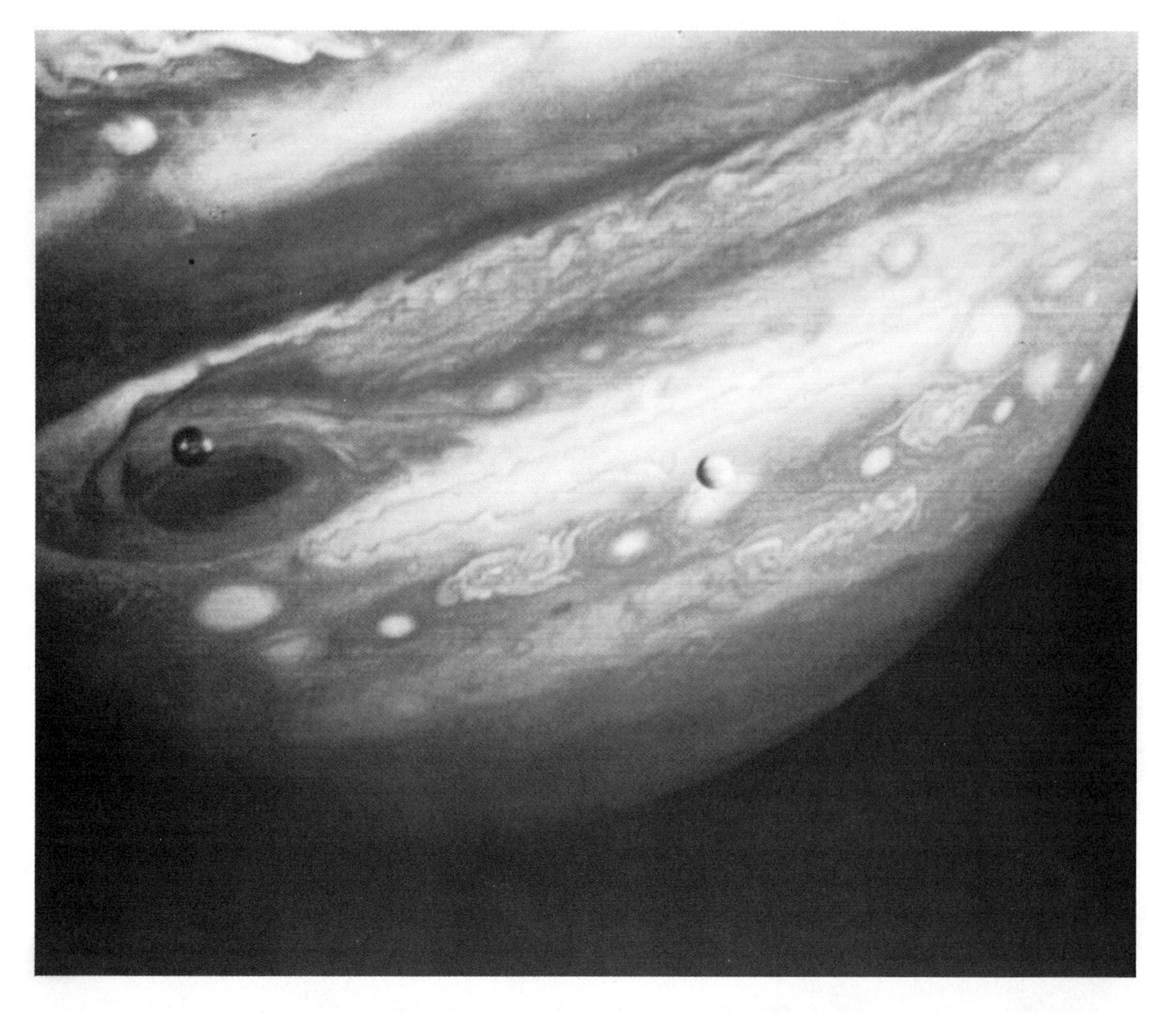

Figure 2-21. Jupiter looms beyond two of its satellites. Europa is the lighter moon (right), and Io the darker moon silhouetted against the Great Red Spot. Numerous other cloud formations on Jupiter are prominent in this Voyager 1 view from a distance of 20 million kilometers. Europa and Io are about the size of our moon. (NASA)

solar system. As shown in Figure 2-21 and later chapters, these vehicles took superb photos of Jupiter and the moons and gathered abundant additional data. In fact, Voyager 1 discovered a ring around Jupiter, fainter and narrower than the famous rings of Saturn and probably composed of fine, rocky particles.

Jupiter has a complex system of at least 16 satellites outside the rings, somewhat resembling a small solar system. There are four large moons, called the **Galilean satellites** because they were discovered by Galileo. The discovery date was January 7, 1610; and on the next night they were independently discovered by the German astronomer Marius, who named them *Io*, *Europa*, *Ganymede*, and *Callisto* (in order outward from Jupiter) after associates and paramours of Zeus, the Greek version of the Roman god Jupiter. Jupiter's moons are also labeled with numerals; the Galilean satellites are J1, J2, J3, and J4 in the order given above. Subsequent numbers are assigned in order of discovery.

Within the orbits of the Galilean moons is the largest remaining moon, Amalthea (numbered J5), about 155 × 270 km in size, dark reddish-gray in color, and heavily cratered. From its surface, Jupiter would present an enormous visage in the sky, subtending an angle of about 46°, as shown in Figure 2-22. The Voyagers discovered three other small moons in this region. Two are about 30 to 40 km across, very dark in color, and located between the rings and Amalthea (Jewitt, 1979). The other is about 75 km across and is located between Amalthea and Io.

Beyond the Galilean moons are two peculiar groups of small moons. The inner group includes J13, J6, J10, and J7, in orbits from about 11 000 to 12 000 km from Jupiter, and all inclined from 24° to 29° to the planet's equator. The second group, including J12, J11, J8, and J9, all lie at distances of about 21 000 to 24 000 km and inclinations of 147° to 164°. (The use of an inclination figure greater than 90° is a convention for indicating a retrograde direction of revolution. Thus, an inclination of 147° is equivalent to an inclination of 180° − 147° = 33° but with retrograde orbital motion.) Why the outer satellites of Jupiter should be arranged in two such tidy groups rather than randomly scattered is a mystery to which we will return in Chapter 5. The surfaces of at least some of these moons are black owing to a surface soil rich in carbonaceous minerals like those in certain dark asteroids and meteorites.

Charles Kowal, the Palomar observer who discovered J13, photographed another apparent outer moon in 1975. The image was rephotographed on several occasions in 1975 by Kowal and by Elizabeth Roemer at the University of Arizona; but the number of images was not sufficient to determine the orbit or even the direction of revolution. Despite further searches, this satellite has not been seen again (Brian Marsden, private communication, June, 1978). Recent orbital calculations (for example, Carusi, Valecchi, and Kresak, 1981) suggest that cometary or asteroidal bodies are sometimes captured by Jupiter into temporary satellite orbits lasting months or years; Kowal's 1975 moon may have been one of these.

Figure 2-22. Jupiter's closest satellite, the small moon Amalthea, should provide a spectacular view of Jupiter. Jupiter dominates the sky, subtending about 46°. (Painting by James Hervat)

Small Worlds of the Outer Solar System: A Simplified Overview

Telescopic views of even the largest satellites of the outer solar system revealed only pinhead-sized disks (less than 2 arc seconds across) on which visual observers reported dusky shadings and spectroscopists usually reported H_2O ice with admixtures of soil. Astronomers tended to assume that these worlds would be rather alike—perhaps resembling icy versions of our own moon. Close-up photos of Jupiter's and Saturn's moons by Voyagers 1 and 2 revealed the opposite—an astounding variety from one world to the next, as if each were attempting to assert its own uniqueness.

The variety might make a theorist throw up his hands and conclude that each world is governed by laws unto itself. But we can make some sense of the pattern by keeping in mind the following simple model. Theories of mineral condensation in the primordial solar system (which we will discuss in more detail in Chapter 5) suggest that the main constituents formed in the cold, outer parts of the system would be frozen water and a type of dark rocky soil rich in black carbon minerals. Spectrophotometric observations seem to confirm that either ice or dark soil, or a mixture of the two, are the main constituents visible on the surfaces of moons and asteroids from the outer asteroid belt outward. Note that a chunk of frozen water is relatively stable against sublimation by solar heating only beyond about 4 AU (Lebofsky, 1975). If mixtures of ice and black soil are finely powdered and mixed in a regolith soil layer, the material will look dark if the soil content is more than a few percent, because the opaque black minerals efficiently absorb the light. Thus, a primitive surface of, say, 50% ice and 50% carbonaceous soil would look dark. Similarly, any surface in which the carbonaceous component had been concentrated would look like dark soil (even though it might have some ice content). On the other hand, if heating occurred, the denser soil would sink and watery "lava" could coat the surface with regions of relatively pure bright ice or ice/soil layers (Hartmann, 1980). If enough heating occurred to evaporate most of the water, components such as sulfur compounds could be left on the surface (Fanale, Johnson, and Matson, 1974).

As we will now see, this "salt-and-pepper" model ("salt"-bright ices; "pepper"-black soils) gives a first-order explanation of the range of surfaces that have evolved in the outer solar system. It is easiest to apply this idea if we describe the four Galilean moons, starting with the outermost and working our way inward.

Callisto

Of the four large Galilean moons of Jupiter, the outermost one, **Callisto,** most closely resembles a terrestrial planet or satellite.* At 5000 km across, it is about 2% larger than Mercury. Voyager photos and Earth-based spectrometric observations indicate a moderately dark surface of silicatelike soil mixed with 30% to 90% H_2O ice by mass (Clark, 1980). The soil may be carbonaceous material that may contain further water chemically bound in its minerals. The Voyager 1 and 2 flybys in 1979 revealed a heavily cratered surface resembling the uplands of the moon or Mercury, as seen in later chapters. The low mean density, 1790 kg/m^3 (compared with 1000 for water and ice, and about 3000 for rock) is lowest of the Galilean moons, indicating that the interior consists of about one-fourth ice by mass. Bright rims and rays of some craters suggest that impacts have blown away a dark soil surface and exposed brighter icy material underneath. The largest feature is an enormous multiring bull's-eye, with outer rings spanning 2400 km. This and other smaller, similar features are probably caused by impacts on the icy crust, later modified by isostatic leveling involving a watery substrate (McKinnon and Melosh, 1980).

*References for the discussion of the various Galilean worlds include Pollack and others (1978) and Voyager 1 and 2 results described by Smith and Voyager Imaging Team (1979), along with other papers in *Science, 204,* no. 4396 (June 1, 1979) and *Science, 212,* no. 4491 (April 10, 1979).

Ganymede

Ganymede, the largest and next moon inward from Callisto, has a quite different surface. With a 5270-km diameter, it is 8% larger than Mercury and 75% the size of Mars. It has a moderately bright surface containing an average of 90% frozen water or frost by mass (Clark, 1980). Voyager close-ups reveal some provinces of old, cratered, dark, Callisto-like terrain. These are broken and cut through by light areas and fracture zones. Offsets have occurred on some fractures, but overlying undeformed craters indicate that the fracture zones are moderately old and inactive. The bulk density is only 1930 kg/m^3. These observations (plus theoretical work—see Chapter 10) suggest that Ganymede once had a thin icy crust floating on a watery interior. Early motions cracked the ice, but later cooling produced today's thicker, more rigid crust.

Europa

Europa, the smallest Galilean moon, is still different. Twelve percent smaller than our moon, it has a bright, featureless surface except for a system of shallow, dusky grooves. It looks like an icy billiard ball with lines drawn on it. The near absence of impact craters suggests that the surface is relatively young and that resurfacing processes may be continuing in relatively recent times. The surface composition is more than 90% ice by mass, but the bulk density of 3030 kg/m^3 implies that rocky material composes most of the interior. The evidence strongly suggests heating by an uncertain mechanism: Eruptions of water formed a surface "ocean" of relatively pure ice, cracked here and there by final adjustments to Europa's shape as the world cooled.

Io

Io, the innermost Galilean moon, is one of the strangest worlds in the solar system. At 3640 km across, it is only 2% larger than our moon. Its density is 3530 kg/m^3, highest of the Galilean worlds, implying an interior of rocky material with very little or no ice content.

A long history of mystifying discoveries indicated Io's strange nature even before the Voyager photos revealed Io's unique properties. In 1964, observers reported Io to be anomalously bright for about 15 min after some, but not all, eclipses (Binder and Cruikshank, 1964). This effect was seen occasionally in later years. Also in 1964, radio observers discovered that bursts of radio radiation from Jupiter were correlated with the position of Io in its orbit relative to Earth (Bigg, 1964). Spectral observers in the next few years found that unlike the other satellites, Io lacks water frost or ice on its surface. High-resolution telescopic photos taken in 1973 show reddish and white patches (Minton, 1973). Observers then discovered that Io is surrounded by a thin, yellow-glowing cloud of sodium atoms (Brown, 1973; Trafton, Parkinson, and Macy, 1974), many Io diameters across and extending along Io's orbit. This cloud is too faint to be seen through telescopes but probably bright enough to be seen as an aurora by an observer on Io. Sulfur and oxygen atoms have subsequently been found associated with Io also. The sodium, sulfur, and oxygen atoms are spread along Io's orbit, and neither Io nor any of the other satellites has any substantial atmosphere of its own. Telescopic observers, some months before Voyager 1 reached Jupiter, noted mysterious flareups of infrared (5μm wavelength) emission, a possible indication of shortlived hot spots, but the observers did not guess the real cause (Witteborn and others, 1979).

Many of these findings have now been clarified. For example, Io orbits within Jupiter's intense magnetic field and is coupled to Jupiter by electric currents through this field, explaining its influence on the directions of radio emissions arising in Jupiter's magnetic field.

Voyager close-ups in 1979 revealed an extraordinary world of mottled patches in colors of red, orange, yellow, white, and black. Reporters compared Io to a pizza. One Voyager investigator quipped that he didn't know what was

Figure 2-23. Photo on which Io's active volcanoes were discovered. The bright hemisphere is lit by the sun, while the dark side is faintly lit by Jupiter. Bright clouds (left and right edges of the crescent) are plumes of debris shot as much as 260 km above Io's surface by active volcanic eruptions and falling back through a near vacuum onto Io's surface. (NASA)

wrong with Io, but it looked as if it might be cured by a shot of penicillin. Linda Morabito, a Voyager flight engineer, processed certain photos and discovered fountainlike clouds of material erupting as much as 260 km above Io's surface, as seen in Figure 2-23. This was the first discovery of active volcanoes on another world and confirms that Io has an intensely heated interior. As we will see later, this heat comes from gravitational forces that cause tidal stresses inside Io. The heating drives the volcanism. The volcanoes are hot (~500K or 440°F) and their outbursts account for infrared flareups seen from Earth. It also has driven off water, leaving a rocky interior composition and a surface dominated by sulfurous lavas, explaining the high density and unusual colors. Ions striking the surface dislodge atoms of sulfur and sodium that are excited by sunlight, explaining the yellow auroral glow around Io. Condensations of compounds such as SO_2 when cooling occurs during eclipses may explain some of Io's reported brightness changes. In short, Io is a world with a colorful red, orange, and white surface, an intermittent yellow-aurora sky, and volcanoes ejecting vast fountains of gas and sulfur-ash debris.

Saturn

The globe of **Saturn** is rather like Jupiter but smaller, more flattened, with less-prominent cloud markings (Figure 2-24). Yellowish and tan cloud belts parallel the equator as seen in Figures 2-24 and 2-25. Occasional bright and dark markings disturb these belts, but a greater depth of overlying haze makes them less visible than on Jupiter. Spectroscopic studies from Earth and Voyagers 1 and 2 identified the main gases as molecular hydrogen (88.2% by mass) and helium (11%). Minor identified components include methane (CH_4), ammonia (NH_3), and ethane (C_2H_6).

Saturn has the lowest mean density of all the planets, 710 kg/m^3, indicating a very hydrogen-rich interior with few rocky materials. Given a sufficiently immense ocean, Saturn would float!

Saturn is best known for its **rings,** shown in Figure 2-26. When Galileo first turned his crude telescope on Saturn in 1610, he could not see the rings clearly. They seemed to be fuzzy appendages on each side of the globe, and he drew Saturn as a triple planet. Saturn's true nature remained controversial until the 1660s, when the rings were clearly seen to encircle the planet (Alexander, 1962). In 1859, the Scottish physicist James Clerk Maxwell showed that they could not be a solid plate but must be made up of innumerable particles, each moving in an independent orbit around Saturn. The American astronomer James Keeler (1895) was the first person to prove this when he detected the varying orbital velocities of different parts of the rings; it was one of the first triumphs of spectroscopic astronomy. Although the rings are 270 000 km (170 000 mi) in diameter, they are extremely thin, probably no more than a few hundred meters thick. Spectra show that the ring particles are com-

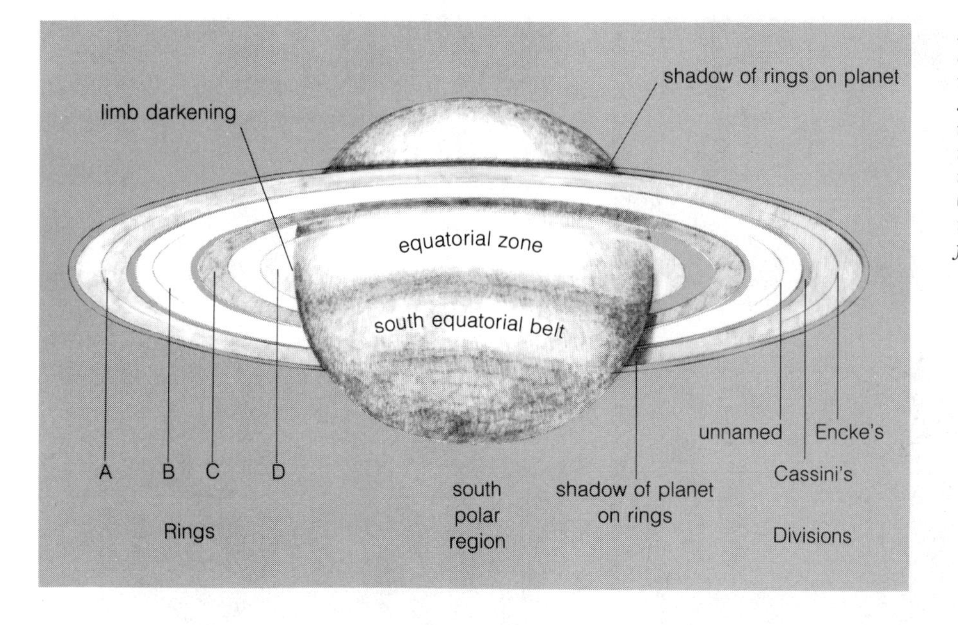

Figure 2-24. Features of Saturn. Telescopes with apertures as small as 5 cm (2 in.) will show the rings; telescopes larger than about 24 cm (10 in.) will sometimes show all these features.

posed of, or covered by, frozen water (Pilcher and others, 1970; Kuiper, Cruikshank, and Fink, 1970), possibly with traces of silicate or carbon minerals (Lebofsky, Johnson, and McCord, 1970; Pollack, 1975; Cuzzi and Pollack, 1978). Radar and photometric studies (Cuzzi and Pollack, 1978), and Voyager radio data show that common ring particles range from a few centimeters to a few decameters across; they are hailstones from the size of a ping-pong ball to the size of a house. A Voyager search for larger ring moonlets yielded negative results at sizes larger than a few kilometers.*

The rings are divided as sketched in Figure 2-24. The dusky outer ring, called ring A, is separated by a gap from the brighter ring B. The gap is called **Cassini's division.** A narrower division, called Encke's division,** cuts ring A. On the inner side of ring B is a very faint, tenuous ring, C. The French observer Guerin (1970) announced a still fainter and more tenuous innermost ring, D, which extends from the planet toward ring C; the Voyagers confirmed a D ring here, but not bright enough to have been seen from Earth (Smith, 1990). Rings A through D, together with the Cassini and Encke divisions, have been called the classical ring elements (Smith and others, 1981); they were charted from Earth. Pioneer 11, flying by in 1979, discovered a narrow ring just outside the A ring; it came to be called the F ring. Beyond it lies a thin G ring and a very diffuse E ring that extends beyond the orbit of Enceladus.

Saturn's Satellites

Saturn's satellite system, like Jupiter's, is extensive and includes exotic worlds. The system includes a host of small moons near the rings, a small outer moon, and a grouping of larger moons at intermediate distance, including the giant moon, Titan. As with Jupiter's Galilean system,

*Important Voyager results on Saturn, the rings, and the satellites have been published in *Science, 212,* no. 4491 (April 10, 1981) and *Science, 215,* no. 4532 (Jan 29, 1982). See also Beatty and others (1981).

**Osterbrock and Cruikshank (1982, Tucson Saturn Conference, May 11–15) show that the actual outer-A-ring gap commonly associated with this name was really discovered by James Keeler in 1888. Encke had earlier described a more diffuse shading in mid-ring A.

Figure 2-25. Varying aspects of Saturn during its 29-y orbit around the sun, as photographed with Earth-based telescopes. Because the rings maintain a fixed relation to the ecliptic, the Earth-based observer sometimes sees them from "above" and sometimes from "below" the ring plane. This series shows half of the 29-y cycle, including two views of the apparent disappearance of the rings as Earth passes through the ring plane. (Lowell Observatory)

we can gain some insights by starting with the outermost moon and working inward.

The outermost moon is the small, dark satellite **Phoebe,** about 220 km across. Its orbit is retrograde and the most highly inclined of the satellite family—leading some theorists to suggest that it may be a captured interplanetary body rather than a native Saturn satellite.

Next inward from Phoebe is the strange moon, **Iapetus.** It is a moderate-sized body, about 40% the size of our moon. Its diameter is 1440 km. Its strangeness was recognized as early as 1671 when G. D. Cassini discovered that he could see Iapetus easily when it was on the west side of the planet, but not when it was on the east side! Larger telescopes revealed that Iapetus keeps one face toward Saturn so that one hemisphere leads and one trails. As seen in Figure 2-26, the leading hemisphere is covered with very dark soil, but the trailing hemisphere (visible when Iapetus is on the west side of the planet) is nearly five times as bright and covered with ice or frost. Voyager 2 photos in 1981 revealed that the boundary between the dark and light sides is irregular and sharply defined, sometimes along contours of crater rims or floors. One theory is that dark dust knocked off Phoebe by meteoroids spirals in toward Saturn and hits Iapetus' leading side, enriching that side's soil with dark material. But the final answer remains uncertain.

As if to reemphasize the rule that each world is unique, the next satellite is again different. **Hyperion** is a biscuit-shaped chunk about 350 km across and 200 km thick. Voyager 2 photos (Figure 2-27) showed its irregular, cratered shape and led to controversy over whether its long axis is lined up with Saturn, as would be expected if gravitational tidal forces had acted for a long time. Perhaps a recent major impact knocked it

Figure 2-26. Saturn's unusual satellite Iapetus has dark soil on its leading hemisphere (left) and bright icy material on its trailing hemisphere (right). A dark crater rim is seen on the right side. Cause of the strange soil/ice distribution is uncertain. (NASA, Voyager 1)

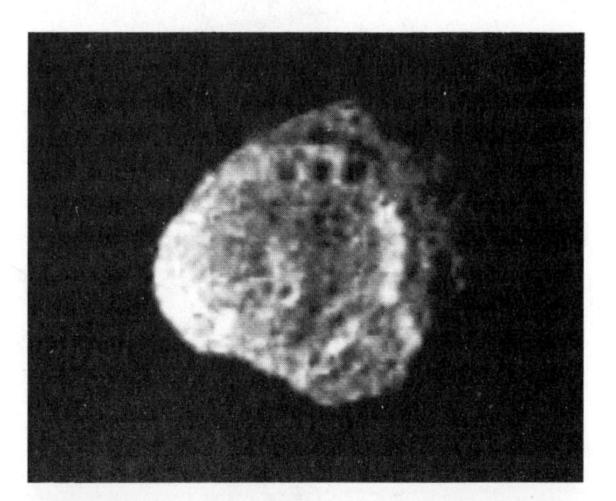

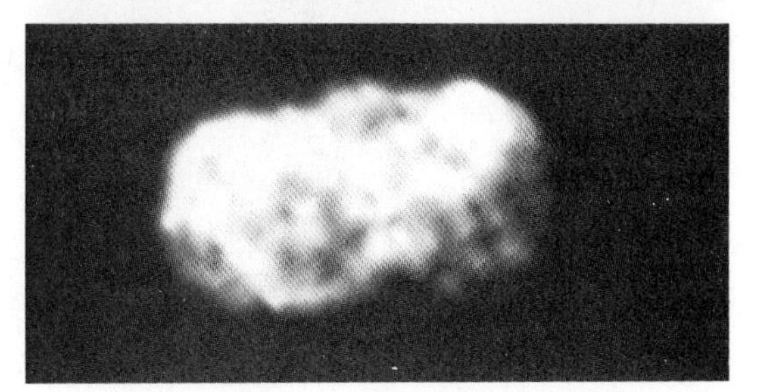

Figure 2-27. (right) Top view and side-on view of Saturn's biscuit-shaped satellite, Hyperion. The irregular shape may have occurred during impact fragmentation. Hyperion is about 360 km long. The side-on view (bottom) was made from a greater distance and is less sharp. (NASA Voyager 2 photo)

out of alignment, where it will wobble until it is eventually pulled so that its long axis again points toward Saturn.

Because the next satellite inward, **Titan,** has a massive cloudy atmosphere and is one of the largest moons in the solar system, we will discuss it separately in a moment.

The next three moons, **Tethys, Dione,** and **Rhea,** are around 1000 to 1500 km across and have bright, icy surfaces. Tethys, Dione, and Rhea have densities of around 1210 to 1430 kg/m^3, and probably have some rocky material mixed with their ice. These worlds are moderately to heavily cratered, as seen in Figure 2-28. They have bright, icy surfaces. However, Tethys' surface is broken by a huge canyon system and Dione's by swaths of lighter-toned material, reminiscent again of Ganymede. Cracking of the surfaces and resurfacing by water (quickly frozen to form bright ice plains and flows) may have occurred due to some source of internal heat.

In 1980, Earth-based observers discovered interesting additional small moons perhaps 20 to 50 km across in the orbits of Tethys and Dione. One orbits 60° behind Tethys; one, 60° ahead of Tethys; and the third, 60° ahead of Dione. At least two others, associated with Tethys and Dione, are suspected from Voyager photos. Gravitational forces make this 60° point a stable location for small objects. Certain asteroids are known in similar 60° points ahead of and behind Jupiter. These discoveries expand the list

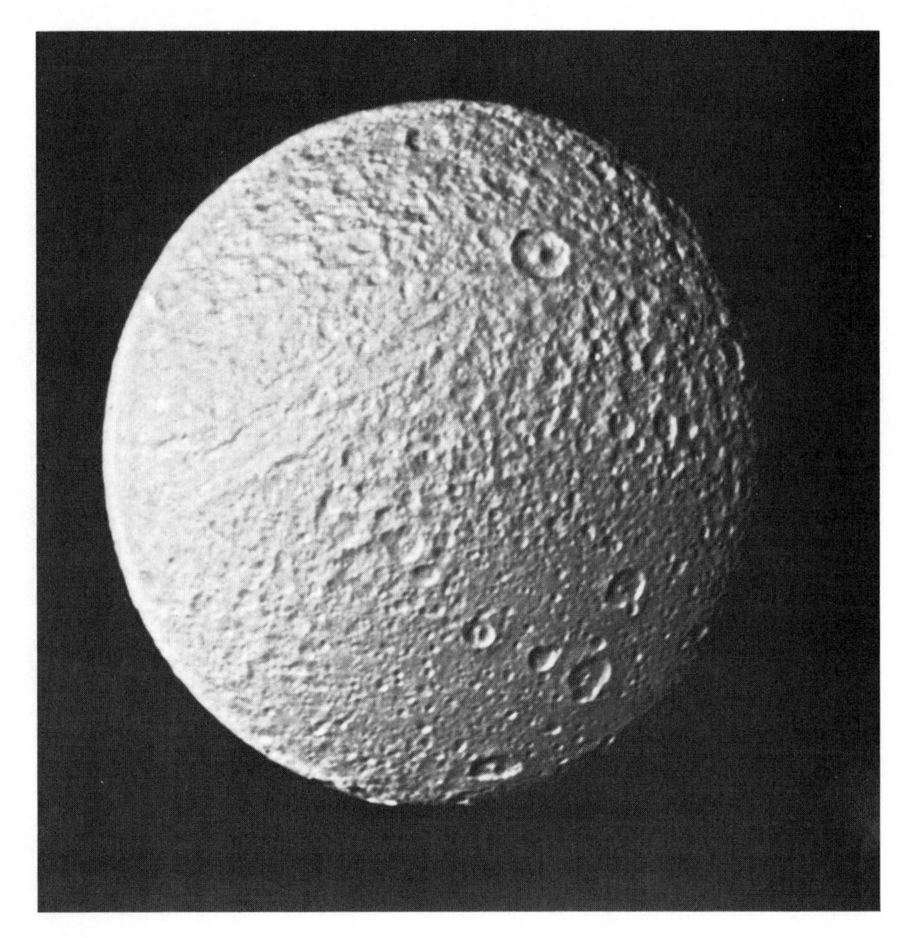

Figure 2-28. Saturn's moon Tethys, about 1050 km across, has a heavily cratered, icy surface. The cratered surface is cut in places by broad valleys or rift systems (upper left center). These suggest internal activity that split the original rigid surface. Several neighboring moons at intermediate distance from Saturn have similar surfaces. (NASA, Voyager 2)

of similarities between satellite families and the solar system as a whole—supporting the concept of satellite families as miniature solar systems.

The next satellite, **Enceladus,** is especially fascinating as a "missing link" between Jupiter's moons Europa and Ganymede. Enceladus (Figure 2-29) has the linear grooves and bright, sparsely cratered, icy surface of Europa, but it also has some moderately cratered, fractured regions that resemble Ganymede. Voyager analysts suspect that Enceladus has a fairly young surface created by eruptions of water and ice from an interior heated by tidal interactions with other Saturnian moons. This mechanism would resemble the heating that keeps Io's volcanoes erupting, but calculations have failed to show how the heating mechanism would be adequate. The bulk density of 1200 Kg/m^3 indicates that Enceladus is mostly icy throughout so that radioactivity from rock minerals (such as heats Earth's interior) also appears minimal. A concentration of material in Saturn's E ring at the position of Enceladus's orbit also leads to suspicions of (volcanic or geyserlike?) eruptions of material off Enceladus. But the mystery of Enceladus's heat sources and degree of geologic activity remains unsolved.

The next moon inward is **Mimas,** 390 km across, round, icy, and heavily cratered. Mimas's bright surface and low bulk density (1190 kg/m^3) indicate that it is composed mostly of ice.

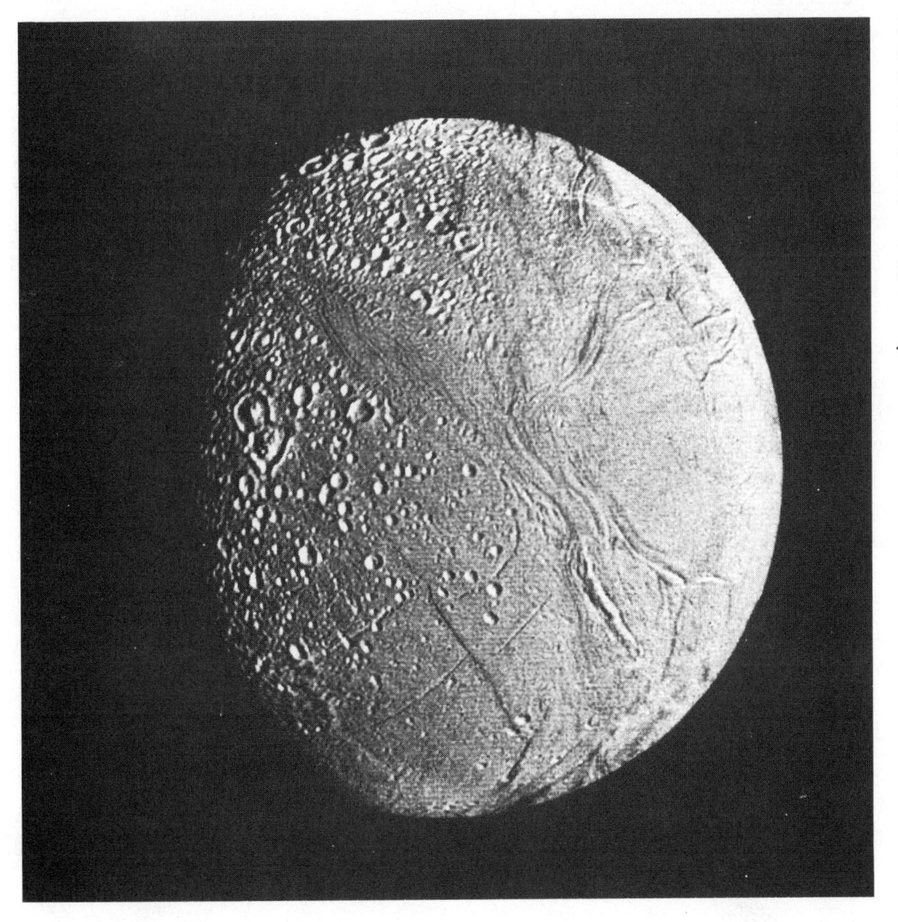

Figure 2-29. The surface of 500-km Enceladus is of special interest because it is markedly less cratered than neighboring Saturn moons. Although Enceladus is much smaller than Jupiter's Galilean satellites, its surface features suggest a missing link between the "icy billiard ball" of Europa, and the fissure-swathed surface of Ganymede. (NASA, Voyager 2)

On the outskirts of the rings are at least five small moons with diameters of 20 to 220 km. Most were discovered by Voyager 1 and by intense Earth-based observations during a period of a few weeks when the rings were seen edge on and thus did not obscure the satellites by their glare. Voyager photos show that at least some of these moons are lumpy and cratered, but their composition is uncertain.

Titan

In the midst of the Saturn satellite system is the giant, **Titan,** with a diameter of 5120 km. It is probably the solar system's second-largest moon, ranking just smaller than Ganymede. Titan rivals Io as the solar system's most bizarre world. It is the only moon with a thick atmosphere. Methane (CH_4) was discovered from spectra of Titan in 1944. Observations in 1973 showed that Titan's sky is not clear,* but rather filled with reddish haze. Later observations showed that this haze is photochemical smog produced by reactions of the methane and other compounds when they are exposed to sunlight—like the smog produced by the action of sunlight on hydrocarbons over Los Angeles. Titan is the smoggiest world in the solar system.

The Voyagers showed that the methane and smog are no more than 10% of the atmosphere. The main constituent is nitrogen; the atmosphere is much denser than realized before the Voyager flights. The surface air pressure was measured by Voyager at 1.6 times that on Earth! Since Earth's air is also mostly nitrogen, Titan's atmosphere offers fascinating comparisons to present or primeval conditions on Earth. The main difference is that Titan is very cold, around 93K (−292°F). Minor constituents detected in Titan's air include ethane (C_2H_6), acetylene (C_2H_2), ethylene (C_2H_4), and hydrogen cyanide (HCN). The abundance of these organic molecules suggests that Titan offers a good natural laboratory for research on the origins of life.

*As fervently wished by a generation of astronomical artists, who loved to depict Saturn hanging in a blue sky over the icy plains of Titan.

Under conditions on Titan's surface, pools of liquid methane may exist. If the methane content exceeds some 8%, methane may also rain out of the clouds and exist as snow or ice, playing the same triple role of gas, liquid, and icy solid as water does on Earth. The smog may even produce gasoline-like compounds that rain out of the hazy clouds. One Voyager scientist characterized Titan as a bizarre murky swamp. . . . The swamp is liquid nitrogen and the murk is frozen nitrogen and hydrocarbon muck." Titan remains a fascinating target for future exploration.

Uranus

Now we come to the outer three planets, which were not known to the ancients. **Uranus** was discovered accidentally on March 13, 1781, by the English musician-turned-astronomer William Herschel. Herschel was observing star fields at the time with his telescope; later studies showed that other accidental observations of Uranus had been made earlier but the observers had mistakenly plotted the planet as a star.

While Uranus is nearly four times the size of Earth, it is only 41% the size of Saturn and so remote that it presents only a tiny, somewhat greenish disk in large telescopes. Some visual observers* have reported bright spots or dusky bands parallel to the equator, as on Jupiter and Saturn, but the best photographs with earthbound and balloon-borne telescopes show only a featureless disk, shown in Figure 2-30. Spectro-

*In astronomy, the term *visual observer* refers to observation by eye through an eyepiece as opposed to observers using photography or other techniques. Visual observers of planets can often perceive more detail than revealed by a photo made with the same telescope, because visual observers can sense detail during near-instantaneous moments of "good seeing" when the image momentarily ceases to jiggle and flicker in response to air currents. Unfortunately, the perception of such detail has a subjective element that has led to errors as well as important discoveries by visual observers of the past.

scopic studies from Earth and later from Voyager 2 indicate a dense atmosphere consisting of about the same mix of hydrogen to helium found on Jupiter and Saturn (Ingersoll, 1990).

Uranus, however, *looks* much different (Figure 2-30). It is a nearly featureless globe about one-third the size of Jupiter, and has a greenish blue color. The featurelessness comes from colder temperatures and lesser inner thermal activity, so that the atmosphere is less restless than those of Jupiter and Saturn. The color comes from Rayleigh scattering (see page 409) of blue light, as in our own atmosphere, plus a strong absorption of red light by methane gas (CH_4), which constitutes a few percent by mass. The net result is a bland, ethereal blue globe, with a slight greenish cast.

Uranus has a peculiar dynamical property. As shown in Figure 2-31, its axis of rotation, instead of having only a slight tip to the plane of the solar system as is true of the other planets, lies almost in the plane of the solar system. This means that sometimes the "north" pole of Uranus points toward the sun, while half a revolution later the "south" pole points toward the sun. The inclination of the axis to the orbit plane is 98° and the spin direction is retrograde. These are more than curiosities; they are properties that must be explained by any theory of the evolution of planets.

The rotation period was uncertain for many years due to the lack of cloud features to observe. Published results ranged from 11 hours in the 1930s to 16 and 24 hours around 1980. Voyager 2 measured the period more exactly at 17.9 hours.

The first edition of this book, in 1972, remarked that Uranus resembles Saturn without rings—a fine analogy until March 10, 1977, when many astronomers watched Uranus pass in front of a relatively bright star. They expected to see the star dim only as it passed behind Uranus' atmosphere, an event that would allow measurement of atmospheric properties. Unexpectedly, they saw the star dim several times on both sides of Uranus' disk, indicating that it had been obscured by rings that go all the way around

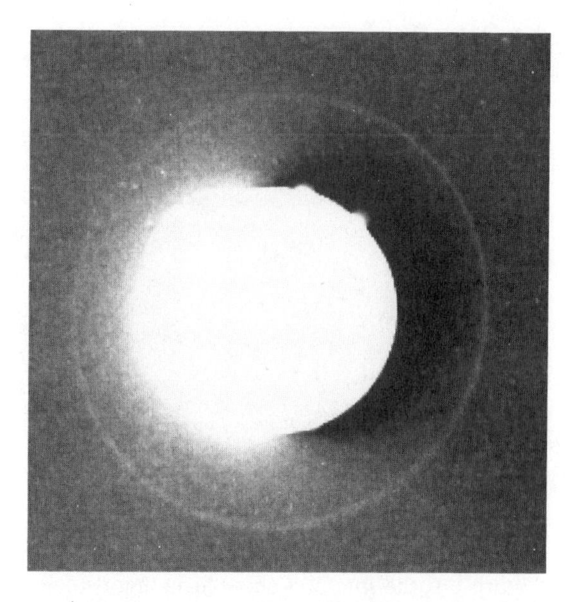

Figure 2-30. Overexposed photo of the globe of Uranus from approaching spacecraft Voyager 2 shows the brightest of the thin rings surrounding the planet. Because of Uranus' high axial tilt, Voyager approached from the poleward direction, giving a "face on" view of the rings. (NASA)

Uranus (Elliot, Dunham, and Mink, 1977, and many other papers published that year). When Voyager 2 arrived, it confirmed an elegant system of narrow, faint rings (Figure 2-30) totally unlike Jupiter's broad, tenuous ring, or Saturn's dramatic system of bright rings and narrow gaps. The faint rings are composed of black dust particles. Between some of the narrow rings are broader, still fainter rings. A flurry of studies followed, trying to understand what keeps each main ring so narrow. Dynamicists believe that the net gravitational actions of small satellites near the rings confine the ring particles into narrow, tightly defined rings.

Uranus has five substantial satellites discovered from Earth, and ten more discovered by Voyager 2. When Voyager 2 approached Uranus on its grand tour of outer planets, scientists thought that surely at this large distance from the sun, the satellites would turn out to be mere cratered iceballs, hardly distinguishable one from another. Once again, Voyager provided a sur-

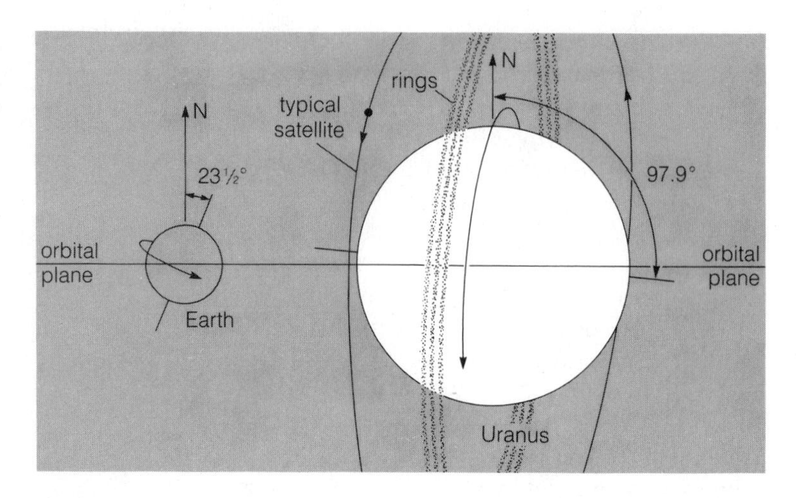

Figure 2-31. Comparison of the sizes and rotations of Earth and Uranus. Earth has an obliquity (or axial tilt) of $23\frac{1}{2}°$ and a prograde (west-to-east) rotation. Uranus has a much steeper obliquity and retrograde rotation.

prise. As shown in Figures 2-32(b) and (c) and Figure 1-1, the satellites show amazing geologic features and individual personalities. The reason for the dark color of 1170-km-diameter Umbriel is unknown, though carbonaceous dust, ubiquitous in the outer solar system, is probably involved. Miranda (Figure 1-1), innermost of the five big moons, is the most notable; it has wide swaths of grooves and faulted cliffs, astonishing for a body only 470 km across. The energy source for this ancient tectonic activity is uncertain. Other satellites, such as 1160-km Ariel, also show tectonic signs of ancient activity (Figure 2-32a).

Neptune

After Uranus was discovered in 1781, celestial mechanicians tried to determine its orbit from the new observations by Herschel and from older observations. They found that the gravitational force of another planet beyond Uranus was affecting Uranus' motions.

An interesting chapter in the history of science ensued (Lyttleton, 1968). Unknown to each other, two astronomers, an Englishman and a Frenchman, set out in the 1840s to predict where the undiscovered planet must be. Based on Bode's rule, which seemed to be confirmed by Uranus, they assumed a solar distance of 38 AU—ironically, since **Neptune** is the one serious failure of Bode's rule. Adams, just finishing his undergraduate work at Cambridge, had trouble getting his professors interested in the search. LeVerrier predicted a position in 1846, and the English astronomers began a desultory search, actually spotting it, but failing to recognize it. Meanwhile, LeVerrier got two young German astronomers interested, and they discovered it within half an hour of starting their search on September 23, 1846.

Adams and LeVerrier are both credited with the discovery, but at the time it was a scandal because of the failure of the British to grasp their opportunity.

Neptune is so far away that little was known of it before Voyager 2 climaxed its flight with a close flyby in 1989. Neptune is slightly smaller than Uranus, shares Uranus' blue color for the same reason, and rotates slightly slower, in 19.2 hours. Voyager 2 revealed a more active atmosphere than on Uranus, with faint blue cloud belts and a large, dark, oval storm system reminiscent of Jupiter's red spot. The atmosphere, as with the other giant planets, is roughly three-fourths hydrogen and one-fourth helium, with minor admixtures of methane (CH_4) and other gases.

Neptune's rings (Figure 1-3b, page 9) were first suspected during stellar occultation measurements in the 1980s (similar to the case of Uranus), but the measurements indicated a strange discontinuous structure to the rings. Voy-

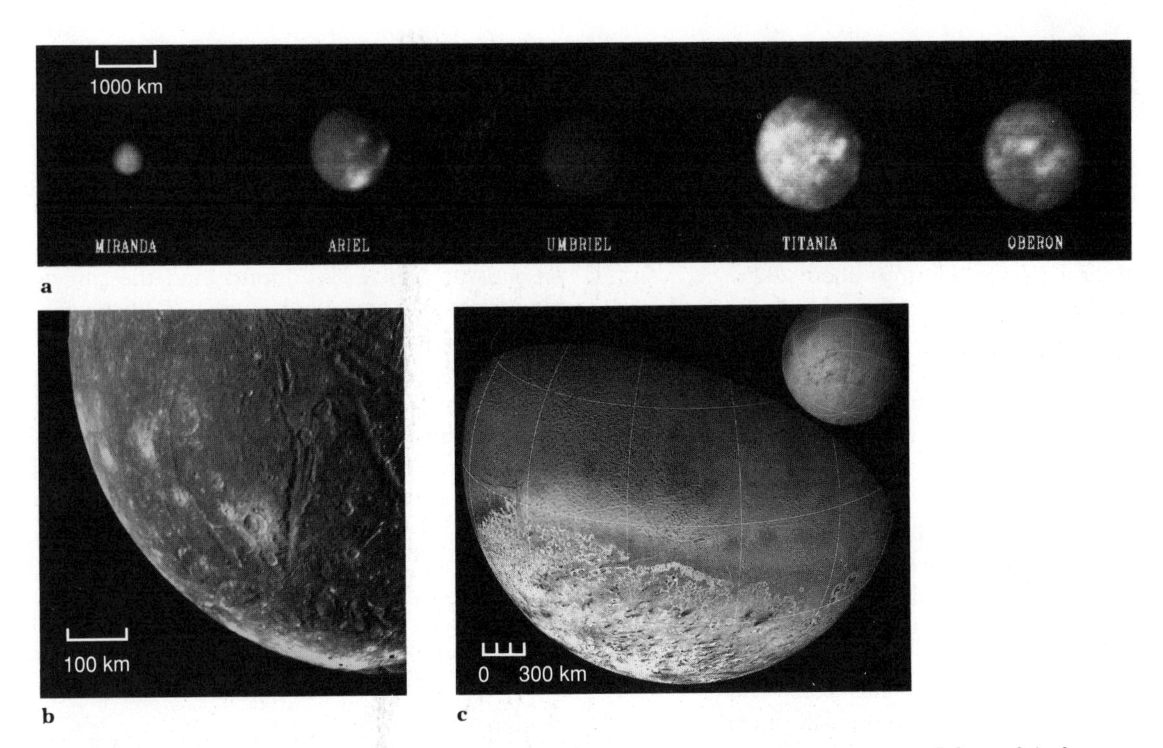

Figure 2-32. Satellites of Uranus and Neptune. (**a**) *The five large moons in order outward from Uranus (left to right). They are reproduced at the same scale and with a uniform brightness scale. The unique, puzzling dark color of Umbriel is well seen, as is the small size of Miranda.* (**b**) *Close-up of part of Uranus' satellite Ariel shows old, heavily cratered surface, broken by numerous tectonic rifts. Compare with image of neighboring moon, Miranda, on page 4.* (**c**) *Two mosaics of Neptune's sparsely cratered, largest moon, Triton. Due to the inclined orbit, lighting was mostly on the southern hemisphere, as shown by latitude/longitude grid. Inset shows a view that includes more of the bright southern polar region. Dark smudges (bottom) are associated with geyserlike vents. (NASA Voyager 2 photos)*

ager 2 confirmed narrow rings that aren't uniform but contain concentrations of particles called ring arcs.

Once again, the satellites provided the big surprise. In addition to seven small moons, about 50 to 400 km in diameter, Neptune has one big moon, **Triton** (Figure 2-32c). Its diameter is 2700 km, about three-fourths the size of our moon. Triton revolves "backwards" around Neptune, in a retrograde, inclined orbit, suggesting a capture origin. During the 1980s, an atmosphere was suspected on Triton. Voyager 2 revealed not only a thin atmosphere of molecular nitrogen (N_2) and methane, and a high haze layer, but also a geologically young, uncratered surface of CH_4 and N_2 ices, along with a seasonal polar cap of transient frost (probably N_2). Most amazing of all were geyserlike plumes of darkish smoke, indicating active internal heating (Stone and Miner, 1989; other papers in that special issue of *Science*).

The energy that drives the "geysers" is uncertain. Triton may have originated as a Pluto-like interplanetary body, been captured by Neptune, and heated by tidal forces (Goldreich and others, 1989, and see next chapter).

Pluto

Irregularities in Neptune's motions suggested at least one more planet beyond Neptune. Percival Lowell began a search for it in 1905. His assistant, Clyde Tombaugh, discovered **Pluto** in 1930 on Lowell Observatory photos. It was named for a god of the underworld, whose first two letters matched Lowell's initials.

A satellite, named **Charon** (KEHR-on, not to be confused with Chiron, discussed on page 14), was discovered later (Christy and Harrington, 1978). Observations of Pluto and Charon's light variations, spectra, stellar occultations, and a fortuitous set of eclipses (1985–1990), established that Pluto rotates every 6.39 days on an axis tipped 122° to the orbit plane, has an atmosphere thinner than Mars', and that Pluto and Charon's mean density is about 2000 kg/m^3, suggesting a composition of around 70% rock and 30% ice (Binzel, 1990). The observations showed further that Pluto is somewhat pinker, and has a surface rich in methane ice, while Charon is more neutral grey and has a surface rich in water ice. The surface temperature of these remote bodies is a frigid 50 K (−369°F).

Though Pluto was hailed as the ninth planet, its status as a planet is questionable. It is much smaller than any other planet and even smaller than our moon; its orbit crosses Neptune's; and the discovery of asteroid/comet 2060 Chiron shows that sizable interplanetary bodies exist. If other Pluto-sized bodies are eventually discovered, Pluto may come to be regarded as one of the largest of the innumerable interplanetary bodies of the outermost solar system.

Occasional searches for other such bodies in the outermost solar system continue (Luu and Jewitt, 1988).

Telescopic Appearance of the Planets

Though much of the knowledge we will discuss comes from spacecraft, it remains interesting to know what can be seen by examining the planets visually with a telescope. Except on rare nights, Earth's atmospheric turbulence blurs details with an angular size of less than 0.5″ of arc. Rare conditions with little turbulence are called good "**seeing**." With even fair seeing, amateur astronomers' telescopes of 15–30 cm (6–12 in.) show most of the details shown in Figure 2-33. Trained observers under good conditions can always see much more detail than beginners; and visual observing generally reveals more detail than can be photographed (see footnote on page 44).

The sizes of the sketches indicate very roughly the apparent sizes of the planets seen from Earth. Note that the inner planets display the largest apparent size when they are in crescent phase between Earth and the sun. A good rule of thumb is that the east–west ("horizontal") thickness of Venus in most of its phases is roughly 10″ (10 seconds of arc) across. Jupiter and the Saturn ring system are typically around 45″. When passing near Earth, Mars reaches as much as 25″ across. Jupiter's four Galilean satellites and some satellites of the giant planets are easily visible in small telescopes; but only the four large moons of Jupiter approach even 1″ in angular diameter, so that vague detail can be seen only with the largest telescopes under the finest conditions.

MISCELLANEOUS BASIC DATA AND TERMINOLOGY

Readers intending to pursue scientific careers would be well advised to memorize a few numerical facts about the planets. In spite of popular arguments against "memorizing mere facts," a small investment in memorization can produce large savings in time spent looking up trivia and allow quick checks of ideas that might otherwise turn out to be blind alleys. Table 2-2 lists some data that are used in a large number of simple calculations. The **solar constant,** listed in Table 2-2, is defined as the rate at which energy is received from the sun by a 1-m^2 surface facing the sun at the mean distance of the earth's orbit (Joules per meter squared per second). In other words, it is the mean flux* of sunlight at the top of Earth's atmosphere. Helpful facts to remember are that the sun's mass is about 1000 times that of Jupiter. Also, the sun's radius is about 10 times that of Jupiter, which is about 10 times that of Earth, which in turn is about 10 times that of the largest asteroid.

*Flux is a general technical term for amount passing through unit area per second.

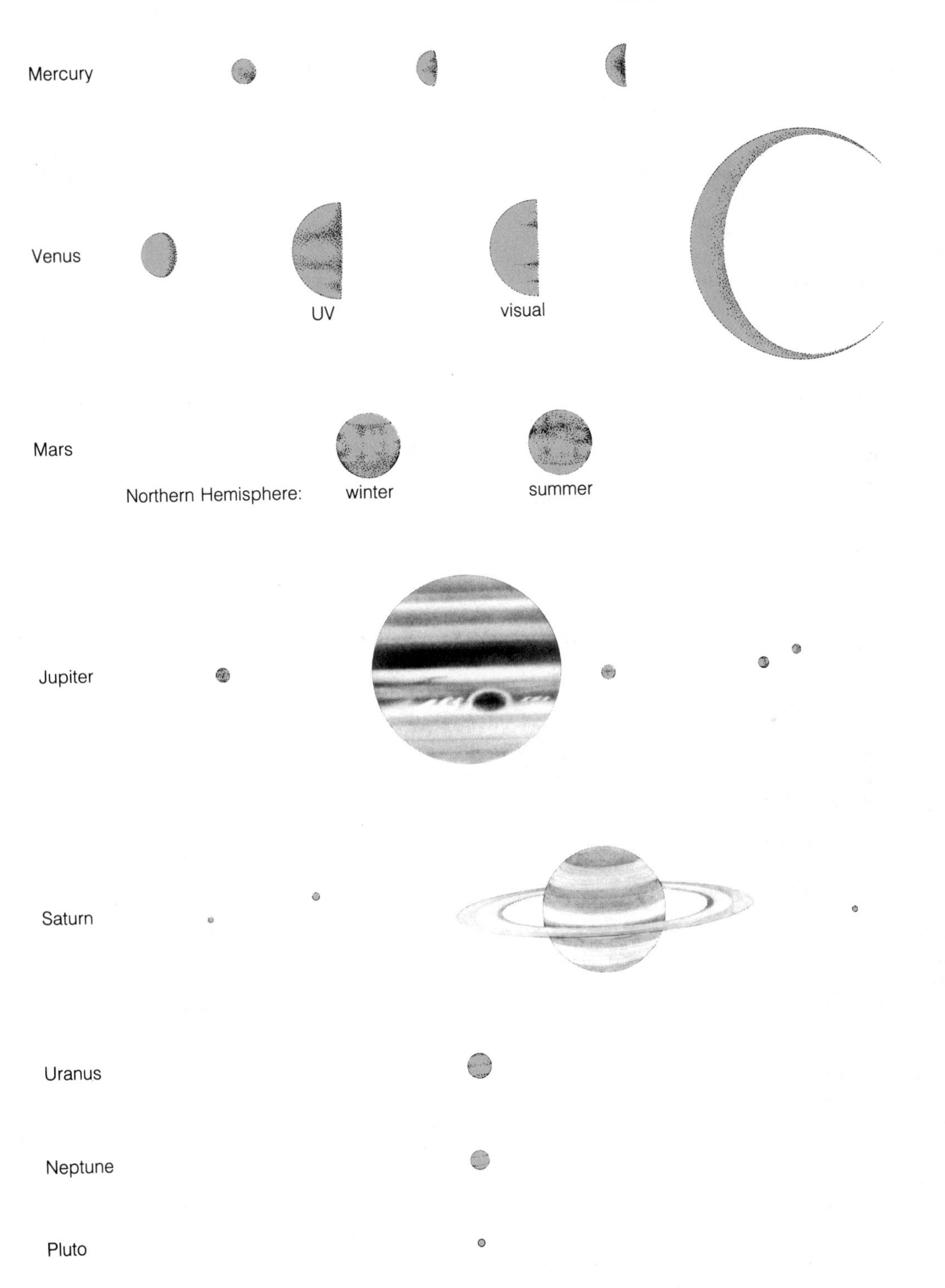

Figure 2-33. Appearance of the planets to a visually experienced observer using a moderate-sized telescope.

Figure 2-34 illustrates a number of terms common to planetary astronomy. The term **apparition** refers to an extended period of visibility of a planet—for example, the 1956 apparition of Mars (when it was on our side of the sun) or an apparition of Venus (when it is within a month or two of elongation and visible in the evening or morning sky). **Elongations** occur for inner planets and can be either eastern (appearing east of the sun in the evening sky) or western (west in the morning). A mnemonic device is the *e* in *east* and *evening*.

As shown in Figure 2-30, an eclipse occurs when one body enters the shadow of another (not necessarily when one appears to pass behind another, which is called an occultation). The shadow of any object (a planet or your hand) has two parts: the umbra, or dense inner region in which the light source is completely covered, and the penumbra, or lighter outer region in which

Table 2-2
Useful Facts to Memorize

Body	Important satellite	Mass (kg)	Radius (m)	No. known satellites
Sun	Jupiter	2×10^{30}	7×10^{8}	thousands
Mercury		—	—	0
Venus		—	—	0
Earth		6.0×10^{24}	6.4×10^{6}	1
	Moon	7.4×10^{22}	1.7×10^{6}	—
Mars		—	—	2
	Phobos	—	—	—
	Deimos	—	—	—
Asteroids		$\leqslant 10^{19}$	$\leqslant 5 \times 10^{5}$	?
Jupiter		2×10^{27}	7×10^{7}	16
	Io	—	—	—
	Europa	—	—	—
	Ganymede	—	—	—
	Callisto	—	—	—
Saturn		—	—	17
	Titan	—	—	—
Uranus		—	—	5
Neptune		—	—	2
Pluto		—	—	1
Comets		10^{15}–10^{16}	—	—

Gravitational constant	$G = 6.67 \times 10^{-11}$ N·m²/kg²
Astronomical unit	1 AU = 1.49×10^{11} m (~150 million km)
Boltzmann constant	$k = 1.38 \times 10^{-23}$ J/deg
Stefan-Boltzmann constant[a]	$\sigma = 5.67 \times 10^{-8}$ J/m²·deg⁴·s
Velocity of light	$c = 3.00 \times 10^{8}$ m/s
Mass of H atom	$M_H = 1.67 \times 10^{-27}$ kg
Planck's constant	$h = 6.63 \times 10^{-34}$ J·s
Mass of sun	$M_{\odot} = 2.00 \times 10^{30}$ kg
Luminosity of sun[a]	$L_{\odot} = 4 \times 10^{26}$ J/s
Solar constant[a]	$F_{\odot} = 1.36 \times 10^{3}$ J/m²·s = 1.36 kw/m²

[a]Note that one Joule/second (J/s) is equal to one watt (W).

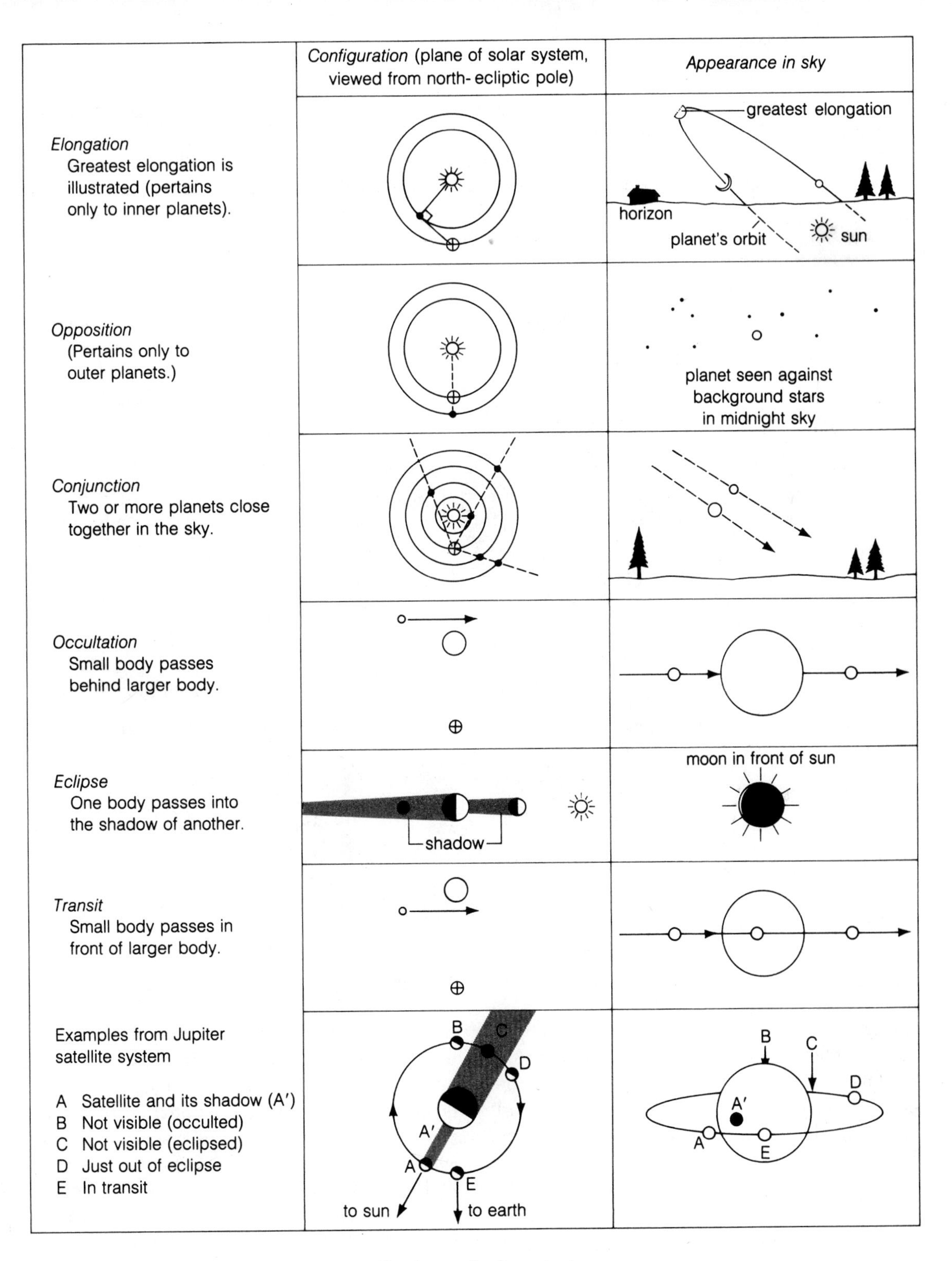

Figure 2-34. Terms used in planetary astronomy. (See the text for discussion.)

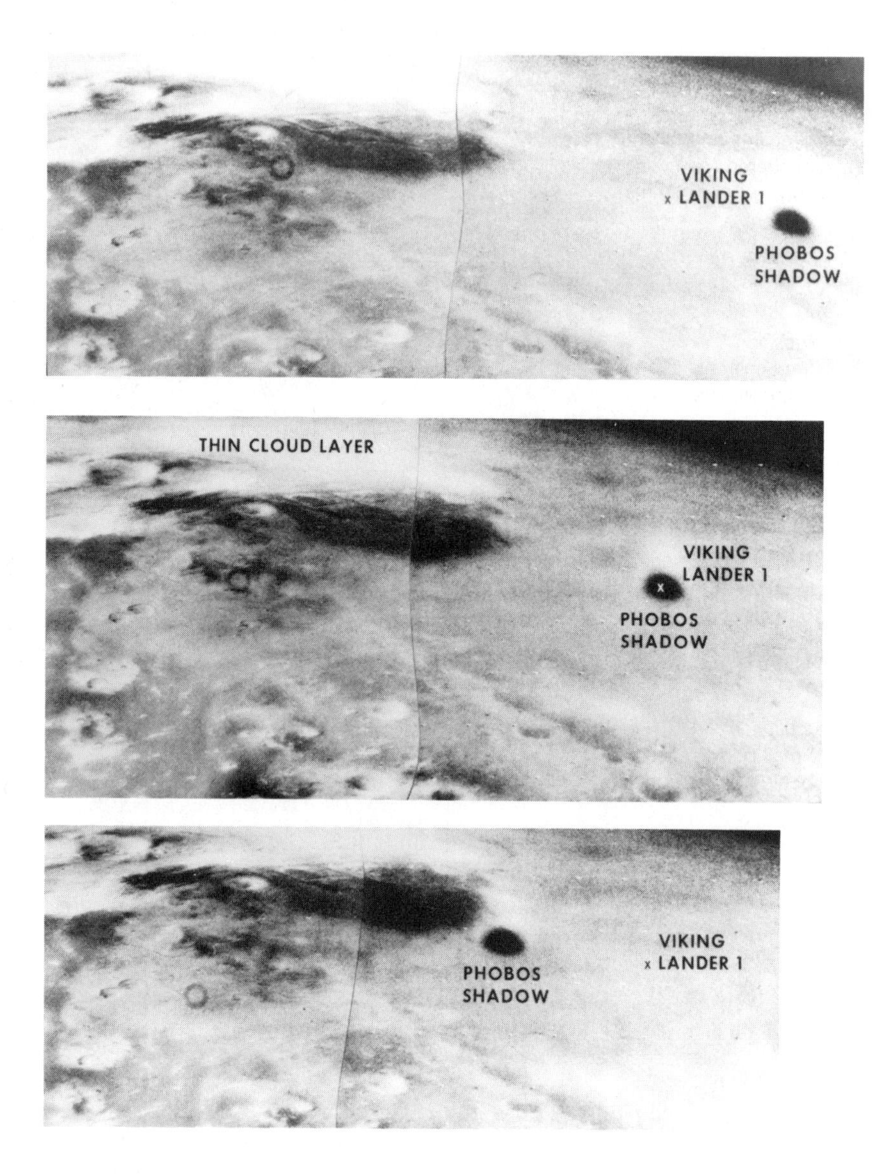

Figure 2-35. Three-minute sequence of pictures shows the penumbral shadow of Phobos passing across the Martian landscape, including the Viking 1 landing site. The shadow is about 90 km (56 mi) long. Timing of the event as observed here by the Viking orbiter and also on the ground by the Viking lander helped locate the lander. Contrast enhancement makes the shadow appear darker than it would otherwise. (NASA, Viking Project; courtesy T. C. Duxbury)

the light source is only partly covered. An eclipse can be either umbral or penumbral, depending on which part of the shadow is occupied by the eclipsed body. Figure 2-35 shows a penumbral eclipse of the Viking 1 landing site on Mars by Mars' satellite Phobos.

Certain dynamical terms should be noted—for instance, the prefixes *peri-* (designating the point when a smaller body is closest to the primary body) and *ap-* or *apo-* (designating the farthest point). If one wants to speak in general terms, without specifying a particular primary body, one uses the suffix *-apsis;* thus, **periapsis** and **apapsis** can be used to designate points in orbits around any planet or satellite. For referring to specific bodies one uses specific suffixes (usually Greek), such as *-gee* for Earth (**perigee** and **apogee**), *-helion* for the sun (**perihelion** and **aphelion**), and *-lune* for the moon (**perilune** and **apolune**). These are the bodies most frequently discussed.

Certain other terms are of very general use.

The **ecliptic** is the plane of Earth's orbit. **Inclination** is the angle between the plane of a planet's or satellite's orbit and the Earth's orbit. Earth's orbit, or the ecliptic plane, was chosen long ago as the standard of reference. Celestial mechanicians sometimes refer to the **invariable plane,** which is defined by the total angular momentum of the entire solar system. The distinction is small, because Jupiter's inclination is only 1°.3. The planets with the greatest inclinations are Pluto (17°.1) and Mercury (7°.0). Sometimes the "inclinations" given for satellite orbits are the angles between their orbit and the planet's equatorial plane, not the plane of the solar system. Thus, a listing under "inclinations" of Uranus' satellites might be given as about 1° or as 98°, and one must be careful to determine which definition is being used.

As shown in Figure 2-36, **revolution** is the motion of one body around a second body, while **rotation** is the spinning motion of a body around an axis within itself. Rotation and revolution are commonly confused, especially in everyday speech (to be entirely proper we should speak of rotating doors and rotators).

Obliquity is the tilt angle between a planet's axis of rotation and the pole of the orbit. For Earth it has the familiar value 23½°. Most of the other planets have similar values, except Venus (which has a value near 180°), Uranus (with the unusual value of 98°), and Jupiter (with the smallest value of 3°). The fact that the largest planet is the most accurately aligned with the plane of the solar system may prove significant to theories of planetary origin. Obliquity is sometimes incorrectly labeled *inclination*. Obliquity is responsible for seasons on the planets, for obliquity causes one hemisphere to be tipped toward the sun at a given point in a planet's orbit. As with inclinations, retrograde obliquity motions are indicated by angles greater than 90°.

Albedo is a measure of the reflectivity of a surface, that is, the percentage of sunlight that the surface reflects. An albedo can be calculated for each color, or an average albedo can be given for all the colors of sunlight (that is averaged over the wavelength range of sunlight), which is the more common practice.

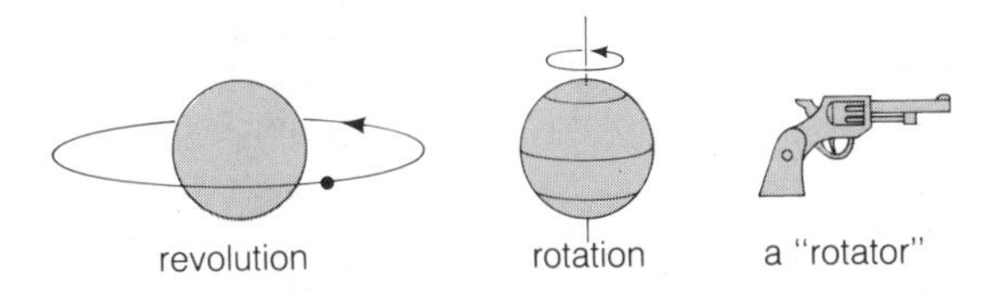

Figure 2-36. Revolution *refers to the orbiting of one body around another.* Rotation *refers to the spinning of a body around its axis. A revolver, which has a rotating chamber, is misnamed.*

A complication arises because of phase effects. **Phase** is defined as the angle between the sun and the observer as seen from the planet. The **Bond albedo,** the most commonly used definition of albedo, refers to the total percentage of sunlight reflected *in all directions*. To determine the Bond albedo, we have to observe a planet over a wide range of phases to determine how the amount of light reflected depends on direction, a dependence called the phase function. Different surface particles, such as fine dust, rocks, or cloud droplets, have different phase functions. For this reason the Bond albedo (A) is broken down into two factors, p and q ($A = pq$). The factor q is known as the phase integral and can be either observed or theoretically calculated. The factor p, known as the **geometric albedo,** gives the percentage of light reflected at zero phase angle.

Both Bond and geometric albedos can be given for specific colors or averaged over all colors. For example, the visual Bond albedo measures the percentage of light reflected in all directions at wavelengths to which the eye is most sensitive. In general, all these albedos are 2% to 10% for dark rocks, 10% to 25% for lighter rocks, and about 40% to 70% for clouds and frosts.

Albedos are thus related to surface and atmospheric properties. Venus, because of its nearly white clouds, has a Bond albedo of 76%. Earth has an intermediate value of about 36% (averaging over clouds, oceans, and land). The moon and Mercury have low values, in the range 6% to 10%, depending on the region. Certain asteroids and satellites whose surface soils are rich in carbon minerals have albedos of only 2% to 5%.

The timescale of planetary evolution is measured in units of 10^9 y, which Americans call billions of years (abbreviated "b.y."). Unfortunately, the British use *billion* to mean 10^{12}. To avoid this and many other problems, the International System of Units (ISU or SI for *Système International*) has been adopted in this book. The SI system employs the following prefixes to indicate various multiples of 10^3:

giga-	means	10^9 (abbreviated G)
mega-	means	10^6 (abbreviated M)
kilo-	means	10^3 (abbreviated k)
milli-	means	10^{-3} (abbreviated m)
micro-	means	10^{-6} (abbreviated μ)

One may thus encounter 1 billion years = 1 b.y. = 1 Gy = 10^9 y. Similarly, 1 million years may be written 1 m.y. = 1 My = 10^6 y.

SUMMARY

Dynamical properties (such as orbit and size) and surface and atmospheric properties (such as composition and temperature) are now known to some extent for all principal planets and many satellites. The frontier of the solar system visited by spacecraft is now beyond Saturn. Earth has turned out to be unique, apparently being the only body in the solar system with substantial amounts of liquid surface water, a temperate climate, a breathable atmosphere, and life. The moon, Mercury, and most satellites present airless, or nearly airless, cratered landscapes. Titan has a cold, dense atmosphere. Venus has a very hot surface with a high-pressure CO_2 atmosphere. Mars has the most earthlike surface, with a thin CO_2 atmosphere, blowing dust, temperatures occasionally above freezing, ice caps, seasonal change, and perhaps had some running water during ancient times. Io has active volcanoes. Europa is an icy billiard ball with scattered cracks. Enceladus is partly smooth, partly cracked, partly cratered. Ganymede has swaths of fissures crossing cratered terrain. Moons and planets have remarkably varied and eccentric "personalities."

CONCEPTS

planet
terrestrial planet
giant planet
comet
asteroid
meteoroid
satellite
Ceres
Chiron
planetary body
principal planet
planetesimal
world
Bode's rule
astronomical unit (AU)
Mercury
mare
Mariner 10
craters
uplands
resonance
Venus
evening star
morning star
retrograde motion
prograde motion
Venera 7
Earth
moon
regolith
Mars
surface coloration (Martian)
polar caps (Martian)
plurality of worlds
canals (of Mars)
Mariner 4
chaotic terrain
resolution
channels (of Mars)
Jupiter
belt
zone
disturbance
knot
festoon
Great Red Spot
Pioneer 10
Galilean satellite
Callisto
Ganymede
Europa
Io
Saturn
Saturn's rings A–G
Cassini's division
Phoebe
Iapetus
Hyperion
Titan
Tethys
Dione
Rhea
Enceladus
Mimas
Uranus
Neptune
Triton
Pluto
Charon
seeing
solar constant
apparition
elongation
periapsis, apapsis
perigee, apogee
perihelion, aphelion
perilune, apolune
ecliptic
inclination
invariable plane
revolution
rotation
obliquity
albedo
phase
Bond albedo
geometric albedo

PROBLEMS

1. Why is the term *planet* difficult to define in a scientifically useful way?

2. How many "planetary bodies" are known in the universe if one defines them as (a) principal planets, (b) nonstellar bodies occupying non-overlapping zones of the solar system (such as described by Bode's rule), (c) nonstellar bodies larger than 5000-km diameter, and (d) nonstellar bodies larger than 1000-km diameter?

3. Why is Bode's rule not a "law of nature" with the same status as, for example, Newton's law of gravitation?

4. Why is Venus the planet most commonly associated with the phrase the "evening star"? Explain by referring to its distance, size, albedo, and angular distance from the sun as seen from earth. What planet would make the most prominent evening star as seen from Mars?

5. Explain why the "canals of Mars" were the most widely discussed Martian surface features at the turn of the century but are not discussed at all now.

6. What satellites have been photographed at a range close enough to reveal surface features, and how do these features compare with those of our moon?

7. Which planet, Earth or Pluto, is usually closer to Uranus?

8. Briefly describe the appearance or properties of each planet and of the satellite systems.

9. If Venus, Earth, Mars, and Jupiter are in a straight line on the same side of the sun, what phenomena does an observer on Earth see? What ones would an observer on Mars see?

PROJECTS

1. Observe as many planets as possible with the largest telescope available. Make sketches showing the appearance and any surface features seen (a disk 5 cm in diameter is often considered standard for sketching planetary images). Observe each planet for at least three nights if possible and observe changes from night to night. (Even experienced observers may take at least three nights before they can see substantial detail on the planetary disks, which usually seem disappointingly small to the inexperienced observer.)

2. By using an almanac, an astronomical guide such as the *American Ephemeris and Nautical Almanac*, or monthly listings in magazines such as *Sky and Telescope* and *Astronomy*, determine which planets will be reaching their best positions (oppositions and elongations) for observing in the next few months and plan observing programs similar to that in project 1 above for a week or two at those times.

3. If the Great Red Spot or other cloud details can be seen on Jupiter, monitor their position for an hour or so until you detect the planet's rotation. Monitor Jupiter's satellites for an hour or so until you detect their revolution.

CHAPTER THREE

CELESTIAL MECHANICS

Celestial mechanics is the science that attempts to describe and predict motions of objects such as trapped asteroids, broken satellites, rings, and other curiosities in space. Archaeological evidence suggests it began with observations of solar and planetary positions in order to calibrate calendars for agricultural and ceremonial purposes. An even greater motivation arose from **astrology,** the belief that planetary positions influence our lives on Earth. Astrologers from about 1000 B.C. to A.D. 1500 created an incessant demand for forecasts of planetary positions. More recently, celestial mechanics has developed through efforts to improve marine navigation, to regulate the world's clocks, and to chart interplanetary flights. This chapter will describe many celestial mechanics phenomena, from basic orbital properties to peculiar effects that help explain some planetary properties.

The system of rings around Saturn, composed of independently orbiting particles, demonstrates the need to use celestial mechanics and dynamical theories of orbits in order to understand planetary phenomena. (NASA, Voyager)

HISTORICAL DEVELOPMENT THROUGH THE RENAISSANCE

As practical observations and astrology developed side by side, the Greeks discovered the **scientific method** of drawing conclusions about nature and making hypotheses on the basis of observations, and then testing the hypotheses by making more observations. In this way, some of the Greeks gained a fair conception of the solar system. Aristarchus of Samos (c. 270 B.C.) is said to have advocated that Earth rotated and revolved around the sun, and by using geometry he estimated the relative distances of the moon and the sun from Earth and the proportions of Earth and the moon. Eratosthenes (c. 200 B.C.), by a famous measurement of the elevation of the sun above the horizon at different latitudes, measured the spherical shape and size of Earth.

The most popular concept, however, was that of Ptolemy (c. A.D. 100), who placed Earth stationary at the center of the planetary system, with the sun and other bodies revolving around it. This **Ptolemaic system** allowed the positions of the planets in the sky to be predicted, and it held sway for 15 centuries. Five names dominate the overthrow of the Ptolemaic system, a revolution that shook science, theology, philosophy, and perhaps our basic psychology. These people and their contributions to this revolution are as follows:

Nicolaus Copernicus	1473–1543	Theory of circular motion of Earth around the sun (1543)
Tycho Brahe	1546–1601	Observations of planet positions (c. 1600)
Johannes Kepler	1571–1630	Analysis, Kepler's laws (c. 1610)
Galileo Galilei	1564–1642	Telescopic observations supporting Kepler (c. 1610)
Isaac Newton	1642–1727	Laws of gravity (*Principia*, 1687)

The work of these five scientists produced a steady advance in our knowledge. During the rise of the spirit of free inquiry, the Polish astronomer Copernicus questioned the Ptolemaic assumption of Earth's central position. The motions of the planets across the sky could be explained much more simply, he found, by assuming that Earth rotated on its own axis and that Earth and other planets revolved around the sun. Copernicus suspected that the planets moved in perfect circles around the sun.

Inspired by the idea of making a new star catalog, measuring the planetary motions, and illuminating the controversy between the Ptolemaic and the Copernican theories, the Danish astronomer Tycho Brahe built a private observatory and started recording planetary positions night by night. There were no telescopes in those days, but Tycho (as he is called) had elaborate instruments for naked-eye observations.

Tycho hired a German assistant, Kepler, who inherited the stacks of observations. Kepler studied the data for years before discovering the empirical rules that describe how the planets move. He found that the planets move in ellipses, not circles. His three basic rules, known as Kepler's laws, are discussed in the next section.

Meanwhile, the Italian observer Galileo became the first to use the telescope (probably invented in Holland) for observing the planets. His discoveries included craters and mountains on the moon and rings around Saturn, and he overturned earlier, idealized conceptions of the planets as perfect spheres. Galileo made the first of several direct telescopic observations supporting the Copernican theory. He found the four large satellites, now called the Galilean satellites, that orbit Jupiter, thus proving that not all bodies revolve around Earth. Legend tells of some defenders of the Ptolemaic theory who refused to view this sight through Galileo's telescope. A black day in the history of formalized religion came when the Church forced Galileo to recant his work and sentenced him to house arrest for the rest of his life. Galileo remarked with just cause, "In questions of science, the authority of a thousand is not worth the humble reasoning of a single individual."

A second telescopic observation supporting the Copernican theory was Giovanni Cassini's 1666 description of a large spot on Jupiter (probably the Great Red Spot), which Cassini followed as Jupiter's rotation carried the spot across the disk. This discovery, showing that other planets rotate, helped convince skeptics that Earth also rotates.

In the year of Galileo's death, the English physicist Isaac Newton was born.* Newton, who said he made his discoveries "by always thinking about them," thought about the meaning of planetary motions. At this time, the motions of

---

*A certain science fiction story has it that Galileo's spirit was transferred and reincarnated as Newton in 1642! This theory may be questionable, but it helps one to remember the historical framework of the scientific renaissance.

the planets were known empirically, but no one knew any theoretical reasons for Kepler's laws (described below) or any reason why planets move in ellipses rather than, say, squares. Newton reasoned from his observations and experiments that masses were endowed with gravitational fields and that this caused bodies to attract each other with a force proportional to the product of their masses but inversely proportional to the square of the distance between them. This discovery is known as **Newton's universal law of gravitation** and is discussed in his book *Principia* (1687), probably the most important scientific book ever written. Through his law, Newton was able to derive all of Kepler's laws.

Newton's accomplishment exemplifies how science works. Science takes a lot of seemingly unrelated observations and combines them into theories that any person can use to predict correct answers to questions about nature. Of Newton, Alexander Pope wrote,

> Nature and Nature's laws lay hid in night:
> God said, "Let Newton be!" and all was light.

KEPLER'S LAWS

Kepler's laws can be stated as follows:

1. Each planet moves in an ellipse with the sun at one focus.

2. The line between the sun and the planet sweeps out equal areas in equal amounts of time (law of areas).

3. The ratio of the cube of the semimajor axis to the square of the period is the same for each planet (harmonic law).

Newton's work showed that Kepler's laws apply to any situation where a small body revolves around a much more massive body. The two pairs of bodies can be a planet and a satellite, the sun and a comet, or the moon and a spaceship, except that the ratio in the third law is different for each pair of bodies, since it depends on the sum of their masses.

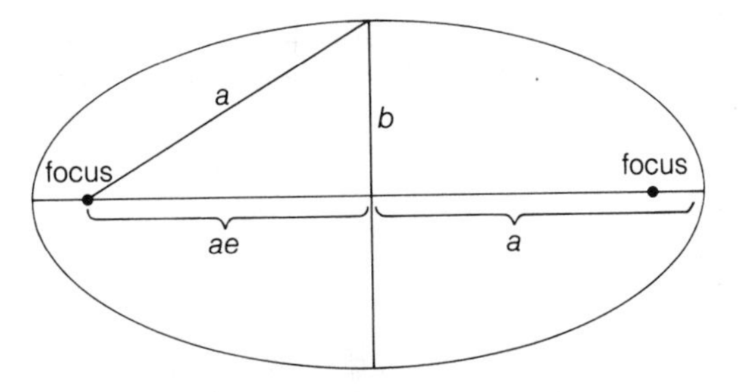

*Figure 3-1. Geometric properties of an ellipse: (**a**) semimajor axis, (**b**) semiminor axis, and (**e**) eccentricity.*

As shown by Figure 3-1, each ellipse has two **foci,*** or points of symmetry (one of which is occupied by the **primary body,** which is the more massive body). Ellipses have different shapes, the circle being a special case where the two foci coincide in the center. The point closest to the primary is called the **periapse,** and the point furthest from it, the **apoapse.**** The properties required to define the shape of a planet's orbit and its position in the orbit are called the **orbital elements.** Five elements define the size, shape, and orientation of the ellipse, and the sixth (the moment when the orbiting body passes through periapse) defines the orbiting body's position in the ellipse. The three most important elements, shown in Figures 3-1 and 3-2, are the following:

a **semimajor axis:** greatest distance from the center of the ellipse to its periphery (for a circle, the radius)

e **eccentricity:** a measure of the departure from circularity

i **inclination:** the angle between the orbital

*The plural of focus is foci.

**If the primary is the sun, these are usually called the perihelion and aphelion. If the primary is the Earth, they are usually called the perigee and apogee. "Periapsis" is an alternate form of the term periapse.

plane and some reference plane, usually Earth's orbital plane (that is, the ecliptic)

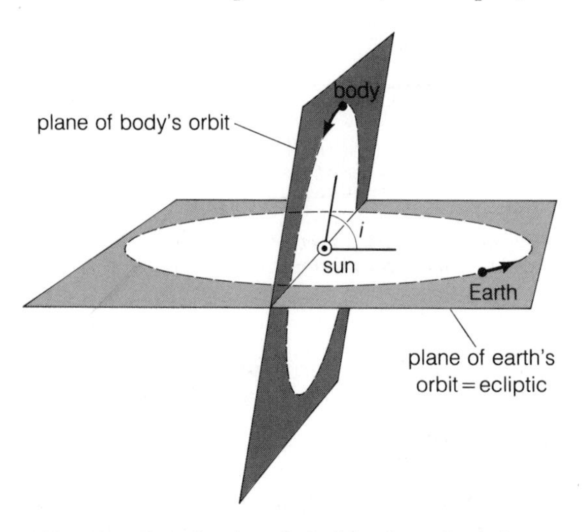

Figure 3-2. The inclination of a body's orbit is the angle between its orbital plane and the ecliptic plane.

Another useful property of an orbit is P, the **period,** or the time required for the body to complete one orbit.

NEWTON'S LAWS

Newton clarified the meaning of the concepts of force, motion, energy, and so on. He realized that if any body is *not* moving in a straight line, some force must be acting on it to deflect it from straight-line motion. In the case of the planets, the principal force is **gravity,** the force by which any mass attracts any other mass. Newton determined that gravitational attraction between the sun and a planet must be proportional to the mass of the sun and to the mass of the planet, and he showed what mathematical form the force law has. This formulation is his **law of gravitation.**

MATHEMATICAL NOTES ON KEPLER'S LAWS

The laws can be simply stated mathematically:

1. $$r = \frac{a(1 - e^2)}{1 + e\cos\theta} \text{ (equation for an ellipse)}$$
2. $$\frac{dA}{dt} = \text{constant}$$
3. $$\frac{a^3}{P^2} = \text{constant}$$

where r = distance from orbiting body to one focus

a = semimajor axis

e = eccentricity

θ = angular position in orbit measured from focus

A = area

P = period

In the most general case, with any small mass m that is not negligible and any large mass M, the third law takes the following form:

3. $$\frac{a^3}{P^2} = \frac{G}{4\pi^2} \cdot (M + m)$$

Because every planetary mass is so much smaller than the solar mass (M here), application of the law to the system of sun and planets converts the right-hand side of the equation to approximately a single constant, $GM/4\pi^2$.

An important exercise in advanced courses is to derive Kepler's laws from Newtonian physical theory. This is very simple in the case of the second law, because the law of areas is a restatement of one of the basic conservation laws that are so important in physics: the law of conservation of angular momentum. Angular momentum is equal to the linear momentum times the

ORBITS

A body anywhere in the solar system is under the gravitational influence of the sun. If the body is not too close to a planet, the sun's influence is dominant. If the body is given a velocity v in any direction, it will start to move in some orbit influenced by the sun. Body A in Figure 3-4 is pushed to the left and starts to curve downward toward the sun. Body B is pushed away from the sun but eventually turns back. Both bodies are in orbit around the sun. Similarly, a football thrown by a quarterback is in orbit around the center of Earth (in this case the orbit intersects the surface of the ground).

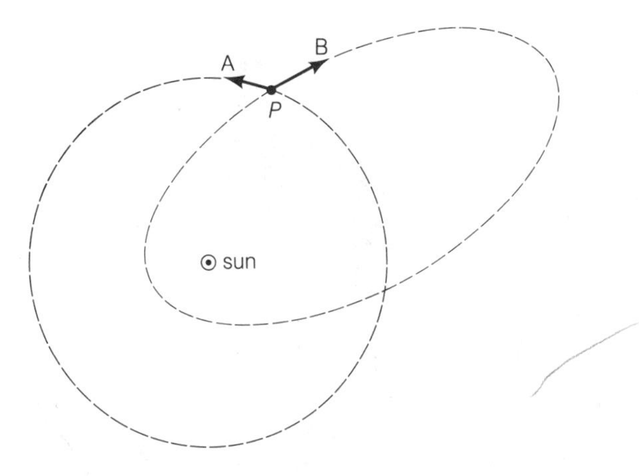

Figure 3-4. A body at point P can go into a variety of different orbits around the Sun, depending on its initial velocity. Two possible orbits are shown.

Whether the object is in orbit around the sun, Earth, or some other body, the principle is the same. The primary body will be at one focus of the orbit. (Detailed analysis shows that the focus lies at the center of gravity of the two bodies, which is near but not exactly coincident with the larger body. The sun is so massive relative to the planets that it virtually coincides with the focus.)

---

radius, or mvr. Thus conservation of angular momentum says that

$$mvr = \text{constant}$$

or, substituting for v,

$$m\left(r\frac{d\theta}{dt}\right)r = \text{constant}$$

But from Figure 3-3 we see that the area swept out in a time dt (being $\frac{1}{2}$ the base times the height of the triangle) is $dA = \frac{1}{2}r\,d\theta\,r$. Thus, substituting in the above equation,

$$m\,2\frac{dA}{dt} = \text{constant}$$

Since the mass of the planet m and 2 are both constants, we now have the law of areas, dA/dt = constant. Another important exercise, beyond the scope of this book, is to derive Kepler's first and third laws from Newtonian physics.

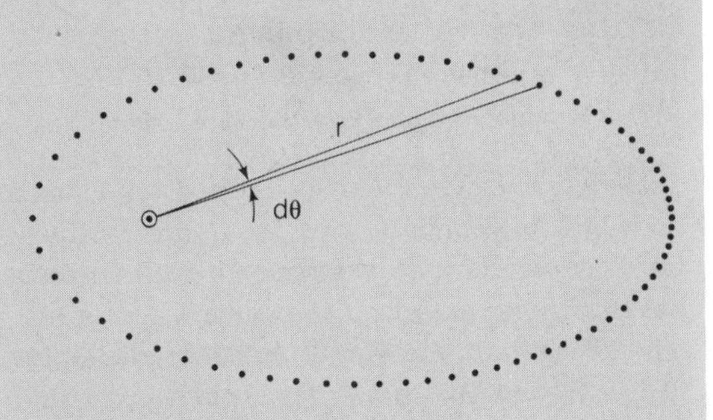

Figure 3-3. Orbital motion in an elliptical orbit. The dots are spaced to show motion during equal time intervals, indicating fastest motion at periapse and slowest motion at apoapse. Area swept out during the indicated interval is $\frac{1}{2}rd\theta r$.

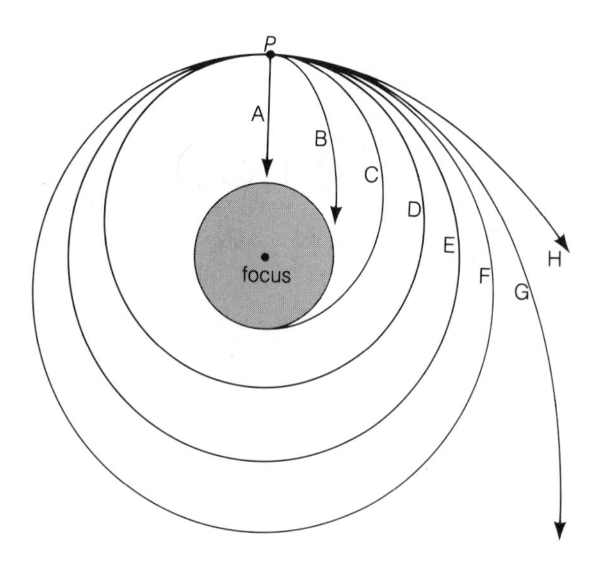

Figure 3-5. A rocket at point P above Earth can go into a variety of different orbits. Cases A through H correspond to firing the rocket parallel to the ground but with different velocities (see the text).

Circular Velocity

Circular velocity is the speed of a satellite in a circular orbit about a primary body. The circular velocity decreases with increasing distance because the pull of gravity declines with distance.

It is instructive to imagine a body being inserted into orbit at various speeds. Suppose that a spacecraft has reached a point *P* above Earth's atmosphere* (see Figure 3-5). Its attitude-control system has aligned it parallel with the ground, but its position is stationary with respect to the center of Earth. If a rocket motor is not fired, the spacecraft will fall directly to the ground on path A. If a short burst is fired, the spacecraft will fall into an elliptical orbit around the center of Earth,

*If the spacecraft were in the atmosphere, atmospheric friction would produce a resistive force called *drag,* which would disturb the motions described here.

MATHEMATICAL NOTES ON THE LAW OF GRAVITATION

The law of gravitation states that

$$F = \frac{GMm}{r^2}$$

where F = force between two bodies
M = mass of larger body
m = mass of smaller body
r = distance between the two bodies
G = gravitational constant (see Table 2-2).

The equation gives the force exerted on a small body by a large body, or vice versa. Earth attracts familiar objects with a force that we call their weight. If you are given the value of the gravitational constant, the mass of Earth M, the radius of Earth, and the mass of a rocket, you could compute the rocket's weight, which equals the force necessary to lift it off the ground. This is why the force exerted by rocket engines is usually expressed by American rocket engineers as pounds of thrust.

■ *Sample Problem*

Prove that a person weighs only 1/6 as much on the moon as on Earth, given that the moon's mass is 1/81 of Earth's mass and the moon's radius is 1/3.7 of Earth's.

Solution

The ratio of weight on the moon to the weight on Earth is

$$\frac{F_☾}{F_⊕} = \frac{GM_☾ m/r^2_☾}{GM_⊕ m/r^2_⊕}$$

where M and r are the mass and radius of the planet involved, and m is the mass of the person. Canceling G's and m's, we have

$$\frac{F_☾}{F_⊕} = \frac{M_☾ r^2_⊕}{M_⊕ r^2_☾} = \frac{1}{81}\frac{(3.7)^2}{1} \approx \frac{1}{6}$$

but the orbit, B, will intersect the ground like that of the football. With a higher speed the rocket could go nearly all the way around, striking the ground on the far side, on orbit C (similarly to an ICBM missile). A still higher speed will put the rocket into a complete orbit D, with the perigee on the far side. A still faster velocity is needed to put the rocket into a precisely circular orbit E. The description remains the same whatever the primary body and whatever the smaller body (Figure 3-6).

Nearly circular orbits were chosen for the first space flights because the rockets were barely powerful enough to climb to the top of the atmosphere, and the space capsules had to skim along at a constant height to avoid atmospheric drag.

Time-lapse movies from space probes dramatically record Keplerian orbital motion among satellites of planets. One of the most interesting cases exists within ring systems, such as the famous rings of Saturn. Each of the millions of small particles pursues its own nearly circular orbit around Saturn. As shown in Figure 3-7, certain swarms of these particles can be tracked in their movement around Saturn on consecutive photos—but the reason for the visibility of these swarms is unknown.

Figure 3-6. The condition called weightlessness in space is not due to any lack of gravitational interaction but merely to the fact that all objects in a spacecraft are co-orbiting together around some distant object (Earth, the moon, the sun, and so on) in nearly identical orbits. The condition is sometimes more aptly called "freefall" (if you jump off a high diving board and release a tennis ball during your fall toward Earth's center, it will appear to float weightlessly in front of you—until you hit the water). Here, astronauts William Pogue and Gerald Carr clown to show the effects of weightlessness inside Skylab 4 during an 84-d flight in 1975. (NASA)

MATHEMATICAL NOTES ON CIRCULAR VELOCITY

It is easy to remember how to derive the circular velocity, because gravity must be just matched by centrifugal force:

$$\frac{GMm}{r^2} = \frac{mv_{\text{circ}}^2}{r}$$

Thus,

$$v_{\text{circ}} = \sqrt{\frac{GM}{r}}$$

Note that this gives v as a function of r and that as r increases, v decreases.

Escape Velocity

Escape velocity is the speed required to project a body completely free of a primary body. Like circular velocity, the escape velocity is less at larger distances from the primary body.

If our imaginary spacecraft in Figure 3-5 is accelerated to an even higher speed than that required for circular orbit E, it climbs into an elliptical orbit F, with the apogee on the far side of Earth. At the critical speed of escape velocity, the apogee is at infinite distance and the rocket never comes back. Instead, it keeps slowing down as it moves out on path G, the speed approaching but never quite reaching zero. Another way of saying the same thing is that the rocket with escape velocity has a kinetic energy just equal to Earth's gravitational potential energy. The orbit, G, has the shape of a parabola and is called a **parabolic orbit.** At each point on G, the speed will just equal escape velocity (sometimes called **parabolic velocity**) at that point. If the rocket is made to move faster than escape velocity, it goes into a different orbit H, with a slightly different shape from orbit G. Again the rocket never comes back, but in this case it has a kinetic energy greater than Earth's gravitational potential energy, and the rocket is always moving faster than the escape velocity at each point in its orbit. This path is called a **hyperbola** or a *hyperbolic orbit.* (Ellipses, circles, parabolas, and hyperbolas are a geometrically related family known as **conic sections,** since they can all be produced by taking cross sections of a cone at various angles.)

Figure 3-7. (left) Three photos of Saturn's rings, taken a half hour apart, reveal motion in circular Keplerian orbits. The rings are made of millions of separate particles (not visible here). Due to unknown causes, swarms of particles may become visible as spoke-like shadings, whose orbital motions can be tracked. In the top view, a double spoke approaches the tip of the rings. In the middle view, it rounds the tip, followed by another spoke. In the last view, this spoke has moved around the tip. As predicted by Kepler's laws, particles nearer the planet orbit faster, causing the spokes to shear from radial to diagonal patterns. (NASA, Voyager 1 sequence)

Another way of viewing escape velocity is to imagine a body dropped from rest at an infinite height. It falls toward the ground with increasing speed, and its speed at any point will be the escape velocity at that height.

ASTROMETRY, ORBIT DETERMINATION, AND THE DOPPLER SHIFT

If you could watch a spaceship returning from the moon night after night from your backyard, how could you determine what its orbit was in three-dimensional space? The task of measuring positions is a branch of celestial mechanics called **astrometry.** In principle, only three observations of appropriately spaced positions are required to determine an orbit, but in practice many observations are used to determine the orbit as accurately as possible. In the 1800s, when many planets and asteroids were being discovered, the laborious computations needed to derive the orbit had to be done by hand and might require months; today they are done by computer in minutes (although the computer programmer may have to work for days to get the computer routine in working order).

The science of **spectroscopy,** or analysis of the distribution of colors in light, is of extreme importance in studying all celestial bodies, clarifying not only their composition but also their motions. Each color corresponds to a certain wavelength. If a source, such as a distant satellite, is moving, then the wavelength of the light coming from it is shifted from that of the light received from a stationary source, and the amount of the shift is proportional to the velocity. This shift in wavelength is called the **Doppler shift,** after Austrian physicist Christian Doppler, who discovered it in 1842 while studying stars. The Doppler shift allows measurement of the velocity toward or away from the observer (but not the component of velocity perpendicular to the observer). Measurements of this **radial velocity,** as it is called, clarify the orbital motions of distant objects. If the object being tracked is a spacecraft emitting signals of precisely known frequency, the Doppler shift measures can be made with great precision and allow incredibly precise tracking.

MATHEMATICAL NOTES ON ESCAPE VELOCITY

It is easy to derive escape velocity, since the gravitational potential energy must just equal the kinetic energy:

$$\frac{GMm}{r} = \frac{1}{2}mv_{esc}^2$$

$$v_{esc} = \sqrt{\frac{2GM}{r}}$$

Note the important and useful fact that

$$v_{esc} = \sqrt{2} \times \text{circular velocity}$$

■ *Sample Problem*

Relative to the center of the sun, Earth's orbital speed is about 29.8 km/s. How fast would a rocket have to leave Earth's orbit in order to reach interstellar space?

Solution

This would be escape velocity. Using the $\sqrt{2}$ rule, we have

$$v_{esc} = \sqrt{2}\,(29.8) = 42 \text{ km/s}$$

THE THREE-BODY PROBLEM

The discussions above exemplify the **two-body problem:** the description of motion in a system of two bodies with no other influences. Because the planets are small and have only minor influences on each other, two-body theory gives a good approximate description of planetary motion and accounts for Kepler's laws. But the real universe is not so simple; in many cases we need to consider more than two bodies in a system.

The **three-body problem** is to describe the motions of three bodies when they are big enough to influence one another or at least when two of them are big enough to influence the third. An example would be a spacecraft moving in the Earth-moon system, where both Earth and the moon are big enough to influence the spacecraft's

MATHEMATICAL NOTES ON THE VELOCITY EQUATION

Suppose that an orbit has been determined for a body and you want to know how fast the body is moving at a particular point relative to the primary body. A useful equation, sometimes called the *vis visa* or **velocity equation,** gives the velocity as a function of position (see Figure 3-8). With semimajor axis a and distance r to the object,

$$v^2 = GM\left(\frac{2}{r} - \frac{1}{a}\right)$$

This equation is a much more general form of our equations for circular and escape velocity, since it allows calculation of velocities required for any desired orbit.

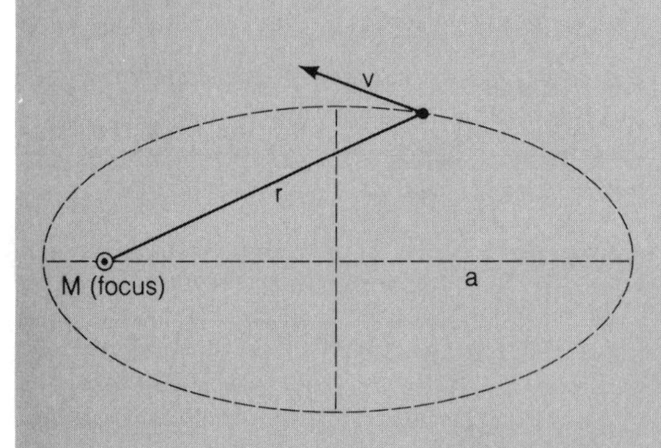

Figure 3-8. Orbital parameters used in the velocity equation.

■ *Sample Problem*

A comet in orbit around the sun is at aphelion 100 AU from the sun, and its perihelion is at Earth's orbit. Find its velocity at perihelion relative to the sun.

Solution

First, remember that all units must be in the same system, such as SI. A sketch of the orbit shows that the semimajor axis $a = 50.5$ AU $= 7.5 \times 10^{12}$ m. Substituting in the other quantities from Table 2-2 (page 50) in SI units, we have

$$v^2 = 6.67(10^{-11})\,2.00(10^{30})\left[\frac{2}{1.49(10^{11})} - \frac{1}{7.5 \times 10^{12}}\right]$$

$$\therefore\ v = 4.21(10^4)\ \text{m/s} = 42.1\ \text{km/s}$$

Note that this is essentially the escape velocity from Earth's orbit, where perihelion is located, as derived in the sample problem in the mathematical notes on escape velocity. In other words, 100 AU is so far out that the comet falls into the inner solar system essentially as if it had fallen in from infinity. Mathematically, we see that the second term in the brackets is almost negligible, reducing the equation to the equation for escape velocity.

trajectory. There is no single, general analytical solution that describes such motion, contrary to the known equations in the case of the two-body problem. Therefore, the only general way to predict motions in a three-body system is by **numerical integration,** the process of solving the problem in small steps. A computer could be set up to start with the spacecraft at position 1 and compute the forces on it from Earth and the moon, and then to let the spacecraft move a small distance under the influence of these forces to position 2, and so on, for hundreds or thousands of positions, depending on the accuracy needed.

The three-body problem applies not only in the Earth-moon-spaceship case, but also in many other solar system cases, such as a sun-Jupiter-comet system. In addition, some special cases cause very interesting effects that help explain some puzzling properties of the solar system, as described in the following sections.

Perturbations and Resonances

Each planetary body in the solar system is affected not only by the sun but also by the other planets (such as massive Jupiter) and other smaller bodies. The sun dominates, but the interplanetary forces cause small effects, called **perturbations.** Because of these perturbations, the planetary and satellite orbits are not perfect ellipses, but ellipses with minor sinuosities. Uranus' motion is perturbed by Neptune, an asteroid's motion by Jupiter, and so on. A satellite of a planet may be perturbed by the sun and also by the mass contained in the equatorial bulge of the planet (produced by centrifugal force due to the planet's rotation).

Effects of perturbations are interesting and varied. The elements of a perturbed orbit change with time. The changes may be **periodic** (varying smoothly between limits) or **secular** (tending to change in a certain direction). Perturbations of satellite orbits inclined to planets' equators typically cause the plane of the orbit and the line from periapsis to apoapsis (the line through the two foci) to swing around the planet on time scales much less than the solar system's age. In the lunar case, the line from perigee to apogee swings around Earth in only 9 y.

In many cases the perturbations cause only

MATHEMATICAL NOTES ON THE DOPPLER EFFECT

The Doppler effect may be expressed as

$$\frac{\Delta\lambda}{\lambda} = \frac{v}{c}$$

where

$\Delta\lambda$ = change in wavelength
λ = normal wavelength
v = velocity along the line of sight
c = velocity of light = 3×10^8 m/s

For example, consider a spacecraft moving away from Earth at 3 km/s, which is 10^{-5} of the velocity of light. All its radio signals would be shifted in wavelength (or frequency) by 10^{-5} of their original amount. Monitoring the wavelength with great care thus gives information on the acceleration of the spacecraft. Such accelerations can be caused by local masses near which the spacecraft is moving. Using this fact, researchers have detected masses of small satellites, mass concentrations on the moon (Chapter 9), and effects of winds as they disturb the motions of spacecraft parachuting into the atmosphere of Venus (Chapter 12).

A body moving away from the observer has its light shifted toward longer wavelengths (the red end of the spectrum), an effect called a *red shift.* An approaching object has a decrease in the wavelengths of its light, called a *blue shift.*

minor fluctuations in the orbital elements that look more or less random. But consider the case of one body whose period is $\frac{1}{2}$, ⅓, ⅔, or some other small-integer fraction of the period of a neighboring larger body, a condition in which the two bodies are said to be **commensurable.** A good example would be an asteroid whose period is $\frac{1}{2}$ of Jupiter's period, as shown in Figure 3-9. On every second trip of the asteroid around the sun, it will find itself again next to Jupiter (positions 1 and 9), and the perturbations of the asteroid by Jupiter will repeat.

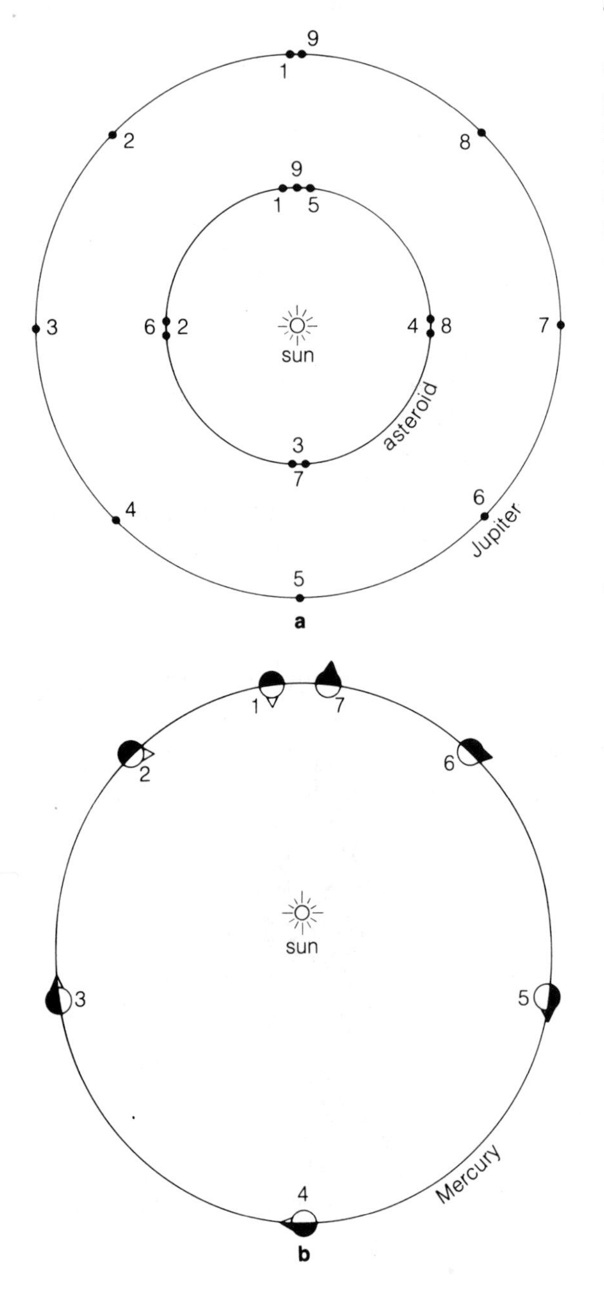

In other words, commensurability causes a condition called **resonance,** the repetition of perturbations. If you push on a swing at random intervals, you may produce a small oscillatory motion, but if you push on the swing in a constantly repeated way that matches the swing's natural frequency, you can "pump up" the motion of the swing to dramatic, wide oscillations. Similarly, some resonances can have a dramatic pumping-up effect on a planetary orbit, causing a drastic change in the orbital elements, usually markedly increasing the eccentricity *e*.

Some asteroids have been forced out of certain zones in the asteroid belt by their resonances with Jupiter, creating gaps called the **Kirkwood gaps** and presumably forcing these asteroids into adjacent parts of the belt. In 1969 dynamicist James Williams discovered another type of asteroid-Jupiter resonance* that can apparently throw asteroids clear out of the belt into orbits that can approach Mars and subsequently Earth, possibly explaining the supply of meteorites that fall on

*This secular resonance is designated the $\dot{\nu}_6$ resonance (read "nu dot six" resonance) after a term in the relevant dynamical equations.

Figure 3-9. (left) Examples of commensurabilities. Numbers give positions of bodies after equal intervals of time. (**a**) *An asteroid in a 1:2 resonance with Jupiter completes two revolutions while Jupiter completes one. At position 9 they repeat configuration 1. Such a repeated perturbation has kicked asteroids out of their orbits, creating Kirkwood's gaps.* (**b**) *A 2:3 commensurability controlling the rotation of Mercury, illustrated with a mountain or tidal bulge on one side. Mercury, which completes $1\frac{1}{2}$ rotations in one trip around the sun (or 1 rotation in ⅔ orbits), is locked into this commensurability by tidal forces strongest at perihelion (positions 1 and 7) in its elliptical orbit.*

Earth (see Chapter 6). Other types of resonances can apparently have a stabilizing effect on orbits (see fundamental review paper by Peale, 1976). Rotations can also be affected by resonance. An example is Mercury, whose rotation is sketched in Figure 3-9b.

Lagrangian Points

As mentioned earlier, predicting positions of bodies in three-body or n-body systems is very difficult. However, a specific analytic solution is known for a certain type of three-body system that corresponds to a condition of 1:1 resonance—namely, a system of two major co-orbiting bodies with nearly circular orbits (such as Earth and the moon), with a third small body nearby having the same revolution period P as the other two. In such systems, there are five points where the gravitational forces of the two bodies plus the centrifugal force just balance, so that a third body put in one of these points would tend to remaiin in a fixed position with respect to the other two.

The five positions are called **Lagrangian points** after the mathematician Lagrange (1736–1813), who first discovered their locations. The Lagrangian points are shown schematically in Figure 3-10. Of the five points, only two, $\boldsymbol{L_4}$ and $\boldsymbol{L_5}$, are truly stable.* The other three are only quasi-stable; if a small body were put at L_1, L_2, or L_3, it would stay only until an outside influence (such as a fourth body) moved it out of position,

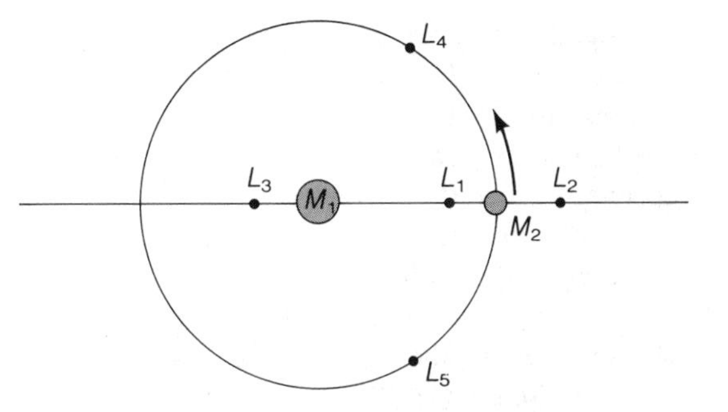

Figure 3-10. Location of the five Lagrangian points in an orbiting system.

whereupon it would drift away. A body put at L_4 or L_5, if slightly disturbed, would oscillate around its original position.*

A key idea in considering Lagrangian points is that the whole system rotates together, because the period of all objects around the center of gravity is equal. It is as if the points in Figure 3-10 were marked on a rotating phonograph record, with the biggest body near the center and the second body partway out. L_2 is then near the rim and L_4 and L_5 are 60° ahead of and behind the second body in its orbit. Objects at a Lagrangian point would seem stationary, and objects near the L_4 and L_5 points would seem to circulate aimlessly around them as seen by an observer rotating with the system (or riding on the phonograph record). Of course, no part of the system is stationary with respect to the surrounding stars.

The Lagrangian points are practical realities. The sun and Jupiter form a system of two massive bodies, with the L_4 and L_5 points 60° ahead of and behind Jupiter in its orbit. Two groups of asteroids have been discovered occupying these positions. They are called the **Trojan asteroids** and are named after the Homeric heroes. Until 1906, when the first of the Trojans was discovered by the German astronomer Max Wolf, the Lagrangian points were mathematical abstractions, but now more than a dozen Trojans are

---

*Greenberg and Davis (1978) review the Lagrangian points emphasizing the important L_4 and L_5 points. They note that many works have erroneously described these two points as stable potential minima, analogous to dimples in a plastic surface in which a steel ball could come to stable rest. (Greenberg and Davis charitably neglect to point out that the first edition of this book was one of those works.) They state that, in fact, the L_4 and L_5 points are potential maxima (analogous to bumps on the plastic surface), but that the Coriolis force is responsible for driving departing particles back toward these points. Greenberg (1978) emphasizes that in the presence of a resisting medium, such as a gaseous or dust nebula, energy loss could reduce the effectiveness of the Coriolis force and cause particles to drift away from the L_4 and L_5 points.

*This stability applies only if the mass ratio of the primaries, M_1:M_2, is less than 0.0385.

cataloged and many more are known to exist. Their orbits are seriously perturbed by other planets, and they may drift as far as 20° out of the L_4 and L_5 positions. Earth-based telescopic observations and Voyager photos in 1980–81 revealed small moonlets near L_4 and L_5 Lagrangian points of Saturn's satellites Tethys and Dione. Searches for objects in the L_4 and L_5 points of Saturn and other planets have so far been fruitless (Gehrels, 1977).

The L_4 and L_5 points of our Earth-moon system have gained attention as a result of proposals by Princeton physicist Gerard K. O'Neill (1974, 1975) and others that they are ideal locations for large, permanent space cities that could utilize at low cost construction materials and other resources from the moon. An advantage of these points is their fixed orientation relative to a lunar base, allowing easy ballistic launching of lunar material to them. These ideas are discussed in more technical detail in a NASA study (Johnson and Holbrow, 1977) and in popular books by Heppenheimer (1977) and O'Neill (1977).

Horseshoe Orbits

An interesting type of three-body orbital relationship, shown in Figure 3-11, has recently been found relevant to several solar system situations. Consider two small moons, A and B, almost in the same circular orbit. At the outset, B lags behind and is slightly closer to the planet. In keeping with Kepler's laws, B catches up to A. Figure 3-11 shows subsequent events as perceived by an observer riding on A. As B approaches on its "inside track," it is accelerated forward by A's gravity. This pushes it into a higher orbit, now farther from the planet than A. In keeping with Kepler's laws, B now moves slower than A and begins to drop behind, in spite of A's gravitational attraction. B thus starts to recede from A on its new "outside track." In this sense, A and B have met, done a little dance, and exchanged orbits. A and B continue to move apart until B drifts all the way around the planet (as seen by A), and a new encounter occurs in which B moves back to an orbit lower than A. The mean positions of the two orbits remain fixed over many encounters. As seen in Figure 3-11, the one moon (especially the smaller one) describes a horseshoe path relative to the other (Dermott and Murray, 1981).

This situation actually applies to two 200-km-scale inner moons of Saturn, 1980S1 and 1980S3, which have orbits only tens of kilometers apart. They don't collide, but approach to within about 16 000 km of each other—where one would loom about 1.5 times the apparent size of our moon in the sky of the other for a short while before shrinking and receding again into the distance (Harrington and Seidelmann, 1981). The encounters of these two moons occur roughly 4 y apart.

TIDAL EFFECTS

Tides

When two bodies orbit around each other, each exerts a force on the other, according to Newton's law of gravity. Consider Earth and the moon. The side of the moon facing Earth has a stronger force on it than does the far side, because the facing side is closer. This means that the body of the moon experiences a **net differential force** acting along the Earth-moon line at each moment. The moon stretches slightly along this line to an extent limited by the elasticity of the solid-rock interior of the moon. The stretching forms approximately symmetric **tidal bulges** toward and away from Earth. These bulges, raised in the solid body of the moon, are called **body tides** (Figure 3-12).

The same situation arises in the case of Earth but with two differences. First, as shown in Figure 3-13, Earth has oceans that are free to flow and form **ocean tides**, bulges A and B. Body tides form as well, but the ocean tides are much easier to observe, as beach people well know. Second, Earth rotates faster than the Earth-moon system revolves, so the side toward the moon is always changing. An observer at a given place on the turning Earth sees the ocean tide rise and then fall as the tidal bulge sweeps past him. In actual fact high tide does not come when the moon is overhead because the shapes of shorelines and other complications (Figure 3-13) modulate the tides.

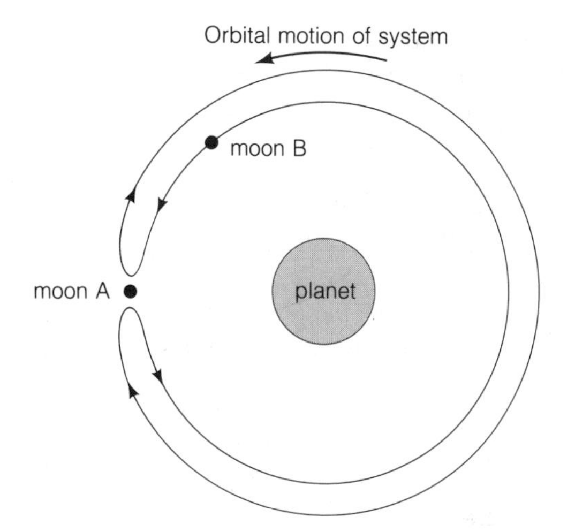

Figure 3-11. A horseshoe orbit: the apparent path of a small moon B as seen by an observer in a rotating coordinate system moving with moon A. (Both moons travel all the way around the planet, with respect to the stars. See text.)

Tidal Evolution of Orbits and Rotation Rates

Because Earth is turning, the tidal bulges A and B are dragged off the Earth-moon line. This happens to both the body and ocean tides, although the amount differs because of the different amounts of friction involved. Bulge A is thus in front of the moon as it moves in its orbit, and bulge B is behind the moon. The bulges are, in fact, concentrations of mass, and so by Newton's law of gravity they exert forces on the moon. Bulge A is closer to the moon and thus stronger; the net effect is to pull the moon ahead. It is as if the moon had a small rocket motor pulling it steadily ahead in its orbit. As a result, the moon is slowly spiraling outward, away from Earth. This effect was actually detected late in the nineteenth century when observers noticed that the moon tended to depart from the position predicted by ordinary, nontidal, two-body Keplerian laws. Stated another way, tidal effects cause a transfer of angular momentum between two orbiting bodies.

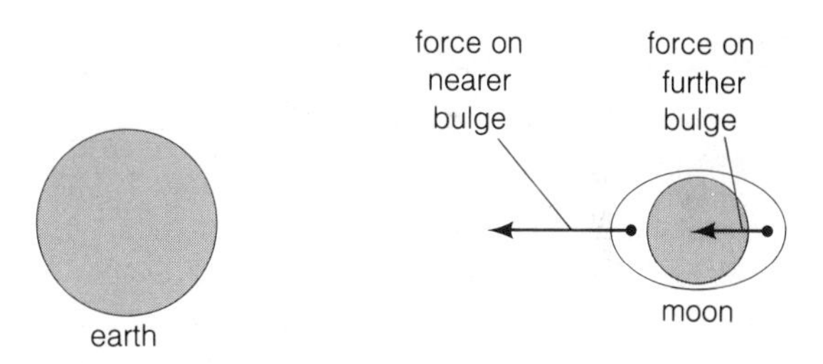

Figure 3-12. Body tides raised in the moon by the differential gravitational attraction of Earth.

In the same way, the moon has a net effect of pulling backward on bulge A, slowing Earth's rotation. Millions of years from now, the day will be noticeably longer than 24 h and the moon will be farther away. The angular momentum gained by the moon is given up by Earth, thus satisfying the law of conservation of angular momentum in the two-body system.

Similarly, Earth has tidally grabbed onto the longest axis of the moon and used it to slow the moon's rotation so that the moon now keeps its longest axis pointed along the Earth-moon line. This state is called **synchronous rotation** of a body—the state where rotation and revolution

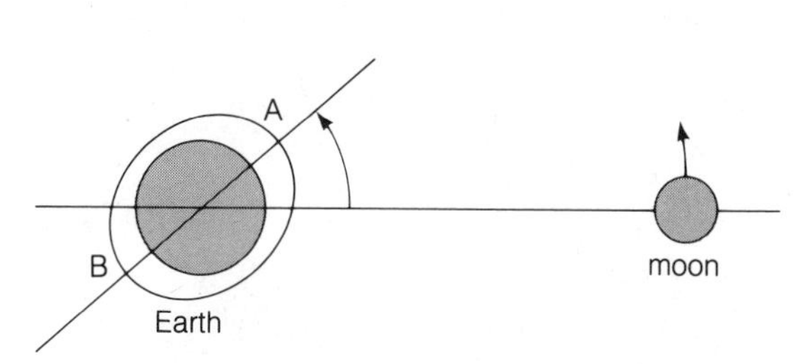

Figure 3-13. Tidal bulges (A and B) in Earth are dragged off the Earth-moon line by Earth's rotation. The effect of bulge A (which exceeds the effect of B) pulls the moon forward in its orbit, hence driving the orbital path slowly outward.

rates are equal so that one side of a satellite always faces the companion body about which it orbits.

In the past Earth turned faster, the moon was closer, and the month was shorter. Actual evidence of this comes from certain marine creatures whose shells have daily and monthly growth bands, allowing biologists to count the number of day bands in a monthly cycle in fossils of different ages. Whereas the present month of lunar phases (synodical month) is 29.5 d long, it was only about 29.1 d long 45 My ago (Kaula and Harris, 1975). Older fossils reveal monthly cycles as early as 2.8 Gy ago, when the month may have been as short as 17 d.

A linear extrapolation of the present recession would put the moon near Earth only 1 to 2 Gy ago, a result ruled out by the old fossils. In practice, it is not possible to compute when the moon was closest from tidal theory, because the extent and depth of oceans, and hence tidal effects, were different in the geologic past. Most researchers believe that the moon's closest approach was several billion years ago, probably during the planet-forming era. Mathematically, we can trace the motions backward and find that when the moon was closest, Earth's day (rotation period) was only 4 to 5 h long.

The same kind of tidal effects have acted on satellites throughout the solar system. The satellites closest to planets have had their long axes aligned toward the planet and are thus locked into synchronous rotation, keeping one side toward

MATHEMATICAL NOTES ON TIDAL FORCES

Newton's law of gravity states that

$$F = \frac{GM_{☾}M_{\oplus}}{r^2} \quad (1)$$

Defining masses as shown in Figure 3-14, we have

$M_{☾}$ = mass of moon
m = mass of tidal bulge
$M_{\oplus}$ = mass of Earth

Thus, the differential force on Earth caused by the moon is

$$dF_{\oplus} \propto r^{-3}\, dr$$

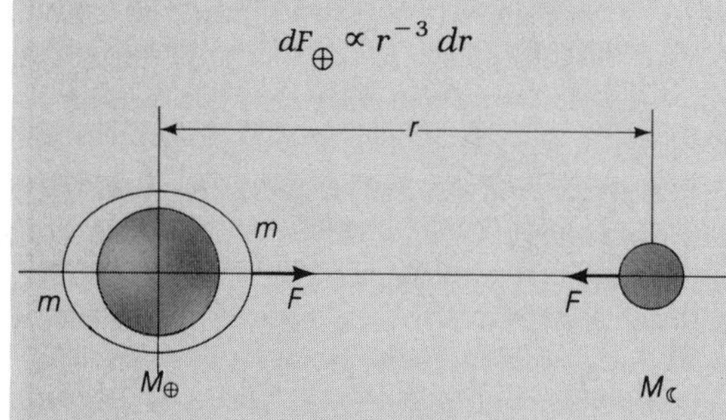

Figure 3-14. Properties related by Newton's law of gravitation in the case of tidal forces. $M_{\oplus}$ and $M_{☾}$ are the masses of Earth and the moon, which attract each other with force F; m's mark masses of tidal bulges.

The mass m of the tidal bulge raised by this force must be proportional to the force, so

$$m \propto dF_{\oplus} \propto r^{-3}\, dr \quad (2)$$

We want to know the net force on the moon caused by the two bulges. This is also a differential force, dF, proportional to the mass of the tidal bulge:

$$dF_{☾} \propto mr^{-3}\, dr \quad (3)$$

obtained by differentiating (1). This becomes

$$dF \propto r^{-6} \quad (4)$$

obtained by substituting (2) into (3). What we have shown is the important fact that *tide-raising forces are proportional to the inverse cube power of distance* and that the net effect of tidal forces depends on the inverse sixth power of distance. These are much stronger dependences on r than the inverse square law of gravity. Therefore, tidal effects can be very strong if the two bodies are close together but weak when the two bodies are somewhat farther apart. For this reason, tidal recession was very fast when the moon and Earth were close together, and is slow now.

the planet. The largest satellites are big enough to raise tidal bulges on the planets and thus are undergoing orbital evolution, usually spiraling outward like our own moon. In multisatellite systems, this has the interesting consequence that the largest satellites arrive in positions that put them in resonances with inner moons. Systems of several major satellites, such as the Galilean moons and several pairs of Saturn's moons, may have arrived in their observed resonant relationships in this way (Weisel, 1981).

Tides raised on the planets by the sun may have slowed the rotations of Mercury and Venus. Mercury is locked into a 2:3 resonant rotation (Figure 3-9b). In the case of Venus, the solar tide was probably raised mainly in Venus' dense atmosphere (Lago and Cazanave, 1979), although a resonant effect of tidal forces raised by Earth may also have been involved.

Roche's Limit

In discussing the mathematical theory of tides, we showed that the force raising a tidal bulge obeys a $1/r^3$ law (whereas the net tidal interaction obeys a $1/r^6$ law). This means that as the distance r decreases, the stretching force gets very large. There exists a critical distance between two bodies within which the tide-raising force on the smaller body is strong enough to tear it apart. This critical distance is called **Roche's limit** after its discoverer Edouard Roche (1850), a French mathematician.

To understand Roche's limit more clearly, imagine the Earth-moon system. Suppose that we represent the moon by two particles just touching each other (Figure 3-15). The only force holding the particles together is their own mutual gravity; the only force tending to separate them is the tidal force. If for any reason this moon approaches Earth, the tidal force will increase much faster than the gravitational force, and at some point it will exceed gravity, causing the two particles to drift apart.

In reality, of course, the moon is made of more than two particles. Its many particles are

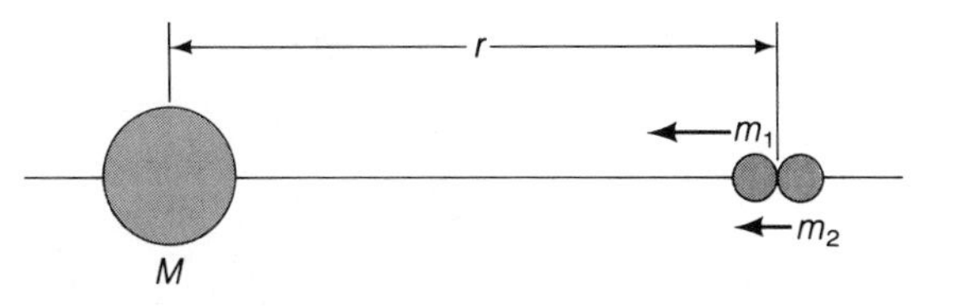

Figure 3-15. If two masses m_1 *and* m_2 *orbit around a planet* M, *Roche's limit occurs where the excess attraction of* m_1 *is equal to the gravitational attraction of* m_1 *and* m_2 *for each other.*

bonded together in the form of rock, with high **tensile strength** (strength against rupture by stretching), but the principle is the same. For mathematical convenience, Roche's limit is defined as the critical distance at which a body with no tensile strength would be torn apart by tidal forces. A solid-rock satellite would have to pass well inside Roche's limit before it would actually fragment.

RINGS

Planetary **rings** are systems of small bodies moving in circular orbits in a planet's equatorial plane. Roche's limit and resonances are believed important in shaping rings (Burns, 1981, provides more detail in a nontechnical review). As discussed in more detail in Chapter 5, the outer edges of ring systems lie near Roche's limit, suggesting that rings are fragments of weak bodies destroyed by approaching too close to a planet or fragments that never coalesced into a moon because they started too close.

As shown in close-up photos of rings and their shadows (Figure 3-16), rings have sharp edges and well-defined gaps. These features are believed to result primarily from resonant perturbation forces acting on the rings from the nearest satellites outside the rings. For instance, the inner edge of Saturn's ring B, the inner edge of Cassini's division, and the outer edge of ring A fall very close to the 1:3, 1:2, and 2:3 resonances of the nearby satellite Mimas (Cuzzi, 1978). Particles

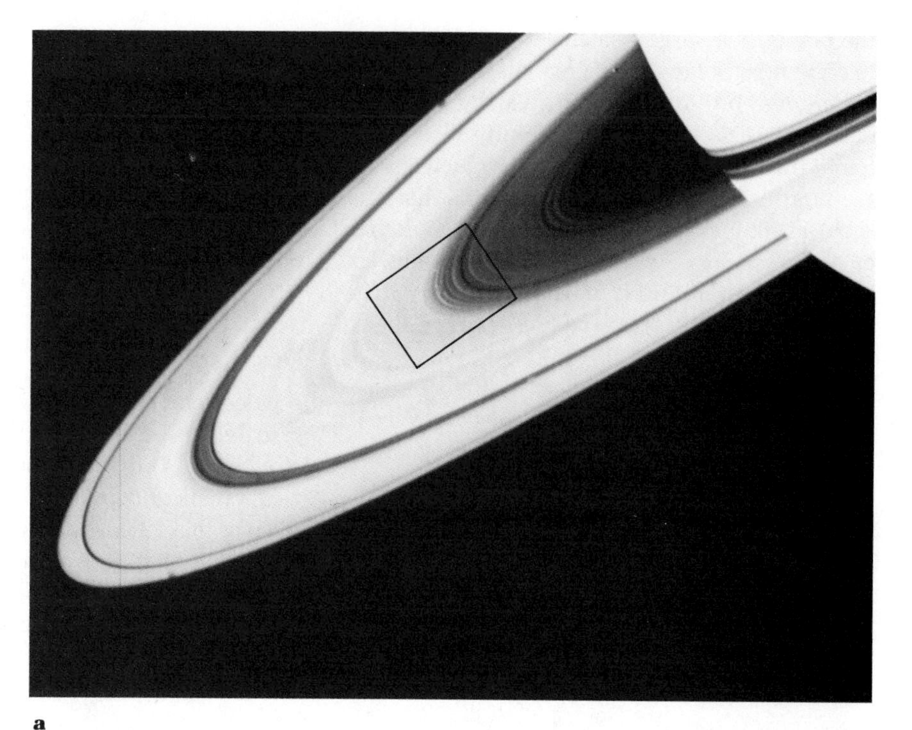

Figure 3-16. ***(a)*** *The rings of Saturn, photographed at intermediate range by Voyager 1, showed much more complex structure than expected from Earth-based observations. Hundreds of fine divisions were found, in addition to the well-known Cassini division (main dark gap) and Encke division (narrow gap, lower left edge). Note shadow of rings and gaps on Saturn's globe, upper right. Box shows region of picture* ***(b)****.* ***(b)*** *Detailed closeups of certain ring regions showed more unexpected anomalies. This picture is a composite of two views of the leading and trailing ansae. Comparison of thin bright ring inside dark gap at right shows that the thin ring is displaced from one view to the next: its position changed. Dynamical forces causing such changes in fine structure are not well understood. (NASA, Voyager photos)*

lying at these resonances are presumably kicked out of the rings, creating gaps and sharp edges.

The dynamics of n-body motions among the ring particles produces fascinating situations. Consider the moonlet in Figure 3-17, which moves in a circular orbit around Saturn. As in any Keplerian motion, particles on exterior orbits move slower, and particles on interior orbits move faster. Thus, small ring particles approach only from the outer leading quadrant (upper left in Figure 3-16) and the inner trailing (lower right) quadrant. But in the strange gravitational field inside Roche's limit, these particles are not pulled directly into the moonlet. As seen by an observer on the moonlet, approaching particles perform odd loops (related to horseshoe orbits) if the moonlet is large enough to have substantial gravity. Relative velocities among smaller interacting particles may be much less.

Because we are inside Roche's limit, you might suppose that there is no chance for one of the incoming particles to come to rest on the moon-

MATHEMATICAL NOTES ON ROCHE'S LIMIT

The mutual gravitational attraction for two equal-sized touching particles is (see Figure 3-15)

$$F = \frac{Gmm}{(dr)^2} \quad (1)$$

The disruptive tidal force is the differential gravity force:

$$dF = \frac{GMm}{r^3} 2dr \quad (2)$$

(see discussion of tides). At Roche's limit r_R, these two forces are equal. Equating (1) and (2) thus gives

$$\frac{Gm^2}{(dr)^2} = \frac{2GMm}{{r_R}^3} dr$$

Therefore,

Roche's limit (for two touching particles)

$$\equiv r_R = \left(\frac{2M}{m}\right)^{1/3} dr.$$

Since the mass of a body M is $4\pi R^3 \rho_M/3$ (where R is its radius and ρ_M its density) and since the radius of each body m is $dr/2$, this expression can be simplified to

$$r_R \simeq 2.5 \left(\frac{\rho_M}{\rho_m}\right)^{1/3} R$$

A somewhat more complex derivation gives the classical Roche expression for the breakup of a single zero-strength (or liquid) spherical body (instead of two touching spheres):

$$r_R = 2.44 \left(\frac{\rho_M}{\rho_m}\right)^{1/3} R$$

Aggarwal and Oberbeck (1974) have studied the case of the breakup of orbiting spheroidal bodies held together by the strength of their rocky or icy material. Generally, for bodies larger than about 40 km in diameter, orbiting icy or stony bodies of modest strength will break up at

$$r_{AO} = 1.38 \left(\frac{\rho_M}{\rho_m}\right)^{1/3} R$$

An incoming body about to impact a planet will get even closer to the planet before breaking up:

$$r_i = 1.19 \left(\frac{\rho_M}{\rho_m}\right)^{1/3} R$$

The actual limits depend on the size and strength of the bodies. Bodies smaller than about 30 to 60 km in diameter can penetrate much closer to a planet and will not break up at all if they are small and strong enough. Of course, aerodynamic stresses (treated by Melosh, 1981) may cause breakup of a different sort if the planet has a dense enough atmosphere.

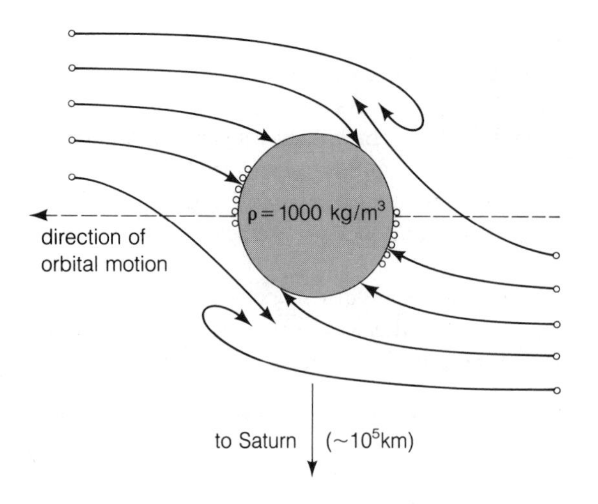

Figure 3-17. Motions of small, equally-spaced ring particles passing an icy, 200-km diameter moonlet in Saturn's rings as seen by an observer moving with the moonlet and looking down on the ring plane. Particles approach from only 2 quadrants and collide or pass by with a velocity of about 60 m/s. They land only in a restricted band. Although this example exceeds the size of likely moonlets in Saturn's rings, it illustrates effects that would occur to lesser extent among other major moonlets in ring systems. (Adapted from Greenberg and others, 1977; see text)

let's surface. But remember that the main Roche tidal disruptive force lies along the moonlet–Saturn line. Thus, at some point inside Roche's limit, a particle placed on the Saturn-facing side would spontaneously levitate and drift away. But a particle that landed 90° from the sub-Saturn point could remain on the moonlet's surface if it did not rebound. Thus, the moonlet could accrete material in a band about 90° from the Saturn direction. It could grow, slowly changing shape and axis of rotation (calculations by D. R. Davis; see Greenberg and others, 1977).

These results may explain the curious discovery that two opposing quadrants of Saturn's A ring (the quadrants preceding conjunction, as defined by the Saturn-Earth line of sight; that is, northwest and southeast of Saturn's disk) are a few percent brighter than the other two quadrants (Rietsema, Beebe, and Smith, 1976). This could arise if the larger ring particles are rotating synchronously and are struck by neighbors primarily only on the two quadrants shown in Figure 3-16, making the light-reflecting qualities of these sides different from those of the other two quadrants (Columbo, Goldreich, and Harris, 1976). Other explanations have been proposed (Franklin and Columbo, 1978). One involves formation of spiral, wavelike patterns in the rings, similar to spiral arms of some galaxies.

Another application of n-body ring dynamics is the attempt to understand unexpected details of ring structure observed by Voyagers 1 and 2 in the Saturn system. For example, Goldreich and Tremaine (1979) and others developed theories in which narrow rings could be maintained by the gravitational action of "shepherding satellites"—small moons in or between rings, which force ring particles into narrow rings like sheepdogs shepherding their flock. This theory gained support when Voyager 1 photographed two 100-km moons on either side of Saturn's narrow, outer F ring (Figure 3-18), apparently acting as shepherds in the way described by Goldreich and Tremaine. The theory received a setback, however, when a follow-up search by Voyager 2 revealed no analogous moons (down to a size of about 5 km) among the inner ringlets.

The detailed structures of rings thus remain puzzling subjects of current research in celestial mechanics. In addition to the question of how thousands of well-defined ringlets and gaps are maintained, scientists are pondering mysteries such as noncircular ringlets (Figure 3-16b) and ringlets with kinks or braids (Figure 3-19). These curiosities may involve disturbances of streams of ring particles as they pass by small moonlets, but the situation is very unclear.

Whatever the solution to ring dynamics, it is clear that ring systems offer extraordinary environments. In Saturn's rings the particles may be relatively close, as we infer from the fact that the rings may be no more than 100 m thick but absorb considerable light. They probably move along together in virtually circular orbits; Goldreich and Tremaine (1978) suggest a random velocity dispersion of 0.2 cm/s or less! In the rings we would be surrounded by a cloud of floating hailstones,

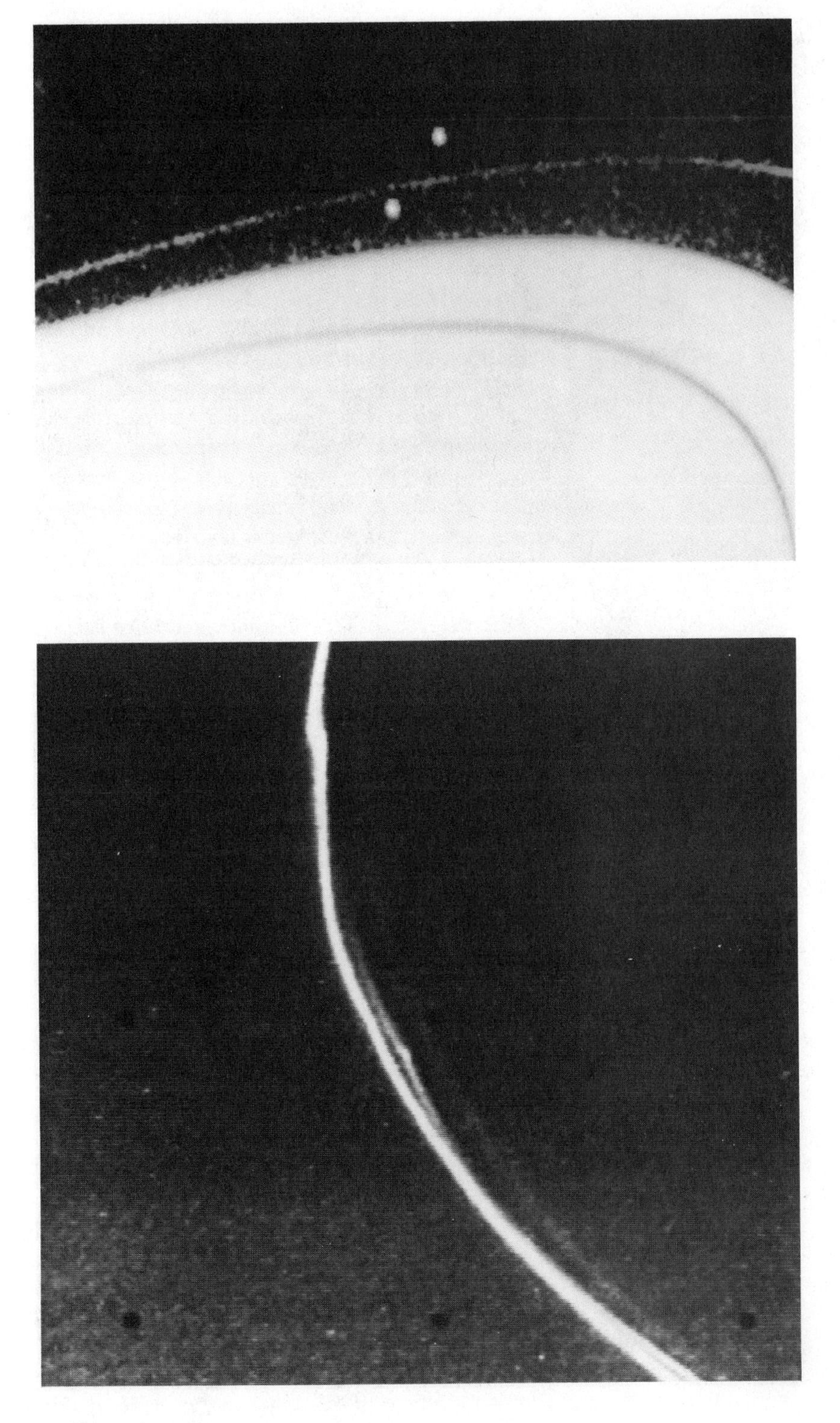

Figure 3-18. Two "shepherding satellites" lie on either side of Saturn's narrow F ring. The lower part of the picture is filled by the outer part of Ring A. In this view the satellites are less than 1800 km apart, and will pass each other in about 2 hours. (NASA, Voyager 2)

Figure 3-19. Unexpected twists or braids in Saturn's narrow F ring were discovered by Voyager 1. The diffuse ring segment to the right is about 35 km across. The knot near the center may be a local concentration of ring particles. (NASA, Voyager 1)

perhaps occasionally jostling into one another, as imagined in Figure 3-20.

DYNAMICAL EFFECTS OF SOLAR RADIATION AND SOLAR WIND

Radiation Pressure

Sunlight, like all light, is a form of electromagnetic radiation. Electromagnetic radiation consists of pulses transmitted through electric and magnetic fields; it has some of the properties of wave motion yet some of the properties of particles. We commonly speak of radiowaves or the wavelength of light, but the particlelike properties are not so familiar. One of these properties is that the pulses of light, called **photons**, carry momentum. Like BBs, they transmit this momentum to whatever they strike. When a photon strikes an object, an impulse is transmitted away from the light source.

Suppose a body is orbiting in space, exposed to the sunlight. The body is being struck by photons that transmit momentum (that is, exert pressure on the body). This phenomenon, called **radiation pressure**, has the effect of a force pushing the body away from the sun. If the cross-sectional area exposed to the sun is very large and the mass of the body is low, the radiation force can exceed the gravitational pull of the sun, causing the body to be "blown out" through the solar system.

One application of this has been the idea of a radiation-pressure-driven spacecraft, which would consist of an enormous thin sail. No such spacecraft has yet been built, although, minor radiation pressure effects have been utilized in navigating existing space vehicles. True solar sailing would allow long solar system voyages with low power and low fuel consumption.

Figure 3-20. (left) An imaginary view within Saturn's ring system shows a sky full of floating hailstones. Saturn's cloudy surface lies below; the sun lies some degrees to the left of the ring plane. The densely populated portion of the rings may be as little as 100 m thick. (Painting by author)

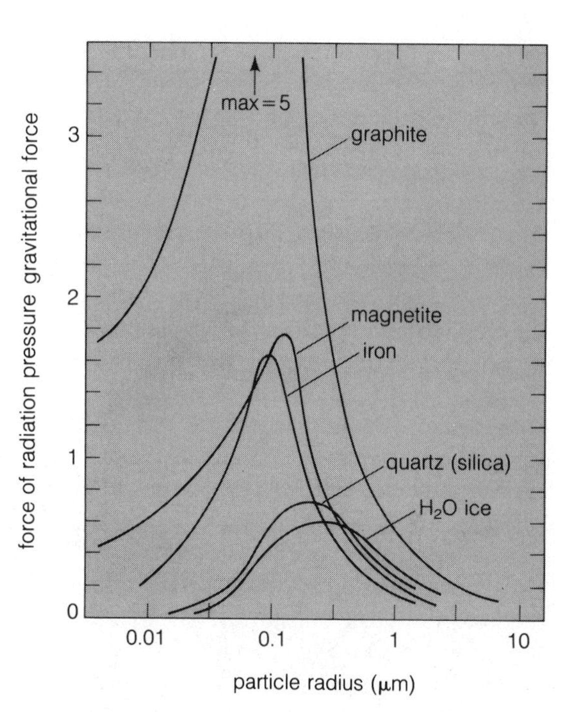

Figure 3-21. Ratio of outward radiation force to inward gravitational force for various-sized particles orbiting the sun. The force due to radiation pressure tends to maximize around sizes of 0.1 to 1 μm, but is greater for darker-colored particles. (Adapted from Burns, Lamy, and Soter, 1979)

Radiation pressure affects interplanetary dust particles in the solar system. Because the ratio of cross-sectional area to mass (r^2:r^3) increases as the radius gets smaller, only small particles are strongly affected. (But if the radius gets much smaller than the wavelength of sunlight [5×10^{-7} m], the interaction with photons changes and radiation pressure again becomes negligible.)

Burns, Lamy, and Soter (1977) describe these effects and derive the results shown in Figure 3-21 for different kinds of particles. Radiation pressure acts strongly on micrometer- and submicrometer-sized particles ($1 \mu m = 10^{-6}$ m). Since dark particles absorb the most light, the effect is strongest for them. Radiation pressure acts so strongly on submicrometer graphite and metal particles that they could be blown out of the solar system even if initially pushed toward the sun. For other submicrometer particles of stone or ice, radiation pressure merely decreases the net force toward the sun.

Solar Wind and Interplanetary Gas Motions

Analyses of comets, especially by Biermann (1951), showed that submicrometer material expelled from them is being accelerated away from the sun faster than can be accounted for by radiation pressure, suggesting that the interplanetary gas itself is moving away from the sun and carrying the cometary material with it. This idea led to the concept of the **solar wind**—an expanding, low-density **plasma** (ionized gas), emanating from the sun. Near Earth, this material is composed of a plasma with average electron densities around 2 electrons/cm^3, temperatures around 200 000 K, and expansion velocities around 600 km/s, ranging occasionally up to 1000 km/s. The

MATHEMATICAL NOTES ON RADIATION PRESSURE

Pressure can be expressed as the rate of transfer of momentum to a unit surface. The momentum carried by an individual photon is $h\nu/c$, where h is Planck's constant (see Table 2-2), ν is the frequency of the light, and c is the velocity of light. The total rate of momentum transfer is this momentum per photon times the number of photons per square meter per second for all frequencies:

$$P = \sum_{\nu} \frac{h\nu}{c}\frac{dn}{dt}\bigg|_{\nu} = \frac{\mathcal{F}}{c}$$

where $\mathcal{F}$ is the total flux of radiation in Joules per square meter per second. The substitution of $\mathcal{F}$ can be made because $h\nu$ is the well-known energy per photon, so the total flux of energy is $h\nu(dn/dt)$ summed over all frequencies. If we want to know the force caused by radiation pressure on a small particle, we recall that pressure is force divided by area. Thus,

$$\text{radiation force on a spherical particle} = F = \frac{\mathcal{F}}{c}\pi a^2 Q$$

where a is the particle radius and Q is a correction factor on the cross section, since small particles may have an absorption cross section different from the geometric cross section:

$$Q = \text{correction factor} = \frac{\text{effective absorption cross section}}{\text{geometric cross section}}$$

Q depends on particle size, particle composition, and wavelength of the light, but typically has values of 0.1 to 1.0.

■ *Sample Problem*

For particles of what size would the outward force of radiation from the sun be just balanced by the inward force of gravity? Assume that the particle is in Earth's orbit at solar distance r and that it is made of stone with density $\rho = 3 \times 10^3$ kg/m^3 and that $Q = \frac{1}{2}$.

Solution

Equating radiation and gravity force from solar mass M, we have

$$\frac{\mathcal{F}\pi a^2 Q}{c} = \frac{GMm}{r^2} = \frac{GM}{r^2}\frac{4\pi a^3 \rho}{3}.$$

Solving for a and substituting the values for the constants (from Table 2-2), we have

$$a \cong 1 \times 10^{-7}\ \text{m}.$$

This is about 0.1 μm. Particles of this size and smaller would thus be blown out of the solar system when exposed to sunlight.

solar wind can thus carry gas and fine dust from the inner to the outer solar system in only a month or so.

Theoretical models treat the solar wind in the first approximation as an expansion of the hot gases of the sun's outer atmosphere, or **solar corona** (temperature about 2 million K at three solar radii from the sun's center).

Spacecraft have returned good observations of the solar wind all the way out to the vicinity of Saturn. A current problem is the nature of the interface of the solar wind with the interstellar gas. One hypothesis places this beyond Pluto's orbit at around 50 AU from the sun. Space probes that leave the solar system, such as Pioneer 10, may clarify the location of this interface.

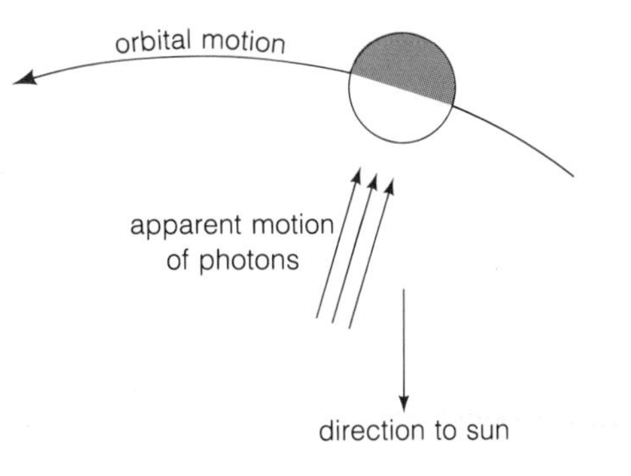

Figure 3-22 The Poynting-Robertson effect is caused by drag due to the apparent displacement of photons from the solar direction (see the text).

Poynting-Robertson Effect

The **Poynting-Robertson effect** is an interaction of light with (roughly) centimeter-scale particles. Unlike the solar wind and radiation pressure, this effect causes the particles to spiral *inward* toward the sun.

Consider a small particle moving in a circular orbit around the sun. Sunlight, which can be thought of as a steady stream of photons, flows outward from the sun, while the particle moves at right angles to the photon stream. Thus the photons strike the particle preferentially on its leading side, just as a car driven through a rainstorm is struck on its front side even though the rain may be falling vertically. This effect is shown in Figure 3-22. The displacement of the apparent source of the light is called the **aberration of light**; the apparent positions of stars seen from Earth are actually displaced up to 20.5″ of arc due to the aberration of light, as discovered by the English astronomer James Bradley in 1727.

The preferential absorption of photon momentum on the leading side is only one small part of a complex relativistic effect, whereby reradiation of absorbed photons causes a net loss of energy, and the particle slowly settles into a smaller and smaller orbit, spiraling in toward the sun.

This effect was first predicted by the British physicist Poynting (1903) and later amended by the American physicist Robertson (1937) to take relativity into account. Wyatt and Whipple (1950)

MATHEMATICAL NOTES ON THE POYNTING-ROBERTSON EFFECT

A concise discussion of the Poynting-Robertson effect (sometimes called the PR effect) is given by Burns, Lamy, and Soter (1979), who discuss several radiation effects. In general, if a particle starts on an eccentric orbit, the eccentricity will be reduced until the orbit is virtually circular. During the rest of its history, the particle will slowly spiral inward toward the sun (though any single trip around the sun will follow a nearly circular path). Most of the time may be spent in a fairly elliptical orbit. Once circularized, the orbit decays into the sun in a time given by

$$t_{PR} = 7.0\,(10^6)\,a\rho r^2/Q$$

where

t_{PR} = decay time (y)
a = particle radius (m)
ρ = particle density (kg/m^3)
r = orbit radius (AU)
Q = correction factor (typically 0.1 to 1.0)

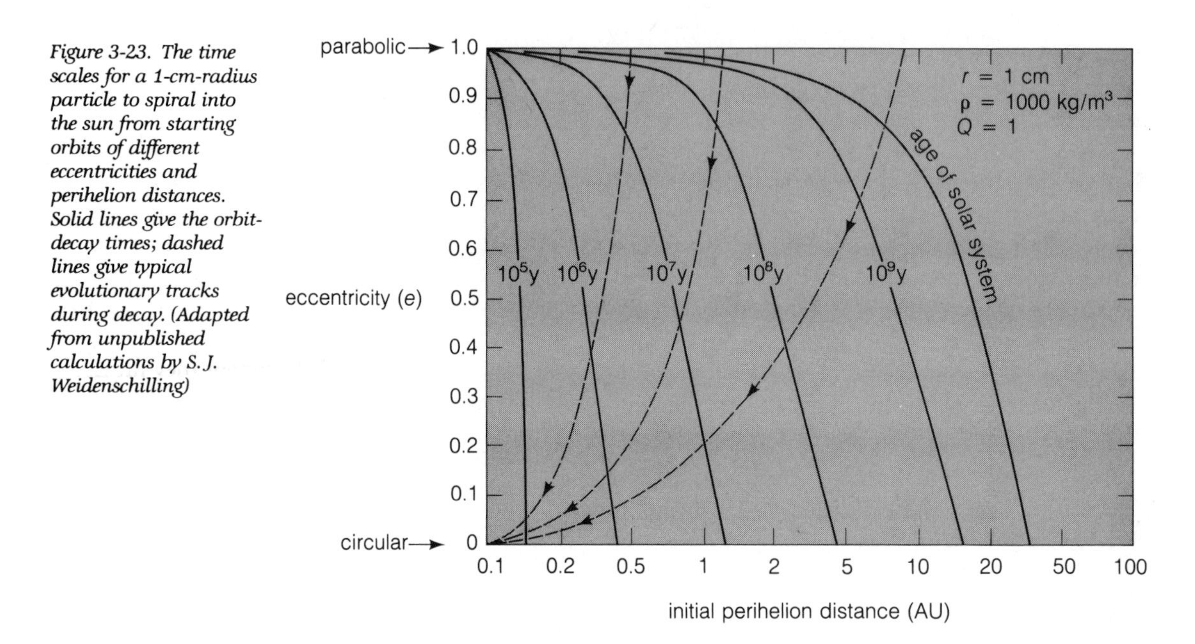

Figure 3-23. The time scales for a 1-cm-radius particle to spiral into the sun from starting orbits of different eccentricities and perihelion distances. Solid lines give the orbit-decay times; dashed lines give typical evolutionary tracks during decay. (Adapted from unpublished calculations by S. J. Weidenschilling)

showed how this effect affects motions of interplanetary debris.

Mathematical theory enables us to calculate the history of a particle of any size, starting in any given orbit. Particles smaller than a few centimeters across will be swept from most orbits into the sun in a time less than the history of the solar system. Figure 3-23 summarizes the situation by plotting the time scale (solid lines) for particles with any initial eccentricity and perihelion distance to be swept into the sun. The dashed lines show the track followed by sample particles as they evolve across the diagram. For instance, a 1-cm particle starting at 1-AU perihelion with an eccentricity of about 0.7 will have a lifetime of a few 10^7 y. In its last million years, the particle will have an eccentricity of only about 0.15 and a perihelion of about 0.4 AU. Particles with a 10-cm radius have lifetimes 10 times as long, particles with 0.1-cm radius have lifetimes 1⁄10 as long, and so on.

Combined Poynting-Robertson and radiation pressure effects may cause a particle as small as 0.1 to 1.0 μm to evolve a quite elliptical orbit, in some cases removing it outward from the solar system by radiation pressure.

Yarkovsky Effect

A curious effect with a curious history is related to the Poynting-Robertson effect. The **Yarkovsky effect** is an orbital change on (roughly) meter-scale rotating particles due to the radiation that they emit after they are warmed by the sun. If there were no rotation, the warmest spot would be the noon area, but rotation carries this spot around to the afternoon side. All bodies radiate photons in proportion to T^4 (T = surface temperature), so that the warmest area radiates the most photons. Thus, the most photon momentum is ejected on the afternoon side, and these photons act like a rocket exhaust, pushing the particle forward into an ever-expanding orbit. But if either the rotation or the orbital motion is retrograde, the effect is reversed and the particle spirals inward.* Thus, if the rotation direction of an asteroid or comet were occasionally changed

*The Yarkovsky effect can also change the inclination of the particle's orbit if the particle's "equator" is not in the orbital plane, though in the presence of planetary perturbations this is probably a small effect (S. J. Weidenschilling, private communication).

by collisions, the planetesimal might wander about the solar system, sometimes moving outward and sometimes spiraling in toward the sun.

The Yarkovsky effect became widely known only in recent years after E. J. Öpik (1951) pointed out that it had been discovered by a Russian engineer, I. O. Yarkovsky, around 1900 and described in a paper that was lost. Subsequently this effect was apparently rediscovered independently by the Russian astronomer V. V. Radzievskii in 1952 and again by the American astronomer C. Peterson in his master's thesis at MIT (Burns, Lamy, and Soter, 1979).

Evaluating the actual importance and time scale of the effect is difficult because of uncertainties in the rotations and changes of rotation of small particles. The consensus seems to be that the Yarkovsky effect is the most important effect in changing orbits of bodies ranging in size from about 1 m up to perhaps tens of meters (Burns, Lamy, and Soter, 1978); other influences offset the Yarkovsky effect on larger and smaller bodies. An example of the application of the Yarkovsky effect is Peterson's (1976) suggestion that asteroid fragments are moved by this effect until their orbits reach configurations where resonances with Jupiter allow them to be thrown onto Earth-crossing orbits, where they hit Earth and become meteorites.

TURBULENCE

Turbulence, the swirling motion of moving gas or liquid, is familiar to anyone who has watched

MATHEMATICAL NOTES ON TURBULENCE

In the mathematical theory of fluid mechanics there is a quantity known as **vorticity**, which is a measure of the amount of rotary circulation of the medium per unit area. The *vorticity equation* gives the rate at which vorticity increases or decreases. This equation has several positive terms that increase vorticity and a negative term that decreases it. The ratio of these terms gives the Reynolds number, which can be expressed as

$$R \equiv \frac{vL}{\nu}$$

where

v = flow velocity in the medium

L = dimension of possible turbulent eddies

ν = kinematic viscocity = $\dfrac{\text{absolute viscosity}}{\text{density}}$

In looking up viscosities, the reader should be aware of the difference between kinematic viscosity and absolute (sometimes called dynamic) viscosity, as given above. Some typical absolute viscosities are

1.0×10^{-3} N·s/m^2 for water near room temperature
1.8×10^{-5} N·s/m^2 for air near room temperature
8.8×10^{-6} N·s/m^2 for H_2 gas near room temperature
5.7×10^{-6} N·s/m^2 for H_2 gas at 160 K
7.6×10^{-6} N·s/m^2 for CH_4 gas at 195 K

■ *Sample Problem*

With normal wind conditions, would you expect turbulence in Earth's atmosphere involving dimensions like those of a thunderstorm?

Solution

We must use SI units. Windy gusts of 36 km/h equal an air speed of 10 m/s for v. If we take 10 km for the storm dimension, $L = 10^4$ m. The kinematic viscosity will be the absolute viscosity of air (above) divided by air's density of about 1 kg/m^3. Thus the Reynold's number comes out to be of the order 10^{10}, which is considerably greater than 1; and we correctly predict turbulence on this scale, as confirmed by satellite photos of storm clouds.

the eddies in a flowing river, or an explosion, or the dust swirling behind a passing car. It is the opposite of streamline flow, or **laminar flow**. In planetary science turbulence is important in a number of cases. The clouds of gas and dust from which stars form are turbulent; turbulence characterizes the motions of clouds in some parts of planetary atmospheres.

The **Reynolds number** is a useful concept in discussing turbulence; it tells whether turbulence will occur on a certain scale in a given gaseous medium. In the Reynolds number, which is a ratio, the numerator represents the forces that promote turbulence and the denominator represents viscous forces that tend to damp out turbulence. If the Reynolds number is greater than 1, turbulence will increase in the medium; if it is less than 1, turbulence will decrease.

SUMMARY

The motions of the main bodies in the solar system, such as planets orbiting the sun or small satellites orbiting planets, are well described by Kepler's three laws. These in turn have been given a satisfactory physical basis by Newton's laws of motion and law of gravitation. Motions of one small body around one large body (an example of the two-body problem) are easy to treat and well understood. The three-body or n-body problems are much harder and result in many interesting phenomena in the solar system, including small perturbations of orbits, resonance, and Lagrangian points. Many other effects cause major or minor alterations of Keplerian orbits. Tidal effects, which change orbits and planetary spins, are examples of motions modified by massive bulges raised in two-body (or n-body) systems by gravity. Tidal effects also create Roche's limit and help explain planetary rings. Orbiting bodies interact with solar radiation and the solar wind in ways that tend to eliminate small particles from the solar system, either by blowing them out or by pulling them into the sun. As a function of size range, the effects are as shown at the bottom of this page.

CONCEPTS

celestial mechanics
astrology
scientific method
Ptolemaic system
Newton's universal law of gravitation
Kepler's laws
foci
primary body
periapse
apoapse
orbital elements
semimajor axis
eccentricity
inclination
period
gravity
law of gravitation
circular velocity
escape velocity
parabolic orbit
parabolic velocity
hyperbola
conic sections
astrometry
spectroscopy
Doppler shift
radial velocity
velocity equation
two-body problem
three-body problem
numerical integration
perturbations
periodic
secular
commensurable
resonance
Kirkwood gap
Lagrangian points
L_4, L_5
Trojan asteroids
net differential force
tidal bulge
body tide
ocean tide
synchronous rotation
Roche's limit
tensile strength
ring
photons
radiation pressure
solar wind
plasma
solar corona
Poynting-Robertson effect

Diameter (m): 10^{-9} 10^{-8} 10^{-7} 10^{-6} 10^{-5} 10^{-4} 10^{-3} 10^{-2} 10^{-1} 10^{0} 10^{1} 10^{2} 10^{3}

Effect:	**Solar wind**	**radiation pressure**	**Poynting-Robertson effect**	**Yarkovsky effect**	**negligible force**
	blows out	blows out	spirals inward	acts in or out	assumes Keplerian orbit

aberration of light
Yarkovsky effect
turbulence
laminar flow
Reynolds number
vorticity

PROBLEMS

1. Why did Kepler's, Galileo's, and Newton's celestial mechanics discoveries have such a profound philosophical impact on humanity?

2. Why are those discoveries called the Copernican revolution?

3. Consider three hypothetical solar systems:

a. the sun with one planet of Earth-like size;

b. the sun with two planets, Jupiter and Earth; and

c. the sun with two companions, a "planet" of 0.1 solar mass and Earth.

In which system would the Earth-sized planet most nearly obey Kepler's original laws? Why?

4. In which of the above systems would the motion of the Earth-like planet be most disturbed and least well obey Kepler's laws? Why?

5. Compare the revolution periods of asteroids to the revolution period of Jupiter and explain why, in some cases, those orbital relations may result in asteroids being kicked out of the central asteroid belt into other parts of the solar system. What features of the asteroid belt's structure may arise from this process?

6. Suppose Jupiter had only one large satellite (as large or larger than Ganymede). Where might you look in that system for evidence of possible smaller bodies or trapped meteoroidal debris? Why?

7. If Jupiter's semimajor axis is 5.2 and Jupiter has a period of 11.86 y, what is the typical semimajor axis and period for a Trojan asteroid?

8. Figure 3-13 shows how the tidal bulge raised on Earth by the moon accelerates the moon forward (like a small rocket pushing forward) and hence drives it slowly away from Earth. (a) Consider a system (such as Neptune's) in which the planet rotates prograde but the slowly orbiting large moon revolves retrograde. What is the result in terms of orbital evolution of the satellite? (b) What is the result if the planet rotates retrograde?

9. Why is tidal evolution ineffective if the satellite is much much smaller than the planet, as in the case of Mars and Phobos?

10. Suppose a comet breaks apart while in an elliptical orbit, producing dark-colored silicate particles of various sizes traveling in the same orbit. The diameters are 0.01 μm, 1 μm, 1 mm, and 1 m. Predict what happens to each particle, and roughly estimate the time scale involved.

ADVANCED PROBLEMS

11. Show that an asteroid at $a = 3.28$ AU (just outside the orbit of the largest asteroid, Ceres) is in a 1:2 commensurability with Jupiter. Are many such asteroids known?

12. Show that the velocity of Earth in its orbit around the sun is about 30 km/s.

13. Show that the velocity equation reduces to the equation given for circular velocity (when $r = a$) and the equation given for escape velocity (under the appropriate condition in that case).

14. If a comet moving on a parabolic prograde orbit passes Earth, how fast would it move relative to Earth? What if it is moving on a retrograde parabolic orbit?

15. Escape velocity from a planet's surface is equivalent to the speed at which a falling object hits the ground if dropped with zero initial velocity from infinity (or very high altitude), neglecting atmospheric drag. What are the fastest and slowest velocities at which a large meteorite orbiting the sun could strike Earth's surface? Explain your reasoning.

16. (a) In the past, when the moon was 0.25 as far from Earth as it is now, estimate how much more massive Earth's average tidal bulge was than it is today. (b) How much stronger was the net accelerating effect of this bulge on the moon?

17. An early measurement of the outer radius of Uranus' rings was about 5.0×10^7 m from the center of the planet. Compare this to the Roche limit for icy bodies near Uranus, using the expression derived here for touching particles. What is the significance of this result? Would you expect Uranus' ring particles to aggregate into a satellite?

18. Show that the ratio between outward radiation force and inward gravity force is independent of the distance from the sun.

19. (a) Consider two particles in circular orbits in Saturn's rings 10^8 m from Saturn's center. One is located 1 m farther from Saturn than the other. By Kepler's laws they have different periods and must occasionally pass each other. How fast do they pass by each other? (*Hint:* You could compute V_{circ} for each particle and subtract, but the difference in orbits is only one part in 10^8, so you would have to maintain accuracy to at least eight decimal places, which is unrealistic. Instead, use calculus and differentiate $V_{circ} = (GM)^{1/2}(r^{-1/2})$ with respect to r, getting

$$dV_{circ} = \text{difference in velocity}$$
$$= \tfrac{1}{2}(GM)^{1/2} r^{-3/2}\, dr$$

where dr is the difference in distance.

(b) What if the particles are about 200 km apart, as in the example in Figure 3-17? Compare your result with the 60 m/s figure given there (based on a more detailed calculation including gravitational attraction of the particles for each other).

(c) How fast would a shuttle in a 200-km-high circular orbit approach a space lab in a 201-km-high circular orbit? (Answer: about 0.6 m/s.)

20. During a Mars exploration expedition, a package is to be launched from Phobos to reach a party of astronauts on Deimos. Relative to Phobos, what launch velocity is needed for the minimum energy orbit, which has periapsis on Phobos and apapsis on Deimos?

21. Two asteroids in the asteroid belt collide. They are on roughly circular orbits whose semimajor axes differ by 0.3 AU. At what speed do they collide?

22. A tidal stretching force of 1 lb is observed across a large space structure constructed in synchronous orbit around Earth. Later, it is decided to bring the structure down to a height of 200 km. What is the new stretching force?

23. During planet formation, a planet has acquired a satellite of rocky or icy material. The planet itself is still growing rapidly, due to accretion of meteoritic material. (a) How does the satellite's orbit change? (b) What will happen to the satellite as the satellite's orbit evolves? Specify distances for various events. (c) Describe the plausible final appearance of the system as compared to the appearance of, say, Saturn.

CHAPTER FOUR

THE FORMATION OF STARS AND PLANETARY MATERIAL

Having surveyed our planetary system and the forces that control the motions of particles in it (Chapters 2 and 3), we are ready to ask how this system formed. This question, as it turns out, boils down to a question of how innumerable small particles around the early sun aggregated into a few big particles—the planets. To see this, it will be helpful to consider the general astrophysical question of how stars form.

EVIDENCE THAT STARS ARE FORMING TODAY

Two proofs that stars have continued to form throughout astronomical time come from a combination of stellar and planetary research. The first involves the relative youthfulness of the solar system itself. Stellar studies show that the universe began in an outrushing of material, called the **big bang**, some 10 to 18 Gy ago. Galaxies, or huge masses of typically 10^{11} stars, aggregated out of the outrushing material; our galaxy formed about 10 to 12 Gy ago. Yet from several kinds of dating methods based on the study of radioactive isotopes in lunar samples, terrestrial samples, and many different meteorites, we know that our solar system formed only 4.6 Gy ago. So our solar system has formed in only the last 38% of our galaxy's history; all stars did not form at the beginning.

The second proof of continuing star activity is based on astrophysical understanding of stellar evolution. A star like the sun takes about 10 Gy to consume its supply of hydrogen fuel. But the more massive a star, the hotter it is and the faster it consumes its fuel. Consequently, a large star has a shorter lifetime as a normal, luminous star. A star of two solar masses ($2\ M_{\odot}$) may last only about 1 Gy; a star of $4\ M_{\odot}$, about 0.1 Gy; a star of $10\ M_{\odot}$, about 0.01 Gy. The most massive stars, of several tens of solar masses, may last only a few million years and then explode in dazzling **supernovas**, throwing much of their interior into neighboring interstellar space.

Many of these massive stars exist, and we have

The Pleiades star cluster illustrates some features of star formation, including formation in groups and association with gas and dust, which form clouds of debris around the young stars. Brighter cores of clouds in this picture are about 20 000 AU across, similar in dimension to the cloud of comets surrounding the solar system. (Haute Provence Observatory)

historical records of supernovas, proving that stars have formed within the last few million years. Still more interesting is the fact that these stars invariably occur in **open star clusters**—groupings of several hundred stars that typically disband in a few hundred million years—and in **molecular clouds**—the densest, cool concentrations of interstellar dust and gas. Open star clusters and molecular clouds are often associated with larger **nebulas**—huge, diffuse clouds of interstellar dust and gas, such as shown in Figure 4-1.

These findings have led to several conclusions about how stars form. First, they form in clusters rather than singly. Second, the clusters form from nebular concentrations of **interstellar material**—huge clouds of atoms, molecules, and dust grains. These findings suggest that interstellar material is the ultimate parent material of stars and planetary systems and that we should look at it more closely.

THE INTERSTELLAR MATERIAL

All evidence indicates that the overall composition of the interstellar material approximately matches that of the sun, supporting the idea that the sun formed from a concentration of such material. This composition is shown in Table 4-1. As shown in the table, spectroscopic studies show that some of the heavier elements, such as Ca, Al, Fe, Mg, and C are strongly depleted in the gas and concentrated in the dust. Provocatively, these are just the planet-forming elements! These elements are especially prone to aggregate into

Table 4-1
Composition of Interstellar Medium in Our Part of the Galaxy

		Composition (% by mass)		
Element	**Atomic no.**	**Interstellar medium[a] (total)**	**Interstellar atomic gas[b]**	**Concentration factor (total ÷ gas)[c]**
H	1	78.3	~78	1
He	2	19.8	~20	1
O	8	0.8	0.2	4
C	6	0.3	0.02	15
N	7	0.2	0.03	7
Ne	10	0.2	0.2	1
Ni	28	0.2	0.1?	2?
Si	14	0.06	0.03	2
S	16	0.04	0.03	1.3
Fe	26	0.04	0.001	40
Mg	12	0.015	0.0007	21
Ca	20	0.009	0.000002	4500
Al	13	0.006	0.00003	200
Ar	18	0.006	0.001	6
Na	11	0.003	0.0006	5
		99.98	~99.6	

[a]Includes atoms, molecules, and dust grains. Based on observed solar composition (which represents all interstellar material from which sun formed). Adapted from Gibson (1973).
[b]Does not include molecules and dust grains. Estimates based on abundances spectroscopically observed in interstellar gas between the solar system and the star ζ Ophiuchi as reported by Field (1974).
[c]Probable indicator of concentrations in minerals of interstellar dust grains (quotient of preceding two columns).

Figure 4-1. Dramatic nebular clouds of dust and gas lie in the star-forming region of the constellation Monoceros. The dark "cone" is a dust-grain-and-gas concentration estimated to be about 50 000 AU in width, silhouetted against more distant glowing clouds. This region lies in the outer part of the young star cluster NGC 2264, which is only a few million years old. Molecules of formaldehyde (H_2CO) have been found in the region. (Lick Observatory)

dust grains. The grains are the first steps on the road to rocky and icy planetary material.

Molecules, Molecular Clouds, and Dust

For many years, the formation of interstellar dust grains was a mystery. Early calculation showed that in most of interstellar space, atoms are too far apart to collide often enough to form molecules (Oort and van de Hulst, 1946). Where, then, could molecules form? Astronomical research in the 1970s showed that molecular clouds are the dense, cool environments where collisions occur frequently enough to allow many molecules and dust grains to aggregate. For example, hydrogen atoms (H) collide and form molecular hydrogen, H_2.

Even larger particles form, as revealed by recent infrared spectroscopic observations (see Figure 4-2). Most molecules and mineral grains tend to absorb certain colors of light, not in the visible parts of the spectrum, but at **infrared wavelengths**, the portion of the spectrum with wavelength too long for us to see. Detectors for measuring infrared light were not developed until about World War II and not widely applied in astronomy until after 1960. An explosion in the discovery of interstellar molecules occurred in the late 1960s. One molecular (hydroxyl, OH) was discovered between 1963 and 1967; 24 molecules were discovered in the next 5 y. These include water (H_2O), ammonia (NH_3), formaldehyde (HCNO), and carbon monoxide (CO). A 1978 detection of methane (CH_4) in the Orion nebula was its first detection outside the solar system. Complex molecules such as methylam-

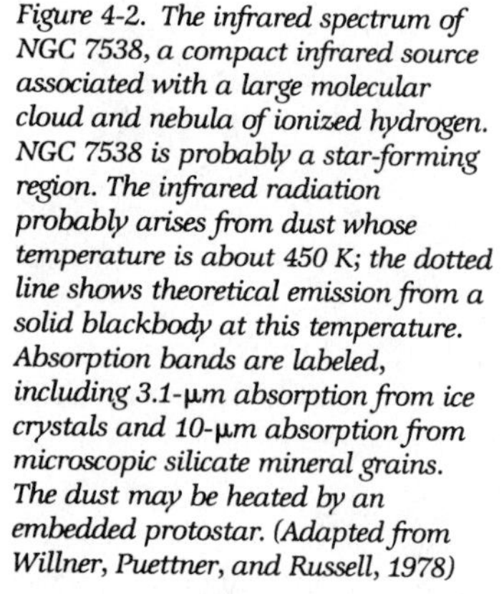

Figure 4-2. The infrared spectrum of NGC 7538, a compact infrared source associated with a large molecular cloud and nebula of ionized hydrogen. NGC 7538 is probably a star-forming region. The infrared radiation probably arises from dust whose temperature is about 450 K; the dotted line shows theoretical emission from a solid blackbody at this temperature. Absorption bands are labeled, including 3.1-μm absorption from ice crystals and 10-μm absorption from microscopic silicate mineral grains. The dust may be heated by an embedded protostar. (Adapted from Willner, Puettner, and Russell, 1978)

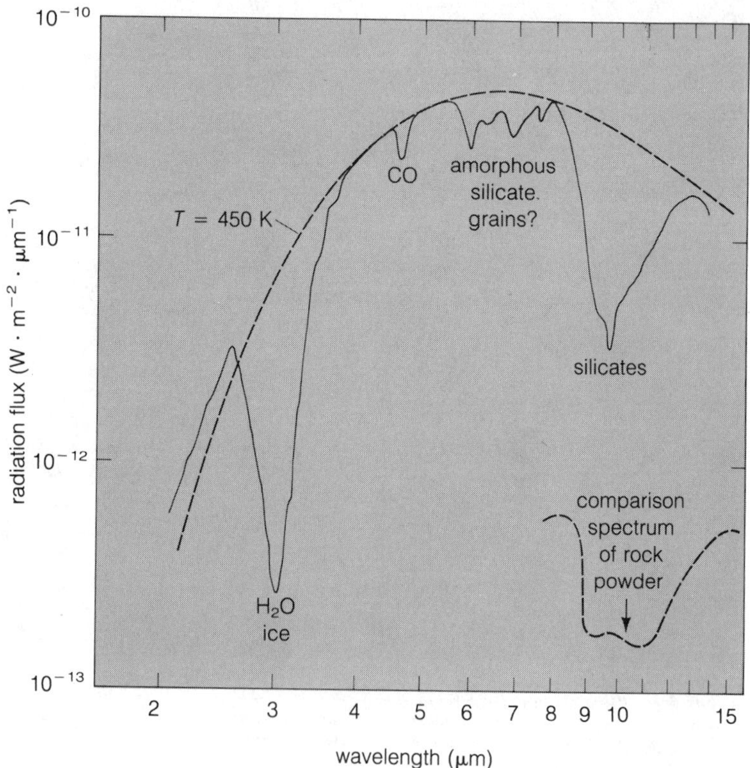

ine (CH_3NH_2) and ethyl alcohol (C_2H_5OH) have also been found in interstellar clouds.

Note in Table 4-1 that the atoms involved in these molecules, such as C, O, and N, are among those most abundant in cosmic material, explaining why they collide often enough to make numerous molecules. Note also from the last column that these atoms are among those depleted from the atomic gas, confirming that most of them have been tied up in molecules.

Heavier atoms have been even more depleted. They are tied up in larger particles, the grains of **interstellar dust**. These are detected by the fact that the dust reddens the light of distant stars, much as dust in our own air reddens the light of the setting sun. The amount of reddening can be used to calculate the size of the interstellar dust grains, which turn out to be about 0.1 to 1 μm in diameter. On the basis of spectroscopic and theoretical work, these grains are now believed to be composed of mineral grains similar to ordinary rock-forming compounds. As shown in Figure 4-2, spectral **absorption bands near 10 and 20 μm**, found in starlight transmitted through dust nebulas, match spectra of light transmitted through ordinary rock powders. In addition to silicate and carbon minerals, the dust probably also contains iron, a conclusion based on evidence that elongated grains, like little compass needles, are aligned along the galactic magnetic field.

As we will discuss in more detail in the next chapter, chemists have calculated what types of grains ought to condense in a cooling cloud of interstellar gas. The process is rather like that by which snowflakes and raindrops condense in clouds in our own atmosphere. Interestingly, if a relatively dense nebula cools sufficiently, iron and nickel tend to form metallic minerals, while silicon, oxygen, magnesium, calcium, and aluminum form rocky minerals. These are the very elements observed by astronomers to be depleted from the interstellar gas clouds! The correlation gives strong evidence that we understand roughly how the interstellar grains form (Donn, 1978). More importantly for our present purpose, we see that planetary material begins to form whenever we get a dense, sufficiently cool concentration of cosmic gas.

The next problem, then, is to understand where dense, cool nebulas are likely to form in the interstellar material. As we will see, one of the most likely places for such nebulas to evolve is alongside newly forming stars.

GRAVITATIONAL COLLAPSE: A THEORY OF STAR FORMATION

The only way for a dense, cool interstellar cloud to contract from nebular to stellar dimensions is to be dense enough so that the gravitational attraction of its particles for each other is strong enough to start it contracting or to keep it contracting if it gets compressed by external forces. The complete contraction process is called by astrophysicists **gravitational collapse** because once the process has begun, the cloud usually collapses to stellar sizes (less than a millionth of its initial size) in a relatively short time. During its collapse, but prior to onset of nuclear reactions, the object is called a **protostar**.

But since all particles have a gravitational attraction for other particles, why shouldn't all nebular clouds collapse and become protostars? The answer is that other forces compete with gravity by pushing outward on the cloud. Probably the most important as well as the most easily understood force is the outward force of gas pressure caused by heat in the cloud. The hotter the cloud, the more it tends to expand. In terms of atomic theory, each atom has a certain mean velocity that increases with the temperature of the cloud (remember that the temperature of a substance is defined as a measure of the mean velocity of its atoms).

In a given cloud at a given temperature, the atoms are darting about, colliding with each other. This exerts an outward thermal pressure. If gravity is inadequate to hold the cloud together, the cloud rapidly expands away into space. The cloud can collapse only if the density of the gas becomes great enough for interatom gravitational attractions to overcome the outward pressure.

In addition to heat, other forms of energy in the cloud tend to affect contraction. Two of these are magnetic fields and turbulence. Detailed theories of star formation must consider them—a difficult consideration that remains an active area of current research. In this introductory discussion we will consider only the dominant influences of gravity versus thermal pressure.

The Virial Theorem

The competition between gravity and thermal pressure is expressed by a crucial astrophysical theorem called the **virial theorem** or *Jeans theorem* (after one of its developers, English astrophysicist Sir James Jeans). This theorem, which can be derived by the methods of statistical mechanics, shows that gravitational collapse of a protostar will begin only if the total gravitational potential energy of the cloud exceeds twice the total thermal energy (neglecting other energy sources, such as magnetic fields).

The fact that the entire interstellar medium is in motion—with turbulence, heated regions, and expanding supernova clouds adding to the chaos—helps assure that some portions of the interstellar gas will be compressed and become dense enough (that is, have enough mass and gravitational potential energy) for collapse to start (for example, see Herbst and Assousa, 1978).

The virial theorem further implies that energy is released in a specific way: About half the gravitational energy released by the infalling material heats the gas of the protostar, and the other half radiates away into space.

To see why gravitational energy gets converted into thermal energy that heats the cloud, try to imagine a collapsing cloud in which all the atoms are moving radially inward. In this case, no gravitational energy is converted into heat. The gravitational energy released (that is, the change in gravitational potential energy) is completely transformed into the kinetic energy of the atoms as they fall toward the center. In real life the atoms do not fall radially inward because of their erratic thermal motions. Throughout the collapse, the atoms collide, slowing the collapse. The atomic collisions guarantee that some gravitational potential energy is converted into heat, and subsequent radiation by the gas ensures that some energy is radiated away into space.

The virial theorem has many important applications. It can be used to estimate the temperature and the amount of radiation coming from a collapsing protostar, and these estimates can be used to identify protostars in space and to calculate conditions during planet formation. We can also use the virial theorem to compute the size and density of interstellar clouds that could collapse to form a star such as the sun. Figure 4-3 shows an example of the application of the virial theorem (in the absence of magnetic fields and turbulence) to predict the masses of clouds that would collapse under different temperature and density conditions. For example, the diagram shows that for cool ($T = 10$ K) interstellar clouds to collapse into objects ranging in mass from the largest to smallest stars (see right-hand scale), the initial cloud densities would have to be about 10^{-9} to 10^{-10} kg/m^3.

Are nebulas of such mass and density common in interstellar space, so as to explain the continuing formation of stars? Paradoxically, the answer is no. Most nebulas are less dense.

How, then, do stars form? To resolve the paradox, remember that stars are observed to form in clusters. Looking along the bottom scale in Figure 4-3 to the densities typical of dense interstellar clouds (about 10^{-18} kg/m^3), we find that such clouds, if reasonably cool, would contract into masses about the size of star clusters rather than individual stars. This explains the origin of the star clusters. A typical contracting cloud, as it collapses, subdivides to make several hundred or a thousand stars. Theoretical studies have revealed that during the collapse of these **protoclusters**, local condensations form within the cloud and begin to collapse as independent entities. For example, Wright (1970) has followed theoretically the contraction of a 500-$M_{\odot}$, 12-pc-

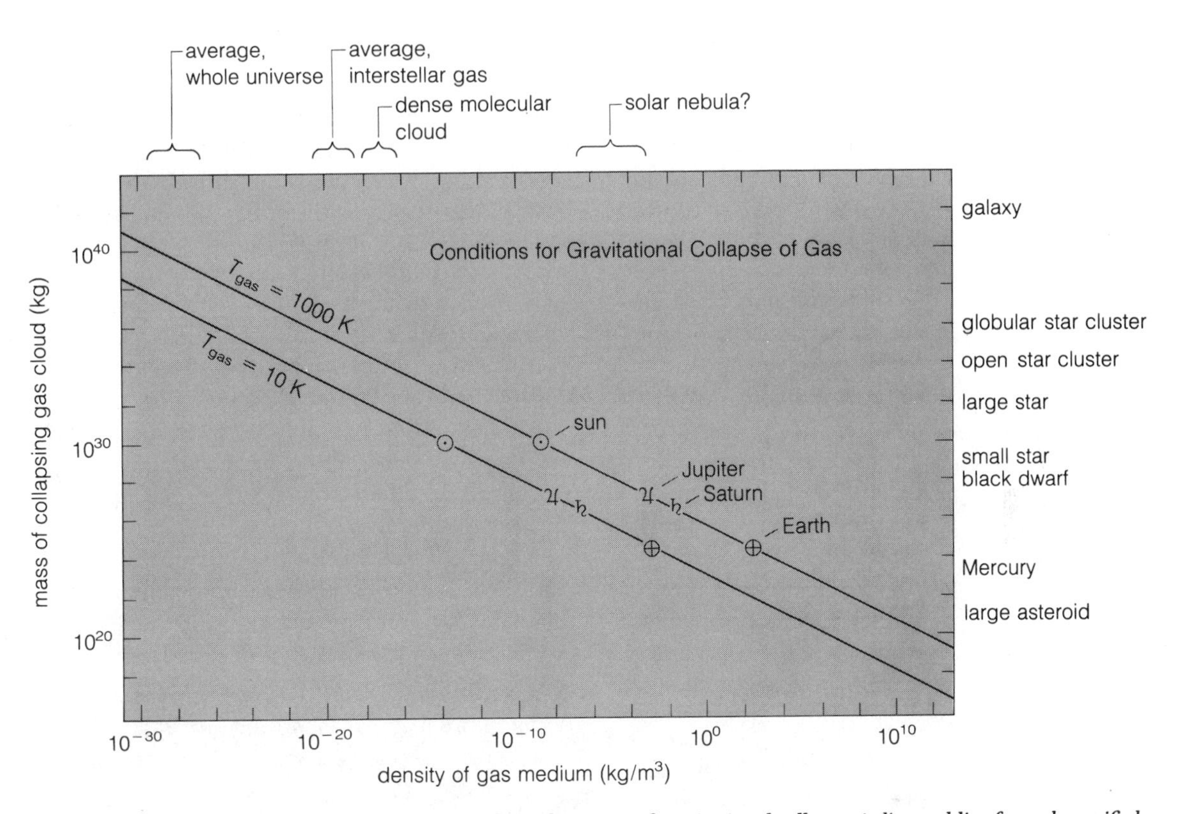

Figure 4-3. A guide to formation of astronomical objects by means of gravitational collapse. A diagonal line for each specified temperature shows the minimum density of gas (bottom scale) at that temperature required to initiate collapse in a gas cloud of specified mass (left scale). Clouds to the right of their appropriate temperature line would tend to collapse; clouds to the left would not.

wide cloud with an initial density of 4×10^{-20} kg/m^3 and found that various-sized subcondensations form throughout most of it. These subcondensations then collapse to form stars, and the whole thing turns into an open star cluster similar to the Pleiades, the Hyades, or the region around the Orion nebula.

Comment on the Formation of Just About Everything

The theory of star formation is not restricted to masses the size of stars or clusters. That is why Figure 4-3 has been plotted to include such a large range of mass and density. It is provocative to consider the evolution of the whole universe from this point of view.

For example, if all the material in the universe were spread out uniformly, a very low density of roughly 10^{-27} or 10^{-28} kg/m^3 would be obtained. Figure 4-3 shows that in a more or less uniform, relatively hot gas of this density, the condensations would have the mass of galaxies, about 10^{11} $M_{\odot}$. In other words, if ever no galaxies existed and only hot gas filled space, it is reasonable that galaxies should have formed. This helps explain our earlier statement that galaxies formed out of the gas expanding from the big bang.

The same process applies to smaller, denser

masses. Much evidence indicates that the newly formed sun, similar to other newly formed stars, was surrounded by a nebula with a density perhaps as high as 10^{-6} to 10^{-3} kg/m^3. Figure 4-3 shows that if this mixture of gas and dust got cool enough, masses the size of Jupiter or even smaller might have formed directly by gravitational collapse. Some theorists believe that planets (or at least the giant planets and their satellite systems, which resemble miniature solar systems), may have formed in this way. Thus our star-forming theory has led us directly to a plausible planet-forming theory, to which we will return in the next chapter.

MATHEMATICAL NOTES ON THE VIRIAL THEOREM

A complete mathematical derivation of the virial theorem from first principles is beyond the scope of this text, but the basic relations can easily be set up from the verbal statement given earlier. According to that statement, a cloud will contract if the gravitational potential energy exceeds twice the thermal energy. The gravitational energy of a homogeneous sphere (which we can assume to represent the protosteller cloud) is

$$\text{potential energy} = \frac{3}{5}\frac{GM^2}{R}$$

where

G = gravitational constant
M = mass of cloud
R = radius of cloud

The thermal energy is the thermal kinetic energy of *one* particle, $\frac{3}{2}kT$, times the total number of particles, $M/\mu M_H$, where

k = Boltzmann constant
T = temperature of gas
μ = mean molecular weight $\simeq 1$
M_H = mass of hydrogen atom

The inequality stated by the virial theorem is thus

$$\frac{3}{5}\frac{GM^2}{R} > 2 \times \frac{3}{2}\frac{kTM}{\mu M_H} \quad \text{for collapse}$$

Canceling common factors and substituting for the mass of a sphere $M = \frac{4}{3}\pi R^3\rho$, we have

$$\rho > \frac{15kT}{4\pi GR^2\mu M_H} \quad \text{for collapse to start}$$

where ρ = density of gas.

Once the collapse of an unstable object gets well under way (so that hydrostatic equilibrium is almost in effect), the inequalities expressed above become approximate equalities, so that

$$4\pi GR^2\mu M_H\rho \approx 15kT$$

or

$$GM\mu M_H \approx 5kTR$$

This result allows approximate evaluation of the temperature inside a collapsing protostar or protoplanet of specified mass once it has reached a specified radius.

■ *Sample Problem*

Find the density required for collapse to begin in an interstellar cloud consisting mostly of atomic hydrogen at 50 K if the cloud has dimensions somewhat bigger than a typical open star cluster, say about 20 pc in diameter. What mass would be involved in this cloud?

Solution

We will substitute into the inequality for density ρ. Since the cloud is essentially hydrogen (like all interstellar clouds), the mean molecular weight $\mu = 1$. One pc is 3×10^{16} m. Other constants appear in Table 2-2. The density is found to be roughly 8×10^{-20} kg/m^3, giving a total mass of roughly 5000 $M_{\odot}$, enough to make quite a respectable star cluster.

NEWLY FORMED STARS: THEORY AND DIRECT OBSERVATIONS

We have now seen evidence that (1) star systems are forming today, (2) planetary material can condense as dust grains in cool, dense environments, and (3) that star formation is theoretically understood (approximately!) in terms of contracting dust and gas clouds. Consequently, we can now examine the actual properties of newly forming stars and see whether they betray any symptoms of planet-forming activity that might let us understand our own solar system or other possible planetary systems.

Cocoon Nebulas

Solar system studies, theoretical astrophysics, and direct observation have all indicated that a newly formed star is immediately surrounded by a relatively dense cloud of gas and dust, some tens or hundreds of astronomical units across. When searches for newly formed stars were pursued in the 1960s, the Mexican astronomer A. Poveda (1965) argued that new stars are likely to be obscured by this envelope of dust and gas. Davidson and Harwit (1967) agreed and coined the term **cocoon nebula** for the dusty cloud predicted to surround newly born stars. The cocoon nebula was predicted to be shed later (probably by strong radiation pressure or solar wind) so that the star emerges into view like a butterfly emerging from a cocoon. (Employing another metaphor, some authors have referred to this as a *placental nebula*, noting that it sustains the growth of the planetary bodies.) We will now review three lines of evidence for the existence of cocoon nebulas, the environments that may produce planets.

Solar System Evidence for Cocoon Nebulas: The Nebular Hypothesis

Even before stellar astronomers began to speak of cocoon nebulas, planetary scientists realized that a cocoon nebula must have surrounded the early sun in order for the planets to have formed and to have obtained their present motions. After Kepler determined planetary motions around 1610, but before Newton published his analyses of the motions in 1687, the French philosopher René Descartes reasoned in 1644 that the initial gas in the universe broke up into rotating "vortices," much as the surface of a stream may develop rotating eddies. Each such rotating vortex produced a rotating star surrounded by gas, and minor subvortices in the gas might have produced a family of planets near each star.

The German thinker Immanuel Kant correctly reasoned in 1755 (without much mathematical support) that gravity would make a circumsolar cloud contract and that rotation would flatten it. Thus the cloud would assume the general shape of a rotating disk, explaining the fact that the planets revolve in a disk-shaped distribution.

This view, that a disk-shaped nebula formed around the early sun, came to be called the **nebular hypothesis**. It was further developed by the French mathematician Laplace, who proposed in 1796 that the rotating disk continued to cool and contract. Due to conservation of angular momentum, the sun would have spun faster as it contracted, just as a figure skater spins faster as she pulls in her arms. Laplace proposed that the outer rim of the spinning disk shed concentric rings of material at various stages of the contraction, with each ring eventually condensing into a planet. In this way Laplace explained the spacing of the planets in circular orbits.

This early work on the nebular hypothesis was important in two regards. First, it moved scientific thinking toward the modern view that a cocoon nebula developed as part of the normal star-forming process. But this movement was of limited value, because it was not supported by either detailed calculations or direct observations. Laplace himself offered his discussion "with that diffidence that always ought to attach to whatever is not the result of observation or calculation."

Second, and more important, these early ideas

helped clear away the tangle of confusion that connected Renaissance physical theories with theology. The work of Descartes, Newton, Kant, and others implied, contrary to prevailing thought, that even if God had created the original universe full of gas by some special, incomprehensible process, He let it evolve by cause-and-effect relationships. Thus physical laws derived from observations today could be used to infer past events or to predict future ones.

The nebular hypothesis has been retained and expanded. Its modern version—that planets formed from dust embedded in a disk-shaped gaseous nebula—explains many planetary properties, as we will see in subsequent pages.

Theoretical Evidence for Cocoon Nebulas

A current area of active research is to investigate the evolution of a contracting protostellar cloud in hopes of understanding not only the birth of the star but also the evolution of the cocoon. Taken into account are various initial properties such as the total mass of the cloud, its temperature, its rotation rate, the degree of internal turbulence, and magnetic fields. Most models are simplified, since it is hard to treat all these variables at once.

The calculations made so far strongly indicate that a star forms as a central condensation in an extended nebula, since the inner part of the cloud collapses faster than the outer part. The outer part remains behind as the cocoon nebula. The same studies indicate that under many conditions of rotation, turbulence, and so forth, the inner star-forming nucleus may divide into two or three objects orbiting around each other, as shown in Figure 4-4a. This may explain why more than half of all star systems are binary (two co-orbiting stars) or multiple (three or more co-orbiting stars) rather than single stars like the sun.

The same process of fragmentation may also produce bodies too small to be true stars. The minimum mass of a star (as defined by the presence of nuclear reactions in its interior) is about 0.08 $M_{\odot}$. Somewhat smaller objects could be large planets such as Jupiter (0.001 $M_{\odot}$) or hypothetical larger planets in other systems. The whole process has been reviewed by Cameron (1975) and Larson (1968); detailed calculations appear in Bodenheimer and Black (1978), Field (1978), and Larson (1972). The calculations also indicate that the cocoon nebula would obscure the new star from outside view. Only the cocoon, a dusty, warm cloud emitting primarily infrared radiation, would be visible.

Telescopic Evidence for Cocoon Nebulas: R Mon and Infrared Stars

Prior to 1966, there was no firm observational proof that cocoon nebulas existed. However, the Mexican astronomer E. E. Mendoza (1966) observed an object cataloged as star R in the constellation Monoceros and discovered that it was emitting large amounts of infrared radiation (peaking at wavelength around 3.4 μm). Called **R Mon** for short, this object lies near the star-forming nebulosity shown in Figure 4-1. As shown in Figure 4-5, R Mon appears as a bright condensation at the tip of a triangular nebula.

Because R Mon lies in a star-forming region, is erratically variable, and has an unusual spectrum, it has long been regarded as a candidate for a newly forming star. Analyzing its infrared properties, American astronomers Frank Low and Bruce Smith (1966) announced that it is probably a newly forming star buried in the center of a 200-AU-diameter cocoon nebula whose dust grains have been heated by the enclosed star to a temperature of 850 K, a temperature calculated from the spectrum of R Mon's infrared radiation (Figure 4-6). The spectrum also indicates that light from any star inside the cocoon is completely obscured by solid material. This has been inferred to be the long-sought dust of a cocoon nebula, since dust is known to exist in the general region, is efficient at obscuring stars, and is predicted to surround new stars. The star heats the

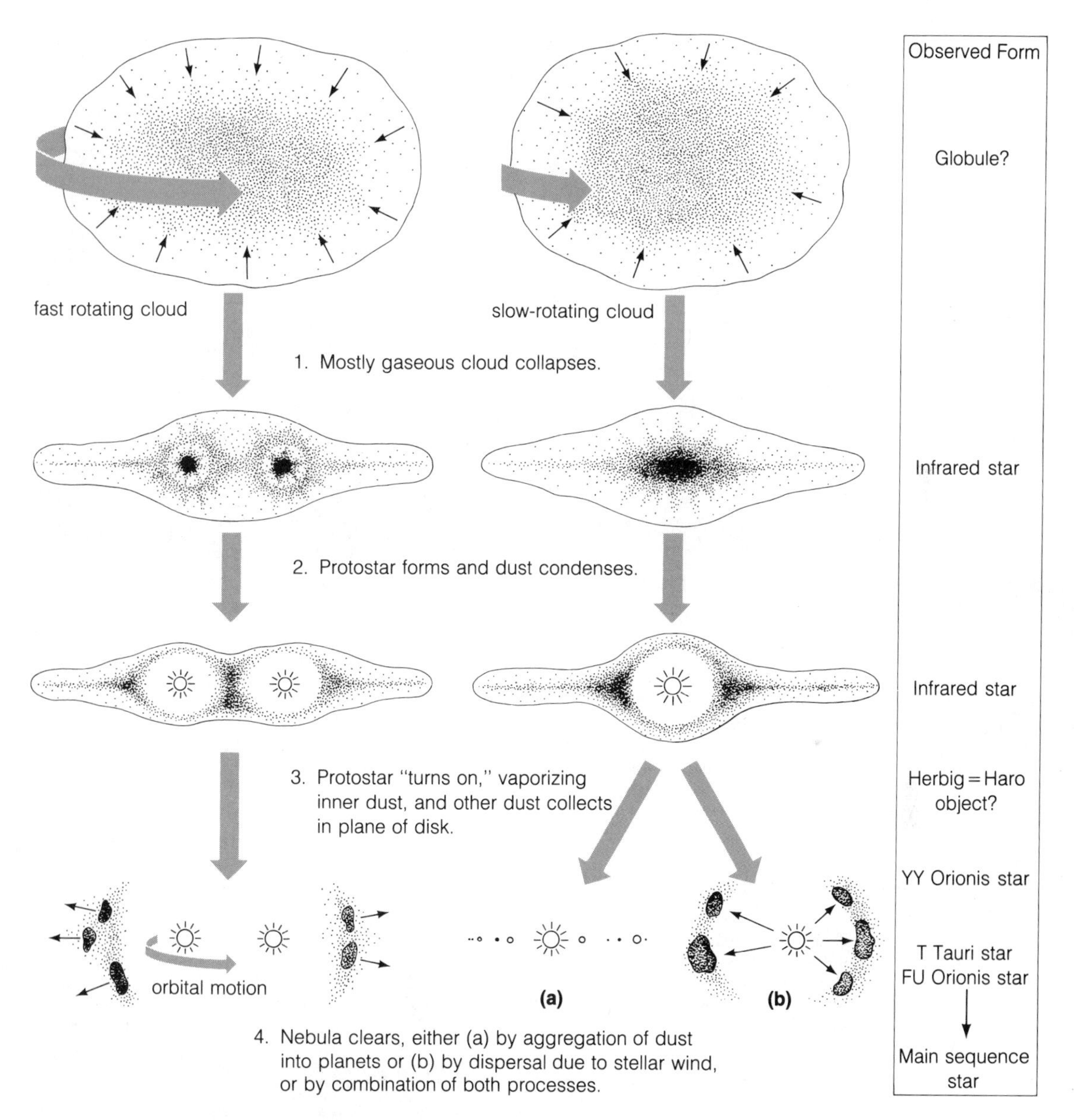

Figure 4-4. Schematic view of the possible sequence of evolutionary processes for a binary star (left side) and a single (nonbinary) star (right side). Depending on the processes that dominate stage 4, the potentially planet-spawning dust may aggregate into planets (case a) or be blown away (case b).

dust that obscures it, and the outside observer sees only the infrared thermal radiation from the warm dust, as shown in Figure 4-7.

From the estimated properties of the system, Low and Smith (1966) calculated that the dust must be orbiting the star at a distance of some tens of astronomical units. They pointed out that the configuration is almost identical to that long

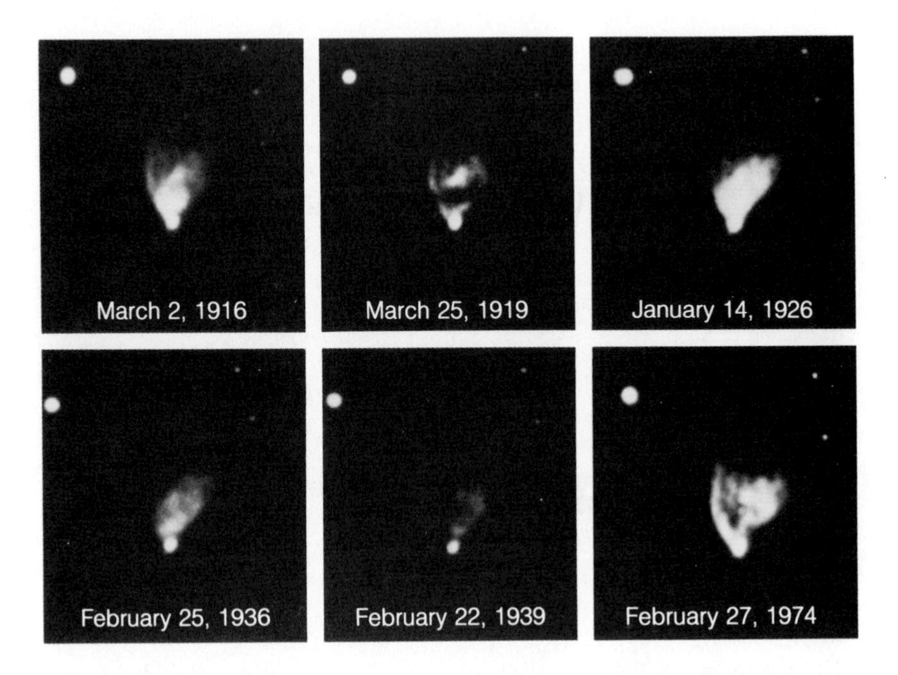

Figure 4-5. The young, starlike object R Monocerotis is the starlike image at the bottom of the triangular nebula shown in these pictures. The form of the nebula changes as shown in this 58-y sequence. The changes may reflect changes in the radiation of R Mon due to its instability. The triangular nebula is roughly 50 000 AU across; the cocoon nebula around R Mon is too small to show here, about 200 AU across. These objects are about 700 pc from the solar system. (Lowell Observatory; 1974 photo courtesy Alan Stockton, Mauna Kea Observatory, University of Hawaii)

Figure 4-6. Spectrum of R Monocerotis. Most of the light is infrared radiation emitted by surrounding dust particles, but some visible light leaks through the nebula.

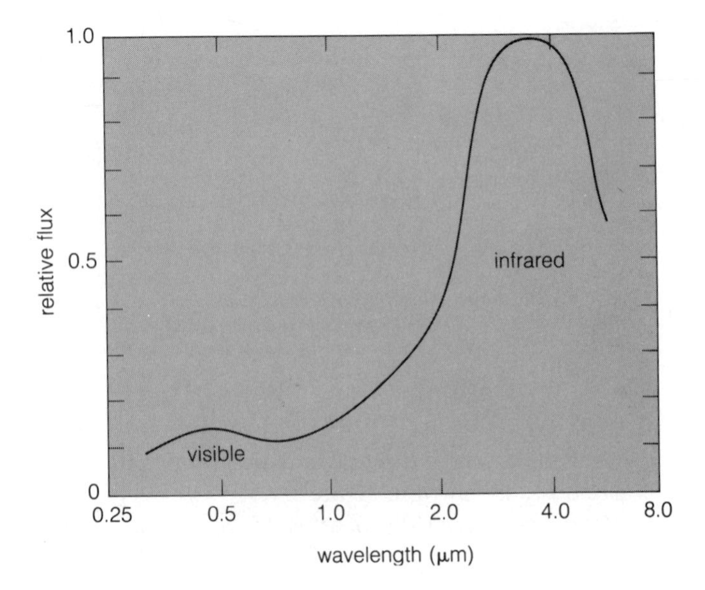

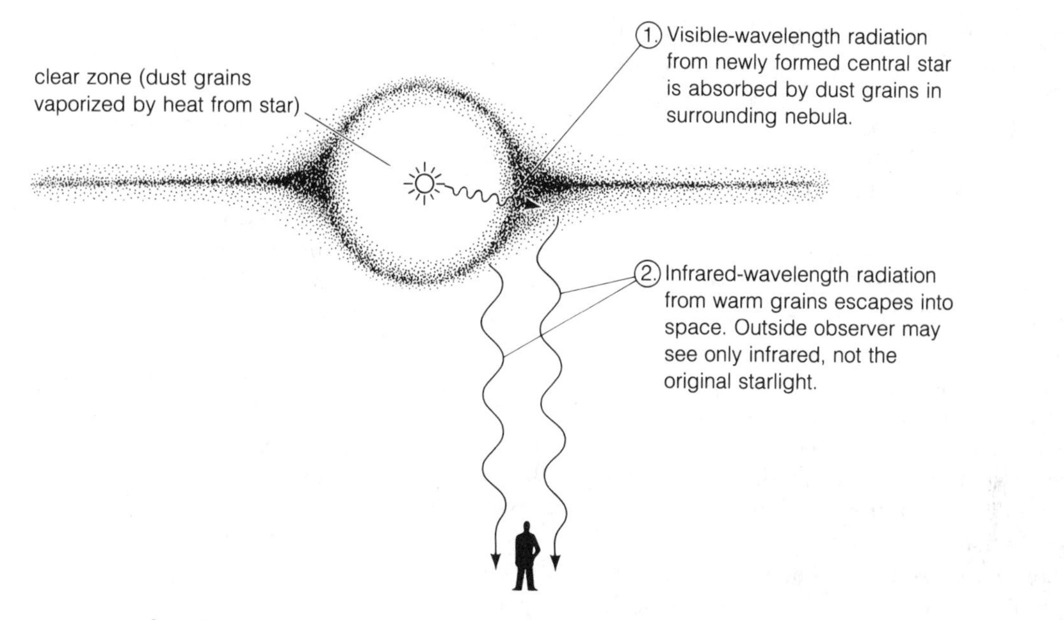

Figure 4-7. Schematic cross section of a cocoon nebula, showing the disk-shaped profile of the dust particles and the conversion of the star's visible radiation to infrared radiation from the heated dust.

MATHEMATICAL NOTES ON WIEN'S LAW AND TEMPERATURE MEASUREMENT

Wien's law tells which wavelength λ corresponds to the maximum amount of radiation at each temperature T. Using SI units, the wavelength is given in meters and the temperature in Kelvin. The law is

$$\lambda = \frac{0.00290}{T}$$

The number 0.00290 is a constant of proportionality and remains the same in all applications of the law. Thus, as T increases, λ decreases, giving shorter wavelengths and hence bluer light.

An example is instructive. The sun has a temperature of about 5700 K. Therefore, we can calculate the wavelength of the strongest solar radiation. Using SI units and powers of 10,

$$W = \frac{2.9 \times 10^{-3}}{5.7 \times 10^{3}} = 0.509 \times 10^{-6}$$
$$= 5.09 \times 10^{-7}\,\text{m}$$

This, of course, is exactly in the middle of the range to which the eye is sensitive. (Otherwise we could not see sunlight!) Light of this wavelength corresponds to a yellow color.

Planetary matter, of course, is not so hot. The wavelength for maximum radiation from dust in a cool (500 K) nebula, for example, would be 5.8×10^{-6} m, or 5.8 μm. This figure is well into the infrared part of the spectrum. The infrared region of roughly 5 to 40 μm is often called the *thermal infrared,* since it is the region of emission caused by the body heat of nonluminous planetary sources.

hypothesized by planetary scientists for the primeval solar system, and they suggested that this might be the first discovery of a planet-forming process outside our solar system.

Further research has strengthened these interpretations. Herbig (1968) studied the structure and spectrum of R Mon and found that the object at the tip of the nebula in Figure 4-5 is not a star but rather a very small nebula whose radiation and spectrum changed from year to year. No normal star has been found; the nebula probably contains a hidden, unstable, young star. Nearby are other infrared nebulosities, such as the so-called Becklin-Neugebauer object, which may be another young star in a dusty shell, and the Kleinmann-Low infrared nebula, which may enclose a whole cluster of newly forming stars. By means of spectroscopic observations such as shown in Figures 4-2 and 4-3, a number of such cocoon nebulas have been identified as containing silicate and/or ice particles, thus confirming that they actually do contain potential planet-forming material similar to that in our solar system.

A small infrared source, NGC 7538E, located in a probable star-forming cloud rich in molecules, is a good example of an object in which such materials have been identified (Willner, Puettner, and Russell, 1978). As shown in Figure 4-2, its infrared spectrum reveals the presence of ice crystals, carbon monoxide, and silicate mineral grains similar to rock-forming material. Such grains are microscopic, probably 0.1 to 1 μm across.

In a 1975 survey of 20 objects considered likely protostars or very young stars (including such objects as R Mon, the Becklin-Neugebauer object, and probable young stars T Tauri and FU Orionis), 19 had excess infrared radiation, presumably caused by circumstellar dust, and 9 actually had silicate-mineral features in their spectra (Terry Teays, 1975, private communication). Many of the objects are detected only or primarily at infrared wavelengths, and hence are called **infrared stars**.

THE H-R DIAGRAM: A TOOL FOR DISCUSSING PROTOSTAR EVOLUTION

We have established that there are a variety of infrared stars (some visually luminous) that look suspiciously like new stars as well as resemble the postulated early solar system. How can we make further sense of these stars? Can we arrange them in order of age? In order of mass? Which ones are really like the early sun?

These are all questions of stellar evolution, a field well developed in astrophysics. Theoretical astrophysicists as well as observational astronomers have long used a diagram called the **H-R diagram** as a tool for clarifying these problems as well as other problems of stellar evolution.

The format of the H-R diagram is based on the format of the first such diagrams, plotted around 1914 by Danish astronomer Ejnar Hertzsprung and American astronomer Henry N. Russell, after whom the diagram was named.* The diagram plots the luminosity of the star (total energy output expressed in Joules per second) against the temperature of the star. By tradition, luminosity increases upward, and temperature increases to the left.

The H-R diagram was a way to begin to make sense out of the observational data accumulating on stars early in this century. One of the first discoveries made with it is that most stars, including the sun, fall in a band called the **main sequence**, as shown in Figure 4-8.** The main sequence contains "normal," middle-aged stars. Physically, they are defined by the fact that they generate their energy in nuclear reactions that convert hydrogen to helium in their central

---

*Although some prefer the term *spectrum-luminosity diagram*, whch describes the two observable quantities, H-R or Hertzsprung-Russell is the more common name.

**The conventional H-R diagram in most astronomy texts is just the upper left third of Figure 4-8, where most observable stars lie, but Figure 4-8 has been extended to include very cool stars and planetary bodies.

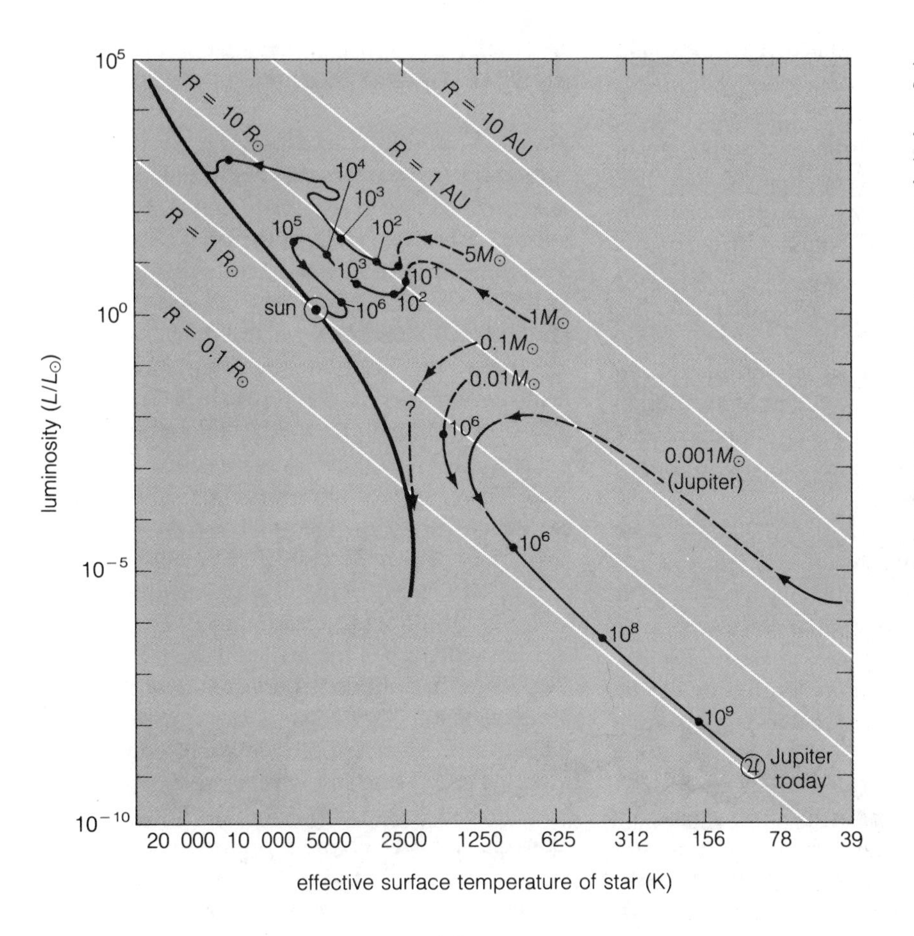

Figure 4-8. Calculated evolutionary tracks in the H-R diagram for stars of different masses. Note collapsing objects smaller than about 0.08 $M_\odot$ do not initiate hydrogen-consuming nuclear reactions; hence, they do not reach the main sequence to become true stars. The planet Jupiter is such an object, yet it can be studied by the techniques of stellar astrophysics, as presented here. (Data for 5 and 1 $M_\odot$ from Larson, 1969; for 0.01 and 0.001 $M_\odot$ from Bodenheimer, 1976; for 0.1 $M_\odot$ interpolated.)

regions. Hot, luminous stars, in the upper left end of the band, are the most massive; cool, faint stars, in the lower right, are least massive. The sun is in the middle of the main sequence.

Stars that do not generate energy in this way fall off the main sequence. These are of two types. Stars in the final stages of evolution run out of hydrogen and begin nuclear reactions that consume other elements, thus becoming **post-main-sequence stars**. These fall in other parts of the diagram and do not concern us here. The others are the **pre-main-sequence stars**, which are the topic of this chapter. These are protostars in which the normal H reactions have not quite started or are just beginning. Thus these stars are just settling onto the main sequence.

Protostars and the H-R Diagram

To predict what a protostar is like—how bright and hot it is—is to predict where it falls on the H-R diagram. One can easily see that a newly forming protostar must begin life far in the lower right corner of the H-R diagram, since it first becomes recognizable as a cool, dark cloud. Detailed astrophysical calculations have followed the evolution of bodies across the H-R diagram (Larson, 1972, 1978; Bodenheimer, 1976; Bodenheimer and Black, 1978; Field, 1978). In general, the results show that once the protostellar cloud becomes gravitationally unstable, it contracts very rapidly to a diameter of some tens of solar radii in only about a hundred years. It then contracts more slowly to produce an inner stellar core only a few solar radii across within about a thousand years. At these rates of evolution, variations should be visible over the few decades that we have been observing suspected protostars; and, indeed, such variations have been reported. Such is the case with R Mon, though these changes may involve obscuration by clouds moving around the star, rather than evolutionary changes in the star itself.

The stages followed by a star as it evolves define an **evolutionary track**. The early, high-luminosity part of the evolutionary track is called the **Hayashi track**, after Japanese astronomer Hayashi (1961), who recognized its properties. The duration and maximum luminosity of the Hayashi stage is controversial and important, since it implies strong heating just before the planets form. Following the decline of the Hayashi track, the evolutionary track tends to move erratically left toward the main sequence before dropping onto it. This erratic path is called the **Henyey track**, after Henyey, LeLevier, and Levee (1955), who first indicated its approximate position by means of simplified calculations.

Evolutionary Track of a Protostar as Observed from Outside the Cocoon Nebula

Most astrophysical calculations of protostar evolutionary tracks, as shown in Figure 4-8, focus on the history of the starlike contracting mass that forms at the core of the larger contracting cloud. But while the inner protostar may radiate large amounts of visible light, this light is absorbed by nebular dust; and it merely heats the nebula to an effective radiating temperature of some hundreds of degrees, so that the object as seen from the outside is an infrared source for 10^5 to 10^6 years (Larson, 1972). The evolutionary track for a cocoon nebula thus lies well to the right of the main sequence, as in the case with R Mon.

T TAURI STARS: A LATER STAGE IN STAR FORMATION

Probably the best-known type of newly formed star is the **T Tauri star**, which sheds light on a postcocoon stage of star formation. T Tauri stars, named after the star T in the constellation Taurus, were first recognized as an unusual group in 1943–1944 in work almost unnoticed because the author, K. Himpel, was publishing inside Nazi Germany. They were later emphasized by American astronomer A. Joy (1945). T Tauri stars probably represent a stage when the cocoon nebula begins to clear or blow away, as shown in Figure 4-9. The star itself draws closer to the main sequence during the Henyey track portion of its evolutionary path across the H-R diagram.

T Tauri stars are highly and irregularly variable in brightness, accompanied by variable nebulas and strong magnetic fields and found in regions with abundant dust clouds and young, massive stars. They rotate fast (as predicted by Laplace's nebular hypothesis about contracting rotating stars) and are enveloped by expanding gas clouds. They occupy a band in the H-R diagram just above the main sequence. This is the region that would be reached by stars that have just evolved to the bottom of their Hayashi tracks or are moving onto their Henyey tracks. For example, the H-R diagram for stars in and near the Monoceros young cluster NGC 2264, shown in Figure 4-1, reveals that many of the stars are T Tauri stars. They lie in a band that indicates an average age of only a few million years.

Many T Tauri stars have infrared excesses, as discovered by the Mexican astronomer E. E. Mendoza (1966), indicating that they probably have warm dust in their expanding nebulas. Doppler shifts in the spectra of most T Tauri stars indicate expansion velocities of 70 to 200 km/s for the nebulas, which may indicate the final outward dispersion of cocoon nebulas by strong stellar winds blowing out from the star. Cocoon nebulas apparently blow off as much as $0.4M_{\odot}$ before reaching the main sequence (Kuhi, 1966). Many T Tauri stars have the telltale 10-μm and 20-μm silicate spectral features seen as emission

Figure 4-9. A view inside a cocoon nebula, viewed from slightly above the plane of the nebular disk. Dust particles have condensed and aggregated. Distant dust clouds partially obscure the central star. Seen from outside, the system would be an infrared star, possibly irregularly variable due to shafts of starlight breaking through the dust clouds. Such a scene may have appeared in the primeval solar system. (Painting by author)

bands, indicating the presence of silicate dust with temperatures on the order of 200 K, but the dust is generally not optically thick (Rydgren, 1978).

Not all T Tauri stars have a smooth outflow of gas. Redward displacements of absorption lines for many T Tauri stars indicate that perhaps a quarter or more of them have an infall of some matter, and some show both infall and outflow. Some that have been observed numerous times have shown an infall about 30% to 50% of the time (Kuhi, 1978). Rather than the spherically symmetric radial outflow that early analysts suggested, T Tauri stars may be disk-shaped rotating systems (Strom, 1972), with outflow in some regions and infall in others. Views of the same T Tauri star at different times or from different angles can give different flow patterns, as turbulence, rotation, and evolution bring different gas-and-dust-cloud patterns into the line of sight.

Observations of T Tauri stars indicate that most of their light gets out through the dust. The dust is apparently too thin to permit planet formation. Recent analysts have viewed the T Tauri stage as postdating planet formation and representing the stage of cocoon nebula dispersal. In order to explain the observed spectra of certain

T Tauri stars, Catherine Imhoff (1978) suggests that some of the infrared radiation comes from 10^4-K gas near the star, and that the infrared thermal radiation from dust can be modeled by optically thin dust within 30 stellar radii of the star, a zone out to about the orbit of Mercury in our system. Thus, the parts of a T Tauri system involved in our observations may be smaller and contain less material than the solar system. Most dust in these systems may have already aggregated into larger (meter-scale to planetary-scale?) bodies. Thus the dust would not be in a dispersed form that shows up in spectroscopic absorptions, excessive infrared radiation, or obscuration and reddening of the star.

Gas near the surface of T Tauri stars contains more lithium than do the sun, the interstellar medium, or the gases farther out in their nebulas. Curiously, the amount of lithium matches that in planetary material such as Earth and the meteorites. How is this extra lithium created? Bonsack and Greenstein (1960) argued that the flares and magnetic fields of T Tauri stars accelerate atomic particles that collide with atoms in the inner nebula, split them, and create the lightweight lithium atoms. The lithium may then be incorporated into nearby planetary material. This theory has several ramifications: (1) It shows how the chemistry of planetary matter can be affected by early stellar conditions; (2) it again shows the relevance of T Tauri stars to planet formation; and (3) it shows how the strong flares and magnetic fields in new stars could play a role in creating unstable isotopes from broken atoms in the inner cocoon nebula, thus adding to the radioactivity incorporated in planetary matter (and thus adding to the heat trapped in any accumulating planets).

T Tauri stars dramatize the importance of astrophysical studies in gathering evidence on the formative stages of possible planetary systems elsewhere in the universe. They also illustrate that common evolutionary processes rather than rare catastrophes accompany the transition from a state where planetary dust obscures the star in a cocoon nebula to a state where most of the planetary matter has become undetectable. But we don't know whether the dust has been destroyed, blown away, or aggregated into planets. Cocoon stars probably exemplify conditions before the dust has aggregated into planetary bodies, and T Tauri stars exemplify conditions shortly thereafter.

Kuhi (1978) sums up the situation as follows:

> Do [T Tauri stars exemplify ongoing] formation of planets? This seems unlikely because the dimensions of all the models (crude as they are) are on the order of a few solar radii to a few 100 $R_\odot$ at most for the line-emitting region. This is considerably smaller than any protoplanetary disk discussed by solar system theoreticians. . . . This all suggests that the planet-forming stage has already passed . . . although the wind may still play a significant role in post-planet-formation cleanup of the interplanetary medium.

OTHER PROTOSTARS AND YOUNG STARS

Infrared cocoon stars and T Tauri stars are the two most firmly established and widely discussed groups of probable newly formed stars, and their properties shed much light on the early solar system. However, a number of other related objects are known.

Bok globules are dark, rounded clouds typically 1000 to 100 000 AU across, often seen silhouetted as dark spots in regions of bright nebulas. They may contain several solar masses of dust and gas and are often millions of times denser than normal interstellar gas. Many may be objects beginning to collapse into cocoon nebulas, but others may be fragments of clouds blown apart by gas outrushing from star-forming regions.

Herbig-Haro objects are small, variable, bright nebulas containing no detectable stars or stellar spectra. They often cluster in groups of several objects, among which any one may brighten without notice, remain visible a few years, and die out again. They may represent evolutionary states between infrared cocoon stars and T Tauri stars. Their spectral emission lines are characteristic of T Tauri stars, and the nebula around T Tauri itself would have been called a Herbig-Haro object had the star not been visible (Böhm, 1978). The distribution of Herbig-Haro objects in space resembles that of T Tauri stars. The variability of both types of object might involve dust concentrations in the cocoon nebula passing either in front of the star as seen from Earth or in a way that allows the starlight to illuminate new parts of the nearby nebulosity. Many Herbig-Haro objects seem to be expanding and have subcondensations estimated to contain only 1 to 10 $M_{\oplus}$ (Earth masses) of dust (Böhm, 1978), again suggesting the breakup of the type of cocoon in which planets could form.

YY Orionis stars are essentially T Tauri stars that show infalling instead of outflowing material. They are classified separately by some authors, who postulate infall of matter during a stage of evolution prior to the terminal flare-ups that blow nebular material outward.

FU Orionis stars are stars exhibiting extreme flare-ups. The prototype, FU Orionis, brightened by a factor of about 250 in luminosity. Later studies show that it has a surrounding "shell" of cooler gas and substantial infrared radiation. About a half dozen such stars (depending on interpretations) have been discovered, at least one of which (V1057 Cygni) was a T Tauri star before it flared up. Depending on interpretation, the FU Orionis phenomenon may be an extreme flaring activity on the surface of T Tauri stars, similar to the flares on the sun that sometimes accompany sunspots. There are suggestions that FU Orionis stars are T Tauri stars that are more massive than 1.5 $M_{\odot}$ and that the flare-ups may help drive away all nebulas from these stars, preventing planet formation around them (Welin, 1978). Herbig (1977) concluded that the FU Orionis phenomenon recurs in T Tauri stars at roughly 10^4-y intervals, but this is uncertain.

As can be judged from our discussion of T Tauri and FU Orionis stars, the dispersal of the cocoon nebula after formation of planetary material is a crucial process. Planetary scientists have attributed to this process all manner of interesting effects, such as stripping of primitive atmospheres from planets; but there is little firm evidence on which to base such models. Ward (1981) points out that typical cocoon nebula models amount to 10 times the total planetary mass, so that the loss of the nebula may seriously alter the gravitational environment of the planet system, may produce resonances among planets, and may affect the orbital arrangement and stability of the planetary system.

BINARY AND MULTIPLE STARS

Our description of star formation has one major flaw: We have hardly considered the fact that at least half of all stars are binary or multiple. A **binary star*** is a pair of stars orbiting around a common center of gravity; a **multiple star** is a system of three or more co-orbiting stars. Since binary and multiple star systems are so common, their formation must be a common process. Their properties, though poorly understood, must contain some messages about the formation of planetary systems, especially since some companions in such systems are nearly as small as planets.

One reason for the lack of understanding of binaries is the lack of good statistics. Selection effects are devastating. If we study only nearby stars, which we can see well, we have too small a number to have a good sample. If we study distant stars, the data are biased in several ways (for example, toward binaries with equal brightness because in very unequal pairs, the faint member is lost in the glare of the bright one). A **visual binary** is a binary with members far enough apart to be seen separately in a large telescope; their statistics are biased toward large separation distances. A **spectroscopic binary**

*The term **double star** applies to two stars at different distances that appear close together in the sky by fortuitous geometric alignment but have no gravitational connection.

has members so close together that their orbital velocities are very high (recall Kepler's third law), causing measurable, periodic Doppler shifts that identify the star as a binary. Statistics of spectroscopic binaries thus favor very close separations. Because of these selection effects in detection and cataloging, no sample can be guaranteed statistically valid for all binary and multiple stars.

A third type of binary, called an **astrometric binary**, is detected by observations of stellar motions having incredible precision. These are called **astrometric observations**. If a star is accompanied by another star or planet, the two bodies orbit each other. Therefore, the star's drifting motion across interstellar space will not be a straight line but will have a "squiggly" shape, because it is always orbiting back and forth around its companion. The companion may be an unseen, tiny body; but if the star is close enough to the sun, the sinuosities will be detectable. It may take decades to get enough good data to detect any periodic sinuous motion.

Fortunately, this type of observational work was started at several observatories by far-sighted astronomers 80 y ago. Analysis of the accumulated observations has shown that a number of seemingly single stars actually have very, very tiny sinuous motions. The sinuosities are so tiny that statistical methods must be used to sort out the unavoidable inaccuracies in measurement. They are also so tiny that the invisible companions must be much, much less massive than the sun. At least one, a companion to Barnard's star (see below), has been claimed to have planetary mass (roughly like Jupiter's), though this result has been questioned.

Rarity of Single Stars

Table 4-2 lists the nearest stars and shows the importance of binary and multiple systems. The largest member of each system is A, the second-largest, B, and the third-largest, C. Six of the 7 nearest stars and 16 of the 32 in the whole list are known or suspected binaries or multiples. Binary and multiple systems are hardly rare!

Several surveys have been made to determine the number of members in star systems. Table 4-3 lists some results. Once again, the selection effects are devastating because the farther away the star, the harder it is to detect small, faint companions by direct observation or astrometric means. Binary statistics suggest that about 50% to 80% of all star systems contain two or more companions (columns 2 through 4 in Table 4-3).

Surveys by Abt (1978) have studied the mass distributions of binaries and multiples, leading to several interesting results. First, the mass distribution of systems with wide separations (about 100 to 3000 AU) are quite different from the mass distribution of systems with orbits about the size of planetary orbits in the solar system. The wide systems seem to follow the size distribution found for ordinary field stars, suggesting that they may be field stars randomly paired in some way. The mass distribution of close binary companions suggests that they formed by a different process, probably a tendency for collapsing protostars to break up into two or more closely spaced, co-orbiting condensations that then became stars (Abt, 1978).

A second finding from recent surveys is that the oldest known stars—which happen to lack heavy elements—have the normal complement of widely spaced companions but do not have close companions! Here again, a separate origin for short-period companions is indicated, as is the possibility that the short-period companions might be related to the dusty planetary material of cocoon nebulas that was absent during the formation of the most ancient stars (Abt, 1978). In addition, Fleck (1978) finds a similarity of angular momentum between close binary systems and planetary systems, and suggests a similarity in origin.*

---

*Note that the most ancient stars are deficient in heavy elements, which were mostly formed by nuclear reactions inside stars and then blown out into interstellar space by supernova explosions of large stars. Star-forming interstellar gas has thus grown richer and richer in heavy elements, which now amount to 2–3% of the total mass.

In summary, the origin of some closely spaced binary and multiple systems may be related to the origin of planetary systems—planets being simply ultra-low-mass companions whose mass distribution is controlled by conditions such as angular momentum, turbulence, and mass distribution in the original collapsing protostellar cloud, as suggested by G. P. Kuiper in 1955.

Table 4-2
Survey of 32 Nearest Star Systems out to 15 Light-years

		Mass of known or suspected components ($M_\odot$)			Approximate semimajor axis of orbit (AU)	
Star or system	**Distance (ly)**	**A**	**B**	**C**	**AB**	**AC**
Sun	0.0	1.0	0.001	0.0003	5.2	9.5
Alpha Centauri	4.3	1.1	0.9	0.1	24	10,400
Barnard's star	6.0	0.15	—	—	—	—
Wolf 359	7.5	0.1?	—	—	—	—
Lalande 21185 = BD + 36°2147	8.2	0.35	0.02??	—	0.07??	—
Luyten 726-8 = UV Ceti	8.4	0.044	0.035	—	10.9	—
Sirius	8.6	2.31	0.98	—	20	—
Ross 154 = Gliese 729	9.4	0.2?	—	—	—	—
Ross 248	10.2	0.2?	—	—	—	—
Luyten 789-6	10.4	0.1?	—	—	—	—
Epsilon Eridani	10.7	0.7?	0.003??	—	—	—
Ross 128	10.8	0.2?	—	—	—	—
61 Cygni	10.8	0.6	0.6?	0.008?	85	?
Epsilon Indi	11.1	0.7	—	—	—	—
Procyon	11.4	1.8	0.6	—	16	—
Σ2398 = BD + 59°1915 = Gliese 725	11.5	0.4	0.4	—	60	—
Groombridge 34 = BD + 43°44 = Gliese 15	11.6	0.4?	0.2?	0.2?	156	?
Tau Ceti	11.7	1.0	—	—	—	—
Lacaille 9352 = CD − 36°15693	11.7	0.4?	—	—	—	—
G51-15	11.9	0.1?	—	—	—	—
BD + 5°1668 = Luyten's star	12.0	0.2?	0.1?	?	—	—
Luyten 725-32	12.3	0.1?	—	—	—	—
Lacaille 8760 = CD − 39°14192	12.5	0.5?	—	—	—	—
Kapteyn's Star	12.7	0.5?	—	—	—	—
Krüger 60 = DO Cep = Gliese 860	13.0	0.27	0.16	0.01	9.5	?
Ross 614	13.4	0.14	0.08	—	3.9	—
BD −12°4523 = Gliese 628	13.7	0.2	0.1??	—	?	—
Wolf 424	13.9	0.2	0.1	—	3.0	—
van Maanen's Star = Wolf 28	14.0	0.3?	—	—	—	—
CD −37°1549	14.5	0.2?	—	—	—	—
Luyten 1159-16	14.7	0.1?	—	—	—	—
Groombridge 1618 = BD + 50°1725	15.0	0.5?	—	—	—	—

Source: Data from Allen (1973; general data); Gatewood and Eichhorn (1973; Barnard's star); Harrington and Behall (1973; Luyten 726-8); Probst (1977; Ross 614); Campbell (1989, Epsilon Eridani and 61 Cygni). Heintz (1978) reviewed data on all reported but unseen companions and concluded that there is no solid evidence for any of them, including companions listed here for Lalande 21185, Epsilon Eridani, and 61 Cygni. Consequently, many reported components have been listed with question marks. Campbell's 1989 data may revise Heintz's conclusion for Epsilon Eridani and 61 Cygni. More data are needed.

Table 4-3
Fraction of Systems Containing *n* Members

No. of members	19 systems within 12 ly	All systems: Average of 6 estimates[a]	All systems: Estimate favored by Batten (1973)[b]	Solar-type main sequence stars[c]
1	0.50	0.40	0.30	≤0.22
2	0.30	0.40	0.53	
3	0.15	0.15	0.13	
4	0.05[d]	0.036	0.03	≥0.78
5	—	0.01	0.008	
6	—	—	0.002	

[a]Estimates by various authors. Data from Batten (1973).
[b]Batten's estimates are averages over all stars on the basis of data from various sources.
[c]Data from Abt (1978). Abt's estimates are based on reviews of several sources, plus several comprehensive observational surveys of various types of stars by Abt and his coworkers.
[d]This figure represents one system, our solar system.

Orbits as a Test of the Origin of Companions

The planets have nearly circular orbits that lie nearly in a common plane, probably because they originated from finely dispersed material in a disk-shaped nebula around the early sun. Collisions would have damped out noncircular motions until everything traveled in parallel, circular, coplanar orbits. These features, then, may be diagnostic of this type of origin process.

Do binary and multiple star systems show these features? No. Binary orbits are frequently much more elliptical than are the orbits in the solar system. As for relative inclinations, in the solar system Pluto's orbit, the most inclined, lies at 17° to the system, and none of the rest is inclined more than 7°. However, in ten triple- and quadruple-star systems reviewed by Batten (1973), no orbit pair is reported at less than 19° inclination, and half are at more than 40°. One other system, BD + 66°34, may have three low-mass stars in coplanar orbits. Furthermore, most multiple systems are not smoothly and concentrically ordered as if by a Bode's rule relation. Instead, they may have two closely spaced members and a very distant member or two co-orbiting, closely spaced pairs.

In general, most binaries and multiples have orbits dissimilar to those in our planetary system, suggesting a different origin. Conversely, should a system be discovered with several small planetary-mass companions, the similarity of its origin to that of the solar system could be tested by determining whether the orbits of the planetary-mass companions are circular, coplanar, and regularly concentric.

Have Extrasolar Planets Been Detected?

What empirical evidence is available as to whether planets exist around other stars? Not much . . . yet. As we will describe in a moment, planets near other stars cannot be seen directly with present-day equipment, but a number of different techniques can be applied to detect them. Generally, massive unseen companions to stars are easier to detect than less massive companions, even if none of them are luminous. Because hydrogen-rich giants larger than 2 M♃ have substantially different internal structure (in terms of hydrogen degeneracy and electron behavior), we might restrict the term planet to objects smaller than 2 M♃ (with the proviso that orbits also need to be measured to affirm solar system–like origin).

Masses smaller than 2 $M_{♃}$ are hard to detect. Nonetheless, several improved systems are going into operation in the 1990s, and statistics about the presence or absence of Jupiter- and Saturn-sized giant planets are likely to accumulate within a decade. Any result will be fascinating—whether such objects are common, rare, or totally absent in a sample of, say, 50 well-observed stars!

Already some statistics are beginning to accumulate, but they are shaky. An interesting social situation compounds the problem. Because the detection of extrasolar planets is philosophically interesting in terms of humanity's place in the universe, some prestige attaches to discovering such an object, and the popular media go into a frenzy any time an astronomer announces even a tentative detection. As a result, perhaps a dozen such detections less than or even greater than 2 $M_{♃}$ have been announced in the press with great hoopla, only to fade quietly when the objects went unconfirmed, or were recognized as too big to be ordinary planets. The press seems also to miss the fact that a continuum of masses runs from stars, through sub-stellar masses, to planets, so that what looks to the press like a black-and-white issue of "discovery of the first extrasolar planet" looks more to an astronomer like the greyer and calmer issue of pushing down the mass frontier to lower limits.

Nonetheless, there are some interesting cases to discuss.

Barnard's star, the fourth-closest star system, caused a flurry of interest in 1963 when Peter van de Kamp announced an unseen companion of about 1.5 $M_{♃}$. As recently as 1981 he augmented his data, indicating two objects of about 0.7 and 0.6 $M_{♃}$ at 2.95 and 3.7 AU from the star. Even four planets were claimed by other authors. However, the more planets one invokes, the easier it is to fit noisy data. Gatewood and Eichhorn (1973) published independent observations indicating no evidence of planets, and today the system is regarded as inconclusive at best.

Other more massive objects are more definite. The binary 61 Cygni, 10.8 light years away, has two visible components and a well-measured unseen companion of about 8 $M_{♃}$. BD + 66°34, about 33 light years away, has two small stars orbited by an unseen object of about 130 $M_{♃}$. The orbits may even be circular.

In 1987–1989, Canadian observers announced that by monitoring radial velocities of 18 stars, they found evidence of unseen companions of 1 to 13 $M_{♃}$ around nine of them (Campbell, 1989). Examples were Epsilon Eridani, 10.7 light years away, with a possible companion of 1 to 5 $M_{♃}$, and Gamma Cephei, with a possible companion having mass as low as 1.6 $M_{♃}$.

Infrared observers have reported small companions in young star systems. The newly forming star T Tauri, with a surrounding nebula of silicate dust grains, has a reported companion of 5 to 80 $M_{♃}$, with a temperature 800 K (Hansen and others, 1983).

Origins of Binary, Multiple, and Planetary Systems

Table 4-4 summarizes current thinking on the origins of binary, multiple, and planetary systems based on astronomical evidence. There are probably several processes at work. The widely separated pairs, with orbits larger than the scale of the solar system, are probably formed by **capture**, a process whereby one star picks up a companion due to gravitational interactions during a close encounter in the presence of a nearby third star or a resisting gaseous medium. Probably this happens just after the star-forming process while the stars are crowded in their initial star cluster. Arny and Weissman (1973) found that half the protostars in a typical cluster undergo collisions or close encounters before the cluster disperses.

Similarly, Hills (1977) finds a high probability that if one already formed binary interacts with a single star, the single star may sometimes steal one member of the binary as its own companion. Hills finds that 1 in 1000 of the field stars of the solar neighborhood have undergone this process, while most binaries with wide separations in open clusters have had one or more such "exchange collisions." Thus, stars engage both in picking up singles and mate swapping. Both processes could drastically affect any chances for growth of planetary progeny.

Another category of binaries, very closely spaced pairs only a few stellar radii apart (or actually in contact in some cases) are probably formed by **fission**, a process whereby a rapidly rotating star splits into two similarly sized objects late in the star's formative process. Some of these close pairs may also result from tidal evolution of more widely spaced pairs (see Figure 3-9). If the satellite star revolves faster than the primary star rotates, it would tend to pull ahead of the tidal bulge (A-B in Figure 3-9), contrary to the Earth-moon case in Figure 3-9; hence the tidal bulge would pull backward on the satellite, causing its orbit to spiral closer to the primary and producing a very close pair.

The large group of remaining binaries, with orbits about the size of planetary orbits, may form from **subfragmentation** of the collapsing protostar, a process in which the protostar divides into several separate condensations, each of which then separately evolves into a star, producing a coorbiting system. Lucy (1977) made a computer simulation of a rotating, axially symmetric, contracting, optically dense protostar, in which the initial protostar divided into a triple system whose components contained 60% of the original total mass, the rest of the mass being ejected. This same process may occasionally produce subcondensations with planetary mass.

We thus tentatively conclude that the last class of multiple star systems—those in which the collapsing protostar breaks into several smaller collapsing objects—could include some systems with small companions that would essentially be planets. However, the next chapter will give evidence that not all planetary bodies in our solar system formed simply by gravitational collapse of part of a protostar, but rather by aggregation of the dust grains.

Furthermore, calculations suggest that planets like those in our system would not form in most systems with two or more stars, because the complicated perturbations caused by the stars prevent the surrounding dust grains from aggregating (Hartmann, 1978; Greenberg and others, 1978). Thus we also conclude that systems of planets are less likely to accompany known binary and multiple stars than to accompany the other, seemingly single, stars. The stars most likely to have planets are the solitary stars of about the same mass as the sun.

BLACK DWARFS VERSUS PLANETS

As shown in Figure 4-8, objects smaller than about 0.080 to 0.085 $M_{\odot}$ contract but never get hot

Table 4-4
Possible Origins of Binary, Multiple, and Planetary Systems

Very closely spaced systems (<< solar system)	Closely spaced systems (≈ solar system): Stellar companions	Closely spaced systems (≈ solar system): Planetary companions	Widely spaced systems (> solar system)
Fission of nearly formed, rapidly rotating protostar Inward tidal evolution of initially closely spaced pair formed by subfragmenting protostar (circular orbits produced by tidal forces)	Subfragmentation of protostar (due to higher-than-average angular momentum?) and subsequent separate evolution by gravitational collapse (orbit types may vary)	Aggregation of dust component of nebula by gravitational collapse and/or collisional aggregation (circular, coplanar orbits)	Capture of adjacent star in original star cluster (elliptical, noncoplanar orbits)

enough to initiate nuclear reactions and become main sequence stars (Hoxie, 1969). Astronomers call them **black dwarfs**, because they do not become visibly luminous and are relatively small.

What is the difference, if any, between black dwarfs and planets? Abt (1978) and others have adopted a clear but arbitrary semantic solution to this problem by simply defining the terms according to mass. According to this usage (with some modifications), stars are objects larger than 0.8 $M_{\odot}$, black dwarfs are objects from 0.002 $M_{\odot}$ (arbitrarily chosen because it is twice Jupiter's mass) to 0.8 $M_{\odot}$, and planets are objects smaller than 0.002 $M_{\odot}$.

Unfortunately, this definition implies that there is a continuum in the formative process as well as in the mass of these objects. We don't know whether this is true. More likely black dwarfs, defined as above, form by the normal process of star formation—gravitational collapse—and when they are companions, they may have eccentric orbits. **Planets** probably often form by processes that (1) separate heavier elements from the stellar material (see next chapter); (2) tend to produce bodies of mass much less than 0.08 $M_{\odot}$; and (3) tend to produce circular, coplanar orbits. In other words, there may be physical as well as semantic distinctions between the larger objects called black dwarfs and the smaller objects called planets. To shed light on this quagmire of problems, we need more data on the existence and characteristics of planet-sized objects near other stars.

DETECTING EXTRASOLAR PLANETARY SYSTEMS

Figure 4-10 illustrates the problem of finding planets near other stars: Planets are extremely faint and stars are extremely bright. If we viewed the solar system from a distance of 10 pc (a moderately nearby star, for instance), the sun would appear as a rather faint star (5th magnitude) but Jupiter would be only 0.5″ of arc away and no brighter than about 26th or 27th magnitude, several times fainter than the faintest stars that can now be routinely photographed. How, then, can we hope to discover for certain the presence of planets outside the solar system? Let us run down a list of techniques:

Figure 4-10. The faint moon almost hidden in the glare of the brilliant sun illustrates the difficulty of detecting planets near distant stars. Light scattered by the optics of the camera system nearly hides the moon in this Apollo 14 view taken in 1971 between Earth and the moon. (NASA)

Photography or imaging: Since the faint planet is overwhelmed by the glare of the star, this technique seems out of the question unless large space telescopes are used, which would remove the glare created by Earth's atmosphere and also use distant dark objects (such as the dark limb of the moon) to occult the star and yet reveal the area within a few seconds of arc. This could reveal large planets near stars within a few parsecs of us.

Astrometric analysis: This method, which detects planets by their effects on the motions of the central star, has yielded positive identification of low-mass objects, of which several luminous examples have been confirmed by direct photography. However, the reported cases with mass in the planet and black dwarf range are controversial. Additional data will help clarify the problem. Large telescopes in space would probably produce sharper images that could be better measured to reveal the sought-for motions produced by planetary companions.

Radio listening: Surveys have been made by radio telescopes to see if intelligent lifeforms might be broadcasting signals from any nearby stars. Through late 1978, such surveys (in the United States and the Soviet Union) have been somewhat spotty but have included at least short listening periods with stars such as Barnard's star, Wolf 359, BD + 36°2147, Luyten 726-8, and Ross 154 (see Table 4-2). Reported results have been negative.

Radial velocity surveys: A new method is being developed by K. Serkowski (1976). He points out that the same back-and-forth motions of the central star around the center of mass, which can be detected astrometrically, would also produce variations in the radial velocity of the star, as detected by the Doppler shift. A Jupiter-mass object orbiting at 1 AU from the sun would produce net variations of some 40 m/s in the sun. Serkowski believes such variations, as seen in distant stars, could be detected by sophisticated applications of the Doppler effect, and he proposes surveys to look for them.

Based on the information discussed in this and the next chapter, most astronomers think that planets are likely to exist around a certain fraction of stars. We have proof that silicate and icy dust exists near many young stars, yet we have no definite proof that such materials have aggregated into planets anywhere but in the solar system. Therefore, we must still accept the possibility that planets may be some rare cosmic accident. Hence, conclusive proof of planets elsewhere would be a major scientific and philosophical discovery. Such proof would increase the probability that environments similar to Earth's have produced life elsewhere in the universe.

SUMMARY

In the complex interstellar environment of gas and dust, stars continue to form today, giving us examples of what the solar system may have looked like during planet formation. These examples confirm the nebular hypothesis: Circumstellar cocoon nebulas of gas and dust hide the stars during their early formation. Such objects have been observed as infrared stars. The cocoon nebulas contain silicate dust and icy grains, exactly resembling the conditions under which planets are believed to have formed. Indeed, planets may aggregate during such a stage. In later stages, such as the T Tauri stage, the nebula thins and disappears, in part by being blown outward. Some of the coarser debris may remain behind to continue aggregating into larger bodies. Such planetary bodies are too small to detect easily, but sophisticated techniques are being developed to search for them.

It is uncertain whether planets *normally* form from this dust and ice, though that is what happened in the case of the solar system. A complication is that most stars form with binary or multiple stellar companions, and the relation of binary star formation to planet formation remains unclear. Some planet-sized masses may have formed in some binary or multiple systems, but planetary systems like ours seem more likely to be found near seemingly single, solar-type stars.

CONCEPTS

big bang
supernova
open star cluster
molecular cloud
nebula
interstellar material
infrared wavelengths
interstellar dust
10- and 20-μm absorption bands
gravitational collapse
protostar
virial (Jeans) theorem
protocluster
cocoon nebula
nebular hypothesis
R Mon
infrared star
H-R diagram
main sequence
post-main-sequence star
pre-main-sequence star

evolutionary track
Hayashi track
Henyey track
T Tauri star
Bok globule
Herbig-Haro object
YY Orionis star
FU Orionis star
binary star
double star
multiple star
visual binary
spectroscopic binary
astrometric binary
astrometric observation
Barnard's star
capture
fission
subfragmentation
black dwarf
planet

PROBLEMS

1. Give some lines of evidence that star formation is continuing at the present time in our part of the galaxy and that we can observe in the sky objects that exemplify the early solar system.

2. How are the Pleiades, Hyades, and other features in the Orion region of the winter sky relevant to star formation? Sketch this part of the sky and identify the Pleiades, Hyades, Orion belt and sword, and the location of R Mon.

3. List four properties of astronomical objects that can be measured by spectroscopic techniques and describe how each measurement is made.

4. A distant star has a spectrum exactly like that of the sun and has no Doppler shift. In addition, infrared observations show emission of infrared radiation peaking near 3 μm. Superimposed on this infrared spectrum are broad absorption bands at 10 μm and 20 μm. Also superimposed are some absorption bands arising in a circumstellar nebular cloud that are slightly blue-shifted from their normal positions. What physical conclusions can you draw about this star? Can you prove from this information whether the star is a newly formed star, a normal, "middle-aged" main sequence star, or a star near the end of its main sequence lifetime?

5. Two infrared stars are observed. Star A has its radiation maximized at wavelength 3 μm, and star B at 6 μm. (a) Compare whatever physical properties of A and B that you can. (b) The two stars are found to have the same luminosity (total Joules radiated per second). What further comparison of physical properties can you now make?

6. How do discoveries of complex organic molecules and grains of rock-forming minerals in interstellar space and in circumstellar clouds support, but not prove, the idea that life might exist on planets outside our planetary system?

7. If a shock wave (high-pressure wave that at least temporarily compresses gas in its path) rushes out from a supernova explosion through neighboring gas clouds and compresses some of the clouds, how might this help initiate star formation in the region near the supernova?

8. Given that supernovas occur in massive stars only about 10^8 y or less after such stars form, would a supernova explosion like that considered in Problem 7 be likely to have gas clouds and a potentially star-forming environment around it, or would it be more likely to occur isolated in deep space? Why?

9. Ejected material from a supernova is likely to contain some radioactive dust grains and gas that decay to distinctive stable isotopes. These stable isotopes are less abundant in ordinary interstellar gas and dust. How might a planetary system blasted during its formation by a nearby supernova be distinguished from a system of the same age never exposed to a nearby supernova?

10. Why are T Tauri stars thought to mark a later stage of formation than infrared stars?

11. (a) What evidence, if any, is there that microscopic rocky material exists in the universe beyond the solar system? (b) What about rocks of, say, 1-m dimension? (c) What about "rocks" of 1000-km dimension, that is, planets?

12. (a) Does the existence of binary and multiple star systems help prove that planetary companions may exist around other stars? Why or why not? (b) Do reports of astrometric binary companions of Jupiter mass prove that planets exist near other stars? (c) If planetary material is prevented by perturbations from aggregating into planets in all binary or multiple star systems, what percentage of other stars could still have planets, according to recent observational estimates?

13. Cite evidence based on observations of suspected young stars that, during the formation of the solar system, the sun may have been much more active in terms of flares, strong solar winds, and similar instabilities than it is now.

14. According to the thermodynamics discussed here, most dust grains vaporize (evaporate into gaseous form) at 2000 K. Suppose the gas in a collapsing protostar, a few astronomical units from the center, reaches 3000 K and then forms a stable, circumstellar disk that slowly cools, so that the same gas remains in a relatively dense state. The gas cools to, say, 1000 K. What would you expect the history of the grains and condensable elements in this nebula to be?

ADVANCED PROBLEMS

15. A T Tauri star has its maximum radiation at 0.5 μm, but also has an infrared excess that peaks at 5 μm. What are the probable sources of these two sets of radiation, and what are their temperatures?

16. (a) If the emission lines near 0.5 μm in the T Tauri star of Problem 15 are blue-shifted up to 0.0005 μm, and the circumstellar absorption lines are blue-shifted by 0.0005 μm relative to the star, what are the motions of the circumstellar material? (b) What would you conclude if a red shift of 0.0001 μm was also seen faintly in some of the stronger absorption lines?

17. Two optically thick infrared stars are measured to be at the same distance from the sun, and both have infrared radiation peaking at 2.0 μm. However, one is 2.56×10^2 times brighter in total luminosity than the other. What is the main difference between them, assuming that they are both blackbodies?

18. Two optically thick infrared stars are at the same distance. Star A has peak radiation at 2.0 μm, and B at 4.0 μm. Star A is 16 times as bright as B. What physical conclusions can you draw about these stars if they both radiate as blackbodies?

19. Once the collapse of a protostar gets well under way, the inequality of the virial theorem becomes an approximate equality, so that

$$GM\mu M_H \approx 5kTR.$$

Assuming that the cloud remains approximately isothermal (equal temperature throughout), calculate a representative internal temperature for the protosun (a) when it reaches the size of the solar system (use Pluto's orbit) and (b) when it reaches its present size. (c) Compare result (a) with the reported temperature for infrared stars and comment on its significance.

20. The present internal (central) temperature of the sun is a few million Kelvin, a temperature sufficient to initiate nuclear reactions that produce the heat and radiant energy to maintain the sun at its present size. Nuclear reactions started in the center of the sun when the center reached that temperature. In view of Problem 19(b), explain why the sun stopped contracting when it reached its present size.

21. A condensation of gas and dust with Jupiter's mass forms within a cocoon nebula whose temperature is 500 K. It has an orbit equal in size to Jupiter's orbit and a radius 0.1 that size. (a) What density must the condensation have in order for it to collapse into a Jupiter-sized planet rather than dissipating if no tidal or other forces act to break up the condensation before it can collapse? (b) Suppose the mean density in this nebula is 10^{-6} kg/m^3. Is such a nebula likely to produce planets of Jupiter's size or *smaller* by purely gravitational collapse?

22. The condensation in Problem 21 is acted on by the central star, which has 1 $M_\odot$ and is 5.2 AU away. Estimate the density that the condensation must have to keep it from being torn apart by tidal forces from the sun. (*Hint:* The condensation would have to be outside Roche's limit.)

23. A distant asteroid has a blackbody spectrum with radiation peaking at 30-μm wavelength. What is its temperature?

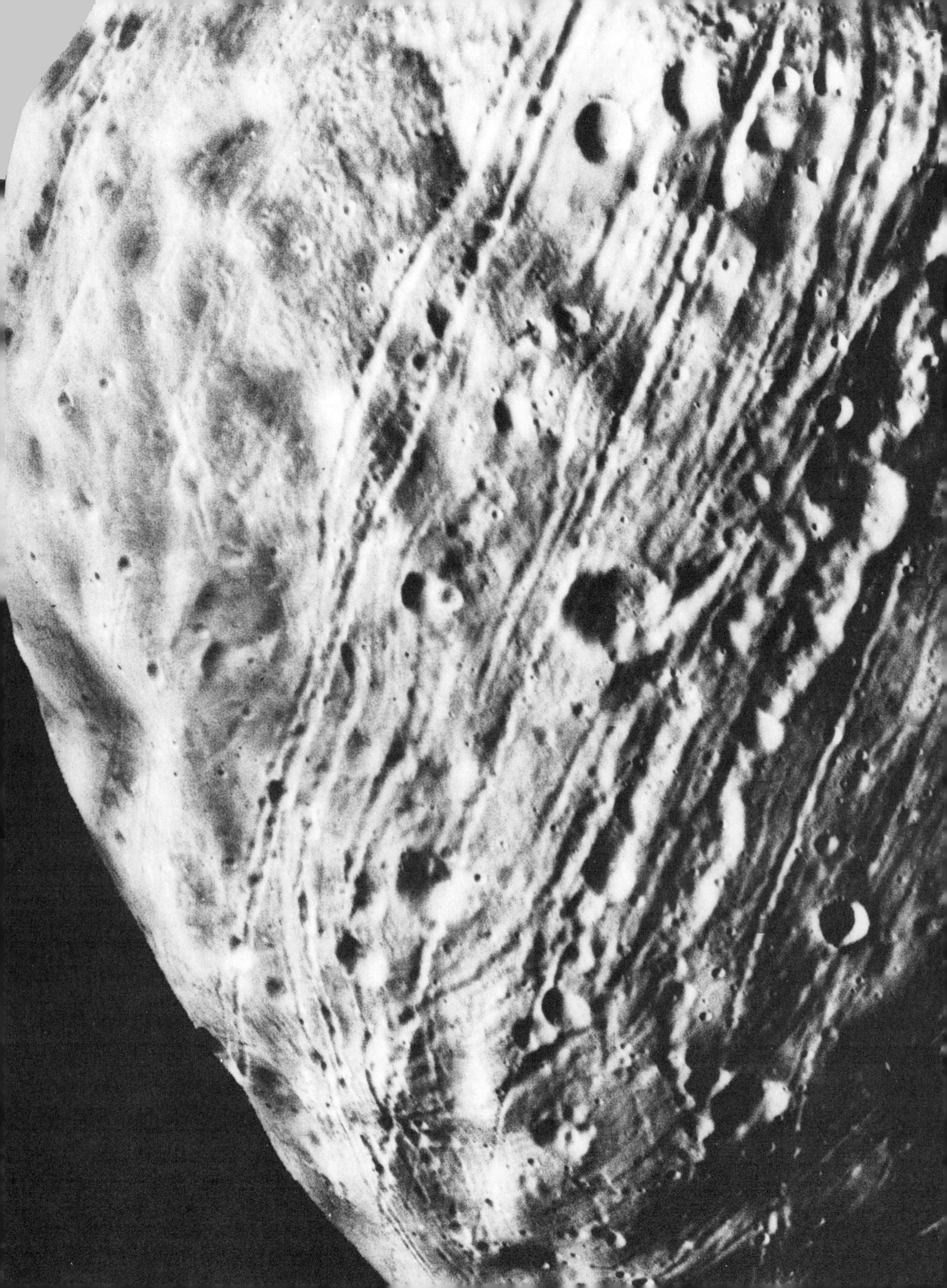

CHAPTER FIVE

THE FORMATION OF PLANETS AND SATELLITES

Chapter 4 emphasized astrophysical evidence about formative conditions among other stars. In this chapter we will emphasize evidence derived from our own star-and-planet system. Both lines of evidence suggest that the sun formed within a surrounding cloud of gas and silicate, icy, and metallic dust. Both lines of evidence also suggest that the dust-and-gas system aggregated into distinct bodies, and we will here attempt to answer when this happened, how long it took, and how it occurred.

DATE AND DURATION OF SOLAR SYSTEM FORMATION

The age of the solar system is 4.6 Gy. This figure has been derived from studies of rocks from three planetary sources: the meteorites (probably representing fragments of asteroids), the moon, and Earth. The derived age comes from **radiometric dating**, a determination of a rock's age by studying the distribution of radioactive and radiogenic isotopes in the rock. **Radioactive isotopes** (**parent isotopes**) are unstable and eventually decay at a known rate into **radiogenic isotopes** (**daughter isotopes**). Since the radioactive decay process proceeds continually, the ratio of daughter to parent isotopes in a given mineral continually increases and becomes a "clock" measuring the age of the mineral. The actual age determination involves a sequence of highly complicated and precise chemical measurements.

Just what do we mean by the age of the mineral? Usually this refers to the time since the mineral solidified from some previous molten or gaseous material. Solidification traps both the parent and daughter atoms in the crustal structure of the mineral grain, thus allowing the radiometric clock to start running. Prior to solidification, when all atoms are in gaseous or liquid form, the parent or daughter atoms can escape from the system.

Consider an idealized example in which crystals of some mineral attract a certain radioactive parent element but not the daughter. As these

Close-up of Mars' satellite Phobos illustrates some properties hypothesized for planetesimals—small bodies from which planets formed in the early solar system. The heavily cratered surface indicates many collisions with other small bodies. Grooves may be fracture patterns from a major collision almost big enough to break Phobos itself. (NASA, Viking)

crystals actually solidify, they incorporate a certain allotment of parent and no daughter atoms, and the radiometric clock starts running. A rock age measured by using this principle is called a **solidification age**, since it indicates when the mineral and rock material solidified into its present form. The actual techniques of measurement are complicated by the need to measure other isotopes as well, which allow estimates of the numbers of parents and daughters (not necessarily zero) actually present in the original rocks.

Even more complex measurements of various isotopes allow estimates of the time when the initial isotopic mixture of a rock's parent material was established—that is, the date of formation of the original planetary material. This is called the **formation age** of the material. Since a given rock might have formed in a recent lava flow or during some other event on the planet, solidification ages may be younger than formation ages for the same material. However, many ancient meteorites have equal solidification and formation ages, showing that they have been little altered since the formation of the solar system. Other types of ages can also be measured, which will be discussed in the next chapter.

Formation Age of the Solar System

Table 5-1 shows ages related to the formation of three types of planetary material. All indicate that planetary matter formed 4.6 ± 0.1 Gy ago. Meteorites give the best determination, because they are abundant and little modified since they formed. The agreement of ages among the various samples not only confirms the date but also shows that the different objects formed at nearly the same time. The planet-forming process was therefore relatively short.

The terrestrial samples require a special comment. Although both the moon and meteorites give samples with solidification ages of 4.4 to 4.6 Gy, no known rocks on Earth solidified earlier than about 3.6 Gy. Older rocks have been eroded, and all rocks have been so heavily modified by chemical and geologic processes that formation ages are hard to estimate. However, Gancarz and Wasserburg (1977) analyzed 3.6-Gy-old specimens from Greenland and found that the initial isotopic chemistry has been little altered. As a result, they could derive formation ages for Earth of 4.47 or 4.53 Gy, depending on the assumptions made about the isotopic history of the material.

All these results are consistent with astrophysical evidence that the sun is several billion years old. The whole solar system apparently formed 4.6 ± 0.1 Gy ago.

Did a Nearby Supernova Explode Just Before the Solar System Formed?

In 1960, J. H. Reynolds discovered that many 4.6-Gy-old meteorites contain radiogenic isotopes that must have come from highly unstable radioactive parents that last only a few millions of years. In particular, he found xenon 129, a xenon isotope that forms by decay of radioactive iodine 129, which has a half-life of only 0.017 Gy, or 17 My. In any radioactive decay process, the **half-life** is the time during which half the atoms present at a given moment will decay. For example, if a billion iodine 129 atoms were initially present, one-half billion would be left after 17 My, one-fourth billion after 34 My, one-eighth billion after 51 My, and so on.

Clearly, because of this relatively rapid decay rate, the original unstable iodine 129 could not have been created too long before the meteorites were created, or else there would have had to be fantastically large amounts of initial iodine 129 in order to have any left to be trapped in the meteorite. What created the iodine 129?

Similar questions have been prompted by other radiogenic material in meteorites. For example, magnesium 26 has been found that formed by decay of radioactive aluminum 26, which has a half-life of only 0.72 My! This suggests that the aluminum 26 must have been created only a few million years before being incorporated in meteorites. What created the aluminum 26?

Growing evidence indicates that the unstable

Table 5-1
Formation Dates of Some Planetary Bodies

Sample	Date of origin (Gy)	Notes
Meteorites:[a] Most primitive available samples		
Guareña, H chondrite	4.56 ± 0.08	Rb-Sr age by Wasserburg and others, quoted by Wasson (1974).
Allegan, H chondrite	4.6 ± 0.1	Ar-Ar age by Podosek, quoted by Wasson (1974).
St. Severin, LL chrondrite	4.6 ± 0.1	Ar-Ar age by Podosek, quoted by Wasson (1974).
All known ordinary chondrites and a group of carbonaceous chondrites	4.6 ± 0.1	Summary by Wasson (1974).
Moon: Rare samples of ancient, little-altered crustal rock		
72417, dunite crustal differentiate (probable fragment of earliest solidified crust)	4.55 ± 0.1	Papanastassiou and Wasserburg (1976). (Albee, Gancarz, and Chodos, 1973, got 4.6 ± 0.07.)
60025, lower limit on lunar age (anorthosite crust fragment)	4.44 ± 0.02	Carlson and Lugmair (1988).
Estimated lunar age from various samples	4.44 – 4.55	Range 4.53 – 4.55 applies *if* moon had chondritic Rb/Sr ratio. Range 4.44 – 4.51 plausible if moon had low Rb/Sr. Carlson and Lugmair (1988).
Earth: Old rock sample		
Amîtsoq 3.59 Gy-old gneiss sample from Greenland	4.50 ± 0.05	Average of U-Th-Pb ages by Gancarz and Wasserburg (1977), based on reconstructions of original chemistry of parent material. This 4.5 result interpreted as age of the Earth.

[a]Meteorite types will be discussed in Chapter 6.

radioactive elements are the type created by nuclear reactions inside stars. Cameron and Truran (1977) wrote a famous paper arguing that a supernova went off near the sun just before it formed, spraying out fresh radioactive isotopes that impregnated the solar nebula. It became widely believed that a supernova happened near the solar nebula, but Cameron (1985) retracted the model, arguing instead that the radioactive nuclei were created in ordinary novas, which are smaller explosions involving binary stars. One line of evidence is that even the ordinary interstellar gas seems enriched in Al-26. In any case, the evidence shows the radioactive isotopes came from material formed inside stars 100 to 400 My before the planetesimals formed, with some late additions no more than 3 My before the planetesimals formed (Wasserburg, 1985).

These conclusions dovetail nicely with Chapter 4, which shows that the primordial sun probably resided in a cluster near many stars, including unstable ones that could have exploded as novas and/or supernovas. Aside from being dramatic, the hypothetical creation of short-lived

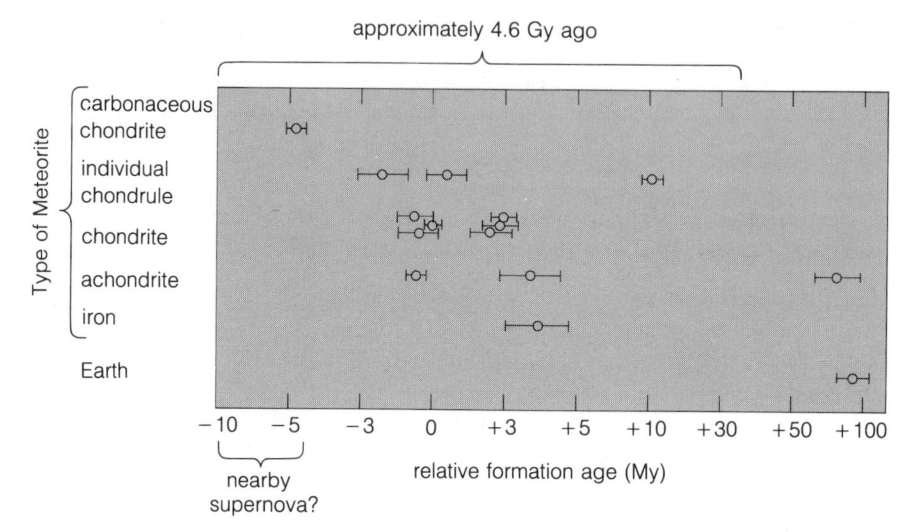

Figure 5-1. The first 110 My of solar system history. Formation ages of Earth and of the several meteorites (identified by type, see also Chapter 6) are not absolute but relative to the formation of a well-studied chondrite meteorite named Bjorböle, whose age is arbitrarily designated as zero. The results show that most meteorites formed during a 20-My period. Dates are interpreted from decay of radioactive iodine 129, apparently produced by a nearby supernova a few million years before the solar system formed. (Adapted from Podosek, 1970; Wasson, 1974; Pepin and Phinney, 1975)

radioisotopes just before the formation of the solar system would give us the opportunity for another type of dating that reveals the *duration* of the planet-forming process as well as its date.

Formation Interval and Duration of the Formative Process

When the short-lived radioactive isotope iodine 129 decays, the inert xenon 129 gas that is formed escapes unless the parent and daughter materials have already been incorporated in a mineral crystal. Suppose a supernova impregnated the solar system material with a certain amount of iodine 129. A meteorite that formed early would have had a certain amount of this iodine, which would have decayed rapidly, leaving xenon 129 trapped in the rock. A meteorite that formed later would have had less initial iodine 129 because the iodine 129 would have been decaying all along; thus, the second meteorite would now have less xenon 129. Therefore, laboratory measurements of the radiogenic xenon now found in a meteorite gives a measure of the time when one meteorite formed relative to another, a measure called the **relative formation age**.

Measurements of this type, together with certain assumptions, lead to estimates of the **formation interval**—the time between the nucleosynthesis (element-forming) event (a supernova?) and the inclusion of the short-lived isotopes into planetary material. Estimates for the formation interval are a few million years, but are model dependent. Rather than reporting model-dependent formation intervals for each meteorite, the current practice is to reference all formation ages to that of the well-studied meteorite Bjorböle, as shown in Figure 5-1 (Wasson, 1974). Figure 5-1 shows that most meteorites formed within about 20 My of each other. Using a similar technique, Pepin and Phinney (1975) found that Earth formed (that is, was big and solid enough to retain its xenon) 96 ± 12 My after Bjorböle.

In summary, we have strong evidence that the planet-forming process occurred 4.6 ± 0.1 Gy ago and was fast. The formation interval between the supernova and the time that solid planetary material formed was only a few million years. In 20 My most of the meteorites had formed, and within about 100 My the planets had developed.

SOLAR SYSTEM CHARACTERISTICS TO BE EXPLAINED BY A SUCCESSFUL THEORY

In constructing a theory of the events of 4.6 Gy ago, we have to choose carefully from among the observational clues. For example, Kepler's laws need not be explained by a theory of origin, because they are required by Newton's laws of gravity. On the other hand, the circularity and coplanar nature of the planetary orbits do need to be explained, because the laws of gravity do not require such conditions.

Nobel laureate Hannes Alfvén, who has spent years researching the solar system's origin, once said, "To trace the origin of the solar system is archaeology, not physics." He meant that our ignorance of the initial conditions forces us to work backward through time, reasoning from whatever clues we can find. The most important clues are *facts about the solar system that have no obvious explanation from present-day conditions or known physical laws*. Table 5-2 gives a list of such facts and we will refer to them in the following discussion, in which we explain them, one by one.

THE SOLAR NEBULA: THE NEBULAR HYPOTHESIS CONFIRMED

Because of a tradition in planetary science that arose before astronomers discovered cocoon nebulas, the sun's cocoon nebula is called the **solar nebula**. The following terms describe objects *in* the solar nebula, to which we will refer in this chapter:

grains or **dust grains:** microscopic particles such as silicates and ices that condensed from the solar nebula or interstellar grains that became trapped in the solar nebula

planetesimals: bodies from submillimeter size up to hundreds of kilometers in diameter (even up to 1000-km-scale bodies in some uses) that formed during the planet-forming process in the solar nebula

protoplanet: (1) any large precursor of a planet, usually referring to an extended mass, including any large atmospheric cloud that may have later

Table 5-2
Characteristics of the Solar System to Be Explained by a Theory of Origin

1. All the planets' orbits lie roughly in a single plane.
2. The sun's rotational equator lies nearly in this plane.
3. The planets and the sun all revolve in the same west-to-east direction, called *prograde* or *direct* revolution.
4. Planetary orbits are nearly circular.
5. Planets have much more angular momentum (a measure of orbital speed, size, and mass) than does the sun, and the sun spins slower than expected if it spun off the solar nebula during its collapse. (Failure to account for this was the great flaw of early evolutionary theories.)
6. Some meteorites contain inclusions of minerals that formed from grains condensed at higher temperatures than were other meteorite materials; these inclusions contain materials with unique isotopic abundances.
7. Planets differ in composition, roughly correlating with distance from the sun. Mercury is dense and metal rich; other terrestrial planets are less metal rich. The giant, hydrogen-rich planets are in the outer solar system.
8. Meteorites differ in detailed chemical and geologic properties from all known terrestrial and lunar rocks.
9. The distances between the planets mostly obey the simple Bode's rule.
10. All closely studied planets and satellites show impact craters indicating collisions with interplanetary bodies whose diameters ranged up to 100 km.
11. Except for Venus, Uranus, and Pluto, all planets rotate prograde with obliquities of less than 29°.
12. Most planets and asteroids rotate with similar periods of about 5 to 10 h, except in cases where obvious tidal forces have slowed them (as is the case with Earth).
13. As a group, comets' orbits define a large, spherical swarm around the solar system.
14. Major planet-satellite systems resemble the solar system on a smaller scale.

dissipated, and usually implying a body as massive as or more massive than the planet today; (2) in theories involving gravitational collapse, the large gravitationally unstable units from the time they first gain an identity by initiating collapse to the time they reach planetary size; initially they may extend across an appreciable part of the orbital circumference of the planet in question.

Origin and Mass of the Solar Nebula

Astrophysical studies of star formation (Chapter 4) suggest that the solar nebula was the sun's cocoon nebula. It consisted of hot gas and dust many astronomical units from the sun by the time the sun began to take on its identity as a central condensation in the gravitationally contracting, prestellar cloud. Stages 2 and 3 of Figure 4-4 showed the development of such a nebula as the star collapses rapidly in the nebula's center. Thus, the solar nebula is not believed to have been material that was ever inside the sun or thrown off its incandescent surface.

Some early theorists incorrectly assumed that the mass of the solar nebula was equal to the total mass of the planets. However, the mass of the nebula must have been many times the planetary mass, since the planets are composed of only residual elements from a nebula originally of cosmic composition. For example, the terrestrial planets are composed mostly of silicon and iron. Table 4-1 showed that only 0.0006 of the cosmic material is silicon, and thus 1700 grams of interstellar gas would be needed in order to obtain even one gram of silicon.

Most of the solar nebula was not condensable material but consisted of gases such as hydrogen, helium, the inert gases, water vapor, and so on. These materials are called **volatiles**, the elements and compounds that, under the given conditions, tend not to form solids. Volatiles have a high vapor pressure, meaning that if they are incorporated into solid or liquid material and find themselves in a vacuum environment, they will readily escape from the solid, atom by atom, and build up the surrounding gas pressure.

A minimum mass for the solar nebula can be calculated if we estimate the total amount of missing volatiles that must be restored to each planet to produce a protoplanet of cosmic composition. Since silicates and iron constitute only a small fraction of cosmic material, the mass of Earth must be multiplied by a large factor to restore the lost mass, as shown in Table 5-3. Interestingly, the masses contributing to the various planets (the last column) were much more nearly equal than the final masses of the planets themselves. The table shows that the minimum material involved in planet formation approached 0.03 $M_{\odot}$. Of course, the nebula might have been still more massive, since we do not know how much extra interplanetary gas and dust may have been blown away without contributing to the planets. The Soviet theoretician Safronov (1966) reviewed solar nebula physics and inferred a mass of 0.15 $M_{\odot}$.

Estimates of the mass of the solar nebula also come from studies of young stars. The young star VY Canis Majoris may have a cocoon nebula of about 0.15 $M_{\odot}$ (Herbig, 1970). T Tauri stars apparently blow away up to 0.4 $M_{\odot}$ of material, and this very strong, outflowing stellar wind also carries off angular momentum that slows the star's rotation from the fast rates observed for T Tauri stars to the slow rate observed for the sun (Kuhi, 1966, 1978). Magnetic fields during the loss of a 0.15-$M_{\odot}$ nebula could have slowed the sun to its present rotation rate (Schwartz and Schubart, 1969).

A much more massive nebula has been postulated (Cameron, 1962, 1975; Cameron and Pine, 1973) on the theoretical grounds that a 1-$M_{\odot}$ star would be unlikely to form in a cloud of only, say, 1.3 $M_{\odot}$, leaving only 0.3 $M_{\odot}$ behind. From models of the collapsing cloud, these authors estimate that about 2 $M_{\odot}$ of gas was left behind in the solar nebula. However, it is difficult to support such a large figure from stellar observations. Perhaps the safest statement is that the mass of the solar nebula was at least 0.2 $M_{\odot}$.

Composition and Shape of the Solar Nebula

On the basis of both meteorite chemistry and the above conclusion that the nebula was a solar remnant, researchers have long argued that the solar nebula had approximately the same composition as the sun, as indicated in Table 4-1.

However, as noted above, variations in isotopic composition among planets and meteorites indicate that the nebula composition was not uniform and that some interstellar (supernova-derived?) material was injected into the nebula.

The solar nebula was disk shaped and rotated in the plane of the sun's equator. As noted in the last chapter, Laplace pointed out this fact in the 1700s. To understand why the initially irregular, rotating protosun collapsed into a star surrounded by a disk, consider a system of particles (gas atoms or dust grains) orbiting almost randomly around the newly formed sun. Because the orbits are nearly random, many will intersect, causing the particles to collide. The net effect of these collisions will be to average out the different directions of motions until most particles are moving in one plane and in nearly circular orbits. This is the same reason why particles near Jupiter, Saturn, and Uranus have formed flat ring systems and why the dust-and-gas complexes of the Milky Way and most other galaxies have formed disks. This also accounts for characteristics 1–4 in Table 5-2, the list of characteristics to be explained by a theory of solar system origin.

Theoretical studies of the collapse process (Tscharnuter, 1978) show that the material "above" and "below" the sun plunges in rapidly toward the equatorial plane, so that a disk shape is attained probably within a million years or so. These results agree with observations mentioned in Chapter 4, favoring a disk shape for some observed cocoon nebulas.

As shown in Figure 5-2, the nebula finally adopts a quasi-equilibrium shape when all forces

Table 5-3
Masses of Nebular Matter Required to Form Planets

Planet	Assumed composition	Present mass (kg)	Estimated ratio required mass: present mass	Estimated nebular mass (kg)
Mercury	Silicates, iron	3×10^{23}	410	1×10^{26}
Venus	Silicates, iron	5×10^{24}	380	2×10^{27}
Earth	Silicates, iron	6×10^{24}	380	2×10^{27}
Mars	Silicates, iron	6×10^{23}	370	2×10^{26}
Asteroids	Silicates, iron (ices?)	$\sim 1 \times 10^{21}$	250	3×10^{23}
Jupiter	Hydrogen, ices	2×10^{27}	10	2×10^{28}
Saturn	Hydrogen, ices	6×10^{26}	16	1×10^{28}
Uranus	Methane, ammonia, water, ices	9×10^{25}	67	6×10^{27}
Neptune	Methane, ammonia, water, ices	1×10^{26}	64	6×10^{27}
Pluto	?	7×10^{23}?	75?	5×10^{25}?
Comets	Methane, ammonia, water, ices	$>1 \times 10^{27}$?	5	$>5 \times 10^{27}$?
			Total nebular mass:	$>5.1 \times 10^{28}$[a]

Source: Adapted from Kuiper (1956a), Cameron (1962, 1975), Hoyle (1963), and Whipple (1964).
[a]Equivalent to $>0.03\ M_{\odot}$.

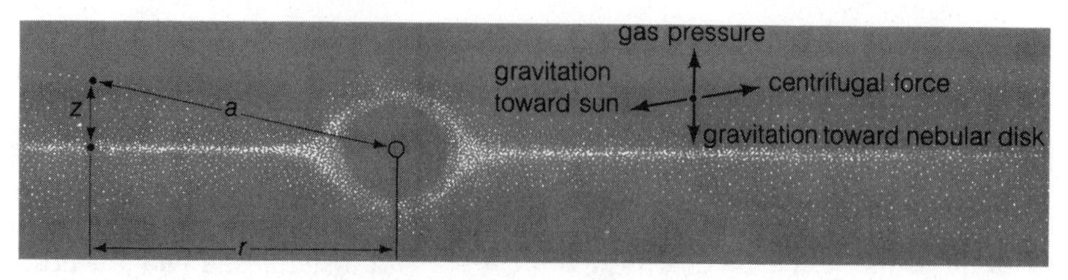

Figure 5-2. Cross section of the solar nebula, showing balanced forces on a particle in the nebula. The disk shape is indicated. Density and pressure are greater in the disk than out of it and increase toward the sun. The outer regions are thicker and less dense than inner regions. Dimensions a, r, *and* z *are employed in the optional mathematical discussion.*

on the particles are balanced—primarily gravity, (directed inward) and gas pressure and centrifugal force (outward). Mathematical models using the **hydrostatic principle**—which assumes balance among these forces—were pioneered by Von Weizsacker (see Chandrasekhar, 1946) and Kuiper (1951). The disk shape can be called only quasi-equilibrium, because as the nebula cools, some of the dust aggregates into planets, and some dust and gas eventually blow away.

MATHEMATICAL NOTES ON THE SHAPE OF THE SOLAR NEBULA

The structure of any gaseous atmosphere or nebula in equilibrium can be analyzed by noting that all the inward forces are equal to the outward forces. In its simplest form, this analysis involves only two equations, the perfect gas law, which merely states that the pressure, temperature, and density are controlled by the fact that the substance is a gas:

$$P = \frac{\rho k T}{\mu M_H} \qquad (1)$$

and the hydrostatic equation, which states that the pressure in the gas responds to a gravitational compressive force:

$$P = -\rho g r \qquad (2)$$

In these equations,

P = pressure
T = temperature
ρ = density
μ = mean gas molecular weight
g = gravitational acceleration at the point in question
r = distance from sun

Other symbols are constants defined in Table 2-2.

We could use these equations to analyze the pressure and density changes along the ecliptic plane in the direction toward and away from the sun. However, the result shows that ρ does not change too much in this direction, and it is more interesting to analyze the structure perpendicular to the ecliptic plane, as shown in Figure 5-2.

For the upward (perpendicular to ecliptic) forces to equal the downward forces, the gas pressure must balance the hydrostatic pressure. The downward force is the z component of gravity:

$$g_z = g\frac{z}{a} = \frac{GM_\odot}{a^2}\frac{z}{a} = \frac{GM_\odot z}{a^3} \qquad (3)$$

Since the variables depend on one another, we set up a differential equation for the change in pressure dP over a small distance dz, equating dP from the perfect gas law (1) with dP from the hydrostatic equation (2):

Density and Pressure in the Solar Nebula

For given temperatures, such as those observed in infrared cocoon nebulas, the density and pressure distributions in the solar nebula can be found from the principles described above. The mean density, of course, is just the total mass divided by the total volume. If we take the mass of 0.2 $M_\odot$ mentioned above and spread it across a 0.5-AU-thick nebula the size of Pluto's orbit, the gas's mean density is only of the order 10^{-7} kg/m^3. Similarly, if we distribute the minimal 380 $M_\oplus$ of cosmic material necessary to make Earth (see Table 5-3) in a toroid 0.1 AU thick around Earth's orbit, we have a mean gas density again around 10^{-7} kg/m^3. Dynamical studies show that the density was greater near the midplane of the nebula and nearer the sun (see mathematical notes on the shape of the solar nebula).

For a hydrogen-rich gas at 800 K and with density of, say, 10^{-6} kg/m^3, the gas pressure would be of the order 10 N/m^2, which is 0.1 mbar, or about 10^{-4} normal atmospheric pressure.

Cooling of the Solar Nebula

The last chapter showed how a contracting gas cloud about to form a star gets warmer as it contracts. This is the source of the heat that warmed the inner solar nebula to temperatures probably exceeding 2000 K (Cameron and Pine, 1973). However, once the nebula assumed a disk shape rotating approximately in hydrostatic equilibrium, its contraction was much slowed

$$\frac{kT}{\mu M_H} d\rho = -\rho g_z dz \tag{4}$$

which assumes that T is about constant at the given distance from the sun r.* Substituting for g_z,

$$\frac{kT}{\mu M_H} d\rho = -\rho \frac{GM_\odot z}{a^3} dz \tag{5}$$

giving

$$\frac{d\rho}{\rho} = -\frac{\mu M_H G M_\odot}{a^3 kT} z dz \tag{6}$$

Integrating this gives

$$\ln\left(\frac{\rho}{\rho_0}\right) = -\frac{\mu M_H G M_\odot z^2}{2kTr^3} \tag{7}$$

where ρ_0 = density in the ecliptic plane, and z is assumed to be not too far out of the plane, so that $a \approx r$, being constant. This gives the density structure

$$\rho = \rho_0 \exp - \frac{\mu M_H G M_\odot z^2}{2kTr^3} \tag{8}$$

This says two interesting things. First, the density decreases above and below the ecliptic plane as z increases. Second, the greater the distance r from the sun, the thicker the disk, as illustrated in Figure 5-2. This can be seen by looking at the exponent. Let us define scale height H as the distance "above" the ecliptic over which the density drops by a factor e. By examining the exponent, one sees that

$$H = \sqrt{\frac{2kTr^3}{\mu M_H G M_\odot}} = 1.11(10^{-8})\, T^{1/2} r^{3/2} \tag{9}$$

The numerical solution is for a hydrogen-dominated nebula. Thus, the scale height increases with r. At 1 AU from the sun in a 1000-K gas, this gives a scale height of about 0.14 AU; at 10AU, it gives a scale height of about 4 AU.

*Incidentally, Safronov (1972, p. 25) gives a similar derivation and this is his equation 2.

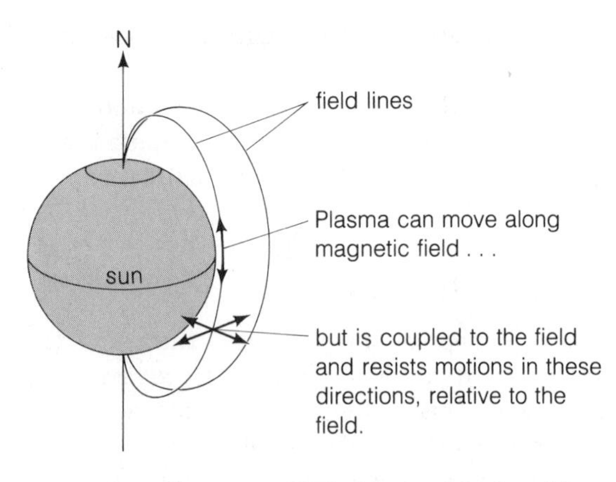

Figure 5-3. In the magnetic field of the sun (represented by two magnetic field lines), neutral gas can move in any direction, but ionized gases (plasma) are deflected if they do not move along field lines. Rotation of the sun would rotate the field lines and cause the plasma to be dragged along in the rotational direction.

and it eventually cooled by means of infrared radiation. Temperatures dropped toward values of around 800 K, as determined for infrared cocoon stars, and eventually reached values as low as a few hundred Kelvin.

MAGNETIC EFFECTS

Our discussion so far has implied that the solar nebula evolved as a neutral gas with some admixture of dust particles. Evolution of such a nebula has been theoretically analyzed in terms of motions controlled by gravity plus additional small forces such as gas pressure or radiation pressure.

However, a different situation exists if there was sufficient energy to keep the nebular gas ionized, as is the solar corona and interplanetary gas today. While a neutral gas can move freely in any direction, a highly ionized gas, or **plasma**, cannot move freely with respect to a magnetic field. As shown in Figure 5-3, a plasma can move freely *along* magnetic field lines, but not *across* them. (**Magnetic field lines** are imaginary lines stretching through space from the north to the south pole of a magnetic source, such as the sun.)

If a magnetic field moves through a plasma, the plasma tends to be dragged along with the field lines and is said to be **magnetically coupled** to the field. If the nebular gas was neutral, therefore, its motions would be analyzed by **hydrodynamics** (the science of fluid and gas motions), but if the gas was a plasma, it would be dragged around by the rotating solar magnetic field (rotation of the sun and field can be visualized in Figure 5-3), and its analysis would require **magnetohydrodynamics** (the science of plasma motions).

Was the Solar Nebula Highly Ionized?

While most theorists have developed models with neutral nebular gas, a few workers have developed plasma models. Nobel laureate Hannes Alfvén (1954) and Alfvén and Arrhenius (1976) have long championed the view that the nebula was largely ionized by its energy of infall, and its motions controlled by magnetohydrodynamics. As Goettel and Barshay (1978) point out, however, the plasma model has not had notable success in explaining the compositions of the planets.

A new reason for proposing at least partial ionization of the solar nebula was suggested in 1978. Consolmagno and Jokopii (1978) noted that if the solar nebula really contained the recently proposed amounts of radioactive aluminum 26 or potassium 40 from a supernova, then the energetic subatomic particles shot out of these atoms during radioactive decay would have ionized neighboring atoms. Their calculations for reasonable models involving aluminum 26 content indicate that ionization would be partial but sufficient to couple the nebular gas to ambient magnetic fields. These calculations suggest that further magnetohydrodynamic studies will be required for precise modeling of the solar nebula's evolution.

Magnetic Braking of the Sun's Rotation

Although opinions on the importance of ionization in the solar nebula vary, most researchers agree on one exceedingly important magnetohydrodynamic effect. The ionized atmosphere, or corona, of the primeval sun merged with the inner nebula, so that at least the inner nebula had a large ion content. As the sun's magnetic field turned through gas at some distance from the sun, the gas would have tended to be dragged along with the field, as can be visualized by imagining the sun and field turning in Figure 5-3.

Initially, as a result of the contraction process, the sun should have been spinning much faster than it does today, as pointed out by Laplace. The big problem with Laplace's original nebular hypothesis is that it did not explain why the sun slowed from this initial fast spin rate. But now we see that the sun's magnetic field would have tended to grip the nebula. The nebula in turn would have caused a drag on the magnetic field, twisting the sun's field and slowing the sun's spin, just as a spinning tennis ball dropped in a swimming pool will rapidly come to rest because of the drag of the water on the fibers in the ball's surface. Thus magnetohydrodynamics explains the fifth important characteristic of the solar system in Table 5-2.

As the sun slowed down, its angular momentum was transferred to the nebular gas. Much of the angular momentum was eventually carried out of the system, as the gas rushed away from the sun in an expanding solar wind. This effect explains not only the sun's spin rate but also the fact that T Tauri stars seem to be rotating much faster than the main sequence stars into which they evolve. Apparently the T Tauri stars will also be slowed by magnetic braking, which occurs between the T Tauri stage (as the nebula begins to clear) and the final shedding of the nebula (Kuhi, 1978).

FIRST PLANETARY MATERIAL: EVOLUTION OF DUST IN THE SOLAR NEBULA

We now have a reasonably good picture of the planet-forming environment. A primeval solar nebula of remnant solar material probably had a mass of some tenths of the solar mass and a disk-shaped volume similar to the present solar system's shape. During contraction much of it reached temperatures exceeding 2000 K, but it was cooling.

The Condensation Process

Suggestions were made as early as the 1940s and 1950s that solid microscopic mineral grains would condense out of such a nebula. Pioneering work in this area was done by the physical chemist and Nobel laureate Harold C. Urey (1952), who discussed the condensation of grains. He showed from meteorite chemistries that the meteorites and planets must have formed from individual solid grains at temperatures as low as a few hundred Kelvin, rather than being directly formed from gravitationally bound, incandescent masses of gas as had once been supposed. Urey was virtually the only physical chemist working on planetary science at the time, and a decade passed before his ideas were fully absorbed and a new generation of chemistry-oriented planetary scientists began to make great strides.

The condensation process is simple enough. If a hot gas contains a condensable substance and starts cooling, eventually a condensation temperature will be reached where the condensable substances change from gaseous form to liquid or solid form—the latter being microscopic dust grains. This is a familiar phenomenon, since we know that raindrops or snowflakes appear when water vapor condenses in air that is lifted to cool regions of the atmosphere.

In the 1960s, Wood (1963), Lord (1965), Larimer (1967), Larimer and Anders (1967), and Blander and Katz (1969) began to sketch out the sequence of mineral types that must have condensed in the cooling solar nebula. In the 1970s, planetary chemists such as Lewis (1972a, b, 1974), Grossman (1973, 1977), and Goettel and Barshay (1978) have brought this work to fruition with models that explain in surprisingly simple terms the gross chemical and mineralogical properties of the solar system. Actual experiments have confirmed the condensation of microscopic smoke particles, such as Si_2O_3 from silicon monoxide (SiO) vapor (Day and Donn, 1978).

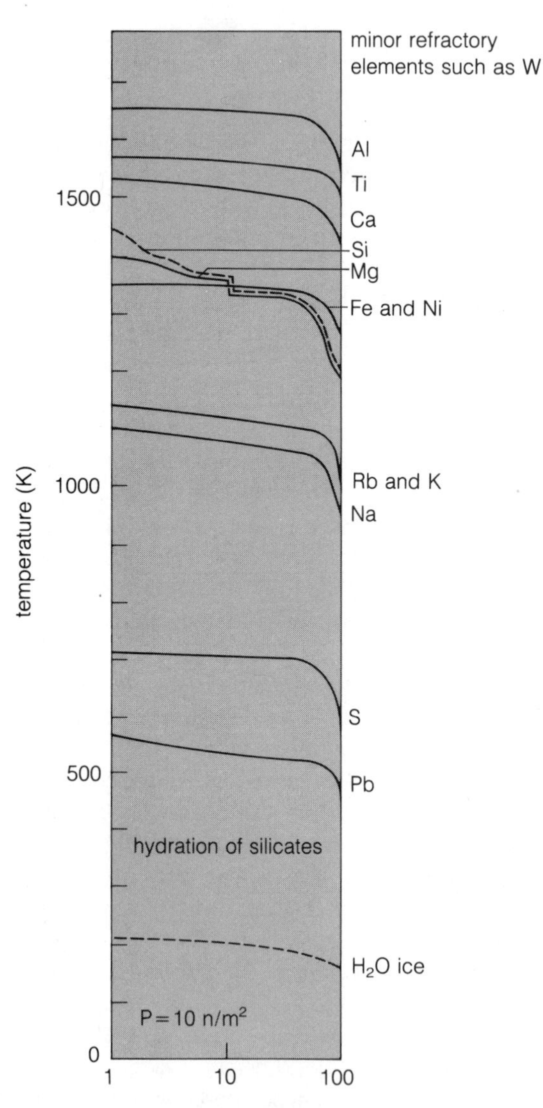

Figure 5-4. Condensation of elements from gas in the solar nebula as temperature drops, calculated by Grossman (1975) at P = 0.1 mbar. Graph does not indicate resulting mineral forms, except to note that water vapor begins to hydrate the silicate minerals below about 500 K and freezes out as ice at about 200 K (Lewis, 1972a). Curves for silicon and water, materials of special planetary relevance, are dashed.

Grossman (1977) studied the process in the solar nebula by considering an idealized system containing the 20 most abundant elements in the solar system (excluding the noble gases since they are chemically inactive). He determined the concentrations of different components at a series of temperatures 25 or 50 K apart as the nebula cooled and as various gases became depleted by condensation of solid grains.

The results of such calculations, seen in Figure 5-4, show the history of condensation as the various elements change from gaseous to solid form. The figure shows only which gases disappeared, not what minerals resulted.

The composition of the mineral dust depends on the cooling rate and the degree to which the early grains remain available to interact with the later material. Fortunately, however, the sequence of events is fairly model predictable. Figure 5-5 gives an example, based on Lewis' (1972a, b) calculations and his assumption that all the dust and gas remains in **chemical equilibrium** with each other—that is, that the minerals present are in equilibrium with the gas and other minerals for the given temperature and pressure. The following sections describe the events of Figure 5-5 in more detail, starting with the initial high-temperature conditions and then following the condensates as the temperature dropped.

Condensation of the High-Temperature Refractories

Refractory compounds are the last to melt during heating and the first to condense during cooling, at around 1500 K. Extreme refractories include relatively rare elements such as tungsten, osmium, and zirconium, which appeared among the earliest crystals condensed. More abundant amounts of refractories appeared as oxides of calcium, titanium, and aluminum.

Proof of the early condensation of refractories appeared when refractory materials were actually discovered in a primitive type of meteorite called a carbonaceous chondrite (see the next chapter

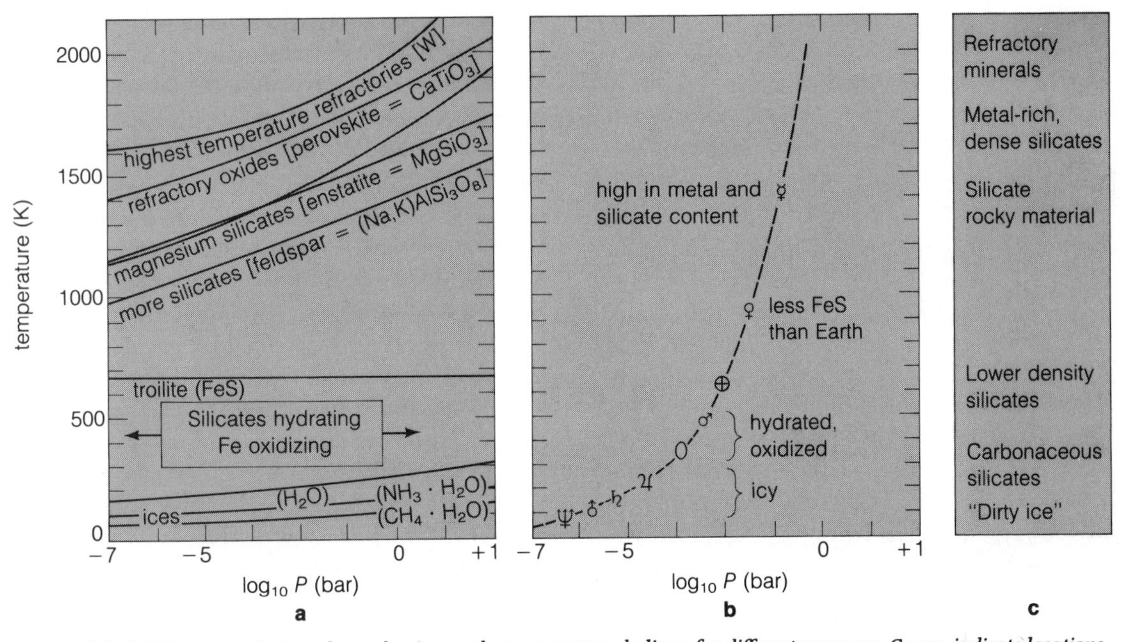

*Figure 5-5. (**a**) The types of minerals condensing as the temperature declines, for different pressures. Curves indicate locations for various mineral groups, and brackets give most prominent examples. (**b**) The dashed curve gives a plausible adiabatic distribution of temperature and pressure in the solar nebula at a given moment, with the relative positions of the planets given. If the gas were suddenly blown away, the mineral grains remaining would have compositions given in part (**a**), which approximately agree with observed planetary properties. (**c**) The bulk composition of solid material that would dominate at various temperatures. (Adapted from Lewis, 1972a, 1972b)*

for more details on meteorite types). **Carbonaceous chondrites** are carbon-rich meteorites that, after being formed, were never strongly heated, thus preserving mineralogy from the days of their formation 4.6 Gy ago. On February 8, 1969, several tons of such a meteorite fell at Allende, Mexico. Samples were rushed to numerous labs, which were just gearing up to analyze the first lunar samples, due later that year. As visible in Figure 5-6, about 5% to 10% of Allende consists of centimeter-scale, irregular inclusions of whitish minerals, which turned out to be calcium-, titanium-, and aluminum-rich minerals such as spinel ($MgAl_2O_4$) and perovskite ($CaTiO_3$). These **Allende inclusions** are believed to be actual pieces of the first material condensed in the solar nebula. They contain little iron and condensed at around 1625 to 1125 K as indicated by their mineral content (Grossman, 1973; Haggarty, 1978). Apparently the Allende meteorite itself formed at a relatively late date but trapped a range of much earlier refractory particles that were floating around in the nebula. This accounts well for item 6 in Table 5-2.

Major Condensates: Nickel-Iron and Silicates

At about 1400 K more abundant materials began to condense (see Figures 5-4 and 5-5). Iron and a fair amount of nickel condensed to form an alloy, and magnesium silicates (very common rock-forming minerals on Earth) condensed as rocky particles. Since these materials are more abundant than the refractories, the dust existing at nebular temperatures around 1300 to 1400 K became mostly a mixture of nickel-iron alloy grains and rocky particles. One of the most common minerals that condensed at this temperature was **enstatite** ($MgSiO_3$).

As the temperature continued to drop, sodium and potassium also formed silicate minerals, especially the feldspars. **Feldspars** are familiar

Figure 5-6. A fragment of the Allende carbonaceous chondrite meteorite, showing the prominent whitish inclusions of refractory minerals, incorporated as the meteorite accreted. (Photo courtesy of R. S. Clarke, Smithsonian Institution)

aluminum-silicate minerals making up as much as 60% of Earth's crust and having formulas such as $NaAlSi_3O_8$, $KAlSi_3O_8$, and $CaAlSi_3O_8$.

Also, as the temperature dropped, chemical reactions occurred between the gas and the dust. In particular, the iron metal began to be oxidized to other mineral forms. Some iron reacted with hydrogen sulfide gas (H_2S) to form the mineral **troilite** (iron sulfide, FeS). Other iron reacted with oxygen to form ferrous oxide (FeO), which in turn reacted with enstatite to form the important mineral **olivine** (a mixture of Fe_2SiO_4 and Mg_2SiO_4). These reactions were relatively complete by the time temperatures reached 500 K, when the dust probably consisted of olivine, feldspars, and other silicate and oxidized iron minerals.

Again, primitive 4.6-Gy-old meteorites little altered by heat or melting actually confirm these chemical models. One type of meteorite, the enstatite chondrite, consists of nearly pure enstatite and 19% to 25% by weight of nickel-iron, as predicted for the dust in the 1300-K nebula (Mason, 1962, p. 71; Wasson, 1974, p. 27). These compose only about 2% of the meteorite population.

Most of the remaining meteorites, called ordinary chondrites and amounting to about 78% of meteorite falls, are rich in olivine, contain troilite, and have about 25% by weight of nickel-iron in the form of pure nickel-iron alloy and oxidized iron minerals.

In other words, these meteorites are aggregates of the dust condensed in the solar nebula; they have not been drastically altered (by melting, for example) since their formation.

The Carbonaceous Condensates

At a somewhat lower temperature, as indicated in Figure 5-5, carbon from the nebular gas began to appear in the form of graphite, combined in complex organic molecules, and also in carbon-rich silicate minerals. Partly because of the complexity of organic chemistry, the mineralogy of this carbonaceous material is not too well understood theoretically. Nonetheless, carbonaceous materials were an important phase of the condensation process. For example, an important type of meteorite, called a carbonaceous chondrite (see next chapter) consists of carbon-rich minerals. Allende is a meteorite of this type, and although it contains high-temperature mineral inclusions, its bulk is low-temperature, blackish carbonaceous material, shown in Figure 5-6. Smith and Buseck (1981) found this black matrix to be rich in poorly crystallized graphite with grain sizes of only 1 to 100 nm.

Telescopic observations reveal this type of blackish material on the surfaces of asteroids and certain small satellites in the part of the solar system ranging from the outermost asteroid belt outward to at least the region of Saturn. Theoretical calculations by Consolmagno and Lewis (1977) suggest that planetesimals in Jupiter's region might have initial compositions of about half carbonaceous material and half ice. Temperatures in our region of the solar nebula, inside the asteroid belt, may never have dropped low enough to have produced abundant carbonaceous minerals.

More Major Condensates: The Ices

Below 500 K, water played an increasingly important role. Between 500 and 200 K, water vapor in the gas reacted with some of the dust grains to form complicated hydrated minerals, such as serpentine ($Mg_6Si_4O_{10}[OH]_8$, formed when olivine reacts with water vapor), tremolite ($Ca_2Mg_5Si_8O_{22}[OH]_2$), and talc ($Mg_3Si_4O_{10}[OH]_2$).

In colder regions at around 200 K, many minerals were hydrated and water began to appear in the form of H_2O ice crystals. In even colder regions, in the outermost parts of the solar system, ammonia (NH_3) and methane (CH_4) also condensed as ices. Probably they were mixed with water, forming hydrated ices such as ammonia hydrate ($NH_3 \cdot H_2O$) and methane hydrate ($CH_4 \cdot 8H_2O$) (Lewis, 1972a).

These ices were so abundant that they swamped the rocky materials that had condensed earlier in these regions. Note that the elements involved, H, O, C, and N, are numbers 1, 3, 4, and 5 in abundance, respectively (see Table 4-1). (Number 2 is helium, an inert gas that does not readily form compounds.) There was more than enough hydrogen to combine with all available O, C, and N, and the amount of the O, C, and N was about 8 times that of the Si, S, Fe, and Mg of the silicate minerals. Thus, planetesimals formed in the coldest regions were mixtures of ice somewhat dirtied by trapped dust grains.

Where were ices stable in the early solar system? As long as a significant fraction of the planetary material existed as subcentimeter dust, the nebular cloud was opaque and shielded from direct sunlight but may have been warmed by infrared radiation within the cloud. Once the dust grains clumped together into meter-dimension bodies, the nebula began to clear (Hartmann, 1970), and sunlight sublimed any unshielded ice crystals in the inner regions. Under such conditions, as at present, H_2O ice is stable only beyond the asteroid belt (Watson, Murray, and Brown, 1963; Urey, 1952).

Observations support these theoretical considerations. The black matrix material of the most primitive, carbonaceous chondrites such as Allende consists of low-temperature hydrated minerals containing as much as 22% bulk water content. The water is chemically bound in the minerals (not in the form of ice) but can be easily driven off by mild heating to a few hundred degrees C. Ice-covered satellites, such as Europa, Enceladus, Rhea, Titania, and Saturn's ring particles are common in the outer solar system but do not exist in the inner solar system. Nuclei of comets appear to be ice chunks containing a peppering of dust grains (see Chapter 8).

GROSS COMPOSITIONS OF THE PLANETS

Lewis (1972b) pointed out an interesting consequence of the condensation theory. Suppose that at some stage in nebular cooling, the nebular gas was suddenly swept away, leaving the dust behind. At each planetary orbit, the dust would have the bulk composition of condensates at temperatures corresponding to that distance from the sun. Refractory and metal-rich dust would lie close to the sun. More silica-rich, lower-density dust would be farther from the sun, and low-density ices would lie farther from the sun. Given a temperature at one point, solar nebula models would determine the temperature at other points, which fall along an **adiabatic curve**, or temperature-pressure relation, affected only by the gas and the sunlight.

Figure 5-5b shows such a curve and the relative positions of planets on it. The result is an amazingly good first-order match to planetary compositions, indicated by Figure 5-5c. It shows that if the gas cleared at the moment when the Mercury material was in the refractory- and metal-rich part of the diagram, then Venus and Earth would be silica rich with lower densities, as observed. Mars would have a still lower density, as observed, being rich in oxidized iron compounds (the rustlike minerals that color Mars' surface red). The outer planets would be rich in ices and have low densities, also as observed.

This first-order explanation of planet compositions accounts for item 7 in Table 5-2. If meteorites are fragments of interiors of bodies formed in the region of the asteroid belt and elsewhere, it also accounts for item 8.

Chemical Complexities

Nature is rarely as simple as a satisfying model. The solar nebula is no exception. While temperature no doubt declined with increasing distance from the sun, producing different minerals in different zones, there may have been some turbulence in the nebula that moved material toward or away from the sun and also "up" and "down" out of the midplane to regions of lower pressure and different temperature. This mixing would complicate Lewis's picture. A major current research problem is whether planet compositions are seriously affected by such mixing.

The Allende calcium-aluminum-rich inclusions, for example, prove that refractory material accreted, but later got embedded in less refractory-rich material.

Also, giant planets' gravities may have widely scattered the volatile-rich carbonaceous planetesimals that formed beyond the outer asteroid belt. Numerous workers require an admixture of carbonaceous material to explain the composition of Mars and other bodies (Pepin, 1991; McSween, 1989).

On the other hand, the total bulk mixing must have been limited, because each zone has preserved a unique composition, as noted above. Even within the asteroid belt such distinct zones exist, from light-toned rocky asteroids in the inner belt to black carbonaceous ones in the outer belt. Wood (1985) notes that inclusions have different properties in different meteorite types, showing that even such small bodies did not mix throughout nebula before being caught up in larger planetesimals; they stayed in their own zones. Finally, a crucial discovery (Clayton and others, 1973) is that materials from different zones, such as Earth/moon rocks, rocky meteorites from the inner asteroid belt, and carbonaceous meteorites from more distant regions, have different ratios of oxygen isotopes O^{16}/O^{18}, showing that the nebular gas did not mix uniformly throughout the nebula.

The rate of accretion vs. the rate of nebular cooling is also critical to planet makeup. At one extreme, if accretion happened faster than cooling, early refractories would accrete into metal-rich bodies before later silicate and carbonaceous dust appeared. Planets might have started with at least partially formed iron cores (Slattery, 1978). This is often called the **heterogeneous accretion model.** Alternatively, in the **chemical equilibrium model,** accretion was slower than cooling, and the dust remained dispersed through the nebular gas, reacting with it as new components condensed. Most workers lean toward some variant of this latter model.

FURTHER EVOLUTION OF THE SOLAR NEBULA DUST

The dust grains must have aggregated. If they hadn't, Earth wouldn't be here and we wouldn't be here; they would have been blown outward into interstellar space eventually or perhaps would have remained behind as a set of rings like Saturn's or as a belt of miniasteroids. The question is, how did they aggregate?

This issue aggravated researchers for some years. Ordinary experience suggests that if neighboring, sun-orbiting rock particles hit at low speeds, they would simply bounce apart without sticking; if they hit at high speeds, they would shatter each other instead of combining. Kerridge and Vedder (1972) designed an experiment with silicate particles hitting each other at speeds of 1.5 to 9.5 km/s (typical of collisions in today's asteroid belt) to test whether any sticking or impact welding occurred. They found none; the particles shattered. Other workers hypothesized electrostatic processes or sticky coatings of organic molecules to explain aggregation.

The dynamics of these particles as they move

around the sun suggests a different approach. Collisions and gas drag would quickly damp them into nearly circular orbits. If the orbits were precisely circular, there would be virtually no collisions, because particles at one distance from the sun would move at the same speed and neighboring orbits would not intersect. Thus, particles would approach each other at virtually 0 km/s rather than several kilometers per second.

But slight eccentricities would develop. For example, particles passing near each other would perturb each other's orbits gravitationally by amounts proportional to the masses of the particles. Thus orbits would become eccentric and collisions would occur. But even if two particles in Earth's orbit developed eccentricities equaling Earth's current eccentricity (e = 0.017), their collision would occur at only about 0.5 km/s, only a tenth the speed of modern asteroid collisions (see formulation by Safronov, 1972, p. 69). And the earliest neighboring particles must have collided even slower than that, since earlier eccentricities were still less. For example, among bodies 0.1 mm to 10 m across, Weidenschilling (1980) described drag effects produced by the nebular gas, giving velocities of only a few centimeters to a few meters per second. Greenberg and others (1978b) suggest collisions as slow as 0.01 to 1 cm/s for the initial microscopic dust particles, about 1 μm across.

In other words, the initial tiny dust grains nestled up to each other like snowflakes falling out of a snowstorm. As with snowflakes, the grains probably clustered together into fluffy clumps—the first aggregated planetary matter. Forces tending to hold them together could have been **electrostatic forces** such as Van der Waals forces (Weidenschilling, 1980). Actual evidence of the proposed fluffy structure comes from microscopic meteoritic particles collected in space and in the uppermost atmosphere (see Figure 8-10).

Further evidence that planets grew from interactions of innumerable small particles comes from the general properties of the solar system. Regularities such as nearly coplanar, circular orbits, prograde revolution, and the usual prograde rotation all suggest that final planetary properties derived from statistical averaging of properties of countless constituent particles.

This was first emphasized as early as the 1940s by the Russian planetary researcher O. Y. Schmidt and his students, even while Western researchers pursued other types of theories involving semi-independent, massive protoplanets. The Russian followers of Schmidt made many of the pioneering studies of the dynamical evolution of swarms of particles orbiting around the sun or planets. These studies were little noticed in the West until V. S. Safronov published a summary of the subject, which reached the West in 1972 as the product of an Israeli translation program. This book has had a major impact on subsequent investigations of the planets' early evolution.

As the dust aggregations grew larger into full-fledged planetesimals, they continued to perturb each other during near collisions. Their motions began to depart from the gas motions, and they approached each other faster than before. Calculations by Safronov (1972) and Greenberg and others (1978a) indicate the important rule of thumb that the approach velocities of the orbiting planetesimals tend to become equal to the escape velocities of the larger, but not the largest, bodies in the swarm, because these are large and abundant enough to have strong, perturbing gravitational effects. The large bodies thus become "gravitational spoons," stirring the smaller bodies in the cosmic mixing bowl of the solar nebula.

By numerical coincidence the escape velocity (V_{esc}) of a planetesimal in meters per second is about numerically equal to the radius of the planetesimal in kilometers.* For example, a 1-km-radius planetesimal would have an escape velocity around 1 m/s. Mars' satellite Phobos, at mean radius 12 km, has $V_{esc} \approx 12$ m/s, the speed of a fast baseball pitch. So the Safronov rule of

*This is exactly true if the density of the planetesimal is 1790 kg/m^3, typical for low-density rock-ice mixtures.

thumb is that if the planetesimals have grown to a meter in size, they interact at speeds around 10^{-3} m/s; if they've reached 1 km, around 1 m/s; and if 10 km or so, around 10 m/s.

Collisional Accretion of Subkilometer Planetesimals?

What do these facts tell us about how the planets formed? To study this, I have tested collisional breakup of various materials in a vacuum (Hartmann, 1978). Weak carbonaceous chondrites (simulated), H_2O ice pieces, and igneous rocks begin to fragment at collision speeds of about 2, 9, and 37 m/s, respectively. At lower speeds, these materials would just bounce apart, and at higher speeds they would break each other up. When clean solids bounce apart after low speed collisions, the rebound speeds are often about half the impact speed, or even less, depending upon their shapes and initial spins.

Thus, the largest objects in a planetesimal swarm (being hit at just above their escape velocities) would be hit by smaller objects that rebound too slowly to escape, fall back onto the surface, and add to the mass of the larger body. Furthermore, the experiments showed that the granular layers thus accumulated would slowly rebound and aid the efficiency of this process, making growth even more rapid. Small bodies in the swarm would not grow by this method, since they would be hit at a speed considerably faster than their escape velocities, and rebounds would occur too fast for accretion.

This method of planetesimal growth, with bodies striking each other one at a time in random collisions and often sticking together, is called **collisional accretion**. Computer models (for example, Greenberg and others, 1978a, b) indicate that primitive dust grains or aggregates of grains could rapidly accumulate into bodies at least several kilometers across by collisional processes alone. Many small planetesimals (such as the refractory-rich, centimeter-scale Allende

MATHEMATICAL NOTES ON COLLISION VELOCITIES

If two bodies are approaching each other from a great distance apart at a speed V_{app}, at what speed V_{imp} will their actual impact occur? Suppose a small body or meteorite m is colliding with a much larger body M. (This is the usual situation, since a wide distribution of sizes makes the collision of two equal-sized bodies relatively rare.) As they approach, the small body accelerates due to the gravity of M (and vice versa, to a negligible extent, since m is much smaller). The collision thus occurs at $V_{imp} > V_{app}$. It is easy to derive V_{imp} in terms of other knowns by the conservation of energy:

$$E_{initial} = E_{final}$$

Total energy E is kinetic energy plus gravitational potential energy. Thus,

$$\frac{1}{2}mV_{app}^2 - \frac{GMm}{\infty} = \frac{1}{2}mV_{imp}^2 - \frac{GMm}{R^2}$$

After cancelling m's, we have

$$V_{imp}^2 = V_{app}^2 + \frac{2GM}{R}$$

$$V_{imp}^2 = V_{app}^2 + V_{esc}^2$$

In other words, the square of the final impact velocity is just the sum of the squares of the initial approach velocity and the escape velocity of the larger body. If the approach is so slow that $V_{app} << V_{esc}$, the meteorite will fall in at escape velocity. If the approach is so fast that $V_{app} >> V_{esc}$, then the meteorite will slam in at its approach velocity without ever "feeling" the gravity of the larger body.

inclusions of early condensates?) may have formed in this way.

Gravitational Collapse of Kilometer-Scale Planetesimals?

But collisional accretion is not the only way to get planetesimals of kilometer scale or even larger. Another way is by **gravitational collapse**, illustrated in Figure 4-3. Gravitational collapse might seem a promising way to make a planet in one fell swoop, without the tedious business of making it one collision at a time, rock upon rock. You might simply imagine a whole piece of the solar nebula becoming dense enough to become gravitationally unstable and then collapsing at once into a planet-sized object.

There are two things wrong with this scenario. First, the nebular gas-and-dust mix is unlikely ever to have become dense enough to make bodies as small as Earth or the parent bodies of meteorites by one-stage collapse of the gas-and-dust mixture (see Figure 4-3). Second, geochemists have long realized (Urey, 1952) that Earth did not form this way, because Earth lacks a full cosmic complement of the inert gases such as neon, argon, krypton, and xenon. Current data show that in Earth, Venus, Mars, the moon, and meteorites, these gases have no more than only 10^{-7} to 10^{-10} of their normal cosmic abundances (Taylor, 1975; Kaula, 1968, p. 377). These gases are heavy and chemically inert, so that if they had ever been gravitationally trapped in Earth's material, they would not have floated off into space (as most of the light hydrogen has done) or have been chemically tied into mineral compounds hidden inside Earth. Therefore, if the terrestrial planets had formed by gravitational collapse of part of the solar nebula, neon and the other inert gases would have been brought along with everything else in their normal cosmic complement and would be prominent in the atmosphere today.

So terrestrial planets did not form by simple gravitational collapse of the gas-and-dust solar nebula.

But what about the dust alone? Important studies by Gurevich and Lebedinskii (1950), Safronov (1972), and Goldreich and Ward (1973)* showed that the freshly condensed dust grains and aggregates, while orbiting around the sun, also tended to settle through the gas toward the ecliptic plane, being attracted not only by the sun but also by the disk-shaped nebula. **Stoke's law**, which gives the maximum velocity at which a particle can move through a gas under the influence of a given force, indicates that the formation of a relatively thin, dense layer of dust in the ecliptic plane was rapid. Safronov (1972, pp. 26–27) estimates that micrometer- and submicrometer-scale particles settled to the plane in only about 10^3 to 10^5 y; larger particles fell faster. Goldreich and Ward (1973) found that early condensates such as iron may have grown to a few centimeters just by further chemical condensation on their surfaces during their fall to the plane (a source for Allende inclusions?).

Once the dust layer formed, the density of material in the dust layer was considerably more than the density in the more dispersed gas. Gravitational instability probably occurred within this dust layer, forming planetesimals with diameters of a few kilometers. Aggregation velocities were very low, so that individual grains, aggregations of grains, and centimeter-scale chunks of early condensates gently accumulated, consistent with Figure 5-6.

Goldreich and Ward (1973) concluded that several stages of gravitational collapse occurred. The first formed planetesimals with diameters up to about 2 km, directly as a result of collapse of the primeval dust. This layer of kilometer-scale bodies was also gravitationally unstable and aggregated into a second generation of 10-km-scale planetesimals—clusters of first-generation

*The Cal Tech researchers Goldreich and Ward began their work without knowledge of the earlier Russian results and independently derived a quite similar model. At about the time of publication, they became aware of Safronov's (1972) book and credited the Russian work in a note added at the end of their paper—an interesting example of the independent confirmation of scientific ideas.

bodies that resembled asteroids. Further growth, according to Goldreich and Ward, was by collisional accretion.

REVIEW OF THE STAGES OF GROWTH

To review, we have discussed four probable stages of planet growth:

1. chemical condensation of dust grains (micrometer scale)
2. electrostatic clumping of grains (submillimeter scale)
3. collisional accretion of clumps (meter scale?)
4. gravitational collapse of grains and clumps (kilometer scale?)

Interestingly, however, steps 2 through 4 may have all happened at once from the point of view of an observer riding on an early dust grain. This observer would have just seen grains and clumps of material getting closer and closer, with collisions, electrostatic attraction, and gravitational attraction all mixed up. What matters is that the dust aggregated into asteroidlike planetesimals and that aggregation continued toward even larger sizes.

COLLISIONAL ACCRETION OF FULL-SCALE TERRESTRIAL PLANETS

In the growth to planetary size, collisional accretion was probably the dominant mechanism, at least among terrestrial planets and the asteroids (Wetherill, 1981).

Perhaps the most detailed models of growth from kilometer-sized to planet-sized bodies are those of Greenberg and others (1978a, 1978b), who began with a computerized celestial mechanical model of a swarm of bodies orbiting around the sun. This model allowed the bodies to collide, assumed collision results as actually determined in the lab, and then recorded the effects on the sizes and velocities of the bodies. These researchers found that for a variety of starting velocities, a swarm of 1-km planetesimals (such as produced by gravitational collapse of the dust) produced a few 500- to 1000-km-diameter "planet embryos" by collisional accretion in only about 10^4 y. As large bodies appeared, they stirred up the velocities among the smaller bodies to the point where collisions began to fragment many of the smaller objects. The size distribution at 10^4 y included a few large embryos and many fragments smaller than 1 km.

Evidence supporting this theory is found in the current asteroid belt. The largest asteroid, Ceres, is about 1000 km across, the next three bodies are about 500 km across, and a host of smaller fragments exists. The size distribution is similar to that calculated by Greenberg and others (1978b). The belt is a fossil tableau of colliding planetesimals where collision velocities are high enough to break up many of the smaller colliding pairs. As seen in Figure 5-7, striking testimony of the collisional history of planetesimals is offered by certain satellites. The grooves that radiate from the region of the largest crater on Phobos (Figure 5-7a) suggest surface expressions of fractures produced in a nearly catastrophic collision.

Growth to the full-fledged planetary state apparently slowed as the largest planetary bodies in each zone began to sweep up the remaining debris, leaving one large body dominating most zones.

Isaacman and Sagan (1977) simulated this growth in a computer model without using specific experimental collision data. Instead, they merely assumed that each collision resulted in particles sticking together. The result was a variety of planetary systems, as seen in Figure 5-8. These included many systems resembling the solar system, with some massive Jupiter-like planets, some Earth-sized planets, and often some leftover asteroidlike bodies. The competition to sweep up the planetesimals leads to a spacing of the large planets in Bode's-rule-like zones, helping to explain item 9 in Table 5-2. Isaacman and Sagan stress that similar results came from a variety of starting conditions, supporting the idea that planet systems have formed elsewhere.

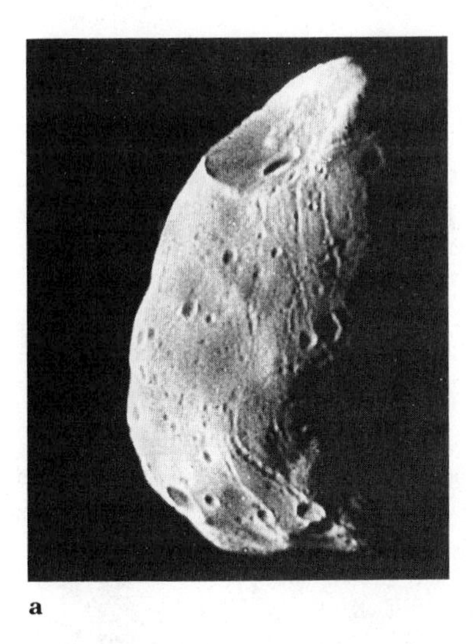

a

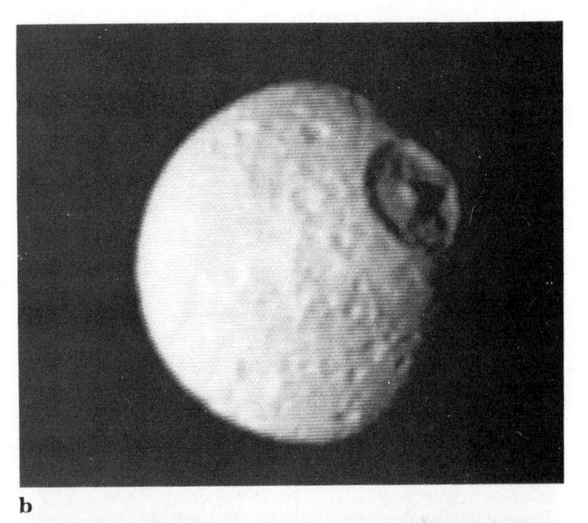

b

c

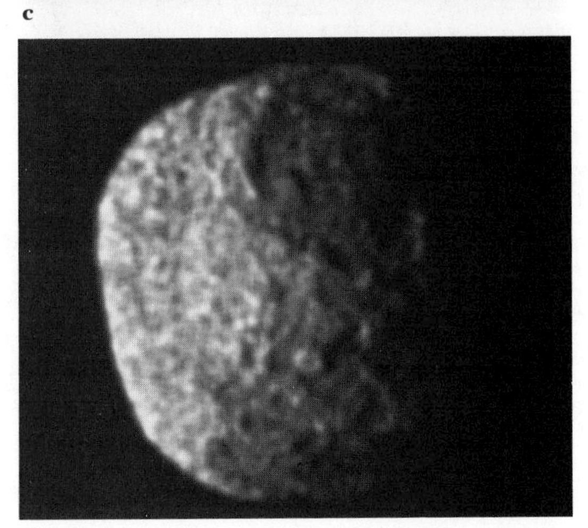

d

Figure 5-7. Several satellites suffered impacts nearly big enough to disrupt them. (**a**) *Mars' satellite Phobos has a crater Stickney (top) from which radiate fracturelike grooves. The crater diameter is 37% of the 27-km long axis of the satellite.* (**b**) *Saturn's 394-km moon, Mimas, has a crater about 34% as big as the satellite.* (**c**) *The nearby 1048-km moon, Tethys, has a similar crater about 40% as large as the satellite.* (**d**) *Neptune's 400-km moon, Proteus, discovered by Voyager 2, appears to have a crater (top right) about 45% as big as the satellite. (NASA Viking and Voyager photos)*

The most complete models of the final stages of accretion have been made by George Wetherill (1990). He suggests that runaway growth of the largest bodies produced a host of Mars-sized planetesimals in the terrestrial zone, spaced about 0.01 to 0.02 AU apart, within as little as 100,000 yr. Further collisional accretion among these eventually produced the terrestrial planets.

FORMING THE GIANT PLANETS

In the outer solar system ices supplemented the planetesimal mass supply, and still larger bodies formed. The process combined collisional accretion and gravitational collapse. Once rocky/ice planetary bodies grew to 10 to 20 $M_\oplus$, they had such strong gravity that they began to pull in

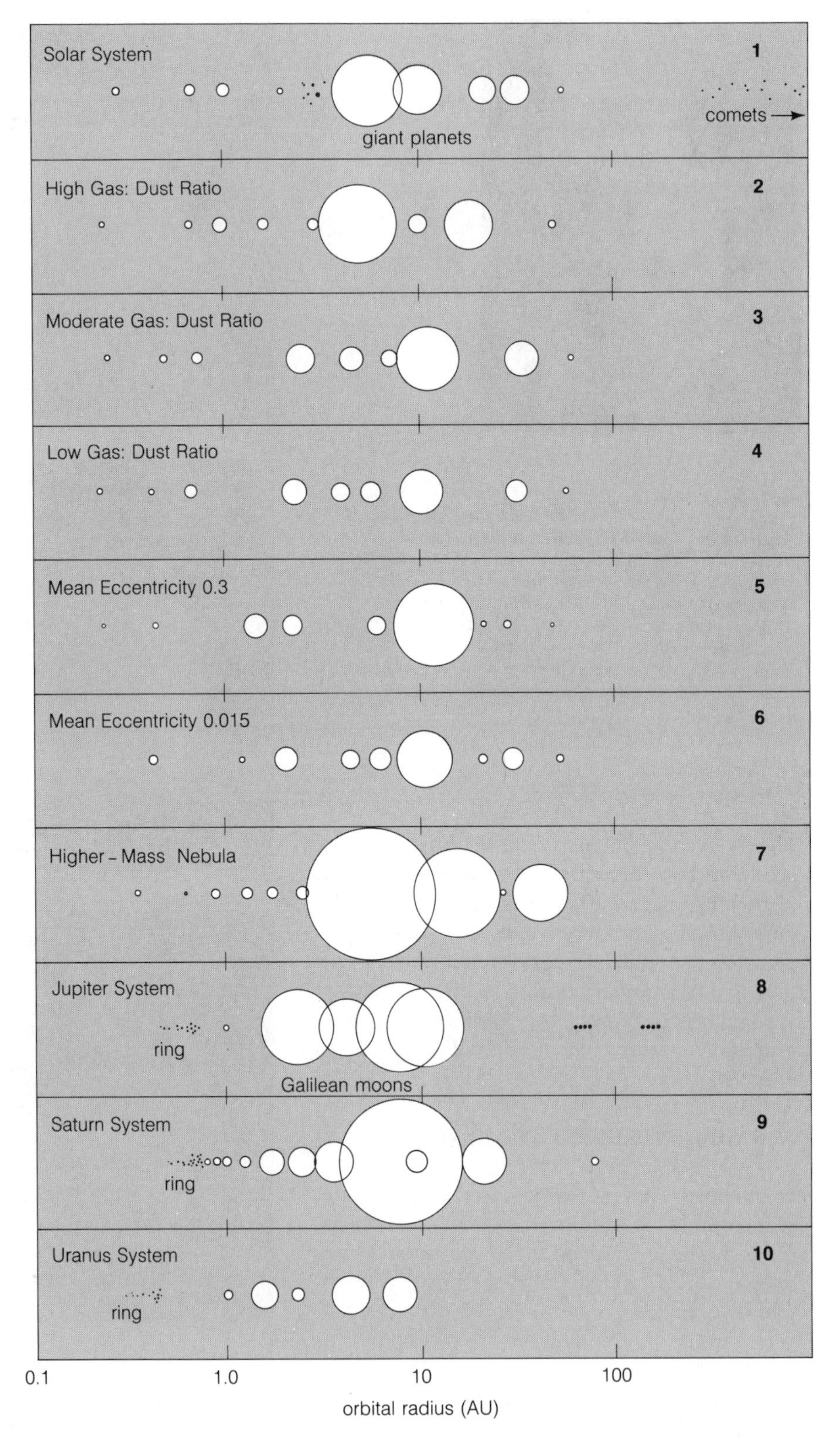

Figure 5-8. Schematic representation of the distribution of bodies in various real and theoretical planetary systems. Orbit scale is given at bottom, and circles represent relative sizes of planets (on a scale different from the orbit scale). Plots 1–7 are after Isaacman and Sagan (1977); plots 2–7 are systems computed by a theoretical model of planetesimals' collisional accretion. Plots 8–10 show satellite systems of three giant planets by a similar schematic method, with one of the inner satellites arbitrarily chosen to represent "Earth." All systems show a tendency to form several bodies, with "giant planets" occupying intermediate orbits.

large amounts of gas directly from the solar nebula, forming a giant planet. In this scenario, the bulk composition of Jupiter might have an intermediate content of inert gases, though the atmosphere, added last, might have the solar complement of inert gases.

A model of this sort is given by Mizuno (1980) who found that as dense, large planetary cores grew past a critical size, the gaseous surroundings collapsed onto these cores, forming giant planets of hydrogen-helium mixtures surrounding the denser cores.

Some support for this intermediate model is found in the computer model by Isaacman and Sagan (1977), mentioned earlier. They found that certain planetesimals, typically at 2 to 20 AU, grew large enough to begin sweeping up the gas, attaining the huge sizes shown in Figure 5-8.

SWEEPUP OF THE LAST PLANETESIMALS

Eventually planet growth reached the stage where the planets had most of their present mass and were sweeping up the last planetesimals. Each collision with a planetesimal left a crater on the planet's surface, accounting for item 10 in Table 5-2.

Craters on the terrestrial planets date back 4 Gy and record the sweepup of interplanetary bodies ranging in diameter up to roughly 200 km. The size distribution of the craters is similar to that of present-day asteroids, and the number of craters corresponds to roughly half the number of asteroids (Hartmann, 1977).

From these statistics of planetesimal sizes, even larger impacts can be inferred for the period before 4 Gy ago. Safronov (1966) pointed out that these large impacts affected planetary obliquities. An important feature of the accretion theory is that statistical averaging of the impact effects of innumerable small planetesimals during planet growth tended to produce planets rotating prograde with periods of about 5 to 20 h (a few times slower than the limiting case of rotational instability, where material would be thrown off the equator) and with rotation axes nearly perpendicular to the ecliptic (Harris, 1977).

The growing planet was only the largest planetesimal in its local swarm, and eventually it had to interact with the 2nd-, 3rd-, 4th- . . . *n*th-largest planetesimals. Some of these interactions may have produced satellites or may have ejected the planetesimals into other parts of the solar system, but some of them resulted in impacts of incredible violence. These largest impacts were statistical flukes. Their effects did not necessarily average out. For example, Safronov concluded that the odd obliquity of Uranus came from the impact of a mass about 0.05 that of Uranus itself. The largest bodies hitting Earth, he found, had a mass about 0.001 $M_{\oplus}$, equivalent to the largest asteroids. Calculations by D. R. Davis and me suggest that impacts of planetesimals with asteroidlike mass distributions ranging up to 0.02 planetary masses could have produced the observed distribution of obliquities and orbit inclinations, in which case the largest planetesimals among the terrestrial planets may have had diameters of 1000 to 4000 km (Hartmann, 1977)! Similarly, Wetherill (1976) predicted that the second-largest bodies in the terrestrial zone grew as large as some 4000 to 6000 km across. The details of the sweepup process thus may account for items 11 and 12 in Table 5-2.

The outer planets were surrounded by many icy planetesimals, and during near-miss encounters between the planets and planetesimals, the strong gravity of the giant planets was adequate to throw many of these icy bodies nearly out of the solar system. Hence they accumulated in a swarm with enormous semimajor axes, spending most of their time on the outskirts of the solar system. This accounts for item 13 on our list.

RING SYSTEMS AS CLUES TO PLANETARY ORIGIN

Chapter 2 described the discoveries of ring systems around Jupiter (Figure 5-9), Saturn (Figure 5-10), and Uranus. Chapter 3 described some of

Figure 5-9. A "close-up" of Jupiter's ring from 1 500 000 km. Photo reveals the relatively sharp outer edge ($1.81R_{♃}$), the less distinct inner edge, and some material inside the ring, apparently extending all the way to Jupiter's upper atmosphere. Photos reveal little structure in the bright ring, though a 700-km wide brightening at 1.79 $R_{♃}$ has been detected. (NASA, Voyager 2)

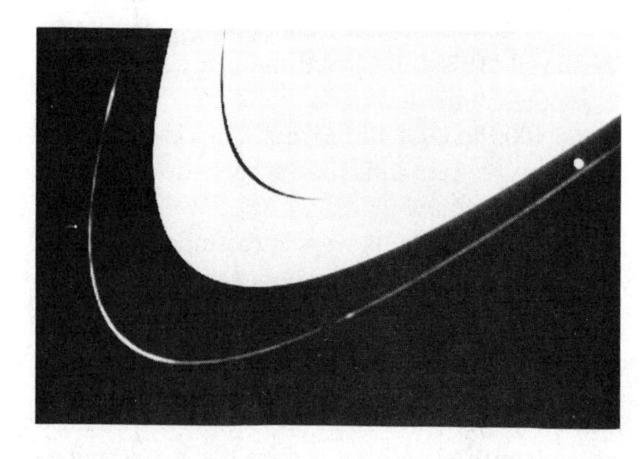

Figure 5-10. Outer tip, or ansa, of Saturn's ring shows the narrow outer "F" ring and the 14th satellite just inside the "F" ring. The "A" ring is overexposed. The "F" ring shows unexpectedly complex structure. The close association of ring material and satellites at their outer boundaries suggests that rings and inner satellites may have evolved in close relationships. (NASA, Voyager 1)

the dynamical properties of the particles in these rings. If we knew more about the composition, sizes, and other properties of the ring particles, we might better understand the **origins of ring systems**, and this knowledge in turn might help explain planetary origins.

Saturn's rings are best understood, and observations indicate that most ring particles are around 1 to 100 cm across, but with a size distribution that may include smaller or larger specimens. The composition of these particles is known to be frozen water from their spectra (Pilcher and others, 1970).* Spectra and radar suggest that the ring bodies are not just ice covered but icy throughout, probably with only a small admixture of silicates, carbonaceous dirt, or other ices (Pollack, 1978).

Uranus' ring particles have an albedo much lower than that of ice and may be rocky or carbonaceous. Jupiter's ring, faintest of all, is probably also rocky dust or dirty ice grains, since its infrared spectrum is inconsistent with clean ices but matches that of dark soil (Neugebauer and others, 1981).

Some of these differences may arise from the source or recent history of the ring particles. For example, the outermost and densest part of Jupiter's ring (Figure 5-9) lies just at or slightly inside the orbit of the innermost known moon, 1979J1, and the ring particles are believed to be μm-scale particles knocked off 1979J1 by impacts. The albedos of 1979J1 and neighboring satellites are believed to be 0.04 to 0.10, implying coloring by dark, carbon-rich soil. Using this model, Jewitt

---

*A story behind this statement illustrates the vicissitudes of planetary research. Kuiper, Cruikshank, and Fink (1970) first obtained the infrared spectrum of the rings. Though Kuiper as early as 1956 hypothesized that the rings are made of water ice, his comparison with lab samples at around −20°C indicated a better match with frozen ammonia, and he published the conclusion that Saturn's rings are made of ammonia ice. Immediately Pilcher and others (1970) and Lebofsky, Johnson, and McCord (1970) pointed out that under the actual ring temperatures, 95 K (or −178°C!), water ice spectra are altered and *do* match the spectra of Saturn's rings. Kuiper and his coworkers at once concurred that the spectra indicate water ice.

and Danielson (1981) calculate a lifetime of 250 000 y for a 5-μm diameter particle to spiral from the satellite orbit into Jupiter. The Jupiter ring is thus analogous to a river, which may always be visible but consists of different water at different times. Just as a river may occasionally flood, a major impact on 1979J1 might send a mass of particles through the ring; the ring might have had ancient episodes of greater prominence. The current total material in the ring would make a ball only 30 m across, and the estimated total material cycled through the ring since the solar system's origin would make a satellite only a few kilometers across (Jewitt and Goldreich, 1980; Burns, 1980).

Whether such a model has relevance to Saturn's rings (Figure 5-10) is problematical. Since the Poynting-Robertson lifetime increases with particle radius and decreases with the solar flux, the larger particles in Saturn's rings would have lifetimes longer than the solar system, although there might still be a population of μm-scale dust particles spiraling inward after being eroded off the larger particles. In the case of Uranus' rings, Goldreich and Tremaine (1979) suggest decay by Poynting-Robertson and collisional effects in only 10^7 to 10^8 y unless the rings are stabilized by small unseen satellites of diameter around 20 km. Further, Uranus' ϵ, or outer, ring is apparently slightly elliptical or inclined to the other rings, suggesting that the Uranus system may be a collisional product where the fragments have not had time to collapse to a circular disk (Mills and Wasserman, 1978; Nicholson and others, 1978). Thus it is hard to determine how much the present ring systems tell us about planet-forming processes during the dawn of the solar system.

Nonetheless, ring positions suggest clues about primeval conditions. How, for example, did the oldest material in Saturn's rings originally arrive there? As shown in Figure 5-11, various authors have defined various types of Roche's limits, within which different types of bodies are pulled apart. Positions shown in the figure are only approximate, since they depend on the size and strength of the bodies in question. However, we see that rings generally lie inside the zone where accretion is hampered. In the regions of rings, centimeter- to kilometer-scale bodies can persist, once formed by accretion of little particles onto surfaces of big particles. Yet particles can easily be knocked off again; velocities of only about 4 to 10 m/s are needed to knock debris off 1979J1 into Jupiter's ring (Burns, 1980). In other words, if material is once distributed in the ring zones, small bodies may aggregate but are unlikely to accrete rapidly into major satellites, as happens outside the ring zones.

Thus, three scenarios are visualized for ring formation. In one, the rings formed from the same miniature solar nebula that produced the satellites, except that the primordial material never accreted inside Roche's limit and was left stranded (Pollack, 1978). In the second scenario, a body passed within Roche's limit and fragmented due to tidal stress, producing ring material. The body might have been a passing planetesimal (Wetherill, 1976) or a satellite that spiraled in under the influence of tides, gas drag, or growth of the planet (Harris, 1978).* The third scenario involves very large scale collisions in which the largest remaining planetesimals in the neighborhood hit either inner satellites or the planet itself and spewed material into the zone within Roche's limit.

ORIGIN OF SATELLITES

Complexities of Satellite Systems

In the 1800s many theorists thought that satellite systems were merely scaled-down versions of the solar system, formed in just the same way and perhaps holding the key to planet formation. Indeed, Figure 5-8 gives considerable support to the idea that similar underlying processes may be involved. The general trend of low inclina-

---

*Recall that under Kepler's laws, as a planet's mass grows, its satellites' orbital radii shrink.

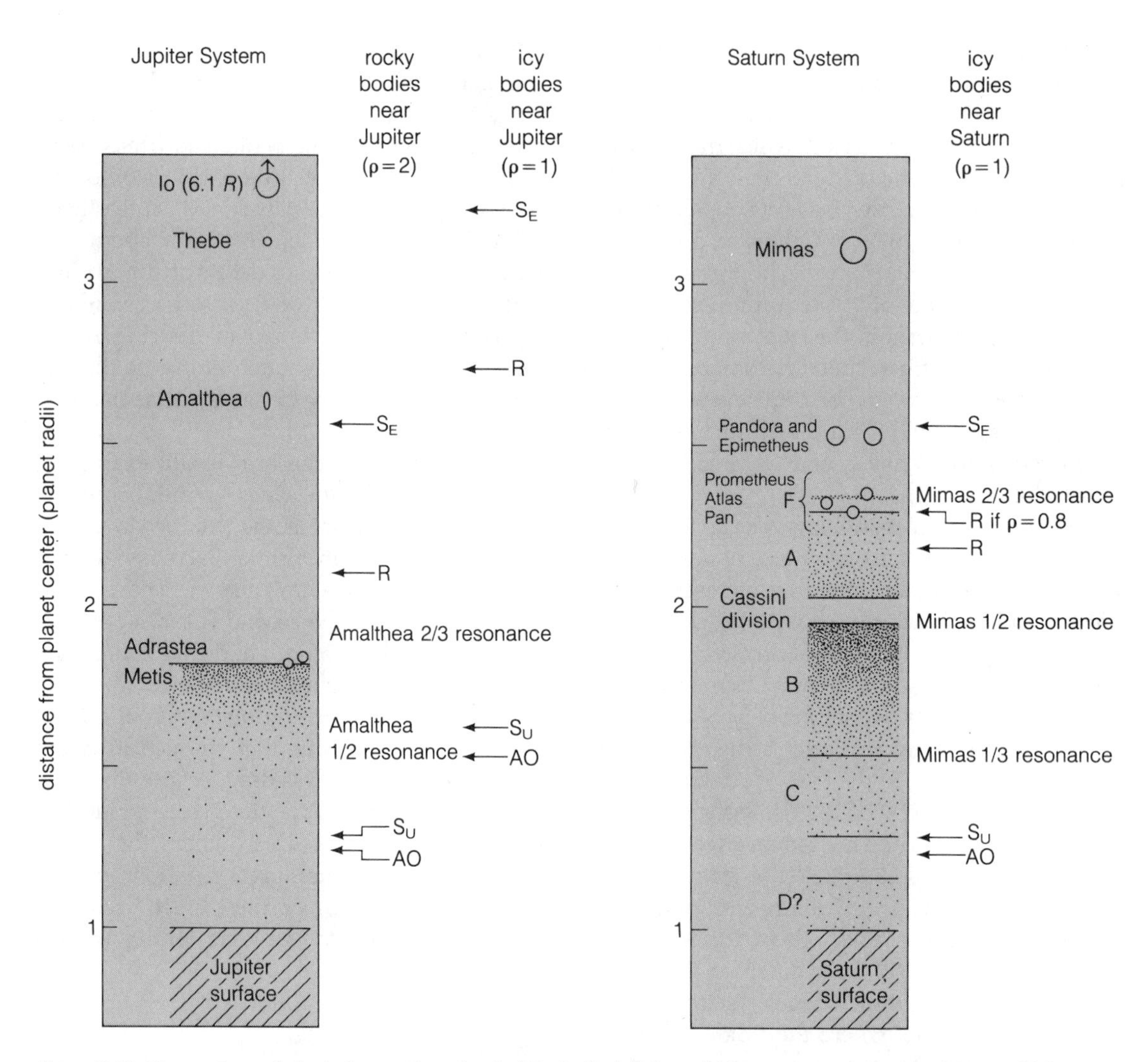

Figure 5-11. Ring systems of giant planets. R = classical Roche limit below which zero-strength liquid sphere would fragment. S_E and S_U = Smoluchowski (1978) limit below which touching equal-size and unequal-size bodies, respectively, would fragment. AO = Aggarwal-Oberbeck (1974) limit below which large ($D \gtrsim 40$ km) solid spheres would fragment. Smaller solid bodies would have still closer AO limits. Rings tend to lie inside R; satellites, outside R.

tions, low eccentricities, and wide spacings support this. Each new satellite system discovered in the past seemed to verify this theory, except for Neptune's largest satellite, Triton, which revolves retrograde.

However, the twentieth century brought the discovery of a number of new satellites that upset this orderly scheme. The new satellites tended to be small and faint (hence their late discovery); many were in orbits outside those already known, and these were often in relatively high inclination orbits (about 30°) or in eccentric or retrograde orbits.

It thus became clear that the satellite systems are somewhat more complex than mere copies of the solar system, although the early-known regularities are undoubtedly significant.

A number of complexities about satellite systems should be noted. First, we cannot trace the past history of satellite orbits by contemporary

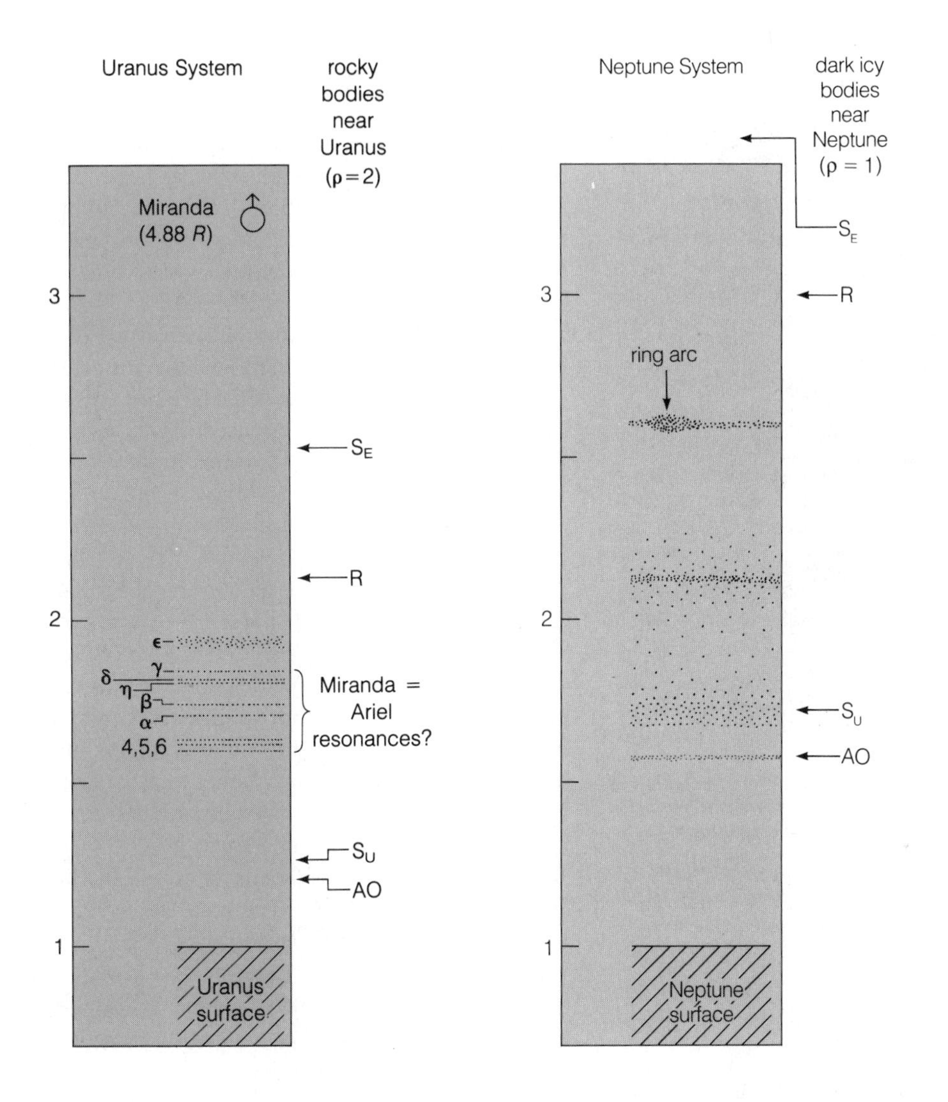

celestial mechanics. The contemporary theory and observational parameters are only accurate enough to trace the orbits through roughly the last 100 My—a few percent of solar system history (Kuiper, 1956b). Thus we cannot rigorously calculate the initial orbital properties of satellites.

Second, some regularities of satellite orbits are associated with resonances between pairs of satellites (such as Saturn's), not with initial conditions. Enceladus is in a 1:2 resonance with Dione, Mimas 1:2 with Tethys, and Titan 3:4 with Hyperion. These satellites may have "fallen into" these stable positions during the tidal evolution of their orbits (Goldreich, 1965; Dermott, 1968; Greenberg, 1977; Peale, 1978).

Third, calculation of the past orbital evolution of several comets revealed seven that spent periods of a few months to a few years as captured satellites of Jupiter, mostly within this century (Carusi and Valsecchi, 1980).

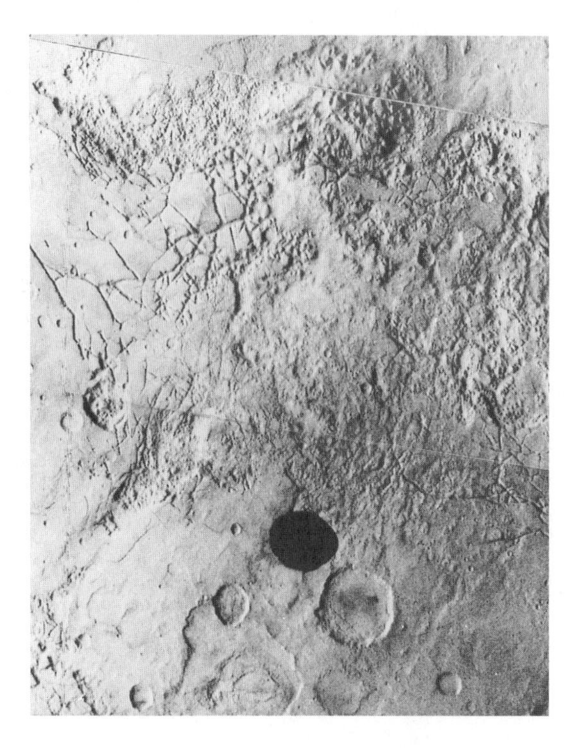

Figure 5-12. Satellite Phobos (dark object) photographed against the surface of Mars, showing its small size relative to features of the fractured terrain in Mars' equatorial region. Because Phobos is covered with carbonaceous-chondrite-like material with only one-fourth the albedo of Mars, its surface is very underexposed in this image, which was exposed for Mars, and is correctly rendered as virtually black in this image. See page 471 for a more detailed image of Phobos. (NASA, Viking Orbiter 1, 1977)

Fourth, tidal transfer of angular momentum from planets to satellites tends to force prograde satellites outward toward escape, as with the moon (Chapter 3). By the same mechanism, retrograde satellites and satellites inside the synchronous point are drawn in toward planets. (Can you confirm this from Figure 3-13, page 71?) Many more retrograde satellites may have existed long ago. Similarly, Triton will approach and crash onto Neptune. Once thought imminent (within 0.1 Gy; McCord, 1968), this crash is more likely some 3.6 Gy in the future (Chyba and Nicolson, 1987). Phobos is expected to crash onto Mars in as little as 50 My (Burns, 1990)!

Capture Origin for Retrograde and Other Small, Outer Satellites

How could retrograde-revolving satellites have been born from prograde-spinning planets? Current results, such as those reviewed above, strengthen the idea that they must have been captured from closely approaching planetesimals, as suggested years ago (Kuiper, 1956b).

As exemplified in Figure 5-12, all well-studied members of the following groups are asteroidlike bodies with very dark surface soil resembling certain black, carbonaceous meteorites: Phobos; Deimos; the small, outer Jupiter satellites J6 through J13; the retrograde, outermost Saturn satellite, Phoebe; the Trojan asteroids caught in Jupiter's two Lagrangian points (see Chapter 3); and most asteroids in the outer part of the asteroid belt (see Chapter 7). Data from the Soviet PHOBOS-2 probe confirmed a Phobos density of 1950 kg/m^3, in the range of volatile-rich carbonaceous chondrite material, which originated primarily in the outer solar system. These relations suggest that these moons are carbonaceous asteroids perturbed into planet-crossing orbits when Jupiter or Saturn grew toward their present sizes and scattered small bodies (Hartmann, 1987, 1990). There are several possible capture mechanisms (Pollack, 1985):

1. Early planets probably had massive extended atmospheres that could have captured passersby and perturbed them eventually into low-inclination orbits. Phobos and Deimos seem especially good candidates for this mechanism of origin, particularly since their carbonaceous-chondrite-like composition suggests an origin in the outer asteroid belt (see Chapter 7; Hunten, 1979; Burns, 1978; Veverka, Thomas, and Duxbury, 1978; Pollack, Burns, and Tauber, 1978).

2. Close approach of a large planetesimal to an existing satellite might have altered the satellite orbit and led to the capture of the planetesimal. The Neptune system, with Triton's weird orbit, is a good candidate. This theory looks especially attractive if we remember that the largest pla-

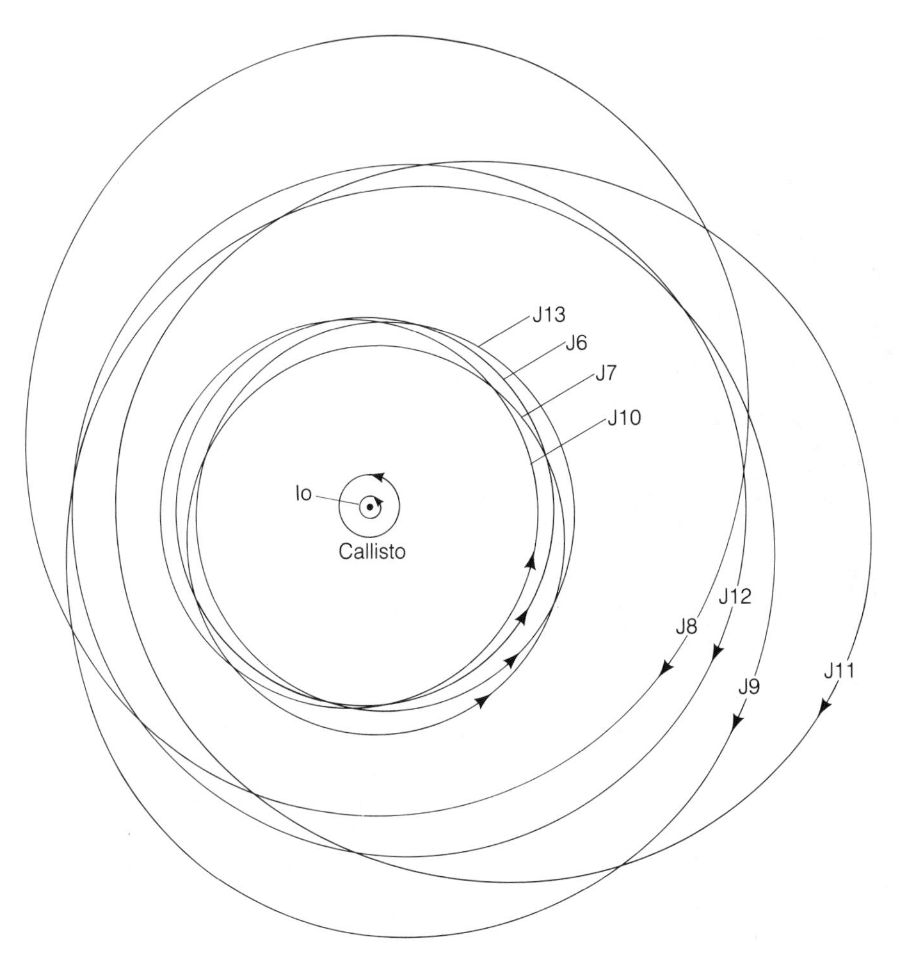

Figure 5-13. Schematic diagram of orbits of the outer Jupiter satellites. Callisto is the outermost Galilean satellite. Satellites beyond Callisto fall into two groups: an inner, direct group with nearly coinciding orbits, and an outer, retrograde group with similar orbits.

netesimals kicked out of the neighborhood of Jupiter and Saturn may have been enormous—possibly larger than Earth. They might easily have disturbed outer satellite systems before colliding with planets.

3. Collision with a satellite might have thrown out debris that could have reaggregated into a new satellite.

A remaining puzzle has been the two groupings of Jupiter's outer satellites—four prograde and four retrograde in clustered orbits—shown in Figure 5-13. Photometric studies of two or three out of the eight suggest that they all have dark stony surfaces but with minor detectable differences. Russian scientists (Aitekeeva, 1968; Bronshten, 1968) suggested that each group might be fragments of a satellite broken by collision. Possibly deceleration of very weak planetesimals during capture could have led to the breakup and formation of each group.

Baily (1971a, b) suggested that the inner direct group resulted from capture of asteroids at Jupiter's perihelion, while the outer retrograde group resulted from capture at aphelion. Davis (1974) found that Bailey's capture mechanism would work but would produce the observed groups of orbits only during the 35% to 40% of the time that Jupiter's slightly changing eccentricity had almost exactly its present value. The chance of all eight capture events occurring independently

Figure 5-14. The strange satellite Io, seen against the backdrop of Jupiter's clouds. Io is almost exactly the size of our moon, and about one-fourth the size of Earth, giving an idea of the immense scale of Jupiter's clouds. This was one of the first photos to show Io's surface detail; the dark spots were later found to be volcanic formations. (NASA, Voyager 1, March 2, 1979)

in this way is less than 10^{-3}, suggesting again that the satellite groups may be fragments produced in the present orbits. Pollack, Burns, and Tauber (1979) suggested that the original bodies were captured by interaction with a massive gaseous envelope around proto-Jupiter.

Origin of Major Prograde Satellites

The large, regularly-spaced satellites with circular zero-inclination orbits (Figure 5-14) are probably products of accretion in "miniature solar nebulas" around planets. This idea would explain item 14 in Table 5-2. For example, the smooth decrease in density and rock:ice ratio as one moves outward among the Galilean satellites mimics the decreasing density and metal:rock:ice ratio in the solar system, suggesting similar sequences of condensation with higher temperatures toward the central "star." The dynamics of giant planet gravitational collapse and the high luminosity of the primordial giant planets kept the central regions of their "miniature solar nebulas" hottest. Chemical studies of the condensation sequence in such nebulas have helped explain satellite compositions. For example, Figure 5-15 shows a diagram of condensation in the Jovian nebula, adapted from Consolmagno (1981). Based on an assumed model of the nebula, Figure 5-15 shows that Io and Europa, in the warm region near Jupiter, would have accumulated mostly from hydrated rocky material (perhaps like carbonaceous chondrites; later internal heating could have driven the water out to form the observed ice on Europa's surface). Ganymede's and Callisto's materials, condensing in a colder region, would have been about one-half water ice, mixed with rocky dust and ammonia ice. Once condensed, the dust and ice grains would aggregate into planetesimals and grow by collisional accretion into major satellites (Harris, 1978; Burns, 1978).

A major satellite, once formed, was not necessarily safe from orbital change or even destruction. Tidal interactions may have been quite effective in removing primeval moons, either by making them spiral into the planet or by making them spiral out, escaping into solar orbit only to collide later with the planet or be ejected from the solar system during a near miss with the planet (Burns, 1973; Ward and Reid, 1973; Harris and Kaula, 1975). As discussed in Chapter 3, observed orbital resonances between the largest moon and other smaller moons might have been established as the larger moons' orbits changed under tidal forces. Harris (1978) noted that surviving satellites may be the last-formed ones, composed partly of late-arriving planetesimals with somewhat different chemistries from those that formed the parent planet. Hartmann (1976) showed that this effect was notable during accretion of our moon, with the late-added material differing in composition not only from earlier material, but also from material being simultaneously added to Earth because of a gravitational sorting process. Thus, because of orbital and chemical evolutionary complexities, discovering the origin of satellites by direct rock sampling remains difficult.

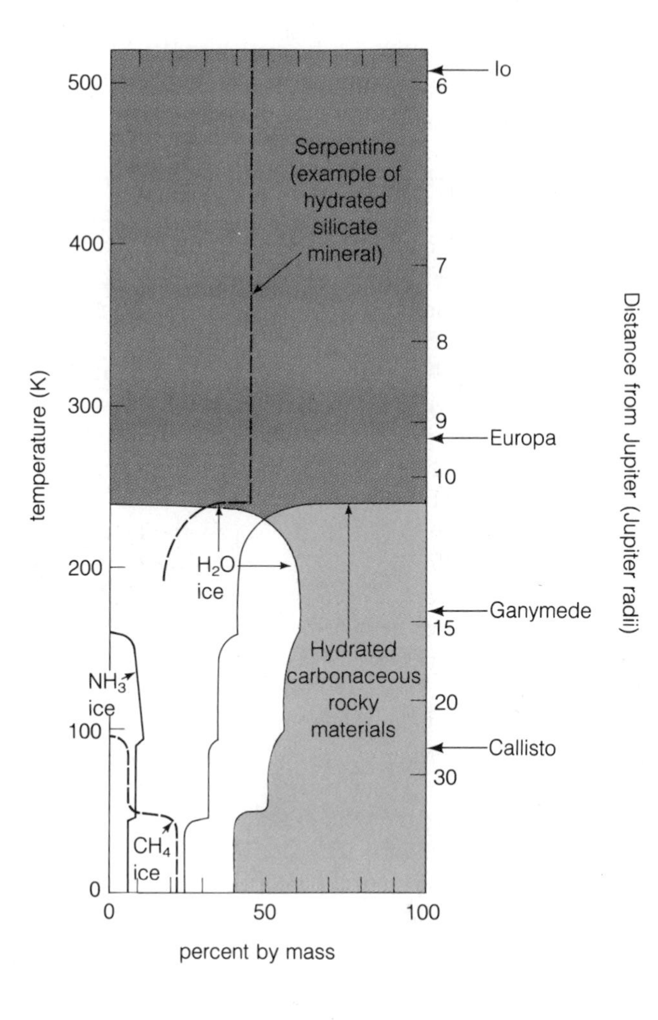

Figure 5-15. A chemical analysis of the condensation sequence among the Galilean satellites, based on an assumed model of the "miniature solar nebula" around primordial Jupiter. The format is a modification of that in Figure 5-4, showing the bulk composition of material condensing at different temperatures (left) and hence different distances from Jupiter (right). For example, materials condensed at more than about 240 K were entirely rocky (about 45% being the strongly hydrated mineral serpentine). At around 200 K, the percent of rock (shaded area) drops to 40% and the rest is water ice. Smaller amounts of ammonia ice and methane ice appear at lower temperatures. (Adapted from Consolmagno, 1981)

Origin of the Moon

Earth's satellite is widely regarded as unique, partly because it has an unusually large ratio of satellite mass: planet mass (0.012, exceeded only by the value of 0.05 to 0.10 for Pluto, which itself is probably anomalous). Prior to Apollo, three general theories about the moon's origin were: (1) the **fission theory**, asserting that the moon broke off Earth; (2) the **binary accretion theory**, asserting that it accreted as an independent protoplanet in orbit around Earth; and (3) the **capture theory**, asserting that it accreted elsewhere in the solar system and was captured by Earth.* Researchers hoped that the Apollo flights would help prove or disprove one of these theories, but they did not produce a final answer.

One of the main problems of lunar origin centers on the difference between the bulk composition of the moon and Earth. For example, the mean densities are 3340 and 5250 kg/m^3, Earth's being higher due to the larger amount of iron and nickel concentrated in its core. From the densities alone, we see that the moon lacks

*NASA geologist Bevan French has called these the daughter, sister, and pickup theories of lunar origin.

much iron. Lunar samples confirm that the moon is deficient in **siderophile elements**—iron, nickel, and other elements with chemical affinity for iron and nickel. Also, the moon is much richer than Earth in refractory elements and depleted in volatile elements and compounds, such as lead and water. These findings suggest that the moon's material was once very hot and that the volatile elements vaporized and disappeared into space during this heating.

The fission theory of the moon's origin has had a long history. It was first suggested in 1898 by George Darwin as a result of his analysis of tides. This analysis showed that the moon was once much nearer Earth, as discussed in Chapter 3. The fission theory was later abandoned when geophysicists pointed out that tides probably had insufficient energy to tear material off Earth and get it outside Roche's limit, where it could reaccrete. Later, Wise (1966) rejuvenated the theory, pointing out that Earth's spin and centrifugal force would have increased when the iron drained to the center and formed a core (as the mass concentrated toward the middle), and this might have given the extra kick needed to throw the lunar mass off. Also, the fission theory explains the moon's lack of iron, since the moon would be made only of Earth-mantle material, of density around 3300 kg/m^3, as actually observed.

Study of samples brought back by Apollo astronauts and Soviet Luna probes has shown that the moon has a bulk composition close to that of Earth's mantle or upper mantle (Wanke and Dreibus, 1986).

An extremely important step in acceptance of this idea comes from work on oxygen isotope ratios. As mentioned above, an important discovery has been that different zones of the solar system have different O^{16}/O^{18} ratios. But lunar rocks have oxygen isotope ratios virtually identical to those of Earth rocks (Clayton and others, 1984). This appears to prove that the Earth and moon formed from material at about the same solar distance, from the same basic kinds of material.

This observation ruled out the capture theory of a moon from some remote part of the solar system. Capture was also shown to be very improbable dynamically. Binary accretion seemed ruled out by lack of an adequate explanation of how Earth could get so much more iron than a moon supposedly accreting in nearby orbit from the same material. The fission theory at least predicted a mantle-like composition, but suffered when dynamicists concluded that not enough energy was available in Earth's spin to throw off the lunar material.

After the lunar rocks had been examined, all three pre-Apollo theories of lunar origin seemed to have failed, and one scientist quipped that since our theories don't explain it, the moon must not really be there.

Ringwood (1970) developed the first major post-Apollo theory. This **precipitation theory** pictured Earth's accretion as so rapid that infalling material heated the surface to around 2000 K. Consequently, a hot, silica-rich vapor cloud surrounded Earth. Already deficient in siderophiles, this cloud lost its volatiles into space, so that the remaining condensates in the cloud had the refractory-rich composition of the moon.

The trouble with this theory was that it was *too* evolutionary. That is, it seemed to predict that each terrestrial planet should undergo such a history and have a large moon.

To meet these problems, Hartmann and Davis (1975) developed a new theory, the **giant impact theory.** We pointed out that the second-largest, third-largest, and other planetesimals in Earth's neighborhood may have grown to very large size before hitting Earth. The fate of the largest subordinate planetesimal could have a strong, but stochastic, effect on the final planetary configuration. The most likely fate would be a close encounter throwing it to another region of the solar system, possibly to a Jupiter encounter that could throw it out of the system entirely. But a few planets might be hit by large planetesimals. A tangential impact could change the spin rate or obliquity, and so on. We suggested that a very large planetesimal hit Earth after the core formed, and blew hot mantle material into orbit where it lost its volatiles and aggregated into the moon. Our paper suggested second-largest planetesi-

Figure 5-16. One-half hour after the giant collision that led to formation of the moon. A Mars-scale planetesimal has hit Earth somewhat tangentially and is shearing off mantle materials from both bodies in a blaze of incandescent material. (Painting by author, based on computer reconstructions of the impact by Jay Melosh and coworkers)

mals 1000 to 6000 km across. Independently, Cameron and Ward (1976) developed almost the same picture from angular momentum calculations, concluding that Earth was hit by a Mars-sized body (6800 km across). Wetherill's accretion models (1976) predicted second-largest bodies 4000 to 6000 km across.

At a conference held in 1984, the giant impact theory became the leading model of how the moon formed. Subsequent computer models of the impact have supported the concept that a fraction of the hot mantle debris could go into orbit, rapidly forming a ring around Earth, where the moon would aggregate close to Earth and then move to its present distance. A number of other studies support the concept (see the collection edited by Hartmann and others, 1986). Currently, geochemical studies of the mantle are seeking evidence of whether such an impact left distinctive chemical traces in the mantle mineralogy. For example, computer models predict transient high temperatures in the mantle that would never have existed without the impact, and these might leave distinctive traces in the mineral chemistry.

SUMMARY

In most current thinking, the solar nebula is pictured as a cloud left after the sun's formation, analogous to other cocoon nebulas, and impregnated by material from a nearby supernova or supernovas. The cloud was heated by its gravitational contraction during solar formation but then cooled. During its cooling, solid microscopic grains formed by condensation. Refractory minerals, followed by metals, ordinary silicates, carbonaceous materials, and then ices, condensed as the temperature fell. The local composition at any time and place was controlled largely by the ambient temperature.

The dust settled toward the ecliptic plane and aggregated into planetesimals, probably largely by gravitational collapse. Collisions among the earliest planetesimals (of kilometer scale or less) led to their accretion into larger bodies. Some collisions led to fragmentation, with fragments later accreted by other bodies. The asteroids probably never completed the growth process.

The largest bodies in each zone of the solar system grew fastest, so that one body eventually came to dominate each zone except the asteroid belt, accreting small planetesimals or scattering them out of their original zones to be swept up by neighboring planets. The giant planets may have formed directly by gravitational collapse or may have accreted to such large size that they gravitationally attracted massive circumplanetary clouds, instead of growing purely by collisional accretion of planetesimals. Giant planets probably developed miniature, disk-like solar nebulas, in which satellites accreted. The extended atmospheres of primeval planets also apparently captured some planetesimals, adding additional outer satellites. Earth's moon probably formed after Earth collided with a giant planetesimal.

CONCEPTS

radiometric dating
radioactive isotopes
parent isotopes
radiogenic isotopes
daughter isotopes
solidification age
formation age
half-life
relative formation age
formation interval
solar nebula
grains
planetesimals
protoplanets
volatiles
hydrostatic principle
plasma
magnetic field lines
magnetic coupling
hydrodynamics
magnetohydrodynamics
chemical equilibrium
refractory compounds
carbonaceous chondrite
Allende inclusions
enstatite
feldspars
troilite
olivine
heterogeneous accretion model
chemical equilibrium model
adiabatic curve
electrostatic forces
collisional accretion
gravitational collapse
Stoke's law
origin of ring systems
fission theory
binary accretion theory
capture theory
siderophile elements
precipitation theory
giant impact theory

PROBLEMS

1. (a) What was the date of solar system origin? (b) How long did the origin process take to create planet-sized bodies?

2. Describe how answers to Problem 1 can be determined by measurements.

3. What is the evidence for a supernova explosion near the contracting solar nebula shortly before the solar system formed?

4. (a) Describe the solar nebula. (b) Summarize evidence that it existed.

5. Assume that you observed the sun and solar nebula from a nearby star during their formation, and compare your observations with modern observations of T Tauri stars and related objects.

6. Calculations indicate that if a sun-sized star formed from a contracting, rotating, interstellar cloud, it would end up spinning much faster than the sun unless slowed in some way. What theory has been advanced to explain why the sun does not spin much faster?

7. (a) You are traveling through space and come to a star of normal solar-like composition but with planets composed of refractory silicate minerals rich in aluminum, titanium, and calcium, and containing no water or ice. What can you conclude about formation conditions? (b) Compare these planets with our own moon and comment on formation conditions of the moon.

8. Citing their distance from the sun, temperature history, and gravity, give two reasons why the giant planets have so much more volatile and icy material and mass than do terrestrial planets.

9. What are the Allende inclusions, and why are they important in interpreting early solar system conditions?

10. (a) By what factor are inert gases depleted in the terrestrial planets? (b) Why does this enormous depletion indicate that terrestrial planets formed by accretion of small, independent planetesimals, rather than by a single gravitational collapse of a whole planetary mass from the solar nebula?

11. (a) Describe some theories of the moon's origin and some problems associated with each. (b) How old are typical lunar rocks? (c) What circumstances prevent us from getting many lunar rocks 4.5 Gy old, which might reveal clues about the moon's origin?

12. (a) Describe the satellite system of Jupiter. (b) What circumstances indicate that the outer Jupiter satellites are fundamentally different from the Galilean satellites and might have formed by capture?

13. Describe ring systems in the solar system and indicate some possible modes of origin for them.

ADVANCED PROBLEMS

14. (a) If a planetesimal has an effective cross-sectional area of πR^2 and is sweeping through a cloud of smaller planetesimals with velocity V, show that the number of collisions per second will be

$$\frac{dn}{dt} = \frac{\pi R^2 V \rho_N}{m}$$

where ρ_N = the spatial density (kg/m^3) of particles in the surrounding nebular cloud, and m = the mass of each small particle. (b) Show that if each collision results in the target mass sticking to (or embedding in) the planetesimal, then the planetesimal will gain mass at a rate

$$\frac{dM}{dt} = \pi R^2 V \rho_N$$

where M = planetesimal mass.

15. Under the assumptions of problem 16 and assuming that R is the actual geometric radius of the body, show that an expression for the time to grow to radius R is

$$T = \frac{4R}{V}\frac{\rho_P}{\rho_N}$$

where ρ_P = density of the planetesimal itself. Assume while solving the differential equation that ρ_N and V remain constant during the growth process (that is, that the nebula acts as a very large reservoir for material to be accreted, rather than be used up).

16. (a) Assuming the equation in problem 15 holds and that the reasonable value for density ρ_N of accretible material in the nebula is 10^{-7} kg/m^3, estimate the time scale to accrete a body 1000 km across in the solar nebula. Assume a reasonable ρ_P. (b) How does the time scale change if the accretion process is inefficient, so that not every collision results in mass gain? (c) Is the calculated result reasonable compared with the measured formation interval?

17. (a) What is the escape velocity of a 1000-km-diameter planetesimal if its density is 3000 kg/m^3? (b) A small meteoroid at a great distance from this planetesimal is approaching it directly at 1 km/s. At what speed will it impact the surface of the planetesimal?

18. If the density of gas and dust in the solar nebula was 10^{-3} kg/m^3 in some region and the temperature was 500 K, what would have been the smallest body that could have formed directly by gravitational collapse?

CHAPTER SIX

METEORITES AND METEORITICS

Between the planets, bits of interplanetary flotsam pursue their orbits around the sun. They range in size from dust particles to substantial worldlets. Among the terrestrial planets, the largest are about 30 km across. Among the giant planets, the largest are hundreds of kilometers across. In the asteroid belt between Mars and Jupiter, where most asteroids are concentrated, the largest object is Ceres, about 1000 km across. The most numerous of these bits are the comets, in the outermost solar system, reaching at least dozens of kilometers in diameter.

All these objects—dust particles, asteroids, and comets—are collectively known as the **meteoritic complex**, from the Greek term *meteōron*, meaning a celestial phenomenon or airborne object.* They are interesting from several points of view. Some collide with Earth, surviving the passage through the atmosphere and hitting the ground. They are called *meteorites* (discussed below). Many more pass nearby. The larger of these may be visited in the near future, and some of their materials may be utilized for space colonization.

The meteoritic complex is also interesting because it sheds light on the past. Some meteoritic bodies survive from the days of planetary formation and have not been much heated or chemically altered since then. They give more clues about planetary origin than most planetary samples, which have experienced severe geologic processing.

One important consequence of meteorites is that they have cratered virtually all known planetary surfaces. Low-angle lighting on this portion of the lunar surface indicates that meteorites of many sizes have fallen and created a surface nearly covered with craters. (NASA, photo by Apollo astronauts orbiting the moon)

FATE OF PLANETESIMALS

To understand the history of meteorites, consider the five possible destinies awaiting planetesimals left over after planet formation.

*Until the 1900s the term was used to refer to rainbows, lightning, snowfalls, and so on, as in "the meteor of the ocean air shall sweep the clouds no more" (Oliver Wendell Holmes). Hence the discipline of meteorology refers to the atmosphere—not meteors.

Ejection from the Solar System

Most planetesimals orbited around the sun, being gradually perturbed by the planets until they made a close encounter with one of the planets. Commonly such close encounters sent planetesimals into the outermost solar system. Many were perturbed onto Jupiter- or Saturn-crossing orbits (that is, orbits crossing the orbit of Jupiter or Saturn), eventually leading to such strong perturbations that the objects were kicked out of the solar system on a nearly parabolic or hyperbolic orbit (Öpik, 1963, 1966; Arnold, 1965; Wetherill, 1976, 1977a, b).

Collision with Planets

In many cases, instead of passing several planetary radii away, the planetesimal actually collided with the planet. Such collisions caused the densely cratered surfaces we see on most planets and satellites studied so far at close range, such as Deimos, Phobos, Amalthea, the moon, Callisto, Ganymede, and Mercury. Even the more geologically active worlds, with their younger surfaces, usually have some impact craters, often eroded. Examples include Europa, Mars, Venus, and Earth, as shown in Figure 6-1. Most planetesimals were swept up by the planets in the first 0.5 Gy after the planets formed.

Capture into Satellite Orbits or Resonant Orbits

As mentioned in the last chapter, primeval planets may have had extended atmospheres or satellites whose interactions with the approaching planetesimals allowed capture of the planetesimal into a satellite orbit around the planet (Hunten, 1979). A variant of this fate would be capture into a dynamically stable configuration, such as the L_4 and L_5 Lagrangian points in Jupiter's orbit, where Trojan asteroids are now located.

Fragmentation

If a planetesimal got perturbed by Jupiter onto an Earth-crossing orbit (with eccentricity 0.67), it would speed by Earth's orbit at about 10 km/s, rather than at the velocity of about 0.7 km/s at which "local" planetesimals (with eccentricity 0.05) move by Earth. (Note that velocities of only about 0.04 km/s are necessary for rocks to fragment each other instead of merely bouncing apart upon collision.) Thus, as planetesimals got scattered around the solar system by the larger planets, encounter velocities between them increased to the point where fragmentation rather than accretion dominated (Kaula and Bigeleisen, 1975; Wetherill, 1977a). An additional planetesimal fate, then, was to be smashed to bits. The fragments, which were smaller than the original, could of course undergo ejection, collision with planets, or further fragmentation.

This effect probably explains why accretion stopped in the asteroid belt before a planet grew there. Calculations indicate that even by the time Jupiter reached only 0.1 of its present mass, it scattered bodies through the asteroid belt at relative speeds as high as 10 to 15 km/s (Ip, 1978), causing much damage.

Preservation Until Today

Those planetesimals and fragments that escaped the first four fates have lasted until today. There are few places in the solar system where bodies could be "stored" with much hope of lasting this long. Most bodies, even if placed on circular orbits midway between two adjacent planets, will get perturbed onto planet-crossing orbits and swept up within about 0.1 Gy. One relatively permanent reservoir of objects is the asteroid belt (see Chapter 7), and another is a swarm of comets in the outermost solar system (see Chapter 8). Wetherill (1977a) describes how some planetesimals that originated among terrestrial planets may have been perturbed into the inner edge of the asteroid belt and preserved there.

Figure 6-1. An ancient meteorite impact crater on Earth: Brent crater, in Ontario, Canada. The circular structure has been eroded and partly filled in by glacial and other erosive activity, forming two lakes in the original crater's interior. Its diameter is 4 km, and it has been dated at 450 ±30My old. Other prehistoric craters up to 65 km across and 485 My old are known in the Canadian Shield, an old and stable region of Earth's crust. (Aerial photo, Earth Physics Branch of the Department of Energy, Mines, and Resources, Ottawa, Canada)

METEORITES AS "FREE SAMPLES" OF PLANETESIMALS

The preserved planetesimals and their fragments are especially interesting because some of them can interact with Earth. Orbits of such bodies are typically inclined to the plane of the solar system and therefore pass through the ecliptic plane at two points, called **nodes**. If either node happens to lie very close to Earth's orbit, then sooner or later the planetesimal and the Earth will arrive at this point at the same time. For a typical planetesimal on such an orbit, such an encounter will occur within about 10^7 to 10^8 y. The most common result of such an encounter is a near miss that gravitationally alters the orbit of the object, just as the orbits of Pioneer and Voyager spacecrafts changed dramatically as they swung close around Jupiter. A fraction of the approaching planetesimals, however, penetrate through the atmosphere and reach the ground, giving us free samples of interplanetary material—meteorites.

On April 10, 1972, a house-sized interplanetary body approached Earth from behind at 10 km/s on a 15° inclined orbit that reached as far as 2.3 AU from the sun. As Earth's gravity pulled it in, it accelerated to 15 km/s and entered the atmosphere, traveling south to north across the western United States. Friction with the atmosphere heated its outer skin to incandescence, and it became visible as a brilliant glowing object over Utah, moving north over Idaho and Montana. Here it dipped as low as 50 to 60 km in the atmosphere, at which point most such bodies would have fallen into the ground with explosive energy. But like a stone slicing into water and skipping back out again, this body skipped off the atmosphere, ascending at a shallow angle over Alberta and exiting into space once again on a different orbit.

Hundreds of eyewitness reports were collected by the American Meteor Society; *Sky and Telescope* magazine over the next few months published 11 still photos and 72 movie frames (a small sample of the total movie footage) of the object, and a classified Air Force satellite detected it from above with an infrared radiometer designed to look for heat emitted from missiles.* These observations permitted considerable analysis of the orbit and the object, which was found to be a rocky body with a diameter between a few meters and 80 m (Jacchia, 1974). This, then, was a rare case of a large interplanetary body that encountered the atmosphere but barely escaped plummeting into the ground. The miss was fortunate, since a hit by this object would have caused a (nonnuclear) explosion of about the magnitude of the Hiroshima bomb, probably somewhere in Alberta, Canada.

METEORITICS AND SOME ASSOCIATED DEFINITIONS

The study of the interplanetary samples is called **meteoritics**. Meteoritics attempts to determine where these samples came from and how they relate to either the primordial or present-day planetary bodies. It is useful to review the related terms that have evolved:

meteoroid: Any small, extraterrestrial, solid body floating in space. Dimensions usually smaller than a kilometer, and frequently dimensions of millimeters, are implied, with larger objects usually classed either as asteroids or comets.

meteor: A heated, glowing meteoroid in transit through the atmosphere, having not yet hit the ground. A meteoroid may enter the atmosphere at 10 to 70 km/s, and friction between it and the atmosphere heats its surface and ionizes atmospheric molecules. Airless planets such as the

*This large body of solid observational reports contrasts strongly with the absence of good observations of UFO cases allegedly involving alien spacecrafts. One might infer that if passages of such spacecrafts through our airspace occur as frequently as proposed by a few UFO researchers, we ought to obtain better observational evidence within a few years, due to the increasing numbers of amateur photographers and surveillance devices.

moon would have no meteors, for an atmosphere is needed to heat the meteoroids and make them visible.

meteorite: Any of those meteoroids or their fragments that strike the ground. There are three broad classes: stones, irons, and stony-irons. We cannot assume that the meteors we see on an average night are physically related to the meteorites we see in a museum. Indeed, there is contrary evidence, which we will describe later. Meteorites are generally named after a city or geographical landmark near where they fell or were found.

fireball: One of the very brightest meteors. These are infrequent and can be much brighter than the full moon, in rare cases rivalling the sun.

fall: A meteorite witnessed during descent. Some 94% of falls are rocks, and hence most meteoroids in space are believed to be rocks. However, a few are iron alloy objects.

find: A meteorite found lying on or buried in the ground. Many finds may have fallen thousands of years ago. The people in locales of some ancient falls recount legends or place names associated with the sky, as at Campo del Cielo, Argentina. Such legends may have persisted since the time of the fall. Most finds are irons, since these resist weathering longer than rocky material.

parent body: A meteorite is probably a fragment of a larger object known as the parent body. The interiors of these hypothetical parent bodies sustained pressures and temperatures suitable for forming the minerals found in meteorites.

HISTORY OF METEORITE STUDIES

Early Chinese, Greek, and Roman writings describe the fall of "stones from the sky." Often these celestial stones were collected and enshrined as sacred objects. The famous black stone in the Kaaba, the sacred shrine of Islam in Mecca, is apparently a meteorite that was enshrined before

Figure 6-2. Burial of the Winona stone meteorite in a crypt by prehistoric Indians in northern Arizona. (Museum of Northern Arizona, Anthropological Collections; photo by Gayle Hartmann and the author)

Muhammad conquered the city around A.D. 600 (Sagan, 1975). The Romans organized an expedition to seize a meteorite venerated in a temple in Turkey in order to worship it themselves and thus ensure victory over the invading armies of Hannibal. The temple and stone were eventually abandoned, and efforts to relocate the meteorite in the 1800s failed (Nininger, 1952). Figure 6-2 shows meteorite fragments buried around A.D. 600 in a stone-slab crypt by Pueblo Indians near Winona, Arizona (Heineman and Brady, 1929).

During the Middle Ages in Europe, fallen meteorites were likely to be brought by incredulous peasants to local priests or city magistrates. Thus a number of meteorites, such as the Ensisheim stone of 1492 (Figure 6-3), were preserved in European village churches.

In 1751 a widely witnessed fall occurred in Zagreb, Yugoslavia. The bishop of Zagreb carefully collected and protected pieces of the iron meteorites and sent them, along with a tabulation of witness testimony, to the Austrian emperor. This was passed on to the Vienna museum but was not treated as important.

Figure 6-3. An old woodcut showing the fall of a meteorite near the town of Ensisheim (now in France) in 1492. Preservation of the stone in a local church played a historically important role in causing some scientists in the 1700s to dismiss meteorites as superstition.

By the 1790s, philosophers and scientists were aware of many allegations of stones falling from the sky, but the most eminent scientists were skeptical. The first great advance came in 1794, when a German lawyer and physicist, E. F. F. Chladni, published a study of some alleged meteorites, one of which had been found after a fireball had been sighted. Chladni accepted the evidence that these meteorites had fallen from the sky and correctly inferred that they were extraterrestrial objects that were heated by falling through Earth's atmosphere. Chladni even postulated that they might be fragments of a broken planet—an idea that set the stage for early theories about asteroids,* the first of which was discovered seven years later. Chladni's ideas were widely rejected, not because they were ill conceived, for he had been able to collect good evidence, but because his contemporaries simply were loathe to accept the idea that extraterrestrial stones could fall from the sky. Gradually the evidence became more compelling. A 1796 paper accepted the celestial origin but argued that the stones had been swept up by tornadoes, only to fall again to Earth (Wood, 1968).

The last holdouts were members of the austere French Academy of Sciences, who were considered the scientific leaders of the day. These gentlemen pooh-poohed the idea of stones falling out of the sky and argued that the whole business was nonsense because the peasant witnesses could (supposedly) not make reliable observations. In retrospect, it appears that part of the academy's reluctance was generated by subtle social pressures. In the years following the French Revolution, there was a philosophic wave of antipathy to religious authorities. The academy frowned on samples that had come to reside in village churches or had been collected by country priests.

As if in answer, a shower of fragments from a large meteorite pelted the French town of L'Aigle on April 26, 1803. The academy sent the noted physicist Biot to investigate. His exhaustive report finally convinced the scientific world that stones do fall from the sky and that Chladni had been right.* A century of further work revealed that meteorites were unlike any terrestrial stones and had very complex histories. This work continues today, as will be seen in this chapter.

In 1891 a naval expedition on the H.M.S. *Challenger* dredged up seafloor sediments containing numerous submillimeter spherules. These have been proven to be extraterrestrial and probably formed as molten droplets sprayed off meteorites burning in the atmosphere; they are probably the most common form of extraterrestrial material on Earth (Ganapathy, Brownlee, and Hodge, 1978). The collection of known meteorites was considerably expanded by the pioneer meteoriticist H. H. Nininger (1952), who toured the American plains states and obtained falls and finds from farmers and other residents.

*This accounts for the once-popular theory that asteroids were fragments of a single planet that exploded for unknown reasons.

*The knowledge only slowly filtered to the fledgling United States. According to an apocryphal story, upon hearing two Yale professors report a meteorite fall, the Virginia naturalist (and politician) Thomas Jefferson remarked, "It is easier to believe that Yankee professors would lie, than that stones would fall from heaven."

In 1969, Japanese geophysicists discovered that Antarctic ice fields are littered with meteorites lying on the surface or preserved in deep freeze in subsurface ice. Japanese and American teams have collected over 1000 new meteorite samples, which, unlike earlier specimens, are being processed through NASA's lunar rock curatorial facility and preserved in sterile surroundings. Pieces of perhaps 3000 different meteorites were collected as of 1979.

Recovery of a new meteorite specimen is a rare privilege and should be reported to the Smithsonian Institution in Washington or to a nearby major university, since each meteorite has its own peculiarities and might provide some major insight into the history and environment of the solar system.

PHENOMENA OF METEORITE FALLS

Sound

The largest meteorites produce energetic shock waves when they enter the atmosphere at supersonic velocities. These "sonic booms" have long been known, being described in terms of contemporary culture: "like distant guns at sea" (England in 1795) and "like a horse and carriage clattering over a bridge" (America in 1913). The Tunguska fall of 1908 in Siberia was a mass of roughly 10^7 to 10^8 kg (see Chapter 8); its explosions were heard over 1000 km (600 mi) from its flight path (Krinov, 1966). Peculiar sounds reported from fireballs brighter than the full moon have been suspected to have an origin involving electromagnetic radiation at very low frequencies (Keay, 1980), but are poorly understood.

Brightness

The larger the object, the brighter it appears. For example, the Sikhote-Alin fall in Siberia was a 70-t iron meteorite; it dazzled the eyes of observers and cast shadows even though the sun was shining (Krinov, 1966).

Train

Many large meteorites leave a dusty **train** made up of debris blown off during atmospheric passage. Many break apart, spreading pieces in a line along the flight path, or detonate with a burst of light. Meteors and meteorites seen at night may leave a faintly luminous train visible for a few seconds to several minutes. This is caused by photochemical reactions among high-altitude air molecules disturbed by the meteor's passage.

Temperature

Passage through the atmosphere is so swift that only the outer surface of a meteorite is heated—the meteorite's interior remains cold. The widespread idea that a meteorite may remain too hot to touch for some hours after its fall is a fallacy, because the outer hot layer, about 1 mm thick, cools rapidly. A thin **fusion crust**, usually black, is produced on meteorite surfaces by atmospheric heating.

Velocity

A typical meteorite strikes the atmosphere at about 10 to 20 km/s. Those smaller than about 1000 t lose a substantial part of their initial velocity due to atmospheric drag, but larger meteorites can strike the ground with most of their initial velocity. Meteoroids with the highest velocities (up to 70 km/s) are most likely to fragment in the air.

IMPACT RATES: METEOROID FLUX

Various studies estimate that the total **meteoroid flux**, or mass of meteorites that falls onto Earth, is about 10^7 to 10^9 kg/y (Dodd, 1981); the lunar figure is about 4×10^6 kg/y (Wetherill, 1976; Hartmann, 1980). The terrestrial rate equals hundreds of tons per day! The number of particles increases with decreasing size. Microscopic particles, called **micrometeorites**, are very abundant; objects large enough to hit the ground

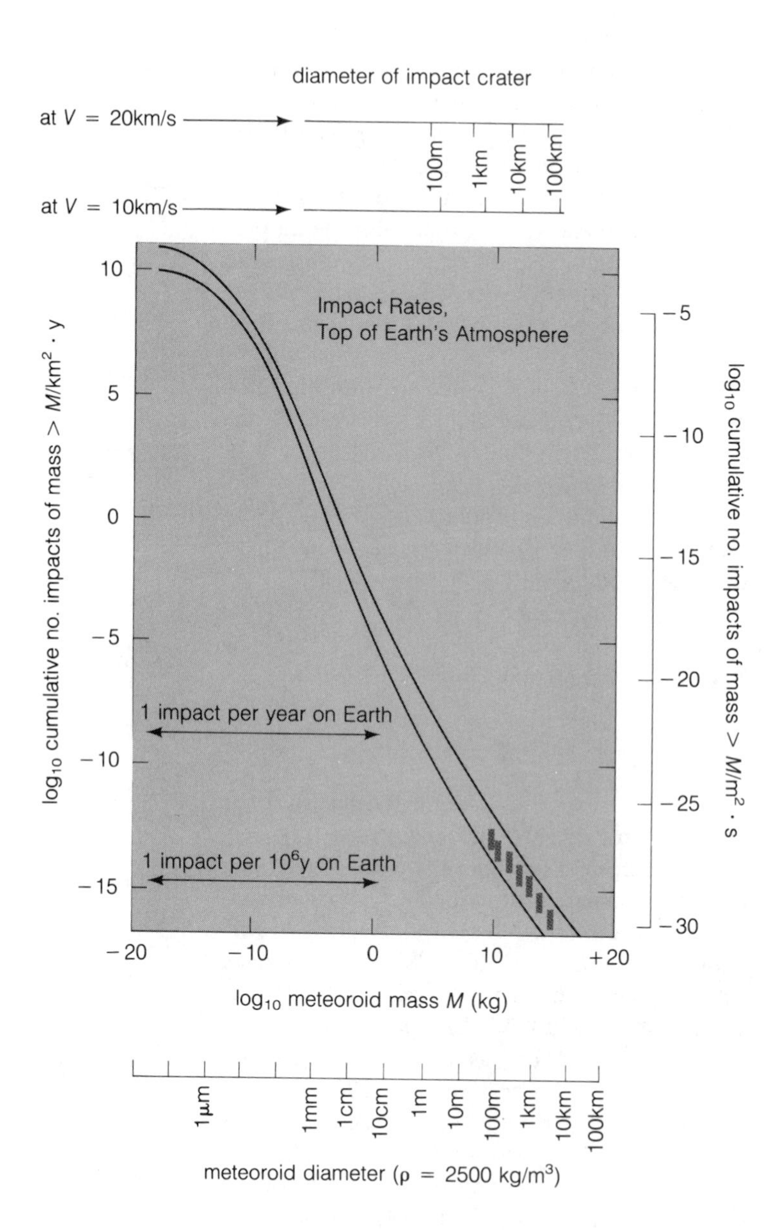

Figure 6-4. The rate of meteoroid impacts at the top of Earth's atmosphere as a function of meteoroid size. Upper and lower curves give estimated limits on meteoroid flux. Scale at top gives size of crater formed for impact velocities of 10 and 20 kn/s; boxes at lower right give crater formation rate derived from craters observed on 3.5-Gy-old lunar maria; box size indicates uncertainty. (See text for further discussion.)

and form substantial craters hit Earth roughly annually; giant explosions happen once or twice per century.

Present-Day Impact Rate

Figure 6-4 shows the numbers of objects of different sizes presently hitting Earth. The curve is based on spacecraft measures of meteoroids and counts of craters formed on Earth in recent geologic time (Dohnanyi, 1972). The scale at the top shows the sizes of craters produced if the objects strike surfaces at 10 or 20 km/s, speeds typical of planetary impacts (see Chapter 10 for further discussion of crater formation). The box-shaped data points in Figure 6-4 show impact rates derived from counting craters on lunar lava plains of known age and multiplying by 1.5 to allow for the fact that Earth's gravity field produces a terrestrial crater formation rate (craters/km^2·y) about 50 percent greater than the lunar rate.

Primeval Impact Rates—The Early Intense Bombardment

As early as 1952, Urey postulated an intense bombardment of planets during the close of planet formation. This occurred because the newly formed planets were efficient at sweeping up debris. Planetesimals on orbits similar to planets' orbits had half-lives as short as 15 to 20 My. By 100 My after planet formation, only the more scattered planetesimals were left, and they had half-lives around 20 to 100 My, according to dynamical calculations (Öpik, 1963, 1966; Arnold, 1965; Wetherill, 1977).

As shown in Figure 6-5, Apollo data confirm these calculations. Dates determined for surface rock layers at different landing sites, combined with crater counts at those sites, give impact rates at different times. The box-shaped data points measured in this way clearly show the decline from the early intense bombardment some 4 Gy ago to the more constant value experienced in the last 2 Gy or so.

One of the controversies about the early intense bombardment is that lunar data give little evidence about conditions before 4 Gy ago except that the impact rate necessary to accrete the moon and Earth is off the scale to the upper left (Hartmann, 1980). One interpretation (curve A) is that the cratering rate simply declined smoothly, with

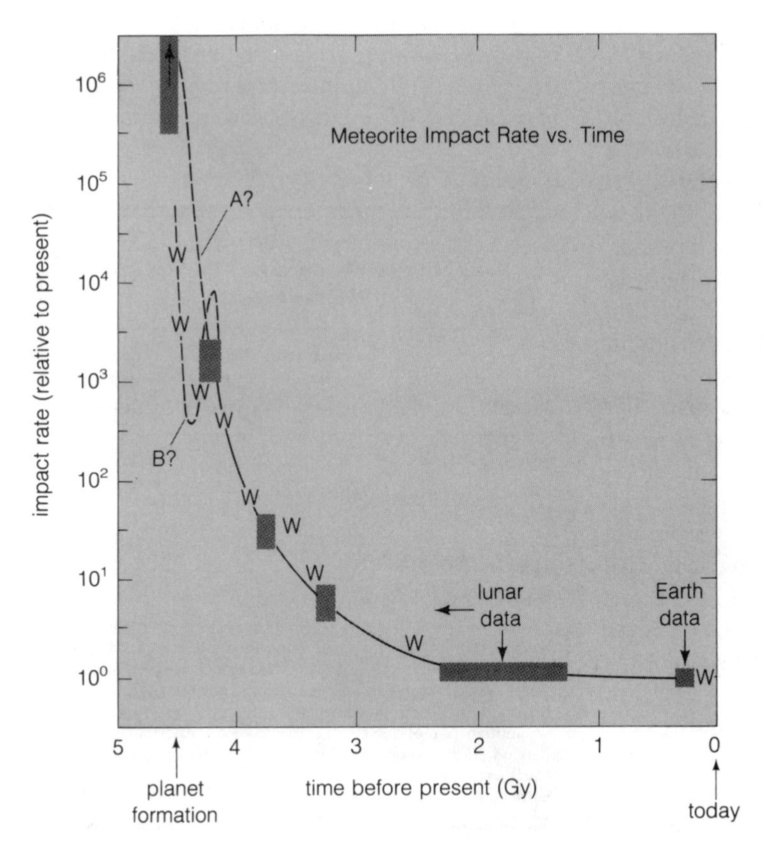

Figure 6-5. The history of impact rates in the Earth-moon system, as derived from terrestrial and lunar data (adapted from Hartmann and others, 1980). Sizes of boxes indicate approximate precision of available data. Two interpretations of earliest cratering, A and B, are discussed in text. Points labeled W mark calculated impact rates at different times, based on an assumed initial swarm of planetesimals with a = 0.9 AU, e = 0.27, and i = 5.7° (from Wetherill, 1977a).

impacts pulverizing most lunar rocks formed before 4 Gy ago. A decline of this sort would result if the Earth and moon simply swept up a population of nearby leftover planetesimals, as shown by the "W" points in Figure 6-5, based on sweepup calculations by Wetherill (1979a). However, some lunar rock analysts believe that the sharp cutoff in numbers of lunar rocks older than 4.0 Gy indicates an intense burst of cratering that destroyed rocks at that time (Tera, Papanastassiou, and Wasserburg, 1974). Such bursts (curve B) could come about if large planetesimals from the outer solar system broke up in the inner solar system (by tidal stresses during close encounters with terrestrial planets?), supplying bursts of meteoroidal fragments (Wetherill, 1975, 1977a, b).

Impact Rates on Other Planets

The impact rates vary from planet to planet. Spacecraft have measured the flux of micrometeoroids, which are abundant enough to be intercepted with high frequency by such vehicles. But fluxes of larger objects are quite uncertain. The inventory of larger objects includes comets and asteroids ejected from the asteroid belt. Table 6-1 lists estimated cratering rates for various planets, assuming a mixture of cometary and asteroidal meteoroids. No comparable data are known for the outer solar system, though Callisto's highly cratered surface shows that impacts have been very numerous in the Jupiter satellite system. This implies that the outer planets have also swept up debris. In their studies of Jupiter and Saturn satellites, Voyager team analysts calculated cratering rates for these moons, based on estimated populations of cometary bodies in the outer solar system. The rates for inner moons, like Io and Mimas, are much higher than rates on outermost moons, because of gravitational focussing toward the giant planets.

Because all planets formed at about the same time and swept up planetesimals at a high rate during early solar system history, the relative rate of cratering through time is believed to resemble Figure 6-5 for all planets, with only the absolute rate varying from planet to planet.

A METEORITE CLASSIFICATION SYSTEM

We now turn from impact effects to the meteorites themselves. In the last chapter, we saw that chemical properties of certain primitive meteorites fit beautifully with the condensation theory of the origin of planetary matter. Other meteorites have been **metamorphosed**, or altered, by heat, pressure, mechanical shock, or other effects. Many have been gently heated so as to alter their minerals slightly. Others have been completely melted, allowing heavy material such as metal to separate from light materials. These materials then resolidified into lavalike rocks or metal.

Any process that segregates chemicals from their original mixture is called **differentiation**. Meteorites have been classified primarily by their degree of differentiation, and understanding the classification scheme helps us to understand what

Table 6-1
Rates of Crater Production of Various Planets Relative to Earth

Planet	Minimum likely value	Most likely value	Maximum likely value
Mercury	0.5	1.3	3.3
Venus	0.5	0.7	1.4
Earth	1.0	1.0	1.0
Moon	0.5	0.7	0.9
Mars	0.6	1.3	2.6
Outer planets	?	?	?

Source: Data adapted from Hartmann and others, 1981.
Note: These values are not quite the same as relative meteoroid fluxes, since a crater on one planet may have been formed by a somewhat different-sized meteorite than the same-sized crater on another planet due to an impact velocity difference. Estimates depend on the assumed distribution of Mars-crossing asteroids, Earth-crossing asteroids, short-period comets, long-period comets, and so on.

meteorites tell us about early conditions in the solar system and about the interiors of other planetary bodies. Unfortunately, the classification scheme is complex and is occasionally revised by researchers. Table 6-2 summarizes the current classifications, of which beginning students should learn at least the "broad classification."

Meteorites are broadly divided into **stony meteorites**, **iron meteorites** (pure metal, essentially nickel-iron alloy), and **stony-iron meteorites** (mixed iron and stony portions). Stony meteorites, often called simply "stones," are by far the most common (94%) and include some that are hardly differentiated—that is, the ones least altered by heating, melting, or other chemical changes. Among the stones, the most numerous are **chondrites**, a class named because they contain BB-sized spherules called **chondrules**, after a Greek term for seedgrains. Chondrules may be melted droplets formed during early planetesimal collisions, and chondrites are believed to be quite primitive objects. Chondrites make up about 80% of all meteorites falling on Earth. A smaller class of stones are **achondrites**, which

Table 6-2
Classification of Meteorites

	Broad classification	Group name assigned by Wasson[a]		Mineralogical-structural name	Percentage of falls[b]
Primitive Bodies ↑	Stones 95% → Chondrites 86% → Carbonaceous chondrites 5%	CI	=	C1 = CI	0.7
		CM	=	C2 = CII	2.0
		CO, CV	=	C3 = CIII	2.0
	Stones 95% → Chondrites 86% → Ordinary chondrites 81%	E	=	Enstatite	1.5
		H[c]	=	Bronzite	32.3
		L[d]	=	Hypersthene	39.3
		LL[e]	=	Amphoterite	7.2
		Others		—	0.3
Increasing Differentiation ↓	Stones 95% → Achondrites 9%	Ureilites	=	Olivine-pigeonite	0.4
		Aubrites	=	Enstatite	1.1
		Diogenites	=	Hypersthene	1.1
		Howardites and eucrites	=	Pyroxene-plagioclase	5.3
		Others	=	Other differentiated stones	1.0
	Stony irons 1%	Mesosiderites	=	Mesosiderites	0.9
		Pallasites	=	Pallasites	0.3
	Irons 4%	IA	≈	Medium and coarse octahedrites	0.8
		IIA	≈	Hexahedrites	0.5
		IIIA	≈	Medium octahedrites	1.5
		IVA	≈	Fine octahedrites	0.4
		Others	=	Ataxites, etc.	1.3
					99.9%

[a]From Wasson (1974).
[b]From Scott (1978) and Dodd (1981).
[c]H = high iron content.
[d]L = low iron content.
[e]LL = very low iron content.

have been heated, melted, and/or shattered enough to destroy the chondrule structures by melting or chemical alterations. Some achondrites resemble volcanic lavas of the moon and Earth.

Another, smaller class is of special interest. These are the **carbonaceous chondrites**, black stony meteorites that have been heated so little that they retain water (bound in the minerals), carbon compounds, and other **volatile elements and compounds**—the ones that would be easily driven off by moderate heating. Carbonaceous chondrites probably include the meteorites least altered since their formation.

An interesting additional complexity of meteorites is that many of them are intensely fractured, and others are *breccias*, or rocks formed of broken fragments cemented together (Wilkening, 1977). These fragments may be pieces of different meteorite types—for example, chips of chondrites interspersed among broken pieces of achondrites!

Current ideas about meteorite origin derive from these general properties. Early planetesimals were made of primordial condensates, as outlined in the last chapter. Some of them may have survived with little alteration to produce meteorites like carbonaceous chondrites and some ordinary chondrites. Others were partly melted by heating processes like radioactivity.* During melting, chemical differentiation occurred. Iron and elements with a chemical affinity for iron, called **siderophile elements**, drained toward planetesimal centers, forming nickel-iron core regions. Lighter, silica-associated elements, called **lithophile elements**, floated toward the surface layers, forming stony mantles. Some of this lavalike material erupted onto some planetesimal surfaces. When these planetesimals were blown apart during collisions, fragments were ejected from different regions—metal objects from the cores, mixed objects from core-mantle interfaces, lavalike rocks from upper mantles and surfaces, and primitive rocks from the least-heated regions. These would explain meteorite types such as irons, stony-irons, achondrites, and chondrites. During violent collisions, fragments from different parent bodies, or from different regions of one parent body, were intimately mixed and buried in ejected rubble on planetesimals, where they were welded together, only to be released again during a later collision. Some planetesimal material may have gone through several episodes of shattering and recementing, thus explaining the complex breccias found in many meteorites.

With these complexities in mind, and with Table 6-2 as a guide, we turn now to more specific properties and types of meteorites.

*Evolution of interiors will be discussed in more detail in Chapter 9.

CHONDRULES

Chondrules are glassy spherules (Figure 6-6) embedded in meteorites. They are composed principally of moderately high temperature minerals such as olivine, $(Mg,Fe)_2SiO_4$, and pyroxenes such as enstatite, $Mg_2(Si_2O_6)$.* They range in size from about 0.5 to 5 mm, and their structure shows that they formed during rapid cooling of droplets of molten material. Chondrules can make up 70% of a chondrite's total mass, in which case the meteorite consists of a mass of chondrules held together by a dark matrix, as shown in Figure 6-7. The matrix typically consists of submicrometer-scale original condensate grains as well as fragments of broken chondrules, as can be seen in the magnified view in Figure 6-8.

How chondrules formed is hotly debated. There has hardly been much advance since the English mineralogist Sorby in 1877 invented the process of making thin sections of rocks for study through the microscope and then first described the chondrules. Sorby inferred that the chondrules formed from "melted . . . glassy globules,

*See preceding chapter's discussion of mineral condensation, but note that minerals and rocks will be discussed in more detail in Chapter 10.

Figure 6-6. A 3-cm portion of the chondritic meteorite Björbole. Protruding from the surface are chondrules of various sizes, including a large one at lower left. (Center for Meteorite Studies, Arizona State University)

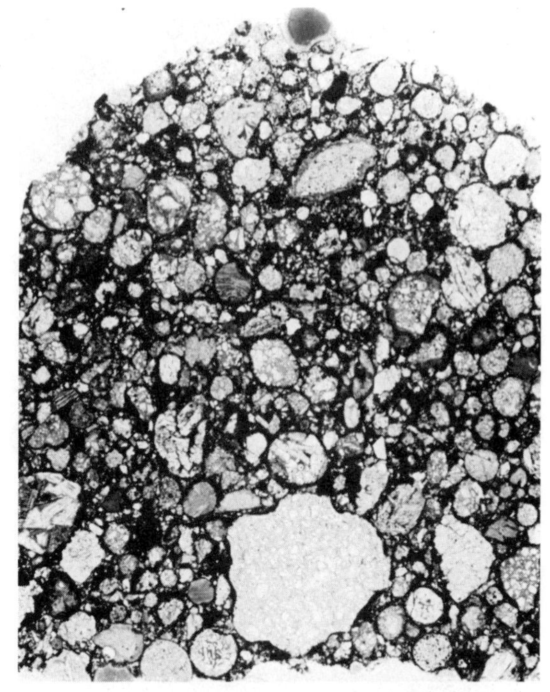

Figure 6-7. A cross section of the highly chondritic meteorite Chainpur. It consists mostly of chondrules embedded in a black matrix. Numbered units on the scale are millimeters. (Courtesy Laurel Wilkening, Lunar and Planetary Laboratory, University of Arizona)

like drops of a fiery rain." A number of chemical indicators suggest that the initial cooling of the chondrules from high temperatures occurred very fast, with a possible drop from 1500 K to solidification within minutes. One bit of evidence is the glassy structure of some chondrules. Properties of metal particles in the chondrules (Figure 6-8) suggest that the final stages of cooling to a few hundred Kelvin was much slower, requiring 10^4 to 10^7 y (Wood, 1968).

Since chondrites seem to be the least differentiated meteorites, chondrules must have formed at about the same time as planetesimal growth. What process in the earliest history of the solar

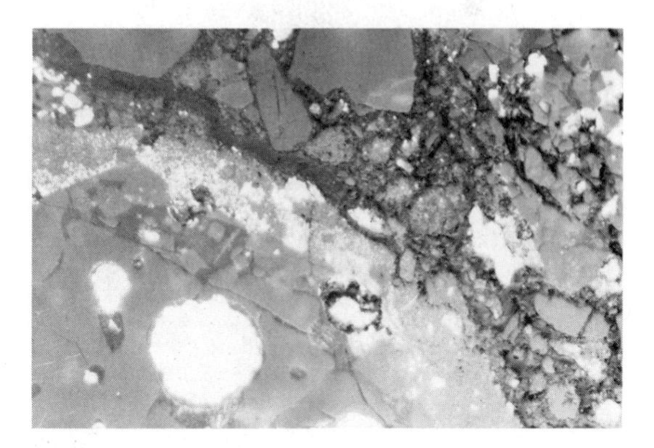

Figure 6-8. This microscopic view of the Chainpur chondrite, shows the interface between a chondrule (rounded body, lower left half) and the dark matrix, which consists of irregular fragments of chondrules and fine grains (original condensates from the solar nebula?) a few micrometers across. The white, circular body in the chondrule is a 50-μm metal spherule. (Courtesy Laurel Wilkening, Lunar and Planetary Laboratory, University of Arizona)

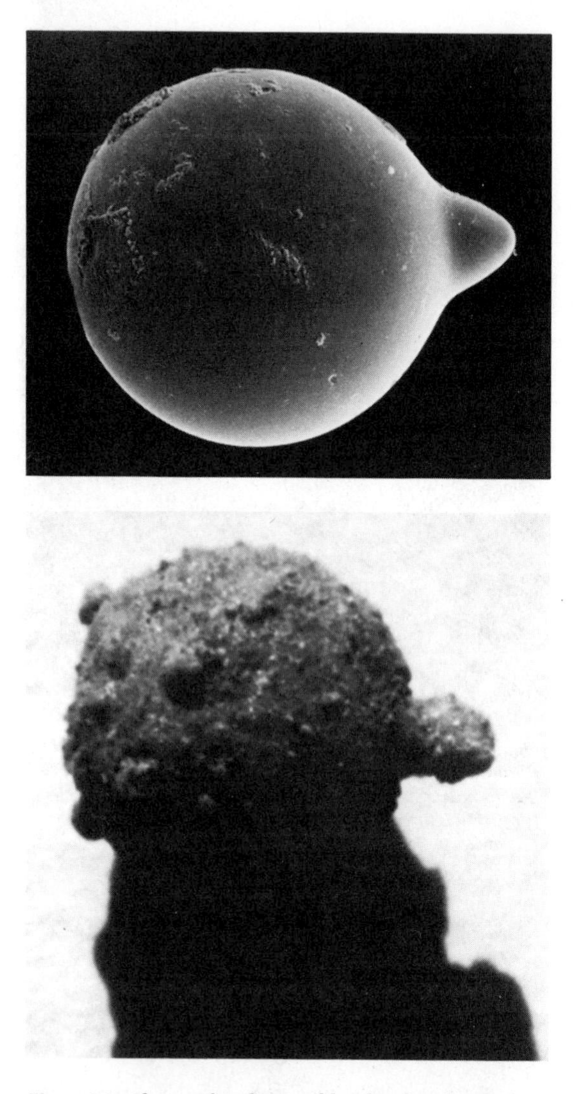

Figure 6-9. Glassy spherules possibly related to chondrule formation. (**a**) *Green glass spherule from lunar soil. Such spherules are probably splashed from lunar impact craters. Diameter, about 0.2 μm. (NASA, courtesy David S. McKay)* (**b**) *Glassy spherules, welded during flight, from ejecta of experimental crater formed on Earth by explosion of 500 tons of TNT. Possibly due to moisture in original soil, interior texture is much more frothy glass than found in chondrules. Diameter about 3.5 mm. (Photo by author)*

system could have produced "drops of a fiery rain"?

One group of theories states that chondrules formed directly out of the solar nebula gas, requiring gas pressures higher than normally estimated for the nebula (Suess, 1949; Wood, 1963; Wood and McSween, 1977). These theories suggest that chondrules condensed during disturbances such as shock waves that boosted local pressures in the nebula to unusually high values. Wood and McSween criticize "a tendency to oversimplify [chondrules], to see all of them as quenched droplets of silicate melt." They attribute the wide variety of forms and composition to different gas pressures and gas:dust ratios in different regions of the nebula. Whipple (1966) and others have suggested that lightning discharges or other electromagnetic effects in the nebula could have produced fused droplets resembling chondrules.

A more widespread idea is that chondrules formed during impacts on surfaces of planetesimals (Fredriksson, 1963; Ringwood, 1966; Dodd, 1971), which may have been meter-sized or smaller (Kieffer, 1975). The idea of "fiery droplets" spraying out of impact craters is buttressed by the finding of glassy spherules around at least one explosion crater on Earth (Hartmann and Wilkening, 1981) and chondrulelike spherules in lunar soil (Figure 6-9). But lunar soil has less than 1% chondrulelike objects and is composed mostly of angular fragments, so the problem is to explain how so many chondrules accumulated on a surface, somewhere, to be aggregated into chondritic rocks. Impact velocities at least 2.5 km/s are needed to melt rock, whereas the escape velocity of a 500-km planetesimal is only about 0.5 km/s, indicating that chondrules would probably be blown clear off of all but the largest planetesimals (Wasson, 1974). Vast numbers of chondrules may have accumulated in space, only to reaccumulate later on planetesimals. Dodd (1965) showed that elongated chondrules often lie with their long axes parallel, typically the result of an orderly deposition process, perhaps during chondrules' reaccumulation on planetesimals' surfaces. Dodd (1981) finds that chondrules in carbonaceous chondrites relate more to an early process of nebular condensation, while those of ordinary chondrites represent later, lower temperature impact melting processes on dusty and rocky planetesimals.

CARBONACEOUS CHONDRITES

Carbonaceous chondrites are samples of unaltered, ancient planetesimal material that condensed at moderate to low temperatures in the solar nebula. There are several indications that they have not been highly heated, compressed, or otherwise altered: (1) They have unusually high abundances of volatile compounds, such as water; (2) they have low densities; (3) they are rich in organic compounds that condensed at relatively low temperatures and would be driven off during any subsequent heating (Anders, 1963); and (4) they contain the heavier elements in nearly their original cosmic proportions.

Carbonaceous chondrites (which we will abbreviate as CC's) have a dark, black color because of finely dispersed black, opaque carbon and magnetite in their fine-grained matrix material between the chondrules. For some years the nature of this matrix was a mystery; detailed studies have revealed carbon in the form of poorly crystallized graphite (Smith and Buseck, 1981); fine-grained clay minerals, organic polymers, magnetite, and iron sulfide (Wilkening, 1978); and perhaps even evidence of primordial interstellar grains trapped in the matrix during its formation (Lewis and others, 1979).

CC's were divided into three types by Wiik (1956) on the basis of chemical analysis. Type I CC's (C1 in Wasson's newer nomenclature; see Table 6-2) are the most extreme in the "primordial characteristics" listed above. They have some 8% to 22% water content (bound in their minerals), a density of only 2200 kg/m^3 compared with about 3600 kg/m^3 for other kinds of chondrites, minerals stable only at low temperatures, and the highest volatile content. Thus it appears that Type I CC's have never been subjected to high heat or high pressure (that is, they have not been buried deep inside planet-sized bodies). They probably formed in a layer of surface soil only 100 to 300 m thick (Heymann, 1978). Material resembling other chondrites could have derived from such material by chemical processing (Ringwood, 1966). Nonetheless, even the Type I CC's have probably experienced some chemical and geologic activity on their parent planetesimals.

Various types of CC's have been exposed to small amounts of liquid water in their histories (DuFresne and Anders, 1963), and microscopic veins running through Type I CC's contain carbonate and sulfate deposits (the type of deposits forming bathtub rings and caliche layers, left after evaporation of mineral-bearing water), apparently having accumulated over several generations of chemical activity as water seeped through the material (Richardson, 1978). Dodd (1981) concludes that the parent bodies were ice rich.

Type I CC's have no chondrules—the nomenclature is misleading! Perhaps their initial parent bodies were so ice rich that impacts did not produce silicate spherules; the present type I CC meteorites may be residual stony material left after the ice sublimed. Alternatively, they may represent planetesimal material from a pre-chondrule-forming era.

Type II CC's (Wasson's CM's and some CV's) do have chondrules, but the matrix between the chondrules represents Type I material with minimal chemical alteration (Wood, 1968). They have 2% to 16% bound water in their minerals.

Type III CC's (Wasson's CO's and some CV's) are yet more modified, but still retain about 1% water.

All the observations indicate that CC's as a class formed at low temperature, perhaps in outer asteroid belt regions or beyond. They underwent slight heating, never to temperatures above 500 K. Ices in the original parent bodies melted and altered some of the minerals; but, on the whole, CC's are very primitive.

Table 6-2 lists CC's as being only about 5% of meteorite falls, with the C1's being only about 1%. These numbers are usually assumed to be the actual percentages of the different types of meteoroids in space. However, several facts suggest that we Earth-bound observers have seriously underestimated the percentage of CC's in space. First, these objects are weak, so they probably break up in the atmosphere more easily than other types, thus explaining why they may be underrepresented in the samples that reach the ground. Second, the dominant type of meteorite in the lunar soil is the C1 carbonaceous chondrite (Ganapathy and others, 1970; Taylor, 1975). Third, photographic studies of meteors and meteorites suggest that a substantial fraction of very weak bodies may burn up in the air, and these may be the "missing" CC's. Furthermore, about 75% of the larger asteroids (especially in the outer asteroid belt) have **spectrophotometric properties** (colors and reflectivities observed as sunlight reflects off the surface) similar to those of CC's. Carbonaceous chondrites might thus be fragments of those types of asteroids or of comets, which most theorists believe are icy bodies with stony inclusions resembling CC material.

Carbonaceous chondrites get their name from their relatively high abundance of carbon. Much of this carbon is in the form of complex organic matter composing about 1% or less of the meteorite, as discovered in 1834 by Berzelius. An exciting area of research has been the relation of these organic materials to the origin of life. By the 1960s various organic compounds had been identified in meteorites, but there was controversy over whether any of them had biological origin. The situation was clarified somewhat when Kvenvolden and others (1970) conclusively identified in the freshly fallen Murchison (Australia) CC amino acids that had not formed on Earth. Amino acids are the class of molecules that join to form proteins and are thus an important paving stone on the road to life. Murchison's amino acids could be identified as extraterrestrial since they had different proportions of "right- and left-handed" molecular structure than do terrestrial amino acids.* Researchers believe that living material did not develop in meteorite parent bodies but that infalling CC's delivered organic material to planets. It might have served as basic material for biological evolution there, even in the unlikely event that such matter did not form on the planets themselves. Experiments (see Wasson, 1974, for review) show that organic compounds in CC-like material can be formed by heating mixtures of carbon monoxide (CO), water (H_2O), and ammonia (NH_3) and catalysts to about 1000 K and then allowing extended cooking at temperatures around 500 K. These ingredients and ovens were undoubtedly common in the great kitchen of the solar nebula.

*The chains of H, H_3N^+, COO^-, and other molecules can join in two different mirror-symmetrical ways to form amino acids. Nearly all living organisms on Earth formed from and replicate the "left-handed" type. This type dominates on Earth but not in space. See Pellegrino and Stoff (1979) for a nontechnical review.

CHONDRITES

Chondrites, the most numerous meteorites, have bulk chemical compositions similar to CC's, and the presence of chondrules shows that they have not been melted. However, they do not have as much carbon, and they have been chemically or geologically processed slightly more than CC's. The chemistry of iron in chondrites is one indicator of some chemical processing. Iron may either be strongly bound to oxygen atoms in rock-forming minerals (**oxidized**) or exist as pure metal (**reduced**). It is found in both states in chondrites. The geochemist G. T. Prior (1920) estimated that all chondrites have about the same iron content, and he developed the so-called **Prior's rules**, which state that as the amount of oxidized iron decreases, the amount of reduced iron correspondingly increases. This relationship indicates some chemical reactions but not enough geologic activity to drain away any iron from the system. As an example, metallic iron could be oxidized (rusted, in fact) by interacting with water or steam in heated, ice-rich planetesimals.

In a now classic review, Urey and Craig (1953) revised Prior's result somewhat by showing that different groups of chondrites have different total iron contents. As shown in Table 6-2, these came to be called the high-iron **H group** (Figure 6-10), the low-iron **L group**, and the very low iron **LL group**. These discontinuities in total iron might be evidence of chondrite origin in several environments, perhaps different parent bodies or dif-

Figure 6-10. The Lost City meteorite, an H-type chondrite. The meteorite was detected by a network of cameras designed to detect meteoritic events and was recovered in Oklahoma 6 days after it fell in 1970. The black "fusion crust" caused by heating during the fall through the atmosphere is typical of freshly fallen meteorites. (NASA)

ferent nebular regions or different eras.

As mentioned in the last chapter, the enstatite chondrites have a composition quite close to that for early-condensed material formed at around 1300 K. Some well-preserved enstatite chondrites may therefore exemplify condensation products in an early, high-temperature nebular regime. But Dodd (1981) notes that others are metamorphosed and chemically related to certain achondrites and irons. He therefore hints that we may someday reconstruct the whole core-to-surface structure of an enstatite parent body!

Another subgroup of chondrites is called the unequilibrated chondrites. As with CC's, much of their iron is oxidized, perhaps by interaction with water or steam, but the iron and other minerals are in mixtures that could not have survived if they had ever been heated to high temperatures. Unequilibrated chondrites may also be primitive objects accumulated at lower temperatures than did enstatite chondrites.

In summary, all chondrites, including CC's, accumulated within parent bodies at different temperatures (different nebular regions?), but none were fully melted by subsequent heating in the parent bodies. Nonetheless, minor heating did affect some of them. All chondrites probably formed near planetesimal surfaces while major heating was confined to the planetesimal centers. Cooling rates can be calculated from properties of individual crystals in the metal particles of chondrites (see Wood, 1968, pp. 29–39, for an introductory discussion). Results (Wasson, 1974, p. 185) show that the chondritic material was typically insulated by overlying material about 2 to 50 km thick. Therefore, the parent bodies must have had diameters at least 4 to 100 km.

Figure 6-11 shows one of many examples of a chondrite that includes an intact, angular fragment of another meteorite (a CC in this case), showing that deposits of chondrules and rock chips were in intimate contact in at least some parent bodies. This supports a model of chondrite origin in the surface layers of fragmental soil in the outer kilometers of planetesimals, similar to the fragmental soil layer on the moon.

Gas-Rich Chondrites

Further evidence for chondrite formation in soil-like layers on the surfaces of parent bodies came in the 1970s, when meteorite researchers realized that many brecciated chondrites consist of grains that were once exposed to cosmic rays and solar wind gases. Microscopic tracks can be seen in these crystals where cosmic atomic particles penetrated into them, and they have accumulated hydrogen and other solar gases. (This phenomenon was clarified by lunar studies that showed that lunar surface soils had the same effect, while lunar buried soils were shielded from such effects.) Detailed studies of both meteorites and lunar soils show episodes of burial, re-excavation and reburial due to impacts. For example, Bogard and Husain (1977) found a gas retention age of 3.7 Gy for the fragments now cemented together in the Plainview chondrite. Since the fragments themselves contain solar gases, they must have lain on some surface before being cemented together. The gases could not predate the impact producing the fragments, because that impact would have driven out the loosely bound solar gases. Thus, the data tell a complete story: the material was broken loose in a collision 3.7 Gy ago, thrown out onto a surface where it was mixed and irradiated by solar wind, and then buried and cemented together into a single rock. In another example, two different pieces of the Djermaia gas-rich chondrite were made out of breccia fragments that showed different sequences of burial and exposure in the upper meter or so of a soil-like layer on some parent body (Lorin and Pellas, 1979). Layers of fragmental debris on the surfaces of cratered planetesimals seem the only plausible setting for such dramas. Important modeling of the evolution of such layers has been published by Housen and others, (1979).

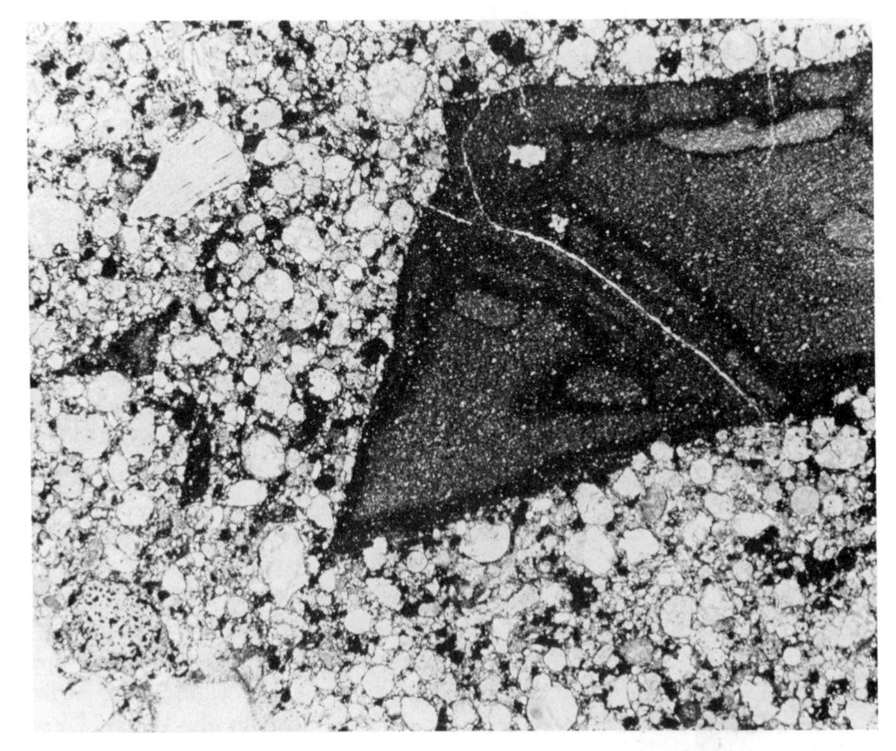

Figure 6-11. An angular CC fragment (dark mass) included inside the chondrite Sharps, showing that mixtures of chondrules and rock chips existed on parent bodies. The white "fish" in the upper corner of the dark mass is a white refractory inclusion, similar to the famous inclusions in Allende (Figure 5-6). Dark rim on the CC fragment is indicative of some heating at the time the entire mass was forming into a chondritic rock. (Courtesy Laurel Wilkening, Lunar and Planetary Laboratory, University of Arizona)

ACHONDRITES

Achondrites are stony meteorites without chondrules. The crystal structure of achondrites is coarser than that of chondrites. Coarse crystal structure is usually an indication of slow cooling in insulated surroundings—for example, it is found in underground intrusions of solidified magma. Of all the meteorites, achondrites are closest to terrestrial rocks. In particular, they are similar to **igneous rocks** (rocks produced from molten material). Some even resemble basaltic lavas and are called **basaltic achondrites**.

These characteristics indicate that the achondrites were produced when some kind of parent material melted and differentiated. The question naturally arises whether chondritelike material might be this parent material. The answer appears to be: probably, at least in part (see discussions of meteorite relationships in Kerridge and Matthews, 1988; McSween, 1987; Wasson, 1985).

Chondrules would indeed be destroyed if chondrites had melted, and chondrites do show a range of chondrule states, from perfect preservation to nearly obliterated, recrystallized examples (see Wood, 1968, p. 43). Also, the iron content has been severely reduced in many achondrites, as we would expect if the heavy iron had drained away from the silicate phase during the melting and differentiation. Finally, study of the cooling and recrystallization process indicates that molten chondritic material could crystallize into mineral groups similar to those observed among the achondrites (Mason, 1962).

The suggestion that achondrites were produced by melting chondrites is qualified, because the parent material of the achondrites was probably not precisely the same as that of the chondrites. Compared with cosmic and chondrite compositions, achondrites show some differences, for example in certain isotope ratios. Recent evidence suggests that the location in the solar nebula may have controlled whether melting occurred and produced achondrites. Furthermore, most achondrites are breccias (Wasson, 1974), so that repeated fragmentation and recementing may be as important as melting in determining their present appearance.

Some types of achondrites stand out as prime examples of lavalike basaltic igneous rock, formed from solidification of molten material. Among these are the **eucrites**, a type of lavalike rock that often displays the bubbly texture of lava flows on Earth. They are believed to be pieces of basaltlike lava flows extruded onto the surface of some parent body in the form of lava billions of years ago. The asteroid Vesta matches the spectrum of eucrites, and for some years there was discussion of how fragments of Vesta reached Earth from the midasteroid belt—a difficult dynamical feat. This problem was lessened when Cruikshank and others (1991) recognized three small Earth-approaching asteroids with spectra indicating basaltic-eucritic composition. Fragments of these may hit Earth, providing a eucrite source close at hand. These three asteroids may be fragments of a Vesta-like parent that was fragmented by a major collision.

STONY-IRONS

These are divided into two main classes, depending on the minerals in the silicate stony part. The mesosiderite silicates are mainly composed of plagioclase and pyroxene minerals and resemble surface lavas or crustal silicate rocks somehow brought into contact with metal. The pallasite silicates are principally olivine and hence resemble mantle material. The pallasites may thus have formed where mantle materials abutted the surface of an iron core or the surfaces of iron blobs near the core. How the more crustlike rocks of mesosiderites came into contact with iron is more problematic. Suggestions include sinking of solid basaltic slabs into a melted interior (C. Chapman and R. Greenberg, private communication, 1982), and intimate mixing during collisions. Figure 6-12 illustrates the intimate contact of the different materials in stony-irons.

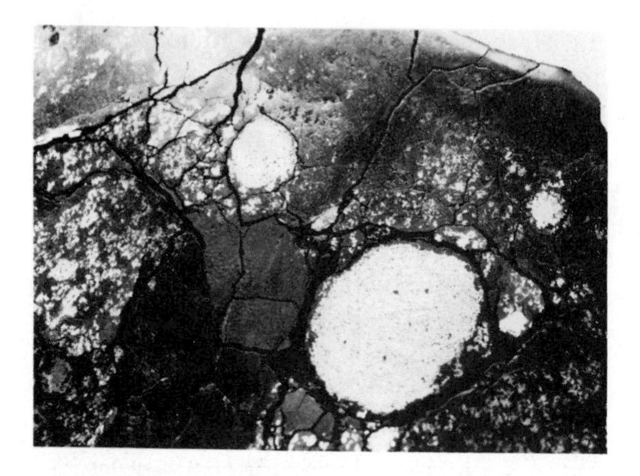

Figure 6-12. Cut and polished section of a stony-iron meteorite from the Bondoc Peninsula, Philippines. The dark, fractured matrix is silicate rock, possibly fractured in collisions. The bright nodules are nickel-iron inclusions a few centimeters across. (Photo by author)

Figure 6-13. A cut and polished section of the Campo del Cielo (Argentina) iron meteorite. Intersecting metal crystals can be studied to reveal environmental conditions inside the parent body. The dark inclusions are stony silicate bodies, the largest being about 4.5 cm across. This pattern of metal crystals is called the Widmanstätten pattern, after its discoverer. (Photo by author)

IRONS

Iron meteorites are the most useful in telling us about the size, number, and history of the meteorite parent bodies. Irons are classed according to nickel content, crystal structure, and related parameters.

Widmanstätten Pattern: Size of Parent Bodies

In many irons the two major minerals, kamacite and taenite, occur in crystals that interlock in an interesting geometry called the **Widmanstätten pattern*** shown in Figure 6-13. The pattern is prominent only after the meteorite is cut, polished smooth, and then etched with acid. Large crystals mean long cooling times, suggesting that these materials were buried deep inside parent bodies, where they would have been insulated against rapid cooling. Measures of the structures of the crystals suggest locations near the centers of bodies about 30 to 400 km across (Wood, 1968; Wasson, 1974; Scott, 1977).

Irons were first classified according to the Widmanstätten pattern characteristics, but Wasson (1974) and other modern workers have reclassified them according to chemical properties. When abundances of certain specific elements such as gallium, germanium, and nickel are considered, certain groups of meteorites are found to have nearly the same amounts of these elements, giving rise first to the concept of four major **gallium-germanium groups** and later to Wasson's groups IA, IB, IIA, and so on. Each of these groups may have come from a separate parent body, suggesting perhaps one-half dozen or dozen parent bodies supplying most of our iron meteorites.

DIAMONDS AND NEUMANN BANDS: EVIDENCE FOR COLLISIONS

Diamonds, the high-pressure form of carbon, are found inside both irons and stones, making early researchers think that lunar-sized parent bodies were needed to generate adequate pressure. More recent work attributes diamond formation to high

*Named after Count Alois von Widmanstätten, director of the Imperial Porcelain works in Vienna and discoverer of the pattern in a meteorite in 1808.

shock pressures produced momentarily during parent body collisions or meteorite impacts (for example, Wood, 1968, p. 53).

Neumann bands are fine striations visible in a polished and etched surface of most hexahedrites and some other irons. They are caused by an alteration of the crystal structure within the kamacite crystals, known as *twinning* (not to be confused with the interlocking Widmanstätten arrangement of the crystals themselves).

Uhlig (1955) showed that Neumann bands are the result of strong shocks applied to the meteorites when they were at low temperatures, not above 900 K and probably below 600 K. Again, the interpretation of this finding is that after the interiors of parent bodies had solidified through cooling, they were subjected to violent impacts, probably collisions with other parent bodies.

BRECCIATED METEORITES: EVIDENCE FOR COLLISIONAL MIXING

We already mentioned that many stony meteorites are breccias, or rocks consisting of cemented fragments, but this fact needs more emphasis. Three categories are recognized: **monomict breccias**, in which all fragments come from the same meteorite type; **polymict breccias**, in which fragments come from very different environments (for example, the chondrite and carbonaceous chondrite in Figure 6-11); and **genomict breccias**, a category coined by Wasson (1974), in which the fragments have similar chemical compositions but show evidence of different geologic processing. These breccias give strong evidence that parent bodies went through complicated collisions and fragmentations that mixed pieces of one or more parent bodies. Monomict breccias might have arisen when rocks were locally shattered (during impact crater formation?) and remixed; polymict breccias, when two different parent bodies contributed their fragments (Wahl, 1952); and genomict breccias, when different regions of the same parent body were mixed by an explosive collision followed by recollapse of the fragments into a new body (Davis and others, 1979; Hartmann, 1979).

METEORITE AGES

The beginning of the last chapter described how formation ages, formation intervals, and solidification ages can be measured from planetary rock materials. The most important concept about radiometric ages is that several different kinds of time intervals can be measured, and the three kinds mentioned above by no means exhaust the list. These different measurements allow reconstruction of many events in meteorites' histories.

If you are told the "age of a rock," you must inquire what kind of age is meant: the time since the rock crystallized, the time since the sample was broken off the parent body, or the time since the sample was last altered by shock or chemical reactions, and so on. Table 6-3 lists the various age relations that can be determined. The story revealed by these ages can be summarized as follows.

Formation Age, Formation Interval, and Solidification Age

As described in the last chapter, studies of **formation ages** indicate that parent bodies, each with its own fixed isotopic composition, formed during the interval 4.55 to 4.65 Gy ago. **Formation intervals** indicate that about 80% of the chondrite parent bodies formed within about 20 My. Most **solidification ages** also fall in this narrow "Genesis interval," indicating that the parent bodies cooled fairly rapidly after they formed and that most of the material underwent no further remelting.

Gas Retention Age

The **gas retention age** is the interval from the most recent degassing of the rocks to the present. The method utilizes gaseous daughter elements, such as argon, which accumulate in the crystal-lattice interstices. An event such as a major collision among parent bodies can cause tiny fractures and heating, which allow the gas to escape. New gas then starts to collect, and measuring this gas dates the time since the shock event. If the rock was never shocked strongly enough to cause heating, the gas retention age may be the same as the solidification age.

Gas retention ages are concentrated at 4.0 to 4.5 Gy ago, but perhaps a third are dotted through more recent time. These almost certainly reflect a long history of collisions among parent bodies, with most major collisions occurring early (in the same 4.0–4.5-Gy period as the intense early cratering of the planets), but others occurring more recently.

Cosmic Ray Exposure Age

Sometimes called a CRE age, the **cosmic ray exposure age** measures the duration of the meteorite's exposure to space. Generally this is the interval from the meteorite's first exposure to space (or when it was first within a meter or less of the surface of a larger body) until its impact on Earth. Some meteorites had a multiphase exposure history, first being just under the surface of the parent body and then being exposed directly to space when the parent body fragmented.

Exposure ages may be measured either by utilizing radioactive and stable isotopes created by nuclear reactions with cosmic rays, which penetrate roughly 1 m into the parent body, or by counting microscopic particle tracks left in the material by the passage of cosmic rays through it. The former method is widely applied to meteorites. Both methods have also been used on lunar samples.

Stones have much lower CRE ages than irons. If they come from the same general source, why have they been exposed for different periods? Probably stones and irons erode at different rates due to their mutual collisions and collisions with

Table 6-3
Types of Meteorite Ages

Type of age and interval measured	Examples of isotopes involved	Meteorite type	Results
Formation age (establishing of isotopic identity)	^{87}Sr–^{86}Sr	Stones	4.55–4.65 Gy
Formation interval (nucleosynthesis to gas retention)	^{129}I–^{129}Xe	Stones and irons	c. 20 My
Solidification age (solidification to present)	^{87}Rb–^{87}Sr ^{235}U–^{207}Pb ^{238}U–^{206}Pb ^{232}Th–^{208}Pb ^{40}Ar–^{39}Ar	Stones and irons	Mostly 4.4–1.6 Gy. A few lavalike meteorites solidified later. Nakhlites (a type of basaltic achondrite) solidified about 1.3 Gy ago.
Gas retention age (onset of retention to present)	^{40}K–^{40}Ar	Stones (irons contain little K)	Often 4.4–4.6 Gy. Some younger, suggested to be due to degassing during shock events.
Cosmic ray exposure age (onset of exposure to present)	^{3}He[a]	Stones	0.1–100 My (young due to erosive impacts in space?)
		Irons	1–1000 My
Date of fall (end of cosmic ray exposure to present)	^{14}C	Stones	c. 0–10 000 y
	^{26}Al	Irons	c. 0–50 000 y

[a]Produced by cosmic rays.

micrometeoroids. If stones fragment more often than do the stronger irons, their outer layers would be exposed more recently, on the average (Dohnanyi, 1970). Since cosmic rays can penetrate only in these outer layers (especially in the outer 20 cm), stones would usually appear to be younger than irons. The history of a meteorite in space might be likened to that of a rock struck by intermittent sledgehammer blows during continuous sandblasting.

Clustering of Ages: Evidence for Specific Collisions

Certain groups of chemically related meteorites have about the same gas retention or CRE age, implying that they all broke loose from some parent body at once. For example, studies of H chondrites reveal strong concentrations of CRE ages at 4.5 and 20 My ago and show that each age group contains H chondrites belonging to various petrologic subtypes (Crabb and Schultz, 1981). These results indicate that two major collisions released most of the fragments now falling as H chondrites and that the parent bodies on which these collisions occurred already had mixtures of fragments. Similarly, a certain group of L chondrites, recognizable by blackening due to mineral changes induced by a strong mechanical shock (impact?) have the same gas retention age of approximately 400 My (Heymann, 1967; Goplan and Wetherill, 1971). These and similar results, tabulated in Table 6-4, show that major collisions somewhere in the solar system created many of our meteorite types in the last few percent of solar system history, even though the first-generation meteorite parent bodies date back to the formation of the solar system.

Table 6-4
Dates of Meteoroid Collisions

Meteorite type	Collision dates (years before present)[a]
H chondrites	4.5×10^6
	2×10^7
Aubrite achondrites	4×10^7
L chondrites	4×10^8
Hexahedrite irons	2×10^8?
Medium octahedrite irons	6×10^8

Note: See Wood (1968), and Wasson (1974) for discussions of the data. The collisions and suspected collisions greater than 10^8 y ago were probably major collisions that fragmented large objects, possibly original parent bodies. More recent collisions may have fragmented only pieces of parent bodies. Taylor and Heymann (1969) have shown that many meteorites give evidence of other major collisions scattered in time between 0.6 and 4.5 Gy ago.
[a]Dates determined by clustered ages. All ages determined from cosmic ray exposure except for L chondrites, the age of which was determined from gas retention.

Date of Fall

This measures the interval from when the meteorite ceased being shielded from cosmic rays (its arrival on Earth) to the present. Stones are not identifiable much longer than about 10 000 y on most parts of Earth's surface because they are subject to weathering and erosion. Four Antarctic stony meteorites have fall ages of 0.03 to 1.5 My, testifying to the long-term stability of the Antarctic ice sheet (Fireman, Rancitelli, and Kirsten, 1979).

TEKTITES

Tektites are small, unique, and rare glass objects rich in silica. The largest have a mass of only 200 to 300 g (about 0.5 lb). Their origin is very mysterious. CRE ages show that tektites differ from meteorites, most importantly in that tektites have not been in space nearly as long. Most iron meteorites have been exposed in space for some 10^8 y, and most stony meteorites for some 10^7 y. But tektite exposure times have been less than 300 y—perhaps only minutes—which is much too short for asteroidal or cometary origin. This restricts their place of origin to the Earth-moon neighborhood.

The high silica content of tektites (about 73% SiO_2) suggests that they come from Earth, where silica-rich sediments abound, and not from the moon, whose igneous rocks are not silica rich. On the other hand, Chapman and Larson (1963) showed in wind tunnel experiments that the tektites had either entered or reentered the atmosphere at hypersonic speeds, causing melting and reshaping. Chapman and Larson's finding was a final proof that tektites were "stones that fell from the sky." This does not prove extraterrestrial origin, however. The most accepted current interpretation is that tektites are bits of fused silicate blasted out of terrestrial meteorite impact craters. They apparently flew outward in huge expanding clouds of debris, solidified, and fell back through the atmosphere (Faul, 1966).

Two lines of evidence in particular support this interpretation. Tektites are found only in restricted areas, each tektite field possibly related

MATHEMATICAL NOTES ON THE RUBIDIUM-STRONTIUM SYSTEM OF ROCK AGE MEASUREMENT

Suppose a rock crystallizes from a uniform molten mass, as shown in the top of Figure 6-14. As it solidifies, different mineral crystals will form, with different affinities for the two chemicals rubidium and strontium. For simplicity, imagine the rock composed of three minerals, A, B, and C, each having increasing initial amounts of rubidium. Initially there is relatively little strontium 87, since that is the daughter element produced by radioactive decay of rubidium 87, with a half-life of 72 Gy. A chemist is now about to measure the strontium 87 and rubidium 87 in each mineral crystal, A, B, and C, but he will express his results normalized to the strontium 86 isotope. That is, rubidium 87 will be expressed not in terms of numbers of atoms, but as the ratio rubidium 87:strontium 86 and strontium 87 as the ratio strontium 87:strontium 86. One convenience of this is that the initial strontium 87:strontium 86 ratio will be the same in all crystals; this is because the crystals grow according to only the chemical nature of the strontium (the outer electron structure of the atom being all that the crystal "sees"), rather than the isotopic nature (determined by the nucleus of the atom). Thus each crystal incorporates the same initial proportions of strontium 87 and strontium 86. This initial abundance is marked in Figure 6-14, along with the initial compositions of the three crystals.

As soon as the rock forms, the rubidium 87 decays, of course. For each atom of rubidium 87 lost, the amount of strontium 87 increases by one atom, which is trapped in the rock. Therefore the compositions of minerals A, B, and C will each evolve to the upper left on the 45°-angle dashed lines, producing points A′, B′, and C′ by the time a chemist makes his measurement a billion years or so after the rock forms. The composition measured at other times would fall along other lines. The line is called an *isochron* (*iso* = same, *chron* = time), since all its points represent the rock at a specific age. The slope of the isochron is a measure of the age.

Note that all the isochrons marking different ages would go through the same point on the ordinate. This point, marked BABI in Figure 6-14, denotes the initial isotopic composition ratio of the strontium, strontium 87:strontium 86, at the beginning of the solar system's history. The peculiar acronym comes from "basaltic achondrite, best initial," meaning the best estimate of the initial strontium ratio, based on studies of a basaltic achondrite, particularly studies by Gerald Wasserburg and his colleagues at Cal Tech.

As the rubidium-strontium dating system was first worked out, planetary chemists assumed that the BABI value was a universal standard for all material in the solar system. However, lower values were found in the Angra dos Reis anomalous

in age and location to an ancient impact crater. The best example is the Czechoslovakian tektite group called moldavites (after the Moldau River). They are 14.8 My old, exactly matching the age of the nearby 24-km-diameter Rieskessel crater in Germany (see Table 10-4). Chemical similarities between tektites and terrestrial soils also support the theory of origin in terrestrial impacts (Taylor and Kaye, 1969); and Faul gave his 1966 paper on the subject the seemingly definitive title, "Tektites Are Terrestrial."

An opposing theory, that tektites were blasted off the moon, possibly in connection with lunar volcanism, was abandoned by most workers when Apollo rock samples revealed no tektites or silica-rich lunar rocks. This theory has been given vigorous, if relatively solitary, support by O'Keefe (1963, 1978).

achondrite meteorite (the so-called ADOR value), an inclusion in the Allende carbonaceous chondrite meteorite. Later studies suggest that there was an evolution of strontium isotope compositions, perhaps different in different parts of the early solar system, culminating in a metamorphosing event that mixed and homogenized the strontium isotopes about 74 My after the time when the BABI value applied (see summary by Wasson, 1974). The clustering of rubidium-strontium formation ages, together with the quick evolution of the original strontium value, gives striking evidence of the rapidity of the planet-forming events that happened in the first 2% of the solar system's history.

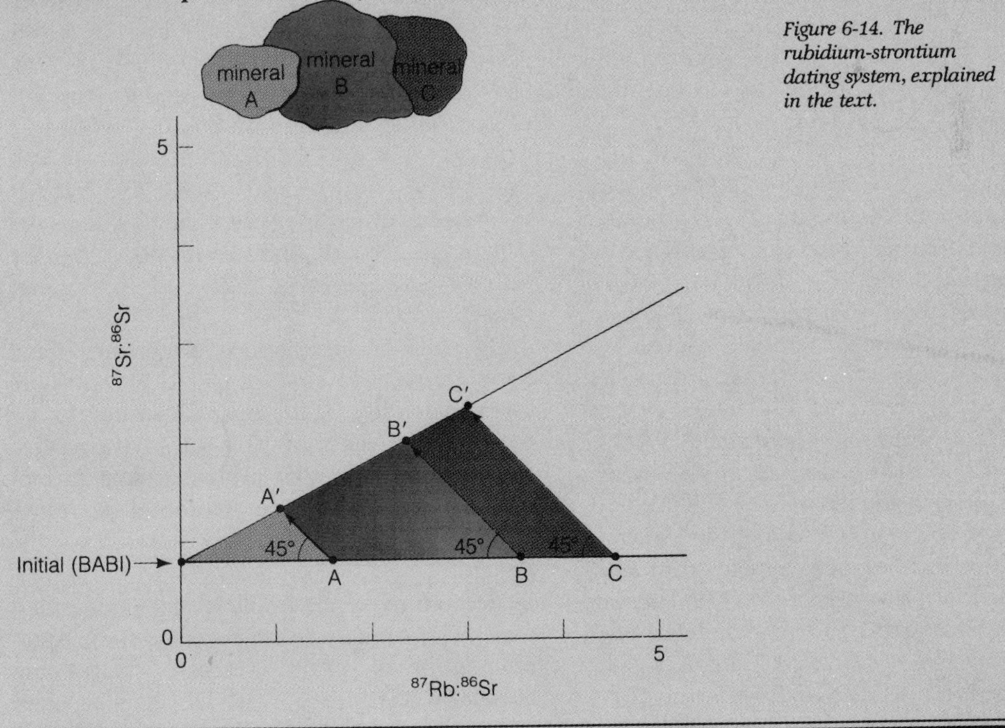

Figure 6-14. The rubidium-strontium dating system, explained in the text.

WHERE DO METEORITES COME FROM?—ORBITS AND ORIGINS

It is a provocative thought that the meteorites in our museums once existed in the dark interiors or on the alien surfaces of bodies hundreds of kilometers wide somewhere in the solar system. These bodies had their own landscapes, lava flows, and geologic processes. Some of these bodies may have been blasted to smithereens by collisions long ago, but others may still survive. Certainly the continuing supply of meteorites and dust particles indicates that new objects are being blasted out of craters on substantial-sized parent bodies somewhere in the solar system today. Obviously, meteorites approach Earth on orbits that intercept Earth's orbit, and these orbits have half-lives against Earth-collision of only 0.01 to 0.1 Gy. Therefore, virtually all meteorites would have been swept up long ago unless they were constantly replenished by a reservoir somewhere in the solar system. Where is this reservoir?

An interesting clue is that the CRE ages are about equal to the orbital lifetimes, meaning that meteorites were placed in Earth-intercepting orbits very soon after being broken into their present sizes.

Another clue is that although many stony meteorites come from a soil layer like the moon's, their CRE ages and the amount of inert gases that their surfaces have absorbed from the solar wind are only around 0.01% to 1% of the lunar values. The relations indicate that the stones come from a region where the impact erosion rate must be around 1000 times that of the moon, but a region that is in our general part of the solar system, about 1 to 8 AU from the sun (Anders, 1978).

Orbits of only a few meteorites are well known, but they all lie in the inner solar system, with aphelia in or near the asteroid belt, as shown in Figure 6-15. The best-determined orbits (Pribram chondrite, Czechoslovakia, 1959; Lost City chondrite, Oklahoma, 1970) were determined from photographs of the meteorite infall (see also the meteorite orbital data in Table 8-5).

All these clues point strongly toward the asteroid belt as the reservoir producing most meteorites. The larger asteroids have the sizes inferred for meteorite parent bodies, and calculations show that they undergo numerous collisions. Ejected fragments would not fly off fast enough to intercept Earth directly but could undergo a two-stage process: (1) The collision would eject fragments onto orbits that could be perturbed by Jupiter resonances (or in some cases close encounters with Mars?); and (2) these perturbations would then send the fragments onto orbits that intercept Earth (Williams, 1973; review by Chapman, Williams, and Hartmann, 1978). The second-stage orbits would resemble those shown in Figure 6-15.

Interplanetary objects having such orbits have been discovered telescopically. They are the **Apollo asteroids**, which are Earth-approaching asteroids a few kilometers across.

A seeming clincher to the argument is that belt asteroids and Apollos have spectra indicating various composition classes close to carbonaceous, stony, stony-iron, and iron meteorite classes. Indeed, the second-largest belt asteroid, 550-km Vesta, matches eucrites and perhaps one or two other subtypes of basaltic achondrite (Consolmagno and Drake, 1977; Feierberg and Drake, 1980), while Apollo asteroid Toro matches the most common meteorite type, L chondrites (Chapman, McCord, and Pieters, 1973). We may actually have pieces of Vesta and Toro in our museums.

In spite of these strong arguments, there remain problems. First, comets may contribute some meteorites. Stony material inside the icy matrix of comets may be weak carbonaceous-chondrite-like material. Thus, a large flux of cometary CC's may enter the top of the atmosphere but disintegrate before reaching the ground.

Second, while the Apollo objects are labeled "asteroids," some researchers think some Apollos may be the stony residues of "burnt-out" comets (Wetherill, 1975; also see Chapters 7 and

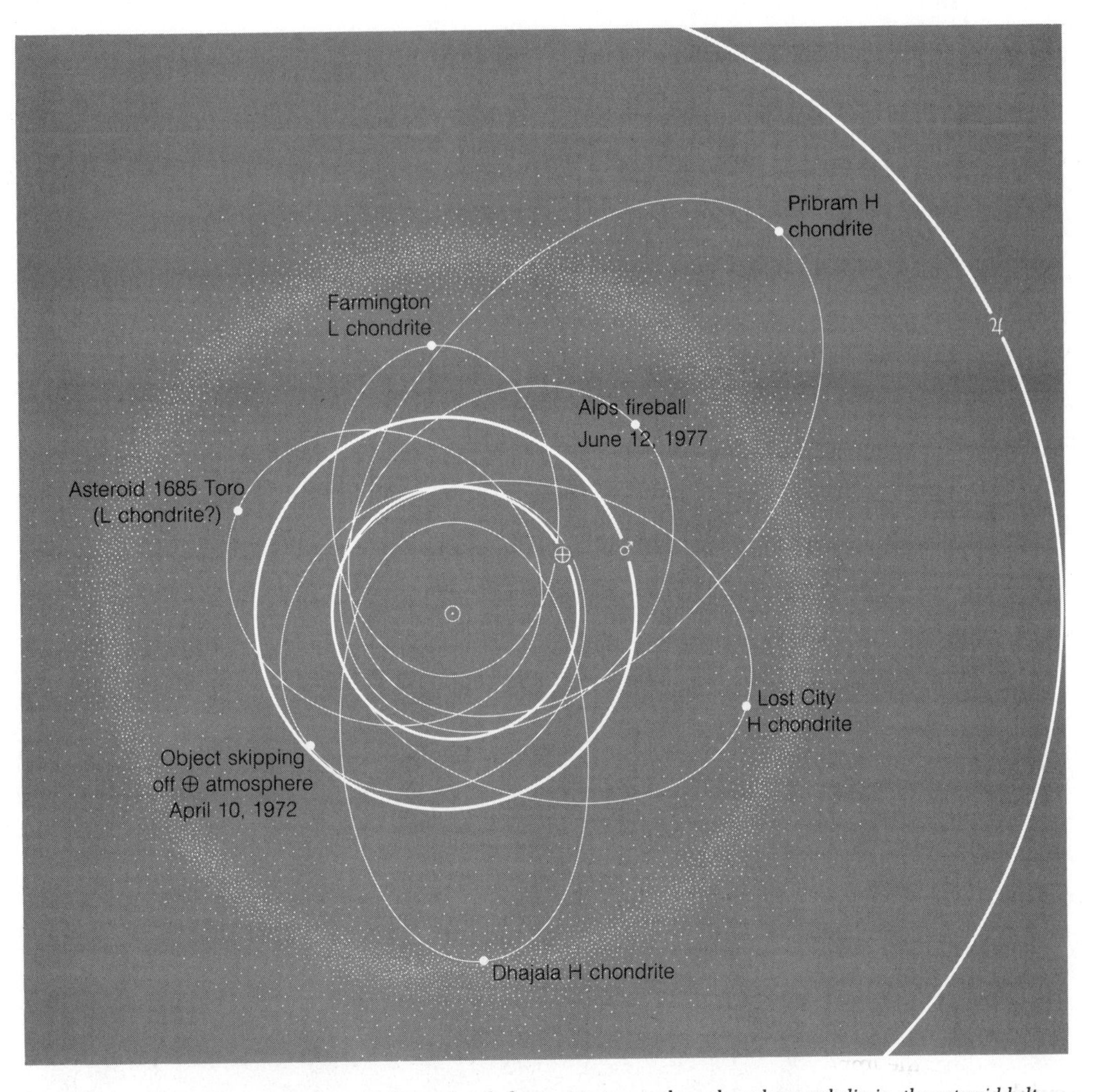

Figure 6-15. Orbits of four meteorites and three related objects. Most cross through, or have aphelia in, the asteroid belt, which is shown stippled. Drawing shows orbit sizes, perihelia, and aphelia to scale, but orientations of semimajor axes are chosen for clarity. (Data from Wasson, 1974; Ballabh and others, 1978; Jacchia, 1974; Clepecha, 1981)

8 for more detail). This idea is not inconsistent with the earlier discussion but confuses the semantics of "asteroidal versus cometary" origins.

A third problem is that since 85% of the meteorites are chondrites, why don't we see more asteroids that are spectral matches for chondrites? A significant fraction (one half) of belt asteroids have spectra like CC's; so, if the belt is the source, why aren't most meteorites CC's instead of chondrites? One answer is that they may be CC's, but because CC's are fragile, they disintegrate. Another answer is that most CC-like asteroids are in the outer part where they cannot easily throw fragments onto Earth-crossing orbits;

virtually no CC's are on the inner edge of the belt. About two-thirds of asteroids on the inner edge of the belt are of a spectral type resembling stony meteorites (labeled "type S"—see Chapter 7), and some investigators suspect that these contain chondrite-like meteorite materials, although the spectral matches are not so good as with other meteorite/asteroid matches (Feierberg, Larson, and Chapman, 1981; Dodd, 1981). Perhaps chondrites are fragments of S asteroids easily perturbed toward Earth from the inner asteroid belt by the action of nearby Mars. A few Apollo asteroids appear chondritic, and they may be sources of some chondrites; but we don't know if they came our way from the inner asteroid belt or as "burnt-out" comets. These questions need more research before we understand the full meaning of meteorites.

An additional complexity is the possibility that the current distribution of meteorite types falling on Earth is not representative of the average over geologic time. This could happen because the meteorite types falling in the last million years are strongly dependent on the last few major asteroid collisions, as shown in Figure 6-16. This means that our meteorite sample depends on the "luck of the draw" among recent large colliding asteroid pairs that may have sent fragments our

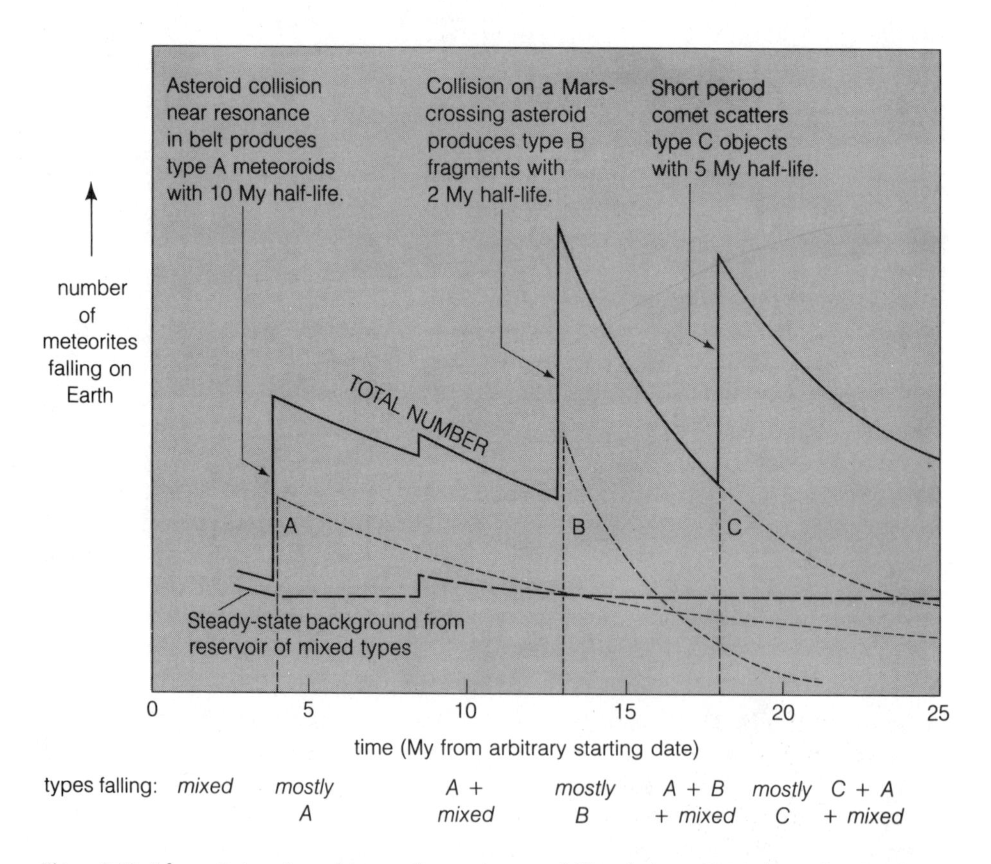

Figure 6-16. Schematic imaginary history of meteorite types falling during a 25 My interval on Earth. Various events (top; subject to random chance) provide bursts of fragments of different compositional types. These have different half-lives against sweepup by Earth, depending on their orbits. The net result is episodic bombardment by different groups.

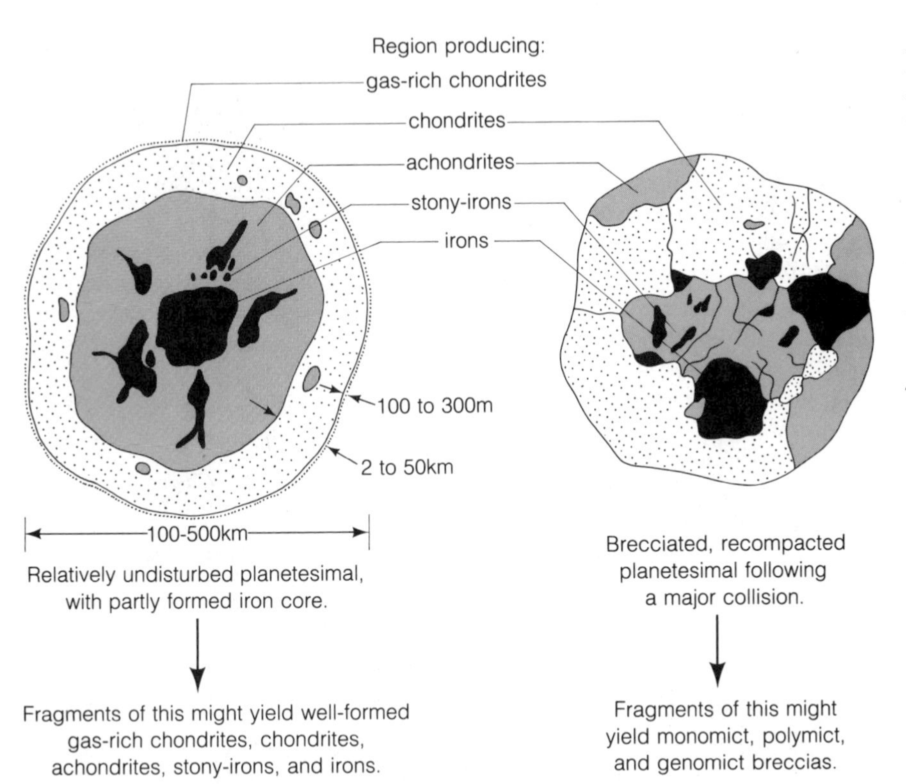

Figure 6-17. Left diagram shows layered structure and dimensions of a hypothetical parent body moderately affected by internal heating and melting. Other parent body (right) shattered and re-aggregated into breccias during large-scale collisions. Still later fragmentation of these two bodies would yield many types of observed meteorites.

way. The meteorite record may bear some evidence of type variations due to this effect.

METEORITES FROM THE MOON AND MARS

An irony followed the massive American Apollo program and Soviet Luna program to bring samples back from the moon: Lunar samples were found here on Earth among meteorites collected in Antarctica! Antarctic ice fields have produced major meteorite collections because glaciers drop them as they melt, and they stand out on the white ice. From 1982 to 1991, 12 lunar rocks were found there and elsewhere. These samples don't negate the value of Apollo and Luna, however, because without lunar flights we would not have known how to recognize lunar rocks, and in any case the Antarctic samples' scientific value is lessened because we don't know the location or geologic context of their sources on the moon.

Their importance is that they give us "free" extra samples, and prove that impacts can blast intact rock samples off a body as big as the moon (escape velocity 2.4 km/s).

This importance was heightened because it led to recognition of even more exotic celestial rocks: rocks that come from Mars! They are rare subtypes called shergottie, nakhlite, and chassignite—collectively called **SNC meteorites,** or SNC's (affectionately dubbed "snicks"). They are all basaltic meteorites only 1.3 Gy old—much younger than other meteorites. Only a large, geologically evolved body could produce such young basalts. Oxygen isotope ratios proved they weren't from Earth or the moon. Where were they from?

As early as 1979, some researchers suggested Mars as the source, but others scoffed that intact rocks couldn't be blasted off Mars, with its 5 km/s escape velocity. (This was based on an erroneous "golf ball" conception of the process, as if the rock were blasted off with an instantaneous impulse; in a large cratering event it is cushioned by the massive ejecta cloud.) The lunar meteorites reversed this argument. Soon the SNC's were found to contain nitrogen and noble gases matching the ratios and isotopic composition of gases found on Mars by the Viking lander (Bogard and Johnson, 1983). Also, the 1.3 Gy age matched estimated ages of widespread basaltic lava flows on Mars, based on crater counts.

Thus, most researchers now believe we have at least a few Martian rocks from unknown Martian sites (see Pepin and Carr, 1991, for review of evidence). The next step is to visit Mars and compare the chemistry and geology of different "known" sites!

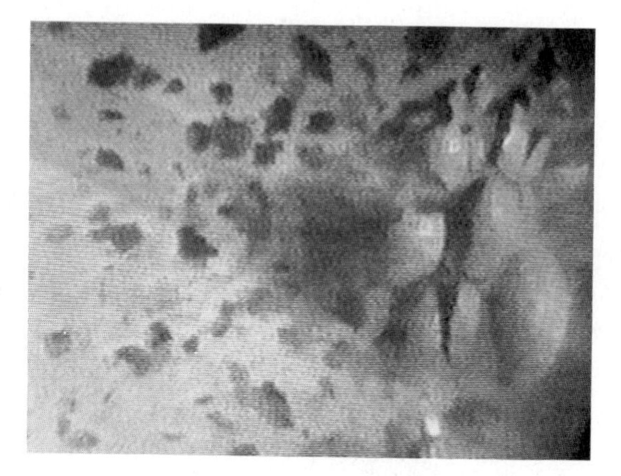

Figure 6-18. Collision of asteroids to produce meteorites is simulated in the laboratory. Here a basalt sphere is shattered by the impact of a smaller projectile from the left at 3.2 km/sec, similar to actual collision speeds in the asteroid belt. Studies of fragments clarify size distribution and other properties of meteorites and asteroids. (Courtesy Akiko Nakamura, Kyoto University, Japan)

SUMMARY

Accretion and chemical fractionation processes in the solar nebula 4.6 Gy ago produced grainlike planetesimals with compositions different from the nebular composition and varying with distance from the sun. Such planetesimals may have ranged in composition from the nearly primordial Type I carbonaceous chondrites (in the colder, outer solar system) to slightly differentiated chondrites. This material was incorporated into parent bodies, or planetesimals of asteroidal dimension, up to at least 200 km in diameter and probably larger (Dodd, 1981; Wood, 1990). Because of orbital data on meteorites and chemical data on their parent bodies, most investigators identify the asteroids as the source of most, if not all, meteorites. The most likely exceptions are the carbonaceous chondrites, which may be associated with comets.

Some parent bodies were heated around 4.5 Gy ago, perhaps by short-lived radioactive isotopes or other mechanisms. Probably the outer layers were least metamorphosed, while the interiors of some bodies got hot enough to destroy chondritic textures, produce achondrites, or melt and produce irons, as shown in Figure 6-17. Other parent bodies (smaller ones?) may have remained cold, preserving chondrite textures and volatiles as in the case of carbonaceous chondrites.

About 4.5 to 4.0 Gy ago and later, collisions among parent bodies affected their structures (Chapman and Davis, 1975). Fragmental layers built up on the surfaces of parent bodies, which later compacted into brecciated chondrites and achondrites. Large-scale collisions (Figure 6-18) may have fragmented entire parent bodies, sometimes scattering the fragments and sometimes allowing them to resettle into confused, brecciated masses. Scattered fragments from both colliding objects sometimes reaccumulated into polymict breccias. Many fragments hit planets and moons, producing craters.

As Figure 6-17 shows, the collisions produced small fragments of widely varying character, many of which became our meteorites. Many meteorites come from objects classifiable now as Apollo asteroids, and some Apollos probably originated in parent bodies in the asteroid belt. Other Apollos may be extinct comets. Specific sources of carbonaceous and ordinary chondrites are still somewhat puzzling.

Figure 6-19. Possible appearance of an asteroid collision spraying out meteoritic fragments. Larger asteroid is light-colored rocky type from inner half of asteroid belt; smaller is black carbonaceous type from outer half of the belt. Fragments are ejected into new orbits. Because of resonances or close passage to Mars, some may eventually be perturbed onto orbits that lead to collisions with Earth, and they may reach the ground as polymict breccia meteorites. (Painting by author)

A collision that ejected many of the L chondrites appears well dated at about 0.4 Gy ago, and other meteorites originated in collisions within the last 0.1 Gy.

Meteorites help confirm the planetesimal theory of planet formation, indicate conditions inside certain ancient planetary bodies, explain craters, and indicate kinds of material available in space for future space explorers.

CONCEPTS

meteoritic complex
nodes
meteoritics
meteoroid
meteor
meteorite
fireball
fall
find
parent body
train
fusion crust
meteoroid flux
micrometeorite
metamorphosed material
differentiation
stony meteorite
iron meteorite
stony-iron meteorite
chondrite
chondrule
achondrite
carbonaceous chondrite (CC)
volatile elements and compounds
siderophile element
lithophile element
spectrophotometric properties
oxidized
reduced
Prior's rules
H group
L group
LL group
igneous rock
basaltic achondrites
eucrites
Widmanstätten pattern
gallium-germanium groups
Neumann bands
monomict breccias
polymict breccias
genomict breccias
formation age
formation interval
solidification age
gas retention age
cosmic ray exposure (CRE) age
tektite
Apollo asteroid
SNC meteorites

PROBLEMS

1. Describe some different groups of small bodies in the solar system. Comment on any relationships, such as meteorite relationships with Apollo asteroids. (These relationships will be dealt with further in later chapters.)

2. (a) Assuming that manned space vehicles continue to evolve, what materials do meteorites suggest might be available in interplanetary space that would be useful in space exploration? (b) Depending on the destination, why might it cost less energy (hence money) to haul raw materials from a near-Earth meteoroid (approaching Earth at, say, 5 km/s) than to haul the materials from Earth's surface?

3. (a) How do we know that the meteoroid flux was hundreds or thousands of times greater 4 Gy ago than it is today? (b) Relate this finding to theories of planet formation.

4. If you saw a meteorite fall, describe some phenomena you might expect to experience and state what you might do with the recovered samples.

5. How many meteorites heavier than 100 kg fall into the atmosphere in a year over an area of a million square kilometers (about the area of Texas plus New Mexico)? Would the full 100 kg reach the ground?

6. Give some lines of evidence that most multikilometer craters on planetary bodies are formed by meteorite impacts rather than, for example, volcanic action.

7. (a) Here is a theory of meteorite origin: Two large parent bodies, each 500 km across, formed. Their interiors melted, and each formed an iron core surrounded by stony-iron material overlain by an achondritic layer and surface layers of chondritic materials. The parent bodies then collided, and their fragments were shot onto Earth-crossing orbits, becoming meteorites and Apollo asteroids. Indicate several lines of evidence showing that this theory is too simple to explain meteorites. (b) How would a similar theory with at least 10 such parent bodies undergoing multiple collisions and partial reaccumulations be more realistic?

8. Describe evidence that meteorites formed inside parent bodies that were about 30 to 400 kilometers across.

9. (a) Must a meteorite have melted 1.5 Gy ago in order to have a gas retention age of 1.5 Gy? (b) What are such young gas retention ages interpreted to mean? (c) Since a number of meteorites have gas retention ages less than 2 Gy, why are the nakhlites, with solidification ages of 1.2 Gy, considered so unusual?

10. Why are CRE ages not equal to gas retention ages of meteorites?

11. List some lines of evidence that meteorites and their parent bodies underwent major high-energy collisions.

12. Why are the Antarctic meteorites especially valuable to planetary scientists?

13. How do we know that tektites did not come from Mars, Venus, or some more distant location that could explain their special chemical properties?

ADVANCED PROBLEMS

14. Convert the velocity equation from Chapter 3 into the form

$$v^2 = k\left(\frac{2}{r} - \frac{1}{a}\right)$$

where r and a are given in astronomical units and v is given in kilometer per second. (Identify k.)

15. (a) Using the result in problem 14 (or the original general form in Chapter 3), calculate the orbital velocity of Earth and the orbital velocity at 1 AU of a prograde meteoroid with its aphelion in the middle of the asteroid belt at 2.8 AU and perihelion at 0.9 AU. (b) Assuming that the meteoroid orbit lies in the ecliptic plane, at what speed does it approach Earth and from what direction? (c) Compare this with quoted meteorite approach velocities. (d) Assuming that the same meteoroid has an inclination to the ecliptic of 30°, at what relative speed would it approach Earth? (*Hint:* Think of the problem as the subtraction of two vectors.) (e) Modest inclinations are typical of meteoroids. Is the result in (d) more typical of meteorite velocities than the result in (b)?

16. (a) How old, in half-lives, is a meteorite whose rubidium-strontium isochron lies at 45° to both axes in Figure 6-14? (b) How old in years? (c) Would this result be likely in a real meteorite?

CHAPTER SEVEN

ASTEROIDS

The popularization of Bode's rule about 1772, followed by its confirmation through the discovery of Uranus in 1781 (see Chapter 2), led to widespread belief that there ought to be a planet at 2.8 AU. In 1800 six German astronomers, under the leadership of Johann Schröter, determined to search out the missing planet. Before these "celestial police" (as they were nicknamed) could succeed, the announcement came that the Italian astronomer Giuseppe Piazzi had discovered an unknown body while making routine stellar observations at the observatory at Palermo, Sicily, on New Year's Day, 1801. The new object moved with respect to the stars and Piazzi called it a new planet, Ceres. By the fall of that year, the famous astronomer Gauss had derived the first general method for determining orbits from observations of celestial bodies, and he computed Ceres' orbit—at 2.8 AU, just as predicted by Bode's rule.

Hypothetical view from the surface of an asteroid passing near the Earth-moon system (upper right) symbolizes the relation between asteroids and other interplanetary material. The sun has set beyond the crater rim, where the glow of the corona (solar atmosphere) and inner zodiacal light (reflection from meteoritic dust) are visible. Above the sunset is a comet—a type of body to which some asteroids may be related. At right is a satellite of the asteroid on which we are standing—a feature recently suggested by some observers and theorists. (Painting by author)

Since Ceres seemed too small to be the sought-for planet, the German astronomers continued their survey program. In March 1802 one of them, Olbers, discovered a second asteroid, which he named Pallas. Pallas was even smaller and fainter than Ceres but was located at the same distance. The addition of a second small body raised the possibility that one normal planet had once occupied the predicted position but had somehow fragmented, producing many small bodies. On the basis of this hypothesis, the search continued. Juno was discovered in 1804 and Vesta in 1807. No more had been found by 1815 and the search stopped until a Prussian amateur named Hencke began a one-man program in 1830 and eventually discovered Astraea. Other discoveries by various workers followed. The bodies came to be called **asteroids** after their faint, starlike images. By 1890, the total known was 300. In 1891 the German astronomer Max Wolf began the first photographic patrol, detecting asteroids by their trails on long-exposure plates guided on the stars, as shown in Figure 7-1. This

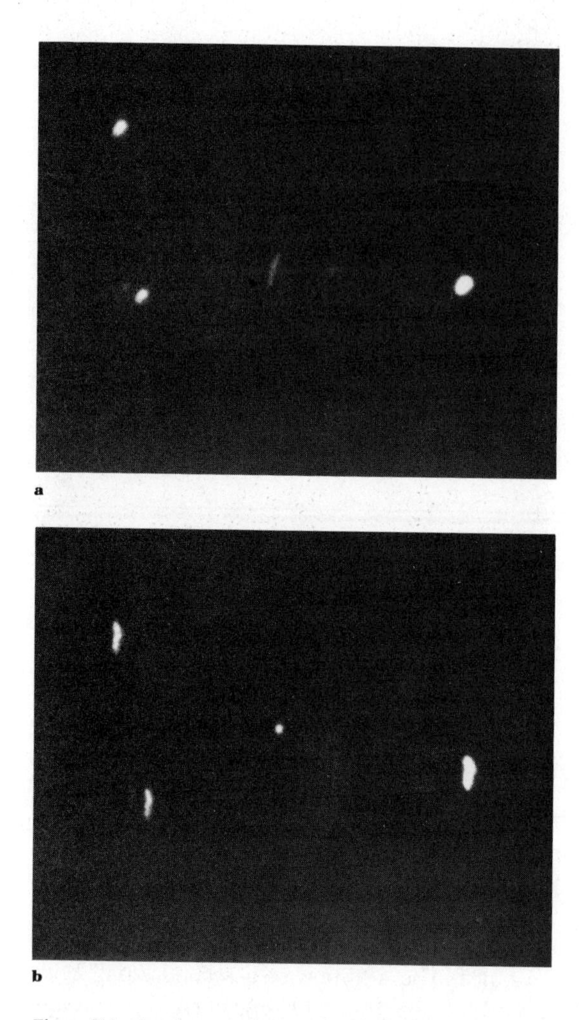

Figure 7-1. Two images of asteroid 887 Alinda. (**a**) *The telescope was guiding on the stars, and the asteroid appears as a streak due to its own orbital motion during the time exposure. Asteroids are often discovered by such trails on stellar photos.* (**b**) *The telescope was guiding on the asteroid, causing the star images to be trailed. (New Mexico State University Observatory)*

mechanization greatly increased the rate of discovery. Keeping track of the many asteroids—no mean task—was an undertaking performed by the Rechen-Institut in Heidelberg until World War II, when it was assumed by Russian astronomers.

The annual Soviet catalog of asteroid orbits, *Ephemerides of Minor Planets*, listed 2095 asteroids* in 1980. Many more have been observed once or twice but lost before enough observations could be made to determine a good orbit.

Asteroids are initially given a provisional number based on the year of discovery, such as 1978 SB. Later they are designated by number when an accurate orbit has been found and the asteroid relocated during a later opposition. The discoverer is also given the right to give a name to a newly numbered asteroid, and the names come from all aspects of human endeavor from mythology to universities to political figures (Hooveria!) to spouses and other lovers.** The full name of a cataloged asteroid is thus a number and name, such as 1 Ceres or 1685 Toro.

GROUPING OF ASTEROIDS BY ORBITS

Asteroids can be divided into a number of groups by their orbits in different parts of the solar system, as shown in Table 7-1.

The Asteroid Belt

The great majority of asteroids are confined to a swarm known as the **asteroid belt**, lying between Mars and Jupiter. Mars is at 1.5 AU and Jupiter at 5.2 AU; the mean distances of most asteroids are between 2.1 and 3.3 AU, but due to their eccentricities, their perihelia and aphelia occupy a wider zone.

Few asteroid orbits are as nearly circular as

*Asteroids are often called **minor planets**, especially by astronomers who specialize in their dynamical properties. They are infrequently called planetoids, although this usage is dying out. Note that the *observational* distinction between asteroids and comets is simply that asteroids do not have a fuzzy, gaseous halo around the starlike image, which comets develop as they approach the sun. Much current research revolves around the *physical* distinction between asteroids and comets.

**I was recently informed of a new amusement: making sentences using only asteroid names. My favorite (not my own creation, I'm sorry to say) is "Rockefellia Neva Edda McDonalda Hamburga."

the orbits of the major planets. Planetary eccentricities are mostly less than 0.1, but in the asteroid belt eccentricities are more typically 0.1 to 0.3. The belt asteroids also have more inclined orbits than do the planets; typical asteroid inclinations range up to 30°, while the planets are mostly inclined less than 3°. On the other hand, the orbits of the asteroids are far more regular than the orbits of long-period comets, which are highly eccentric and almost randomly inclined. Asteroids and short-period comets have some orbital similarities, which we shall discuss later. All the asteroids in the belt have prograde revolutions, as do the planets.

The belt contains at least the 4 largest asteroids, ranging from the 443-km object, 10 Hygiea, to the 914-km world 1 Ceres. Thousands more range from tens to hundreds of kilometers across.

Kirkwood Gaps

We have already seen that Jupiter, because of its large mass, is the most important perturbing influence in the solar system; gravitationally, it dominates a large region. Thus, it has a strong perturbing influence on the asteroids, which are its nearest neighbors.

As explained in Chapter 3, if an asteroid is in a **commensurate orbit**, with a period such as one-half, one-third, two-thirds, or one-fourth of Jupiter's period, it undergoes a repeated perturbation, or **resonance**. In many cases a resonance causes a rapid change in the orbital elements and kicks the asteroid out of that particular commensurate orbit. This creates a narrow unpopulated range of semimajor axes, called a **Kirkwood gap**. This does not mean there is a noticeable empty zone in the actual belt, because eccentric asteroids with noncommensurate a's pass through these zones. But strong depletions mark orbits such as a = 3.28 AU (resonance 1:2 with Jupiter) or 2.50 AU (resonance 1:3).

However, under some circumstances, commensurabilities can actually act to stabilize asteroids. Thus, in the outer belt at the 2:3 resonance (a = 3.97 AU), there is a group called the Hilda asteroids, named after the first-known member, 153 Hilda. Precise mechanisms to explain when resonances eject asteroids and when they stabilize them are still under study.

Jupiter and the Trojan Asteroids

The **Lagrangian points** of Jupiter's orbit, 60° ahead of and behind the planet, are sites of stable orbits (Chapter 3), which mark the 1:1 resonance. Members of these two groups have been named after the Greek and Trojan heroes of the Homeric epics (although names from the two warring sides are unfortunately mixed—there is a Trojan among the Greeks and a Greek spy in the Trojan camp!). These asteroids are thus called the Trojan asteroids. Dozens are cataloged, but recent surveys indicate so many fainter ones that the two Trojan clouds approach the main asteroid belt in numbers of smaller objects! The Trojans are not tightly clustered at the two Lagrangian points, but oscillate back and forth around these points, drifting 20° away.

The largest Trojan, 624 Hektor, is about the 21st-largest asteroid and unusually elongated in shape, being about 150 × 300 km. From rotational lightcurve statistics, Hartmann and others (1988) found that highly elongated shapes are more common among Trojans than in the belt. Binzel and Sauter (1991) confirmed this among the larger Trojans with additional statistics. The Trojans are dynamically isolated from the belt, and this finding may indicate a different history, or possibly an effect of ice sublimation.

Important analogs of Trojans are the small moonlets occupying Lagrangian points in the orbits of Saturn's satellites Tethys and Dione.

Mars-Crossing and Amor Asteroids

A few asteroids on the inner fringe of the asteroid belt cross inside Mars' orbit, and are called **Mars-crossing asteroids**. Shoemaker and his co-workers use the term **Amor asteroid** (after a prominent example) for a subgroup of Mars-crossers that come within 1.3 to 1.0 AU of the

Table 7–1
Properties of Asteroids

Number and name	a (AU)	e	i	Diameter (km)	Rotation period (h)	Geo-metric albedo	Spec-trum[a] type	Est. composition	q^b (AU)	Q^c (AU)
Inner solar system										
Aten type ($a < 1$ AU; $Q \geqslant 0.983$ AU; Earth-crossers interior to Earth's orbit)										
2100 Ra-Shalom	0.83	0.44	16°	3.4	19.8	—	C	Carbonaceous	0.47	1.20
2062 Aten	0.97	0.18	19°	1.0	—	—	S	Silicates	0.79	1.14
2340 Hathor	0.84	0.45	6°	0.2	—	—	—	—	0.46	1.22
Apollo type ($a \geqslant 1$ AU; $q \leqslant 1.017$ AU; Earth-crossers exterior to Earth's orbit)										
1866 Sisyphus	1.89	0.53	41°	10	—	—	—	—	0.87	2.92
1685 Toro	1.37	0.44	9°	4.8	10.2	0.03	S	L chondrite	0.77	1.96
1974 MA	1.78	0.76	38°	6?	—	—	—	—	0.42	3.13
1973 NA	2.39	0.63	68°	6	—	—	—	—	0.88	3.91
1863 Antonius	2.26	0.61	18°	3	4.0	—	S?	Silicates?	0.89	3.63
1947 XC	2.25	0.63	1°	3?	—	—	—	—	0.83	3.67
1864 Daedalus	1.46	0.62	22°	3.4	8.6	—	SQ	Silicates	0.56	2.36
1620 Geographos	1.24	0.34	13°	2.0	5.2	—	S	Silicates	0.83	1.66
Amor type ($a > 1$ AU; $1.017 < q < 1.4$ AU; Mars-crossers)										
1036 Ganymede	2.66	0.54	26°	30?	10.3	0.17	S	Silicates	1.22	4.10
433 Eros	1.46	0.22	11°	7 × 19 × 30	5.3	0.18	S	Silicates	1.13	1.78
1474 Beira	2.73	0.49	27°	11?	—	—	—	—	1.39	4.07
1974 UB	2.12	0.36	36°	8?	—	—	—	—	1.36	2.89
1975 AD	2.37	0.38	20°	8?	—	—	—	—	1.48	3.26
1139 Atami	1.95	0.26	13°	7?	—	—	S	Silicates	1.45	2.44
1963 RH	2.38	0.38	21°	7?	—	—	—	—	1.48	3.28
1580 Betulia	2.20	0.49	52°	6	6.1	—	C	Carbonaceous	1.12	3.27
1627 Ivar	1.86	0.40	8°	6?	4.8	—	S	Silicates	1.12	2.60
Inner Edge of Belt ($a < 2.5$ AU)										
4 Vesta	2.36	0.09	7°	500	5.3	0.24	V	Eucrite basaltic achondrite	2.15	2.57
7 Iris	2.39	0.23	6°	204	7.1	0.16	S	Olivine, nickel-iron	1.84	2.93
6 Hebe	2.43	0.20	15°	192	7.3	0.16	S	Silicates, nickel-iron	1.94	2.92
19 Fortuna	2.44	0.16	2°	190	7.4	0.03	C	Carbonaceous	2.05	2.83
8 Flora	2.20	0.16	6°	153	13.6	0.12	S	Pyroxenes, nickel-iron	1.85	2.55
11 Parthenope	2.45	0.10	5°	150	10.7	0.12	S	Silicates, nickel-iron	2.20	2.70
18 Melpomene	2.30	0.22	10°	150	11.6	0.14	S	Silicates, nickel-iron	1.79	2.81
20 Massalia	2.41	0.14	1°	131	8.1	0.16	S	Silicates, nickel-iron	2.07	2.75
12 Victoria	2.34	0.22	8°	126	8.7	0.11	S	Silicates, nickel-iron	1.83	2.85
21 Lutetia	2.43	0.16	3°	115	6.1	0.09	M	Metallic	2.04	2.82

Mid Belt (2.5 < *a* < 3.1 AU)										
1 Ceres	2.77	0.08	11°	914	9.1	0.05	C	Carbonaceous	2.55	2.99
2 Pallas	2.77	0.23	35°	522	7.9	0.08	B	Altered carbonaceous?	2.13	3.41
704 Interamnia	3.06	0.16	17°	334	8.7	0.04	C	Carbonaceous	2.57	3.55
15 Eunomia	2.64	0.19	12°	272	6.1	0.17	S	Silicate, nickel-iron	2.14	3.14
16 Psyche	2.92	0.14	3°	264	4.3	0.09	M	Nickel-iron, enstatite?	2.51	3.32
3 Juno	2.67	0.26	13°	244	7.2	0.15	S	Olivine, pyroxene, nickel-iron	1.98	3.36
324 Bamberga	2.69	0.34	11°	242	8	0.03	C	Carbonaceous	1.78	3.60
451 Patientia	3.06	0.07	15°	230	7.1	0.03	C	Carbonaceous	2.82	3.30
13 Egeria	2.58	0.09	16°	214	7.0	0.03	C	Carbonaceous	2.35	2.81
45 Eugenia	2.72	0.08	7°	214	5.7	0.03	C	Carbonaceous	2.50	2.94
Outer Edge of Belt (3.1 < *a* < 4.1 AU)										
10 Hygiea	3.15	0.10	4°	443	18	0.04	C	Carbonaceous	2.84	3.46
511 Davida	3.18	0.17	16°	336	5.2	0.03	C	Carbonaceous	2.65	3.73
87 Sylvia	3.48	0.10	11°	272	5.2	0.04	P	Silicate?	3.13	3.83
31 Euphrosyne	3.15	0.22	26°	248?	5.5	0.07	C	Carbonaceous	2.46	3.84
65 Cybele	3.43	0.12	4°	246	6.1	0.02	C	Carbonaceous	3.02	3.84
107 Camilla	3.49	0.07	10°	236	4.6	0.04	C	Carbonaceous	3.25	3.73
24 Themis	3.14	0.12	1°	228	8.4	—	C	Carbonaceous	2.76	3.52
94 Aurora	3.15	0.09	8°	212	7.2	0.03	C	Carbonaceous	2.84	3.47
702 Alauda	3.19	0.03	21°	202	8.4	0.06	C	Carbonaceous	3.09	3.29
165 Loreley	3.13	0.08	11°	202	7.0	0.07	C	Carbonaceous	2.88	3.38
Extreme Outer Belt (4.1 < *a* < 5.1)										
279 Thule	4.26	0.03	2°	60?	—	0.03	D	Reddish, carbonaceous	4.12	4.40
Trojans[d] (Lagrangian clouds in Jupiter's orbit; 5.1 < *a* < 5.3 AU)										
624 Hektor (P)	5.15	0.02	18°	150 × 300	6.9	0.02	D	Reddish, carbonaceous	5.0	5.2
911 Agememnon (P)	5.19	0.07	22°	148?	7?	0.04	D	Reddish, carbonaceous	4.8	5.5
617 Patroclus (F)	5.21	0.14	22°	140	—	0.02	P	Carbonaceous	4.5	5.9
1437 Diomedes (P)	5.08	0.05	21°	130?	18?	0.03	C	Carbonaceous	4.8	5.3
1172 Aneas (F)	5.17	0.10	17°	125	—	0.02	P	Carbonaceous	4.4	5.9
588 Achilles (P)	5.17	0.15	10°	118?	—	0.03	D	Reddish, carbonaceous	4.4	6.0
1143 Odysseus (P)	5.21	0.09	3°	118?	—	0.04	D	Reddish, carbonaceous	4.7	5.7
659 Nestor (P)	5.26	0.11	4°	102?	—	0.04	C?	Carbonaceous?	4.7	5.8
1208 Troilus (F)	5.17	0.09	34°	98?	—	0.04	C?	Carbonaceous?	4.7	5.6
1583 Antilochus (P)	5.28	0.05	28°	98?	—	0.05	D	Reddish, carbonaceous	5.0	5.6
Outer Solar System (*a* > 5.3 AU)										
944 Hidalgo	5.82	0.66	42°	39	10.1	—	D	Reddish, carbonaceous	2.00	9.64
2060 Chiron	13.7	0.38	7°	200	5.9	0.1?	C?	Carbonaceous?	8.5	18.9
5145 1992 AD	20.6	0.58	25°	150?	12?	0.08?	Z	Very red, organics?	8.7	32.4

Source: Various references have been used to construct the table, including Asteroids II (1989, University of Arizona Press); Wetherill (1979); Chapman, Williams, and Hartmann (1978); Kowall (1988); Beatty and Chaikin (1990); Binzel and Sauter (1992). *Note:* Diameters from various sources. Question marks indicate estimates; other values measured with fair precision. [a]Spectrum lists the type classification; U indicates an unusual spectrum; B is a variant of class C; SQ is probably a variant of S. [b]q = perihelion distance. [c]Q = aphelion distance. [d](P) or (F) indicates preceding or following Trojan cloud.

sun and thus approach, but do not cross, Earth's orbit. (Sometimes all Mars-crossers have loosely been called Amor asteroids.)

Dozens of Mars-crossers are known, and at least 20 of the Amor subclass are cataloged (Shoemaker and others, 1979). From discovery rates, Helin and Shoemaker (1979) estimate a population of 10 000 ± 5000 Mars-crossers and 1500 ± 500 Amors larger than about 1 km diameter. The largest Amors are as large as 30 km across.

Calculations indicate that the **half-life** (time for half of the population to be destroyed by collisions with Mars or perturbation on to very different orbits) is a few billion years.

Earth-Crossing Asteroids: The Apollo Group

More than 100 objects are now known with orbits that come in far enough to cross Earth's orbit. Helin and Shoemaker (1979) estimate a population of 700 ± 300 larger than about 1 km diameter. These are called **Earth-crossing asteroids** or **Apollo asteroids** after a prominent example. Most are only a kilometer or so across, with the largest having an estimated diameter of 8 km (object 1978 SB).

Twelve Apollos were known before 1960, 15 by 1970, and 60 by 1990. The high current rate of discovery is due mainly to special search programs run in recent years by Carolyn and Eugene Shoemaker, by Eleanor Helin, and by Tom Gehrels.

There are several reasons for special interest in Apollos. One is that they are the closest planetary objects beyond the moon and would require the least energy to reach. An Apollo asteroid might be the next planetary body to be explored by humans, as suggested in Figure 7-2. Such a trip would be scientifically interesting, because Apollos are probably sources of meteorites. In fact, an expedition might merely discover a huge example of a meteorite type already in our museums! Nevertheless, the structure and surface of such a body (unaltered by atmospheric entry) would be of interest, and any major differences from meteorites would be thought-provoking indeed. Further, an Apollo's materials might be of use in constructing large space habitats.

Another reason for interest in Apollos is their potential for hitting Earth. Most will end their days in this way many million years in the future. A 100-m object capable of making a 1-km crater hits Earth every 3000 y or so (70% falling into oceans and not leaving permanent craters). Such impacts would devastate areas comparable to moderate-sized cities, send out shock waves causing injury many kilometers away, and throw dust into the high atmosphere, coloring sunsets around the world for years.

A famous example of a close approach by an Apollo capable of making a 10-km crater (!) was the approach of 1566 Icarus on June 14, 1968. It passed only 6 million kilometers away, only about 16 times the distance to the moon. Though this approach was predicted well in advance, the incident touched off a number of scare articles in tabloids by sensationalistic "reporters" who allegedly did not trust the assurances of astronomers that Earth would be spared.

As is the case with meteorites, the fact that Apollo and Amor asteroids have half-lives only a few percent the age of the solar system tells us something important: Either they are the remnant of an enormous initial population or they are being replenished. The first possibility was ruled out by Öpik (1963), who showed that an absurdly high initial mass of asteroids would be required (100 $M_{\odot}$!). The question, then, is, if Apollos are being replenished, where do they come from?

One obvious answer is the asteroid belt. Of Wetherill's (1979) list of 31 Apollos and related objects, 81% have aphelia in the belt region from 1.6 to 4.3 AU. One possibility is that asteroids on the inner belt fringe were converted to Apollos by near-miss encounters with Mars. Computer simulations (Arnold, 1965) show how this can occur, and some 29% of the objects on the Wetherill list have aphelia in this inner belt fringe, 1.6 to 2.5 AU, a fact that is consistent with this idea.

Figure 7-2. Many dozens of small Apollo asteroids probably approach Earth, and may become targets of future exploration. Asteroids may offer not only interesting samples of planetary matter, but also metal and mineral deposits of economic interest. In this imaginary view, a post-Shuttle era exploration vessel approaches an asteroid about 100 m long. Bright objects in the lower right are Earth and the moon, about 0.1 AU away. (Painting by author)

Another possibility is that Jupiter resonances throw objects out of the main belt (see review by Chapman, Williams, and Hartmann, 1978).

A second view, pioneered by Öpik (1963) and others, is that Apollos may not really be asteroids at all, but burnt-out comets. Some comets have Apollo-like orbits (see the next chapter). After all, the term *asteroid* used to describe Apollos is merely an observational descriptor implying a lack of any surrounding, glowing gas that would warrant a comet designation. Certain sequences of perturbations can send bodies from cometlike orbits into Apollo-like orbits where their ices would eventually sublime, leaving a rocky core as a residue. Such a core might or might not have a composition similar to that of belt asteroids; its structure would probably be more granular and more fragile. It might be carbonaceous.

Various early workers proposed that a certain subset of Apollos (generally those with eccentric, inclined orbits) are burnt-out comets, while others are not. Shoemaker and others (1979) identified two Apollos having orbits related to short-period comets. The large Apollo (diameter ~10 km), 1978 SB, has an orbit quite similar to that of short-period Comet Encke (Wetherill, 1979). Such work was done before spectral classes were known for most of these objects. Hartmann and others (1987) reviewed spectra and found that of ten asteroids with cometlike orbits (including some Apollos), all have spectral properties similar to comets and the black, outer-solar-system classes D, P, or C. Of 13 Earth-approachers with noncometary orbits, eight or nine were of spectral class S, probably native to the inner belt, and only one was of the very low albedo D-P-C classes. These data support the idea that some high-eccentricity Apollos of very low albedo are black, "burnt-out" comets from the outer solar system while (most?) others are displaced main-belt asteroids.

The Aten Group of Earth-Approachers

In 1976 Shoemaker and Helin defined a new category, **Aten asteroids**, with semimajor axes *less* than Earth's, so that they lie primarily *inside* Earth's orbit but may cross outside near aphelion. The class is named after the first example, 2062 Aten. They are probably similar to Amors and Apollos, but with orbits even more modified by planetary perturbations. A dozen or so larger than 1 km may exist.

Asteroids of the Outer Solar System

Three "asteroids," 944 Hidalgo, 2060 Chiron, and 5145 1992 AD have orbits that lie mostly or entirely beyond Jupiter. Their respective semimajor axes are 5.8, 13.7, and 20.2 AU, compared to 5.2 AU for Jupiter and 19.2 AU for Uranus. Hidalgo has long been considered a prime suspect for a burnt-out comet nucleus. It has reddish-black spectral class D, consistent with Trojan asteroids and some comets.

2060 Chiron, after its 1977 discovery, was considered a possible candidate for an icy cometlike body. Sure enough, in 1987–88 it flared up in brightness and developed a coma (Hartmann and others, 1990), so that it is now a verified comet (presenting a problem for semanticists, since it is still cataloged with its asteroid number, 2060). It has a neutral black color like C-class asteroids and is less reddish than most comets.

Dynamicists using computers have found that Chiron passes close to Saturn every few 10^4 years (Scholl, 1979). It may have been thrown into the solar system from the Oort cloud within the last 100,000 years, and its present orbit cannot be very old. Similarly, it will be radically perturbed by Saturn into an unknown new orbit within 100,000 years or so. Many such **chaotic orbits** exist in the outer solar system. These are often short-lived; their past and future histories are hard to discover because they depend strongly on initial conditions. Such objects may be tossed back and forth among the giant planets like lacrosse balls among planetary players.

Asteroid 5145 1992 AD, discovered in 1992, was found to have colors entirely unlike Chiron and redder than known asteroids and comets. (To the eye, it would look redder than Io but not quite as red as Mars.) Some observers have suggested that the red colors may come from concentrated exposures of organic materials (Mueller and others, 1992).

"Asteroids" of the outer solar system are intriguing objects; more may be soon discovered.

Hirayama Families

In 1918 the Japanese astronomer K. Hirayama published the first of a series of papers pointing out that various asteroids clustered into groups with similar orbital elements. The groupings are called **Hirayama families**.

Each Hirayama family probably contains the fragments of an asteroid that suffered a collision and fragmented long ago. The pieces are spread out along the orbit of the original body in much the way meteor streams spread along comet orbits. When the original asteroid collided with another (smaller) asteroid, it was blown apart and the pieces separated. Brouwer (1951) shows that the differences in velocity among family members are about 0.1 to 1.0 km/s—values that must approximate the original ejection velocities. Such speeds are only a small fraction of orbital velocity; hence all the fragments remain in approximately the initial orbit. The speeds are also consistent with average ejection velocities of fragments from laboratory impacts at about 2 to 10 km/s (Gault, Shoemaker, and Moore, 1963).

Hirayama family velocity distributions are similar to those of fragments of exploded artificial satellites, again suggesting that the families (particularly the Themis, Eos, and Coronis families, named after prominent members) are fragments from asteroid collisions (Wiesel, 1978).

The family involving the 160-km asteroid 8 Flora is especially interesting. It has a bimodal distribution of semimajor axes among its fragments, possibly reflecting the two colliding bodies. The members would combine to make a spherical parent body estimated at 165-km diameter. But Flora itself, if added to the second-largest body (78-km asteroid 43 Ariadne), would require a long axis of 240 km for the parent. Also, Flora is spherical and rotates slowly. These properties suggest Flora itself is not a fragment. Tedesco (1979c) suggests that proto-Flora was a binary asteroid and that only the satellite was broken in the collision.

Chapman (1979) notes that the spectra of some families show uniform composition among all members, strongly supporting the theory of origin by collisional breakup of a parent body. But other families show mixed compositions. Perhaps their fragments came from different regions in the interior of two colliding bodies that were already nonhomogeneous, or perhaps some families were not created by collisions at all.*

Brouwer (1951) described 9 families and 19 additional "groups" identified by similar orbits. The number of asteroids in the families ranges from 9 to 62, and in the groups from 4 to 11. Arnold (1970) reviewed Brouwer's and Hirayama's data, confirmed the existence of families 1 to 9, and added a number of new families. As many as half of all asteroids have been associated with 100 proposed Hirayama families, but only about a dozen families are populous and well established.

Of a sample of 34 Mars-crossing asteroids, 20 can be identified with four Hirayama families. These 20 contain 98% of the mass of the Mars-crossers (Anders, 1964). Possibly the eccentric orbits of the Mars-crossers increases their probability of disruption by collision, accounting for the large percentage in the form of Hirayama families. Nonetheless, in the crowded central part of the asteroid belt, the percentage of Hirayama families may be higher than we now recognize, simply because they are harder to measure because of the large number of objects.

It is intriguing to think that many Hirayama families may be drifting debris from the titanic, catastrophic collisions that have sporadically marked solar system history.

REMOTE SENSING OF ASTEROIDS AND OTHER BODIES

The various astronomical measurements that can reveal asteroid properties—such as polarimetry, radar, and spectroscopy—have come to be called

*H. Alfvén, 1970 Nobel Prize winner, introduced the term "jet stream" to refer to tightly knit familylike groupings that he believed were brought into similar orbits by accretional processes that damped out initial orbital differences. Alternatively, however, they could be jets of debris from specific fragmentation events, and some workers ascribe them merely to observational selection effects.

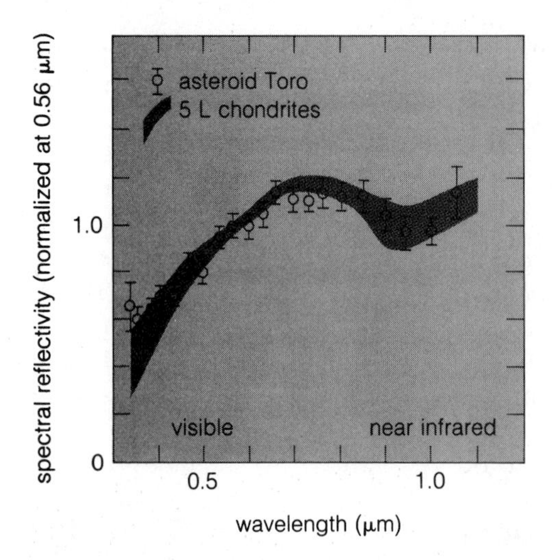

Figure 7-3. Reflectance spectrum of Apollo asteroid 1685 Toro, from blue to near infrared, showing close match to L-type chondrite meteorites, whose minerals are predominantly iron-magnesium pyroxenes and olivine. Other meteorite types lack the absorption near 1.0 μm or have absorptions at different wavelengths. Carbonaceous chondrites and C-type asteroids, for example, roughly match the left third of this spectrum but are nearly flat in the right two-thirds. (After Chapman and others, 1973)

techniques of **remote sensing**. The amount of polarization, for example, and its relation to the **phase angle** (angle sun-object-observer) discriminate between bare rock and a surface layer of fine dust. Such measures have indicated at least thin layers of dust on many asteroids.

The strongest technique applied to asteroids is **infrared reflectance spectrophotometry**, the study of infrared absorption bands produced when sunlight passes through the outer micrometers of certain minerals in the rocks, before reflecting back to the observer. Important examples are the **pyroxene absorption bands** near 1 and 2 μm, which are indicative of pyroxene, a silicate mineral family important in basaltic and some meteoritic rocks. Its formula is $XY(Si_2O_6)$ where X and Y may be the same or different elements. An example is enstatite, $Mg_2(Si_2O_6)$, mentioned in Chapter 5. The pyroxene band appears in Figure 7-3, which displays the close match between the spectrum of Apollo asteroid 1685 Toro and the most common type of meteorite, the L chondrites. As mentioned in the last chapter, many L chondrites may actually be pieces knocked off Toro or the same parent body as Toro. Figure 4-2 shows longer-wavelength examples of mineral infrared spectra.

Absolute Magnitude and Size

For most asteroids the property simplest to measure from Earth is brightness. Generally, the greater the brightness, the larger the asteroid. The apparent brightness depends on the distance from both the sun and Earth. A standardized way to express it is as follows: The **absolute magnitude** of a solar system body is its brightness at zero phase angle (sun-object-observer) when the object is 1 AU from the sun and 1 AU from the observer. The astronomer's stellar magnitude system is used, in which the brightest stars are around magnitude 0 and the faintest visible with the naked eye are around 6. Depending on whether visual (yellow) wavelengths or blue wavelengths are used, the absolute magnitude is designated $V(0,1)$ or $B(0,1)$, where the "0" implies zero phase angle, and the "1" implies the distance of 1 AU. Sometimes $V(0,1)$ is simply designated g.

Diameters can be estimated simply by knowing an asteroid's absolute magnitude, but since albedos vary considerably from 2% to 24%, the diameter will be rough unless the albedo is known. Certain techniques involving observation of polarization or the ratio of thermally-emitted radiation to reflected sunlight allow direct estimation of albedo and hence diameter.

ASTEROID CLASSIFICATION BY SURFACE COMPOSITION

Remote sensing tools have been applied to more than 500 asteroids. The asteroids have been found to have compositions similar to meteorites and have been classified into several groups related to composition (Bowell and others, 1978; Gaffey and McCord, 1978; Zellner, 1979; Tedesco and

Gradie, 1981). There is some confusion in the literature, since various observers have proposed different levels of subclassification and have occasionally suggested new or revised classes. The major types widely recognized are listed in Table 7-2. These so-called taxonomic classes are defined purely in terms of spectral features but are believed to be loosely related to certain meteorite types (based on comparisons of the asteroid spectra and meteorite spectra).

Astronomers have discovered a striking zonal structure in the asteroid belt, as indicated in Table 7-2, according to which the various asteroid types have preferred locations in certain parts of the belt (Gradie and Tedesco, 1982). The rare, high-albedo E type (possibly related to enstatite chondrites) and the very common S type (possibly related to ordinary chondrites) are, for example, mostly on the inner edge of the belt. Examples of the unusual E-type asteroids are 44 Nysa and 434 Hungaria (albedos around 45%!) whose orbits are well placed to produce Earth-impacting meteorites (Zellner and others, 1977). S-type asteroids continue into the mid belt, where M-types (possibly stony-irons?) also peak in abundance.

The most abundant asteroid type is probably the C-type of the outer belt. The C's are thought to be carbonaceous chondrite material, whose strikingly low albedos of 2% to 7% are attributed to opaque minerals (graphite, other carbon-based minerals, and perhaps magnetite) disseminated through their matrix. Remember that carbonaceous chondrites include low-temperature condensates. They contain large amounts of chemically bound water along with fine-grained clay minerals like montmorillonite, as well as organic polymers, magnetite, and iron sulfide. The relation between C-type asteroids and carbonaceous chondrites has been strengthened by Lebofsky and his co-workers (1977, 1982), who found spectral evidence for chemically bound water and montmorillonite clay minerals in the surface material of the largest C-type asteroid, 1 Ceres. Some asteroids may even contain H_2O frost buried beneath their surfaces. Such findings—together with the fact that only a few percent of carbon minerals mixed with pulverized ice can lower the albedo from the pure ice value around 70% to a value around 5% (Clark, 1980)—suggest the likelihood that many blackish C-type asteroids of the

Table 7-2
Selected Asteroid Compositional Types (in Order Outward from the Sun)

Spectral type	Albedo (%)	Composition (meteorite type)	Location[a]
E	>23	Enstatite chondrite?	Concentrated near inner edge of belt
S	7–23	Stony material; possibly chondrites?	Major class in inner to central belt
M	7–20	Stony-iron or iron	Central belt
V	38	Basalt	Vesta and a few Earth-crossers
A	~25	Olivine rich; pallasite	Main belt (rare)
C	2–7	Carbonaceous chondrite	Dominant type in outer belt beyond about 2.7 AU
P	2–7	Uncertain; spectrum resembles M but albedo is lower	Outer and extreme outer belt
D	2–7	Dark reddish-black organics including kerogens and carbonaceous material?	Extreme outer belt and Trojans; about two-thirds of Trojans are type D.
Z (provisional)	4–10?	Organics? Extremely red	Chironlike orbit

[a]After Gradie and Tedesco (1982).

outermost belt or Trojan regions may contain large amounts of ice. The presence of the ice is disguised by the opaque carbonaceous minerals.

The same is true of another important type of the outer belt and Trojans, the D (originally called RD) asteroids, first noted and defined by Degewij and van Houten (1979). D's are as dark as C's, but much redder (RD indicating "reddish, dark"). D materials are probably an even lower temperature condensate than C's, containing a reddish organic sludge that condensed just as the ices were beginning to condense. They are unrepresented among meteorites because their source regions are too far from Earth, but they constitute about two-thirds of the Trojans. Gradie and Veverka (1981) matched their spectra with mixtures of coal-tar residues, magnetite, carbon black, and montmorillonite. A principal coloring material in the coal-tar residues are the organic compounds called kerogens, also found to some extent in carbonaceous chondrites.

The planetesimals of the outermost belt and outer solar system may thus have formed from intriguing mixtures of organic substances, carbonaceous rocky material, and ices.

Calculations of the orbital evolution of early planetesimals in the inner solar system show that many were thrown by Earth outward onto orbits in the inner fringe of the belt. Thus the inner fringe may be enriched in (stony?) volatile-poor asteroids that formed in the inner solar system. These may include many E- and S-types in the inner belt. Further calculations show that some of these would have been thrown rapidly back onto Earth-crossing orbits, while others would have been preserved until today (Wetherill, 1977). Thus, some meteorites may come from E- and S-type asteroids that were originally Earth-crossers but were "stored" in the belt.

ROTATION AND SHAPES OF ASTEROIDS

Many asteroids vary in brightness periodically, as found by measuring the **light curve**, or brightness as a function of time. This variation can have two causes:

1. The asteroid is elongated in shape and tumbling end over end. In this case the true rotation period would include two maxima in the light curve.

2. The asteroid has dark and light markings. Most likely, one hemisphere would average darker than the other (as with the moon), and the true rotation period could include only one cycle of the light curve.

Observations show that the first case is much more common: asteroids are often elongated. There are several observational tests. Apollo asteroid 433 Eros came close enough in 1931 (23 million kilometers) that astronomers could watch the 7 × 19 × 30 km elongated object tumble end over end every 5.3 h. Elongated objects have more symmetric light curves than spotted objects do, and searches for spectrophotometric differences from one side of an asteroid to another during rotation have generally been negative or inconclusive.

Another way to distinguish elongation from spottedness is to compare light curves of reflected solar and emitted thermal infrared radiation. If an asteroid is elongated, both curves will have simultaneous maxima; but if it is spotted, the dark side will be warmer and emit maximum thermal infrared radiation at the same moment that the reflected visual light curve is at minimum due to the low albedo. This test was used to prove that the peculiar Trojan, 624 Hektor, is unusually elongated (Hartmann and Cruikshank, 1980).

Many elongated asteroids vary in brightness by a factor as much as 3 or more. Light-curve statistics reveal fewer kilometer-scale elongated asteroids in the belt than among similar-sized Mars-crossers, possibly due to the higher rate of collision and erosion in the belt (Degewij, 1977).

Photometry indicates not only rotation, nonspherical shapes, and markings but also the obliquity of asteroids. To see how, suppose an irregular asteroid can be seen pole-on from Earth during one apparition; some months later we would see it at some quite different aspect angle, because of our relative motion. When the asteroid is seen pole-on, its light curve would have zero amplitude, but later the light variations would become appreciable. By studying variations in the light curve with aspect angle, we can derive the position of the asteroid's axis of rotation. Taylor (1979) reviews pole positions published for 17 asteroids with obliquities ranging from 30° to 90°. The high, seemingly random obliquities are consistent with spins imparted during major collisions.

Feasibility of Irregular Shapes

If Earth, moon, and other planets are almost exactly spherical, why are most asteroids highly irregular in shape? The asteroids are in a sense large chunks of rock (or stony-iron material); obviously rocks can take a variety of shapes. Perhaps, then, the question should be rephrased: *Why are the larger planets round?*

It is a question of strength of material. The larger the asteroid or planet, the greater the pressure at the center. If the central pressure exceeds the strength of the rocky material, the material will deform, either by plastic flow or by fracture, as a result of failure of the normal elastic properties of the solid rock. Higher internal temperatures also favor deformation toward an equipotential shape. In the case of an ideal nonrotating body, this shape will be perfectly spherical.

As calculated in the accompanying optional mathematical notes, the diameter at which the central pressure begins to exceed the strength of the planetary material is around 440 to 700 km. Table 7-3 tabulates observed irregular shapes among asteroids and satellites; and, as plotted in Figure 7-4, we find a strikingly sharp cutoff in irregular shapes at diameters around 360 to 600 km. As shown in Figure 7-5, planetary bodies smaller than about 400 km often are noticeably irregular; worlds larger than 1000 km never are.*

Centrifugal Force Versus Gravity

Because the asteroids spin rapidly, with periods as short as a few hours, a frequent question has been whether they spin so fast that any loose material would be thrown off their equators by centrifugal force. The answer is no. An asteroid would have to rotate in less than about 2 h in order for centrifugal force to exceed gravity. Even the fastest-rotating asteroid, the Apollo object Icarus, has a period 2.25 h, so loose debris could exist on its surface. Polarimetric data support this conclusion by indicating a dusty surface on many asteroids.

Asteroid Rotations as a Clue to Planet Evolution

The planets' rotation rates, with periods ranging from 0.4 d for Jupiter to 243 d for Venus, might appear at first glance to be random. But if we add data on asteroid rotations, a remarkable trend emerges, as shown in Figure 7-6. Though the masses of asteroids and planets range over 15 orders of magnitude, the rotation periods are relatively constant. Of a sample of 137 planetary bodies, 83% rotate with periods between 4 and 16 h.

The general relation appears to apply to planets as well. (Recall that Mercury's rotation is tidally locked by a resonance with the sun, and Venus' is probably similarly influenced by solar tides or resonance with Earth, so they do not count.) Earth and Pluto have probably been slowed by tidal interaction with their relatively large satellites. Restoring the Earth-moon and Pluto-Charon systems to initial, faster rates based on conservation of angular momentum, we see from Figure 7-6

---

*See Chapter 9 for more detailed discussion of planetary oblateness caused by rotation.

that the planets (except for Mercury and Venus) lie in the general band defined by asteroids.

Why shouldn't planetary bodies rotate with widely scattered periods, such as 1 h, 100 h, or 3 mo? A clue comes from considering centrifugal force versus gravity, as mentioned above. Asteroids spinning with periods faster than about 2 h would throw material off their equators and hence be unlikely to form in the first place. Weidenschilling (1981) considers the further complexity that a weak asteroid spinning up toward this rate will begin to deform, first into a Maclaurin spher-

Table 7-3
Selected Elongated Asteroids and Satellites

Name	Est. maximum diameter (a)	Est. minimum intermediate or diameter (b or c)[a]	Diameter ratio (a/b or a/c)[a]
Asteroids			
1620 Geographos	6.5	1	6
433 Eros	30	7	4
1864 Daedalus	5.5	2.5	2.2
1977 RA	3.3	1.5	2.2
753 Tiflis	34	16	2.1
1685 Toro	9.5	4.5	2.1
624 Hektor	300	150	2
1207 Ostenia	53	28	1.9
1789 Dobrovolsky	9.5	5	1.9
43 Ariadne	99	55	1.8
1580 Betulia	7.7	4.8	1.6
Light-curve averages	11		1.62
	50		1.37
	94		1.24
	164		1.18
	335		1.18
Satellites			
1980S28	40	20	2.0
1980S27	140	80	1.8
J5 Amalthea	270	155	1.7
S7 Hyperion	360	210	1.7
1980S26	110	70	1.6
M1 Phobos	27	19	1.4
1980S3	140	100	1.4
M2 Deimos	15	11	1.4
1980S1	200	150	1.3

Source: Data from Tedesco (1979c), Degewij and others (1977), Harris and Burns (1979), Voyager Imaging Team reports on Jupiter and Saturn satellites and other sources.

[a]The asteroid data are primarily from light-curve amplitudes, which tend to give diameter ratio a/b. The satellite data are primarily from direct imaging, from which minimum diameter c can sometimes be determined, so that the satellite data give a better estimate of a/c. In practice, the data in the last column may be between a/b and a/c, and these two ratios may often be similar.

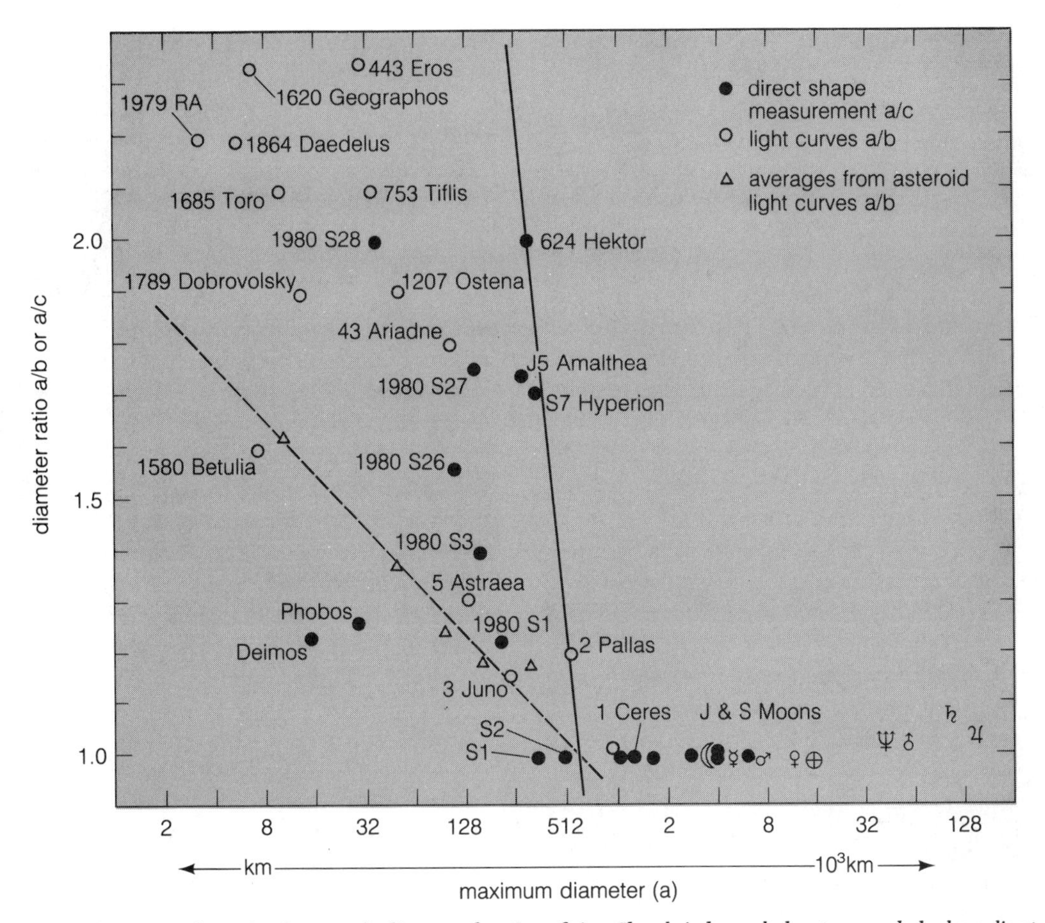

Figure 7-4. Shape irregularity in planetary bodies, as a function of size. Closed circles and planetary symbols show direct measures of a/c, where a and c are maximum and minimum diameters. Open symbols are based on asteroid lightcurves, which reveal the lower ratio a/b, where b is the intermediate diameter. Both the trend of the a/c and average a/b ratios indicates that the maximum diameter of the largest irregular bodies is about 600 km. Larger planetary bodies have spheroidal shapes because planetary interiors are not strong enough to support irregularities of such large scale.

oid (a figure with an equatorial bulge), then into a Jacobi ellipsoid (an elongated, football-like figure), and finally into a body that fissions into two parts. The elongation inhibits spin-up, so that the periods of weak asteroids do not get shorter than about 4 h. Note that if an asteroid once melted and then solidified into strong igneous rock or metal and if it escaped later weakening by impact-induced fractures, it could spin faster without deforming. Weidenschilling (1981) suggests that 321 Florentina is the best candidate; its period of 2.87 is the shortest known in the main belt.

The cumulative effects of collisions during accretion tended to spin up planetesimals toward these limiting rates of 2 to 4 h, but the increased loss rate of ejecta from fast-spinning bodies and the tendency to elongate were probably compensating influences that left nearly all asteroids with periods between 4 and 16 h, as seen in Figure 7-6 (Harris, 1977). Long-period exceptions, such as 128 Nemesis and 393 Lampetia, may have

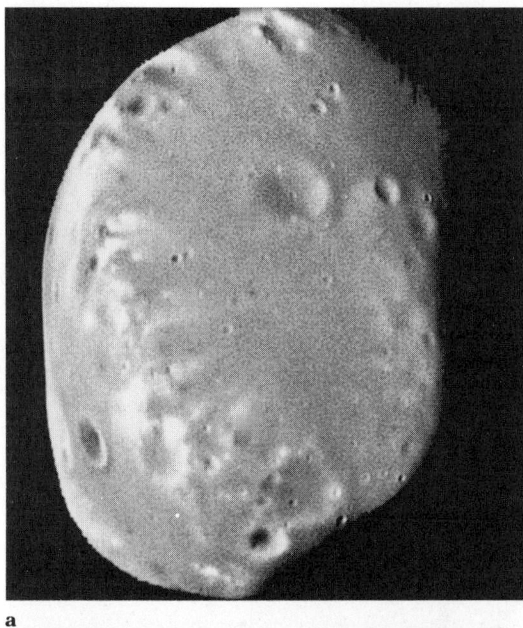

a

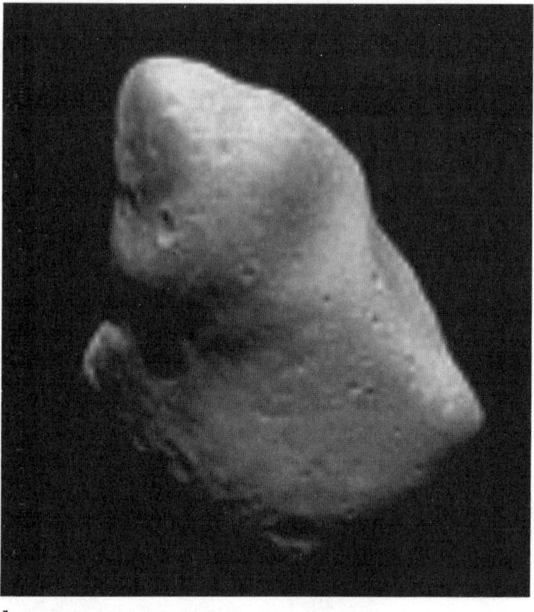

b

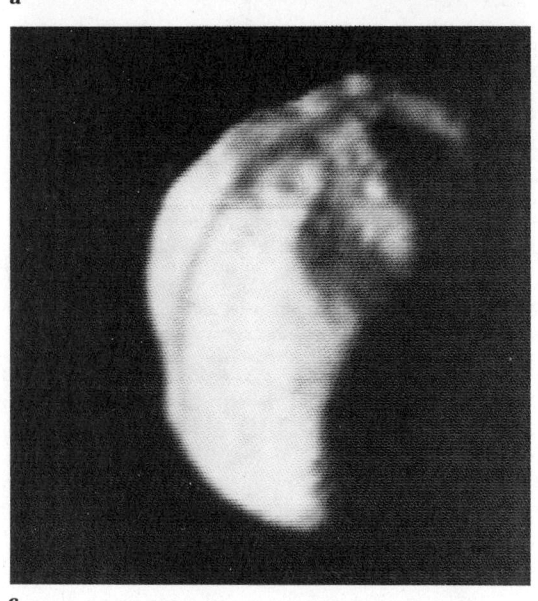

c

Figure 7-5. An asteroid and examples of similar irregularly shaped bodies. (**a**) *Martian satellite Deimos (10 × 16 km), showing a ridgelike crest running down left center side.* (**b**) *Asteroid 951 Gaspra (12 × 16 km).* (**c**) *Saturn's inner satellite S11 Epimetheus (about 140 × 100 km). Dark line along upper left edge is the shadow of Saturn's rings across the satellite. (All NASA:* **a** *courtesy J. Veverka, Viking project;* **b** *courtesy Michael Belton, Galileo Project)*

been struck late in their careers by large-scale retrograde planetesimals, slowing their rotation.

SATELLITES OF ASTEROIDS AND COMPOUND ASTEROIDS?

In the late 1970s, the art of predicting when an asteroid would occult a star reached the point where the track of the "shadow" of the asteroid (cast by the star onto Earth) could be roughly predicted. Teams of observers set out to watch these events because the duration of the "blink-out" is a measure of the size of the asteroid. If blink-outs were measured by several observers at different locations, the asteroid's shape could be determined. On June 7, 1978, 532 Herculina passed in front of star SAO 120774, and a photoelectric record made in Arizona showed not one but two blink-outs, while a photoelectric record in California suggested six events, including a possible match to the Arizona event. A photoelectric record and visual observations later in 1978 gave two blink-outs during occultation of a star by asteroid 18 Melpomene.

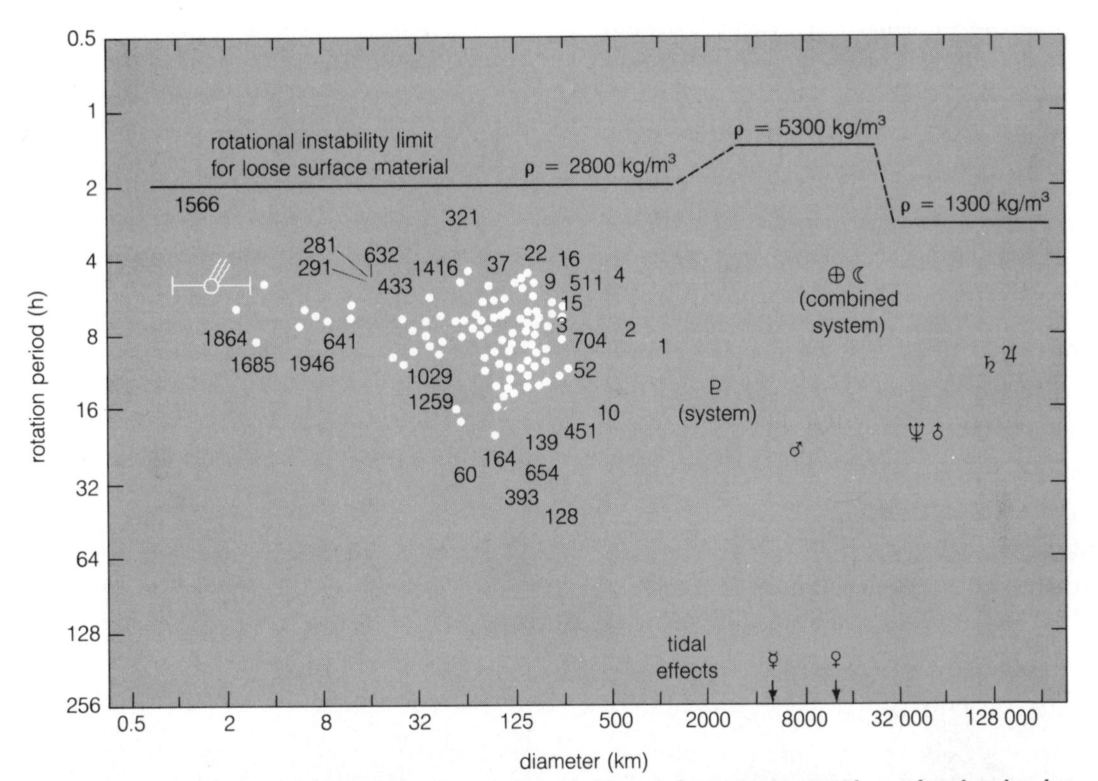

Figure 7-6. A plot of rotation periods versus diameter for asteroids and planets. Most asteroids are plotted as dots, but interesting ones are numbered. The line across the top gives the minimum period (fastest rotation rate) that can be reached without throwing material off the equator. Though the diameters span five orders of magnitude, nearly all planetary bodies seem to have spun up to a rotation rate about 2 to 5 times slower than the instability limit except for those with tidal or resonance effects. Earth-Moon and Pluto-Charon are plotted as initial systems with planet and satellite very close. The comet symbol shows data reported for the comet d'Arrest, whose diameter is uncertain (see Chapter 8).

MATHEMATICAL NOTES ON MAXIMUM ROTATION RATES

It is reasonable to suppose that the fastest rotation rate that could be acquired by an object during its formation by accretion would be the rate at which loose material would be thrown off its surface by centrifugal force. Setting gravity equal to centrifugal force, and expressing the velocity in terms of rotation period, we have

$$\frac{GMm}{R^2} = \frac{mv^2}{R} = \frac{4\pi^2 R^2 m}{P_c^{\,2} R}$$

where P_c is the critical rotation period at which material will be thrown off the equator.

This reduces to

$$P_c = \sqrt{\frac{3\pi}{G\rho}} = \frac{104}{\sqrt{\rho}}\text{hours}$$

For material with densities ranging from 2200 to 3800 kg/m^3 (typical of carbonaceous chondrites and chondrites; Wasson, 1974), this formulation gives critical periods from 1.7 to 2.2 h.

These and subsequent similar observations led to proposals that at least a few asteroids have small satellites (Binzel and van Flandern, 1979; Tedesco, 1979). Dynamical studies of the evolution of orbiting pairs indicated they could not last long, and, if real, must require an ongoing creation mechanism. Hartmann (1979a, b) suggested that during an asteroid fragmentation event, in the chaotic cloud of expanding gas from volatiles, and jostling neighboring fragments, pairs might end up orbiting, or neighbors of comparable size might collide at low speed and come to rest on each others' surface in dumbbell-like configurations he called **compound asteroids.** Tidal forces between a satellite pair would lead to rapid orbit evolution in only some 10^5 y (Binzel and van Flandern, 1979). Satellites outside the synchronous point would evolve outward, and at about 100 primary diameters would escape the primary's sphere of influence, drifting off as independent asteroids. Satellites inside the synchronous point would evolve inward and come

MATHEMATICAL NOTES ON MAXIMUM SIZES OF IRREGULAR PLANETS

In Chapter 9 we will derive the following equation for the pressure of overlying material, measured at some distance r out from the center of a planet with uniform mean density $\bar{\rho}$ and with radius R:

$$P_r = \frac{2\pi G\bar{\rho}^2}{3}(R^2 - r^2)$$

Now consider a planetesimal in whose central region the strength of the material, S, is greater than the pressure exerted by the overlying material. (Note that strength S and pressure P_r have the same units, newtons/meter^2.) As long as S exceeds P_r in the central region, the planetesimal will be strong enough to retain any irregular shape it might acquire, say, by collisional fragmentation. But if the planetesimal grows to a size such that P_r begins to exceed S, then the central material will be crushed and the planet can begin to assume the spheroidal shape of hydrostatic equilibrium. If we set P_r equal to S for material 20% of the way out from the center (an arbitrary choice of distance), the equation reduces to the following condition for the largest planetesimal radius R_{Max} that can sustain substantially irregular shape

$$S = 1.34\,(10^{-10})\,\bar{\rho}^2 R_{Max}^2$$

(assuming SI units)

We might expect the inner regions of a reasonably large planetesimal to be made either of moderately strong rock (like chondritic meteorites) or of iron (like iron meteorites). For these two cases we have the following conditions:

(1) For a *rocky planetesimal*, strengths of chondritic stone meteorites range from about 6×10^6 to 4×10^8 N/m^2 (Wasson, 1974), and corresponding densities range from 2500 to 3800 kg/m^3, respectively. Using a representative strength and density of 2×10^8 N/m^2 and 3500 kg/m^3, we find

$$R_{Max} \simeq 349 \text{ km}$$

(2) For an *iron-cored planetesimal*, using an iron meteorite's strength of 4×10^8 N/m^2 (Wasson, 1974) and the iron density of about 7870 kg/m^3, we find

$$R_{Max} \simeq 220 \text{ km}$$

These calculations suggest that the largest irregular planetesimals might be expected to lie in the *diameter* range of 440 to 700 km. This prediction, though based on a simplified analysis, is in excellent agreement with the empirical data in Figure 7-4, where the largest irregular asteroids and satellites show a well-defined upper diameter limit in the range of about 360 to 600 km.

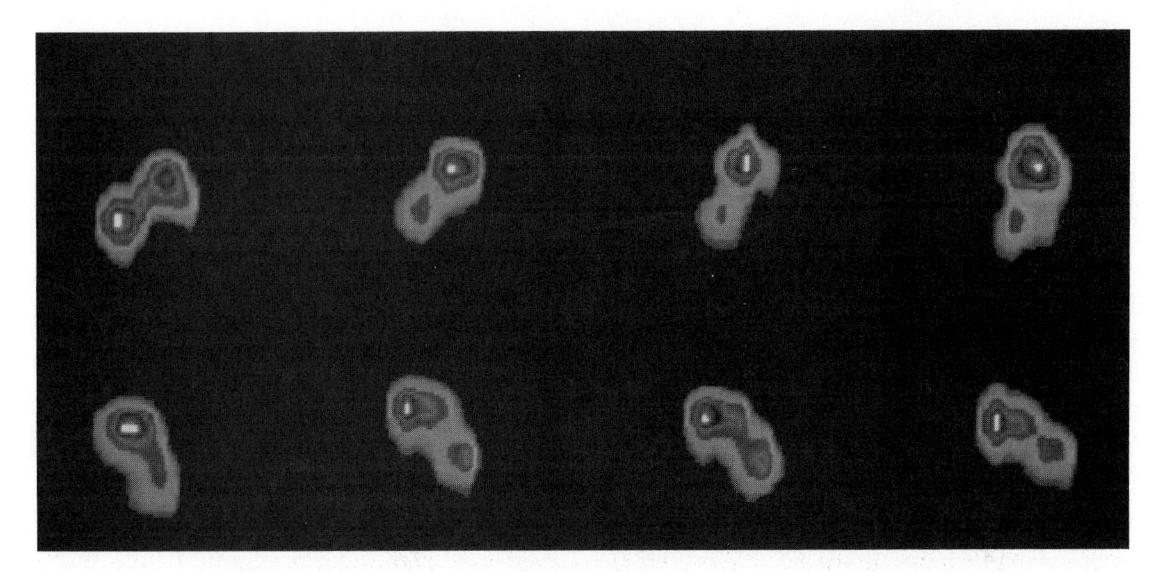

Figure 7-7. Strange shape and rotation of Earth-approaching asteroid 1989 PB, as revealed by radar imagery. Radar signals (here contoured by brightness) revealed two reflecting lobes. The asteroid is about 1 × 1.5 km and apparently has a compound structure of two spheroids barely touching or with a narrow neck between them. (Arecibo Observatory and Jet Propulsion Lab, courtesy Steven Ostro)

to rest on the surface of their primaries, making compound asteroids or at least irregular shapes spinning faster than the asteroid average period of 6.5 hours.

Some support for the compound asteroid idea came from asteroids such as 1580 Betulia (diameter 6 km, period 6.1 h), with a 3-km "bump" on one side (Tedesco and others, 1978), or from 624 Hektor with its roughly 2:1 axis ratio (Hartmann and Cruikshank, 1978; Weidenschilling, 1980). Such shapes were measured from each asteroid's light variations as it rotated.

In spite of all this interest, later observations did not confirm the earlier occultation claims for asteroid satellites, which were often made under poor conditions. Therefore, most observers eventually dismissed supposed asteroid satellites as a figment of faulty data.

But there are some additional haunting observations. For instance, some impact craters apparently formed in pairs. On Earth, for example, the Clearwater Lakes crater pair, 32 and 22 km across, were created by a double impact 290 My ago.

A new, related episode came a decade later. Radar observer Steve Ostro (1990), who had already bounced radar signals off several asteroids and found evidence for occasional nonspherical, unusual shapes, reported that the 1.0 × 1.7-km asteroid 1989 PB had the shape of two touching spheroids of almost comparable size—the classic shape for the hypothetical compound asteroid (see Figure 7.7).

The current research questions are (a) what controls asteroid shapes, (b) whether compound asteroids and asteroid satellites really exist, and if so, (c) how they form. Perhaps asteroid collisions are involved, but there may also be a link to cometary phenomena (Hartmann and Tholen, 1990). Comets apparently gain elongated shapes and sometimes seem to break spontaneously into two or more pieces as a result of their ice loss (see next chapter, pages 231–233). Perhaps some of the elongated or compound "asteroids" (and the double craters) were produced in similar events as larger bodies lost volatiles and broke apart.

NUMBER AND MASS OF ASTEROIDS

To estimate the "total number" of asteroids, we must specify a minimum size, since the smaller the size, the more asteroids there are. Only 3 are bigger than 500 km across, but 12 are bigger than 250 km, and hundreds exceed 100 km. The relation between number and size is called the **size distribution** (or **mass distribution** if mass instead of size is the variable). The estimated size distribution is shown in Figure 7-8.

Mass Distribution as a Clue to Collisional History

Figure 7-8 contains clues about the history of asteroids. In the first place, the trend roughly resembles a power law of the form $N \propto D^{-2}$, where N is the number in a given diameter (D) interval on the logarithmic scale. This type of power law is the size distribution produced when rocks smash against each other in nature (Hartmann, 1969b), thus supporting the idea that asteroids have undergone mutual fragmentation. Figure 7-8 demonstrates this point. The size distributions of shattered rock fragments closely resemble those of asteroids.

Ceres is somewhat larger than expected from the trend predicted by the power law and is nearly twice as large as the second-biggest asteroid. Ceres, at about 1.4×10^{21} kg, contains nearly half the mass of all asteroids. Vesta, the second largest, has a mass probably around 2×10^{20} kg. The total asteroid mass, around 3×10^{21} kg, is only about 4% the mass of the moon.

In 1958 Kuiper and others first determined this distribution and pointed to the hump in the curve at about 60 km. Anders (1965) suggested that the hump might be a remnant of an original size distribution of more Gaussian shape, with most objects 50 to 120 km across. This is not proven. Another idea is that the hump represents strong internal cores that did not shatter.

Chapman and Davis (1975) asserted that the initial belt could have been several times more massive than it is today. This massive belt would have rapidly ground itself down to the present belt, with the smallest dust fragments being removed by Poynting-Robertson and radiation forces. In this view, the hump might involve a difference in strength among different asteroid types, rather than be a fossil remnant of an initial population (Davis and others, 1979; Ryan and others, 1991).

In spite of differences in interpreting the detailed shape of the size distribution, analysts agree that the general form is probably the result of collisional evolution among a swarm of uncertain initial number. Typical encounter velocities in the belt are 5 km/s, quite adequate to fragment many asteroids (Piotrowski, 1953). Modeling of the collisions also suggests that the excess size of Ceres relative to the smaller objects may be due to the beginning of the normal runaway accretionary growth of the largest object in the initial swarm of asteroids (see Chapter 5).

HEKTOR: AN UNUSUAL ASTEROID

Many asteroids could be singled out as having unusual compositions, light curves, orbits, or some other properties, but Trojan 624 Hektor serves as a good example of the questions that arise when certain interesting asteroids are investigated as unique planets in their own right. Hektor is roughly 150×300 km. It is the largest Trojan, about twice as large as the next-largest Trojans, which are not nearly as elongated. This hardly fits the conventional theory that elongated asteroids are splinter-shaped fragments of larger parent bodies. No larger parent is available in the Trojan clouds; and if one did break up, we would hardly expect it to have made a huge elongated fragment and a group of smaller spheroidal fragments.

Seeking an alternative explanation, Hartmann and Cruikshank (1978) suggested that Hektor may be a unique dumbbell-shaped, com-

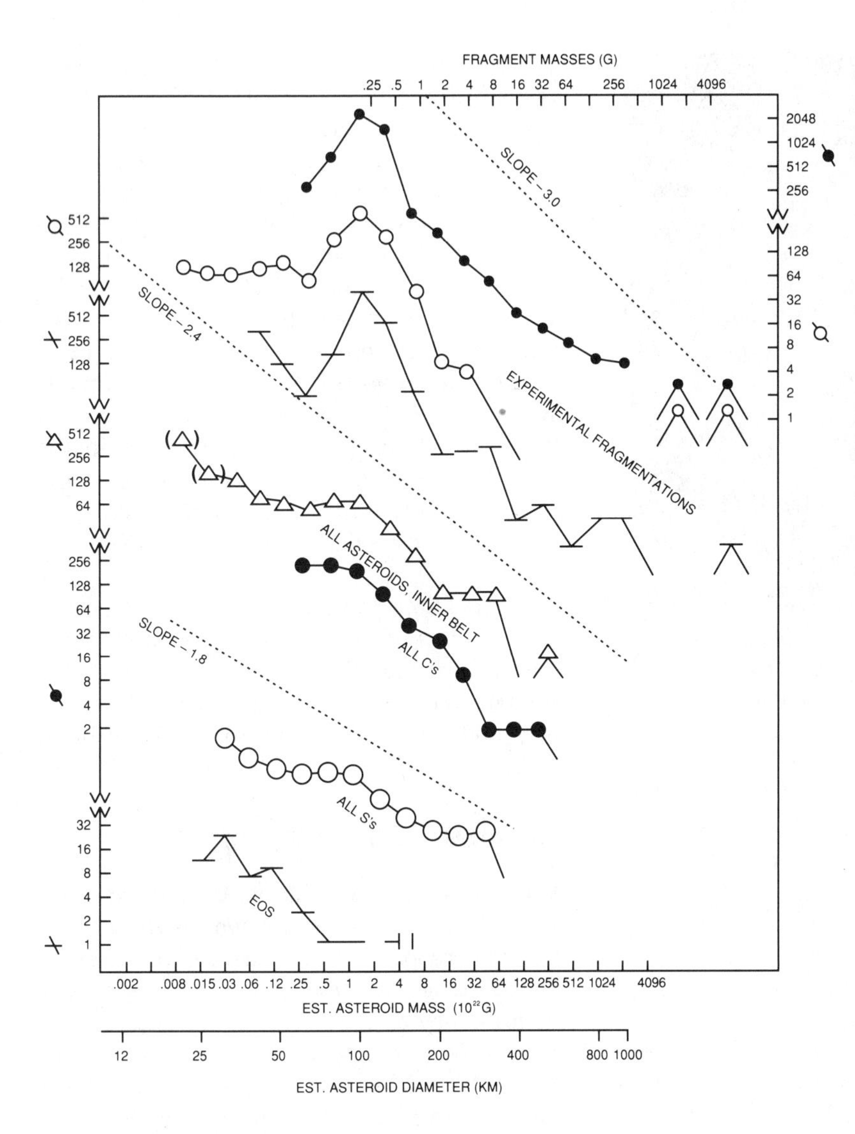

Figure 7-8. Comparison of size distribution of impact-shattered rock fragments and asteroids, showing similar structure, approximating a power law. Dotted lines are reference power laws with indicated exponents in equation of N vs. D (see page 208). Top three curves show fragments of artificial aggregate targets of rock. Bottom four curves summarize actual asteroids: in inner half of belt (a < 2.8 AU), C-class, S-class, and members of Eos Hirayama family, which may be fragments of a single asteroid collision event. "Bump" in asteroid size distribution, at diameters 50–120 km, is seen in curves for all asteroids and C-class asteroids (see text).

pound asteroid formed when two of the larger round Trojans collided at too low a speed to cause complete fragmentation and disruption of the fragments. Weidenschilling (1980) extended these ideas by noting that such a collision could pulverize Hektor's material, creating a weak body that would reaggregate into a shape determined by hydrostatic equilibrium. Calculating such a shape for Hektor's size, lightcurve amplitude, and rotation period, Weidenschilling proposed that Hektor may consist of two elongated ellipsoidal bodies nearly in contact, with a mean bulk density of around 2500 kg/m^3.

THE SURFACE OF AN ASTEROID

No asteroids have yet been photographed or studied at close range, but close-up photos of the asteroid-sized bodies Phobos, Deimos, Amal-

Figure 7-9. Possible analogue of an asteroid surface: a landscape on Phobos. Rolling terrain, bowl-shaped craters, and one of the major grooves were photographed from 120 km away by the Viking Orbiter. Grooves may be surface expressions of regolith-covered internal factures caused by large-scale collisions. Area is about 2 km across, and smallest details about 12 m. (NASA: Viking orbiter photo slightly blurred by spacecraft motion)

thea, and Mimas indicate several interesting features suggesting asteroid configurations (Figure 7-9). First, they are peppered with craters, indicative of the innumerable collisions. Second, as seen in Figure 7-4, they are irregularly shaped but rounded and without ragged fractured surfaces, suggesting that small-scale collisions have rounded off their surfaces. Third, though Phobos and Deimos have a similar composition and environment, they have somewhat different decimeter-scale surface structures. This suggests that surface gravity and time since the last major collisions (both controlling the amount of boulder-sized rubble on the surface) may be important factors in determining surface appearance.

Direct polarization and radar measures of asteroids have been interpreted as indicating dust-covered surfaces, but the depth of the dust layer is uncertain. Some authors suggest that the soil layer may be thin on the smallest asteroids (whose low gravity inhibits soil buildup) and thicker on larger asteroids (Chapman and others, 1978).

Following lunar terminology, a soil layer of dust and rock fragments up to a few meters deep, accumulated by repeated cratering, has come to be called a **regolith** (rocky layer). A computer model of regolith buildup (Housen and others, 1979) suggests that **megaregoliths**, or regoliths a few kilometers deep, may accumulate on the larger asteroids, while regoliths may reach only a few centimeters depth on asteroids 1 to 10 km across.

As mentioned in the last chapter, properties of gas-rich meteorites suggest that they developed in such regolith soil layers. Phobos and Deimos both have regoliths, though Deimos has more large boulders (Veverka, Thomas, and Duxbury, 1978).

THERMAL HISTORIES OF ASTEROIDS

Aside from a complex history of collision, fragmentation, and surface erosion, asteroids have also undergone different histories of internal heating, which melted portions of some parent bodies. Many analysts used to assume that the internal parts of parent bodies were the most strongly heated, but even this is not certain. Direct observations of asteroid surfaces show that some surface layers are not primitive material. While 1 Ceres, the largest asteroid, has a primitive carbonaceous-chondrite-like surface, complete with water of hydration (Lebofsky and others, 1977, 1982), second-largest, 4 Vesta, has a basaltic achondritic surface that resembles lava. Vesta is round and apparently not a fragment exposing a once-melted interior surface. Feierberg, Larson, and Chapman (1981) conclude from the distribution of spectral types that most other observed asteroids have primitive unmelted surfaces, where lava did not erupt in quantity.

Why would some bodies have acquired lava-like surfaces while the surfaces of both larger and smaller bodies remained pristine? The answer is not known, though some models have been suggested and will be discussed in Chapter 9 (see especially Figure 9-11).

MINES IN THE SKY

Because asteroid research has revealed the likelihood of large objects containing nickel-iron alloy and other rare elements, scientists have come to realize that asteroids may have enormous economic value (Gaffey and McCord, 1977). Unlike lunar igneous rocks, in which many elements melted and blended into forms difficult to retrieve, some asteroids (judging from meteorites) contain elements in the form of primitive condensates, and others contain samples of nearly pure nickel-iron cores. Even those with minimal metal content may contain water, hydrogen from the solar wind, and building materials useful to future colonists in space.

Consider a 1-km-diameter asteroid, of which dozens approach Earth's orbit (not to mention the thousands still in the belt). If it contains typical chondritic stony material, it has around 7% to 17% by weight free nickel-iron alloy, whose market value at 1982 rates would be of the order $100 to $300 billion. If such an asteroid had a substantial regolith layer, the iron flecks in that layer could conceivably be harvested with simple magnetic rakes! Gaffey and McCord (1977) discuss the case of a similar nickel-iron asteroid, whose value would be closer to $4 trillion!

As human capability in space increases, we will inevitably head toward the time when the cost of going after these resources will be less than their value. Some writers have raised the specter of humanity despoiling the solar system, in the same manner that overindustrialization is beginning to despoil Earth's environment. But to this writer the prospects seem just the opposite. With a careful balance of research and exploitation, we could learn from and process materials in space in a way that would begin to take the pressure off Earth's ecosystem. A transition from Earth-based manufacturing to interplanetary manufacturing could eventually reduce pollution and ravaging of Earth by an Earth-based society bent on ripping the last dwindling resources from the land.

For these reasons, NASA-sponsored conferences have been devoted not only to the scientific exploration of asteroids (Morrison and Wells, 1978) but also to missions that would visit asteroids and bring small ones back to Earth orbit for exploitation of their materials (Criswell, 1978).

SUMMARY

Asteroids are important to planetary science in a number of ways. As a swarm of small bodies they give us a tableau resembling the conditions of collision in the early solar system, albeit at higher speed. They exemplify bodies that could form craters on planets, and they tie in beautifully with meteorite evidence about parent bodies' orbits, sizes, compositions, and surface regoliths. Certain meteorites are likely to be pieces of certain (known?) asteroids. Asteroids, particularly Apollos, may provide water, hydrogen, metal, and silicate materials for future solar-powered, space-borne manufacturing centers that might alleviate pollution on Earth.

Most asteroids can be viewed as debris left over after planet formation and trapped in long-lived reservoirs such as the main belt and Lagrangian clouds. Many asteroids have gone through complex "processing," however, including extensive collisions and heating.

Different types of asteroids formed in different parts of the solar system, a fact consistent with the condensation sequence theory of planetesimal origin and with the zonal pattern of composition types in the belt. Asteroids formed beyond the belt may have been more ice rich, while asteroids formed within the inner edge of the belt may have been primarily stony or stony-iron. Certain planetesimals originally formed among the terrestrial planets were probably perturbed during close passes to Earth and then Mars in such a way as to have ended up being "stored" on the inner edge of the belt. Such bodies, if identified by composition, might be exciting targets for early asteroid exploration, since they might serve as "Rosetta stones" that preserve records of early terrestrial planet formation.

Perturbations, primarily by Jupiter, have moved some asteroids from the main belt onto Apollo-like orbits that intercept Earth. Other objects on such orbits classified as asteroids may once have been comets, but spent so much time in the inner solar system that their icy volatiles evaporated and left only rocky (carbonaceous?) cores. As will be clearer from the next chapter, comets and asteroids may be more closely related than their seemingly dramatic differences in appearance might imply.

CONCEPTS

asteroid
minor planet
asteroid belt
commensurate orbit
resonance
Kirkwood gap
Lagrangian points
Mars-crossing asteroids
Amor asteroids
half-life
Earth-crossing asteroids
Apollo asteroids
Aten asteroids
chaotic orbit
Hirayama families
remote sensing
phase angle
infrared reflectance spectrophotometry
pyroxene absorption band
absolute magnitude
light curve
compound asteroids
size distribution
mass distribution
regolith
megaregolith

PROBLEMS

1. If Jupiter were magically moved a few percent closer to the sun, describe how the Kirkwood gaps' structure would change.

2. Searches have been made for "Trojan" asteroids at Lagrangian points of planets other than Jupiter, but none have been found so far. What location in the solar system offers the best prospects for such objects, and why?

3. (a) How might an Amor asteroid be converted to an Apollo asteroid? (b) Would the most likely time scale for such conversion of an Amor be 10^3, 10^7, or 10^9 y?

4. Suppose members of a certain Hirayama family were created by breakup of a large asteroid by high-speed collision with a much smaller one. Why might the family members have different compositions (and hence spectral properties) even though mostly originating in a single parent body?

5. (a) Describe three major spectral classes of asteroids and their likely composition. (b) How does 4 Vesta differ from these, and what might its composition be? (c) How does 1685 Toro differ, and what might its composition be?

6. Name and describe the largest asteroid. How does it stand out from the others as a group?

7. Describe some possible solutions to the paradox that most belt asteroids are C type, while only a small fraction of meteorites are carbonaceus chondrites.

8. (a) Describe the range of shapes of asteroids, including the most common shapes and some of the more irregular examples. (b) How are these shapes deduced?

9. Give some lines of evidence that asteroids have regoliths.

10. Give some lines of evidence that many asteroids melted very early in their histories but that the melting was not smoothly correlated with asteroid size.

11. Give lines of evidence that asteroids as a group have undergone many collisions.

ADVANCED PROBLEMS

12. What change of velocity would be required to move an asteroid fragment from a circular orbit on the inner edge of the belt (say 2.0 AU) to an Earth-crossing orbit? (Note: Most analysts believe this change is too high to occur in a single acceleration during a collision or fragmentation—an argument against direct delivery of belt asteroids to Earth.)

13. What change in velocity would be required to move an asteroid fragment from the innermost belt at 1.8 AU onto a Mars-crossing orbit?

14. Asteroid 1566 has an unusual orbit that reaches perihelion only 0.19 AU from the sun. It has a visual albedo of about 0.18. Estimate the maximum likely surface temperatures to be found during perihelion passage.

15. An asteroid 100 km across is suspected of having a satellite 10 km across and 500 km away. (a) What revolution period would the satellite have? (b) Which piece of information is irrelevant? (c) What other numerical datum is needed in part (a)? Defend the value you assumed. (d) Observers monitoring the asteroid's rotational light curve claim to see an event confirming the satellite's existence. What might their evidence look like and what might the event be? (Note: There are four different types of such events!) (e) Describe changes that would be expected in the light curve evidence over a period of a few months after the event was detected.

16. (a) Confirm the derivation of the relation between strength S and maximum size R_{Max} of an irregular body given in the math notes in this chapter. (b) Suppose primordial planetesimals were loosely aggregated objects with the low density of 2300 kg/m^3 and the weak strength of 2×10^6 N/m^2. Describe the shapes that might have existed among populations with a wide range of sizes.

17. (a) Confirm the derivation in the mathematical notes on maximum rotation rates. (b) You are captain of a research vessel approaching a coherent, rocky, 4-km-diameter Apollo asteroid rotating with a period of 1.5 h. Describe a strategy for landing your ship or crew on this asteroid in order to carry out surface investigations.

CHAPTER EIGHT

COMETS

Comet West, photographed in March 1976, displays many features of comets. The faintly glowing tail consists of dust and gas released by the comet nucleus as it is heated by the sun while it passes through the inner solar system. (Steven M. Larson)

A large nearby comet can stand out as a strange, ethereal banner poised in the night sky (Figure 8-1), and thus comets have a long history of affecting both the superstitious and the scientific mind (see review by Brandt and Chapman, 1981). Halley's comet, which returns every 76 or 77 y and is expected in the inner solar system in 1986, is one of many comets regarded as omens. Its appearance in A.D. 66 was said to have heralded the destruction of Jerusalem in A.D. 70. Five circuits later, in 451, Europeans thought it marked the defeat of Attila the Hun. In 1066 it presided over the Norman Conquest of England. In 1456 its appearance coincided with a threatened invasion of Europe by the Turks, who had already taken Constantinople three years earlier. Pope Calixtus III ordered prayers for deliverance "from the devil, the Turk, and the comet." But modern history teaches us that most cultures are not without portentous events that can be handily associated by astrologers with normal celestial events. Today we (some of us, anyway) reject astrological mumbo jumbo.

Truths about astronomical events are more interesting than fictions about them. On June 30, 1908, a mysterious explosion occurred over the Tunguska region of Siberia (Figure 8-2). English observatories 3600 km away noted unusual air pressure waves, and seismic vibrations were recorded 1000 km away. At 500 km, observers reported "deafening bangs" and a fiery cloud.

Some 200 km from the explosion, the object was seen as "an irregularly-shaped, brilliantly white, somewhat elongated mass . . . with [angular] diameter far greater than the moon's." Carpenters were thrown from a building and crockery knocked off shelves. An eyewitness 110 km from the blast reported:

> Suddenly . . . the sky was split in two and [about 50°] above the forest the whole northern part of the sky appeared to be covered with fire. . . . I felt great heat as if my shirt had caught fire. . . . There was a . . . mighty crash. . . . I was thrown onto the ground about [7 m] from the porch. . . . A hot wind, as from a cannon, blew past the huts from the north. . . . Many panes in the windows [were] blown out, and the iron hasp in the door of the barn [was] broken.

Figure 8-1. Comets are a common spectacle in the sky. This view of Comet Ikeya-Seki, a 20-s exposure at f 1.9 on Tri-X film, was made in 1965 with a stationary 35-mm camera. (Steven M. Larson)

Figure 8-2. Devastation caused by the great Tunguska impact event in Siberia in 1908. This view, more than a decade later, shows trees blown outward from "ground zero" by the force of the explosion. (E. L. Krinov, 1966)

Probably the closest observers were some reindeer herders asleep in their tents about 80 km from the site. They and their tents were blown into the air and several herders lost consciousness momentarily. "Everything around was shrouded in smoke and fog from the burning fallen trees."

The probable cause of the **Tunguska explosion** was a collision between Earth and a piece of a comet, about 20 to 60 m in diameter. There are several lines of evidence. The impact left no crater, producing only an atmospheric explosion equivalent to a 10- to 15-megaton bomb. Contrary to ordinary meteorite impacts, no substantial meteorite fragments could be found in the vicinity. (Even the 20 000-y-old Arizona crater is surrounded by scattered iron meteorites; the meteorite did not entirely vaporize when forming the crater.)

These facts suggest that the body was not a firm stony or metallic object, but rather a weak object, perhaps rich in volatiles (ice?) that expanded explosively when it hit the atmosphere without scattering solid fragments. Observations of the object passing through the atmosphere indicate that it encountered Earth in a retrograde orbit at high speed, perhaps 50 km/s, typical of comet orbits but not of ordinary meteorites (Krinov, 1963). For weeks afterward, the night sky in Europe and Russia was anomalously bright, possibly due in part to fine dust blown from the comet explosion into the upper atmosphere and in part to comet-tail material enveloping Earth. A 1961 expedition recovered soil samples containing small spherules, and a later report (Ridpath, 1977) indicated that the soil contained abundant microscopic particles of carbonaceous chondrite.*

Whether the Tunguska object was truly a

---

*Pseudoscientific literature popularized occult explanations of the Tunguska explosion, including the idea that it was a nuclear explosion caused by the crash of an alien spaceship. Among the alleged evidence was a report of radioactivity at the impact site and elsewhere, suggesting a nuclear rather than chemical explosion. As reviewed by Oberg (1977), the increased radioactivity during the year of the blast was actually within normal statistical fluctuation and hence not considered significant. Indeed, physicist Willard Libby pointed out that the levels of radioactivity in his measurements rule out the possibility of a nuclear blast as large as 10 megatons. Furthermore, the identification of carbonaceous chondrite dust in the soil at the site appears to clinch the identification of the Tunguska object as some kind of interplanetary debris.

A thought-provoking novel about the consequences of a larger comet impact on Earth is *Lucifer's Hammer* (Niven and Pournelle, 1977).

comet or an unfamiliar type of outer-solar-system debris is uncertain, but its properties do fit our ideas of a comet's nature. To see this, we need to back up and review observational evidence about comets.

APPEARANCE AND NOMENCLATURE

A **comet** is a diffuse bright object that moves slowly among the stars and may become visible to the naked eye for some days or weeks. Most are quite faint. Although most people know something about comets, many have erroneous conceptions. For example, one sometimes finds references to "a comet flashing across the sky." Such a statement is totally misleading, because comets are almost always visible for several days at a time and are so far away that their motions cannot be detected by the naked eye except by watching their progress among the stars for hours. Authors who speak of comets "flashing across the sky" probably confuse them with meteors, atmospheric phenomena that we shall discuss later in this chapter.

Seneca, the Roman contemporary of Jesus, wrote that "someday there will arise a man who will demonstrate in what regions of the heavens the comets take their way." That man was Tycho Brahe, who in 1577 deduced from the slow movement of comets and their lack of parallactic shift when observed from different locations that they are more distant than the moon, and not in Earth's atmosphere, as some people had supposed.

The word comet comes from the Greek word *komē* ("hair"). The name describes the long wispy **tail** for which comets are famous. Orientals knew comets by the equally descriptive term "broom stars." Comet tails generally extend away from the sun, as shown in Figure 8-3. In addition to the tail, a comet has a **head**, composed of the **coma**, a bright, diffuse region immediately surrounding the starlike **nucleus** (Figure 8-4). The head is the source of most of the light. Among the less spectacular comets, the inner coma is all that can be seen with the naked eye. With a large telescope, the true nucleus of even a spectacular comet appears as no more than a starlike point, indicating that cometary nuclei must be very small, much smaller than a planet. Certain measurements indicate sizes on the order of a few kilometers (see below). Yet these small nuclei are able to produce glowing gas comas that expand to sizes bigger than planets, and dusty tails that span an astronomical unit.

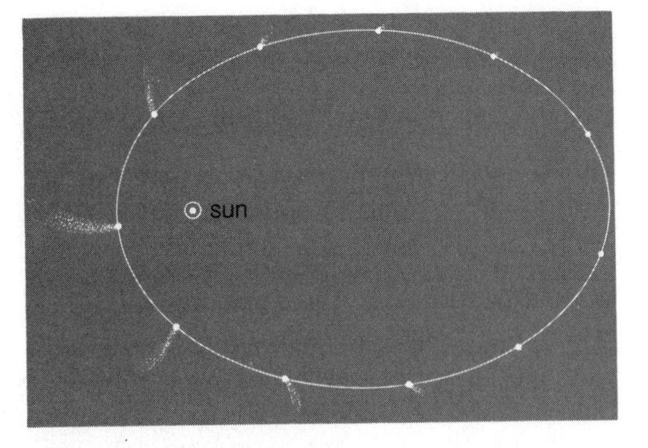

Figure 8-3. Orbit of a typical comet, showing development of the tail as the comet approaches and recedes from the sun.

Comets are usually discovered when they are faint and distant, approaching the sun at a distance of several astronomical units. In general, they brighten and their tails develop and lengthen as they approach the sun. There is a tendency for their comas to reach a maximum diameter at roughly 1.5 to 2.0 AU from the sun. Evidently, this maximum occurs because no gases are released at greater distances, while the molecules that make up the coma are readily disassociated by solar ultraviolet radiation when the comet nears the sun.

As the comet swings around the sun, it is very bright and has a well-developed tail, as shown in Figures 8-1 and 8-3. Often the comet is not visible from Earth at this stage, however, because it is lost in the glare of the nearby sun. As the comet moves away from the sun, it becomes visible once more, and its coma and tail shrink again until the whole comet appears only as a faint, distant, starlike point, even on the best photographs. Most comets then continue their out-

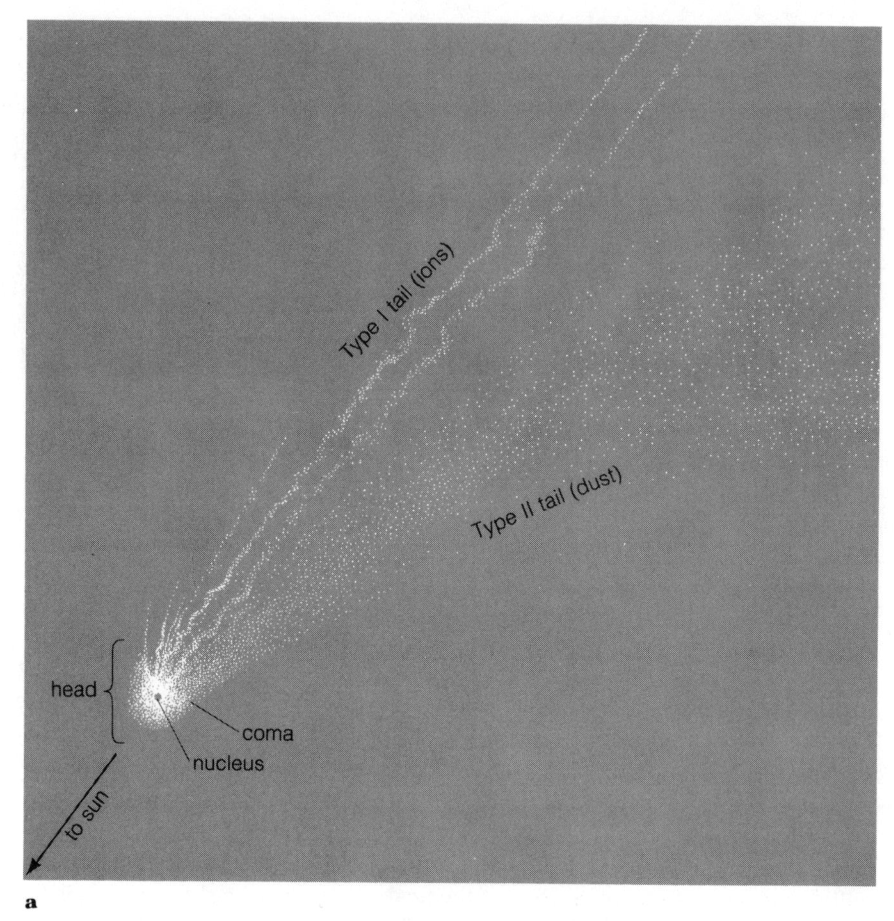

Figure 8-4. (**a**) *Nomenclature of a comet.* (**b**) *Comet Kohoutek, showing the same features, with minimal separation of Type I and Type II tails. (NASA, Lunar and Planetary Laboratory, University of Arizona)*

ward journey, reaching the extreme outer edges of the solar system and returning toward the sun only after many thousands of years.

DISCOVERY

Roughly a dozen comets are observed each year, but only about one per year reaches naked-eye status. About half of the dozen are new discoveries. The other half are known comets on return trips around the sun, which are known as **recovered comets**. Most of the recoveries are first detected on astronomical photographs taken specifically to locate them. Many new comets are found by amateur astronomers who make a hobby of scanning the skies with modest telescopes, checking any suspicious fuzzy objects against the catalogs of known nebulas (which can easily be mistaken for comets in small telescopes) and looking for slow movement against the starry background. In recent years, several comets have been discovered by airline pilots, who are favored by the very clear skies at high altitudes and by their alertness during hours when most people have no time to watch the sky.

When a newly discovered comet is confirmed, an announcement is sent to astronomers around the world by the International Astronomical Union, and the comet is assigned a designation consisting of the year and a letter in order of discovery or recovery: 1973a, 1973b, and

b

so on. Comets are also popularly named after their discoverers, such as Comet Burnham and Comet Ikeya-Seki (which had two independent discoverers). After a year or two, when observations of the comets have been collected and reliable orbits determined, they are assigned new, permanent designations with Roman numerals for the order in which they passed their perihelion points: 1973 I, 1973 II, and so on.

TAILS AND COMAS: APPEARANCE AND COMPOSITION

As shown in Figure 8-4, comets have two types of tails, Type I and Type II. **Type I tails** are essentially straight, often structured by fine, linear streamers or "rays." Their spectra show emission lines indicating a composition of mostly ionized gas. As a mnemonic aid, it is helpful to think of the *I* in Type I: This Roman numeral is linear and can also stand for *i*onized. The principal light-emitting molecules are CO^+, N_2^+, and CO_2^+. Table 8-1 lists additional ions and neutral atoms and molecules that have been identified.

Type II tails are usually broad, diffuse, and gently curved. Their spectra show only the reflected spectrum of the sun. This indicates that instead of being composed of gas atoms and molecules that absorb or give off their own characteristic light, they are composed of grains of dust that merely reflect (scatter) sunlight. They may show delicate rayed structure, seen in Figure 8-5, and caused by individual dust jets from the nucleus.

Some comets have Type I tails, some have Type II, and some show both. The appearance of tails is affected not only by type, but also by aspect angle, as shown in Figure 8-6.

Occasionally a comet displays a seemingly anomalous tail that appears to point toward the sun; Comet Arend-Roland in 1957 was a widely publicized example. This is only a projection effect.

Table 8-1
Chemicals in Comet Tails and Comas

Gases	
Inorganic:	H, NH, NH_2, O, OH, H_2O
Organic:	C, C_2, C_3, CH, CN, CO, CS, HCN, CH_3CN
Metals (and silicon):	Na, Ca, Cr, Co, Mn, Fe, Ni, Cu, V, Si
Ions:	CO^+, CO_2^+, CH^+, CN^+, N_2^+, OH^+, H_2O^+
Dust particles	
Silicates inferred from infrared reflection spectra	

Source: Data adapted from a table by Delsemme (1977).

Figure 8-5 Linear gaseous streamers in the head are shown in this telescopic photo. The nucleus and central coma are much brighter than the outer structure, and are consequently overexposed. (University of Michigan Observatory/Cerro Tololo)

Figure 8-6. The configuration of a comet's tail is affected by the direction from which it is seen. In this view of Comet Humason, the wispy tail was directed almost directly away from Earth, adding to its irregular appearance. This 60-min exposure tracked the comet's motion, blurring background star images. (Naval Observatory, official U.S. Navy photograph)

As we look along the line from Earth to the comet, most of the tail may be on the antisolar side, but a portion of the curved tail may appear in projection against the sky to be on the sunward side. In the case of Arend-Roland, the "solar-directed tail" was part of a Type II tail seen in projection. It appeared as a narrow spike only as Earth passed through the comet's orbit plane. This indicated that the dusty Type II tail was highly confined to the plane of the comet's orbit.

Comet tails, then, are composed of gas and dust emitted by the comet's nucleus and directed more or less away from the sun, with Type I being most nearly antisolar. Thus, during the part of its orbit when a comet moves out away from the sun, the tail leads; it is false to picture the tail trailing behind the comet like a woman's long hair as she speeds down the street on a bicycle.

Why does the tail point away from the sun? The tiny dust grains that compose the Type II tails are repelled from the sun by radiation pressure (see Chapter 3), and each particle follows an orbit determined by gravity and radiation. According to Kepler's second law, the particles farther from the sun revolve more slowly and so they lag behind, giving the Type II tail its characteristic curvature. But this does not explain all properties of Type II tails, not to mention Type I tails. Observations of comets showed that knots of material in Type I tails were accelerated away from the sun faster than could be explained by radiation pressure. This led Biermann (1951) to the concept of the solar wind streaming out from the sun (Chapter 3). Magnetic interactions couple the ions of the Type I tails to the outward-expanding ionized gas, or **plasma**, of the solar wind. To correct our analogy with the bicyclist's hair, we would have to say that she is riding in a fierce wind, sometimes into it with her hair streaming back, and sometimes with it, her hair streaming out in front of her.

The next step beyond identifying gases in comets' tails is to derive from the spectra the relative amount of each gas and the production rate at which the gas is being emitted by the comet nucleus. Table 8-2 gives recent estimates, from which Delsemme (1977) infers that water (H_2O) ice must be the dominant constituent of the nucleus. Methane (CH_4) and ammonia (NH_3) ices are also probably present in the nucleus. Spectral studies (Table 8-1) have not detected CH_4 or NH_3 gas in the comas or tails, although CH- and NH-based compounds are common (Delsemme, 1977).

ORBITS

Determining the orbit of a comet is a complex process requiring the best fit of many observed positions to an elliptical orbit. Study of the computed orbits quickly shows that comets fall into two rather distinct classes: long-period (or nearly parabolic) comets and short-period comets, illustrated in Figure 8-7 and in Table 8-3. **Long-period comets** have periods greater than a few hundred years and move in extremely eccentric orbits. Aphelion distances may reach as much as 40 000 to 50 000 AU, a sixth of the way to the nearest star. In these cases, intervals between solar approaches range from 100 000 to 1 million years. These aphelia are so far away that we can observe the comet in only a small portion of its orbit near the sun, where the comet is brightest. The record distance at which a comet has been seen is 11.5 AU, although few are observed beyond 3 or 4 AU.

Consider the orbits of long-period comets. If you were handed a 1-cm arc of a 1000-cm ellipse and of a similar parabola, you would have difficulty distinguishing them. This indeed is what happens when we try to derive orbits of long-period comets: We see only the small tip of the orbit near the sun, and it is very difficult to tell whether the orbit is very elliptical, parabolic, or even hyperbolic. This is a critical decision, because an elliptical orbit implies that the comet is a member of the solar system, while the last two cases imply that it comes from among the stars. (If you cannot justify these statements, recheck the discussion of orbits in Chapter 3.) Long-period comets are often loosely referred to as **parabolic comets**.

Evidence suggests that most nearly parabolic comets have barely elliptical orbits and thus are members of the solar system. As shown in Table 8-3, their eccentricities tend to cluster near 1.000, whereas interstellar comets would have larger eccentricities (hyberbolic orbits). No definitive comet orbit has been observed to have an eccentricity higher than 1.004; all cataloged hyperbolic orbits probably represent either observational error or perturbations of originally elliptical orbits. Also, inclinations of long-period orbits approach randomness; many are retrograde.

The **short-period comets** have moderately eccentric elliptical orbits, more eccentric than

Table 8-2
Emission Rates of Major Gases from the Nuclei of Four Comets

Chemical	Emission rate[a]
Hydrogen	4–65
Hydroxide	1–30
Oxygen	3–23
Carbon	1–6
Carbon monoxide	~8

Source: Adapted from a table by Delsemme (1977).
Note: The comets measured are Tago-S-K (1969 IX), Bennett (1970 II), Kohoutek (1973 XII), and West (1975n).
[a]Units of 10^{28} molecules/s at 1AU from the sun.

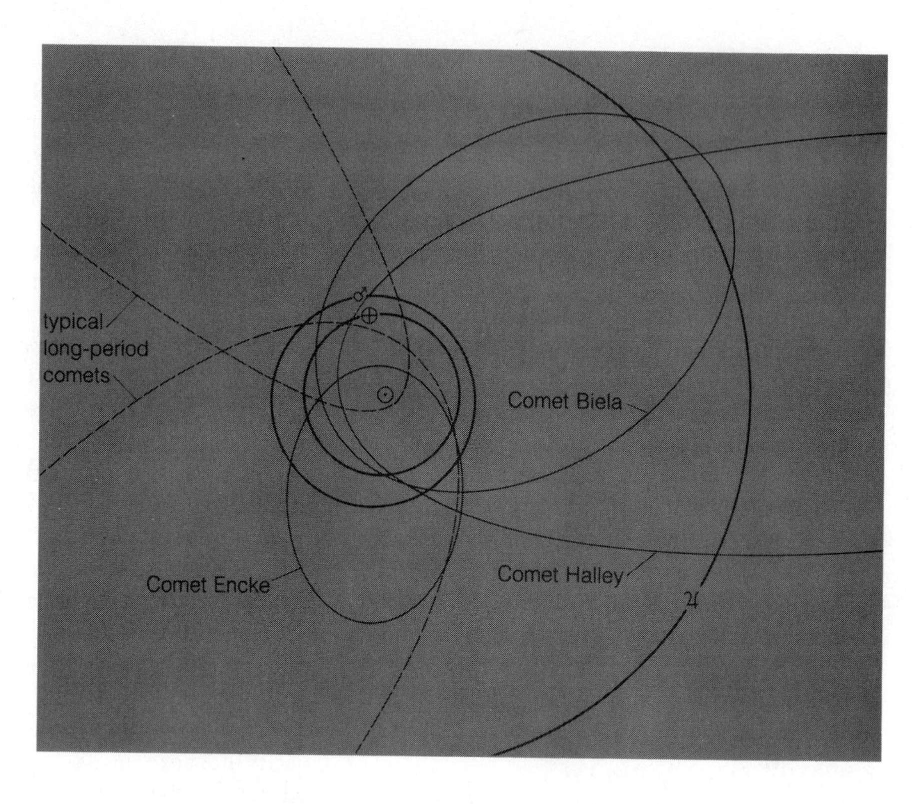

Figure 8-7. Orbits of typical comets, showing three short-period comets (Encke, Bield, and Halley) and two long-period comets.

planetary orbits but distinctly less eccentric than those of long-period comets. Typical aphelion distances are 4 to 7 AU, and inclinations are somewhat compressed towards the plane of the solar system, but with a substantial fraction of inclinations over 45°. About 7% are retrograde (inclinations over 90°). The short-period comets include many bodies with orbits like those of Amor and Apollo asteroids, with an average distance from the sun of a few astronomical units, modest inclinations (perhaps 30°), and perihelia of 1 to 2 AU. This similarity is one of the reasons for suspecting that some Amors and Apollos are burnt-out cores of comets. Short-period comets are often called **periodic comets** and designated by "P/" in front of the comet name.

A sampling of 583 comets cataloged in 1965 gives the following breakdown:

56 short-period ($P < 200$ y), seen in a total of 343 apparitions

43 short-period, seen in only one apparition each

130 long-period elliptical orbits

284 long-period "parabolic orbits"

70 long-period "marginally hyperbolic orbits"

Since hyperbolic orbits may result from observational error and since parabolic orbits ($e = 1.000$) are often assumed in order to make the calculations easier, we might regroup the breakdown more realistically as follows:

99 short-period comets

484 long-period (nearly parabolic) comets

A few comets may actually depart the solar system on hyperbolic orbits. Work pioneered by Stromgren (1914) has indicated that some comets, while approaching the sun on initially elliptical orbits, are perturbed by planets (especially

Jupiter) into hyperbolic orbits. We thereby gain a clue to the history of comets, as seen in the next section.

ORBITAL CLUES TO COMET HISTORIES

The Oort Cloud

The American astronomer H. N. Russell (1920) first pointed out that the distribution of semimajor axes of comets is not what would be expected if the comets were a modern family of objects supplied from *within* the planetary system and randomly perturbed by planets. Further study of the best-determined comet orbits showed a concentration of very large average distances from the sun (the long-period comets). Finally, Dutch astronomer Jan Oort (1950, 1963) showed that these statistics indicate that most comets belong to a vast swarm *surrounding* the solar system out to about 50 000 AU, far beyond the orbit of Pluto (39 AU). This swarm is called the **Oort cloud.***

In other words, the Oort cloud is a reservoir of long-period comets. The inclinations are nearly random; thus, the Oort cloud does not lie in the plane of the solar system, but surrounds it in the

*"Cloud" may convey an inappropriate picture since we refer to a swarm of comets on independent orbits. Maybe we should call it the "Oort crowd."

Table 8-3
Orbits of Selected Comets

Comet	*a* (AU)	*e*	*i*	*P*	Date of next or last perihelion	*q*[a] (AU)	*Q*[b] (AU)	Remarks[c]
Short Period								
Encke	2.2	0.85	12°	3.3	12/80, 3/84	0.34	4.2	D = 0.6–3.5 km
Schwassmann-Wachmann II	3.4	0.39	4°	6.4	3/81, 8/87	2.1	4.8	
Grigg-Skjellerup	3.0	0.67	21°	5.1	5/82	0.99	6.0	
d'Arrest	3.4	0.62	20°	6.4	9/82	1.3	5.5	
Swift-Tuttle	~24	0.960	114°	~120	~1982	0.96	~47	Perseid meteor source
Tempel II	3.1	0.55	12°	5.3	5/83, 9/88	1.4	4.8	
Arend-Rigaux	3.5	0.60	18°	6.8	11/84	1.4	4.9	D = 0.5–3 km
Giacobini-Zinner	3.5	0.71	32°	6.6	9/85	1.0	6.1	Draconid meteor source
Halley	17.9	0.97	162°	76	~2/86	0.59	28.4	
Schwassmann-Wachmann I	6.1	0.04	10°	15	10/89	5.8	6.7	D = 6–38 km; sporadic outbursts
Long Period								
Donati	~157	0.996	117°	~2000	1858	0.58	~313	
Humason	~204	0.990	153°	~2900	1962	2.13	~400	
Morehouse	Large	1.00	140°	Large	1908	0.95	Large	
Burnham	Large	1.00	160°	Large	1960	0.50	Large	
Kohoutek	Large	1.00	14°	Large	1973	0.14	Large	

[a] q = perihelion.
[b] Q = aphelion.
[c] D = diameter

form of a spherical halo where comet motions are half prograde and half retrograde. Comets *in* the Oort cloud are far too far from the sun to have comas or tails and are unobservable.

How many comets are in the Oort cloud? If we recall that roughly six new ones are discovered every year and that these may return only every 100 000 or more years, we could roughly estimate that at least 500 000 comets are available in the cloud. Actual estimates are much higher, since there must be many comets coursing through the solar system that we cannot see. Oort discussed populations as high as 10^{11} comets in the cloud, a number consistent with the possibility that comets are planetesimals formed among the planets and kicked outward by early perturbations into the cloud (similar to Pioneer 10's ejection from the solar system by its close encounter with Jupiter). More recently, it has been suggested that many comets also formed at great distances (Cameron, 1973) and may exist on the inside of the Oort cloud at 100 to 20 000 AU in relatively permanent orbits; estimates as high as 10^{14} such comets have been given (Whipple, 1975)! The 1977 discovery of unique "asteroid" 2060 Chiron, which orbits at 18 to 19 AU, raised the question of whether it is just one of the innermost members of such a swarm of comet nuclei, too far from the sun to show coma or tail activity.

Dynamical Evolution: From the Oort Cloud to the Inner Solar System

Comets in the Oort cloud are so far from the sun that they move very slowly by cosmic standards (on the order of 0.1 km/s relative to the sun) and are perturbed by the nearer stars. Such perturbations occasionally send one into the inner solar system, where it may be discovered by terrestrial observers. This adequately explains the long-period comets, with their mixture of retrograde and direct orbits that are highly inclined. But how can we account for short-period comets with orbits like Amor or Apollo asteroids, entirely within the planetary realm?

As pointed out long ago by Fayet (1910) and Russell (1920), Jupiter is the dominant perturber of incoming comets, and most short-period comets have aphelia near Jupiter's orbit. Usually one of the nodes, or points where the comet passes through the plane of the solar system, lies close to Jupiter's orbit. These facts suggest that incoming long-period comets often pass close enough to Jupiter to be thrown onto short-period orbits.

This explains why most short-period comets are prograde: The retrograde long-period comets move past the planet so fast that there is little time for radical perturbation, while prograde comets move "with" the planet, giving more opportunity for interaction. Nevertheless, early calculations of comets' interactions with Jupiter showed that while aphelia could frequently be reduced to below 10 000 AU, over 96% of the orbit changes would simply lead to reejection from the solar system in about 1 to 100 My (Oort, 1963; Öpik, 1963, p. 232; Öpik, 1966). Only a tiny percentage could evolve directly into short-period comets.

More recent calculations have shown how additional comets may evolve stepwise into the inner solar system, one planet at a time. Everhart (1977) started with a group of comets assumed to have a mean distance from the sun of 50 000 AU, an inclination of 6°, and perihelia in the range of 30 to 34 AU, close enough to interact with Neptune. He found that about 0.15% of those interacting with Neptune were perturbed inward, with about 0.02% eventually ending up as short-period comets. Of such comets interacting initially with Uranus, Saturn, or Jupiter (near perihelion), he found 0.03%, 0.14%, and 0.77%, respectively, ending up as short-period comets.

In summary, for every short-period comet that we see, thousands of comets may have dropped into the outer solar system and been reejected into the Oort cloud.

Orbit Irregularities: Nongravitational Forces

Close tracking of some comets has revealed that their orbits sporadically depart from Keplerian

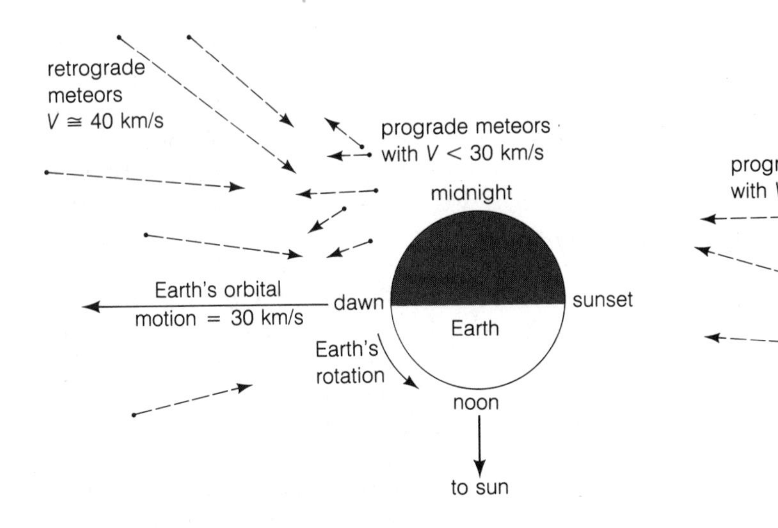

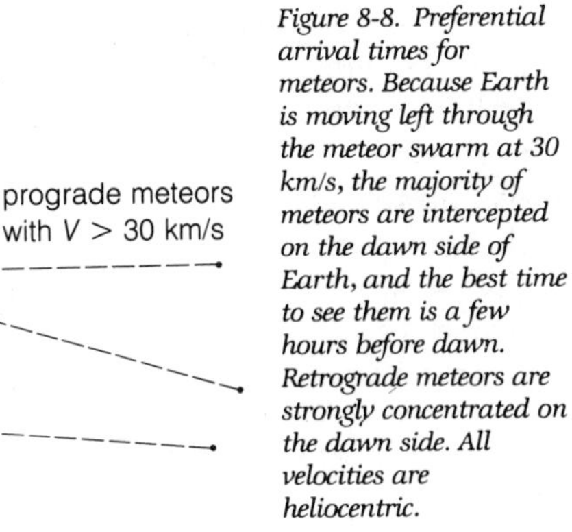

Figure 8-8. Preferential arrival times for meteors. Because Earth is moving left through the meteor swarm at 30 km/s, the majority of meteors are intercepted on the dawn side of Earth, and the best time to see them is a few hours before dawn. Retrograde meteors are strongly concentrated on the dawn side. All velocities are heliocentric.

ellipses! These irregular paths hardly indicate a breakdown in Kepler's laws or Newtonian gravitation, but they do indicate that the comets are being acted on by some other forces not influencing planets and asteroids. The origin of these **nongravitational forces** is believed to be the jet effect of bursts of gas off the surface of the comets. Some of the bursts may be short-lived eruptions, perhaps when pockets of gas break loose from a certain site. But even steady gas sublimation due to solar heating may produce an off-center force, since the maximum heat would occur on the afternoon face of a rotating comet nucleus.

METEORS AND COMETARY METEOR SHOWERS

Meteors, or "shooting stars," are familiar to anyone who has watched the night sky for more than a few minutes. They are small particles that heat up due to friction with air molecules when they plunge into the atmosphere, just as a reentering spacecraft heats up and creates a spectacular fireball. As early as 1861, Kirkwood (of Kirkwood gap fame) argued that meteors are debris of comets, and this has been confirmed through orbital studies.

Velocities and Orbits

Trajectories derived from photographic and radar observations of meteors show that they travel in orbits around the sun, similar to those of comets. Some of these orbits are directly related to specific comets, and most reach farther into the outer solar system than do typical meteorite orbits.

Could any meteors be interstellar debris? Photographic data on 2587 meteors and radar data on some 39 000 indicate no orbits from beyond the solar system (Millman, 1979).

Meteor Showers and Comets

On a typical night, most meteors are sporadic and quasi-randomly directed on downward paths across the sky at a rate of roughly eight per hour. The number seen usually increases toward dawn, for the reason diagramed in Figure 8-8. However, on the nights of about the same dates each year, one can see a **meteor shower**, a large number of meteors falling within a few hours or few days and appearing to come from the same direction in space.

Showers are named for the constellations that lie in the direction from which they come. One of the most famous showers is the Perseid shower, which comes in mid-August each year. The **radiant**, or direction of its source, lies in the constellation Perseus. Some other well-known meteor showers are listed in Table 8-4. During a shower the rate may rise to 60 meteors per hour (typical of the Perseids) or more. The most spectacular shower of recent times occurred on November 17, 1966, when the observers counting Leonid meteors in the predawn hours in the western United States noted a dramatic rise in the rate that continued until they could no longer count fast enough. For about half an hour meteors fell like snowflakes in a blizzard, and the estimated rate exceeded 2000/min! A time exposure of part of this shower is shown in Figure 8-9a. A similar shower of Leonid meteors fell in 1833.

How can we explain such observations? Obviously the earth is encountering a cluster of particles called a **meteoroid swarm**. In the case of the Leonid swarm, the duration of the shower indicated that the inner core was some 32 000 km across (McLean, 1967). Similarly, Davies and Lovell (1955) determined a diameter of 100 000 km for the inner core of the Draconid (Giacobinid) shower. In 1866 the Italian astronomer G. V. Schiaparelli (later more famous as the discoverer of the "canals" on Mars) showed that the Perseid meteoroids were clustered along the orbit of Comet Swift-Tuttle. Within a few years it was found that several shower meteoroids lay in swarms that coincided with comet orbits. The inference was unmistakable: Meteors and meteor swarms are associated with comets.

Their orbits are understandable in terms of celestial mechanics. Should a group of meteoroids become detached at low speed from a comet, they would initially lie in a swarm moving in the same orbit. But perturbation theory shows that the swarm would soon stretch out along the same orbit so that eventually the meteoroids would be more or less smoothly distributed all the way around the ellipse. If Earth's orbit intersects the cometary orbit and intercepts a relatively young, undispersed cluster of meteoroids, a spectacular shower like that of November 17, 1966, ensues. Older swarms are more dispersed.

Physical Nature of Cometary Meteoroids

Carefully calibrated photographs of meteors show how they decelerate as they hit the atmosphere. From this we can calculate the drag due to atmospheric friction and hence estimate the den-

Table 8-4
Dates of Prominent Meteor Showers

Shower name (constellation)	Date of maximum activity[a]	Associated comet
Lyrid	April 21, morning	1861 I
Perseid	August 12, morning	Swift-Tuttle
Draconid	October 10, evening[b]	Giacobini-Zinner
Orionid	October 21, morning	Halley
Taurid	November 7, midnight	Encke
Leonid	November 16, morning	1866 I
Geminid	December 12, morning	?

[a]Showers can appear several days before and after the peak activity on the listed date. Observations are best when the constellation in question is high above the horizon, usually just before dawn.
[b]The Draconids are now weak because their orbits have been disturbed by the gravity of planets, but future perturbations may again strengthen the shower.

a

b

Figure 8-9. Rare intense displays of the Leonid meteor shower. (**a**) *The shower of November 17, 1966. The rate of meteors visible to the naked eye was estimated to exceed 2000/min. This exposure of a few minutes' duration was made with a 35-mm camera. Meteors are the longer streaks. (D. R. McLean)* (**b**) *The intense Leonid shower of November 12, 1833, drawn by a contemporary artist. The meteors were described as "falling from the sky like snowflakes." (NASA)*

sity of the particles. Most cometary meteoroids are very low density, fragile objects. In fact, no known cometary *shower* meteor has been big enough or dense enough to reach the ground. An investigator of meteoroids has compared some of them to bits of burnt newspaper or the balls of fluff that collect under beds!

Various studies (Jacchia, Verniani, and Briggs, 1967; Verniani, 1969) also show that different showers (that is, different comets) produce meteoroids of different character. Most are very fragile and fragment upon hitting the atmosphere, but short-period comets, which have been exposed to the sun more often than long-period comets, tend to produce higher-density, stronger particles. There are exceptions to this tendency; the lowest-density meteors are the Draconids, associated with periodic Comet Giacobini-Zinner. The average densities of photographed meteors range from less than 10 kg/m^3 (Draconids) to 1060 kg/m^3 (Geminids, comet association unknown but possibly associated with a prehistoric, burnt-out comet).

Collections of microscopic interplanetary meteoroid fragments have been made by rockets and high-altitude aircraft. Micrometer-sized particles of this type are strewn off larger meteors in the atmosphere or may even enter the stratosphere without being severely destroyed. One study collected 300 particles, typically some tens of micrometers across but composed of aggregates of micrometer- and submicrometer-scale dust grains, as shown in Figure 8-10 (Brownlee, Rajan, and Tomandl, 1977). The composition of these particles most closely resembles C1-type carbonaceous chondrite material. However, differences from any known meteorites suggest that these meteoroids are a unique material distinct

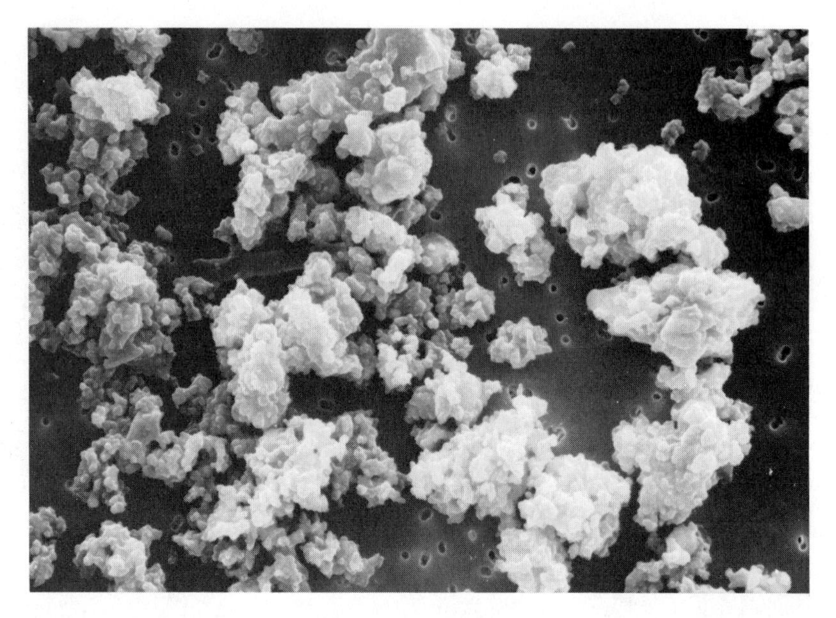

Figure 8-10. A cluster of microscopic chondritic meteorite particles (probably carbonaceous chondritic) fragmented into hundreds of pieces when it hit a collector plate on a U-2 aircraft at an altitude of about 20 km. It is spread across 100 μm, but if reconstituted it would be a fluffy object about 10 μm across. Its fragile, low-density structure as clumps of 0.5-μm scale particles shows clearly. Numerous such particles have been collected. (Electron microscope photo courtesy D. E. Brownlee, University of Washington)

from the normal meteorite population. All these observations are consistent with the idea that comets contain and discharge a carbonaceous-chondrite-like fine dust.

In addition to the dominant carbonaceous chondritic dust aggregates, other denser particles have been found in micrometeoroid collections. Nickel-bearing iron sulfide and crystals of olivine and pyroxene have been reported (Brownlee, Rajan, and Tomandl, 1977). Among microscopic particles netted in one collection experiment was a possible Perseid with mass of 5×10^{-10} kg, density of 7700 kg/m^3, and a magnetic composition presumably mostly iron. Geminid micrometeoroids were found to have densities of 2000 to 2600 kg/m^3, and an apparently glassy, not crystalline, structure. Such particles must therefore also be contained in the icy material of comet nuclei.

Meteoroid Ejection Velocities

The weak structure of most suspected cometary particles indicates that they are gently separated from their parent material. This is consistent with the idea of dust emitted from evaporating ice in a comet, but inconsistent with violent or high-temperature ejection from meteorites. From statistics of Draconids, Davies and Turski (1962) found that these meteoroids separated from their parent comet with velocities less than 10 m/s, again indicating that the particles are not ejected with any great force.

ZODIACAL LIGHT

With microscopic dust grains being spewed out of comets in the form of Type II tails, as well as being blasted off asteroids in numerous large and small collisions, it is hardly surprising that interplanetary space contains much fine dust.

You can satisfy yourself that this is true by observing the dust with your naked eye. It can be seen in the form of the **zodiacal light**, a faint reflection of sunlight off dust grains concentrated in the ecliptic plane. The zodiacal light is visible shortly after sunset and before sunrise (Figures 8-11 and 8-12). In the evening, just after the stars have come out, the zodiacal light is a

a

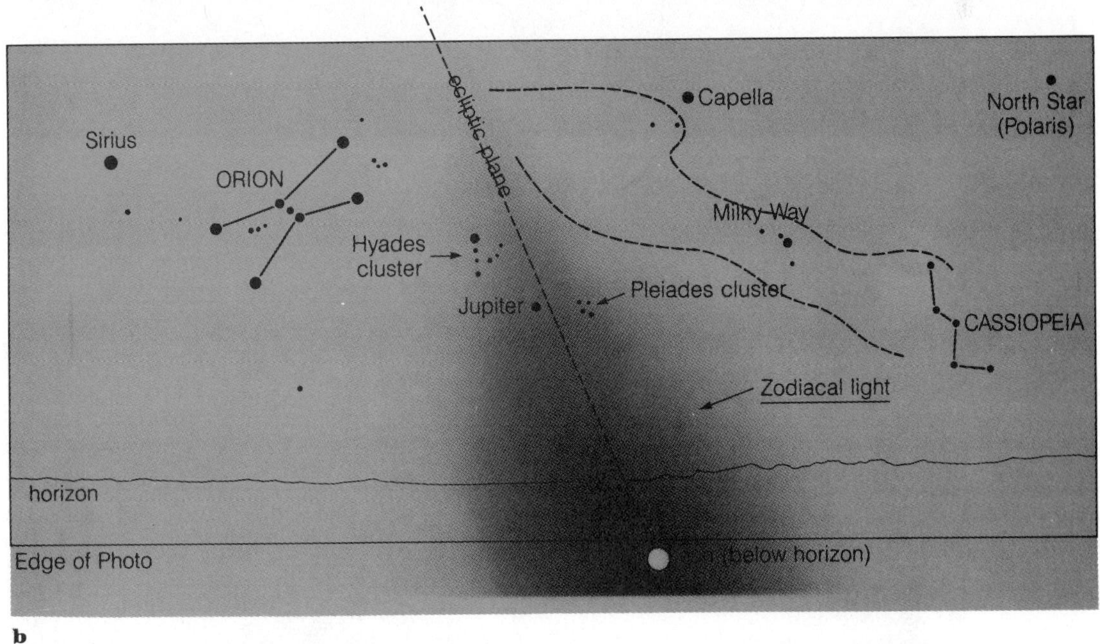

b

Figure 8-11. ***(a)*** *Fisheye panorama of the western horizon on a spring evening in Arizona shows the zodiacal light extending about 40° along the ecliptic plane. Field of view is about 120° × 50°. (Photo by author)* ***(b)*** *Key to features.*

faint band of light, widest at the western horizon, and tapering upward to a width of only 5° or 10°. Seen from mid-northern latitudes, it slants upward to the left, following the ecliptic. In other words, it lies along the plane of the solar system.

The zodiacal light is about as bright as the Milky Way and under good conditions can be traced at least 45° from the sunset point. At a point 180° from the sun is a faintly glowing patch, not as bright as the zodiacal light, called the

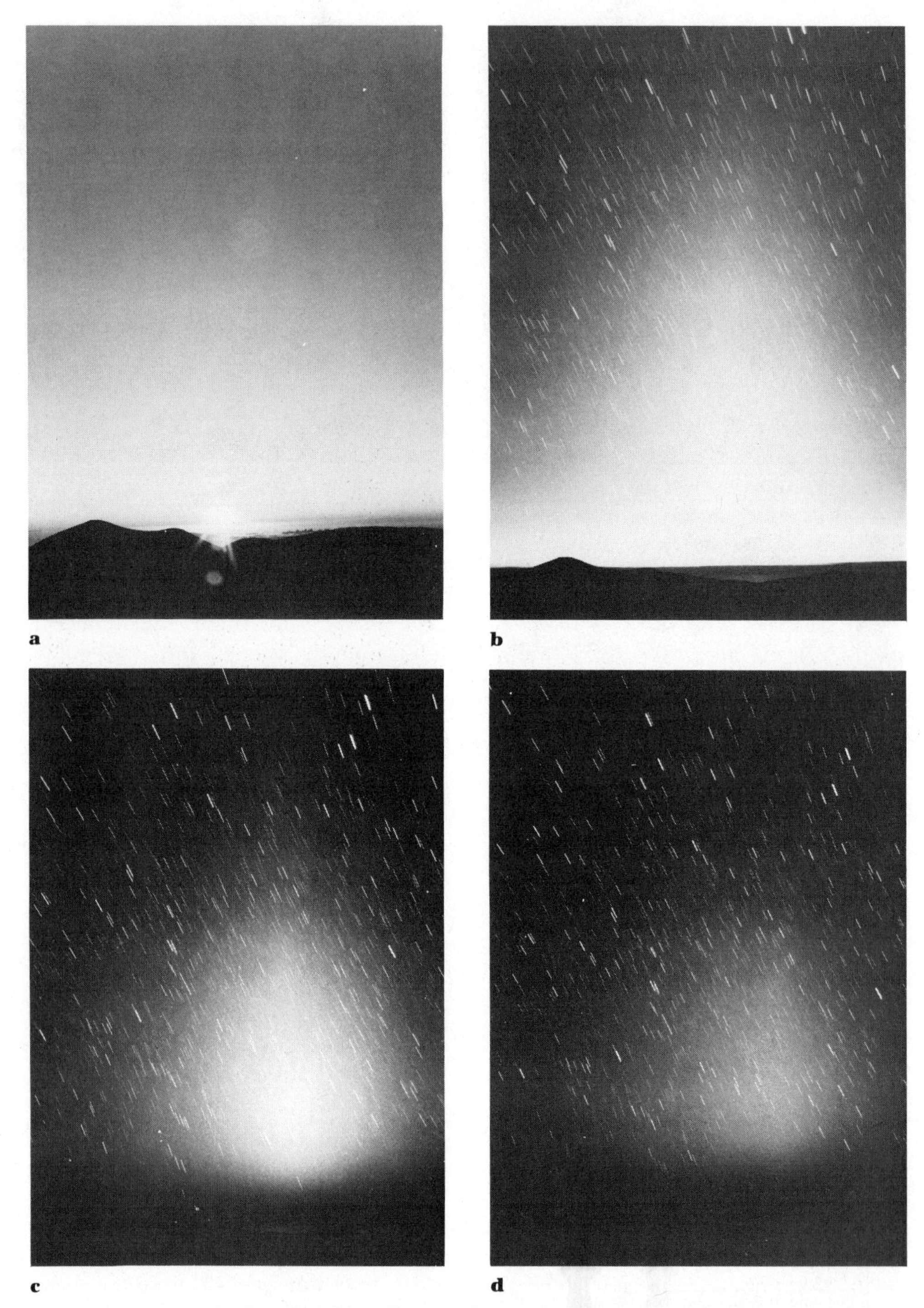

Figure 8-12. Sequence of wide-angle (72° × 52°) views of winter sunset from Hawaii, showing emergence anu setting of the zodiacal light. (**a**) *Sunset.* (**b**) *1 h after sunset, with zodiacal light extending about 49° above horizon.* (**c**) *$1\frac{1}{2}$ h after sunset.* (**d**) *2 h after sunset. (Photos by author; altitude 4 km; Mauna Kea Observatory)*

gegenschein (German for "counterglow"). It is sunlight reflecting off particles at the antisolar point.*

Zodiacal light is seen only from sites with dark skies. One must be several miles from dense population and several hundred meters from streetlights. Averted vision and a technique of scanning—swinging the head back and forth—may be necessary at first to notice the zodiacal light.

Measurements of the brightness, distribution, color, and polarization of the zodiacal light have helped to clarify the nature of the interplanetary dust. The particles responsible for most of the zodiacal light are tiny dust grains about 1 μm in diameter. The total mass of this dust is about 2.5×10^{16} kg (Whipple, 1967), a mass less than that of a single moderate-sized asteroid. Most of this mass is concentrated among the smallest particles.

The shape of the zodiacal dust system can be partly inferred from the narrow width of the zodiacal light; our cross-sectional view shows that the system is flattened toward the plane of the solar system. A total width of the zodiacal light of about 30° near the sun would imply its extension only about 0.2 AU above or below the ecliptic plane, for example. Spacecraft have measured the radial distribution of the meteoroid dust outward through the solar system both by optical sensors and by impacts on the vehicles themselves.

Figure 8-13 compares distributions of different sizes of particles, measured by different means. The largest (asteroidal) bodies are concentrated in the asteroid belt, but subcentimeter pieces are redistributed by Poynting-Robertson and radiation effects. The smallest pieces and the zodiacal glow are concentrated near the sun. The outer luminous atmosphere of the sun, called the **corona**, grades into the interplanetary medium and has two components, the K-corona and the F-corona. The **K-corona** is the gaseous portion, and the **F-corona** is the dust, mixed with the gas. This dust can be traced to within 4 solar radii of the sun by its infrared radiation.

For the zodiacal particles larger than 1 μm, the mean lifetime in the solar system is only about 10^5 y (Whipple, 1967), while still smaller particles are blown away (as in comet tails) in about a year (Biermann, 1967). Whipple concludes that the zodiacal dust cloud is completely replenished in about 2×10^5 y, requiring about 2.5×10^{11} kg/y, or 8 metric t/s in the form of particles ranging from many microscopic dust grains to a few boulders. Most of this material comes from comets, particularly in the inner solar system, where the comets erode as their dust-laden ices sublimate (Millman, 1979); additional particles come from the asteroids colliding in the asteroid belt.

FRAGMENTATION AND OUTBURSTS OF COMETS

So far we have seen that comets lose gaseous material (which contributes to Type I tails) and dust particles. The small particles are driven away from the sun in Type II tails, and the larger ones spread along the orbit in meteoroid swarms.

More catastrophic mass loss can also occur. Some comets, for example, suddenly flare up in brightness, suggesting sudden releases of gas. Comet Humason (Figure 8-6) is known for its peculiar variability in brightness and the unusual intensity of CO^+ emission lines in its spectrum. Such changes have been seen as much as 6 AU from the sun, considerably farther than the limiting distance for the activity of most comets (about 3 AU). Some comets return earlier and some later than predicted. As early as 1836, Bessel attributed this to accelerations produced by the loss of mass from the nucleus. Whipple (1950) showed how the jet action of pockets of escaping gas in

*This phenomenon of brightening at the antisolar point can be seen around the shadow of your own head on dusty, leafy, or other complex surfaces, particularly if your shadow is cast over a long distance. It is also often seen around the shadow of an airplane in which the observer is riding. In all cases, it lies 180° from the sun. This general phenomenon, partly caused by the absence of visible shadows at the antisolar point, is called *heiligenschein* (German for "holy glow").

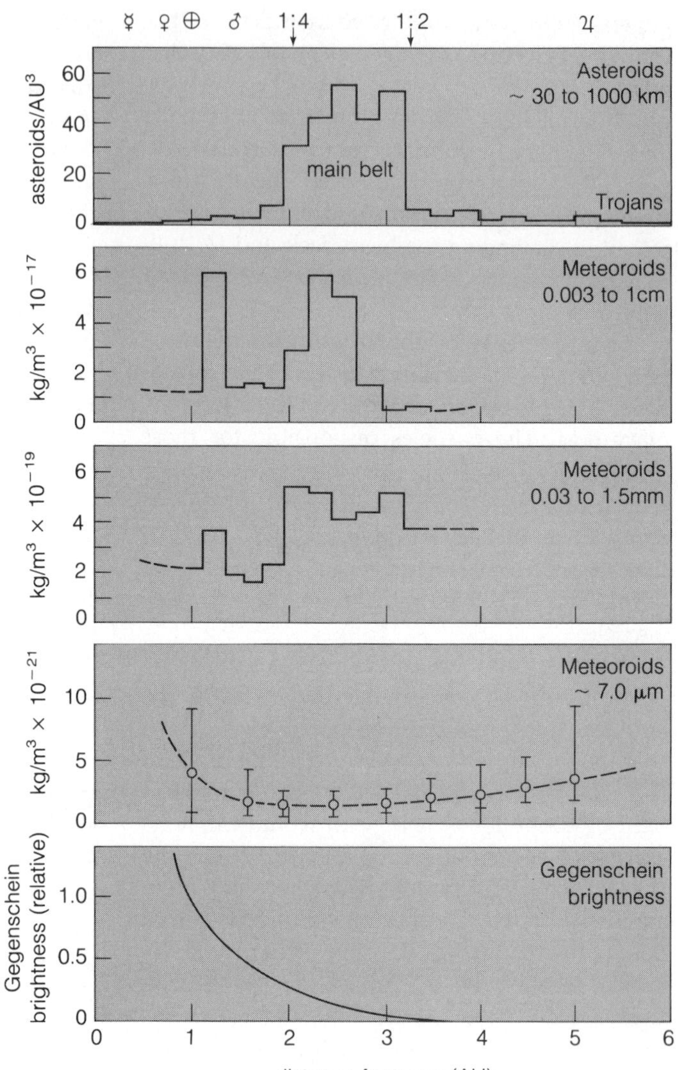

Figure 8-13. Measured distributions of small bodies in the solar system. Asteroids are almost all concentrated in the belt between Jupiter resonances 1:2 and 1:4 (see top). Millimeter-scale asteroid fragments spiral inward due to the Poynting-Robertson force; Mars and Earth sweep up many of them, but a spike appears to exist between Mars and Earth. Other small particles are added by comets, and all are redistributed by Poynting-Robertson and radiation forces. The smallest particles and those involved in zodiacal light are concentrated toward the sun. (Lower four diagrams adapted from Pioneer 10 data of Humes and others, 1974; Soberman, Neste, and Lichtenfeld, 1974)

a rotating nucleus could either accelerate or decelerate a comet. Indeed, thin, curved streamers of luminous material emanating from many comet nuclei in the inner coma strongly suggest jets of material erupting from the nuclei surfaces.

Most catastrophic of all is the fragmentation or complete disruption of comets. Certain comets, known as sun-grazers, pass only about 0.005 to 0.007 AU from the sun at perihelion. Many of these break apart at this time, as shown in Figure 8-14. The breakup is probably caused by heating, gas jetting, and spinning of the nucleus. Sungrazer 1882 II broke into several pieces during perihelion passage, and the 1965 sun-grazer Ikeya-Seki also broke into two pieces that separated at about 12 m/s (Pohn, 1966). Separation velocities of only around 20 m/s have been found for a dozen other disintegrating sun-grazers, suggesting an upper mass limit of 10^{13} kg and a diameter <1.3 km for icy nuclei (Sekanina, 1977; Brandt and Chapman, 1981). (Note that the velocities found for the separation of comet pieces are comparable to those found for the separation of meteoroids from comets, mentioned above.) The motions of the fragments are often affected by strong gas jetting as they break apart, perhaps from sublimation of gas off freshly exposed icy surfaces, as suggested by Figure 8-15.

Comet Biela split in 1846 and was seen as a pair of comets when it returned in 1852, but it was never seen again. This again suggests that breakup and exposure of fresh surfaces encourages the final disintegration of the nucleus. However, not all comet disruptions are due to solar heating in sun-grazers. Strangely, Comet Wirtanen (1957 VI) broke into two pieces near the orbit of Jupiter for no apparent reason.

As seen in Figure 8-16, sun-grazing Comet 1979 XI actually had its perihelion within the outer atmosphere of the sun, spewing debris into the solar corona and disappearing!

AN UNUSUAL COMET: SCHWASSMANN-WACHMANN 1

An example of an unusual comet is **Comet Schwassmann-Wachmann 1** (1925 II), which has a nearly circular orbit between Jupiter and Saturn. Normally its brightness is only 17th to 18th magnitude, and it is visible only in very large telescopes as a stellar image. But every year or so it undergoes sporadic outbursts—even though it never comes into the inner solar system—and can increase 600 times in brightness in 3 days, as shown in Figure 8-17. The outbursts may give us clues about the "geologic" processes of activity in comet nuclei. Hartmann, Cruikshank, and Degewij (1982) noted that the infrared colors during both quiescent and eruptive states resemble not clean ice, but D-like asteroidal soil or dirty ice colored by such soil. They inferred a surface layer of weakly bonded dark-colored dust or low-volatility debris in which gas pressure accumulates as the more volatile-rich interior ices sublime. When the gas pressure exceeds the strength of this weak surface layer, an eruption blows out a transient coma of dust and gas. Whatever the nature of events on Comet

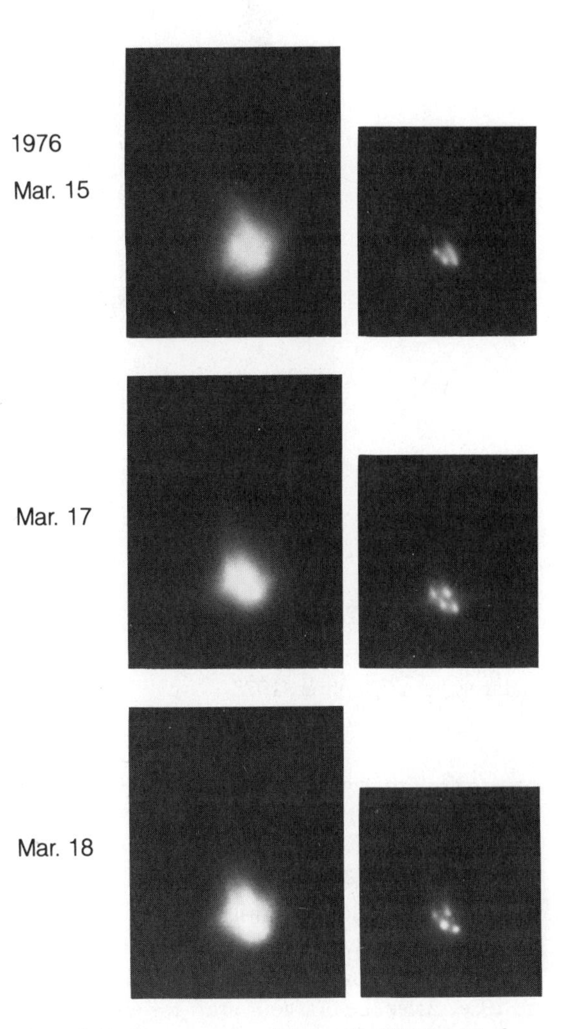

Figure 8-14. Fragmentation of Comet West (1975n) into four pieces in March, 1976. Left-hand pictures show longer exposures dominated by coma. Right-hand pictures show shorter pictures revealing four fragments separating at speeds of about 1 to 5 m/s. One piece was seen for only two weeks; others were followed for more than six months. (Lunar and Planetary Laboratory Mt. Lemmon 1.5 m telescope photo, University of Arizona, courtesy S. M. Larson)

Figure 8-15. The best closeup photo of the nucleus of a comet is this view from the Giotto spacecraft. It shows the black, peanut-shaped comet nucleus with bright jets of gas streaming out of localized vents on the surface. Nucleus is 15 km long. Because of journalistic tradition of printing only "today's news," fuzzier initial pictures were widely printed, but this much improved image has rarely been published in popular media. (Giotto photo courtesy Harold Reitsema and Alan Delamere, Ball Aerospace; copyright 1986 MPAE)

Schwassmann-Wachmann 1, it remains "one of the most interesting and puzzling objects of our solar system" (Richter, 1978).

THE HALLEY FLYBY AND THE PHYSICAL NATURE OF THE COMET NUCLEUS

The climax of comet studies in this century came when an international flotilla of five spacecraft (two Japanese, two Soviet, and one European) flew past Halley's comet during its solar approach in 1986. After the first four craft tested the environment at moderate distance, the European probe, Giotto, flew by closest, only about 600 km from the nucleus.

The heart of Halley's comet was a hazy environment where the probes found more fine dust than expected. Giotto was hit by a dust grain about 960 km from the nucleus, knocking it partially out of commission for 32 minutes during the closest approach. Nonetheless, it obtained extraordinary photos of the black, peanut-shaped nucleus emitting jets of gas (Figure 8-15), and measured the coma gases. The nucleus was found to be 15 km long and 7 to 10 km wide. Its rotation, in keeping with its irregular shape, is also irregular, probably with a basic 2.2-day period and a 7.4-day nodding motion. The rotation state may vary continuously due to the jet effects of the gas plumes.

The dark, fairly neutral-grey color and the reflectivity of only 4% surprised some observers, who had expected brighter ice, but confirmed predictions by other observers that comets are closely related to the black, volatile-rich, carbonaceous asteroids residing beyond the outer asteroid belt. Since the Halley encounter, other comets have also been measured to be just as black (Campins and others, 1987).

Giotto and the two Soviet VEGA probes, together with Earth-based data on CO, give this

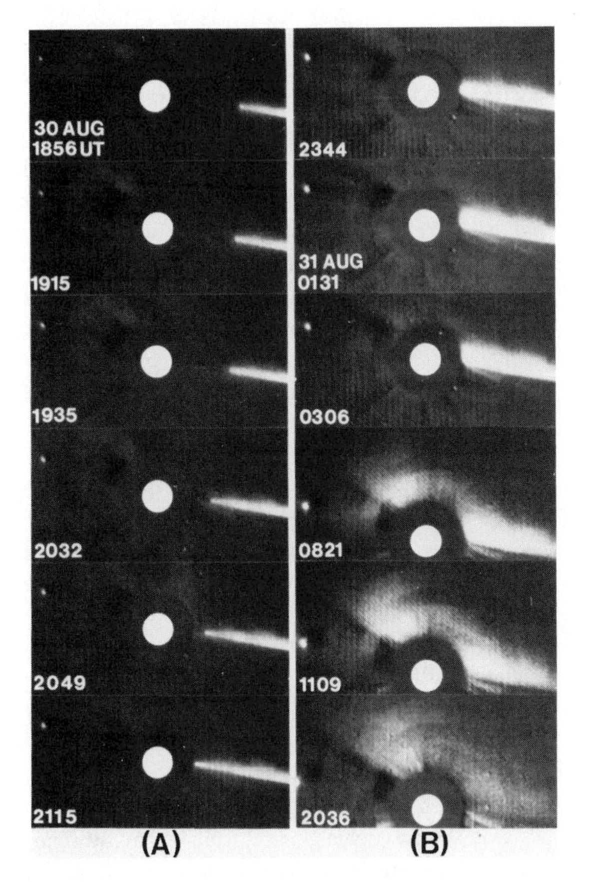

Figure 8-16. The probable collision of comet 1979 XI with the sun on August 30, 1979, photographed with an orbiting coronagraph telescope. The white disk shows the size of the sun, which was occulted by a larger disk (grey, out to 2.5 $R_\odot$) to reduce glare. Orbital elements indicate closest approach to sun occurred August 30 at 22:39 UT ± 20 min. In post-impact series (B) rapid sublimation of the nucleus has added material to tail (exaggerated by contrast enhancement of series B). Material from impact or breakup of nucleus streams radially outward in several directions after 8:21 UT. Comet was not seen to re-emerge from sun. Venus is at left. (LOLWIND satellite photos from Naval Research Laboratory, courtesy D. J. Michels)

February 1

February 10

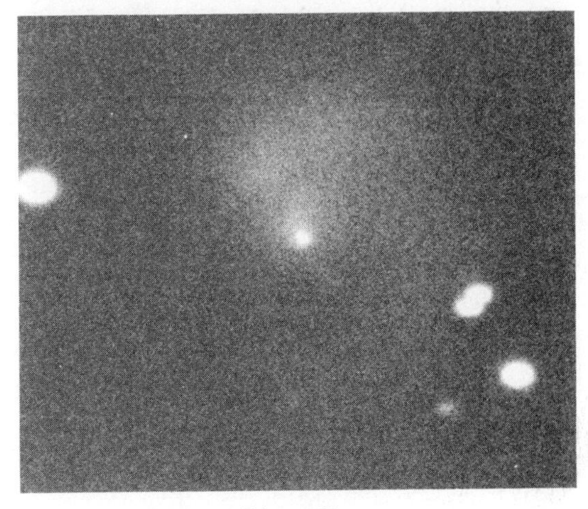

February 28

Figure 8-17. An outburst of peculiar comet Schwassmann-Wachmann 1. The February 1 photo shows the comet in its normal quiescent state, at about 18th magnitude. On February 10 it has brightened by a factor of 100 to the 13th magnitude and developed a spiral tail structure that Whipple has attributed to an eruption of material off one side of a rotating nucleus. Some comet researchers dub this characteristic spiral structure a "ring-tailed snorter." By February 28, the tail structure has expanded and faded. (Harvard-Smithsonian Center for Astrophysics, courtesy Cheng-Yuan Shao)

ice composition by numbers of molecules (Brandt, 1990):

water ice (H_2O)	80%
carbon monoxide (CO)	10%
carbon dioxide (CO_2)	3.5%
formaldehyde polymer $(H_2CO)_n$ and/or other organic compounds	few %

This confirms abundant water ice in the nucleus. The dust particles were found to be rich in C, H, O, and N, somewhat resembling carbonaceous meteroid particles collected on Earth. They are called **CHON particles** and differ from familiar terrestrial dust, which is richer in silicon, iron, other metals, and their oxides. The latter type of dust constituted a second type of particle measured in Halley, and a third, most abundant type, seemed to be a mixture of the two, similar to carbonaceous chondrites. The CHON particles indicate that the nucleus soils are rich in carbon and probably organic compounds (Sagdeev and others, 1986), though no life is expected to have evolved there.

A typical comet nucleus is now believed to be mostly ice by composition (frozen H_2O and other volatiles), but impregnated with black sooty carbonaceous and CHON particles. Only a few percent of the sooty material, scattered through the comet ice, could make even the "fresh ice surfaces" appear very black. As the ice sublimes at the surface of a comet, during passes near the sun, the ice/soil ratio in the surface layers probably decreases, leaving an even darker surface regolith of carbonaceous soil. Gases from subliming ice may build up pressure and eventually break through in certain areas, explaining the sporadic eruptions of comets and the well-defined jets photographed on Halley's surface.

The Dirty Iceberg Model

All these observations support the classic theory of comet structure, known as the **Dirty Iceberg Model.** It was developed by Harvard astronomer Fred Whipple (1950, 1963), who combined all observations available at that time and concluded that the nucleus must be a several-kilometer-diameter mass of ices—water ice, methane ice, and ammonia ice—mixed with dust particles. As the comet approaches the sun and is heated, Whipple theorized, gas sublimes and produces Type I tails. At the same time, dust particles are exposed and loosened, drifting away to form Type II tails and meteor swarms.

Probably the main revision of the model has been post-Giotto acceptance that comets are black and sooty ice, not bright like icebergs.

Another revision of the original iceberg model is that gases don't seem to come off the nucleus in the order or relative abundances predicted for a hunk of ice and dust simply sublimating. More complex models suggest that many of the volatiles appear not in pure ices, but in **hydrates** (compounds with water molecules bound in the crystal lattice or adsorbed on surfaces), and in **clathrates** (compounds with small molecules, such as sulfur dioxide [SO_2] or carbon dioxide [CO_2] trapped within crystal lattices of other compounds, often ices such as H_2O).

Even before Giotto met Halley, such models suggested that chemical reactions might occur among the particles as materials fly away from the nucleus and interact on their way into the surrounding coma (Oppenheimer, 1978); studies of the CHON particles at Halley seem to confirm that. Studies of production and reactions of organics throughout the solar system are an increasingly important area of research.

Rotation and Strength

Comets' rotation rates are harder to measure than asteroids'. As an irregular asteroid rotates, presenting first a broad side and then a more nearly end-on view, the brightness changes in a regular way that reveals the period of rotation. But a comet nucleus is usually buried in an obscuring coma when well placed for observation. Halley's nucleus rotates slowly, with a period of several days and confusing nodding or nutational motions. A few other comets display regular light variations that reveal rotation. For example, Comet d'Arrest has a 5.17 ± 0.01-hour period (Fay and Wisniewski, 1978); Donati has a 4.6-hour period, and Encke, 6.5 hours (Whipple, 1980).

The low rates of separation of split nuclei imply that if pieces are thrown off by rotation, rather than being blown off by gas jetting, the rotation periods must be no more than 2 to 3 hours in those cases (Sekanina, 1977).

Comet nuclei are probably very weak. As ice sublimes and the diameter decreases, the shape may become more elongated (an ellipsoid losing equal amounts off all its surface has its axis ratio depart more and more from unity). Also, fractures that originally extended only partway through the body may now reach from one side to the other of the smaller body. This may explain how an evolving, rotating comet eventually breaks into two or more pieces along a zone of weakness or a narrow neck. Sekanina (1977) noted that if rotational centrifugal forces dominate in causing the breakup, the strength of the nucleus must be about 10^3 N/m^2 at the time of breakup—weaker than concrete and suggesting a crumbly consistency not unlike some carbonaceous chondrites.

Evolution: Fresh Versus Old Comets

The more times a comet goes around the sun, the more it changes. Long-period parabolic comets, which are likely to be dropping in for their first or second visit from the Oort cloud, tend to show features of **fresh comets:** higher rates of dust and gas production, with high dust/gas ratios. At the other extreme, periodic comets that have spent much time in the inner solar system are **old comets:** they have lower rates of dust and gas production, and lower dust/gas ratios.

"Aging" can even be observed with some individual comets during perihelion passage. Comet Arend-Roland, for example, decreased its dust/gas ratio by nearly a factor of 8 during a period of 18 days centered on perihelion passage (Delsemme, 1977). Presumably, incoming comets have a layer of dust-rich ice (perhaps formed as dust fell back during the last departure from perihelion), and this is blown off to expose less dusty, fresher ice beneath. This idea is supported by the fact that Jupiter's satellite Callisto, which is about a 50–50 mix of ice and soil, has a dark, dust-rich surface layer, evidenced by crater punctures into brighter, icier material beneath (see Figure 9-29, page 288).

ORIGIN OF COMETS

Comets are among the most interesting objects to planetary scientists because they have preserved, in deep freeze, some of the most primitive materials in the solar system. But where did they come from?

Many exotic possibilities have been suggested, such as eruption from volcanoes on giant planets or their satellites (Vsekhsvyatsky, 1977, before discovery of the Io and Triton volcanoes!), breakup of a 90 $M_\oplus$ planet in the asteroid belt (van Flandern, 1977), or capture from interstellar material (Van Woerkom, 1948). Most of these ideas have been rejected by the comet and dynamical community. A more widely accepted **planetesimal theory of comet origin** comes from our ideas of solar system origin. Chapter 5 showed that the most abundant solids in the outer solar nebula were water, methane, and ammonia ices with an admixture of carbonaceous and silicate dust. Chapters 6 and 7 showed that carbonaceous materials apparently condensed beyond a "soot line" at 2.7 AU, and ices dominated beyond a frost line at perhaps 4 AU. Watson, Murray, and Brown (1963) and Lebofsky (1975) showed that ices sublime rapidly if closer to the sun than about 4 AU. Thus, comets are believed to be simply the planetesimals that formed among the giant planets, especially Uranus and Neptune (Weissman, 1985).

Öpik (1963, 1966) showed that the history of such planetesimals would be dominated by their encounters with the giant planets. Only a small percentage would actually hit a planet. Most would have near misses and be flung into successively more eccentric orbits. Uruguayan dynamacist J. A. Fernández (1978) presents more detailed

Figure 8-18. Relationships among interplanetary debris are symbolized in this image which captures a comet and an asteroid in a single photo. This 1982 photograph was tracked on Comet DuToit 2–Hartley, which had not been seen since 1945 and had split since then. The second fragment of the comet was recorded on the same photo, about 5 field widths out of this field to the upper right. Coincidentally, asteroid 1982 DB was discovered on this plate (upper right). Star images were trailed during tracking on comet, but asteroid is revealed by its motion, which leaves a trail with a different slant. (1.2 m Palomar Schmidt photo, courtesy Eleanor Helin)

modeling of this process and finds that during encounters with Jupiter, Saturn, Uranus, and Neptune, the percentage of planetesimals ejected into the Oort cloud would be 2%, 7%, 30%, and 40%, respectively. The rest would be permanently ejected into interstellar space. Multiplying these percentages by estimated reservoirs of planetesimal mass in each planet's zone, he concludes that most cometary material in the Oort cloud originally came from the vicinity of Uranus and Neptune, which might explain why carbonaceous-chondrite-like dust in comets is apparently unlike that sampled from meteorites presumably originating closer to home.

Once in the Oort cloud, the comet drifts until some perturbing influence directs it back into the solar system. If the orbit remains highly eccentric, the comet may loop close to the sun and be observed by us as a long-period comet. If it happens to encounter one of the planets, especially Jupiter, in just the right way, it may be deflected into a short-period orbit and become a periodic comet, visible from Earth. Eventually, if it remains near the sun long enough, its volatile ices become exhausted. Then we may see it as an Apollo asteroid (Figure 8-18).

Giant Comets?

Most well-known comets are 5 to 20 km in diameter; some reach 50 km. However, "asteroid" Chiron, which turned into a comet, has a diameter of about 200 km. If comets are planetesimals ejected by giant planets, the Oort cloud may preserve their original diameter distribution, and thus may contain even larger bodies.

The sizes of the largest such planetesimals are unknown, but calculations and observations of planetary obliquities (see Chapter 5) suggest that planetesimals grew to diameters as large as 0.1 or more of the diameter of the planet in their zone before being expelled by the planet or colliding with it. This could mean that among the millions of comets in the Oort cloud, many are hundreds of kilometers across and some perhaps 1000 km or more! Chiron is probably such a body. This suggests that the comets we have seen are just the tip of the cometary iceberg. Perhaps occasionally during planetary history the inner solar system is visited by a world-scale body that blazes to life with a brilliant, immense coma and tail before vanishing again into its long night in the Oort cloud.

SUMMARY

Comets are icy planetesimals usually from 1 to 50 km across and containing bits of fragile dust resembling carbonaceous chondrite material. They probably formed among the outer planets and were ejected into the Oort cloud, from which they occasionally reenter the inner solar system. When a reentering comet reaches a distance of about 2 to 4 AU from the sun, material begins to sublime off its surface, producing a gaseous coma and a gas-and-dust tail that is blown outward by the solar wind. The solid debris become meteoroids strewn along the comet's orbit, since they leave the comet at too low a speed to change orbit very much. These particles are affected by Poynting-Robertson and radiation forces and occasionally hit Earth, producing meteors or their larger cousins, fireballs.

Comets age slowly as more and more gas and dust are lost. Some of them may burn out, leaving inert residues of stony material—possibly cataloged in some cases as Apollo asteroids of spectral class C, P, or D. Comets in their various states would provide interesting targets for close investigation by spacecraft.

An active area of research is the relation between comets and other small bodies, which we have pigeonholed as asteroids, Apollo asteroids, meteorites, and so on. Our terminology (based on observational properties) may divide these bodies more distinctly than is warranted by their chemistries or points of origin. Probably they are all examples of debris from early planetesimals forming in different (adjacent or nonadjacent?) zones of the solar nebula. Comets are surely the more volatile-rich icy bodies formed beyond the asteroid belt, while most asteroids are the more ice-free stony or metallic bodies formed in the belt or even closer to the sun. Some other poorly known bodies may fall in between (Chiron, for instance?).

Table 8-5 illustrates these points by comparing orbits of different bodies in the meteoritic complex. Notice the transition from large aphelia, large eccentricities, and large inclinations to low values as we go down the table. Clearly, the top entries, such as long-period comets and high-velocity fireballs, can reach or nearly reach the Oort cloud; they show its mixture of direct, retrograde, and highly inclined orbits. At the bottom of the table, Apollo asteroids and meteorites are strong objects with mostly direct orbits and low inclinations, seemingly distinct from the long-period comets.

But at some unknown location in the table is an interface between objects that must have been stored in the Oort cloud and those stored in or near the asteroid belt. Kresak (1977) argues that one-eighth of the near-Earth bodies with diameters greater than 1 km are active comets, and that the fraction declines for smaller bodies. Are some Apollos burnt-out comets? Wetherill (1979) says yes, Kresak no. Do some carbonaceous chondrites come out of comets? What is the exact chemistry of material in comet nuclei? Do large numbers of fragile carbonaceous bodies explode in the atmosphere like the Tunguska object and never reach Earth's surface? These questions illustrate the current frontiers of research on the meteoritic complex of dust grains, meteoroids, comets, and asteroids.

CONCEPTS

Tunguska explosion
comet
tail
head
coma
nucleus
recovered comets
Type I tails
Type II tails
plasma
long-period comets
parabolic comets
short-period comets
periodic comets
Oort cloud
nongravitational forces
meteors
meteor shower
radiant
meteoroid swarm
zodiacal light
gegenschein
corona
K-corona
F-corona
Comet Schwassmann-Wachmann 1
CHON particles
Dirty Iceberg Model
hydrates
clathrates
fresh comets
old comets
planetesimal theory of comet origin

PROBLEMS

1. (a) How is a comet distinguished from an asteroid observationally? (b) By physical properties of the body itself? (c) By origin?

2. Assume that you had free access to satellite

Table 8-5
Orbits of Meteors, Meteorites, and Related Objects

Object	a (AU)	e	i	q (AU)	Q (AU)	Remarks
LONG-PERIOD COMET[a]						
Comet Winnecke	∞	1.00	132°	0.58	∞	
HIGH-VELOCITY FIREBALLS[b]						
Nov. 30, 1965	∞	1.00	132°	0.27	∞	Long-period comet debris?
July 31, 1966	32	0.98	42°	0.44	64	Long-period comet debris?
SHORT-PERIOD COMETS[a]						
Swift-Tuttle	24	0.96	114°	0.96	47	Perseid meteor source
Giacobini-Zinner	3.6	0.71	32°	1.03	6.1	Draconid meteor source
SHOWER METEORS[b]						
1966 Perseid	40	0.97	114°	0.96	79	From Comet Swift-Tuttle
1965 Draconid	3.5	0.71	32°	0.99	5.9	From Comet Giacobini-Zinner
1965 Leonid	53	0.98	162°	0.98	104	
1965 Taurid	2.2	0.80	2°	0.42	4.0	
LOW-VELOCITY FIREBALLS[b]						
Feb. 13, 1965	2.3	0.69	13°	0.69	3.9	
May 31, 1966	3.0	0.79	9°	0.60	5.3	
Oct. 27, 1966	2.1	0.53	26°	0.98	3.3	
Apr. 10, 1972	1.7	0.39	15°	1.01	2.3	Wyoming fireball (Chap. 6); probable stone
1974 Leutkirch	1.6	0.40	2°	0.96	2.2	European fireball; probable stone; unrecovered
METEORITES[c]						
Pribram[d]	2.42	0.67	10°	0.79	4.0	H chondrite
Lost City[e]	1.66	0.42	12°	0.97	2.3	H chondrite
Dhajala[f]	1.8	0.59	28°	0.74	2.9	H chondrite
Farmington[g]	1.3	0.55	7°	0.6	2	L chondrite
APOLLO ASTEROIDS[h]						
1685 Toro	1.37	0.44	9°	0.77	2.0	L chondrite meteorite spectrum; source of L chondrites?
1862 Apollo	1.47	0.56	6°	0.65	2.3	
1566 Icarus	1.08	0.83	23°	0.19	2.0	
1620 Geographos	1.24	0.33	13°	0.83	1.7	

[a] For more comet orbits, see Table 8-3.
[b] One hundred fireball orbits are given by McCrosky (1967).
[c] For further discussion of meteorite orbits, see Chapter 6.
[d] Ceplecha, Rajche, and Sehnal (1959).
[e] Wasson (1974).
[f] Ballabh, Bhatnagor, and Bhandari (1978).
[g] Levin, Simonenko, and Anders (1976).
[h] For orbits of other Apollos, see Table 7-1. Further discussion of asteroids appears in Chapter 7.

and planet surfaces and to equipment monitoring interplanetary bodies. Suggest some ways to determine whether impact craters on the moons and planets are predominantly from cometary, stony asteroid, or metallic asteroid origin.

3. If you are scanning the skies with a small telescope and find a faint, fuzzy object not marked on your star map as a nebula, what could you do to test whether it might be a newly discovered comet? (Assume no spectroscopic equipment is available.)

4. Name some important gases seen in comets. How do they suggest that comets contain—in the form of ices—water, methane, and ammonia, which are common in other primitive solar system regimes, such as the atmosphere of Jupiter?

5. (a) How do shapes, statistics, and dynamics of comet orbits imply a vast number of unseen comets in the Oort cloud beyond the planetary system? (b) For every short-period comet we see, roughly how many comets made it from the Oort cloud in as far as Neptune's orbit, only to be ejected again into the Oort cloud or out of the solar system entirely?

6. Give some lines of evidence that meteors are associated with comets and are not simply examples of small ordinary meteorites.

7. What conclusions can we draw about the physical nature of comets: (a) from meteors and the Tunguska event; (b) from spectral observations of comet comas; (c) from fragmentation and outbursts of comets?

8. Give evidence that comets contain some carbonaceous chondrite material. How might this help solve the mystery that the dominant meteoritic component in the lunar soil seems to be carbonaceous chondritic, while only 5% of meteorites reaching Earth's surface have this composition?

ADVANCED PROBLEMS

9. (a) Show that the maximum possible encounter velocity between Earth and a member of the solar system is about 72 km/s. (b) Why can a meteor observed to hit Earth at this speed be associated with long-period comets and not any known asteroids?

10. A water-icy body with albedo 70% and emissivity 0.2 falls toward the inner solar system. (a) What is its surface temperature at Jupiter's orbit? (b) At what distance from the sun does its temperature reach 0°C = 273 K?

11. The same body as in problem 10 is thinly mantled by carbonaceous dust that reduces the visual albedo to 3%, while leaving the emissivity at 0.2. Answer problem 10(a) and 10(b) for this body.

12. (a) A particle of radius 1 cm breaks loose from a short-period comet having a nearly circular orbit at semimajor axis 3 AU. What will be its orbital history and roughly how long will it last in the solar system? Answer the above for a particle with radius (b) 100 μm (0.01 cm); (c) 0.1 μm.

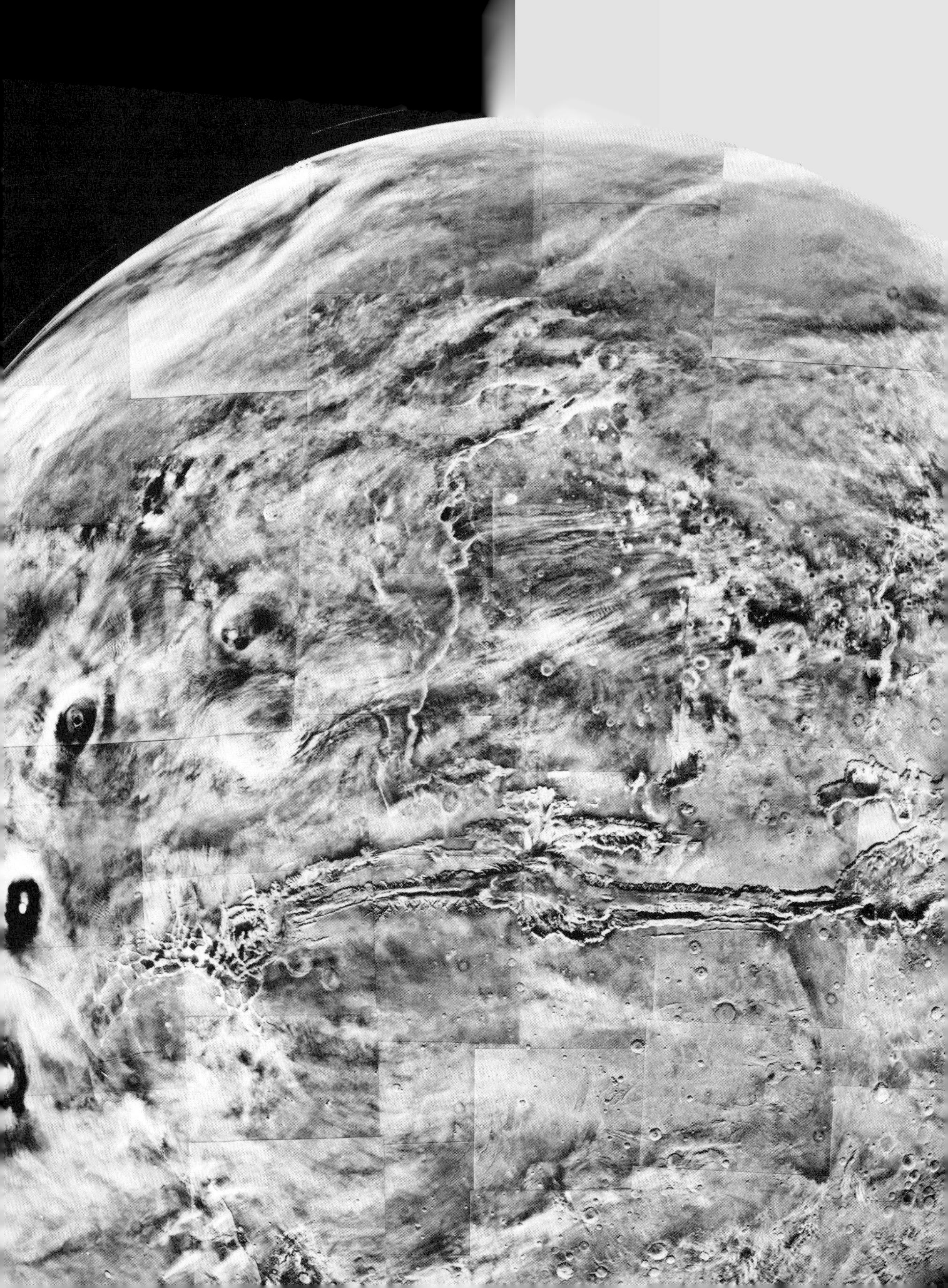

CHAPTER NINE

PLANETARY INTERIORS

Let's take stock of our present position in our solar system survey. Chapters 4 and 5 described the origin of the system. Chapters 6 to 8 described the various surviving representatives of the original planetesimals. The meteorites in Chapter 6 represent physical specimens of the interiors of other, albeit small, planets. Thus the stage is set for a direct confrontation with the problem of planetary structure—the topic of this chapter.

Planetary interiors is the term applied to this topic. This field of research attempts to determine not only what the compositions, pressures, temperatures, densities, and dynamical activity are inside the planets today, but also how these conditions have changed through time. This knowledge will show us how planets evolve.

The first portions of this chapter will be devoted to determining the current conditions inside planets. We can't slice planets open, and so the conditions have to be deduced through a careful combination of observation and theory. Our discussion of meteorites and asteroids has already introduced some of the important ideas. In a sense, nature has sliced some asteroids open for us and sent us samples in the form of meteorites, which suggest that planets have a zonal structure. In planets, the dense materials such as metals tend to drain toward the center while light materials float. The densest, innermost region is called the **core**; its surroundings, a broad zone of intermediate density, are called the **mantle**; and the thin, low-density "scum" on the surface is called the **crust**.

A portion of the planet Mars, exhibiting major features related to the internal evolution of planets. The major canyon system, Valles Marineris (bottom) may be a rift system associated with expansion and crustal tension. Three of Mars' large volcanoes (dark spots with summit craters at left) imply a period of subsurface heating. (NASA Viking photo mosaic, Jet Propulsion Laboratory; courtesy Conway Snyder)

FOUR BASIC OBSERVATIONS AND A BASIC CONCEPT

Imagine that we are approaching a planet in a spacecraft. How can we diagnose its internal properties? Does it have an iron core? Is the interior cold? Warm? Partly melted? Is the planet cooling and contracting? Or heating or expanding?

Before answering these questions, one should determine four basic properties: the mass, diameter, mean density, and surface rock properties. The mass can be found by measuring the planet's gravitational influence on our spacecraft (or on a natural satellite, or even on a distant planet) and applying Kepler's laws.

Measuring the diameter is easy if the disk can be resolved, but is difficult for remote objects with small images, such as Pluto, satellites, and asteroids. One fruitful method is to time the duration of a star's occultation by the planet. Observations from several sites on Earth allow observers to estimate the diameter and rough shape of the planet. One problem when measuring diameters of planetary disks is that planets with thick atmospheres have indistinctly defined limbs. If the atmosphere absorbs sunlight, it produces **limb darkening**, a darkening of the planet's edge that may cause the diameter measurement to be a bit too small.

The third basic observation is not really a separate observation but a combination of the first two. The **mean density**, which is simply the mass divided by the volume, immediately begins to tell us something about the planet's gross composition. For instance, a mean density of 1000 kg/m^3 is typical of ices; values of 2800 to 3900 are typical of volcanic rocks and stony meteorites; values of 5000 to 6000, of iron-rich minerals; and values around 7900, of iron to meteorites.

We can certainly do better than to characterize the planet by its mean density if we add the fourth observation, obtained by sending down a lander and sampling the surface rocks. If the "planet" we are approaching is a small one, say a 500-m asteroid, we will probably find that the surface rocks have a density value equaling the mean density of the planet, perhaps 3500 kg/m^3, which is typical of chondrites. Then we could reasonably conclude that the asteroid is relatively uniform in composition, being composed of the surface material throughout.

If we examine a bigger world, the moon, we find that surface rocks typically have a density around 2800 kg/m^3, while the mean density is 3300 kg/m^3. Now we must conclude that the internal material is somewhat denser than the crustal material. Perhaps a small iron core is hidden in the center, or perhaps the higher central pressure simply compresses the rock material to higher density.

In the case of Earth, we have a striking discrepancy of 2800 kg/m^3 for typical surface rocks and 5500 for the mean density. The reason turns out to be twofold: The mantle has high-density, iron-rich rock, and there is an iron core.

A basic concept evolving from this discussion is that we can use the surface materials as an observable boundary condition and then reason our way into the interior by learning as much as possible about how materials behave under the higher pressures of the interior.

THEORETICAL TECHNIQUES FOR CALCULATING INTERIOR CONDITIONS

The Equation of State

The pressure (P), density (ρ), and temperature (T) in any material are called the **state variables** because they determine the state of the material. For instance, if the pressure gets too high or the temperature too low, molten material may change to the solid, rather than liquid, state. As we go down into a planet, P increases, causing material to compress and reach higher ρ's. For gases, there is a single, simple equation that relates the state variables (called the universal gas law), but for the rocks of a planet, the relation is much more complex. This complex relation can be measured to some extent in the laboratory. Any such relation between P, ρ, and T is called an **equation of state** for the given material, and it tells what density is attained for a specified pressure and temperature.

Computer Models of Planets

Given the equation of state of a planet's inner material, it is easy to see how a theoretical model of the planet's interior can be made. The **hydrostatic equation**

$$dP = \rho g\, dz$$

says that the increase in pressure dP as we descend through a layer is equal to the density ρ of the material in that layer times the gravitational acceleration g times the layer thickness dz (with z increasing downward).

A computer program can be devised that divides a planet into a thousand layers (or some other large number of layers) and then starts iterating its program down from the surface values. At the surface, the pressure is zero (we can neglect the atmospheric pressure in most cases). We know by direct measurement the density ρ of rocks at the surface, at least in the case of Earth and the moon, and we can measure or compute the gravitational acceleration g. The computer starts with these figures and multiplies them by the layer thickness dz to get the pressure at the bottom of the first layer. Then the computer inserts that pressure into the equation of state to compute a new density for the rocks at the bottom of the first layer. Computing the new gravitational acceleration that applies at that depth, it then performs a new multiplication to get the pressure at the bottom of the second layer, and so on.

There are two obvious problems in this sort of scheme. For one thing, the equation of state involves the temperature, and we have not yet supplied the temperature to the computer. Usually a theoretical model of the temperature profile in the planet must be supplied, based on measured temperature gradients at the surface and estimates of interior heat sources. Fortunately, however, the density of a rock does not change very much as the temperature changes (unlike the density of gases). A second problem is that the equation of state must be accurate, and no one yet knows precisely the composition of any planet's interior, so the equation of state has to be guessed at.

In spite of these difficulties, a number of observational facts guide us in constructing our models. One is that we know the total mass of each planet. Therefore, when the computer reaches the bottom of the last layer (that is, the center), it must have "used up" precisely the mass of the planet. If mass is left over, the densities computed were systematically too low; if the allotted mass is used up in the 999th layer or sooner, the densities were too high (and the model would predict a hollow cavity at the center that a sudden collapse would immediately fill in in a real planet). To solve this problem, an adjustment is made to the equation of state, and the computer program is run again. A number of other similar checks can also be made—we shall discuss these later. (Alternatively, the computer program can be designed to start by guessing central conditions and then work outward. It can then check to see if the correct radius and mass have been reached at the level where the pressure equals zero, that is, the surface.)

PLASTIC FLOW INSIDE PLANETS

In Chapter 7, we noted that asteroids smaller than a few hundred kilometers across are often irregular in shape. Worlds larger than 1000 km are round because they obey the hydrostatic equation. This is because the pressures and temperatures in their interiors are great enough to cause failure of the normal elastic properties of rock. Under these conditions the material deforms to meet the equilibrium shape under the influences of gravity and rotational forces.

This shape is called the **figure** of the planet, and because of centrifugal force, the figure of an isolated, fluid, rotating planet is not a sphere but a flattened shape called an **oblate spheroid**. Additional effects, such as tides, interior processes, and irregular mass distribution, cause small departures from the oblate spheroid shape. An example is Earth's slightly pear shape, noted by the first artificial satellites in 1957–1958.

Figure 9-1 dramatizes the remarkable ability of planetary material to deform plastically. This fluidlike deformation of rock is called **plastic flow**, or **rheidity**, and is the property that allows planets to deform into spheres or spheroidal

Figure 9-1. Folded chalky strata at Lulworth Cove on the south coast of England. Examples of folded strata around the world show that rocks subjected to stresses over long periods of time can deform plastically, thus allowing the principle of hydrostatic equilibrium to control planetary configurations. Note figures (top) for scale. (Photo by author)

shapes. A **rheid** is a substance that has brittle properties on a short time scale but fluid ones over long times. Pitch or asphalt is a famous example; if you strike it with a hammer it chips, but if you place a cannonball on the surface of a large vat of pitch, it will eventually sink out of sight.

These properties are believed to arise at least in part by small cracks or imperfections in the crystal lattices of the material. A sudden force may fracture the crystal entirely, but a prolonged gentle stress may cause imperfections (such as "holes" due to missing atoms in the lattice) to propagate through the material, so that on a macroscopic scale the material appears to flow.

It is thus rheidity, as much as melting or crushing, that allows planetary rock to adjust to the hydrostatic equation, bringing planets into round shapes.

For this reason, a planet with a dense crustal layer overlying a lower-density layer is unlikely because such a combination is unstable; given enough time, the dense material is likely to flow downward while the lighter material rises, just

MATHEMATICAL NOTES ON PRESSURES INSIDE PLANETS

Two expressions can be combined in a rather simple exercise to derive the pressure inside planets and to indicate the somewhat secondary role of temperature as a state variable. Let r be the radius to a point in the planet. Then in any thin layer of thickness dr, the pressure P changes by

$$dP = -\rho g\, dr$$

from the hydrostatic equation. The minus sign arises because P increases as r decreases.

If we want to integrate this to obtain P at different r's, we note that both the density ρ and the gravitational acceleration g are functions of r. However, ρ is almost a constant, changing by less than a factor of 2 in many planetary objects. The more important change is in g, which is given by our second expression

$$g = \frac{GM}{r^2}$$

from Newton's gravitational law. The mass of the planet M has to be written in terms of r and is given by

$$M = \frac{4}{3}\pi r^3 \bar{\rho}_r$$

where $\bar{\rho}$ is the mean density out to point r. Substituting all this back into the hydrostatic equation, we have

$$dP = -\rho \frac{4\pi G\bar{\rho}_r}{3} r\, dr$$

To simplify matters, let us assume that the density really is some constant (which is quite true in the outer layers and less true throughout), so that

$$\rho = \bar{\rho} = \bar{\rho}_r$$

where $\bar{\rho}$ equals the mean density for the whole planet. Now we can integrate:

$$\int_{P_{surf}}^{P_r} dP = -\frac{4\pi G\bar{\rho}^2}{3}\int_R^r r\, dr$$

P_{surf} is the pressure at $r = R =$ the surface, which is assumed to be zero (the pressure of the atmosphere is relatively small). Thus,

$$P_r = \frac{2\pi G\bar{\rho}^2}{3}(R^2 - r^2)$$
$$= 1.4(10^{-10})\bar{\rho}^2(R^2 - r^2)$$

in SI units.

This expression gives a central pressure of about 46 kbar (4.6×10^9 N/m^2) in the moon, which is about right. Because of our cavalier assumptions about ρ, this expression yields an answer that is about a factor of 2 too low in Earth, where it gives a central pressure of 1700 kbar, compared to more sophisticated estimates of about 3700 kbar.

Our expression can be improved by keeping the two densities ρ and $\bar{\rho}_r$ separated in the integration, but it has much illustrative value even in this form. For instance, we find that you have to go down 1000 km in the moon to get to a pressure of 40 kbar (4×10^9 N/m^2), while this is reached at only about 100 km down in Earth.

Note also that we derived a planet's pressure structure without any reference to temperature T. So where is T needed in understanding interiors? One place is in understanding the true value of ρ as a function of r. As pointed out in the text, this utilizes the equation of state, and ρ values in each layer are adjusted according to T. Also, as we will see in the text, T controls the dynamics by determining whether rock is molten, in convective motion, and so on.

as the cannonball sinks in the pitch. To the first approximation, then, planets are likely to have their densest materials in the center and lightest materials near the surface, even if they have never been completely molten. Local dense concentrations, such as lava-filled impact basins, may exist on planetary surfaces, however.

CHANGES OF STATE

Minerals, Rocks, and Ices

Planetary samples from Earth, the moon, and meteorites show that terrestrial planets are composed principally of silicate and metallic compounds. Such compounds solidify into different forms known as **minerals**. Assemblages of different minerals make up **rocks**. These traditional geologic definitions need to be broadened for application to outer planets and satellites, where much of the solid bulk is ice. Thus, *minerals* may also refer to frozen water (H_2O), methane (CH_4), or ammonia (NH_3), and the geologic role of "rocks" may be played by both ices and silicates. Nonetheless, we usually use the term *rocky* to mean largely silicate. Chapter 10 will give a more complete review of various mineral and rock types.

To deduce the conditions inside planets, we must have equations of state that describe not only relations among pressure, density, and temperature for each rock type, but also any *changes* in the state of the constituent minerals. The two most important **changes of state**, melting and solid state phase changes, are described next.

Melting

Measurements in drill holes show that temperature increases with depth inside Earth. The same probably applies to all other planets, since planetary interiors contain primordial heat and radioactive atoms that produce heat as they decay. If the interior temperature is high enough, rocks or metallic materials could be molten. In the case of Earth, for example, seismic evidence indicates that the **outer core** is liquid, probably liquid iron. On the other hand, the high pressures at planetary centers may compress material that would otherwise be molten, forcing it back into the solid state. The innermost part of Earth's core, called the **inner core**, has been found to be solid.

Any equation of state used in modeling planets must thus take into account possible sudden transitions between solid and liquid states at various points in the interior. In other words, given the planet's composition, we must know at just what pressure, density, and temperature the material will melt so that the computer can be instructed to change to the liquid state if the critical conditions are encountered while "building" the planetary model.

Solid State Phase Changes

A different and more vexing problem arises because rock-forming chemical compounds may exist in different solid forms at different pressures and temperatures. That is, if the temperature and pressure increase, a mineral with a certain density and crystal structure may change phase to become another mineral with the same composition but denser crystal structure. An example is carbon's existence as graphite and diamond.

Such phase changes obviously affect our efforts to determine the structure of a planet. They could cause a systematic layering—a structure consisting of shells of different mineralogy with discontinuities at the shell interfaces. Thus, a planet could have a complicated internal structure even if it were chemically uniform.

Some of the layered structure in Earth is almost certainly due to phase transformations. The best example is a transition of the mineral olivine, $(Mg,Fe)_2SiO_4$, a mineral common in Earth's mantle, to a denser form called a spinel polymorph, at pressures around 140 kbar.* This **olivine-spinel**

*Note that 1 bar = the pressure exerted by the atmosphere at sea level = 1.01×10^5 N/m^2 = 1.01×10^6 dynes/cm^2.

transition causes a 10% jump in density at a depth of about 400 km in Earth. Other phase transitions may also be important.*

Laboratory Experiments Versus Theory of State Changes

How can we determine accurate equations of state describing the phase changes among planetary materials? If we could simulate conditions inside planets, we could directly measure how various rocks behave. Many tests of this kind have been done, but the highest pressures obtained in *most* laboratory studies are around 20 to 40 kbar, which equals the pressure at depths of only about 63 to 125 km inside Earth, less than 2% of the way to the center (Cole, 1978; Wyllie, 1971).

Higher sustained pressures have been reached in laboratories—around 1700 kbar, which corresponds to pressures in Earth's core (*Geotimes*, August 1980, p. 25). However, because temperatures are also high at this depth, and because it is hard to produce high temperature and pressure at the same time in the lab, we lack accurate simulations of rocks at such depths. Better experimental data should be forthcoming. Still higher transient pressures can be reached during instantaneous shock or explosion tests, but these give limited insight into equilibrium conditions in a planet. The pressure at Earth's center, around 3700 kbar, is well beyond any value sustained in the lab.

*Perhaps the reader has already realized that the study of planetary interiors is laced with uncertainty and difficulties. If not, the following note by the geophysicist Birch may make the situation clearer:

Unwary readers should take warning that ordinary language undergoes modification to a high-pressure form when applied to the interior of Earth. A few samples of equivalents follow:

High-pressure form	*Ordinary meaning*
certain	dubious
undoubtedly	perhaps
positive proof	vague suggestion
unanswerable argument	trivial objection
pure iron	uncertain mixture of all the elements

PRESSURE IONIZATION AND METALLIC STATES IN GIANT PLANETS

In still larger planets, the pressures increase to values beyond 3 Mbar, where the electron swarms of the individual atoms are squashed together and the atomic structure breaks down. Since ordinary chemical properties depend on the interactions of the electron swarms, the matter loses these properties and is said to exhibit **pressure ionization**—a high-pressure state different from those of familiar solids, liquids, or gases. Many or all of the electrons may be freed to move around (as happens in a metal), while remaining ions attempt to locate themselves in some sort of crystal lattice. Since the state of the electrons is similar to their state in a metal, an element in this pressure-ionized form is sometimes said to be in a metallic state,* even if it is not a metal under ordinary pressure. In the interiors of the giant planets, for example, we encounter **metallic hydrogen**, a high-pressure form of hydrogen.

Because the crystal lattices have broken down in metallic matter, many of the complexities of minerals' chemical interactions are absent, and the equations of state are easier to predict than for rocks under intermediate pressure. Also favorable is the fact that hydrogen, a major constituent in the giant planets, is the simplest element and relatively easy to treat theoretically.

Much of the early work on pressure-ionized materials in planets was done by the geophysicist W. C. DeMarcus (1958), whose thesis in 1951 first detailed the structure of the hydrogen planets.

A DENSITY-MASS DIAGRAM FOR PLANETS

Many of the principles just discussed are clearly displayed on a simple diagram that plots two of

*This state is also sometimes referred to as a degenerate state, though this description is not quite correct. Degeneracy in physics involves the role of quantum statistics in determining the condition of the electrons.

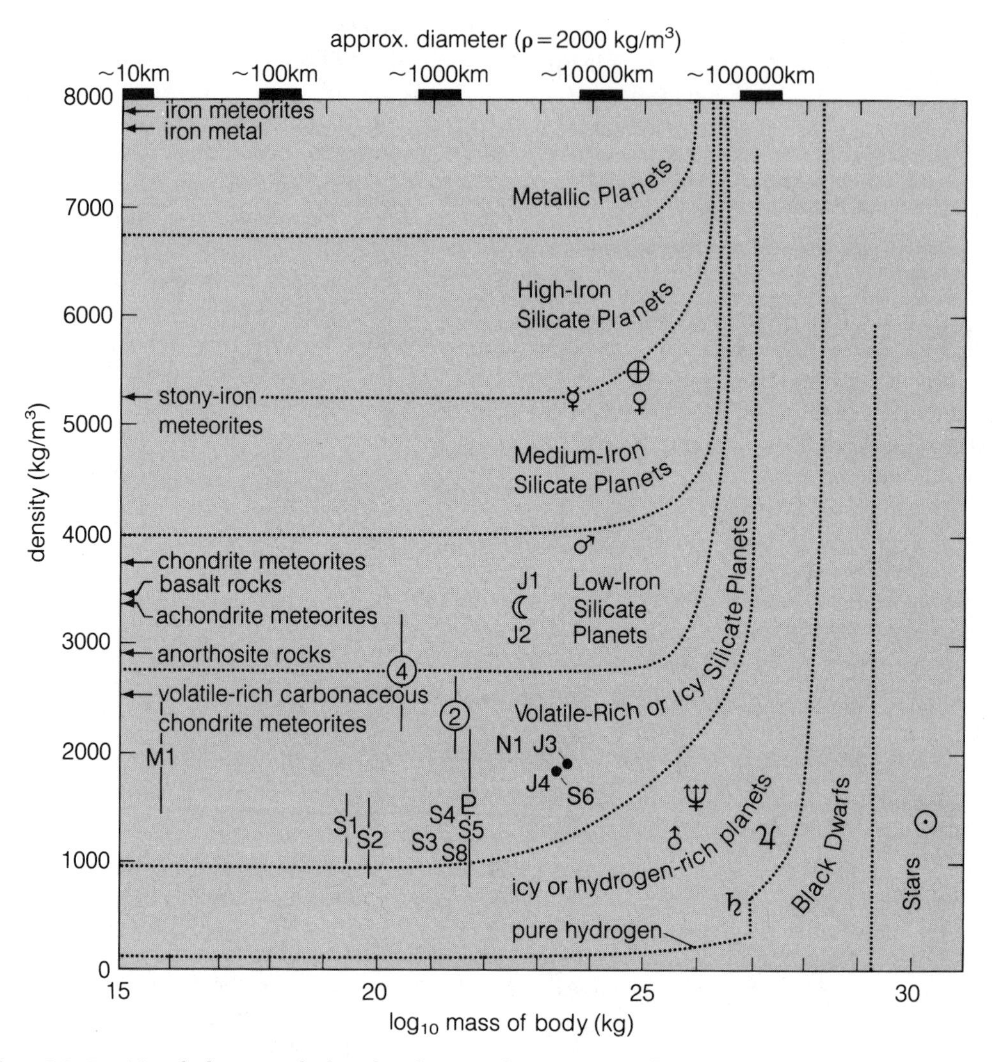

Figure 9-2. Densities of solar system bodies plotted against their masses. Satellites entered by number (S1 = Mimas). Entries along the left edge show densities of common planetary materials, and dotted lines are schematic lines of constant composition. Upward curvature at right is due to matter compressing to higher density and then turning degenerate (shapes based on Wildt, 1972; Lupo and Lewis, 1979). It is evident that planetary bodies evidence a wide range of compositions. Those closer to the top of the diagram have typically been closer to a heat source that excluded accumulation of volatiles.

the most basic measurements that can be made on planets, density and mass. This is shown in Figure 9-2. Suppose we picked some standard material, such as a volcanic rock, and began constructing worlds of it. A small planetesimal would have just the density of the uncompressed rock, around 3300 kg/m^3. As we get into worlds in the range of 10^{22} to 10^{25} kg (typically 1000 to 10 000 km across) the rock substantially compresses and may change phase, increasing the density as shown by the dotted lines. At somewhere around 10^{26} or 10^{27} kg, the central material begins to be pressure ionized and the density shoots up.

The planets, satellites, and asteroids do not lie along a single such dotted line. They scatter widely up and down the diagram, which means that they must be made of different materials. The diagram labels the different general types of materials that fall in the different regions. Bodies near the top have been exposed to heat sources that drove away the volatiles and left high-temperature (usually denser) condensates, as in the case of the inner planets and Io. On the other hand, bodies near the bottom are either cold, icy worlds, such as Saturn's moons, or the giant planets that have gravitationally trapped low-density volatiles such as hydrogen and ices.

DIFFERENTIATION

As mentioned in Chapter 6, **differentiation** is any process by which initially homogeneous material gets divided into masses of different chemical composition and physical properties. Figure 9-2 shows that strong differentiation has occurred among the planets: The initially uniform solar nebula has produced planets of remarkably different compositions.

Differentiation can also happen inside a planet. For example, if a rock mass melts and forms a homogeneous **magma** (molten rock), different minerals may crystallize at different temperatures as the mass cools. As they form, the heavy minerals may sink and light ones rise to the surface, just as happens in the smelting of metals. Thus, the final mass may be differentiated into regions with mineral mixes of different compositions.

Complete melting is not necessary for differentiation; if even partial melting occurs, fluids can move from one region to another, altering compositions. For example, if you heated a mixture of ice and rock, the ice would melt first, while the rock would not melt but settle out of the slushy mixture. Europa's smooth, virtually pure ice surface may have formed in this way. Similarly, even in a layer of melting rock, some mineral crystals melt at a lower temperature than others. **Partial melting** is the condition where some mineral components have melted and others have not. During partial melting, differentiation can occur not only by gravitational settling of the heavy minerals but also by solution of some minerals into the hot fluid component, which may then flow and carry them away from their original locations.

In studying meteorites (Chapter 6), we saw that certain groups of elements have a chemical affinity for each other and cluster together during differentiation processes. The most important examples are **lithophiles**, the elements that chemically bond into low-density silicate minerals and hence tend to float, and **siderophiles**, the elements that chemically bond into higher-density iron minerals and hence tend to sink.

Figure 9-3 is another way of showing that large-scale differentiation of this kind has occurred in planets. The figure plots abundances of some of the heavy elements, ranked from strongly lithophilic to strongly siderophilic properties. The carbonaceous chondrites have solarlike abundances of these elements and have not been strongly differentiated; they establish a reference line. Stony-iron and iron meteorites are, of course, extremely enriched in siderophiles and depleted in lithophiles. They seem to be samples from well inside a melted parent body, where iron and siderophiles have collected near the center. On the other hand, the surface rocks of the moon are enriched in lithophiles and depleted in siderophiles; the moon must have once been at least partially molten, allowing the lighter materials to rise and carry lithophiles toward the surface. Rocks from Earth's granitic crust are even more lithophilic. Probably Earth's differentiation is continuing even today as molten materials flow and volcanic magmas erupt, allowing light materials to accumulate on the surface.

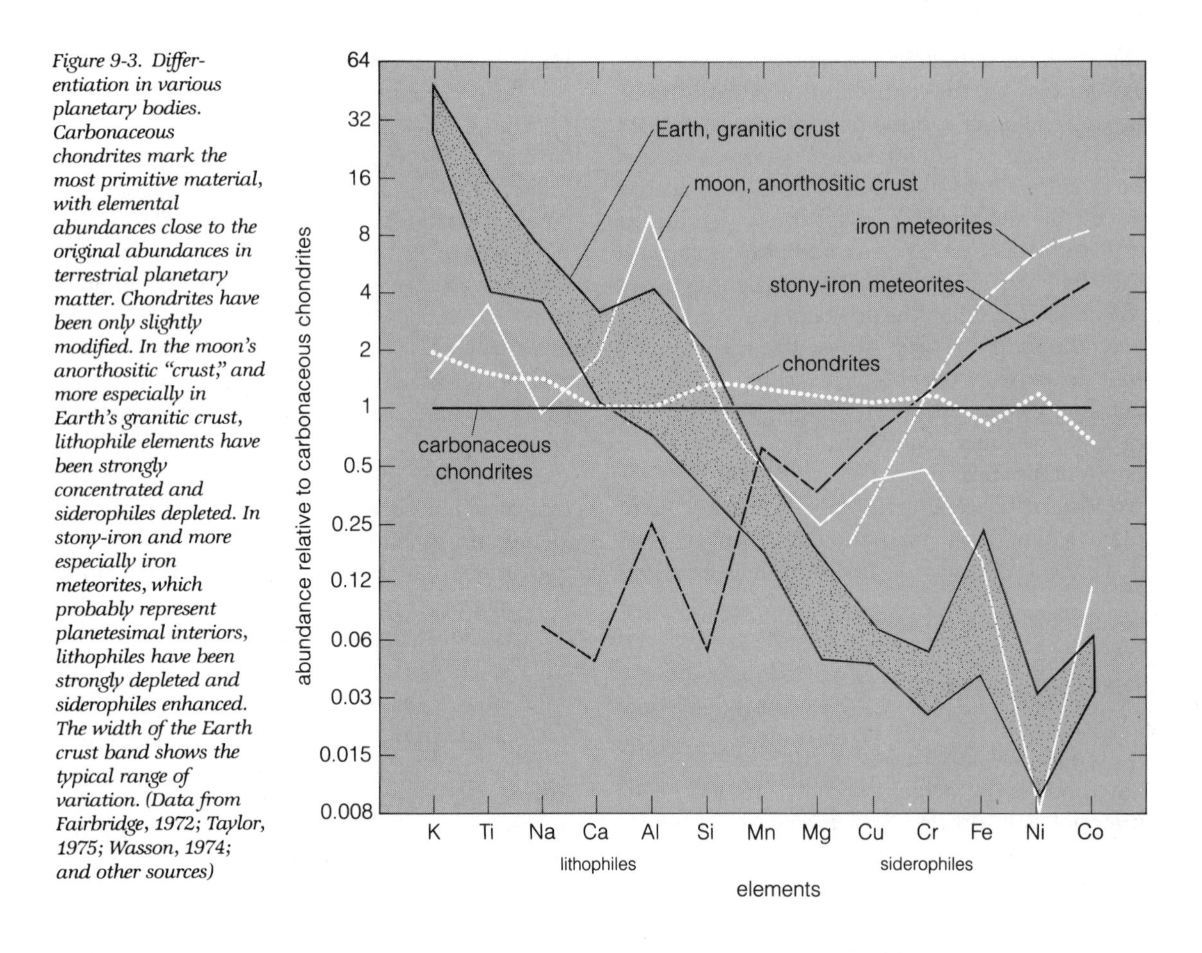

Figure 9-3. Differentiation in various planetary bodies. Carbonaceous chondrites mark the most primitive material, with elemental abundances close to the original abundances in terrestrial planetary matter. Chondrites have been only slightly modified. In the moon's anorthositic "crust," and more especially in Earth's granitic crust, lithophile elements have been strongly concentrated and siderophiles depleted. In stony-iron and more especially iron meteorites, which probably represent planetesimal interiors, lithophiles have been strongly depleted and siderophiles enhanced. The width of the Earth crust band shows the typical range of variation. (Data from Fairbridge, 1972; Taylor, 1975; Wasson, 1974; and other sources)

ADDITIONAL OBSERVATIONAL CHECKS ON PLANETARY MODELS

Given a model of a planet computed according to the scheme outlined earlier, we could use a variety of direct observations to check its accuracy. The first four basic measurements have already been described: (1) the planet's total mass, (2) its diameter, (3) its mean density, and (4) the composition of the surface rocks. We will now discuss the following additional observational checks: (5) the planet's moment of inertia, (6) its geometric oblateness, (7) the form of its gravitational field, (8) its rotation rate, (9) its heat flow, (10) the composition of neighboring planets and meteorites, (11) the form of its magnetic field, (12) drilling and direct sampling, and, most important, (13) seismic properties.

Moment of Inertia

The **moment of inertia** of an object is a measure of the degree of concentration of mass toward the object's center. This is important in planets, since we expect the heavier materials to have concentrated toward the center. For example, if a heavy mass M is whirling around on the end of a long (massless) wire of length R, the moment of inertia of this configuration is MR^2. In planets and other objects of geometrically regular shape, the moment of inertia is said to be kMR^2, where k is called the coefficient of the moment of inertia. For spheroidal objects, k must lie between 0.0 and 1.0, and the numerical value tells how concentrated the mass is. Examples are shown in Table 9-1.

The moment of inertia of a spinning object can be determined by its response to an external torque, which causes it to wobble. For instance, a spinning top wobbles when gravity tries to pull it down. Torques on planets are exerted by the sun and large satellites. For example, the sun and moon produce Earth's precessional wobble, which takes 26 000 y and indicates a moment of inertia coefficient of 0.33. As shown in Table 9-1, this value indicates considerably denser material in Earth's core than in its outer regions. Table 9-1 also shows that the giant planets have dense cores and vast, low-density envelopes.

Geometric Oblateness

A rotating planet is an oblate spheroid. Its equatorial diameter a exceeds its polar diameter b. The oblateness, or flattening, is a measure of this difference:

$$\textbf{geometric oblateness} = \epsilon = \frac{a - b}{a}$$

This ratio can be measured on photographs of the planets or (once we reach the planet) by the more accurate methods of geodesy.

Oblateness is a useful quantity because it reflects the planet's internal properties, such as mass distribution or departure from hydrostatic equilibrium. It is possible to calculate the value of ϵ that will characterize a planet in equilibrium, given the planet's mass, size, rotation period, and its mass distribution. We can calculate values that would correspond to idealized cases, such as a homogeneous sphere, and then list the observed values to see how the planets compare.

Gravitational Field and Dynamical Ellipticity

If a planet departs from spherical symmetry, either by being oblate, having an asymmetric mass distribution, or in some other way, its gravitational field will not be spherically symmetrical. A nearby satellite (either natural or artificial) will respond to the asymmetries by precession of its orbit and by departing slightly from a true Keplerian orbit. By measuring the satellite's motions, therefore, it is possible to estimate the departure from spherical symmetry of the mass inside the planet. These departures are often given in terms of a series of numerical coefficients known as J values, which describe greater degrees of complexity than the mere central concentration of mass described by the moment of inertia coefficient k.

The simplest way to estimate the planetary structure from gravity measures is to assume that the planet's mass is distributed in smooth concentric, equipotential shells that have various departures from sphericity, and then to calculate

Table 9-1
Values of Moment of Inertia Coefficient k for Various Bodies

Body	Moment of inertia coefficient k
Idealized cases	
Hollow sphere	0.667
Homogeneous sphere	0.400
Sphere with core having half the total radius and twice the mantle density	0.347
Mass concentrated entirely at center	0.000
Actual cases	
Moon	0.40
Mars	0.37
Earth	0.33
Neptune	0.29
Jupiter	0.26
Uranus	0.23
Saturn	0.20
Sun	0.06

Source: Data from Cole (1978); Mars value 0.3654 ± 0.001 from Reasenberg (1977).

the planetary figure that would produce the observed J values. The amount of flattening derived in this way is called the **dynamical oblateness**. If it equals the observed geometric oblateness, this suggests that the concentric shell assumption is fairly accurate.

If not, it may indicate "lumps" of higher density scattered here and there. An example appeared in the lunar gravity field, where measures of space vehicle motions near the moon led to the discovery of **mascons**, or mass concentrations in the lunar surface layers (Muller and Sjogren, 1968). These turned out to be large lava flows that fill ancient impact basins, forming rock sheets denser than their surroundings. Since the moon's outer layers formed, they have been too rigid to allow the mascons to achieve hydrostatic equilibrium by sinking into the moon's interior.

Another type of inconsistency between the geometric figure and mass distribution has been found in the case of the moon and Mars. In both bodies, the geometric center is offset from the center of mass. This shows that the mass is not distributed symmetrically around the geometric center and that full hydrostatic equilibrium has not been attained.

In the moon, the center of mass is about 2 km closer to Earth than the center of the figure. This asymmetry is believed caused by the low-density crust being several kilometers thinner on the Earth-facing side than on the far side. This may have resulted from the crust being blasted away during the formation of the largest impact basins on the Earth-facing side. In Mars, the center of mass lies north of the center of the figure. This is associated with the southern cratered highlands, standing about 4 km higher than the northern volcanic plains, a fact that may also be associated with an asymmetry in crustal thickness (Wu, 1978).

Rotation Rate

The rotation period must be known in order to interpret the moment of inertia, the geometric oblateness, and the dynamical oblateness. The telescope reliably reveals actual surface markings of only Mars and the moon. Radar, spectroscopic Doppler shifts, and radio results have been used to confirm planetary rotation rates for all planetary bodies, and even today there is some question about the precise rates for Uranus and Neptune (see Chapter 2).

Surface Heat Flow and Temperature Gradient

The rate at which heat escapes from the interior of a planet and the rate at which temperature increases as we go down into the planet (the **temperature gradient**) are both important parameters for judging the interior conditions. They can be measured in the following way on a planet's surface. First, a hole is drilled and then a thermal probe is lowered into the hole. The hole must be deep enough to get away from **diurnal** (day and night) surface changes of temperature. The probe is several meters long and has a thermometer to measure the temperature at each end. Next, we must wait till the temperatures in the hole reach equilibrium values, because the drilling of the hole created heat by friction. After some hours the frictional heat will have been dissipated and we can measure the environmental temperatures in the hole. The difference between the temperature at the top and bottom of the probe divided by the length of the probe gives the temperature gradient. For example, the difference in temperature might be 0.006 K and the probe 2 m long, giving a temperature gradient of 0.003 K/m, or 3 K/km.

The surface heat flow (that is, the amount of energy coming through a square centimeter of the surface in 1 s [J/m^2·s]) is directly proportional to the temperature gradient and to the conductivity of the rock. If we take a sample of rock from the drill hole and measure its thermal conductivity in the laboratory, we can multiply the conductivity times the temperature gradient and get the heat flow coming out the region of the hole. Such experiments have been done all over Earth. In the outer planets, heat flow is sufficiently large to be measured by infrared detectors. Values are summarized in Table 9-2.

Suppose that on a certain planetary surface we measured a certain rate of heat flow from the interior. Furthermore, suppose that upon measuring the radioactivity of the surface rocks, we found that a layer of those rocks only 30 km thick would provide sufficient heat from radioactive decay to account for the measured heat flow. We would have to conclude that the radioactive elements were strongly concentrated near the surface. This is just the situation on Earth's continents. Heat flow on the terrestrial continents and the flat ocean floors averages about 0.04 to 0.08 J/m^2·s.

But over certain volcanically active areas, especially ridges running along portions of the mid-ocean floors, heat flow ranges between 0.08 and 0.32 J/m^2·s. Apparently, ascending currents inside Earth are bringing hot material upward in these regions. Because radioactive elements are lithophilic, they ascend with the lithophilic chemicals and end up concentrated in the lithophilic surface rocks, especially in the 30-km-thick continental blocks.

Another example is the moon. Before Apollo, many researchers thought the moon might have a chondritelike composition, with a sufficiently low content of radioactive elements to give a low heat flow. However, measurements at the Apollo 15 and 17 sites gave unexpectedly high values, around 0.03 J/m^2·s, approaching Earth's continental values. Since the lunar surface rocks are not as differentiated as Earth's, they don't have as much radioactivity as the very lithophilic continental rocks; and the radioactivity levels measured in the lunar surface rocks must extend to deep levels in order to provide the observed heat flow. These findings show that the moon's overall material is more lithophilic and more differentiated than chondrites, but not so lithophilic as Earth's continents. The moon's bulk is similar in general composition and radioactivity to the subsurface layers in Earth's upper mantle, as noted in the discussion of lunar origin in Chapter 5.

Table 9-2
Heat Flows in the Solar System

Object	Measured heat flow (J/m^2·s)	Energy generated per kilogram per sec (J/kg·s)
Sun	6.6×10^7	2×10^{-4}
Earth	0.055	4.7×10^{-12}
Moon	0.03	1.5×10^{-11}
Jupiter	7.6	2.4×10^{-10}
Saturn	2.8	2.2×10^{-10}
Uranus	<0.18	$<2 \times 10^{-11}$
Neptune	0.28	2×10^{-11}

Note: Giant planet data from Hubbard, 1980 preprint, *Intrinsic Luminosities of the Jovian Planets.*

Composition of Neighboring Planets and Meteorites

Since the gross chemistry of planets was controlled by condensation sequences during the cooling of the solar nebula, there is a relation between gross chemistry and distance from the sun (see Chapter 5). Bulk compositions of planets appear to be consistent with the theory of the condensation sequence. Also, the composition of meteorites, particularly irons and stony-irons, showed that nickel-iron alloys formed cores inside at least some extraterrestrial bodies.

Magnetic Field

The presence of a strong planetary magnetic field probably indicates the presence of a fluid core in the planet, as we will see in more detail later in this chapter. Even if the planet lacks a strong field, magnetic observations near the planet can reveal secrets of its interior. In this case, the magnetic field associated with the solar wind moves through the planet, since the planet's lack of magnetic field causes no deflection. Deflections in the magnetic field around the planet are controlled by the conductivity inside the planet, and a dry, poorly conducting rock would produce a different result than would a conductive interior.

Deep Drilling and Direct Sampling

The deepest drill holes penetrate about 6 km into Earth, only 0.1% of the way to the center, so one might conclude that drilling could hardly reveal any important properties of planetary interiors. This is not so, however, because important layering effects occur within a few kilometers of the surface. The base of Earth's crust, the surface layer of relatively low-density rock, is only about 5 km below the ocean floor.

A national effort to drill through the crust and directly sample the underlying mantle, the **Mohole Project**, was launched in the 1960s but eventually cancelled in a political–scientific wrangle. However, that project was replaced in the 1970s by the building of the extremely successful research ship **Glomar Challenger**, which has sailed all oceans and drilled as much as 1.8 km into the seafloor, including 0.6 km into the basaltic layer at the base of the crust under the ocean sediments. For example, its finding that the oldest known seafloor rocks are only 160 My old proves that the present-day ocean configurations are very young features of a very active terrestrial crust (Nierenberg, 1978).

As for the mantle material, some free samples may be provided by volcanoes. Certain deep-rooted volcanoes erupt nodules of dense rock relatively rich in iron-bearing minerals such as **olivine** (the mixture of Fe_2SiO_4 and Mg_2SiO_4 mentioned in Chapters 5 and 6 and on page 248). Note that olivine is made of the four most abundant elements in Earth: iron, oxygen, silicon, and magnesium, in that order.

Especially important volcanic vents of this type are the **kimberlite pipes** in South Africa, Wyoming, and elsewhere. These are root structures of volcanoes exposed by erosion (Cox, 1978). They contain kimberlite, a dense rock rich in olivine, and even denser nodules composed of olivine and **pyroxene** (a mixture of $MgSiO_3$ and $FeSiO_3$). Kimberlite pipes also contain the well-known high-pressure form of carbon, diamond, and hence have received much (not wholly academic) attention, which has illuminated mantle composition. On other planets, volcanoes and deep meteorite craters are areas where mantle rocks might be exposed.

Seismic Properties

Seismic observations give the most useful experimental data regarding Earth's interior. During the first Apollo flights in 1969, the first seismometers were deployed on another planetary body, opening the way to planetary seismology. The Viking 2 lander carried the first successful seismometer experiment to Mars in 1976. Seismology permits us to measure positions of layers inside a planet, determine their densities, and look for special structures such as cores or mass concentrations nearer the surface. Such measures sketch the geometry of the planet's interior and limit possible compositions. Seismology is considered in more detail in the next section.

SEISMOLOGY AND EARTHQUAKES

The Seismometer

An earthquake is a series of vibrations, swayings, or sudden jiggles of the ground. The jiggles can be so small as to be scarcely detectable; rarely, they can knock over buildings, as shown in Figure 9-4. Imagine that we wanted to record such jiggles; how could we do it? Suppose we suspended a 100-kg weight on the end of a 10 m wire in a quiet room and attached a scribe to the bottom of the weight so that it just touched a tabletop. If Earth suffered a sudden horizontal jolt, the scribe on the bottom of the pendulum would mark a line in the dust (which probably would have accumulated on the table by the time of the first major earthquake). This would be a record of the earthquake; the length of the line would measure the violence of the jolt.

This is basically the principle of the **seismometer**, except that the pendulum must be designed so that its own natural swinging frequency will not be confused with the earthquake's vibrations. Of course, modern seismom-

Figure 9-4. Destructive effects of the San Francisco earthquake of 1906, caused by movement along a major active fault that passes near the downtown center. (Courtesy Steven M. Larson)

eters are very sophisticated and much less bulky than the apparatus suggested above. Television witnesses of the moon landings will recall that the seismometers deployed by the astronauts were relatively small.

Wave Types and Early Seismic Observations

The vibrations felt during an earthquake are waves propagating along Earth's surface. The nineteenth-century physicists Poisson (1829) and Rayleigh (1887) predicted theoretically the main types of seismic waves that could be propagated by earthquakes. The most important of these are the **P waves** and **S waves** predicted by Poisson (Figure 9-5). P waves are *p*ressure waves, such as sound waves, where the motion of individual particles is along the direction of the wave's motion (as a mnemonic aid, think of P waves as *push–pull* action). S waves are *s*hear waves, where the motion of individual particles is at right angles to the wave's motion; an example of such motion occurs in a wave along a rope or garden hose.

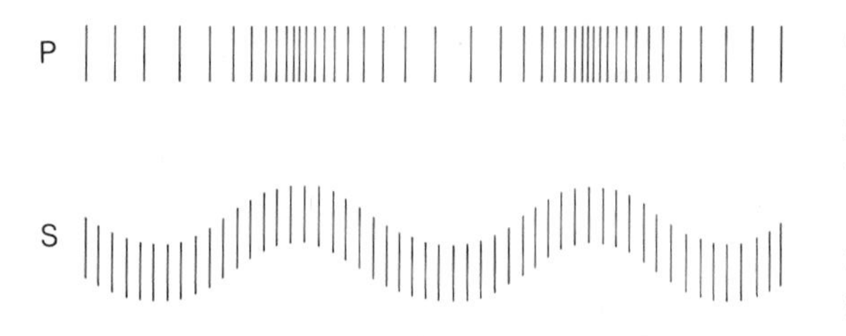

Figure 9-5. Pressure (P) and shear (s) seismic waves.

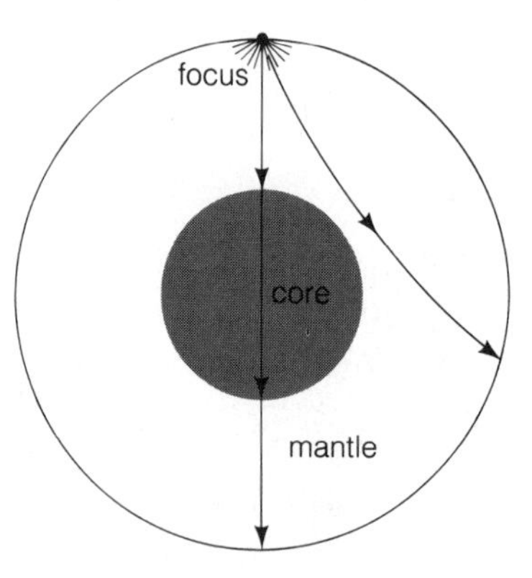

Figure 9-6. Paths of seismic waves through Earth. Waves that arrive on the far side of Earth from an earthquake focus are altered by the core, a fact that led to the discovery and size measurement of the core.

Because P and S waves have different velocities, they reach the seismic observer at different times. If the observer is located close to the **focus,** the earthquake's point of origin, the interval between the P and S arrival times will be short, but if he is farther away, it will be longer. The time interval between the P and S waves is thus a measure of the distance to the focus.

A minimum of three observing stations in a triangular array is needed to locate an earthquake. Each observes the **P-S arrival time interval** and thus computes the radial distance to the focus. On a map, a circle of this radius can be drawn around the station to indicate possible locations. If three stations provide three circles, there will usually be only one mutual intersection. This marks the earthquake's **epicenter,** or surface point directly above the focus. Thus an array of three simultaneously operating seismometers on any planet is far more valuable than a single seismometer.

English geologist R. D. Oldham (1900) was first to distinguish the P and S waves in seismic records, and he first constructed useful tables of travel times for the two waves. Seismologists quickly discovered that they could even observe earthquakes that occurred on the other side of the globe. In 1906 Oldham found that waves arrived at the anticenter (180° from the earthquake) later than would be predicted by his travel-time curves. This meant that the center of Earth contained a low-velocity region, as shown in Figure 9-6. This was the first evidence of a distinct core.

Another famous discovery was soon made by the Yugoslavian seismologist Mohorovičić (1909). In studying shallow earthquakes, he found that there were two sets of P waves and two sets of S waves. One set of each type traveled at a high velocity and was traceable over large distances; the second set traveled at a low velocity and was best observed near the epicenter. Apparently the shallow rocks transmitted waves slower, as shown by Figure 9-7. The interpretation was that a discontinuity in density and wave velocity—hence rock type—occurred at a depth of some 30 km. This is called the **Mohorovičić discontinuity** (sometimes called the "Moho" for short), and its discovery was the first proof of a distinct crust overlying the interior of Earth. Seismic and geological properties showed that the crust was composed of a lower-density rock type than the underlying mantle. Seismology thus first defined the three major compositional regimes in Earth: the core, mantle, and crust. Various minor layerings also appear in the mantle.

In 1926, the German-American geologist Beno Gutenberg pointed out an upper mantle layer, where seismic wave velocities were lower than normal. This zone, which lies about 100 to 350 km deep, was for many years called the **low-velocity zone**. There was no evidence for major rock density change in this zone, and it was con-

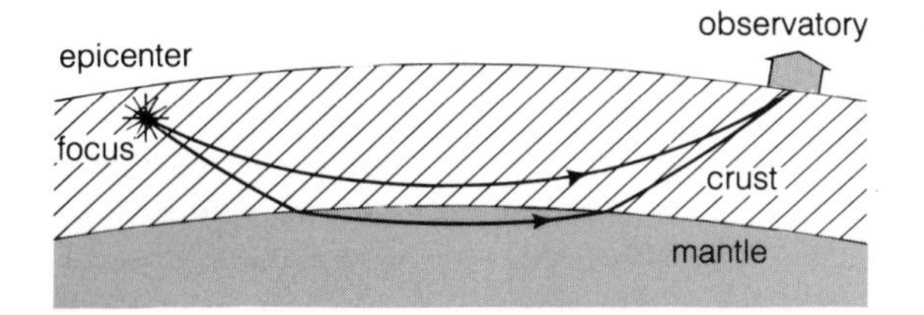

Figure 9-7. Discovery of the Mohorovičić discontinuity by observation of two sets of seismic waves with different velocities coming from a single, shallow earthquake.

sidered poorly defined and unimportant. However, in recent years it has been identified as a weak layer of partially melted material within the upper mantle and is now regarded as very important in interpreting both the dynamics of the upper mantle and the surface geological features of the crust, as we shall see. In this zone some of the lower-temperature minerals have melted, but higher-temperature minerals remain suspended as solid crystals, rather like a slushy mixture of water, ice, and soil. This layer is now called the **asthenosphere** (from a Greek word for "weak").

In 1937 a Danish seismologist, I. Lehmann, detected structure inside the core. P waves traveled faster in the interior part of the core than in the outer part. Studies of the outer part of the core revealed that it did not transmit any S waves, indicating that it has zero rigidity. Therefore, the outer core must be liquid, while the inner core is solid.

Origin of Earthquakes

What causes earthquakes? Earth is a dynamic body, even though its evolution is very slow. To drive an evolving system, energy is required, and in the case of Earth this energy is thermal energy (heat). Earth continually tries to readjust its temperature distribution: Radioactivity produces new heat, while the interior as a whole tries to cool.

Earth might be compared to a house cooling at night; occasionally it creaks or pops. Because of contraction, expansion, or slow movements of materials, stresses build up in the solid outer part of Earth. The rock may stretch elastically, but if the stress is too great, the rock may fail, like a twig bent too far. The sudden splitting and slippage of the rock to a new position of less stress is an earthquake. This process is called faulting, and any rock fracture along which there has been an offset motion is called a **fault**. Many of California's earthquakes are caused by slippage along the San Andreas Fault.

A less important source of earthquakes is volcanic activity at Earth's surface. The spewing forth of billions of tons of lava during a volcanic outburst may be locally accompanied by vibrations and shocks that can literally shake the ground to pieces.

A typical large earthquake may release 10^{17} J of energy. The annual release of earthquake energy is estimated to be about 5×10^{17} J, most coming from a very few large quakes. Table 9-3 lists some of history's most energetic earthquakes and volcanic eruptions.

Not all earthquakes are disastrous. The lower the energy, the more common the earthquake. "Garden variety" earthquakes cause a slight vibration of the ground or a rattling of windows and walls. In addition, seismographs detect vibrations that range in energy all the way down to tremors produced locally by wind, traffic, and so on.

Studies that utilize the seismic signals from naturally occurring earthquakes are called **passive seismology**. In order to better analyze the probing seismic signals, it is necessary to know the exact time of the initial shock and the exact location of the focus; for these purposes, manmade explosions are used as sources. Such studies are called **active seismology**.

Earthquake Distribution: Plates, Asthenosphere, and Lithosphere

Earthquakes are not distributed randomly, either geographically or in terms of depth. Figure 9-8 shows the geographic distribution of shallow earthquakes, deep earthquakes, and volcanoes. This map is full of clues concerning relations between Earth's surface layers and interior, but geophysicists took many years to understand these clues. The map shows that Earth's surface is divided into **plates,** or coherent blocks bounded by zones of earthquakes and volcanism. Some of the plates contain continents, but others are huge segments of the seafloor. Earthquakes are concentrated at the margins of plates.

A great deal of geological evidence, as well as seismological, has revealed the fundamental importance of the plates and their linkage with the asthenosphere, or weak, partly melted zone mentioned above. The combination of heat flow and fluidity within the asthenosphere produces slow currents that drag along the overlying layers. These layers are solid and more brittle than the aesthenosphere and are called the **lithosphere** (from the Greek root for "rock"). The 100-km-thick lithosphere has probably been broken into plates because of movements in the asthenosphere. Earthquakes tend to occur near plate margins, where collisions between plates occur. The motions of plates are very slow—only a few centimeters per year—but they are sufficient to build up great stresses in the rock, which eventually are released as faults and accompanying earthquakes.

The depth distribution of earthquakes, shown in Figure 9-9, reveals more information. Earthquakes are most concentrated in the 100-km-thick lithosphere, because it is continually being cracked by the asthenospheric currents. The number of earthquakes drops sharply as we proceed down through the more plastic asthenosphere, from about 100 to 350 km in depth. Other concentrations may mark alternating fluid layers with

Table 9-3
Destructive Earthquakes and Volcanic Eruptions

Date	Location	Event[a]	Log energy (J)	Deaths (approx.)[b]
Aug. 24, 79	Pompeii, Italy	V	—	15 000
Feb. 2, 1556	Shensi, China	E	—	800 000?
1737	Calcutta, India	E	—	300 000
Nov. 1, 1755	Lisbon, Portugal	E	~17.5	60 000
1815	Sumbawa Island, Indonesia	V	—	75 000
Mar. 26, 1872	Owens Valley, California	E	⩾16.8	~36
Aug. 26, 1883	Krakatau, Indonesia	V	—	37 000
May 8, 1902	St.-Pierre, Martinique	V	—	29 000
Apr. 18, 1906	San Francisco, California	E	16.7	600
Dec. 16, 1920	Kansu, China	E	17.1	~180 000
Sep. 1, 1923	Kwantō, Japan	E	16.8	~143 000
Feb. 29, 1960	Agadir, Morocco	E	13.0	~12 000
May 31, 1970	Peru	E	16.0	~70 000
Feb. 9, 1971	San Fernando, California	E	~14.0	62
Feb. 4, 1976	Guatemala City, Guatemala	E	—	22 000
July 28, 1976	T'angshan, China	E	—	700 000
May 18, 1980	Mt. St. Helens	V	16.7?	65
Aug. 6, 1945	Hiroshima	Atom bomb	13.9	66 794
1969–1970	Moon	Total lunar seismic activity	~ 7	—

[a]V = volcanic eruption; E = earthquake.
[b]Various sources; see Tazieff (1964) and 1978 reference book *Disaster!* prepared by editors of Encyclopaedia Britannica. Figures include all related destruction. Mt. St. Helens energy from *Volcano News*, July 1980.

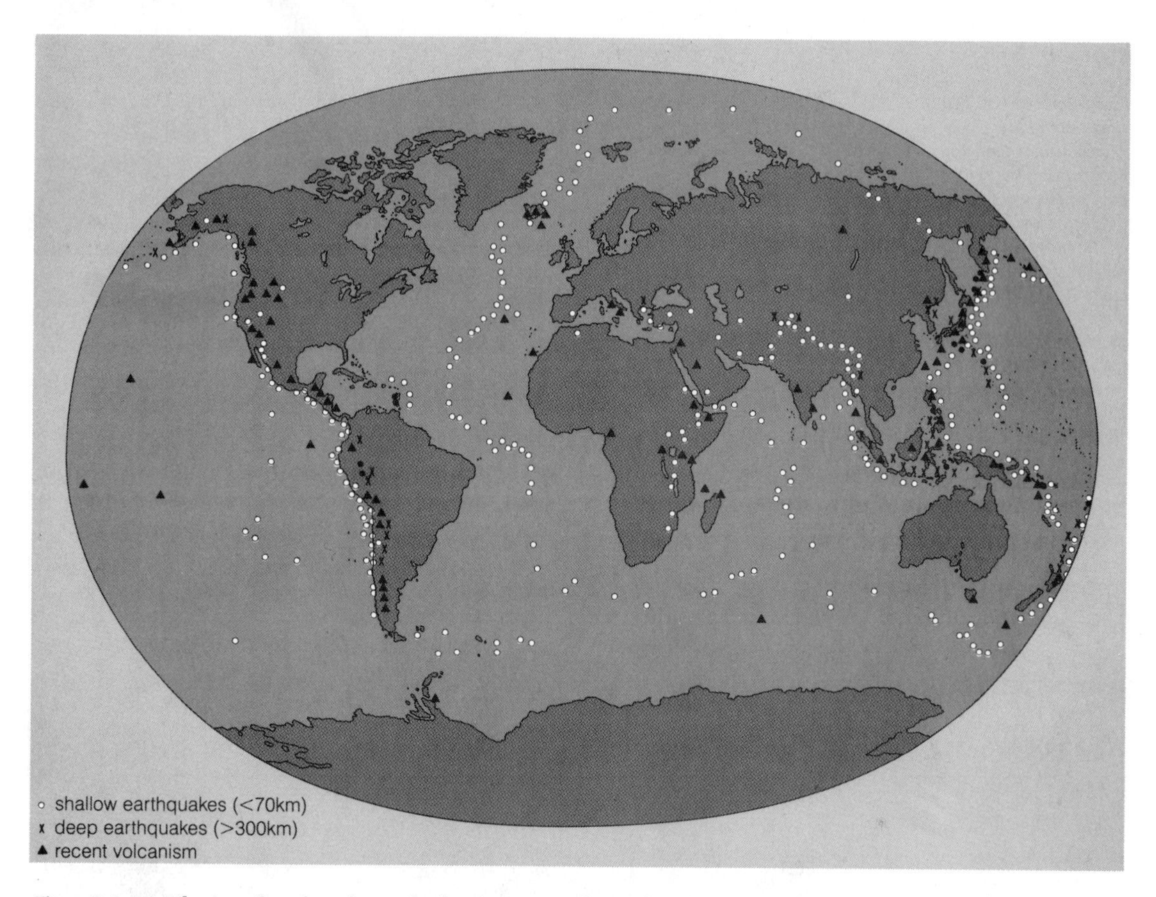

Figure 9-8. Distribution of earthquakes and volcanism on Earth. Earthquake and volcanic activity occur along margins of plates, where plates abut and move relative to each other. Earthquake zones tend to slope downward under continental margins of plates, as shown by the pattern of shallow versus deep earthquakes.

strong, brittle layers. The number drops to near zero at depths below about 700 km, implying that the lower mantle below 700 km is too plastic—too near hydrostatic equilibrium—to build up stresses that cause earthquakes.

"Moonquakes" and "Marsquakes"

Similar concepts apply to other planets. Small planets cool more rapidly than large planets and thus have developed thicker lithospheres. This is clearly shown in the moon, which is seismically much quieter than Earth. During the first year of lunar observations, lunar quakes rated only 1 to 2 on the well-known Richter scale, contrasted with ratings of 5 to 8 for major terrestrial quakes. The total seismic energy released annually in the moon is estimated at about 10^7 J (Latham, 1971), contrasted with 5×10^{17} J in Earth.

As shown in Figure 9-9, the quakes occurring inside the moon occur in a band around 800 to 1000 km in depth, which is interpreted as the base of the lunar lithosphere. Various geophysical measures, such as seismic data and estimated temperature profiles, suggest a partially molten asthenosphere below that level (Taylor, 1975).

The situation on Mars appears similar, according to results from one relatively crude

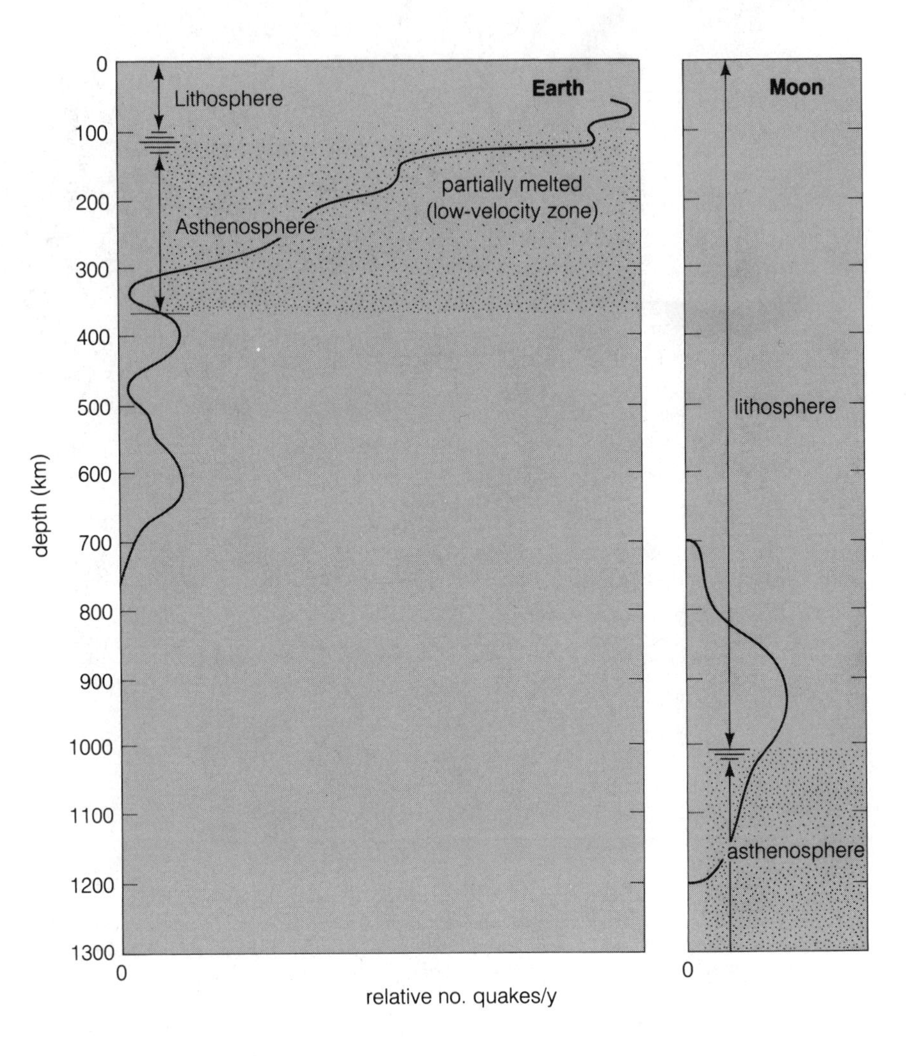

Figure 9-9. Depth distributions of "earthquakes" on Earth and the moon. Earthquakes tend to occur along the bottom edge of the solid lithosphere and the top of the more plastic asthenosphere, where currents build up stresses in the rocks. Moonquakes number nearly 3000/y but are much less energetic than larger earthquakes, so that the two scales are not directly comparable.

seismometer placed by the Viking 2 lander (a duplicate seismometer on the Viking 1 lander failed to be deployed). During 5 mo of operation, no large earthquakes were detected, though one possible event of Richter magnitude 3 was recorded (Anderson and others, 1977). Martian winds, shaking the lander that contained the seismometer, provided a high "seismic" noise level during much of the period. Analysis indicates that Mars is quite likely less seismically active than Earth and perhaps close to the moon in its level of activity.

THERMAL HISTORIES OF PLANETS

Definition of the Problem

Most of this chapter so far has dealt with ways of determining the **statics of planets**, that is, descriptions of their interiors in their present-day states. But planets are evolving systems. We must therefore also come to grips with the **dynamics of planets**, that is, the evolution of their internal properties through time.

The heating or cooling of a planet is one of the driving factors in its evolution. The problem in the study of thermal histories is to discover the temperature and related properties (motions, liquid or solid state, and so on) of planetary interiors as a function of time. This is a more grandiose goal than modeling the present-day interiors.

Theory of Heat Transport

Heat is one form of energy. If a body is hot, its atoms and molecules have a higher amount of energy than if it is cold. The atoms and molecules of a hot body move around and vibrate faster, and hit each other harder, than those of a cold body. **Temperature** is a measure of this thermal energy.

Why are the interiors of planets hot? Some of the heat is original; the material from which the planet formed had a certain temperature, and more heat was generated as the planetesimals crashed and transferred their kinetic energy into the growing planet. Other heat is added to the planet over the course of time by radioactivity, whereby atoms emit small particles. These particles speed out into the surrounding atoms, hitting them and thereby adding to the total thermal energy.

Heat moves naturally only from hot regions to cold regions—never from cold to hot. **Heat transport**, the process of transferring the heat and equalizing temperatures, can occur in three ways. First is **conduction**, the process by which the atoms or molecules in a substance strike each other, preferentially transferring energy from the fast-moving atoms in the hot region to the slow-moving atoms in the cold region. Second is **radiative transfer**, the process by which high-energy atoms emit radiation that travels through the surrounding material and is absorbed by low-energy atoms in a cooler region. Finally, **convection** is the process by which an entire hot region—because it has expanded to a lower density than a cold region and is therefore lighter—rises upward as a unit through the colder material while nearby colder material sinks. An example of this is the hot smoke from a burnt match rising through the air; a more impressive example from a geological point of view can be seen by heating a shallow pan of cooking oil, which will set up a cellular pattern of convection currents reminiscent of the ascending and descending currents postulated by the theory of plate tectonics.

To apply these concepts to the problem of planetary thermal histories, we must have a great deal of quantitative information. First, we must have an adequate mathematical theory describing each of the three modes of heat transport. While conduction and radiative transfer have long been well understood, an adequate theory describing convection is still being developed. Also, we need to know how much heat is being created at each point by radioactivity. Because of their lithophilic chemistry, the main radioactive elements are concentrated in the crust, but we need to estimate their abundance throughout the planet.

Next, we must determine which mode of heat transport is most important in the given planet. If heat transport is by conduction, we need to know the rock conductivity at different depths; if by radiative transport, we need the transparency of the rock to the kind of radiation involved; if by convection, we need to know the mobility of the material. These properties are estimated from lab measurements. Furthermore, at different points and different times inside the planet, different modes of heat transport may dominate.

Convection

To understand the importance of convection, consider some parcel of the fluid or plastic material inside a planet (or, for that matter, in a planetary atmosphere). Suppose some random perturbation starts this material moving upward. As it ascends to a new level, it immediately adjusts its pressure to the surrounding pressure, which is lower. (*Immediately* means at the speed of sound if the medium is truly fluid, since this is the speed at which pressure inequalities transmit themselves.) Thus, the parcel expands and becomes less dense as it rises.

Since any heat transfer between the parcel and its new surroundings takes much longer than the pressure adjustment, the pressure adjustment takes place with virtually no external energy input. This is called an **adiabatic change**. An adiabatic change creates a new density and temperature in a parcel, and if the new density is less than that of the surroundings, the parcel will be buoyed up and will continue to rise automatically. The resulting motion is called convective motion.

Steep temperature gradients favor convection. That is, convection proceeds only if there is a large temperature difference in the material over a short distance. Convection can begin only if the actual temperature gradient exceeds the **adiabatic temperature gradient**, which is that experienced by an ascending parcel that rises adiabatically (that is, without external energy input).

This is one reason that convection may develop in an asthenosphere: The material is plastic and the temperature drops rapidly from the heat of the interior to the cold of the surface in a short distance. If the temperature gradient is great enough, and if the material is fluid enough, convection currents are sure to arise, creating **convective cells** in which hot material rises, spreads outward, and descends along the edge of the cell. If the lithosphere is thin enough, these convection currents may drag on it strongly enough to disrupt it.

Initial Thermal State of Planets

A model of the thermal evolution of a planet gives the calculated temperatures at all points in its interior as a function of time. Once the theory of heat transport is available, calculating such a model requires only that we specify the materials composing the planet and, of course, the equations of state of those materials. Also, we need to specify conditions throughout the planet at some specific time. Therefore, modelers pay special attention to initial conditions and usually calculate forward from the initial state, using the present conditions as a check.

Following the chemical equilibrium theory discussed in Chapter 5, most modelers have assumed that each planet formed as a nearly homogeneous ball, without a core or crust. As an alternative, the inhomogeneous accretion theory of Chapter 5 predicts that the iron-rich early condensates accrete first, forming a ready-made iron core later overlaid by later silicate condensates (Slattery, DeCampli, and Cameron, 1980).

In any case, geochemical evidence from the moon and meteorites indicates that planets (or at least their surface layers) were partly molten by the time they finished forming (Fricker, Reynolds, and Summers, 1974). Lunar rock chemistries and ages seem to require that the moon were initially (4.5 Gy ago) covered by a "**magma ocean**"—a molten layer at least several hundred kilometers deep—while meteorite parent bodies apparently also melted and resolidified 4.5 Gy ago.* So most models begin at 4.5 Gy ago with the surface layers already molten to great depths.

Several heat sources are possible for this initial melting:

1. The impacts of the planetesimals during accretion provided heat. If significant growth occurred in only a few thousand years or less, this heat may have been trapped and helped to melt planets. Slower accretion would have allowed this heat to dissipate.

2. Today's important radioactive elements, uranium, potassium, and thorium, produced more heat at the beginning, as shown in Figure 9-10, but not enough to melt planets in the first few million years.

---

*Note that this doesn't mean that planets grew from incandescent material, as thought half a century ago. Urey (1952) proved from meteorite chemistries that planets formed from cold dust, typically at only a few hundred K. They apparently then melted (at least partially) after reaching dimensions around 500 to 1000 km.

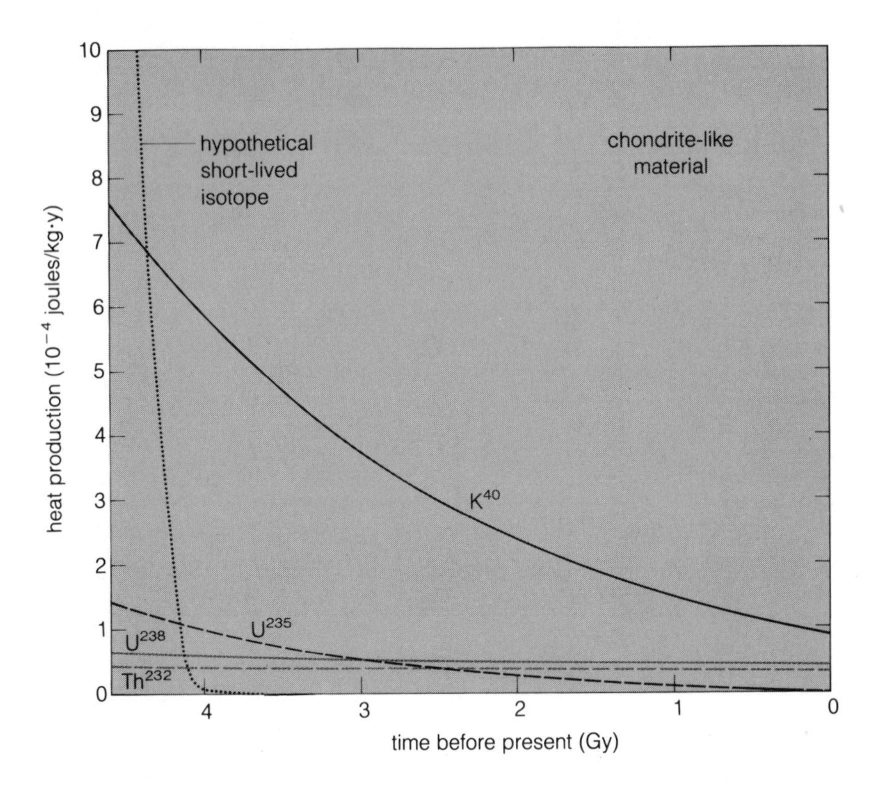

Figure 9-10. Relative heat production in chondritic material as a function of time for important radioactive elements, showing that planetary materials produced several times more heat during the planet-forming period than they do today. Thorium has such a long half-life that it produced little more then than now. On the other hand, nucleosynthesis events may have produced short-lived isotopes that were the dominant heat source during the first few million years (dashed line). (After data of Kaula, 1968, pp. 110–111)

3. Therefore, another possibility is that the short-lived isotopes (dashed line in Figure 9-10), such as aluminum 26 and iodine 129, described in Chapter 5, were so abundant that they produced enough heat to melt planets a few million years after they formed.

4. During the first few million years after the planets formed, the T-Tauri-like sun probably had strong solar winds. As these dragged the solar magnetic field past the planets, they induced electrical currents in the planets, whose strengths and patterns depended on the conductivity in the different parts of the planets. Familiar experience (with toasters, for instance) reminds us that such currents produce heat. This **electromagnetic heating** may have been strong enough to melt planets (or portions of them) initially (Sonett and others, 1970; Herbert and Sonett, 1979). This theory had the remarkable success of showing how asteroids of certain sizes and distances could have been melted while others were unmelted, as shown in Figure 9-11. For instance, 4 Vesta's basaltlike surface indicates that it was melted while larger and smaller neighbors were not. However, many of the parameters in this theory are poorly measured, and results are therefore highly model dependent. Since T Tauri conditions rapidly died out, the theory would explain only why the planets were strongly heated at the beginning.

PRINCIPLES OF CRUST AND LITHOSPHERE FORMATION

Any initial magma ocean cooled rapidly and solidified not only because the heat sources declined rapidly but also because the ocean was at the surface, where its heat radiated into space. However, in the interior, which was insulated by the overlying layers, heat accumulated from the long-lived radioactive uranium and potassium, trapped in the planet's silicates. This heat began

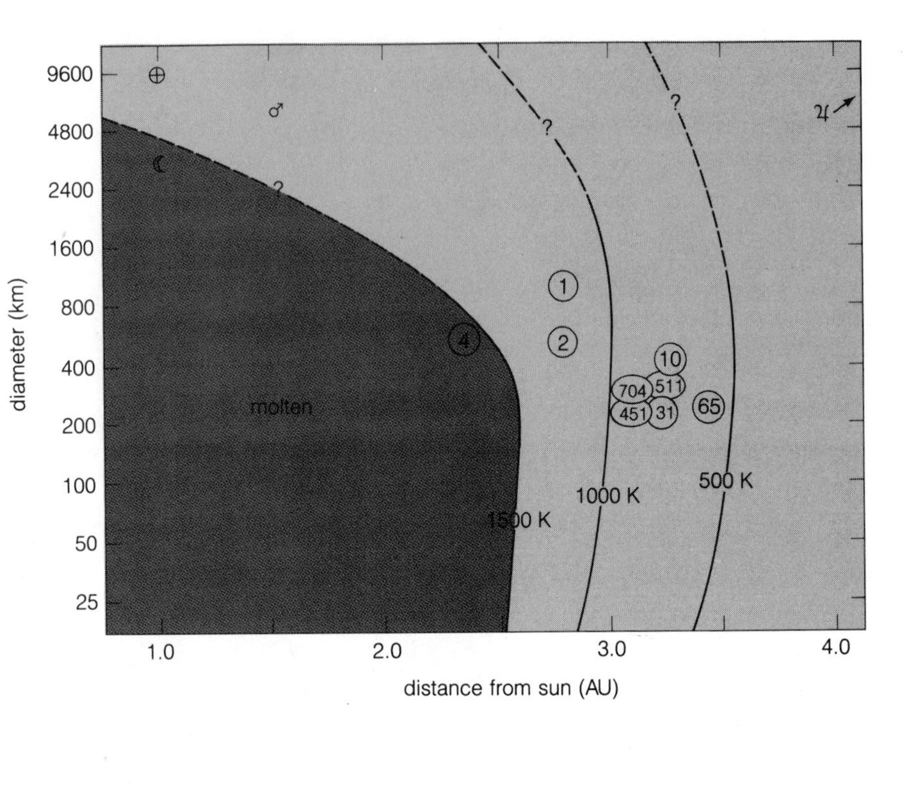

Figure 9-11. Temperatures reached in initially carbonaceous chondritic planetary bodies of different sizes as a function of distance from the sun, under an assumed model of electromagnetic heating by electric currents induced by early, strong solar winds. The model is consistent with observations of all asteroids larger than 300 km: 4 Vesta has a basaltic surface that was apparently once molten; the surface materials of 1 Ceres and 2 Pallas are chondritic and were never molten; and the surfaces of more distant asteroids resemble lower-temperature carbonaceous chondrites. (After Herbert and Sonett, 1979)

to melt the deep interiors of planetary bodies larger than perhaps 1000 km, allowing differentiation and sinking of the iron component. Settling of the metals into a core produced still more internal heat, acting like gravitational contraction. Thus, while a magma ocean rapidly disappeared from the surface, a molten or partly molten asthenospheric zone grew at some depth.

During the asthenosphere phase, the lower-density magmas could work their way to the surface, either during normal volcanic eruptions (see Chapter 11) or during large impacts, which penetrated into magma-rich layers.

Recall that while the lithosphere-asthenosphere boundary is defined by *physical state* (solid versus liquid), the crust-mantle boundary is defined by *chemical composition*. The crust, which may involve some of the same layers as the lithosphere, is merely the surface collection of distinctively low-density materials. Now we can see why **feldspars**, one of the lowest-density types of silicate minerals (2600 kg/m^3), are the most common minerals in Earth's and the moon's crust: In silicate-rich planets, the crusts are likely to be feldspar-rich rocks because feldspars float to the top of magma oceans or subsurface melted zones. Examples of feldspar-rich rocks are **basalts**, a commonly erupted type of volcanic lava, and **gabbros**, which are similar to basalts but have coarser crystals and form when magma solidifies at depth instead of on the surface. Much of Earth's crust is basalt, and the lunar crust is anorthositic gabbro, a type of gabbro composed almost entirely of feldspar. Furthermore, Viking landers discovered basaltic boulders and soil on Mars (see the next chapter for more details).

The lithosphere thickness is defined by the physical state of solidity, which in turn depends on the chemistry, since chemistry determines the melting point. For dry silicates (as in the moon) this is about 1100 K, which may be fairly deep in the planet. But with a few percent water (as in Earth) this is reduced to about 900 K, which may occur at a shallower depth, producing a thinner lithosphere (Warner and Morrison, 1978). Similarly, for a water-rich icy body (such as a satellite in the outer solar system) the melting temperature may be only 200 to 300 K, requiring minimal heating to produce a "molten" zone.

HEAT BALANCE OF PLANETS

If a planet were in equilibrium with its surroundings, it would radiate just as much heat as it receives from the sun. If it radiated less than it received, it would be heating up; if more, cooling down. We have just pointed out, however, that in addition to sunlight, a planet contributes a certain amount of heat of its own from radioactivity and possibly from early initial heat.

In the case of Earth, the solar heat flux coming in (that is, the solar constant) is about 20 000 times the terrestrial heat flux going out. Thus, at Earth's surface, sunlight is the dominant energy source in maintaining the temperature.

In the case of the giant planets, Kuiper (1952) speculated that the Jupiter atmosphere was heated primarily from internal sources, not sunlight, because of the spectacular cloud activity. The internal heat can be measured by infrared detectors on telescopes or spacecraft. Öpik (1962) and Taylor (1965) made the unexpected discovery that Jupiter and Saturn radiate *more* heat than they receive from the sun! Some approximate current figures are (Hubbard, 1990; Voyager team press releases; Stone and Miner, 1989):

	heat generated (watts) / 10^{12} kg of total mass	heat emitted / heat absorbed
carb. chondrite	3	—
Earth	4	0.00005
Jupiter	210	2.5
Saturn	150	2.3
Uranus	~10	~1.1?
Neptune	20	2.7

The first two lines show the heat generated from radioactive isotopes in chondritic meteoritic material, and show that Earth generates about this much heat per kilogram. The next lines show that at least three of the giant planets have additional, stronger heat sources. Their heat is believed to be the last stages of the long, slow gravitational contraction that brought these planets from their initial protoplanetary size down to their present size (see Chapter 4, especially Figure 4-8, page 103).

MAGNETISM OF PLANETS

A compass is a magnetized needle, and if a magnetic field is present, the needle aligns itself in what we call the direction of the field. Arbitrarily we define a "north-seeking" end of the needle, and then say that that end points toward the north direction of the magnetic field. The stronger the field, the stronger the tendency of the needle to align itself. Magnetic fields, then, have both strength and direction.

To make matters conceptually easier, physicists often speak of imaginary "**lines of force**," which are lines paralleling the direction of the field (the direction in which the compass needle comes to rest). This makes it possible to represent a magnetic field in a drawing such as the inset in Figure 9-12, showing the field around a bar magnet. As shown by Figure 9-12, a planet's field resembles a bar magnet's field in having two well-defined poles; such a field is called a **dipole field**. The north and south magnetic poles are defined simply by the direction in which a north-seeking magnet would point when placed near the planet. (By contrast, the north rotational

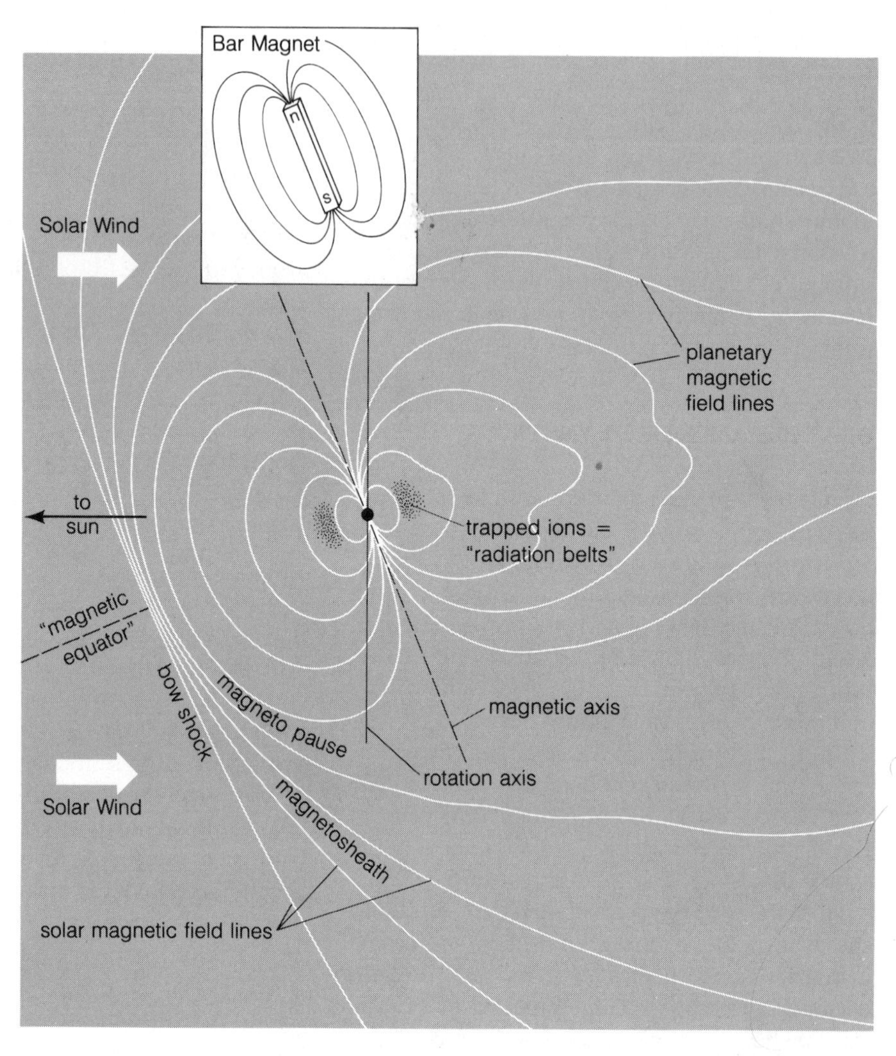

Figure 9-12. Schematic cross section of the magnetic field around a planet with an intrinsic magnetic field. The magnetic field of the passing solar wind is disturbed as it hits the magnetic field of the planet, forming a bow shock analogous to that formed in the wave pattern of lake waters moving past a speedboat. Spacecraft-borne magnetometers can map field configurations near planets and reveal internal planetary properties. The inset illustrates the resemblance between the dipole field of a bar magnet and an inner planetary field.

pole is defined as the pole on the north side of the ecliptic plane.)

The magnetic poles do not necessarily coincide with the poles of rotation. In Earth's case they depart by a slowly changing angle, which is now about 11°. This explains why surveyors and hikers must apply a **magnetic correction** to directions determined with a compass. Furthermore, there are local, complicated nondipole components of the field, about one-tenth as strong as Earth's main field.

Magnetic Interactions with the Solar Wind

Figure 9-12 shows the interaction between a planetary magnetic field, like Earth's, and the field being carried along by the ions of the solar wind. There are several characteristic features. The **magnetosphere** is the volume of space in which the magnetic field is essentially that of the planet. This is bordered by a **magnetopause**, or region where the planet's field is in equilibrium with the impinging solar wind field. Because the solar wind rushes by the planet at hypersonic speed, a **bow shock** builds up where the solar wind strikes the planetary field, and the planetary field is said to be encased in a **magnetosheath**, as shown in Figure 9-12.

Changes in solar activity alter the solar wind and the configuration of these boundaries, while the interior state of the planet controls the strength of the planetary field itself. Spacecraft magnetic measurements to map the magnetic field near a planet are thus important in deducing the planet's properties. Interestingly, Voyager measurements showed that the solar wind drags Jupiter's magnetic field into a tail at least 3 AU long and possibly extending far enough to interact with Saturn (Scarf and others, 1981)!

Changes in a Planet's Magnetic Field

One might suppose that a characteristic as major as a planetary magnetic field would be relatively permanent. On the contrary, measurements show that Earth's field is continually changing strength, drifting so as to change its pole positions and even reversing direction! About 250 My ago, Earth's north magnetic pole was located in the present north Pacific. Measurements show that the present field is fluctuating at a rate of about 0.1%/y, so that major changes in strength are expected in about 1000 y. Also, Earth's field in 500 B.C. was 50% stronger than it is now (Kaula, 1968).

Most mystifying is the fact that the direction of the field has reversed itself at sporadic intervals. No one yet knows why this happens, although evidence for reversed magnetism has been known since 1906. In the last few million years these **magnetic field reversals** have occurred every few hundred thousand years, but there have been intervals as long as 50 My. The process of reversal takes less than 10 000 y, and some believe that the field may reach nearly zero strength for a short time during this interval. There has been speculation that without a strong field to deflect cosmic ray particles, the incidence of cosmic rays at the surface goes up during the times of reversal, increasing biological mutation rates and affecting the course of biological evolution.

Paleomagnetism

The determination of past magnetic fields is the subject of a scientific field called **paleomagnetics**. Whenever a rock crystallizes from molten lava in the presence of a magnetic field, the magnetic elements in the rock, like compass needles, are frozen into position aligned with the field. This happens not at the freezing (that is, melting) point, but at a lower temperature, called the **Curie temperature**. As long as the rock remains below the Curie temperature and fixed in position, it retains a record of the strength and direction of that original field.

If a geologist locates an ancient, undisturbed outcrop of such rock and carefully removes a sample, noting its original orientation, he can take the sample back to the lab and measure the slight magnetism of the rock itself, which is called **remanent magnetism** (not *remnant!*). This gives a measure of the direction and strength of the ancient magnetic field.

Origin of Planetary Magnetic Fields

Earth's magnetic field cannot be a frozen remanent field derived from some strong field in the early solar system, because a remanent field would have decayed in around 10^4 y (Cole, 1978). Further, it cannot be due to the presence of some giant, permanent dipole magnet such as a magnetized iron core because (1) the interior temperature is far above the Curie temperature and (2) the field is too variable. These observations show that planetary magnetic fields must arise from some dynamic process in the interior.

The **dynamo theory** is the most widely accepted explanation of planetary magnetic fields. This theory notes that if a conductive fluid (such as a molten iron or metallic hydrogen core) moves (by means of convection currents or planetary rotation) through an external magnetic field (such as that of the solar wind or a random initial planetary field), electrical currents are induced that create their own local magnetic fields. This causes a feedback effect that further alters the motions of the fluid medium. The effect is that employed in a dynamo motor. The upshot of such an effect is that a partitioning of energy would occur between the kinetic energy of fluid motion and the magnetic energy of the magnetic field being created. An initially small field may thus be converted to a strong field at the expense of some of the motion.

Analysis shows that production of a strong field with good dipole symmetry requires two factors: a large volume of conductive fluid (that is, a large core region) and a rapid enough rotation to create a single symmetric pattern of motion (Cole, 1978). Internal irregularities, such as growth and decay of convective cells due to radioactive heating, could cause magnetic field variations such as polar wandering and field reversals. The patterns in Earth's field suggest that the liquid outer core is convecting with a set of perhaps 10 to 20 convection cells.

Table 9-4
Surface Planetary Magnetic Fields

Object	Magnetic field (nT)[a]	Approx. inclination of dipole to rotation axis
Sun	200 000[b]	6°
Mercury	220	<10°
Venus	<30	?
Earth	30 500	11°
Moon		
3.3 Gy ago	2 000	—
Today	10	—
Mars	40	—
Jupiter	420 000	9.°5
Saturn	20 000	<1°
Uranus	23 000	58°.6
Neptune	~100 000	46°.8

Note: Field values are variable from place to place, typically by about a factor of 2. Values from Taylor (1982); Van Allen, 1990; Hubbard, 1990.
[a]The tesla is the preferred SI unit for magnetic field strength, though values have traditionally been given in gammas or gauss until quite recently. 1 nT = 10^{-9} tesla = 1 gamma = 10^{-5} gauss. For comparison, the interplanetary field in the inner solar system is typically a few (~3) nanoteslas.
[b]Highly variable, up to 10^8 nT in sunspots.

Strength of Planetary and Solar Magnetic Fields

Table 9-4 summarizes available data on magnetic fields of various bodies in the solar system. The sun has a magnetic field in some ways similar to Earth's except with a shorter time scale. The sun's magnetic field reverses every 11 y, in step with a cyclic behavior in the number of sunspots. Every 22 y the sun goes through one complete magnetic cycle and two sunspot cycles.

Spacecraft measures near Venus, the moon, and Mars indicate that none of them have significant magnetic fields comparable to Earth's. This supports the theoretical ideas explained above, since Venus rotates very slowly and the moon and Mars are so small that they have probably cooled to a point where they no longer have large enough molten cores. Mercury's substantial field may relate to its high density, which suggests a large iron core. The strong fields of Jupiter and Saturn are believed to be related to their rapid rotations and the presence of metallic hydrogen inner mantles.

SYNTHESIS: INTERIORS OF SPECIFIC PLANETS

Having reviewed measurement techniques and some scattered data on the present state of planetary interiors and their long-term thermal evolution, we will now describe what is known about some specific planets.

Earth: Core-Mantle-Crust Evolution

The core of Earth is believed to be metallic, about 80% to 90% iron. Its material is an iron-nickel

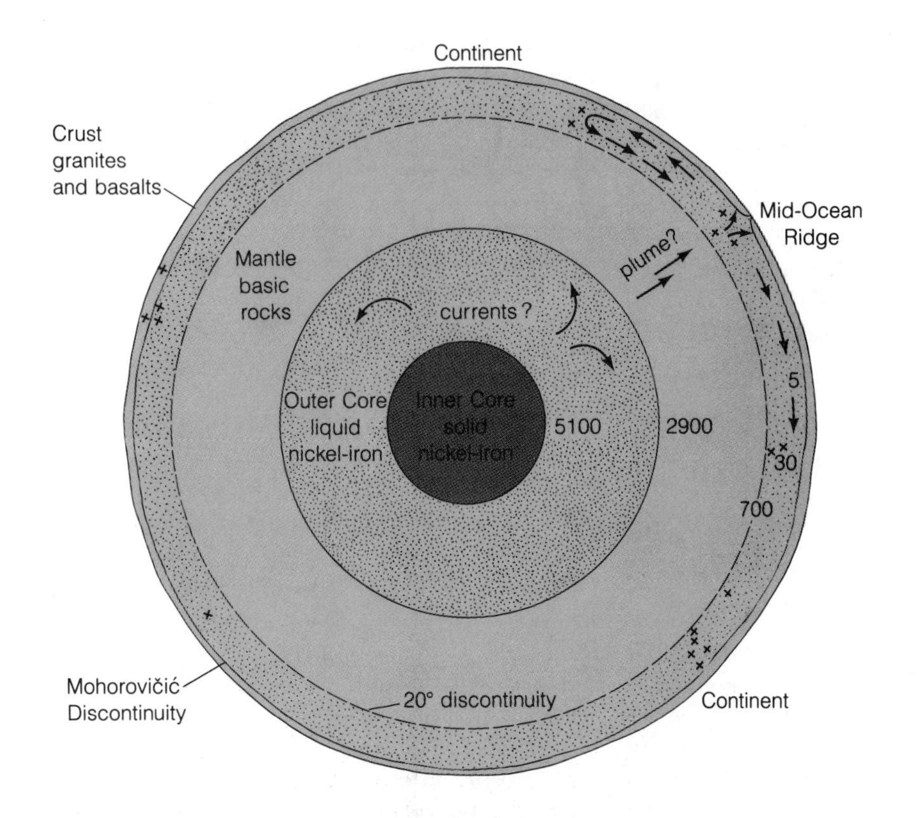

Figure 9-13. Structure of Earth's interior as revealed by seismology, with compositions inferred from additional evidence. Earthquake positions are indicated by X's. Stippling indicates molten material in the outer core and partially molten material in the asthenosphere. Numbers indicate depth in kilometers.

alloy with an admixture of up to 20% sulfur, cobalt, and other minor materials. Iron meteorites (92% iron, 7% nickel) support this. No volcanoes tap the core, but iron-nickel nodules erupted from deep volcanic sources may be corelike material that was trapped in the mantle. Though the inner core is hottest, the high pressures there make it solid while the outer core stays liquid, as shown in Figure 9-13. Circulation at an estimated 0.02 cm/s in the outer core helps produce Earth's magnetic field.

Some analysts have suggested that the core formed intact when iron accreted directly from the solar nebula and before silicates condensed (Slattery and others, 1980; Stevenson, 1981) though most believe it formed somewhat later during melting of Earth's interior. Since remanent magnetism appears in 3.5-Gy-old rocks, the core must have formed within the first 1 Gy. From radioactive isotope abundances in mantlelike rocks, Murthy and Hall (1972) conclude that the core formed by very early melting of the interior with subsequent drainage of iron metal to the center. In addition, they conclude that the major light element in the core is sulfur, in the form of iron sulfide (FeS), which also appears in iron meteorites. Similarly, Basu and others (1981) find from chemical abundances in 3.8-Gy-old East Indian rocks that at least some parts of the mantle had already begun to differentiate from the core and crust by that early era.

The mantle is siderophile-rich silicate rock, which probably has layers of different mineral phases and different rock types, such as the layering caused by the olivine-to-spinel phase transition at about 400 km deep. The heat flow through the top of the mantle under continents is substantially less than the heat flow out of the mantle under the seas, and hence there must be lateral as well as vertical variations in mantle structure. Aside from the olivine-rich rock types, pyroxene-rich and other rock types are suspected to exist in the upper mantle; all these are denser and more iron rich than common rocks on the surface, but they are sometimes found in material erupted from deep volcanoes (see also the discussion of rock types in the next chapter).

Some long-lived regions of high heat flow on Earth's surface, especially on the seafloor, are thought to mark columns of hot or molten material ascending from fairly deep in the mantle. These are called **mantle plumes**. Whether they exist is controversial, but they could explain certain features of Earth's surface, such as the Hawaiian Islands. Here, a lithosphere plate has apparently slid to the northwest across the top of a plume. The plume heated the crust and produced magmas that occasionally erupted to form volcanoes. Instead of piling up into one enormous volcano (as happened on Mars, for example), each intermittent eruption formed a volcanic mass—an island—on the part of the crustal plate occupying the "hot spot" over the plume at that time. As the plate slid by, various islands formed.

Strongly supporting this theory are the facts that the northwest islands in the chain are the oldest, the only active volcanoes are on the largest, youngest island at the southeast end, and hot spots are known off the southeast coast of that island, where new eruptions may soon begin.

The crust is the outermost rock skin—the lowest-density rocks that have accumulated on the surface after aeons of recycling mantle materials through volcanic and erosive processes. In the seafloors, the crust is thinnest—about 5 to 10 km of basaltic rock. In the continents, it has piled up into thicker blocks—about 20 to 60 km thick. These continental blocks consist primarily of granitic rock (more silica rich than basalts; see the next chapter's discussion of rock types) overlying a more basaltic layer 5 to 10 km thick and resembling the ocean-floor crust.

From geochemical abundance patterns in crustal rocks, Jacobsen and Wasserburg (1979) found evidence that a rapid cycling and differentiation of mantle and crustal materials occurred in the period from 4.5 to 3.6 Gy. They found that the present crust began to form at the end of that period, that today's continental crustal rocks have a *mean* age of 1.8 to 1.5 Gy, and that the rate of continental crustal rock accumulation in the last 0.5 Gy is less than that earlier.

Earth: Unraveling the Dynamics of the Interior

For centuries, naturalists have pondered features of Earth's surface. How do immense mountain ranges form? Why do certain regions sink and get invaded by shallow seas? Why are some areas shaken by earthquakes while other areas are quiet?

One fact early recognized was that the continental masses have a lower density than do the heavier mantle rocks. Furthermore, the crust is thickest under high mountain ranges, which thus have low-density roots supporting their weight. An analogue is an iceberg, which has a large volume of ice below sea level to support the small volume that protrudes.

These facts show that the continental masses, which rise above the mean crustal surface, are not piled on a rigid substrate, but rather are very nearly in **isostatic equilibrium**—that is, floating in equilibrium on the denser mantle. Another observation is that the cores of continents, the so-called **continental shields**, are ancient, flat-lying regions not recently disturbed by mountain building. They contain the oldest rocks, about 3.5 Gy old. An example is the Canadian Shield, surrounding Hudson's Bay.

Major mountain ranges, as shown in Figure 9-14, are the youngest, most active, and most distinctive features of Earth. They are vast crumpled crustal masses (Figure 9-15) usually lying around the outer edges of continental shields. Many of them formed in recent episodes such as the "Laramide revolution" about 70 My ago, which saw much of the warping and thrusting that led to the Rocky Mountains.

Interestingly, studies of continents did not solve the problem of the origin of Earth's dynamic processes, because the continents are so complex, contorted, and active that the geological

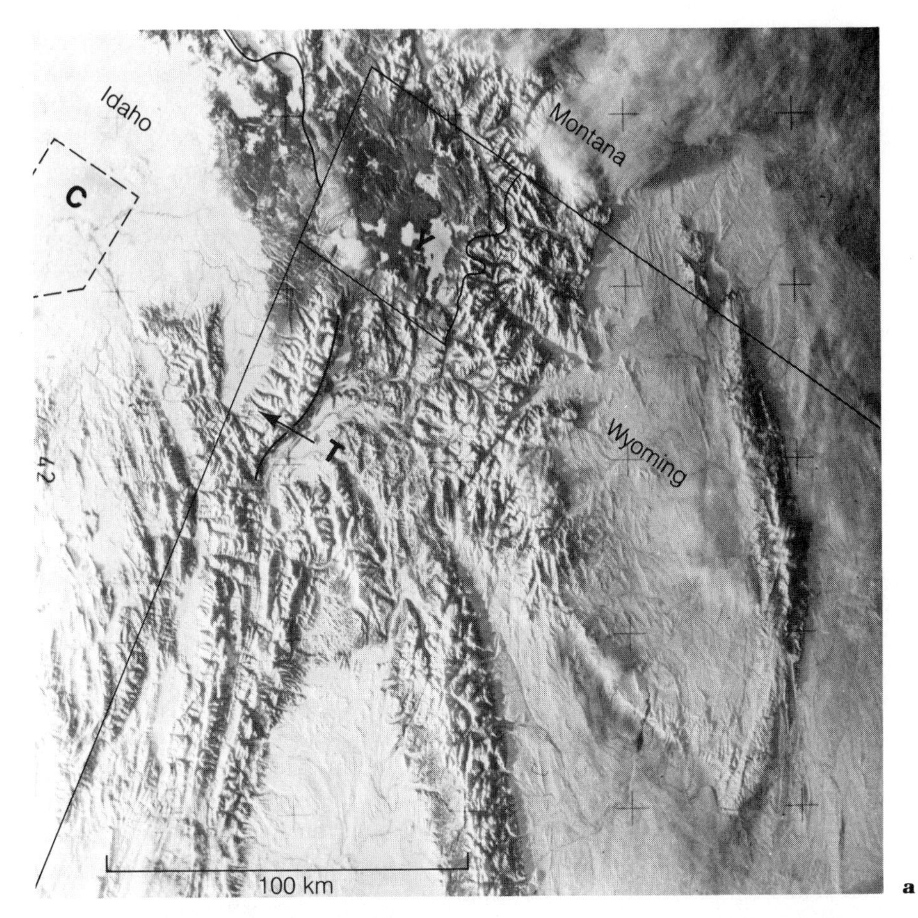

a

b

Figure 9-14. Major folded mountain chains are unique to Earth in the solar system. (**a**) *A portion of the Rocky Mountains photographed with a hand-held camera by a Skylab astronaut. Y = Yellowstone Park, a crustal hot spot with active geysers and fumaroles; C = an area of recent cinder cones and lava flows near Craters of the Moon National Monument; T = a large fault that exposed the east face of the Grand Teton Mountains. Arrow marks position and direction of view in part* (**b**). *(NASA)* (**b**) *East face of the Grand Tetons (Middle Teton Peak), showing cliffs exposed by faulting and a massive dike of basaltic rock. This basaltic rock apparently intruded as magma into a large fracture when the original rock mass was underground, prior to faulting. The dike, partly covered by snow, runs several hundred meters from the summit to a point where it disappears behind loose talus at the cliff base. (Photo by author)*

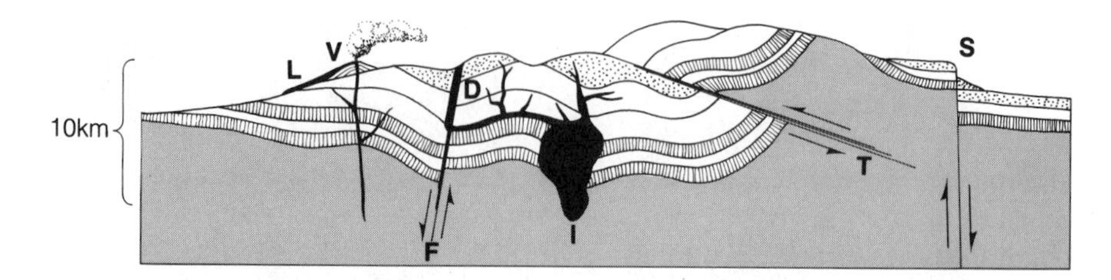

Figure 9-15. Schematic cross section of a terrestrial mountain belt, showing folds and tectonic structures. L = lava flow; V = volcano; F = normal fault; D = dike intruded along the fault; I = intrusive body; T = thrust fault; S = scarp (cliff) formed along fresh normal fault.

messages are lost in the chaos. Historically, it has been the 71% of Earth's surface that is hidden under the sea that has provided the key to the dynamic evolution of Earth's crust.

As early as 1620 Francis Bacon noted that the east and west coastlines of the Atlantic could fit like jigsaw puzzle pieces. He made the remarkable suggestion that the Americas may once have been in contact with Europe and Africa. By the late 1800s paleontologists had noted the similarity of 300-My-old fossil plants from opposite shores of the Atlantic, and the famous Austrian geologist Eduard Suess had noted jigsaw puzzle fits of ancient geological formations across the Southern Hemisphere. Suess hypothesized a primeval southern continent, which he named Gondwanaland (from a key geological province in India). It was the parent of the present southern continents, which took shape as Gondwanaland broke up.

This idea became the theory of **continental drift**, that continents broke apart and drifted. In 1908 the American geologist F. B. Taylor discussed possible mechanisms to keep the continental masses moving. Continental drift was then championed by the German meteorologist Alfred L. Wegener in 1922. He pointed to cross-oceanic fits of geological structures and fossil flora and fauna, indicating that matched coastlines had actually been in contact as recently as a few hundred million years ago. In the 1920s and 1930s, South African geologists added maps of distinctive glacial deposits in South America, Africa, Australia, India, and Madagascar, indicating that these areas were once closer together and grouped near Earth's southern rotational pole. Also in the 1930s, geophysicist F. A. Vening Meinesz made gravity measurements that led him to suggest that convection currents in the upper mantle (that is, asthenosphere) drag the floating continental blocks along, causing drift.

But the continental drift theory remained extremely controversial, and as recently as the early 1960s most geologists simply did not believe it. Aside from an inbred belief in terra firma, many argued that the drag on continental blocks moving on a mantle sea would be enormous, that the mantle was relatively stable, and that no energy sources in Earth were sufficient to push the continents along.

Direct mapping of the ocean floor led to a revolution in geological thinking, in which the theory of continental drift was somewhat modified and finally accepted. In the late 1940s, oceanic expeditions discovered that the ocean floor was not simply a submerged continental landscape; instead, it was marked by major fractures, winding **oceanic ridge** systems marked by shallow earthquakes (see Figure 9-8), and sediments rarely older than 100 My. This showed that the ocean floors are young. At the same time, geophysicists found that heat flow in the oceanic ridge systems was several times that measured elsewhere.

The next advance came from radiometric dating of seafloor rocks on either side of oceanic ridges. These studies showed that rocks along the ridges are new, often currently being erupted,

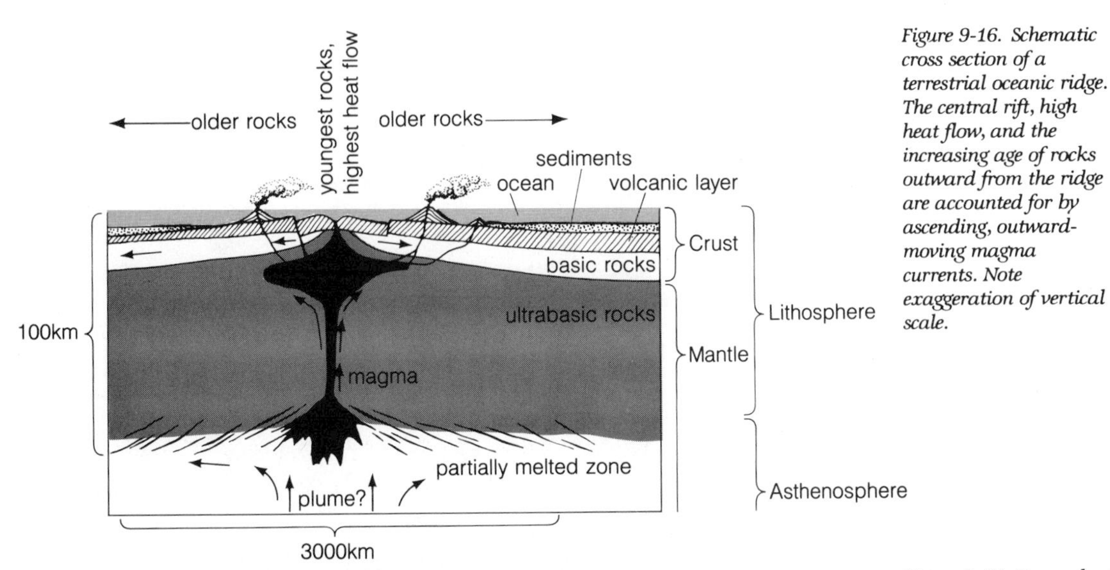

Figure 9-16. Schematic cross section of a terrestrial oceanic ridge. The central rift, high heat flow, and the increasing age of rocks outward from the ridge are accounted for by ascending, outward-moving magma currents. Note exaggeration of vertical scale.

Figure 9-17. Example of rift faulting. The Mid-Atlantic Ridge is exposed in Iceland, which is being split apart by parallel fissures. This view looks across one such fissure, about 11 m wide, that exposes layers of basaltic lava formed by recent eruptions along the ridge. (Photo by author)

while rocks further away are older, as indicated in Figure 9-16. There could now be little doubt that magma is ascending from the mantle beneath oceanic ridges, erupting, and forming new seafloor, while old seafloor is being dragged away from the ridges in both directions, apparently by subcrustal currents. Islands such as Iceland, which are bits of mid-oceanic ridges protruding above the water, dramatically show the evidence of being split. **Rift valleys**, opened by splitting of the landscape due to horizontal stretching, are common, as shown in Figure 9-17.

In 1963, these facts were organized into the new concept of **seafloor spreading**, demonstrating that the width of the Atlantic is growing as new material wells up along the Mid-Atlantic Ridge and as the continents move apart (Vine and Matthews, 1963). A meeting of the Geological Society of America in San Francisco in 1966 was the final turning point in accepting these ideas (Hurley, 1968).

As summarized in Figure 9-18, various lines of evidence indicate that the present continents were associated in a protocontinental land mass

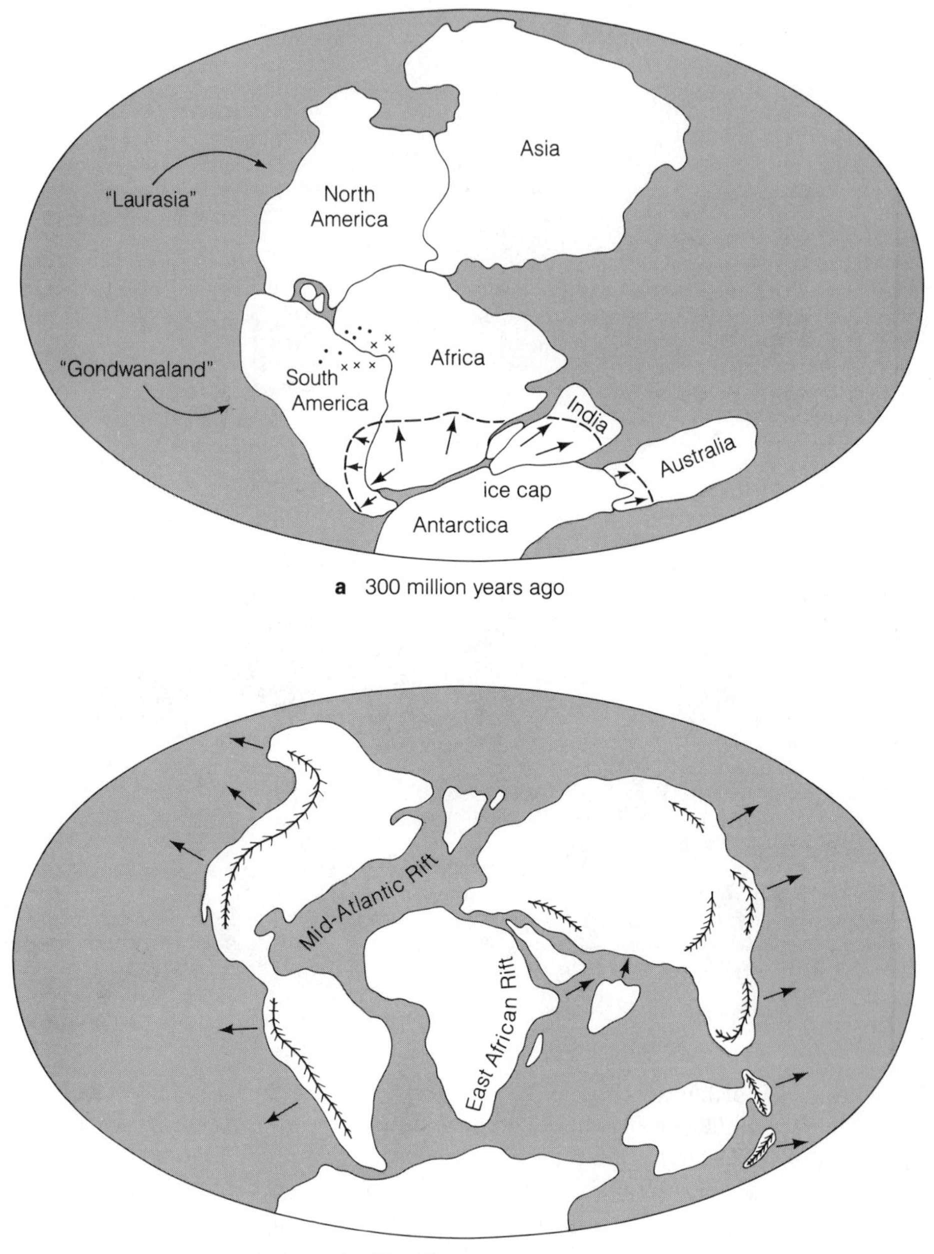

Figure 9-18. (**a**) *Configuration of continental masses roughly 300 My ago at the beginning of the breakup of Pangaea. See text for further discussion.* (**b**) *Configuration roughly 50 My ago as continental masses were drifting toward their present positions. Mountain masses are shown along leading continental edges, at collision zones between plates.*

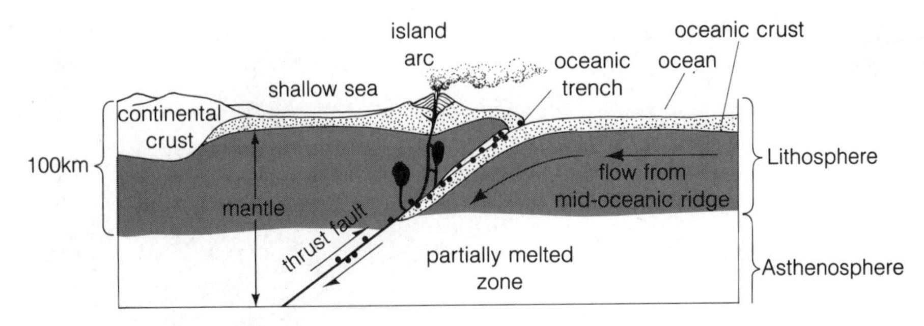

Figure 9-19. Cross section of the collision zone between continental and oceanic plates. The oceanic lithospheric plate, moving outward from the mid-oceanic ridge, slides under the lighter continental plate. Blobs of heated oceanic crustal material rise as magma and produce island arc volcanism. Dots indicate earthquate foci. (After Marsh, 1979)

known as **Pangaea** (pan-GEE-a) as little as 300 My ago. Two major portions were Gondwanaland in the south and **Laurasia** in the north. Evidence includes glacial striations marking the south polar ice cap in Gondwanaland, and jigsaw fits of different-aged provinces (X's and dots in Figure 9-18) in South America and Africa. Pangaea was rifted apart when spreading started at sites such as the modern Mid-Atlantic Ridge.

Similar spreading is still rifting modern continents, as documented in East Africa and along the Sea of California. As continental blocks move, their leading edges tend to ride over seafloor crust, which plunges under the continents to create crumpled mountain belts, shown in Figure 9-18.

Figure 9-19 shows a cross section through such an area. The types of mountains produced depend on the collision speed of the continental and seafloor units, which range from about 4 to 10 cm/y. Among features produced are **trenches**, the lowest suboceanic spots on Earth's surface, which are created when the seafloor is dragged downward as it descends under a continent. The fact that deep earthquakes are landward of shallow quakes (Figure 9-8) is also explained by the angle of descent of the seafloor crust under the continental crust (Menard, 1969; Dewey and Bird, 1970; Marsh, 1979).

The concepts of continental drift and seafloor spreading were merged into a grand, new planetary-scale perspective on the dynamics of Earth processes—the **theory of plate tectonics**. Tectonics is the field of geology that deals with movements in planetary crusts and interiors, from local faults to large-scale crustal deformations.

The theory of plate tectonics views Earth's lithosphere as analogous to a thin scum of wax floating on heated water. The water is not boiling but is stirred by convection currents that break the scum into units, or **plates**, that move independently. On Earth, the plates are large units of the lithosphere. Some of the plates are continental masses, but others are large seafloor units. Rifting of plates occurs along mid-oceanic ridges and other related faults (Figure 9-20). Collision of plates produces mountain ranges and trenches at continental margins. The lithospheric plates are probably driven by convection currents in the asthenosphere or by larger-scale currents in the mantle, which deforms by plastic flow over long periods of time.

In Earth, the adiabatic temperature gradient is about 0.2 K/km, while the observed temperature gradient in the crust is about 3 to 30 K/km. While rocks seem to have a great deal of strength, we have already seen that if they are stressed long enough, they can deform by flowing. Thus it appears theoretically likely that slow convection

Figure 9-20. Example of large-scale plate tectonics influencing the landscape of Earth. Fault boundaries of the Pacific and American plates are mapped, including the San Andreas and other well-known faults that endanger California cities as plates slip by each other; arrows indicate motions. Peaks indicate current or recent volcanoes. Baja California is being rifted off the continent by faults on the floor of the Gulf of California. The Gulf averages about 150 km wide. (NASA, Apollo 16)

currents exist in the relatively fluid asthenosphere. Seafloor spreading rates suggest that the currents have velocities of a few centimeters per year.

Many details of these processes remain to be worked out. For instance, researchers are uncertain about the long-term history of the plate motions. Did convection get more vigorous in the last few hundred million years as a result of thermal evolution and thus cause a sudden breakup of a previously intact Pangaea? Or was Pangaea just a temporary conglomeration of earlier continental blocks that had crashed together? Some 3.7- and 2.9-Gy-old belts of "greenstone" granitic crust in Canada and Australia have been interpreted as evidence of ancient plate collisions (Windley, 1976), and other eroded mountain chains have been interpreted as island arc or mountain zones at the sites of ancient plate collisions. Ben-Avraham (1981) describes bits of old continents embedded inside today's continents and seafloors. Thus there were probably some plate motions long before the "recent" continental breakup 300 My ago.

Going back even further, several researchers have suggested that the primeval intense cratering of Earth, 4.5 to 4.0 Gy ago, may have been responsible for "punching holes" in the crust at large impact sites and piling up crustal ejecta in other regions. This activity established the initial crustal heterogeneity that led to well-defined continental blocks (Goodwin, 1976; Frey, 1977).

Venus: Earth's Enigmatic Sister

Because Venus is about the size of Earth and located close by, we would expect it to have a roughly similar radiogenic heating history, similar bulk thermal and insulating properties, and a similar thermal evolution (Figure 9-21). The theory of the condensation sequence suggests less sulfur in Venus than in Earth, implying that Venus' core might have less iron sulfide and might have formed later (Toksöz and Hsui, 1977). This and Venus' slow rotation might account for its weak magnetic field.

Although the planet was expected to have an internal history similar to Earth's, the nature of the surface geology, as radar-mapped by Venera, Pioneer, and Magellan probes, shows some dramatic differences. Most important, there is little evidence for plate tectonics, large crustal motions, or continental-scale compressional mountain ranges formed by plate collisions. Instead, 60% of Venus is rolling plains with only about a kilometer of relief. About 24% of the surface contains a few Australia-sized continentlike raised plateaus standing a few kilometers above the rest of the surface, and about 16% consists of high mountains including volcanic peaks towering to 10.6 km (34,500 feet). Unlike Earth's mountains, these mountains are believed to be too high to be supported isostatically by low-density roots, because Venus' lithosphere is warm and plastic (Phillips and others, 1991).

Venus' surface is mostly volcanic. Soviet landers Venera 8-10 and 13-14, and VEGA 1 and 2 made rock composition tests of varying quality on the plains near the equator, some closer to uplands than others. Generally, Veneras 9, 10, 13, and 14 and VEGA 1 and 2 found roughly basaltic compositions. Venera 8 found a more granitic composition. (See Chapter 11, p. 398, for more detail.) The general feeling is that some differentiation has occurred, but not as much as on Earth's surface.

Various workers have tried to interpret the interior from these facts, but one gets the impression from the literature that we need more information, such as seismic probes, to make much progress. For example, some workers believe that the lack of water in Venus' atmosphere is partly due to its nearness to the sun—that it got less water to start with (see commentary by Grinspoon, 1987). Warner and Morrison (1978) concluded that the lack of water in mantle rocks would raise their melting temperature, leading to a lithosphere twice as thick as Earth's. On the other hand, Kaula (1990) and Phillips and others (1991) speak of a lithosphere only about 100 km thick, similar to or thinner than Earth's. They reason that Venus' high surface temperature keeps the rocks near the melting point, preventing a thick lithosphere from forming.

If the volcanic peaks are too high to be supported by the lithosphere, they must be recent, temporary lava accumulations that will eventually subside. This implies active volcanic upwelling and eruption to build the mountains.

One thing that most have agreed upon in the wake of Magellan imagery, aside from lack of active plate tectonics, is that plumes of hot material are probably ascending from Venus' mantle. The plumes are thought to hit the underside of the lithosphere, causing uplift. High volcanic mountains, accumulated from sustained eruptions, are a result of long-lived plumes. Shorter-lived plumes may explain coronas, the unique circular features that may mark uplift and subsequent collapse (Figure 9-22). Coronas are often filled with lava. Belts of chaotic, folded, and faulted mountains around some of the high areas may be the result of compression caused by slumping of material off uplifted areas over ascending plumes. Showing the perversity of Venus studies to date, some workers assert that the same

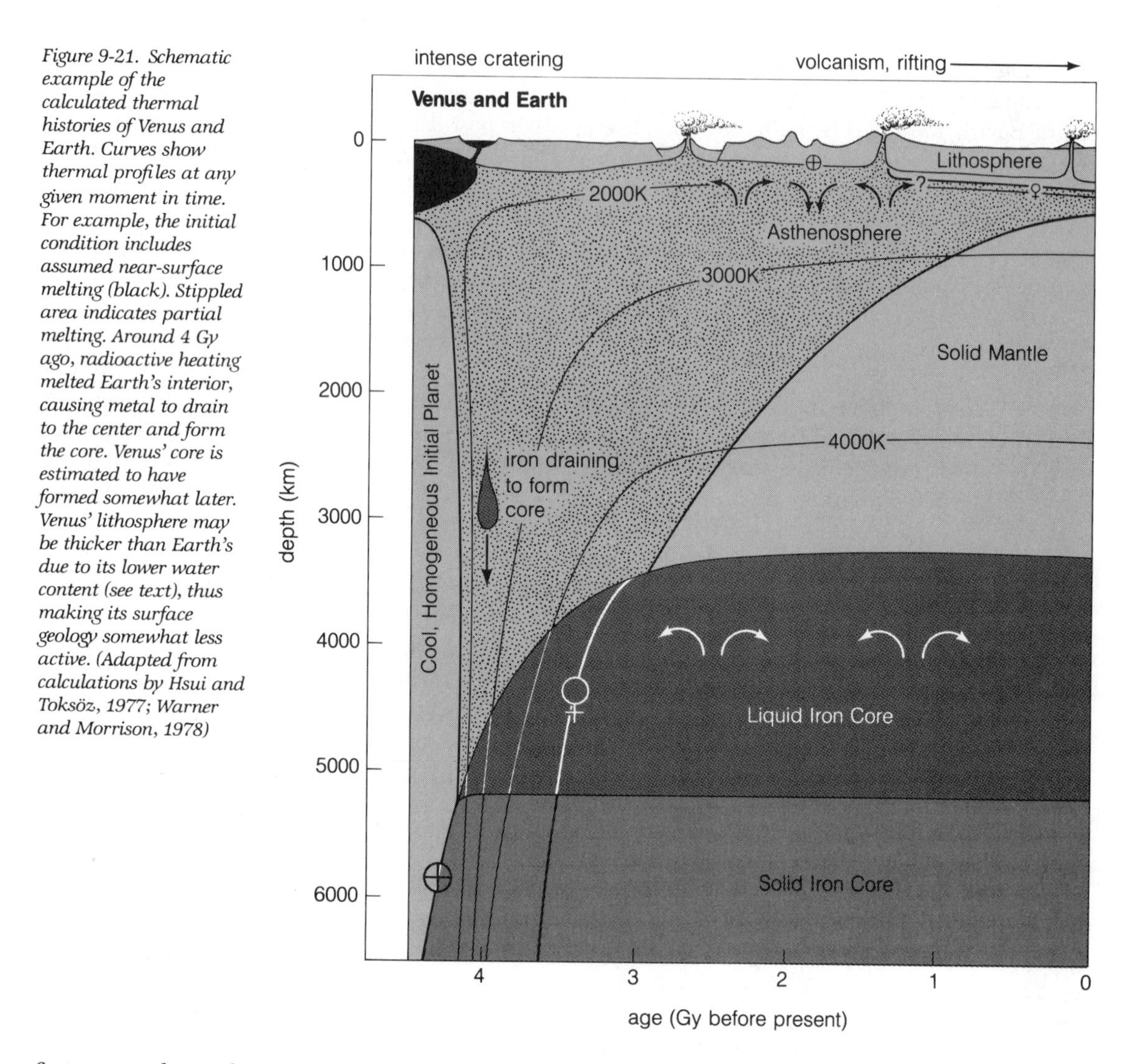

Figure 9-21. Schematic example of the calculated thermal histories of Venus and Earth. Curves show thermal profiles at any given moment in time. For example, the initial condition includes assumed near-surface melting (black). Stippled area indicates partial melting. Around 4 Gy ago, radioactive heating melted Earth's interior, causing metal to drain to the center and form the core. Venus' core is estimated to have formed somewhat later. Venus' lithosphere may be thicker than Earth's due to its lower water content (see text), thus making its surface geology somewhat less active. (Adapted from calculations by Hsui and Toksöz, 1977; Warner and Morrison, 1978)

features are located over areas where convecting mantle material flows *downward.*

Widespread fractures show that Venus' crust and lithosphere are stressed, even if not as mobile as Earth's. Geophysicists Kaula (1990) and Phillips and others (1991) agree that while subduction drags segments of Earth's crust back into the mantle for further differentiation, Venus lacks subduction and hence, instead of a few very differentiated, granitic continental blocks as on Earth, it has accumulated a relatively uniform global crust without much plate motion, composed of mildly differentiated basalts. The proof of the pudding will be in future measurements.

Moon: A Low-Density Planet That Cooled Rapidly

The moon's interior lacks the clear structure of Earth's interior because most of the moon resembles Earth's upper mantle rocks and has only enough iron to form a small metallic core. Seismic evidence is inconclusive, but the lunar moment of inertia and the small, ancient lunar magnetic field suggest a small core that formed 3.5 to 4.5 Gy ago. A core of nearly pure iron may have a radius of 300 to 400 km; a core rich in iron sulfide could reach 500 to 600 km (Solomon, 1979).

Seismic records (Figure 9-23) indicate that

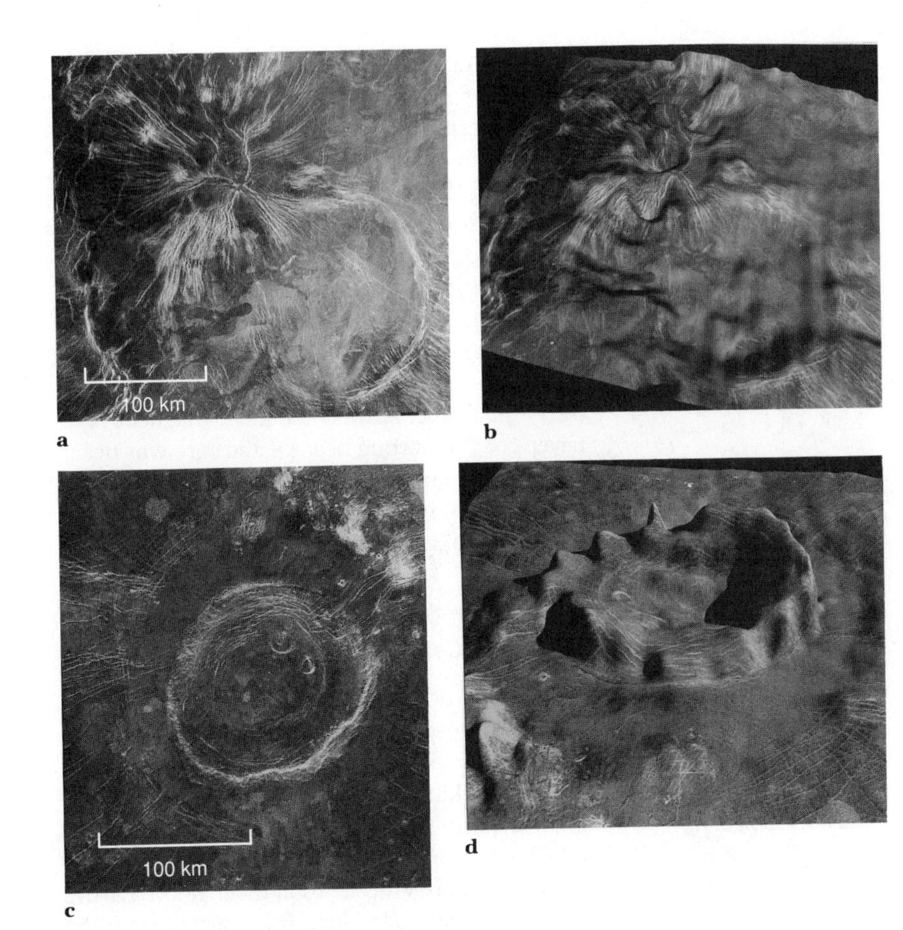

*Figure 9-22. Three examples of coronas in two regions of Venus. Each region is shown in a near-vertical radar view and a computer-constructed oblique view showing exaggerated relief (using radar altimetry data). Some scientists believe these coronas show an evolutionary sequence. Rachel corona (upper left of [**a**] and [**b**]) may represent an early stage. It is an uplifted volcanic complex, radially faulted, with collapse in the center. Bhumidevi corona ([**c**] and [**d**]) may be an intermediate case, where the whole central area has collapsed, leaving a raised ring with circular and radial fractures. Rebecca corona (lower right of [**a**] and [**b**]) may be an old example, where relief almost totally relaxed back into the rolling terrain. (NASA Magellan radar images, courtesy of Daniel M. Janes, Cornell University, and Jet Propulsion Laboratory)*

the lunar crust is about 60 km thick and composed of anorthositic gabbro, which is composed mostly of low-density feldspar minerals. As shown in Figure 9-23, seismic wave speeds in layers shallower than 25 km suggest that the near-surface rocks are intensely fractured by impacts. From the 60-km base of the crust to the core, at about 1300 km down, the rock probably resembles a dense gabbro rock, rich in olivine and similar to Earth's mantle rocks.

These data suggest that the moon started out with an Earth-mantle-like composition and then differentiated during very early melting or partial melting. However, it never underwent such efficient churning and differentiation as Earth.

Lunar researchers quickly recognized that the 4.4-Gy-old lunar crustal rocks are the low-density feldspars that would be expected to float to the surface if the outer layers of the moon had been molten initially. This conclusion led to the concept that the moon started with a magma ocean several hundred kilometers deep, as shown in the calculated thermal history of Figure 9-24. Consistent with this, Warren and Wasson (1979) made a special search among lunar samples for "pristine" crustal fragments, and found eight that indicated an initial lunar crust composed of feldspar-rich gabbroic rocks: namely, anorthosites, troctolites, and norites (see next chapter for further discussion of rock types).

According to the magma ocean model, the outer skin down to perhaps 400 km was melted by some process such as magnetic induction or impact heating as the moon formed. It quickly cooled, forming an impact-shattered, rocky skin that thickened until the whole ocean had dis-

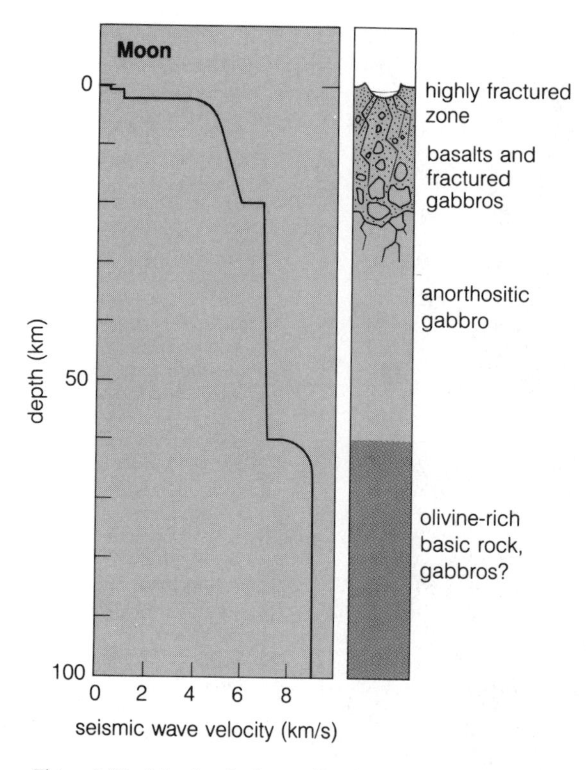

Figure 9-23. Seismic velocity profile of the moon, measured by Apollo equipment. The interpretation is shown at right. (After data from Taylor, 1975)

appeared in perhaps 400 My or less. At the same time, radioactive atoms inside the moon released heat that accumulated until partial melting occurred, probably about 400 km down, around 3.5 to 4 Gy ago. So the magma ocean gave way to a deeper source of molten material, which is believed to have been the source of the 3.5-Gy-old mare lavas.

Since the moon is quite small, the ratio of heat-radiating surface to heat-producing volume is large, and the moon cooled much faster than Earth. The lithosphere thus thickened to its present 1000-km depth, too thick to allow plate tectonic activity. The primeval surface features such as craters and lava plains were thus preserved and not offset by mountain ranges or rift fractures.

The lunar surface lacks obvious compressional or extensional faults, which would indicate major expansion or contraction. This constrains thermal models and precludes an initially molten moon that cooled and contracted since formation (Solomon, 1979).

Interestingly, although the paleomagnetic studies indicate a magnetic field when the lunar maria formed as early as 3.8 Gy ago, the thermal models together with geological data suggest that the moon's interior did not melt much before that. Thus the core could not have formed much before that time. Possibly the lunar magnetic field never got very strong because the core was never large or hot enough to develop vigorous convective currents, which are needed to drive the electromagnetic dynamo.

Mercury: A Moonlike World with a Molten(?) Core

Mercury is moonlike in its cratered surface and lack of atmosphere, but it has a much greater density. Seismic data and moment of inertia figures are lacking, but four lines of evidence help us to hypothesize about the interior (Solomon, 1976): (1) The high density indicates a large iron content of about 60% $\pm$ 10%, much above Earth's and the moon's. This idea is consistent with the condensation sequence in Mercury's zone, close to the sun. (2) The magnetic field suggests that the iron has differentiated into a hot, convecting core. (3) Radioactivity consistent with the condensation theory would also require that Mercury have melted and differentiated. (4) Surface features indicate extensive volcanic flooding and a type of compressional faulting not seen on the moon. This suggests planetary contraction.

Figure 9-25 shows a theoretical model of the thermal history of Mercury. The time of core formation depends strongly on the assumed conditions. The draining of such large amounts of iron to the center released substantial potential energy in the form of heat, causing expansion and also adding to the radiogenic heat that produced the initial internal melting. Thus the whole mantle became molten (Solomon, 1976). This

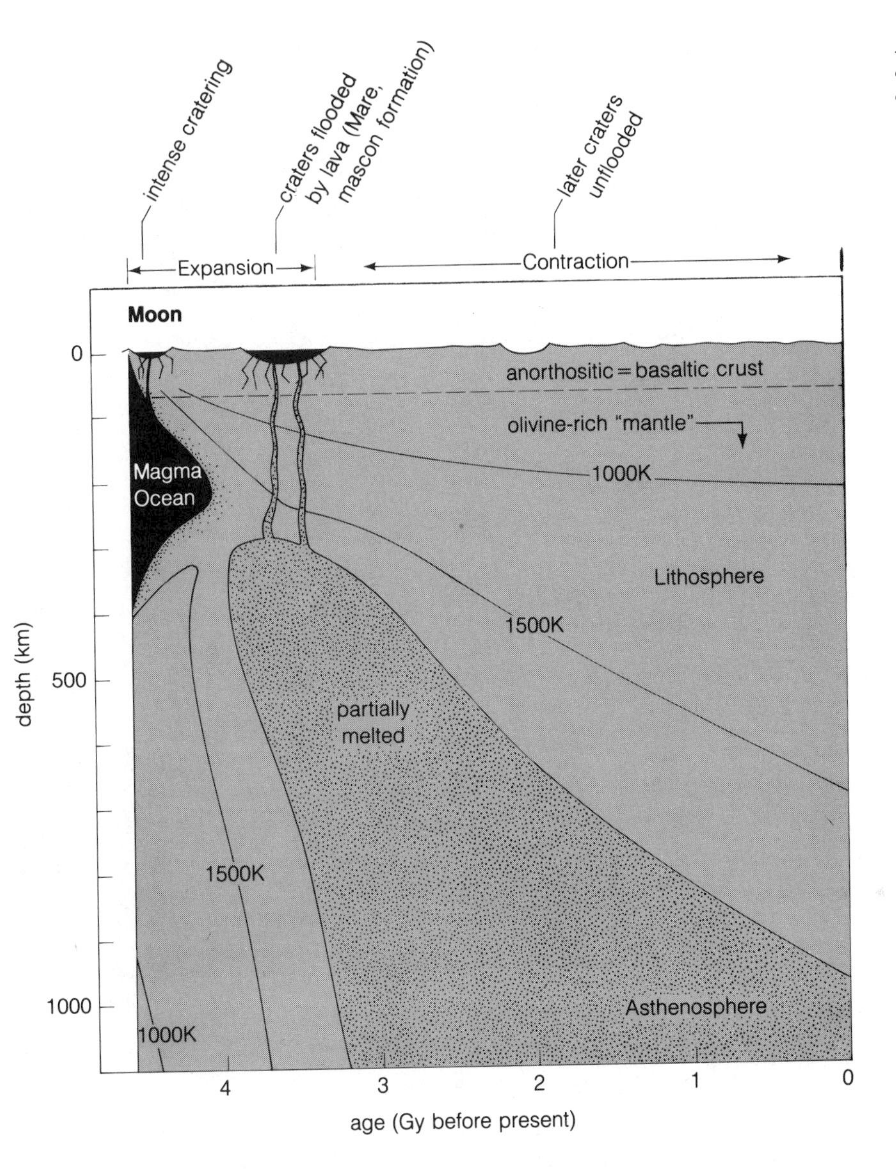

Figure 9-24. Schematic example of the calculated thermal history of the outer part of the moon. As the magma ocean cooled, radioactivity in the interior caused temporary partial melting at greater depths. Due to the moon's small size, rapid cooling produced a thick lithosphere (see text). (Adapted from calculations by Solomon and Head, 1979; Hsui and Toksöz, 1977; Herbert, Drake, and Sonett, 1978; and others)

cannot have happened after the end of the early intense cratering, because we would otherwise see expansion fractures and many well-preserved lava flows on the surface, all of which are absent.

Because of Mercury's small size, the asthenosphere cooled rapidly, and the lithosphere thickened rapidly. According to some models, the asthenosphere has disappeared entirely, so that Mercury's lithosphere reaches hundreds of kilometers down to the iron core, a condition that would be consistent with the absence of recent eruptions or plate activity.

Contraction began as the planet cooled after core formation, and the models predict a decrease in radius of around 2 km. This would compress the surface and cause **thrust faults**, a type of fault in which one rock unit slides up over another. As shown in Figure 9-26, such faults have actually

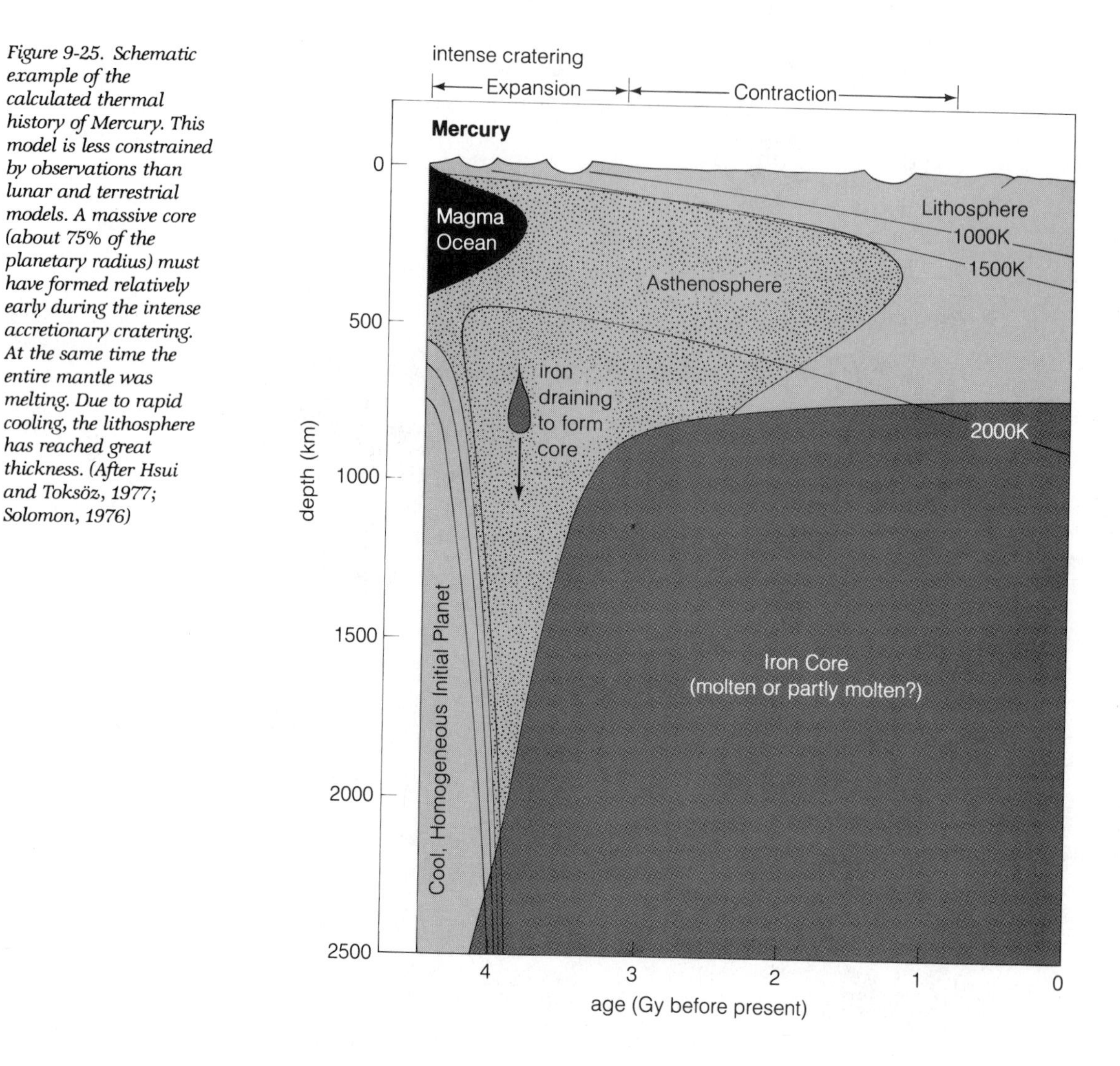

Figure 9-25. Schematic example of the calculated thermal history of Mercury. This model is less constrained by observations than lunar and terrestrial models. A massive core (about 75% of the planetary radius) must have formed relatively early during the intense accretionary cratering. At the same time the entire mantle was melting. Due to rapid cooling, the lithosphere has reached great thickness. (After Hsui and Toksöz, 1977; Solomon, 1976)

been discovered, and indicate a contraction by about 2 km (Strom, Trask, and Guest, 1975). Thus the calculation and observation agree!

Mars: The Intermediate Case

Mars is intermediate in size between the small terrestrial worlds (the moon and Mercury) and the large ones (Venus and Earth). It appears to be intermediate in surface and internal properties, too, thus shedding important light on the relationships among the planets. One hemisphere is dominated by ancient, eroded impact craters, and the other by sparsely cratered lava flows, relatively recent volcanoes, and huge rift valleys.

Data on the moment of inertia have led to general agreement that Mars has a sizable dense core that is about 40% the radius of the planet (Wood, 1978, p. 74). Depending on the efficiency of the differentiation process, this core may be either smaller and nearly pure iron or larger and rich in low-density contaminants, such as iron sulfide and magnetite. Martian seismic data from

Figure 9-26. (**a**) *Cross section of a large thrust fault produced by compression of the planetary lithosphere (open arrows) accompanying substantial planetary contraction, as is believed to have occurred on Mercury.* (**b**) *A portion of the "Discovery Scarp" on Mercury, a feature believed to be created by compressive thrust faulting, described in part* (**a**), *during planetary contraction after core formation. The portion of the fault shown is about 360 km long, and the largest crater through which the fault passes (upper center) is about 55 km across. (NASA)*

Viking are inadequate to tell much about the deep interior. Calculated thermal models differ, reflecting uncertainty about the internal properties.

Figure 9-27 is an example consistent with known facts. The planet may have started with a magma ocean, consistent with the appearance of several ancient, seemingly lava-flooded impact basins. The mantle reached the melting point of iron well before the silicates melted; sluggish iron drainage probably formed a core before a well-developed asthenosphere of molten silicates could appear. Most calculations indicate that molten silicates appeared no less than 400 km down about 0.5 to 3 Gy ago. This magma may have been the source of the vast, sparsely cratered lava plains

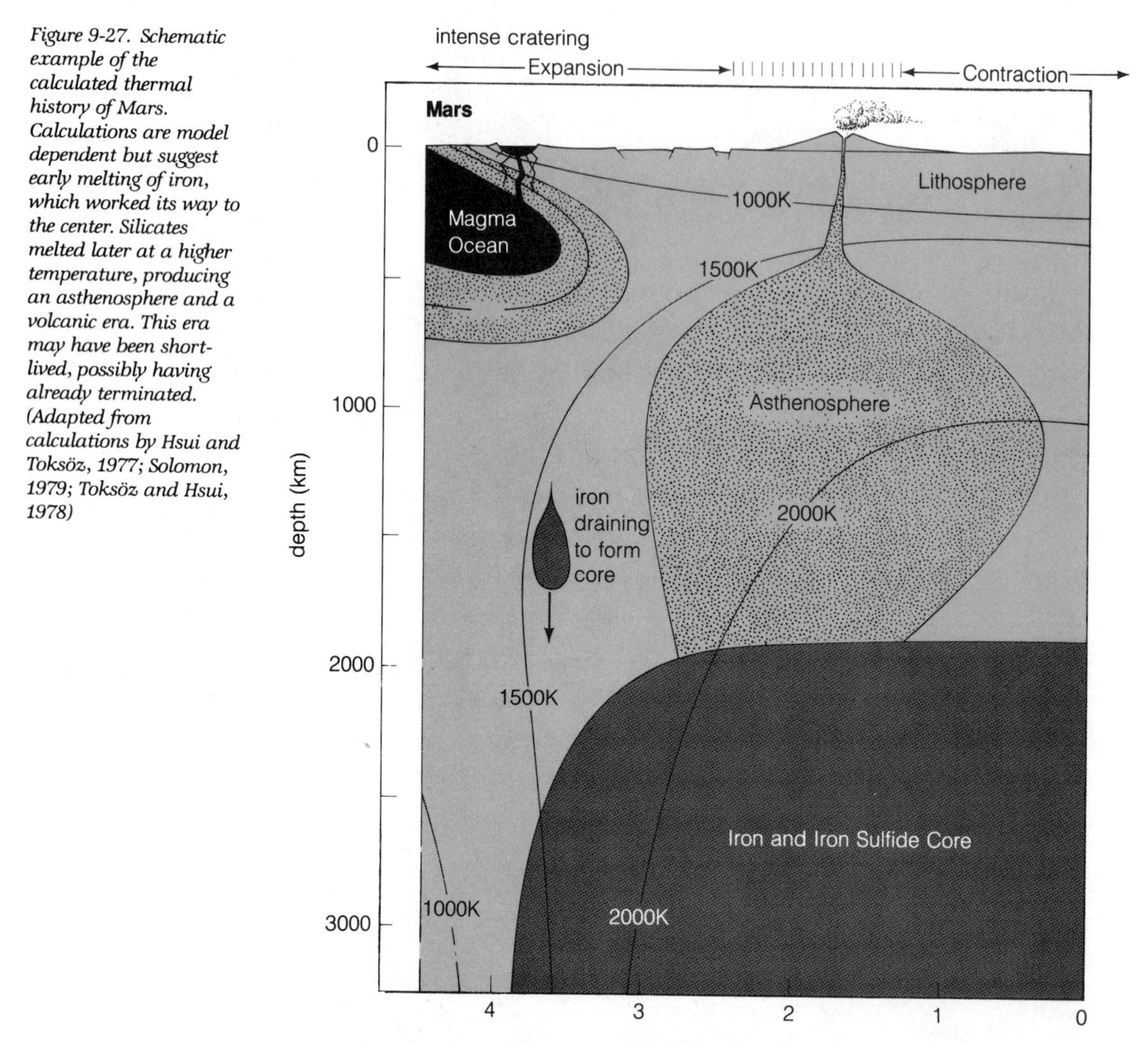

Figure 9-27. Schematic example of the calculated thermal history of Mars. Calculations are model dependent but suggest early melting of iron, which worked its way to the center. Silicates melted later at a higher temperature, producing an asthenosphere and a volcanic era. This era may have been short-lived, possibly having already terminated. (Adapted from calculations by Hsui and Toksöz, 1977; Solomon, 1979; Toksöz and Hsui, 1978)

and huge volcanic cones (see Figure 2-15), which do not appear very old. The models suggest that this asthenosphere has largely disappeared, leaving a lithosphere at least hundreds of kilometers thick.

The suggestion that the lithosphere never got very thin explains why well-developed plate tectonics did not appear. Nonetheless, the calculations suggest a long period of expansion by 5 to 10 km during the heating (Solomon, 1976), explaining the many fractured areas and the appearance of an enormous rift canyon system, shown in Figure 9-28, that is known as Valles Marineris (Valley of Mariner, named after the spacecraft that discovered it).

The huge volcanoes of Mars lead to an interesting suggestion about Martian mantle plumes, related to our discussion of the Hawaiian Islands. On Mars there is virtually no plate tectonic motion, as evidenced by the lack of folded mountain ranges. The accumulating volcanic cones sat directly over the source as long as the source was active, instead of moving over the source as on Earth. This may explain why the volcanic cones of Mars are so much larger than those on Earth. In support of this idea, the total volcanism of the

Hawaiian Islands, if accumulated into one mountain, would make a cone comparable to the Martian examples. The concentration of large Martian volcanoes in one region near the west end of Valles Marineris supports the idea that a Martian mantle plume may have formed under that area.

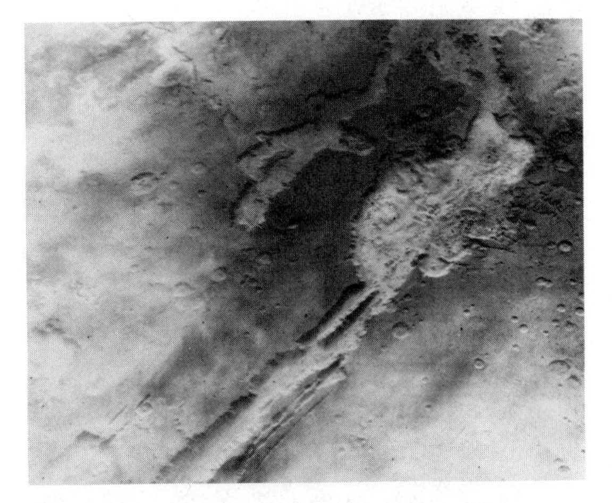

Figure 9-28. The Valles Marineris, or "Grand Canyons of Mars," appear to be a rift system enlarged by splitting, collapse, and erosion. This view shows chaotic (collapsed?) terrain and linear faulted canyons at the east end of the system, whose total length would cross the continental United States. The linear canyons are about 150 km wide. Clouds obscure parts of the surface (left and lower right). This chapter's opening photo shows a global view of the whole system. (NASA, Viking Orbiter 1)

Satellites of the Outer Planets

The American geochemist John Lewis began studies of the interiors of the outer planets' satellites in 1971. From their low mean densities and from condensation chemistry, he predicted that they consist not of iron cores and silicate mantles but primarily of ices and silicates. Lewis (1971a, 1971b) suggested that with even minimal heating, such satellites would form rocky or "muddy" cores and icy surface crusts. The mantles, composed mostly of watery material, might be either frozen or liquid, depending on the thermal state.

Spectral evidence supports this picture, revealing absorption bands of water frost on major satellites such as Europa, Ganymede, Rhea, Dione, Tethys, and one side of Iapetus.

Thermal history calculations for such worlds (Consolmagno and Lewis, 1977) show that the degree of melting was strongly dependent on size and composition. A typical bulk composition for a Jupiter satellite condensing at 160 K would be about half C1 carbonaceous chondrite and half water by mass, which would lead to substantial internal melting at diameters larger than 2000 km and some melting at diameters of 1300 to 2000 km. The presence of ammonia ices would reduce the melting point.

Theoretical models of such worlds suggest a core of hydrous silicates, an extensive asthenosphere of ammonia-rich liquid water, and a thin "lithosphere" (if we may apply that term with its Greek root meaning "rock") of ices. In 1979, California researchers S. J. Peale, P. M. Cassen, and R. J. Reynolds added an interesting wrinkle by pointing out that the hulking presence of a giant planet near such a satellite cannot be ignored: The satellite will be slowly flexed by tides as its orbit goes through subtle variations due to perturbations by other satellites. Just as repeated flexing of a tennis ball makes it warm because of friction, tidal flexing of satellites is an important heat source. Peale and his colleagues calculated that tidal flexing could be significant on Europa and enough to maintain molten conditions inside Io.

Two Voyager flights through the Jupiter satellite system in 1979 beautifully confirmed these predictions about the Galilean worlds. Callisto has the darkest surface (but still twice the albedo of the moon), with bright craters (Figure 9-29). The dark surface is probably a residue of silicate dirt, possibly accumulated after sublimation of some ice (during impacts?) from a poorly differentiated crust. The craters may punch through to cleaner ice below, explaining their bright floors and rays. Since Callisto is farthest from Jupiter and lowest in density, it probably has the least heating from tides or radioactive elements and thus retains the highest ice content of Jupiter's

Figure 9-29. Callisto, seen in a composite view utilizing an ultraviolet image that enhances contrast variations in surface materials. The dark surface soil is believed to be silicate rich, but the craters eject brighter, underlying icy material. (NASA, Voyager 2)

satellites. Schubert, Stevenson, and Ellsworth (1981) propose a never-melted, virtually undifferentiated interior, though earlier models call for differentiation.

Ganymede is entirely different and shows remarkable evidence for incipient plate tectonic activity. As shown in Figure 9-30, Ganymede has broad, Callisto-like regions of dark soil pocked by bright craters with rays of dark and light ejecta. These regions are cut by strange bands of fractures or grooves that may be formed by the jostling of icy plates, one against another. The bright, ridged bands may have formed as water erupted or fresh ice squeezed up between plate margins. Figure 9-30a even shows an offset fault similar to those that occur where Earth's plates are offset. The chaotic fracture systems shown in Figure 9-30b are not unlike the fracture patterns observed in the ice pack "plate" that floats on the Arctic Ocean, as shown in Figure 9-31. Impact craters superimposed on both the older regions and the fracture zones suggest all these features are fairly

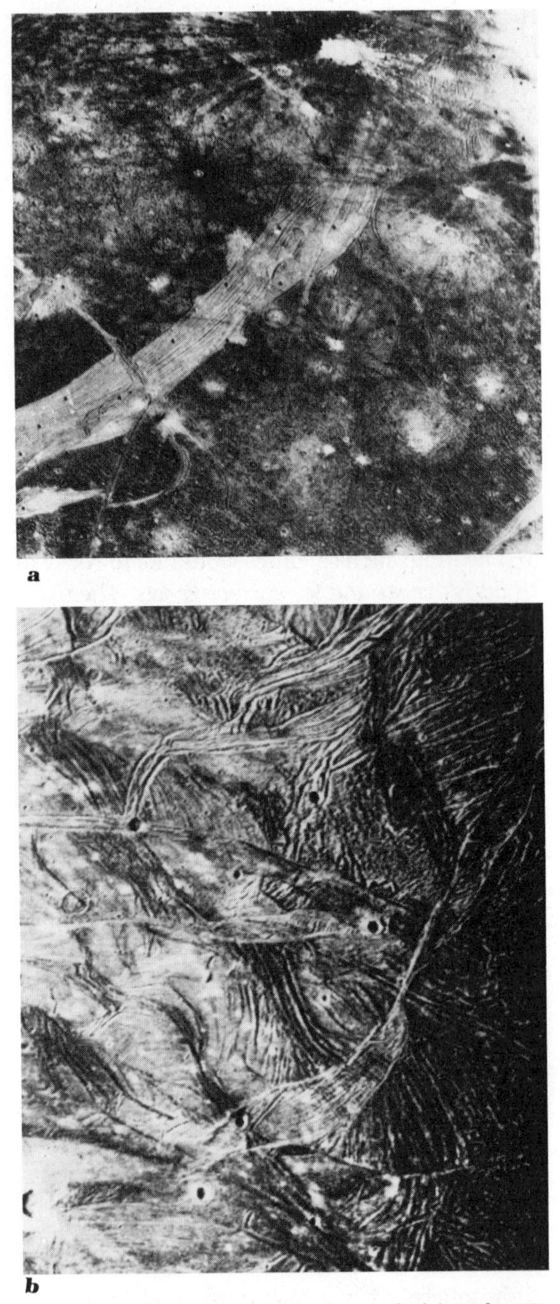

*Figure 9-30. Fracture systems on Ganymede. (**a**) Bright, 150-km-wide band of fractures crossing an older Callisto-like cratered terrain. Note offset in broad fracture band by a narrower fracture band in lower left. (**b**) Close-up of fracture patterns. The frame width is about 580 km, and the smallest features are about 3 km across. (NASA, Voyager)*

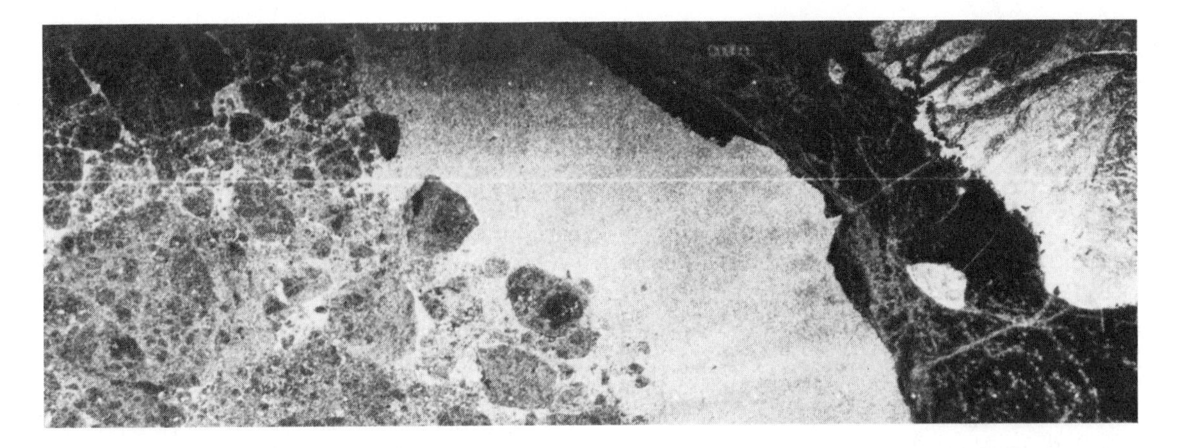

Figure 9-31. Fractured ice pack (left) in Arctic Ocean displays fracture patterns similar to those seen on Ganymede and Europa, which may also involve icy slabs floating on a watery substrate. The land mass of Banks Island, Canada, with stream channels, is at right. Dark lane in the right center is the open sea. Length of the image is 120 km. (NASA, Seasat satellite radar image)

old (possibly 3 to 4.5 Gy, according to preliminary Voyager team analysis), so that Ganymede's geological activity or "plate jostling" may have been confined to a short period of maximum heating, when a thin ice "lithosphere" overlay a watery asthenosphere.

Europa shows a still different "lithospheric" structure—water ice plains with so few craters that they must have been resurfaced by erupted water well after intensive cratering ceased. The age of the surface features is very uncertain (0 to 4 Gy!). High-resolution photos (Figure 9-32) reveal little relief, though ice (or some material) has apparently squeezed up through the cracks, which, again, resemble cracks in the terrestrial ice pack (Figure 9-31).

The Voyager imaging team concluded that the major crack systems could be due to stresses from either tidal forces as they slowed the satellite's rotation or tectonic activity driven by a hot interior. They found that as much as 20% of Europa might be ice in a crust or lithosphere as much as 100 km thick, which is perhaps floating on a layer of liquid water of uncertain depth (Smith and others, 1979). The ice and water are a thin coating on a silicate-rich interior, required by the bulk density of 3000 kg/m^3.

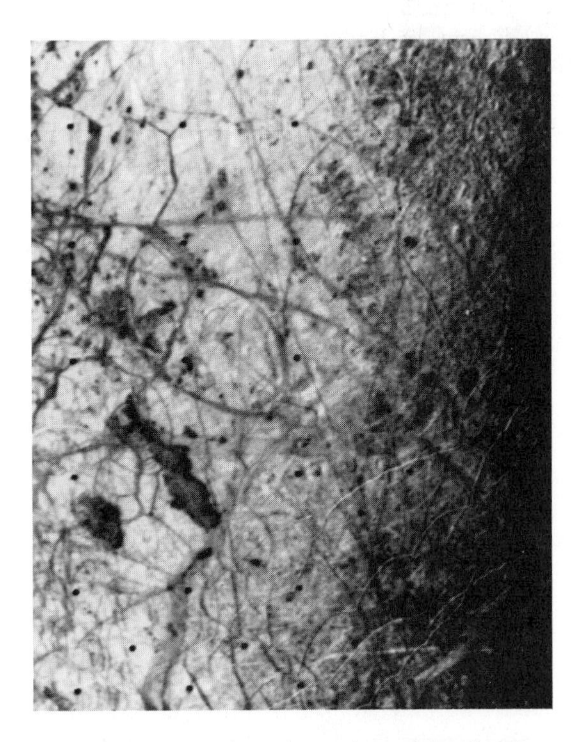

Figure 9-32. Fractures near the terminator (edge of sunlit region) of Europa. The area is about 600 × 800 km, and the smallest features about 4 km across. Low evening light reveals a lack of strong relief but indicates low ridges along many fractures, suggesting material welled up and frozen at the surface. The absence of impact craters indicates that the surface is not very old. (NASA, Voyager 2)

The extraordinary inner satellite Io is most strongly heated by Jupiter's tidal forces and must have an active interior, to a large extent molten. The volcanic activity predicted by Peale and others was dramatically confirmed by Voyager photos of nearly a dozen active volcanoes.

Little or no ice is present in Io. The primordial Io material may have been deficient in ice because of the high temperatures in the proto-Jupiter nebula at Io's distance (see Figure 5-15). Whatever ice may have been incorporated into Io initially has been heavily depleted by recycling through the volcanoes and hence evaporating into space. Preliminary Voyager team models of Io's interior have a thin sulfur crust overlying a silicate mantle. In one model published soon after the Voyager flybys, the sulfur crust has its own solid surface layer, or "lithosphere," perhaps only a kilometer thick, riding on a fluid asthenosphere of molten sulfur and sulfur dioxide. This in turn rides on a solid silicate lithosphere that overlies a deep molten silicate mantle (Smith, 1979). Schubert, Stevenson, and Ellsworth (1981) calculate that convection in the solid interior was able to carry away most of the internal heat, so that only a 100-km-thick asthenosphere is molten in their model. It is underlain by a solid, thick mantle and core. An Io model by Consolmagno (1981) treats both the thermal and chemical evolution. It concludes that Io started with approximately C2 or C3 carbonaceous chondrite composition. Heating and eruptions exhausted water from the crustal layers, leaving a crust rich in sulfur and oxidized iron; a core rich in iron sulfide (FeS) contains about 20% of Io's mass.

Figure 9-33 gives an overview of Galilean satellite models, and following chapters discuss the relation of interior and surface features further.

Saturn's moons (except for huge, haze-covered Titan) are smaller than the Galilean satellites. They have icy surfaces and icy interiors. The moons Mimas, Enceladus, Tethys, Dione, Rhea, and Iapetus have densities about 1200 to 1400 kg/m^3. Even giant Titan has a density of only 1880 kg/m^3. These densities are so low that they imply large proportions of ice in the interior. For example, models of Rhea and Titan by Lupo and Lewis (1979) called for about 66% ice by mass and small rocky cores with about half the radius of each moon.

What is the state of internal heating in such ice-balls? Peale, Cassen, and Reynolds (1980), who successfully predicted the tidal heating and volcanism of Io, concluded that "tidal heating is not an important contributor to the thermal history of any Saturnian satellite." The small sizes of most of these icy moons also suggest that internal radiogenic heat would be inadequate to produce geologically active worlds. For example, models of icy satellites by Consolmagno and Lewis (1978) and Lupo and Lewis (1979) generally predicted that bodies smaller than 1000 km in diameter would never melt; bodies smaller than 1500 km would have had some early melting but would be refrozen by now; and even bodies as big as 3000 km across would not have surface eruptions of water as a result of endogenic (that is, internally generated) heating. Recall that, except for Titan, the largest Saturn moon (Rhea) barely exceeds 1500 km across.

These predictions at first seem to be borne out by Voyager photography of the heavily cratered surfaces on many of Saturn's moons that imply lack of resurfacing after the end of the early intense bombardment. The heavily cratered surfaces are prominent in Figure 2-28, a view of Tethys.

But this is not the whole story. Yoder (1979) made a remarkable prediction, if judged in light of the above work. He noted that of all Saturn's satellites, the Enceladus-Dione pair* offers a tidal resonance which provided the greatest chance for tidal heating. He wrote, "It is possible that originally [the eccentricity of Enceladus] was pumped up to some critical eccentricity which caused catastrophic fracturing. . . . The Voyager 2 flyby of Enceladus may reveal another curious satellite whose past and present state is controlled by tidal friction."

*Note that the resonance associated with tidal heating does not have to be with the neighboring satellite.

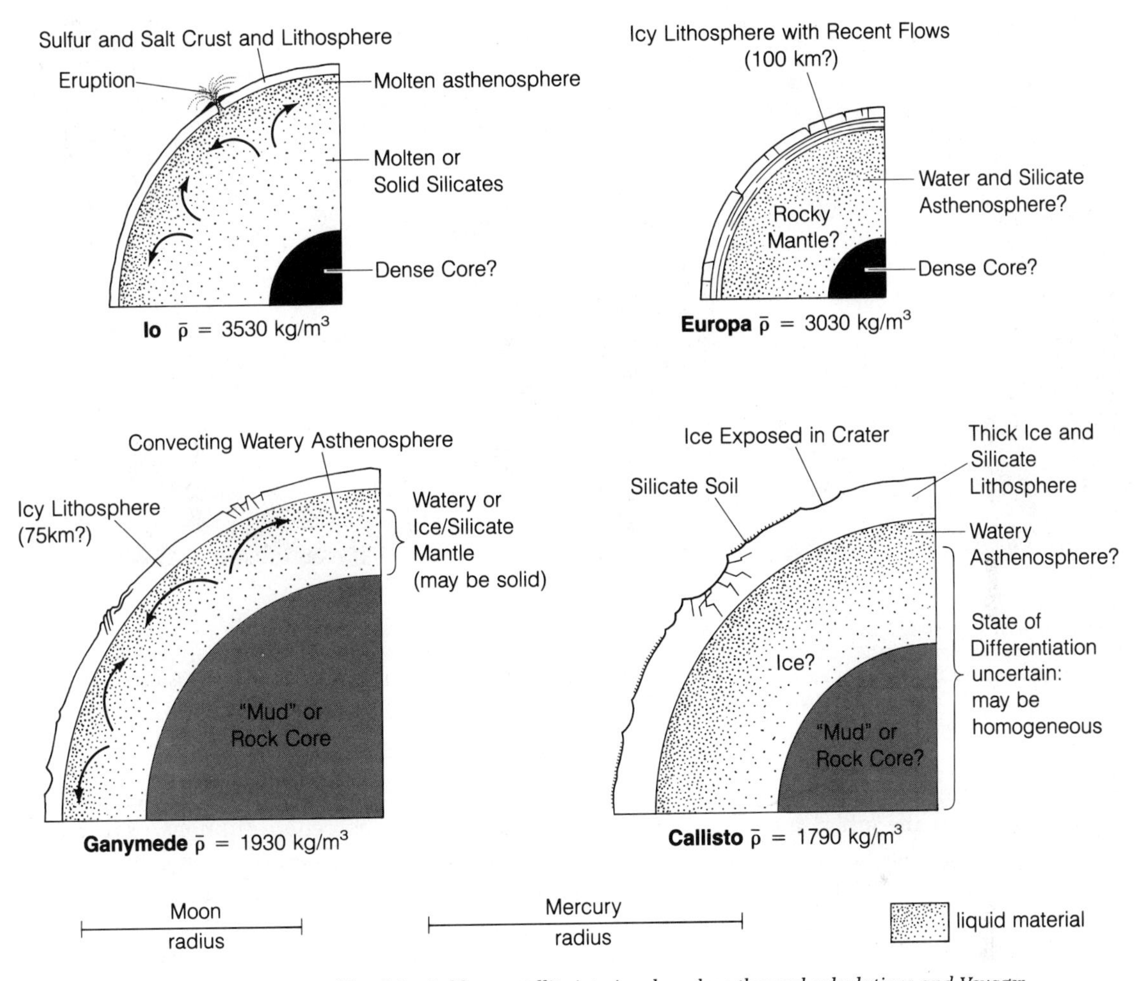

Figure 9-33. Possible schematic models of the Galilean satellite interiors based on thermal calculations and Voyager discoveries. (Adapted from model summary diagrams by Torrence Johnson, plus recent references mentioned in text)

As shown dramatically by Figure 2-29, the Voyager 2 photos did reveal that Enceladus is a "curious satellite." Unlike its neighbors, it is only moderately cratered, and the cratered regions are broken by swaths of smoother plains. The surface material is much brighter than on the other moons, with nearly 100% reflectivity. The appearance of these plains suggests that water erupted and formed fresh ice flows on the surface. From counts of craters, Voyager team members (Smith and others, 1982) concluded that the younger plains units are less than 1 Gy old. Combining this result with the observed presence of debris in a thin circum-Saturn ring spread along Enceladus' orbit, some theorists have suggested that Enceladus may be geologically active even today, occasionally erupting debris off of its surface! Puzzles remain, however, because further calculations of the tidal heating rate suggest that it is inadequate to melt an H_2O ice body (Poirier and others, 1983). Some alternate suggestions are that an admixture of ammonia ice (NH_3) lowers the melting point or that impacts have created localized fracture zones where the tidal heating

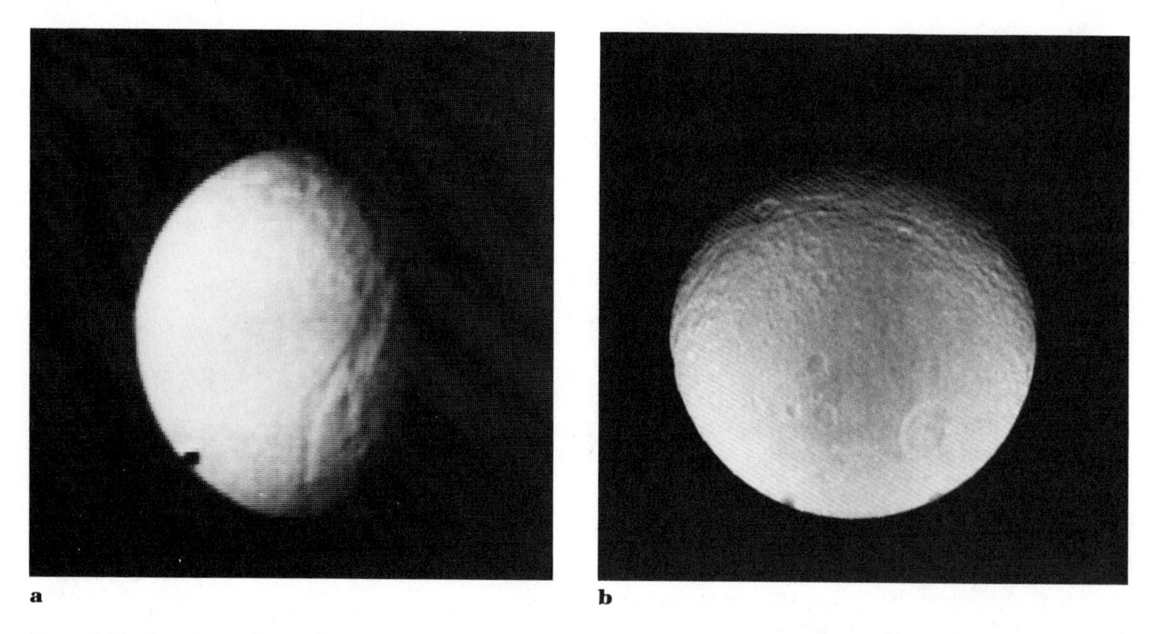

a b

*Figure 9-34. Two views of Saturn's 1060-km moon, Tethys, showing the rift-like trough, Ithaca Chasma. (**a**) North-south segment of the trough near the center of the Saturn-facing hemisphere. (NASA, Voyager 1) (**b**) Ithaca Chasma crossing the north polar region. The portion shown in (**a**) is at the extreme left of (**b**). Crater in upper left and craters in lower left center can also be seen in Figure 2-28. (NASA, Voyager 2)*

process was maximized (Smith and others, 1982).

Even some of the other satellites show features suggesting internal activity. Cratered terrains with different crater densities, hence different ages, have been found on several of the moons. As shown in Figures 2-27 and 9-34, Tethys has a globe-girdling rift or trough. It is an estimated 3 to 5 km deep, 100 km wide, and has been mapped to extend around three-fourths of the circumference. The trough has been named Ithaca Chasma. Calculations suggest that if Tethys had once been melted—a sphere of liquid water with only a thin icy lithosphere—the subsequent expansion during the freezing of the water could have produced a crack with the area of Ithaca Chasma (Smith and others, 1982). Why the expansion would produce one major rift instead of a network of smaller fractures is unknown.

As seen in Figure 9-35, even on the small satellite Mimas, some linear riftlike features can be seen in the heavily cratered terrain. Perhaps a number of Saturn's moons underwent an interior melting at some time in their history; and perhaps circulating currents, eruptions, or freezing-induced expansion modified some of the surfaces. More modeling of the interiors and thermal histories of Saturn's moons is in order, now that the Voyagers have clarified the moons' physical properties.

The Giant Planets

In the early 1900s, it was thought that the giant planets might be gaseous throughout. However, the English geophysicist Jeffreys (1924) showed that they surely would have cooled to a nongaseous state, and he suggested that Jupiter was a small, solid planet surrounded by a vast hydrogen-rich atmosphere. Wildt (1938) computed a more realistic Jupiter model that had an inner core of mean density ~5500 kg/m^3 (the same as for Earth as a whole), an inner mantle of ice, and an outer mantle of frozen hydrogen. Jupiter was realized to have such a strong gravitational

Figure 9-35. South polar part of 392-km Mimas is heavily cratered. Linear valley suggest fractures due either to large impacts or rifting. (NASA, Voyager 1)

field that it could retain all its original complement of even the lightest gas, hydrogen. Since hydrogen is by far the most abundant material in the universe, attention focused on it.

DeMarcus (1958) used theoretical equations of state for pressure-ionized hydrogen and helium to calculate models of Jupiter and Saturn. He found that Jupiter consists of at least 78% by weight of hydrogen, distributed in various pressure-modified forms. The exact distributions depend considerably on the initial and later thermal states (Podolak, 1978; Hubbard, 1981, 1990) and on the equation of state of the material.

The discovery of the large amount of internal heat radiated by Jupiter gave new clues. Hubbard (1970, 1990) concluded that most of the interior must consist of convecting liquid metallic hydrogen, as shown in Figure 9-36, in order to supply the observed heat flow. This coincidentally accounts for the strong magnetic field. Such models also give a cooling rate consistent with the type of evolutionary track shown in Figure 4-8 and reaching today's observed state after about 4.5 Gy (Hubbard, 1977).

Calculations of Jupiter's interior conditions are limited by uncertainties about the equation of state of the hydrogen-helium-trace compound mix. In the atmosphere, hydrogen forms a molecular gas, H_2, similar to our own gaseous atmosphere, with higher density, pressure, and temperature at lower depths, following an adiabatic relation. If the temperature were cool enough, an ocean surface of liquid hydrogen would appear at a depth where the pressure exceeded 13 bars, perhaps 50 to 100 km below the clouds. But the temperature is too high, and the gas simply gets denser and denser, turning into a mush resembling a hot liquid at depths where the pressure exceeds 0.1 mbar, but without a well-defined "ocean surface." At a pressure of 4 mbars, hydrogen atoms' electron structure breaks down, and the material becomes liquid metallic hydrogen, with protons surrounded by loose electrons. This point is about 20 000 km below the clouds (Hubbard, 1981, 1990).

The disposition of silicates and metals in Jupiter is uncertain. Most analysts conclude that Jupiter has lost some volatiles in spite of its strong gravity, so that the relative abundance of heavy elements is perhaps 3 times that in solar material, or the equivalent of roughly 15 $M_\oplus$ of silicates, metals, and so on. A few models suggest that these heavy elements could be dispersed throughout the deep interior, but more likely differentiation has allowed them to accumulate in the center along with some helium. Probably buried deep in the gaseous and liquid hydrogen bodies of Jupiter and of other giants are cores that look strangely like huge terrestrial planets (Figure 9-36). A typical Jupiter model calls for a density of 1000 kg/m^3 at a depth 14 000 km below the visible clouds, a dense core with a mass of 14 $M_\oplus$ compressed to 22 000 kg/m^3 and 1.5 $R_\oplus$, and a temperature around 19 000 K at the core-mantle boundary (Slattery, 1977).

The situation in Saturn is quite similar but with lower pressures, temperatures, and densities prevailing. There is probably a more complete differentiation of heavy elements into a core, perhaps due to less initial heating and less con-

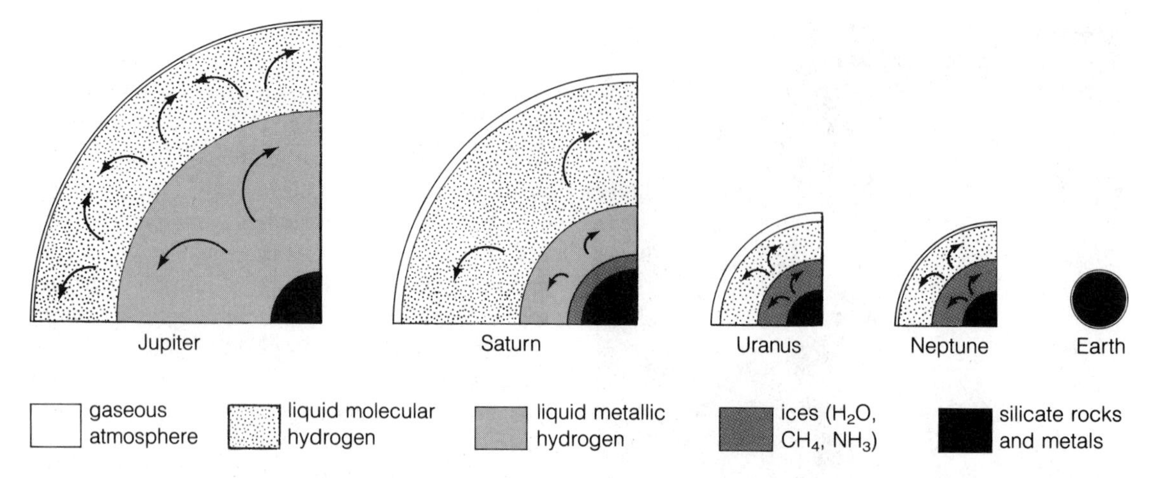

Figure 9-36. Schematic models of the giant planets. Some variations can be found, depending on model assumptions. (Adapted from calculations by Slattery, 1977; Podolak and Cameron, 1974; Hubbard, 1981, 1990)

vection in Saturn (Podolak, 1978). A massive fluid hydrogen ocean with an ill-defined surface overlies a liquid metallic hydrogen region, containing a core with a mass of 17 $M_\oplus$ compressed to 10 000 kg/m^3 and 2.1 $R_\oplus$, with a temperature around 11 000 K at the core-mantle boundary (Slattery, 1977). Ices form on the core surface, but the outer portions of Saturn are so hydrogen-rich and distended that the mean density of the planet is only 690 kg/m^3—the lowest of any planet, and low enough to float in water!

Uranus and Neptune are much denser than theoretical pure hydrogen planets of the same mass, as shown in Figure 9-2. They are richer in heavy elements than Jupiter or Saturn. Of the following two factors, one or both may have been involved: (1) They never accreted such big rocky cores, and thus did not gravitationally pull in such large H/He atmospheres from the nebula. (2) Accretion was slower at Uranus and Neptune, because of the slower orbital velocities, and the nebula had partially cleared by the time large rocky cores formed, resulting in less atmosphere captured. Because Uranus and Neptune are so much smaller than Jupiter or Saturn, the pressure at the base of the H-rich layers does not exceed 4 mbar, and so no metallic hydrogen layer forms (Hubbard, 1981, 1990).

Typical models of Uranus and Neptune call for rocky cores ranging from 4 to 15 $M_\oplus$, with central temperatures around 7000 K, and convecting icy mantles (see Figure 9-36). Pollack (1985) infers that all giants have rocky cores with masses within a factor of 2 of 15 $M_\oplus$.

SUMMARY

The concept of differentiation is helpful in understanding planetary interiors. Differentiation includes any process that arranges elements in different abundance patterns than they had originally, whether in the nebular gas, inside a planet as a whole, or even inside a single geological uni, such as a volcanic magma chamber. Figure 9-2 shows that differentiation in the solar nebula led to planets with different initial compositions. Planets near the sun are composed of dense, iron-rich refractories, and outer planets of low-density volatiles.

Further differentiation occurred inside the planets because of heating. The smallest bodies, such as some asteroids, radiated away this heat before enough accumulated to allow even partial melting. But larger planets were better insulated by their own bulk, which prevented heat from escaping. They melted or partially melted, and their dense materials drained to the center to form cores. In the terrestrial worlds, iron cores formed, and lower-density, more silica-rich materials floated to the surface to form crusts of feldspar-rich rock such as basalt, gabbro, anorthosite, and, on Earth, granite, the most silica rich of all. Icy satellites formed rocky cores and icy crusts. Giant planets formed rocky cores overlain by icy layers and liquid hydrogen seas.

Because of the heating and melting, an understanding of heat sources is crucial to understanding planet interiors. Lunar rocks show that some of the heat was original and produced an initial magma ocean hundreds of kilometers deep. This initial heat may have been common to other planets, also. Magma oceans on other planets may have also cooled and solidified within the first 0.5 Gy. By this time, radioactivity was substantially heating the deep interiors of those planets and creating iron cores and melted mantles in the terrestrial planets.

Cooling of surfaces led to the formation of solid, ever-deepening lithospheres, which overlay molten or partly molten asthenospheres. The thickness of the lithosphere, especially during maximum heating, had a crucial influence on the development of surface features. Small planets tended to cool fast and form thick lithospheres. Large planets and planets heated by tides acquired only thin lithospheres.

Planets that developed early, thick lithospheres preserved intensely cratered surfaces. If the lithosphere was a moderately thick, planet-wide, single "plate," its surface character was influenced by whether the planet underwent significant expansion or contraction during its thermal history. Planets with moderately thin lithospheres developed more modified surfaces, with more fracturing and volcanic activity, and hence fewer surviving craters. Planets with thin lithospheres and convecting asthenospheres (Earth being the prime example) had their lithospheres broken into plates, with resulting major deformation of the surface.

CONCEPTS

planetary interior
core
mantle
crust
limb darkening
mean density
state variables
equation of state
hydrostatic equation
figure
oblate spheroid
plastic flow
rheidity
rheid
mineral
rock
change of state
outer core (of Earth)
inner core (of Earth)
olivine-spinel transition
pressure ionization
metallic hydrogen
differentiation
magma
partial melting
lithophiles
siderophiles
moment of inertia
geometric oblateness
dynamical oblateness
mascon
temperature gradient
diurnal
Mohole Project
Glomar Challenger
olivine
kimberlite pipes
pyroxene
seismometer
P waves
S waves
focus
P-S arrival time interval
epicenter
Mohorovičić discontinuity
low-velocity zone
asthenosphere
fault
passive seismology
active seismology
plates
lithosphere
statics of planets
dynamics of planets
heat
temperature
heat transport
conduction
radiative transfer
convection
adiabatic change
adiabatic temperature gradient
convective cell
magma ocean
electromagnetic heating
feldspars
basalts
gabbros
lines of force
dipole field
magnetic correction
magnetosphere
magnetopause
bow shock
magnetosheath
magnetic field reversal
paleomagnetics
Curie temperature
remanent magnetism
dynamo theory
mantle plume
isostatic equilibrium
continental shield
continental drift
oceanic ridge
rift valley
seafloor spreading

Pangaea
Laurasia
trench
theory of plate tectonics
plate
thrust fault

PROBLEMS

1. Why is understanding a planet's interior helpful in understanding its surface?

2. (a) Does rock have to be molten or partly molten to deform like a fluid over long time periods?
(b) Would such flow be reduced or enhanced if all the crystal lattices in all minerals were "perfect," with no microfractures or missing atoms? Why?
(c) Why is flow enhanced if the rock is heated, even if not completely melted?
(d) Does a rock with remanent magnetism have to be melted in order to destroy the remanent magnetism? Why or why not?

3. The highest lab pressures are around 40 kbar, a pressure reached near a depth of 100 km in Earth, but at a depth of nearly 1000 km in the moon. Explain why this means that knowledge of lunar and Martian mantles could be better than knowledge of Earth's mantle if we had equally accurate information on the chemical composition of the rocks.

4. (a) Compare current seismic knowledge of the mantles of Earth, the moon, and Mars. (b) When better Martian data are available, do you think Mars will be found to display more, less, or the same amount of internal seismic activity as the moon?

5. How does a crust-mantle interface differ from a lithosphere-asthenosphere interface regarding (a) definition and (b) implications for surface structure?

6. (a) What is pressure-ionized matter? (b) In which planets is it important?

7. Explain why the soil in your backyard is not chondritic, even though Earth was made from chondritelike planetesimals and chondrites are the dominant type of "soil" falling out of the sky.

8. Spacecraft are placed in orbit around two identical-looking planets with no natural satellites. The first planet exhibits perfect Keplerian motion. The second exhibits nearly Keplerian motion, but with minor accelerations and decelerations. What do you infer about the interiors of the two planets?

9. Since the moon and Earth may have formed from very similar material, why doesn't the lunar crust have the same composition as the terrestrial crust?

10. Pretend that a sheet of paper is a map. Place several dots labeled A, B, C, and so on at random to represent seismometers. Mark an X at random to represent the epicenter of an earthquake.
(a) How do observers at A determine how far the earthquake was from them?
(b) With only their own data, can these observers locate the epicenter? Mark the points where their data indicate that the earthquake could be located.
(c) Show that a minimum of three stations are needed to locate the epicenter precisely.

ADVANCED PROBLEMS

11. A satellite orbits at a distance of 100 000 km from a planet of 22 000-km radius. The satellite has a mass of 10^{19} kg, a circular orbit, and a period of 1 d.
(a) Find the velocity of the satellite.
(b) Find the mass of the planet.
(c) Find the mean density of the planet.
(d) Which solar system planet does this most resemble?
(e) Describe briefly some plausible characteristics of this planet's interior. Which piece of information is irrelevant (assuming that we can see at first glance that the planet is much larger than the satellite)?

12. At what depth in the ocean does the pressure equal Venus' surface pressure of 90 bars?

13. Make a diagram showing the pressure as a function of depth inside the outer 3000 km of the moon, Mars, Earth, and Jupiter. Contrast the depths at which the pressure reaches the maximum laboratory pressure of roughly 40 kbar.

14. The specific heat c of basaltic rock is the amount of energy involved in a temperature change of 1 K for a cubic meter of rock; it is $c = 2.4 \times 10^6$ J/m^3·K. Suppose a magma ocean starts at time $t = 0$ with mean temperature 1800 K and cools with no significant input of additional heat from below or above. The ocean is 300 km deep and has a surface crust that radiates at a mean temperature of $\overline{T} = 900$ K. Would you expect a significant cooling in 1 y? 10^3 y? 10^9 y?
(a) Make an estimate of the time scale for cooling of the mean ocean temperature by 400 K.
(b) Give some physical complications that require a more exact calculation in order to derive the complete cooling history to 800 K.

CHAPTER TEN

PLANETARY SURFACES 1: PETROLOGY, PRIMITIVE SURFACES, AND CRATERING

Surface environments of other planets are among the most intriguing topics in planetary science. What is it like to stand on another planet? What processes formed those strange landscapes? Our formal goals in studying planetary surfaces are to describe the present-day conditions (such as rock types, surface structures, temperature) and to understand the evolution of the surface by processes such as impact, volcanism, and erosion.

Now and again during the last few chapters we have confronted (and sidestepped) the details of rock chemistry. Now it is time for a more detailed review of minerals and rock types. As the Reverend John Fleming noted as early as 1813, "He who has the boldness to build a theory of the earth without a knowledge of the natural history of rocks will daily meet with facts to puzzle and mortify him" (quoted by Geikie, 1905).

An astronaut's sampling of a large boulder on mountain slopes at the edge of the moon's Sea of Serenity symbolizes the exploration of planetary surfaces to learn about early surface-forming processes. Foreground soil is powdery dust and rock fragments created by meteorite bombardment of the lunar surface. (NASA, Apollo 17)

PETROLOGY

Petrology is the study of rocks—of their description, classification, evolution, and origin. This vast subject is reviewed in many geology textbooks. The present introduction is intended only as a guide for students with little geological background.

Minerals

Minerals are solid inorganic substances that compose solid planetary material. They are characterized by forms determined by their constituent elements or compounds, and they are defined by composition and structure. The crystals visible in rocks are usually individual minerals. **Rocks** are merely assemblages of different minerals, usually in the form of crystals, but sometimes in glassy or amorphous forms that may occur under conditions such as rapid cooling.* Many minerals

*Glasses are noncrystalline; that is, the atoms and molecules have not been able to arrange themselves in the characteristic orderly, geometric manner of a crystal. The atoms may occur in tangled molecular chains or mixed, incomplete crystal lattices. The difference between a crystal and a glass, as George Gamow once remarked, is like the difference between a carefully built brick wall and frozen caviar.

are defined by simple, fixed atomic composition (such as SiO_2, which is called silica or quartz), but they do not all have such simple formulas. For example, olivine has the compositional formula $(Mg,Fe)_2SiO_4$, which means that ions of either magnesium (Mg) or iron (Fe) may *substitute* in the crystal lattice. The governing factor in substitution is often the size of the atom or ion; the situation is analogous to a structure made of packed Ping-Pong balls in whose interstices BB's, but not golf balls, can fit. The number of substituting ions is, of course, constrained by the requirement of electrical neutrality. Many minerals thus cannot be defined in terms of one fixed composition, but rather must be defined in terms of ranges of composition.

Although they are defined chemically, minerals can be visually identified in the field by a series of semiquantitative tests of such properties as hardness (on a 0–10 scale called Moh's scale), streak (color of the powdered form produced when a sample is rubbed against a small, hard tile, or "streak plate"), shape, luster, and density.

In typical rock specimens, the mineral crystals range from easily visible to microscopic. In the latter case the rock seems homogeneous but is really a mass of fine mineral crystals. Microscopic examination is often required to identify the various minerals present.

Minerals and rocks are sometimes loosely classified according to composition, ranging from **acidic** or **siliceous** (high silica content) through **basic** or **mafic** (high content of heavy elements such as iron or of elements with chemical affinity for iron) to **ultrabasic** or **ultramafic** (very rich in heavy elements and low in silica). A very rough rule of thumb is that the darker or denser the rock type, the more basic it is. In a planet that has been at least partially molten, crustal minerals and rocks tend to be siliceous, erupted lavas tend to be basic, and deep-seated minerals and rocks tend to be ultrabasic.

The vast majority of rocks on terrestrial planets are composed mostly of a few important minerals. The most important minerals include the following:

Feldspars: $(K, Na, Ca)AlSi_3O_8$. Density 2600 to 2800 kg/m^3. Feldspars are one of the most important groups of minerals in planetary sciences and by far the most common mineral in terrestrial surface rocks, making up about 60% of the crustal minerals (Mason, 1966). They are silicates of aluminum (Al), with a variable admixture of potassium (K), sodium (Na), or calcium (Ca). The formula shows how K, Na, or Ca may substitute for each other in the aluminum silicate lattice, providing a range of compositions. Note that feldspars contain a majority of the most abundant elements in Earth's crust (and probably other terrestrial planet lithospheres): O, Si, Al, Fe, Ca, Na, K, and Mg. One important property of feldspars is that they are among the lowest-density silicate minerals. The density of about 2700 kg/m^3 contrasts with densities of 3300 to 4300 kg/m^3 for common volcanic silicate minerals, such as olivine, spinel, and garnet, and values such as 5200 kg/m^3 for the iron-rich mineral magnetite. A consequence is that feldspars tend to float in magmas and accumulate closer to planetary surfaces than do the denser minerals. Because feldspars are so common they are intricately subdivided. **Orthoclase** feldspars are the K-rich group, tending to occur in lighter-colored and more silica-rich rocks. **Plagioclase** feldspars are the Na- and Ca-rich group, tending to occur in darker-colored, less silica-rich rocks, although they are nonetheless common in most igneous rocks. The orthoclase feldspars are still further subdivided according to the Na/Ca ratio. The sodium-rich end of the spectrum is a mineral called albite (sometimes abbreviated Al), and the Ca-rich end, anorthite (sometimes abbreviated An). We will not emphasize these subtypes, though their names are as likely to be encountered in rock analyses as the general term *feldspar.*

Quartz, a form of **silica:** SiO_2. Density 2600 kg/m^3. If a cooling magma has silicon left after feldspars have formed, silica is likely to form. The most common form is the familiar mineral quartz. Because of its low density, quartz is also likely to float in the magma and accumulate in surface rocks. If there is very active geological processing and differentiation, quartz may become very concentrated at the surface, because its density is even less than that of many feldspars. These conditions are especially prevalent on Earth, where many surface rocks are much richer in quartz crystals than are surface rocks sampled on the moon and Mars.

Pyroxenes: A group of Mg, Fe, Ca, Na, Al, and Ti silicates, including specific minerals such as augite (probably the most common), enstatite, and hypersthene. The density is high, ranging from 2800 to 3700 kg/m^3. These minerals are common in meteorites as well as in basic planetary igneous rocks. They constitute roughly 10% of Earth's crustal minerals.

Amphiboles: A group of Mg, Fe, and Ca silicates with different crystal structure from the pyroxenes. The densities are only slightly less than those of pyroxenes. The amphiboles make up perhaps 7% of Earth's crustal minerals, being common in basic igneous rocks.

Micas: K, Al, and Mg silicates, with intermediate densities ranging from 2760 to 3200 kg/m^3. They are common in igneous rocks and compose roughly 4% of Earth's crustal minerals.

Beyond about 4 AU from the sun (the outermost fringe of the asteroid belt) H_2O ice constitutes as much as 60% by mass of the material that condensed in the solar nebula. Hence, H_2O and other ices play the role of major "rock"-forming minerals on worlds in the outer solar system.

Water ice: H_2O. Density 932 kg/m^3 at 100 K and 1 bar (917 at 273 K, 1 bar). Water ice dominates the surfaces of many distant moons, such as Europa, Ganymede, Enceladus, Rhea, Titania, Charon, and Saturn ring particles (Cruikshank and Morrison, 1990), and also appears in the polar caps of Earth and Mars.

Carbon dioxide ice: CO_2. Density 1560 kg/m^3 (216 K, 5 bar). This ice sublimes at 194 K at 1 bar pressure, and is thus unstable up to about 10 AU from the sun. It might be an important volatile powering comet eruptions from 4 to 10 AU.

Ammonia ice: NH_3. Density 817 kg/m^3 (194 K, 1 bar). Ammonia ice has a low melting point, 195 K at 1 bar, and may be a partial constituent of some outer planet satellites. Because it melts at a lower temperature than does H_2O ice, it could play an important geological role as a fluid or gas in slightly heated bodies.

Methane ice: CH_4. Density 415 kg/m^3 (109 K, 1 bar). Methane ice is very volatile because of its low melting temperature. It should be stable in sunlight only on the most distant bodies. It is present on Pluto (discovered by Cruikshank and Silvaggio, 1980) and Triton (discovered by Cruikshank and Silvaggio, 1979), probably mixed with frozen N_2 ice (Stone and Miner, 1989). It might be among the first gases to sublime as a comet approaches the sun from the Oort cloud.

Some other types of minerals are commonly encountered in planetological literature:

Olivine: $(Mg,Fe)_2(SiO_4)$. Density 3300 to 4400 kg/m^3. Because of its high density, it sinks to the lowest parts of magma volumes. It is thus important in materials formed at depth such as some meteorites and planetary lavas. It is believed to be a major component of terrestrial planet mantles. It is usually a greenish crystal and is often found as inclusions in basaltic lavas. Because the Mg/Fe ratio varies, olivine has a variety of subtypes. A-class asteroids are very rich in olivine.

Iron oxides: Important examples are **magnetite** (Fe_3O_4) and **maghemite** (an unusual form of Fe_2O_3), which are magnetic and likely candidates for the 1% to 7% of loose Martian soil that adhered to magnets on Viking landers (Hargraves and others, 1977). Other nonmagnetic forms are **hematite** (the more common form of Fe_2O_3), **goethite** (pronounced GHER-tite;

$HFeO_2$), and **limonite** ($FeO[OH] \cdot nH_2O$). These are oxidized iron minerals, cousins of ordinary rust, and have reddish colors ranging from yellow to brown. They are believed to be the minerals that redden the soil of Mars, where they may occur as fine coatings on dust grains of other compositions. Limonite is the hydrated form, of variable composition, caused by alteration of other iron minerals, often by intermittent exposure to water.

Troilite: FeS. Density 4600 kg/m^3. An accessory mineral in most meteorites, typically 5% to 6% by weight of chondrites.

Graphite (C) and **carbonaceous minerals:** Density 2200 kg/m^3. Various forms of carbon-based minerals are low-temperature condensates that appear to play a very important role in the outer solar system, creating the blackish and reddish-black (albedo 2% to 8%) surfaces common there. Organic compounds are common among these materials, sometimes causing some coloration.

Clay minerals: Example: **montmorillonite**, $(Al,Mg)_8(Si_4O_{10})_3(OH)_{10} \cdot 12H_2O$. Clay minerals are essentially hydrous aluminum silicates. They play two important roles in the solar system. First, they are major minerals in the erosion products of Earth and Mars. Montmorillonite and **nontronite** are considered to compose roughly 20% and 50%, respectively, of the fine dust of Mars; they are frequent products of weathering of volcanic lavas (Toulmin and others, 1977). Second, they are major minerals in the carbonaceous matrix that condensed among C- and RD-type asteroids and outer solar system objects. Spectra suggest the presence of montmorillonite on Ceres, the largest asteroid. The open crystal structure of clay minerals allows them to absorb and hold large amounts of chemically bound water.

Rocks and Rock Types

Rocks are traditionally divided into three groups. **Igneous rocks** are rocks formed directly from the cooling of molten magma. **Sedimentary rocks** are formed by deposition and cementing of small particles, either of biogenic or nonbiogenic origin. **Metamorphic rocks** are rocks originally formed in either of the above modes but changed to a new rock form by high pressure, high temperature, or the addition of new chemicals.

Igneous rocks. Igneous rocks have a variety of compositions and textures, depending on the composition and cooling history of their parent magmas. Terrestrial magmas typically have temperatures of 1300 K and may solidify underground. If they work their way upward and extrude onto the surface, they are called **lavas**. Rock formed on the surface from cooling lava is also loosely called lava or lava rock.

Igneous rocks are the most important rock type in planetary surface studies. Some 95% of Earth's outer few kilometers is composed of igneous rocks. The high percentage of sediments we see around us is misleading, because sediments and sedimentary rocks form on the surface, accounting for much area but little depth. The moon's rocks are all igneous. Lavas abound in the dark plains, but in the uplands, which are older regions, intense cratering has shattered most igneous rocks. There they have recemented into breccias, or rocks formed of welded angular fragments. Many Martian and Venusian rocks and soils are also igneous, as determined from their appearance and soil analysis by several landers, though sedimentary layers also exist on Mars.

If we broaden our horizons to include the predominantly nonsilicate (that is, icy) crusts of the outer solar system, we can anticipate that sulfur compounds are common igneous "rock-forming" materials on Io and that ices are such materials on many other satellites.

Igneous rocks are subdivided into compositional classes and subclasses, as shown in Figure 10-1. Arranged down the figure are compositions ranging from silica rich to silica poor.

The classifications shown in Figure 10-1 reflect the eruptive history of the magma. If the magma remains underground, it is called **intrusive**. There

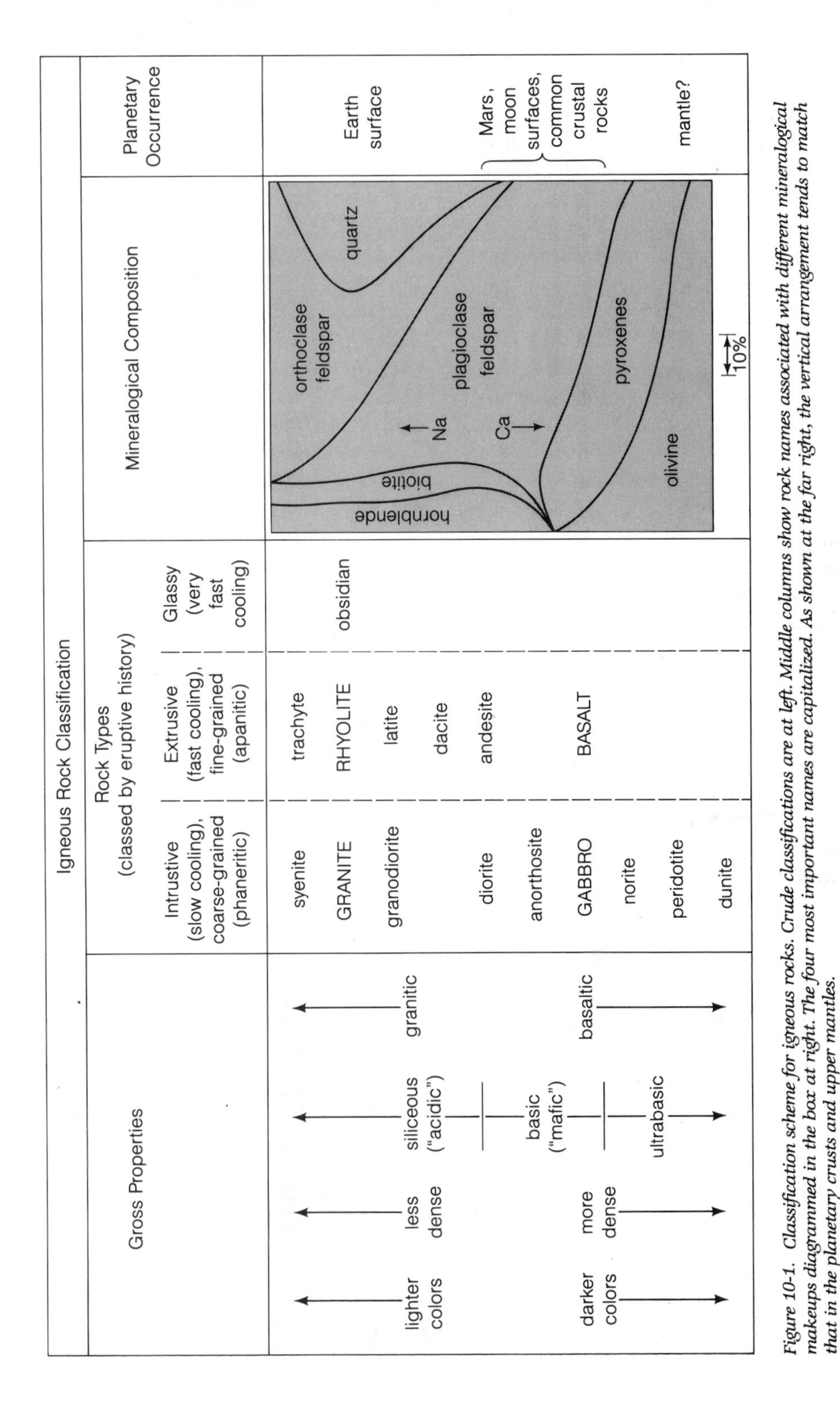

Figure 10-1. Classification scheme for igneous rocks. Crude classifications are at left. Middle columns show rock names associated with different mineralogical makeups diagrammed in the box at right. The four most important names are capitalized. As shown at the far right, the vertical arrangement tends to match that in the planetary crusts and upper mantles.

it is well insulated, cools slowly, gives crystals lots of time to grow, and hence produces coarse-grained, or **phaneritic**, rock. Such rocks may be exposed on the surface by later erosion. If the magma erupts onto the surface, it is called **extrusive**; it cools rapidly and produces fine-grained, or **aphanitic**, rock. Extrusive materials may cool so fast that no crystals grow because the molecules do not achieve a crystal lattice structure; in this case the material is called **glass**.

In planetary science, the most important group of igneous rocks are the **basaltic rocks**. These include **gabbros** (the coarse-grained, intrusive type) and **basalts** (fine-grained, extrusive). These are very common because they are formed from the feldspars and other materials common in terrestrial planet crusts. While basaltic rocks were once known only at scattered volcanic sites, they have now been identified as major surface rocks covering the dark lunar plains and much of Mars, Venus, and asteroid Vesta. The lunar uplands have many gabbros, some erupted basalts, and a gabbroic rock made largely of plagioclase feldspar, called **anorthosite**, or anorthositic gabbro.

The next most important group is the **granitic rocks**: **granite** (coarse-grained, intrusive) and **rhyolite** (fine-grained, extrusive). On Earth, granites are common. Rhyolite flows, the result of eruptions of granitic magma, were once thought relatively uncommon but are increasingly being recognized.

The importance of basaltic and granitic rocks on Earth is shown by the distribution of silica contents in random rock samples. There are two peaks: a basaltic group averaging 53% SiO_3 and a granitic group averaging 73%. Thus, basaltic and granitic rocks, with their distinct colorations, often appear as distinctive outcrops in terrestrial landscapes, as shown in Figure 10-2.

Granitic rocks are much less common on other terrestrial planets (with the possible exception of Venus—one out of five Soviet landers, Venera 8, measured radioactivity levels comparable to a granitic rock called granodiorite, while the other landing sites yielded chemistry typical of basalts).

The kind of rock produced by a given magma depends on the magma's composition and on what happens during the cooling. As the temperature drops, minerals form and react chemically with each other and with the remaining magma. Therefore, the composition of the resulting rocks depends on the rate of the cooling process and on the degree of **fractionation**—the differentiation process by which early-crystallizing minerals separate from the rest of the magma. Bowen (1928) expressed this concept in his **reaction series**, the sequence of minerals that crystallize from a given melt as the temperature drops. Interestingly, the basic minerals tend to crystallize first (at higher temperatures), and the more silica-rich ones last (at lower temperatures).

The significance of the reaction series can be seen better by considering the example of a melt of basaltic composition. Among the first minerals to crystallize are olivine and calcium-rich feldspars. If there is no fractionation, these crystals remain in the melt and eventually react with the hot fluid magma. The olivine is converted into pyroxenes. The whole mass cools. The final rock is basalt (if the magma erupts on the surface and cools fast) or gabbro (if the magma does not erupt and thus cools slowly), because that is the composition of the initial melt.

But fractionation is likely, and the sequence is likely to change. The olivines are denser than the melt and tend to sink to the bottom of the melted region, if there is enough time. This stops the reaction of the olivine with most of the melt, so that the melt itself evolves differently. Amphiboles and more potassium-rich feldspars may form. Feldspars, too, may fractionate by floating to the top of the molten material and forming feldspar-rich rocks there. If fractionation continues, the last stages of crystallization will produce quartz, often in a watery solution of hydrothermal minerals (minerals affected by the presence of hot water and associated dissolved material).

Conversely, if igneous rocks are heated, the first stages of melting may produce a hydrothermal mineral solution (which might drain off),

Figure 10-2. Contrast of basaltic and granitic rocks on Earth is prominent in this Mexican range, where basalt lavas broke through a light-toned granitic ridge. The dark basaltic cap has eroded and produced basaltic talus spreading down the granitic hillside. (Pinacate volcanics, Sonora; photo by author)

then melting of quartz, and lastly melting of certain basic minerals.

Sedimentary rocks. Sedimentation is essentially a planetary skin effect, "the interaction of the atmosphere and hydrosphere on the crust of the planet" (Mason, 1966). Chemical and mechanical **weathering** constantly attacks rocks exposed at the surface of a planet with an atmosphere. Chemical weathering involves chemical reactions that break down the rocks; mechanical weathering involves nonchemical breakdown, for example by abrasion from blowing sand or moving water. The rock particles are transported away in suspension and in solution to form new rocks of different compositions and textures in other places.

The most important type of sedimentary rock on Earth is **shale**, a rock composed of very fine grains derived from consolidated beds of clays, silts, or mud. Shales make up an estimated 4% of the upper 6 km of Earth's crust (Mason, 1966). **Sandstone**, a coarser-grained rock, is composed of cemented sand grains (often of quartz but sometimes of other minerals). On Earth (and perhaps Mars) shale and sandstone are usually composed of rounded grains, because the grains have been abraded by being tumbled in wind or water.

If rock fragments were violently broken and recemented, they will be highly irregular and angular; a rock made of such fragments is called a **breccia**. Breccias are important in planetary science. As mentioned before, terrestrial breccias are considered sedimentary rocks because the fragments have usually been transported in water or air and then deposited. But 85% of the moon is ancient highlands where virtually all rocks are breccias composed of gabbroic and basaltic fragments blasted out by meteorites. Similar conditions probably exist on Mercury and perhaps some asteroids. Since impact breccias on other planets have usually not been transported or sorted in air or water, most geologists do not consider them sedimentary, though some authors have done so. I prefer not to say that the moon is 85% covered by sedimentary rock!

Another type of sedimentary rock important on Earth, Mars, and perhaps elsewhere is a type known as **evaporites**—solid materials left behind when solutions evaporate. Salt and sulfate minerals often result. For example, gypsum ($CaSO_4 \cdot 2H_2O$) is a major mineral that is left behind as moisture evaporates from soil. Salt (NaCl, sometimes called rock salt) is another. Evaporites often bond other rock sediments together into a loose, crumbly rock. A crumbly or flaky crusty material on Mars, called **duricrust**, was found among the dust deposits by Viking, and is believed to be evaporite material.

Significant geochemical differentiation can occur during the formation of sedimentary rocks. Certain minerals, especially quartz, are hard—resistant to solution during low-temperature weathering processes—and are dropped as fragments; sandstones may result. Other minerals are broken down; the aluminosilicates give rise to clays, muds, and shale; ferrous iron (+2 charge) is oxidized to the ferric state (+3 charge, usually producing red mineral colors), which may produce hydroxide precipitates abundant enough to form ore deposits; calcium minerals produce calcium carbonate solutions, which may yield limestones. Certain elements that remain in solution are carried preferentially from the land to ocean basins; sodium, the best example, is constantly being added to oceans.

Metamorphic rocks. Metamorphism is the net effect of all processes (short of complete remelting) acting to alter and recrystallize solid rock material beneath a planet's surface. These processes are a result of changes in pressure, temperature, and chemical environments. A given environment *and a given composition* imply equilibrium only for a certain assemblage of rock minerals, and these particular minerals may or may not be present. If they are not present, adjustments in the existing rock will tend to produce them.

The presence of chemically active volatiles is perhaps the most important factor in metamorphism. They may carry off or introduce new material to the system, a process called **metasomatism**. Examples of other metamorphic processes are an increase in pressure and the physical rearrangement of minerals (for example, along parallel bands by vertical pressure or shearing), both of which create high-density minerals.

Some common metamorphic rocks are listed in Table 10-1. Pressure plays an important role in producing the banding and cleavage characteristic of the first three entries.

A SURVEY OF PLANETARY ROCKS

Many principles of planetary geochemistry become clearer if we review the compositions of planetary samples collected to date. The most basic way to look at rock compositions is to look at elemental abundances, as shown in Table 10-2. However, analysts more commonly report compositions in terms of oxides, as if the available elements were all combined into compounds with oxygen. This method of presentation is shown in Table 10-3. Since most of the minerals in rock are likely to be oxides, this type of table comes close to indicating the mineral makeup of the rocks, especially in the case of igneous rocks. But in the real rock, many of the atoms may be bound up in minerals more complex than the simple oxides indicated in Table 10-3. We will discuss information to be gleaned from both modes of presentation.

Table 10-2 starts with the most primitive solar system materials, such as the solar gas from which the planets ultimately derive. Note the tremendous amount of differentiation implied by going from the primeval gas to the composition suspected for even the primitive, icy comets, not to mention the carbonaceous meteorites, ordinary chondrite meteorites, and iron meteorites, with their 91% iron. The material remaining in the terrestrial and lunar mantles, judged from deep-seated samples, is somewhat depleted in siderophiles and correspondingly enriched in lithophiles, such as silicon (Si), magnesium (Mg), and aluminum (Al), relative to meteorites. The processes that produced crusts, such as floating of low-density minerals and chemical affinity of the lithophiles for silica-rich minerals, produced complex patterns in crustal rocks. Crustal rocks are generally enriched in oxygen (O) and Si and depleted in iron (Fe). The moon's crust of anorthositic rocks is notably rich in Al. Earth's crust, as a result of more complete differentiation, is notably rich in Si. The planetary basalts, which are postcrustal lavas erupting from the upper mantles of the planets, are remarkably similar.

More data are available in Table 10-3, and some of the same patterns can be seen. Primitive meteorites are even richer in iron than planetary mantle rocks; they represent planetary material before the iron differentiated into cores. The crustal rocks, such as lunar highland materials and Earth's crust, are depleted in dense iron-rich minerals. Earth's continents are the most silica rich, while the oceanic crust resembles the lunar crust and basaltic materials.

The numerous basaltic samples are similar, but show a range of silica contents. They may show anomalous enrichments of one mineral or another. Many lunar mare basalt flows are unusually high in titanium, (Ti), while so-called **KREEP basalts**, often found in the lunar uplands, are enriched in potassium *(K)*, rare-earth elements *(REE)*, and phosphorus *(P)*. Though Martian rocks were not analyzed by Viking, Martian soils were analyzed and were found to be derived from basalt, but with an excess of sulfur compounds and related materials left as evaporites. In some places, the soil is welded together into duricrust by these evaporite materials.

Among the nonigneous rocks, Earth's sediments represent an averaging of different rock types, and the limestones are noteworthy for having a biological origin that (presumably!) never occurred on other solar system worlds. The sea creatures whose shells produced limestone rep-

Table 10-1
Common Metamorphic Rocks

Metamorphic rock	Common parent rock	Description
Gneiss	Granite	Marked parallel bands, coarse grained
Schist, phyllite	Shale, granite	Banded, fine grained
Slate	Shale	Fine grained, splits into thin sheets
Quartzite	Sandstone	Massive quartz rock, breaks through quartz grains
Marble	Limestone	Familiar in architecture

Table 10-2
Elemental Abundances in Cosmic Materials (percent weight)

Material	Element[a]														Total	Primary reference
	H (V)	He (V)	O (L)	Fe (S)	Si (L)	Mg (L)	S	C	Ca (L)	Ni (S)	Al (L)	Na (L)	Ti	K		
Primitive Materials																
☉ ("cosmic" abundance)	78	20	1	0	0	0	0	0	0	0	0	0	0	0	99	Gibson (1973), p. 72
Comets	7	?	76	?	6	?	?	3	?	?	?	?	?	?	97	Delsemme (1977), p. 9
C1 carbonaceous chondrite	0	0	46	18	11	10	6	3	1	1	1	1	1	1	100	Wasson (1974), p. 78
Meteorites (avg.)[b]	0	0	33	29	17	14	2	0	1	2	1	1	0	0	100	Mason (1962), p. 151
Planetary Interiors: Ultrabasic Rocks																
Iron meteorites	0	0	0	91	0	0	0	0	0	9	0	0	0	0	100	Wyllie (1971), p. 97
⊕, core (est.)	0	0	0	85	5?	3	4	0?	0?	3	0?	0?	0?	0?	100	Wyllie (1971), p. 104
⊕, mantle (est.)	0	0	44	10	23	19	0?	0?	2	0?	2	1	0	0	101	Wyllie (1971), p. 104
☾, bulk (est.)	0	0	42	8	20	18	0	0	5	0	4	0	0	0	97	Taylor (1975), p. 316
Crustal Materials																
☾, highland soils	0	0	45	5	21	4	0	0	11	0	13	0	0	0	99	Taylor (1975), p. 64
⊕, crust (avg.)	0	0	47	5	28	2	0	0	4	0	8	3	0	3	100	Mason (1958), p. 44
Basaltic Materials																
☾, mare basalt	0	0	40	11	21	6	0	0	8	0	8	0	6	0	100	Lofgren and others (1981), p. 239
☾, mare soils	0	0	43	11	21	5	0	0	8	0	8	0	2	0	98	Taylor (1975), p. 64
♂, Chryse soil	0	0	42?	13	21	5	3	0	4	0	3	0	1	0	92	Clark and others (1977), p. 4588
♂, Utopia soil	0	0	42?	14	20	?	3	0	4	0	?	0	1	0	84	Clark and others (1977), p. 4588
⊕, basalts	0	0	43	9	23	5	0	0	8	0	8	2	1	1	99	Lofgren and others (1981), p. 14ff
Other Material																
♂, duricrust	0	0	42?	13	21	5	4	?	4	?	3	?	1	0	93	Clark and others (1977), p. 4588

[a](V) = volatile (easily lost by heating); (S) = siderophile (tending to associate with iron minerals); (L) = lithophile (tending to associate with silica minerals).
[b]Essentially ordinary chondrites.

resent an extreme geochemical differentiation favoring calcium.

EARLY EVOLUTION OF LITHOSPHERE MATERIALS

Now we have the tools to discuss in more detail the formation of the primeval surfaces of planets. We have emphasized that feldspar crystals, because of their early appearance during solidification, their low density, and their abundance, tended to float in hypothetical magma oceans and accumulate near surfaces. This explains why the crustal layers of Earth, the moon, and Mars appear to be anorthositic, gabbroic, and basaltic in composition, since these are the rock types associated with abundant feldspars. On Earth, of course, the primitive crust has been nearly obliterated by geological activity.

As the lunar magma ocean solidified, the Ca-rich feldspars (that is, plagioclase feldspars) fractionated from the magma so efficiently that many lunar highland samples are classed as anorthosites (rocks defined by very high plagioclase content). Related, specialized rock types also commonly described among primitive crustal samples from the lunar highlands are **norites** (hypersthene-plagioclase-rich gabbros) and **troctolites** (olivine-plagioclase-rich gabbroic rocks).

Some researchers have attempted theoretical extensions of these principles to characterize other planets' primitive crusts. Geochemist J. Wood (1977) emphasized that feldspars crystallize from magma readily at pressures less than 12 kbar but are replaced by denser minerals at pressures greater than 12 kbar. This means that in any planet, feldspars would not float upward from below depths where the pressure equals 12 kbar, thus limiting the thickness of feldspar-rich crusts. Morrison and Warner (1978) carried this a step further and calculated the following maximum thicknesses for feldspar-rich crusts that extend to the 12-kbar level in terrestrial planets:

Earth	40 km	Mercury	108 km
Venus	45 km	Moon	250 km
Mars	105 km		

These depths would be reached only if there were enough feldspathic material to fill such zones.

Did the primitive Earth therefore have a thin anorthositic crust resembling the moon's? Anorthosite masses are found in some of the primitive shield areas of Earth, but they are thought to be products of fractional crystallization of large, localized magma masses rather than relics of a primeval crust.

Aside from the thinness of the feldspathic layer available for Earth-sized planets, Warren and Wasson (1979) raise another reason for concluding that "no substantial anorthositic crust is to be expected on Earth-sized planets": Much of the magma ocean crystallization would have taken place on the bottom of the ocean layer, which would have been beneath the 12-kbar level on the larger planets. The denser minerals crystallizing there in place of plagioclase (namely garnets and spinels) would rob the remaining magma of much of the calcium and aluminum needed to make the plagioclase feldspars before the feldspar-rich crust could form on the top of the ocean. With Earth's surface materials thus seriously depleted in plagioclase relative to the moon's crust, a more granitic primitive composition would be favored. Venus' crustal composition will provide an interesting test of these ideas, since Venus is similar in size to Earth.

We have been discussing initial surface formation as if it happened without external effects. Had that been the case, primitive worlds like the moon ought to display nicely stratified sequences of accumulated crustal rocks, with numerous 4.5-Gy-old "**Genesis rocks**" lying around, waiting to tell us their stories of how it was when the planets had just finished forming. Lunar astronauts were trained to look for such rocks. They weren't there, at least not in any great number. Primeval rocks were mostly pulverized by the intense meteoritic bombardment associated with the sweepup of the planetesimals (see Figure 6-5), as shown in Figure 10-3.

REGOLITH AND MEGAREGOLITH

Prior to the 1960s, many scientists and artists pictured the moon's surface as incredibly rough, craggy, and fissured. Yet the first close-up photos, obtained as Ranger 7 crashed into the moon in 1964, revealed a smooth, rolling topography, with only scattered rocks. The explanation came at about the same time. McCracken and Dubin (1964) pointed out that the amount of meteoritic material that has accumulated on mare surfaces in the last few billion years amounts to a layer a few centimeters deep. Since each meteorite dislodges several hundred times its own mass (Gault, Heitowit, and Moore, 1964), this means that the meteorites would have created a layer of rubble a few meters deep. Summarizing a 1963 conference on the lunar surface, Salisbury and Smalley (1964) wrote: "It is concluded that the lunar surface is covered with a layer of rubble of highly variable thickness and block size. The rubble in turn is mantled with a layer of highly porous dust." A more accurate description could hardly have been written, as seen in Figure 10-4.

Following the first Ranger photos, the continuous layer of rubble and fine dust came to be called the **regolith** (Greek for "rocky layer"). Shoemaker (1965) considered more carefully the statistics of postmare impact craters and the amount of debris thrown out by each one and confirmed depth estimates of about 10 m for the regolith in the maria.

The first lunar astronauts remarked on the dust kicked up by their rocket exhaust during their landings and by themselves during their hikes. Their photos of a valley hillside (Figure 10-4b) show a rare example where the regolith and underlying rock beds can be glimpsed in cross section. The great amount of fine dust shows that most surface rocks have been pulverized by impacts, but the scattered boulders and subsurface rock layers show that bigger impacts penetrate into rock layers and throw out intact rocks. Apollo data show that in the lunar maria, the regolith is typically about 2 to 10 m deep, grading into broken or coherent lava flow layers below.

Table 10-3
Compositions of Planetary Rocks (percent weight)

Rock	Oxide										Primary reference
	SiO_2	Total iron[a]	MgO	Al_2O_3	CaO	Na_2O	TiO_2	K_2O	H_2O	Total	
Primitive Materials											
Carbonaceous chondrites	28	27	19	2	2	1	0	0	13	92	Mason (1962), p. 74
Enstatite chondrites	38	32	22	2	1	1	0	0	0	96	Mason (1962), p. 74
Hypersthene chondrites	40	27	25	2	2	1	0	0	0	97	Mason (1962), p. 74
Ultrabasic (Mantlelike) Igneous Rocks											
⊕, deep-source nodules	43	9	43	2	1	0	0	0	0	98	Wyllie (1971), p. 112
⊕, dunitic peridotite, Switzerland	42	8	42	2	2	0	0	0	4	100	Rittmann (1962), p. 105
⊕, mantle (est.)	44	8	40	3	2	0	0	0	?	100	Wyllie (1971), p. 114
☾, bulk silicate portion	43	9	28	11	9	0	0	0	0	100	Warren and Wasson (1979), p. 2054

"Crustal" Igneous Rocks											
☾, breccias, Descartes highlands	44	3	3	30	17	1	0	0	0	98	Taylor (1975), p. 216
☾, anorthosites, highlands	44	1	1	35	19	1	0	0	0	98	Taylor (1975), p. 234
☾, gabbroic anorthosites, highlands	44	3	3	31	17	0	0	0	0	98	Taylor (1975), p. 234
☾, highland rocks and soils (avg.)	45	6	7	25	15	0	1	0	0	99	Taylor (1975), p. 81
⊕, oceanic crust (avg.)	48	10	7	16	12	3	2	1	1	100	Wyllie (1971), p. 154
⊕, crustal composition (avg.)	57	8	5	16	8	3	1	2	1	101	Wyllie (1971), pp. 153, 155
Basaltic Materials											
☾, high titanium mare basalts, Tranquillitatis, and Serenitatis	39	20	8	9	10	0	12	0	0	98	Taylor (1975), p. 136
☾, mare basalts, Imbrium, Procellarium	45	20	9	9	10	0	3	0	0	96	Taylor (1975), p. 136
☾, KREEP basalt, highlands	49	8	10	18	12	0	1	1	0	99	Taylor (1975), p. 228
♂, Chryse soil[b]	45	18	8	6	6	0	1	0	2?	86	Toulmin and others (1977), p. 4629
♂, Utopia soil[b]	43	20	?	?	5	0	1	0	2?	71	Toulmin and others (1977), p. 4629
⊕, olivine basalt, Mauna Loa, Hawaii	46	13	24	6	6	2	2	0	1	100	Rittmann (1962), p. 105
⊕, basalt, Kilauea volcano, Hawaii	50	11	7	14	12	2	3	1	0	100	Rittmann (1962), p. 105
⊕, plateau flood basalts (avg.)	49	13	7	14	9	3	2	1	2	100	Rittmann (1962), p. 105
⊕, oceanic basalt layer (avg.)	50	9	7	17	12	3	2	0	1	101	Wyllie (1971), p. 153
⊕, continental basalt layer (avg.)	58	8	4	16	6	3	1	3	1	100	Wyllie (1971), p. 153
Siliceous Igneous Rocks											
⊕, continental granitic layer	64	5	2	15	4	3	1	3	2	99	Wyllie (1971), p. 153
⊕, dacite, Krakatoa 1883 explosion	68	4	1	15	3	6	1	2	1	101	Rittmann (1962), p. 105
⊕, rhyolite, New Zealand	72	2	0	12	1	3	0	4	4	98	Rittmann (1962), p. 105
⊕, rhyolite, Sicily	75	1	0	13	1	4	0	5	1	100	Rittmann (1962), p. 105
Nonigneous Rocks											
♂, duricrust pebbles[c]	44	18	9	6	5	0	1	0	3	86	Toulmin and others (1977), p. 4629
⊕, sediments, all types (avg.)	58	5	3	13	6	1	1	3	3	93	Mason (1958), p. 147
⊕, shales (avg.)	58	6	2	15	3	1	1	3	5	94	Mason (1958), p. 147
⊕, sandstones (avg.)	78	1	1	5	6	0	0	1	2	94	Mason (1958), p. 147
⊕, limestone (avg.)	5	1	8	1	43	0	0	0	1	59	Mason (1958), p. 147
Tektite											
Australite (avg.)	73	5	2	12	4	1	1	2	0	100	Taylor (1975), p. 81

[a]Iron may exist in various states of reduction and oxidation. Primarily, these are Fe, FeO, Fe_2O_3, and FeS. For example, iron is all in the form of FeO in carbonaceous chondrites, and mostly in the form of Fe metal in enstatite chondrites. In the basaltic samples, it is almost always FeO, except in the Martian soil samples, where it is more oxidized and listed as Fe_2O_3.

[b]The Martian soil is believed to be derived mainly from weathered basaltic material similar to the boulders scattered at the two Viking landing sites. It has an additional component of weathering products, such as 7% SO_3 and 1% Cl.

[c]These are fragments of crumbly material believed to be evaporites. They have a high content of sulfur and other weathering products: 10% SO_3 and 1% Cl.

[d]Limestone has a very high carbon content, listed by Mason as 42% CO_2. The terrestrial sediments listed above this entry also have 3% to 5% CO_2.

Figure 10-3. Meteorite impacts continually churned the early surfaces of planets during the sweepup of planetesimals prior to about 4 Gy ago. This view shows the impact of a meteorite fragment after its breakup in a thin atmosphere. (Painting by Ron Miller)

The regolith is quite variable, and depths as much as 36 m were estimated at some points near the Apollo 17 site in Mare Serenitatis (Taylor, 1975; Short, 1975).

The term regolith, originally applied to the lunar maria, has been extended to dust layers created on other planetary surfaces by meteoritic bombardment.

The ancient lunar uplands are much more heavily cratered than the lunar maria. Hence, they are believed to contain a much deeper regolith than the maria. Short and Forman (1972) and Hartmann (1973) independently used crater statistics and other data to estimate that the densely cratered highlands contain a roughly 2-km layer of pulverized material, which was termed the **megaregolith**. Still earlier craters, now obliterated by later craters, may have created even greater amounts of debris, possibly tens of kilometers deep. However, the deeper layers of debris were probably bonded into coherent breccia rock, rather than remaining in the form of loose regolith.

PRIMEVAL LITHOSPHERIC EVOLUTION: ROCK FORMATION VERSUS ROCK DESTRUCTION

Some consequences of early intense cratering for the moon's primeval lithosphere structure can be seen in Figure 10-5, which shows depth cross sections through time. The curves involved are based on quantitative calculations, but are model dependent (Solomon and Longhi, 1977; Herbert and others, 1977; Hartmann, 1980a). Thus the diagram should be considered a semischematic view of the evolution of early lunar and planetary surface layers. The solid curve LS shows the rate of lithosphere solidification due to cooling of the magma ocean—thus it represents the tendency to form coherent igneous rock. The dashed curve MG shows the amount of megaregolith generation produced by cratering in any interval of about 100 My—thus it represents the tendency to destroy coherent igneous rock by grinding it into dust and debris.

a

b

Figure 10-4. Views of the lunar regolith. (**a**) *Regolith surface texture. The Surveyor spacecraft bounced before coming to rest, leaving a footprint in the flourlike soil.* (**b**) *Regolith cross section in the wall of Hadley Rille. Among the loose boulders on the slope, layered outcrops of rock can be seen some tens of meters below the surface. These layers are probably intact portions of lava successive, bedded lava flows. (NASA)*

These two processes, magma solidification and cratering, competed. Initially (the first 10^8 y or so) cratering was so intense that impacts frequently broke through any primitive lithosphere, splashing magma about and completely pulverizing the earliest rock materials. Eventually (at about 4.3 Gy on Figure 10-5) curves LS and MG crossed so that rapid pulverization no longer occurred at the level where new rock was forming. Therefore a true lithosphere, intact except for the sites of the deepest late impacts, began to evolve at that depth, overlain by megaregolith. The rubble just above this level, which had been rapidly churned before, was now less disturbed because the cratering rate had declined, and impacts rarely penetrated to this depth. With its long exposure to moderate pressure from the overlying material and heat from below, this material may have welded into the breccias so commonly found in the uplands. Note that any lava flows erupting at this time (such as the one indicated at about 4.12 Gy) would have been rap-

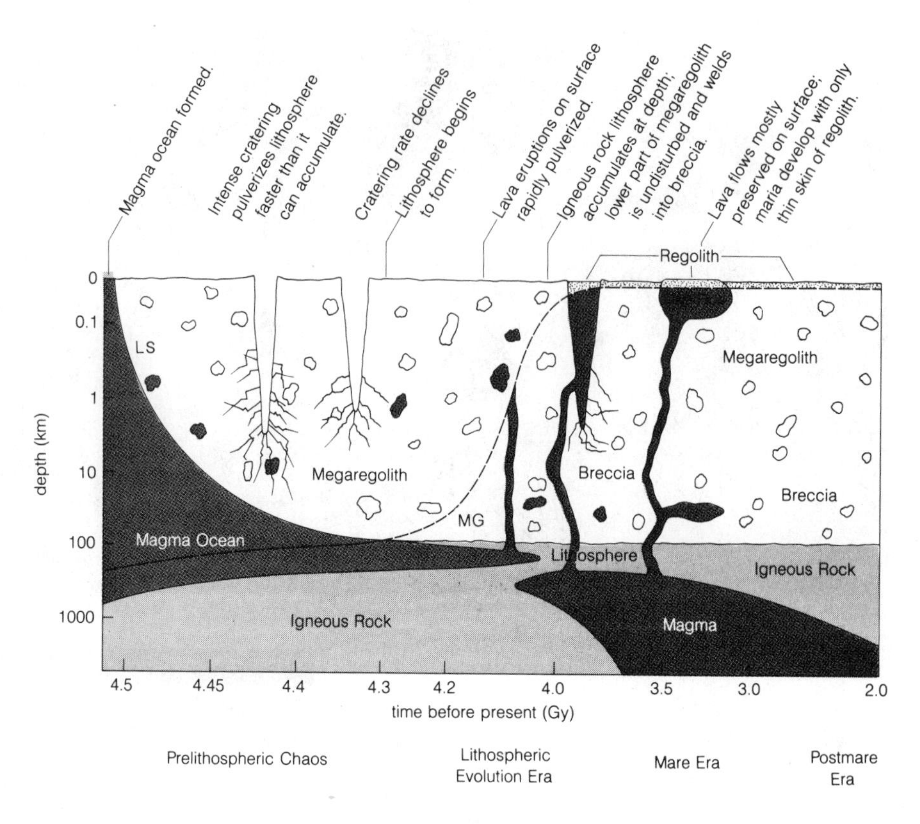

Figure 10-5. Schematic cross section of the moon's surface layers, showing evolution during early lunar history. See text for further description.

idly eroded by cratering. A new era began later, around 3.9 or 4 Gy ago, when lava flows such as the 20- to 50-m-thick flows measured in Mare Imbrium could partially survive, with only a thin layer of regolith on the top. The lava flows of this era became the familiar lunar maria.

This scenario of early evolution divides lunar history into four phases. In the first phase, marked **prelithospheric chaos** in Figure 10-5, no substantial layers of igneous lithosphere formed above the magma ocean. In the second phase, the **lithospheric evolution era** (perhaps 4.3 to 4.0 Gy ago), intact lithosphere formed at depth but no flows were well preserved on the surface. In the **mare era** (4.0 to 3.0 Gy), flows erupted from the deep-seated magma and surface structures were preserved. The moon, being small, cooled quickly enough so that much of its history involved a fourth phase, the **postmare era**, when eruptions had virtually ceased.

If some other planets lacked magma oceans initially, cratering would have simply pulverized their surface layers until a decline in cratering allowed preservation of surface features and undisturbed welding of lower layers into coherent breccias. Also, the larger planets maintained hot mantles and eruptive activity longer, so that their "postmare" periods would have been shorter or nonexistent. Various cratering rates in different parts of the solar system competed with various thermal histories of different-sized planets and led to different patterns of surface evolution.

It is clear from Figure 10-5 that cratering is an exceedingly important effect in the development of planetary surfaces. On some small, rapidly cooling planets, eruptive activity never occurred, and the surface landscapes were entirely produced by cratering effects. Due to the importance of cratering, most of the rest of this chapter will be devoted to consideration of cratering effects.

DISCOVERY OF METEORITE IMPACT CRATERS

Galileo first turned his telescope to the moon on November 30, 1609, and within a few months he announced that circular pits pockmarked the moon's surface (Whitaker, 1978). Galileo used the Greek word **crater** ("cup") to describe these. Until the middle of this century, a debate raged about the origin of craters. Some scientists thought they were impact features, but others thought that they were volcanic or formed by giant gas bubbles that rose through a molten primeval moon and broke on the surface.

Not until this century were meteorite craters, complete with meteorite fragments, firmly identified on Earth. Further geological fieldwork revealed terrestrial **astroblemes** (from Greek words for "star wounds"), or eroded circular structures that have turned out to be remnants of ancient, eroded meteorite craters (see Table 10-4 and Figure 6-1). These results favored the meteorite theory of planetary craters. Further evidence came when astrophysicist Ralph Baldwin (1949) showed that the properties of lunar craters matched those expected for impact explosions. Next, fieldwork and space probes of the 1960s revealed compelling similarities between known meteorite craters on Earth (Table 10-4) and the ubiquitous craters on other planetary bodies. These similarities included hummocky rim forms and patterns of ejected rubble. Furthermore, fresh craters on various planets have size distributions consistent with size distributions of the interplanetary meteoroids, as noted in Figure 6-4. So the majority of craters larger than a few kilometers across on planetary bodies are now believed to have been formed directly by impact of interplanetary bodies.*

*Unfortunately, astronauts did not investigate lunar craters in enough detail to recover meteorite specimens or specifically prove their meteoritic origin. Also, they did not explore any multikilometer-scale fresh craters.

Craters are interesting in several regards. They are fascinating as weird landscapes and scenes of ancient cataclysms.* Their structures reveal subsurface properties. And their numbers reveal the age of the surface, since the more that craters have accumulated, the longer the surface has been exposed.

MECHANICS OF IMPACT CRATER FORMATION

Due to the combination of orbital speed and planetary gravity, meteoroids typically strike planets at 10 or more kilometers per second. If the planet has an atmosphere, the smaller meteoroids are slowed and do not form craters.

The kinetic energy of high-speed meteorites is converted upon impact into thermal, acoustic, and mechanical energy that distorts, fractures, and ejects rocks. The result is like an explosion centered a few meteorite diameters below the ground, and meteorite impact craters are thus somewhat like bomb craters. When the meteorite enters the ground, it is usually **hypersonic** (that is, moving faster than the local speed of sound), since seismic waves (sound waves in rock or soil) typically move at 1 to 4 km/s. Thus, the rock materials cannot dissipate the impact energy by seismic waves until the meteorite has penetrated and slowed. A **shock wave**, or highly compressed zone in front of a supersonic body, carries a high density of energy and matter and builds up around the impact point, like a shock wave around the front of a supersonic aircraft. The explosion is the spreading of this shock wave, which compresses the rock and initially makes

*A meteoriticist friend tells of visiting Meteor Crater, Arizona. As he stood on the rim, looking at the contorted strata, he imagined the thunderous explosion of the impact, the fiery burst of ejecta shooting upward, and the shock wave racing out across northern Arizona, devastating life throughout the region in a matter of minutes. His mood was broken when a woman approached and remarked, "Oh, my! It's nowhere near as big as the Grand Canyon!"

it deform almost like a fluid around the impact site. The rock bends backward, upward, and outward, excavating a volume of material much larger than the meteorite itself. A relation exists between the kinetic energy of the meteorite and the size of the crater (Figure 10-6), though the crater size also depends to a lesser extent on the nature of the surface materials. Since the explosion center

Table 10-4
Selected Meteorite Impact Craters on Earth

Name	Location	Est. original diameter (km)	Est. age (My)
Sudbury	Ontario	140	1840 ± 150
Vredefort Ring	South Africa	140	1970 ± 100
Popigai	Russia	100	39 ± 9
Lake Manicouagan	Quebec	100	212 ± 2
Acraman	Australia	90 (or 160?)	~600
Puchezh-Katunki	Russia	80	183 ± 3
Siljan	Sweden	52	368 ± 1
Kara	Russia	50	57 ± 9
Charlevoix	Quebec	46	360 ± 25
Araguainha Dome	Brazil	40	<250
Carswell Lake	Saskatchewan	37	117 ± 8
Clearwater Lakes	Quebec	32, 22[a]	290 ± 20
Manson	Iowa	32	< 70
Slate Island	Ontario	30	350
Lake Mistassini	Labrador	28	38 ± 4
Teague	Australia	28	1685 ± 5
Rieskessel	Germany	24	14.8 ± 0.7
Gosses Bluff	Australia	22	142 ± 5
Wells Creek	Tennessee	14	200 ± 100
Sierra Madera	Texas	13	<100
Deep Bay	Saskatchewan	12	100 ± 50
Bosumtwi	Ghana	10.5	1.3 ± 0.2
Kentland	Indiana	9	~300
Serpent Mound	Ohio	6.4	~300
Decaturville	Missouri	5.6	320
Crooked Creek	Missouri	5.6	320 ± 80
Brent	Ontario	3.8	450 ± 30
Flynn Creek	Tennessee	3.6	360 ± 20
Steinheim	Germany	3.4	14.8 ± 0.7
New Quebec	Quebec	3.2	<5
Meteor Crater[b]	Arizona	1.1	0.05

[a]Two craters lie almost tangent to each other. They may indicate fracture of a meteorite into two pieces in the atmosphere or impact by a body and its satellite.
[b]Meteorites are known at this location and at many lesser sites in various locations, ranging down to clusters of pits caused as a meteor broke up in the atmosphere. The smallest example is an 11-m single pit at Haviland, Kansas.
Source: Based mainly on Grieve, R., and four others (1988): "An Astronaut's Guide to Terrestrial Impact Craters" (Houston: LPI Tech. Report 88–03).
Note: Table lists probable meteorite impact craters, with known evidence of mineral alteration due to shock wave metamorphism. List is complete down to 28-km diameter, with selected smaller examples.

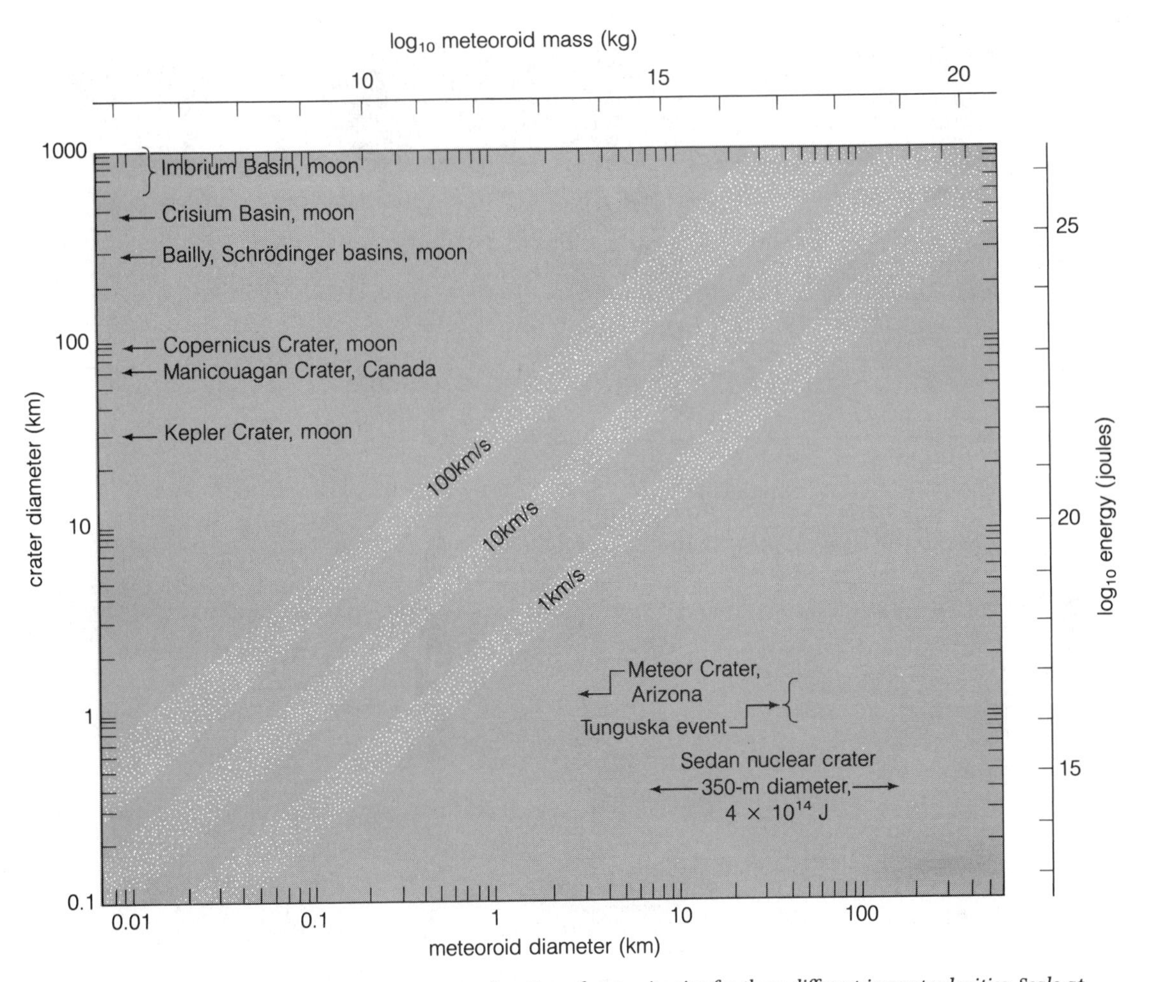

Figure 10-6. The diameter of an impact crater as a function of meteorite size for three different impact velocities. Scale at right gives the estimated energy required to make each crater. Documented craters on Earth and the moon are shown. The assumed meteorite density is 3 g/cm^3. (Crater energy and diameter data from Baldwin, 1963; Wasson, 1974, p. 145; Vortman, 1977)

is below the ground, strata that initially lay flat are heaved upward and outward, bent back, and even overturned in big flaps like petals of a giant flower opening. The rim is built partly from this upthrust rock and partly from excavated debris dumped on the crater edge.

Features of Impact Craters

Craters formed by meteorite impacts are often called **primary impact craters**. Figure 10-7 shows some common features. The rim has a hummocky structure grading outward into a thinner layer of debris. All this externally deposited material is called the **ejecta blanket**. Discrete blocks or clumps crashing down at high enough speed form **secondary impact craters**. Powdered and melted material resolidified as glassy beads is thrown out at very high speed (as much as 1 km/s or more) and leaves long, bright, linear deposits called **rays**. The rays radiate from the primary crater, often with secondary craters clustered along them, as shown in Figure 10-8b. The famous ray system of the 90-km crater Tycho

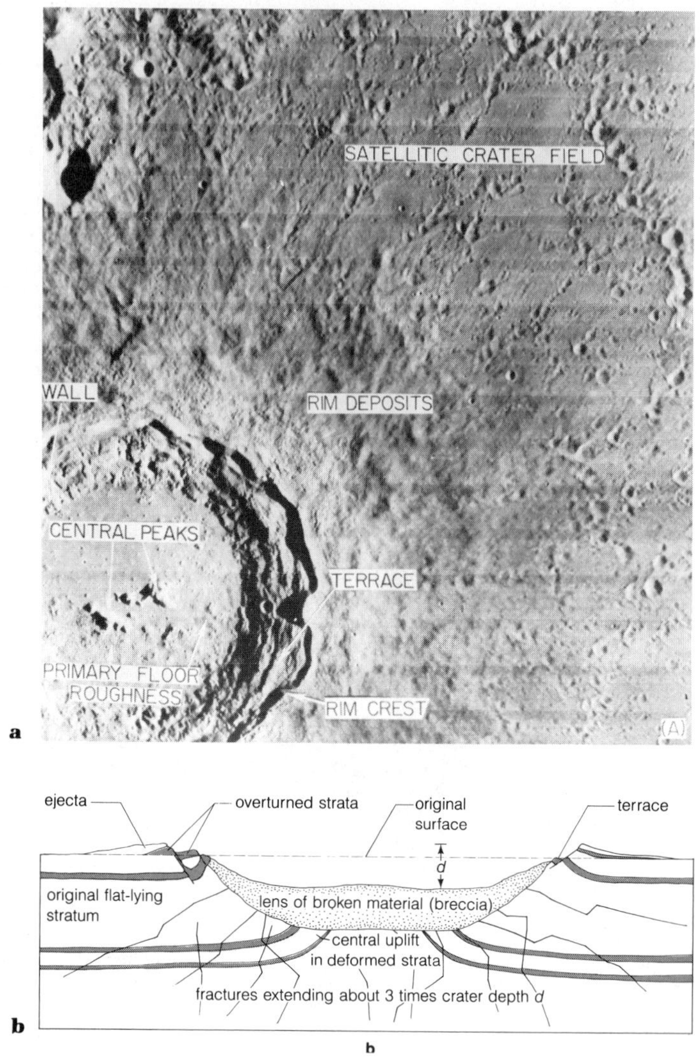

Figure 10-7. Features of impact craters. (**a**) *Vertical view of the relatively fresh 90-km lunar crater Copernicus. (Courtesy James Head, Brown University)* (**b**) *Cross section of a typical large crater showing additional features (see text).*

stretches more than 1000 km over much of the moon, as seen in Figure 10-8c.

Drilling reveals other features. The floor is typically a lens-shaped mass of breccia, rubble, and small amounts of lava produced by melting during the impact. This is ejecta that has fallen back into the crater's original cavity. Below this is highly fractured **bedrock** (rock not moved out of its original position). The fractures typically penetrate about 3 times deeper than the depth of the crater itself.

Some idea of the origin of crater features can be gleaned from Figure 10-9, showing the explosion of 91 t of TNT in a cratering experiment. The turbulent cloud around the base of the fireball is expanding, soil-laden gas called a **base surge**, which deposits some of the hummocky, dunelike ejecta around the crater rim. The high-speed jets angling upward may be analogous to the spurts of material that created ray systems.

Visibility of crater features from above depends strongly on the lighting angle, as shown by Figure 10-10. Rays are prominent under high light but disappear under low light. But low light brings

a

b

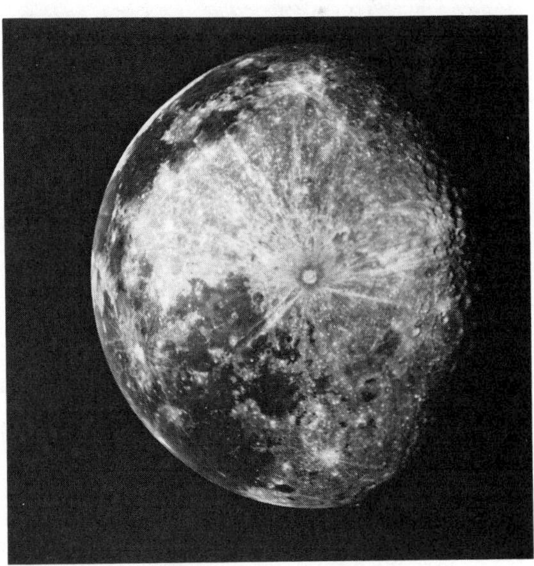

c

Figure 10-8. Features of impact craters. (**a**) *View along inner wall of Meteor Crater, Arizona, showing upended strata. Original beds (as found nearby) lay flat. (Photo by author)* (**b**) *View toward the lunar crater Copernicus (see Figure 10-7a) on the horizon. Ejecta has been thrown over the Carpathian Mountains (the Imbrium rim, middle distance), littering mare surface with secondary craters and bright rays. (NASA; National Space Science Data Center)* (**c**) *View of the moon, centered directly above the young crater Tycho, showing the Tycho ray system. Note many rays are tangent, not radial, and begin outside a dark nimbus around the crater's rim. This photo was made by projecting an Earth-based photo on a globe and rephotographing the projected image from above Tycho. (Lunar and Planetary Laboratory, University of Arizona)*

out relief and allows rim and mountain heights to be measured from the lengths of shadows.

Surface views of modest-sized craters dramatically reveal many features, as shown by Figure 10-11. Hummocky rims and scattered rock fragments are prominent in young impact craters but become muted with time as material is deposited or eroded. Atmospheres inhibit the formation of small craters. For example, Martian craters smaller than about 50 m are absent due to atmospheric breakup of meteoroids as well as erosion (Binder and others, 1977).

Simple Craters, Complex Craters, and Multiring Basins

The structural features of craters change with increasing size. Small craters (about 1 km across) tend to have smooth, bowl-shaped interiors and are called **simple craters**. At larger sizes, the floor flattens. At still larger sizes, the floor develops a **central peak**, or mountain mass, such as shown in Figure 10-12. Central peaks are probably formed by a rebound phenomenon like the rebound of a droplet in a coffee cup. Terraces may also appear on the inner walls, apparently due to slumping of the rim inward. Craters with such features are called **complex craters**.

Figure 10-9. Test explosion of 91 metric tons of TNT illustrates features of crater formation (see text). This 1972 test in Colorado produced a crater 39 m across and 7 m deep with secondary impact craters up to 110 m away and ejecta as much as 201 m away. (Photo by author)

The transition from simple to complex occurs at smaller sizes on larger planets. For example, lunar central peaks are most common in young craters larger than 60 km in diameter; Martian central peaks, above 10 to 30 km; and Earth's, above 1 to 3 km. The difference probably relates to the wall height that can be sustained without slumping, given the gravity of the planet.

Although classic lunar central peaks occur only in large craters, some lunar craters as small as 100 m have rough central mounds (Figure 10-13) or terraces (Figure 10-14). This is caused by shallow layers of resistant rock, which disturb the otherwise smooth bowl shape. Experimental impact craters in layered targets show similar effects.

Still other features occur at larger crater diameters. For example, at about 100 to 300 km on the moon, Mars, and Mercury, a rare transitional shape develops, in which the central peak broadens and turns into a ring of hills, or **peak ring** (Hartmann and Wood, 1971), as shown in Figure 10-15.

Finally, the largest impact features are huge systems of concentric rings, called **multiring basins** (or just **basins**). Figures 10-16 and 10-17

Figure 10-10. Three views of the young, 35-km lunar crater Timocharis under different lighting. Low lighting exaggerates relief, while high lighting brings out ray material and bright crater walls. (NASA, National Space Science Data Center)

a b

Figure 10-11. Surface views of impact craters. (**a**) *North Ray crater, near Apollo 16 site in the lunar uplands. Crater is about 900 m across and about 50 My old according to rock sample dates. (NASA)* (**b**) *Interior of Meteor Crater, Arizona, about 1100 m across and 20 000 y old. (Wide-angle photo by author)* (**c**) *Martian crater rim on the horizon, about 2.5 km southwest of Viking 1 site. Crater diameter is about 400 m. (NASA)*

c

show spectacular examples on three planets. Some of the rings, especially the inner ones, may be only roughly defined circles of hills, sometimes partly flooded by lavas that have covered the basin's inner floor. Other rings, often outer rings such as the Appennine arc around the lunar Imbrium basin, have a well-defined crest resembling the rim of an ordinary crater. These outer rimlike rings may be the true rim of the original impact crater, with other rings being rebound or slump features. Curiously, the rings are often distinctly spaced at intervals of about $\sqrt{2}$ times the inner ring radius (that is, 1.0, 1.4, 2, 2.8 . . . radii from the center). Multiring basins are the largest individual geological structures in the solar system. Some examples on the moon had been recognized before Apollo. Concentric, multiring lunar structures were described by Baldwin (1949, 1963), but they were more clearly recognized as a class after a number of others were discovered by "rectified" photography (projecting ordinary lunar photos on a globe and then photographing selected regions from "overhead." The great Orientale system (Figure 10-16) was discovered in

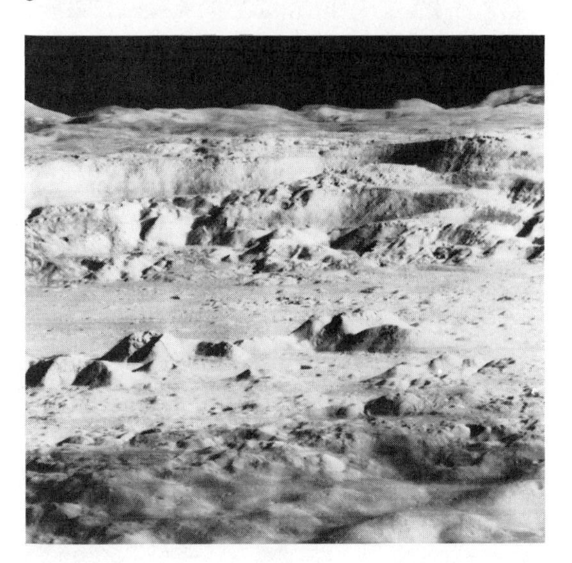

Figure 10-12. Oblique view of the central peak complex on the floor of the lunar crater Copernicus (compare Figure 10–7a). The near rim of Copernicus is at bottom foreground, the peaks in lower center, and the far rim just below the horizon. Boulders and outcrops appear to be present on the peaks. (NASA, Viking Orbiter 2)

a

b

Figure 10-13. Central mounds in two small craters. (**a**) *Terrestrial test explosion crater about 30 m across, formed by 109 t of explosive. Central mound may be related to a resistant caliche layer below dusty sediments in which the crater was formed. (Miser's Bluff, Arizona; photo by author)* (**b**) *Lunar crater adjacent to Apollo 11 site. Central mound may be related to a resistant lava layer below surface regolith. (NASA)*

this way by Hartmann and Kuiper (1962) prior to lunar mapping by spacecraft.

Discovery of additional multiring basin systems on Mercury, Mars, Callisto, and Ganymede spurred interest in understanding their formative processes and the roles of giant impacts in establishing crustal heterogeneity on the planets. They appear to provide fracture systems allowing lava to gain surface access (Hartmann and Wood, 1971), and their ring spacings may be indicators of subsurface layering (Wilhelms, Hodges, and Pike, 1977). In addition, they are the centers of vast systems of radiating valleys and ridges that suggest profound fractures radiating from the impact sites (Figure 10-18).

UTILIZING IMPACT CRATERS TO LEARN ABOUT PLANETS

Impact craters give several types of evidence about planets. They excavate and expose material. Their central peaks expose material originally about one-tenth of the crater diameter below the surface. For example, Pieters (1982) found that the central peak of the lunar crater Copernicus has a spectrum not seen in other lunar areas, and interpreted as olivine-rich, upthrust material from a 10 km deep crustal layer. As indicated in Figure 10-19, the structure of an ejecta blanket may indicate properties of the material at the impact site. Figure 10-19 shows one of many Martian ejecta blankets with lobate structure (lobe-shaped sheets extending from the crater). These are unknown on the moon and Mercury. Such craters have been termed **rampart craters**; they are attributed to impacts into soils containing large amounts of ice or water, which formed muddy ejecta flows with entrained gas (Carr and others,

Figure 10–14. (left) Lunar crater about 150 m across in Oceanus Procellarum shows terraced rim and ejecta blanket of boulders averaging about a meter in size. These features may indicate that the crater intruded a layer of intact lava below the regolith surface. (NASA, Orbiter 3)

a

*Figure 10-15. Craters with peak rings, a transitional form between central peaks and multiple concentric rings. (**a**) Lunar crater Shrödinger, about 320 km across. (**b**) Mercurian craters Ahmad Baba (top) and Strindberg (200-km diameter, bottom). (NASA)*

b

1977). Rampart craters are the most common craters in many areas of Mars, though other areas contain primarily the lunar type of crater, probably caused by impacts into drier soil.

On icy satellites such as Ganymede, Callisto, Mimas (Figure 5-7b), Tethys (Figure 5-7c), and Dione (Figure 10-20a) many of the larger craters (diameter $\gtrsim$ 40 km) show shallow profiles, domed floors sometimes showing central pits, and central peaks that (in a few cases) protrude above the crater rim, a situation unprecedented in the inner solar system. The original floors appear to have been pushed upward. These properties tell us something about the properties of the material in which the craters formed. The material is probably predominantly ice, possibly underlain by more fluid (watery) or rigid (rocky) layers. The surface ice layer would be more fluid than the surface rock layers of terrestrial planets, although the viscosity of H_2O ice at the temperatures of the satellites' subsurface layers is uncertain.

Parmentier and Head (1981) calculated profiles of craters in relaxing viscous ice layers and matched observed craters with models in which the craters formed in a deep ice layer whose viscosity was constant or increased with depth. Large craters (with their greater initial elevation differences) flatten quicker than small craters do, as measured in proportion to their initial depths.

a

b

Figure 10-16. Multiring basins on two planets. The systems nearly match, with outer rings being about 1300 km in diameter, or capable of stretching from the Great Lakes to the Gulf of Mexico. (**a**) *Orientale Basin, on the limb of the moon, discovered in 1962 by Earth-based photos. (NASA)* (**b**) *Caloris Basin on Mercury, discovered in 1974 by Mariner 10. (NASA)*

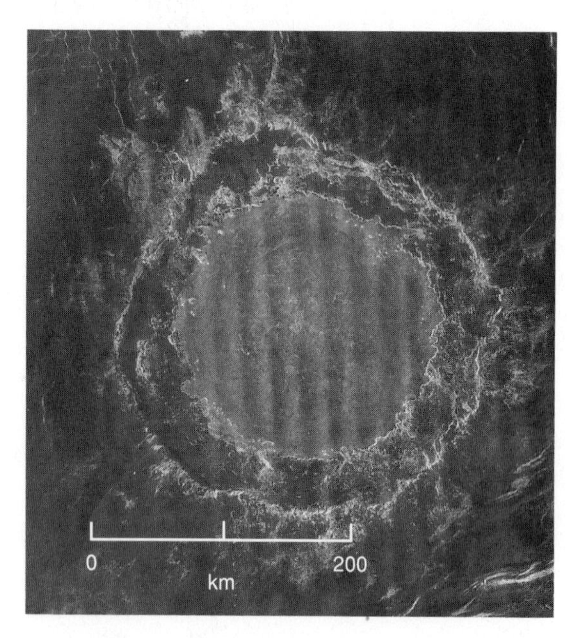

Figure 10-17. Radar mapping of Venus by the Magellan spacecraft in 1990–91 revealed the solar system's best examples of transitions between craters with peak-rings and full-fledged multiring basins. This crater, named Mead (after anthropologist Margaret Mead), is the largest impact feature discovered to date on Venus, and shows beautiful rings with the $\sqrt{2}$ *diameter ratio typical of multiring basins. (NASA Magellan radar image, Jet Propulsion Laboratory)*

Figure 10-18. Portion of radial pattern around the moon's Orientale Basin. Some larger striations may be faults, while the finer structure suggests a turbulent flow of material across the surface (upper left to lower right) as ejecta was deposited. Picture width, 275 km. (NASA, Orbiter 4)

Figure 10-19. Lobate ejecta around the 25-km Martian crater Arandas. This pattern, unique to Mars, is interpreted as due to impact into water- or ice-bearing soil. Patterned ground to the right is also possibly related to the freeze-thaw cycle of ice in the soil. (NASA, Viking 1)

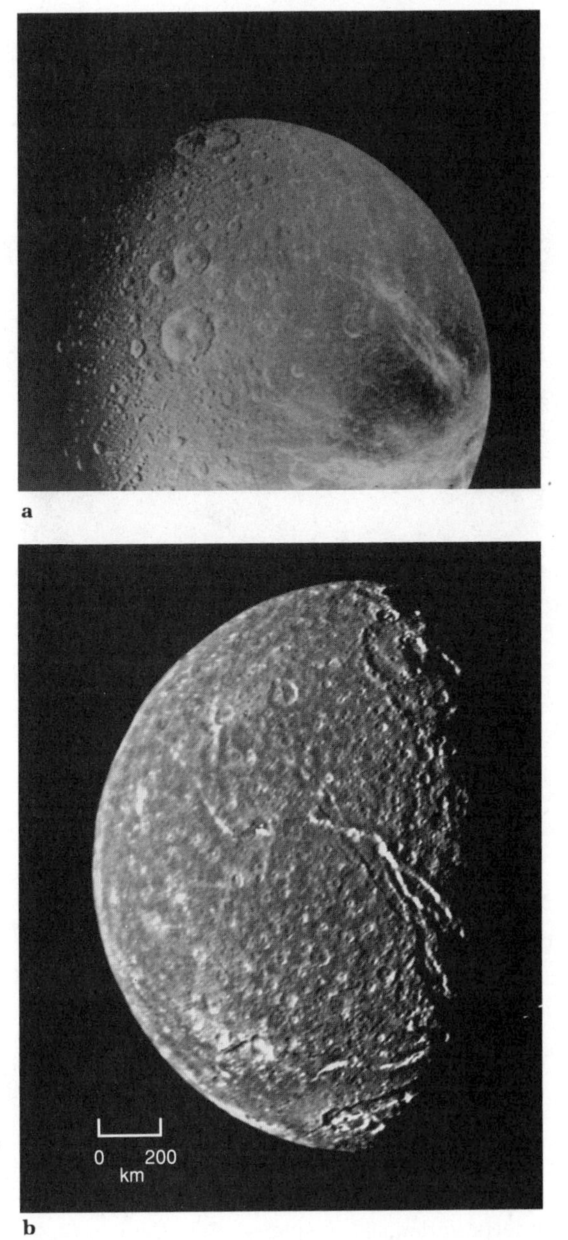

Figure 10-20. Examples of densely cratered surfaces on icy satellites. Some researchers find that broad regions, though crowded with craters, have fewer large craters than on the moon, possibly due to viscous relaxation. (**a**) *Saturn's 1120-km moon Dione. Origin of dark regions and bright swaths on the trailing hemisphere (right) is uncertain. North at top.* (**b**) *Uranus' 1580-km moon, Titania, with cratered areas cut by moon-wide fracture systems. (NASA Voyager 1* (**a**) *and 2* (**b**) *images, respectively)*

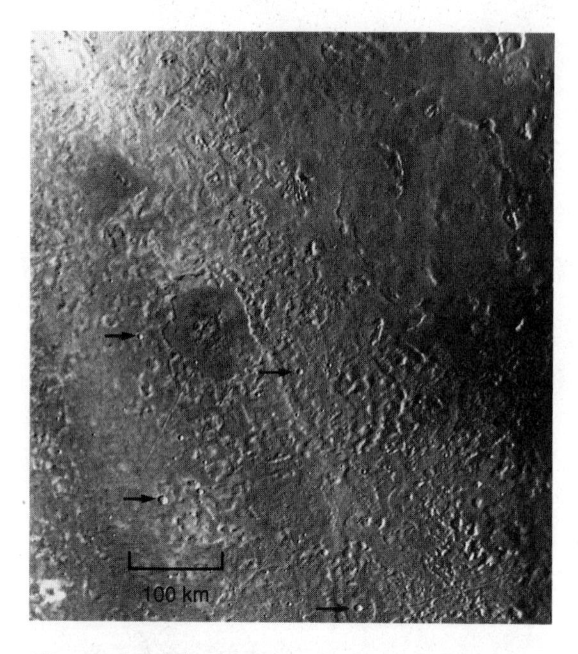

Figure 10-21. Lack of abundant impact craters is a sign of a youthful surface, recently resurfaced. This part of Neptune's icy satellite, Triton, shows rolling hills similar to "cantaloupe terrain" found in other areas. Very small impact craters (arrows) are sparsely scattered through this region—similar in number to those of lunar maria. In upper right are darker, smooth plains with still fewer impact craters. Narrow linear fractures and smooth "ice lake" (left center) can also be seen. (NASA Voyager 2 photo)

Craters 10 and 100 km across were found to relax in 30 Gy and 30 My, respectively. Thus, large craters can be subdued or obliterated while small ones survive. These data may explain why some heavily cratered regions of some satellites (Figure 10-20) seem to have strangely flattened intercrater plains and seem to lack large craters.

On the other hand, Voyager analysts (Smith and others, 1982; Strom 1987) have a different interpretation. They suggest that these remote worlds have been hit by meteoroids (comets? intersatellite debris?) with size distributions different from those that hit the moon. In this view, differences in crater size distributions reflect differences in impactors more than differences in obliteration processes. The problem deserves more study.

The most important use of craters has been to date planetary surfaces. On an erosion-free surface, such as a lava flow on an airless world, meteorite craters accumulate with time. Thus two types of dating are possible.

The simpler but less informative type of dating measures the **relative age** of geological provinces by merely measuring their crater densities (craters per square kilometer). On a given planet, a less-cratered province is younger than a more-cratered province. **Stratigraphic relationships**, that is, young formations overlapping older formations, assist in establishing the relative ages.

The more exact type of dating determines the **absolute age** in years. This can be estimated if the rate of formation of craters is known, as from Figure 6-4 and Table 6-1. Absolute ages allow comparison of events on different planets. Unfortunately, crater formation rates outside the Earth-moon system remain quite uncertain—errors in age may be a factor of ± 3. These rates could be improved by better knowledge of the numbers and lifetimes of asteroids and comets or by datable rock samples from certain geological provinces on other planets, which could then be used to calibrate the relationship between crater density and relative age for each planet.

Figure 10-22 shows examples of this technique, comparing observed numbers of craters with predicted numbers for surfaces of various ages on 11 planetary bodies. The small, airless worlds have retained the most craters of various sizes. Large, geologically active worlds have younger surfaces and fewer craters. Diagrams such as Figure 10-22 contain key data for understanding the evolution of planetary surfaces in the solar system. Such data are reviewed in further detail by Hartmann and others, 1980.

Stratigraphic Studies of the Earth-Moon System

Once we know the relative ages of features on a planet, we can construct a **stratigraphic column** for the planet, an imaginary vertical col-

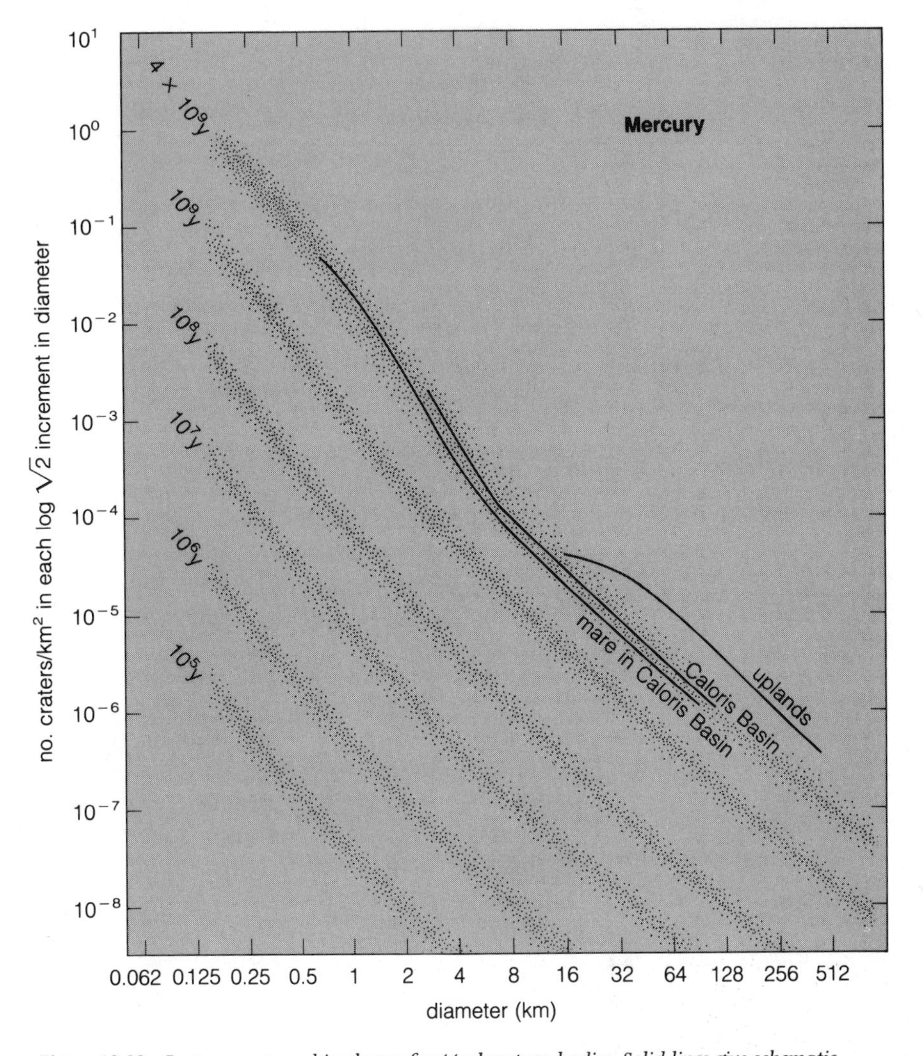

Figure 10-22a.

Figure 10-22. Crater counts and isochrons for 11 planetary bodies. Solid lines give schematic observed diameter distribution of craters. Dotted isochrons are estimates of numbers of primary impact craters expected for surfaces that have preserved all craters since the times of formation listed at left. Estimates are based on data such as Figure 6-4. Data show that Phobos, Deimos, the moon, Callisto, Ganymede, and Mercury are relatively primitive bodies with surfaces that are a few billion years old in most regions. Mars has many old surfaces but also some extensive younger volcanics, perhaps a few hundred million years old, and evidence of erosive loss of smaller craters in some regions. Venus (as mapped by radar; Phillips and others, 1991) and Earth have regions that have preserved large craters for about 800 My, but smaller craters have been eroded on Earth and perhaps on Venus. Venus' thick atmosphere prevents formation of craters <3 km in diameter. Europa's surface (based on the detection of three 20-km craters) is perhaps 0.1 to 2 My old. Io is being resurfaced, probably in less than 1 My, by volcanic eruptions. (Galilean data based on Smith and others, 1979)

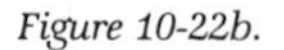

Figure 10-22b.

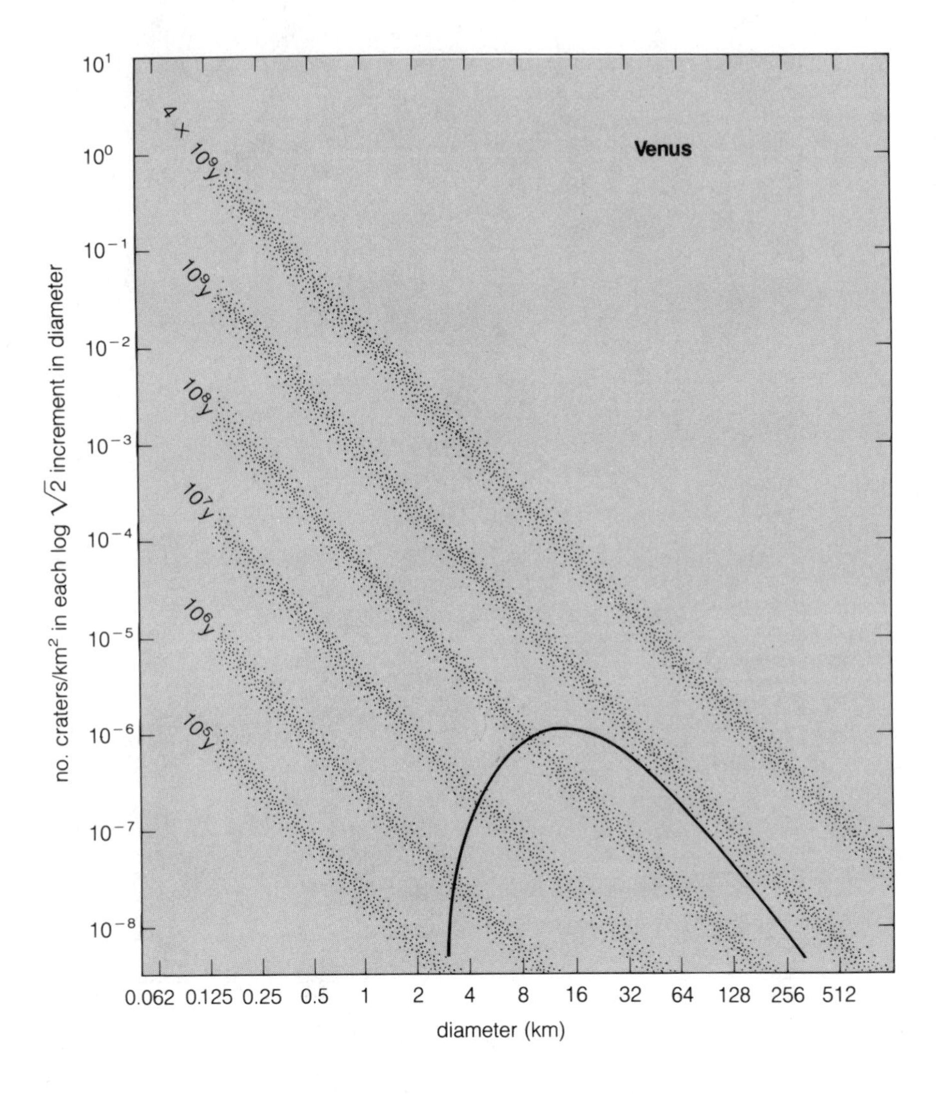

umn showing the sequence of events in different geological periods and resembling a cross section through an idealized portion of the planet's lithosphere.

Systematic work of this kind was begun when U.S. Geological Survey researchers Shoemaker and Hackman (1962) laid out the beginnings of a stratigraphic system for the moon. Just as terrestrial geologists developed a system of names such as Cambrian and Jurassic (often from locales) to designate periods whose absolute dates were uncertain but whose relative ages were established from fossils, Shoemaker and Hackman chose names such as Imbrian and Copernican (from lunar locales) to designate periods whose relative ages were established from crater counts and overlap relations.

In the 1960s, Geological Survey geologists preparing for the Apollo missions mapped much of the moon's stratigraphy in this way (Mutch, 1970). The Apollo missions then gave absolute calibration for certain geological provinces,

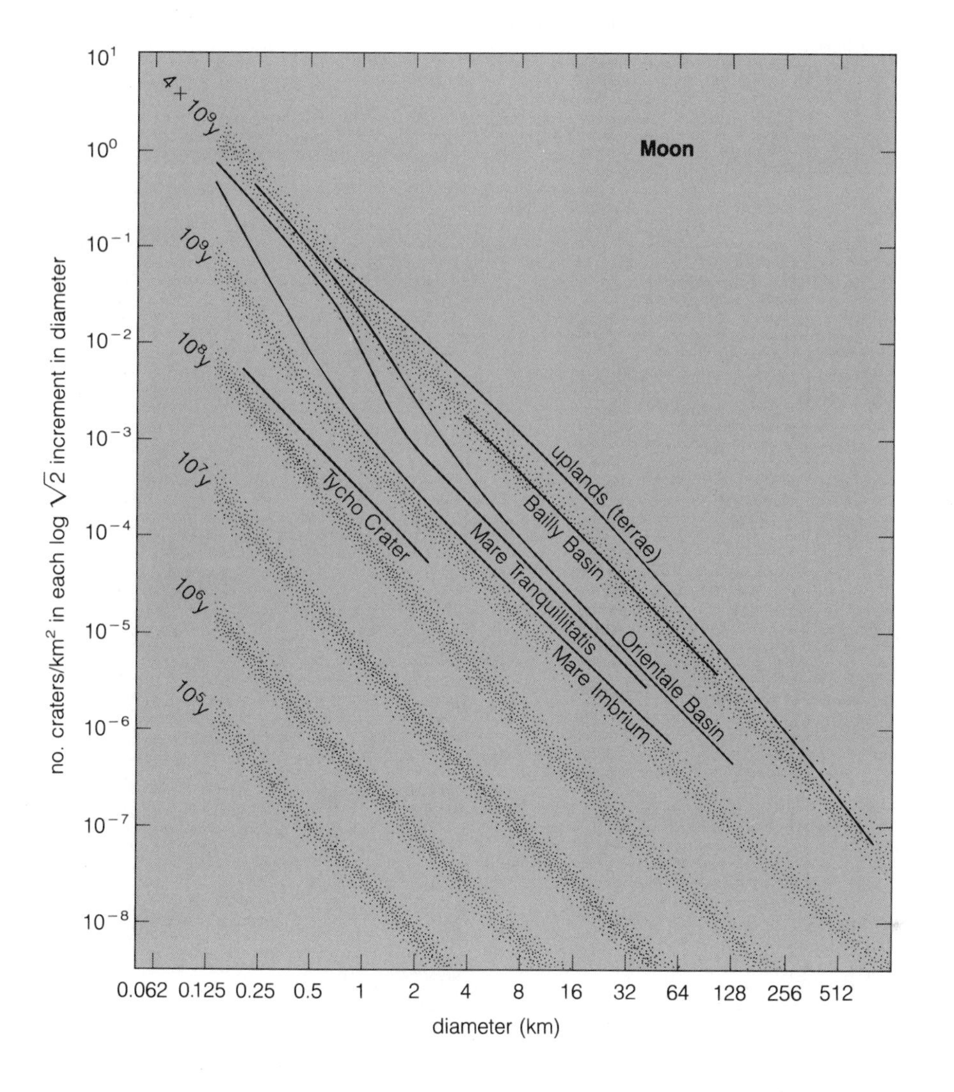

Figure 10-22c.

allowing the time scale based on lunar crater counts to be calibrated globally. Absolute dates can now be estimated for all major lunar features.

Table 10-5 compares the stratigraphic systems that have been built up for the Earth-moon system in this way. Earth's stratigraphic column has a hierarchy of divisions including broad eras and finer periods (and even finer subdivisions). Note that fine detail appears in the terrestrial system only in the last 10% of solar system history (the only interval with a good fossil record and abundant rock samples), while fine detail appears only for the earlier history of the moon. Though stratigraphic names such as Cambrian or Imbrian have been used to indicate dates of geologic provinces, advances in absolute radiometric dating of planetary rock samples have led to widespread quoting of numerical ages (for example, 3.63 Gy) in much of the lunar and general planetary literature. This trend seems likely to continue, since a single system of well-deter-

Figure 10-22d.

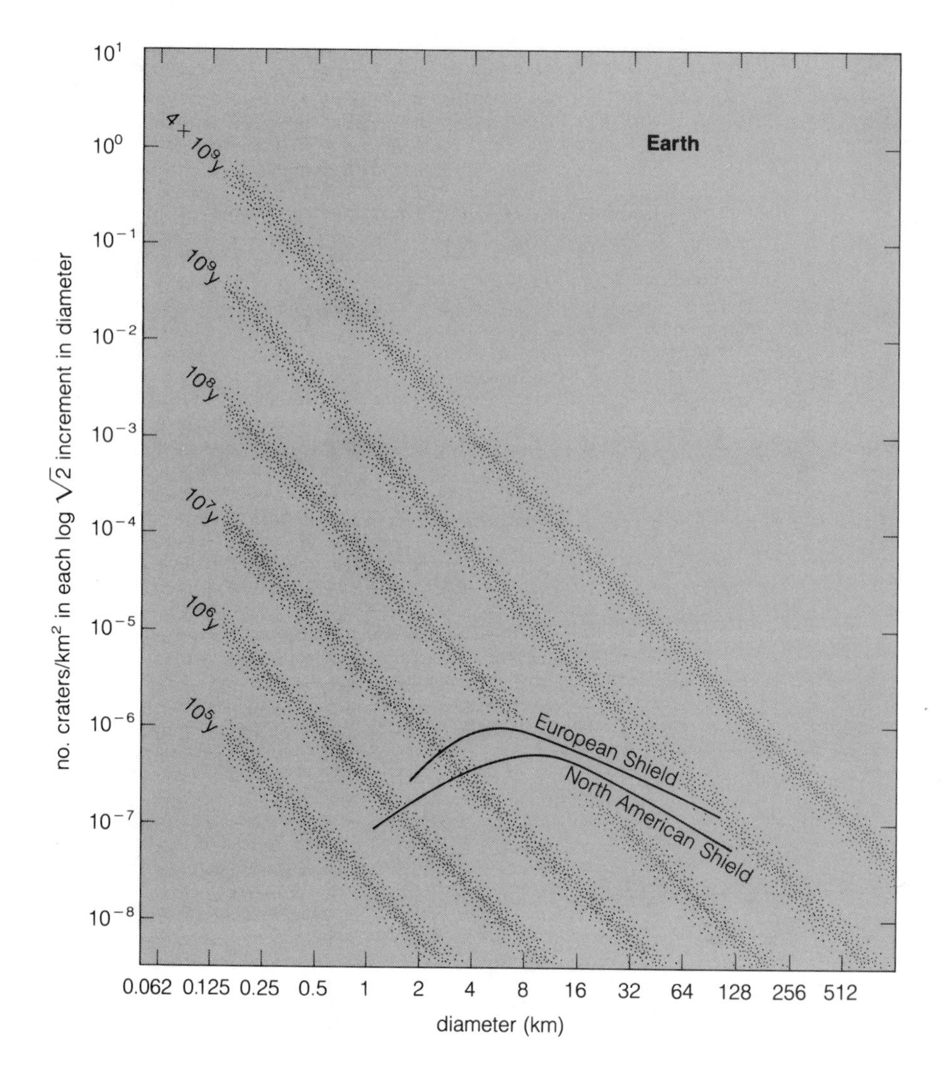

mined dates is much preferable to separate stratigraphic terminologies, one developed for each planet.

Crater Retention Ages

What age is actually measured when we count primary impact craters? If the surface is a sparsely cratered, thick lava flow on an airless world, the number of craters measures the time since the flow's origin. But if the flow is on a planet such as Mars, where dust blows into craters and other erosion occurs, small craters may last only (say) 1 My and large ones, 100 My. Then the number of craters of a certain diameter measures not the age of the surface but the length of time a structure with that dimension can withstand erosion. Thus, ages measured by crater counts have been termed **crater retention ages** (Hartmann, 1966). With care, these ages can be used not only to learn about ages of certain features but also to interpret rates of erosive activity (for example, Chapman and Jones, 1977).

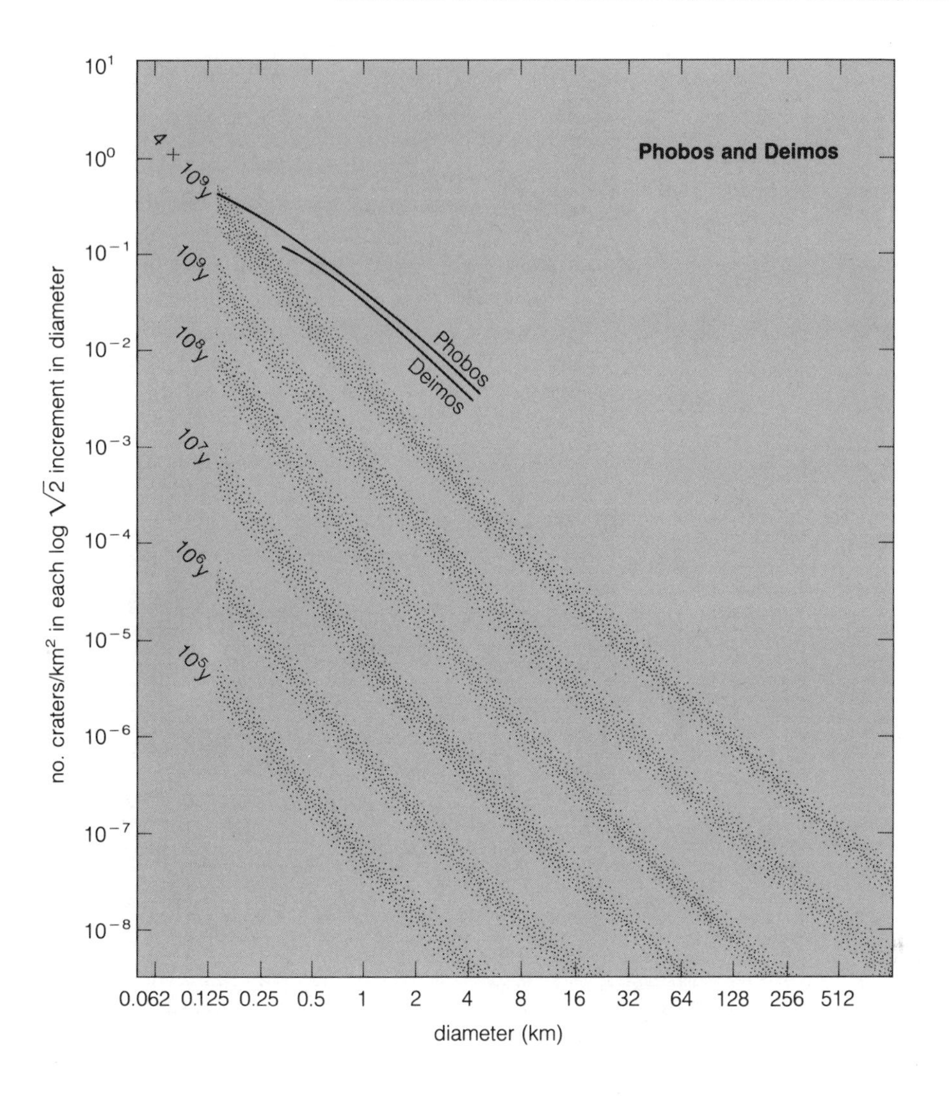

Figure 10-22e.

Steady State Cratering

Consider a fresh lava flow on a planet. As time passes, the surface will accumulate craters. Eventually a point is reached where craters are so crowded that new impacts destroy old craters as fast as they make new ones. The surface thus reaches a condition called **steady state cratering** (sometimes loosely called **crater saturation**), beyond which the observable crater density does not rise. Figure 10-23 shows the most profusely cratered surfaces in the solar system all have a similar density of cratering, roughly 30 times that found among the multikilometer craters on the lunar maria. The similarity of these curves suggests that these surfaces actually reach or approach a saturated steady state, though computer simulations (Woronow, 1978) have led to suggestions that crater densities could attain somewhat higher values.

A surface that is nearly saturated with craters must be ~4 Gy old, because only surfaces dating back to the era of early intense bombardment acquired enough hits to approach a steady state.

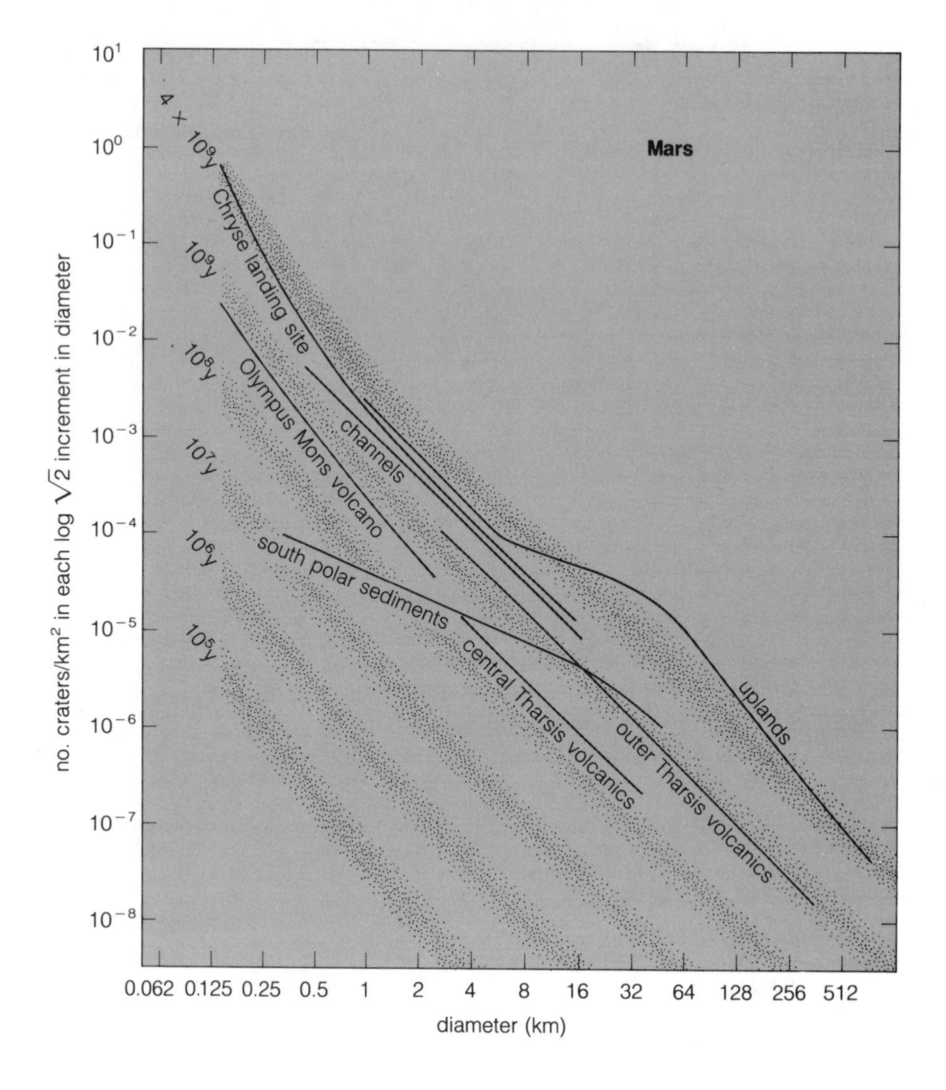

Figure 10-22f.

We can't tell from photos whether a given saturated surface has been exposed long enough to absorb enough hits barely to achieve saturation or long enough to absorb many times more hits. Surfaces anywhere in the solar system younger than 3.5 Gy should not approach the crater densities shown in Figure 10-23.

Steady state cratering is also important because it leads to the formation of deep megaregoliths on planets, as described in our previous discussion of megaregolith formation (see Figure 10-5). While cratering has covered only about 3% of the lunar maria with craters larger than 4 km and made a regolith in those areas about 10 m deep, cratering has covered about 100% of the highlands with craters larger than 4 km and thus stirred the surface to a depth of at least 1 or 2 km.

Problems with Small Craters

In order to get an adequate statistical sample of primary impact craters for dating purposes,

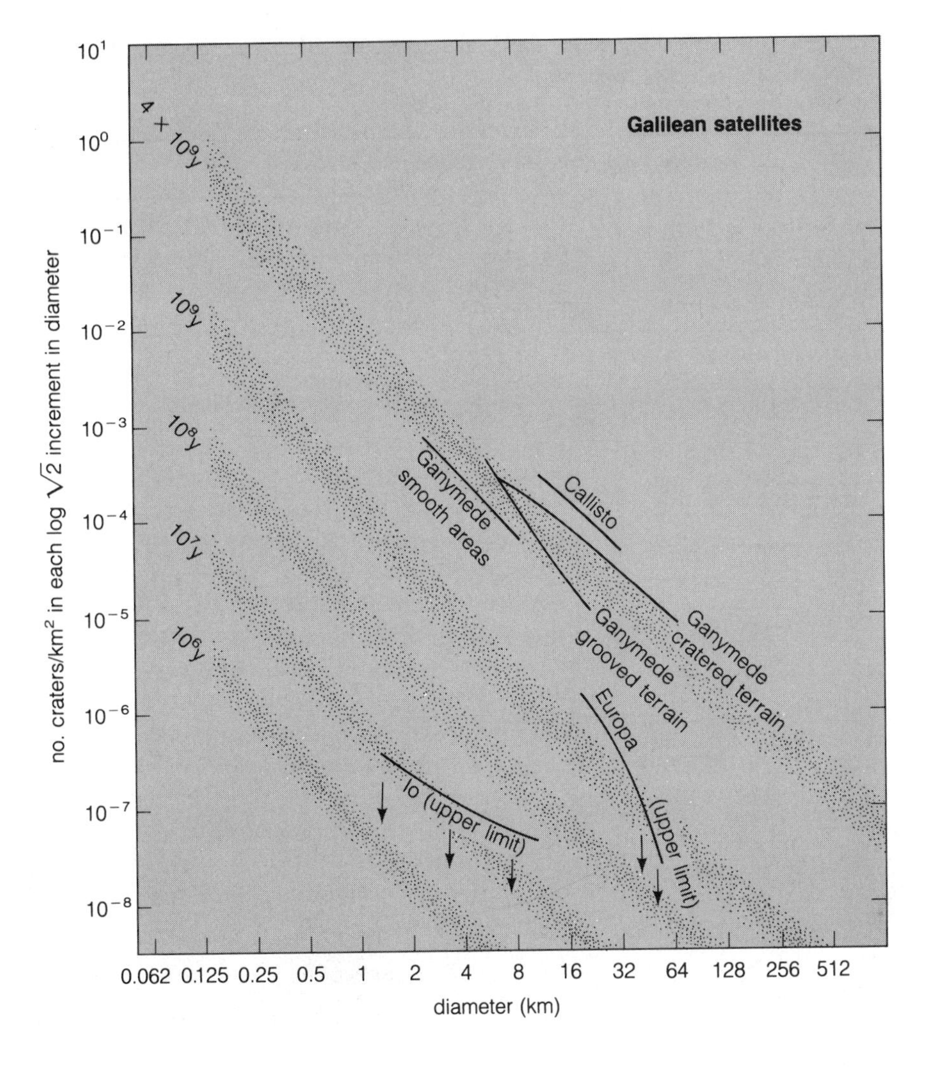

Figure 10-22g.

multikilometer craters must be counted over mare-sized areas hundreds of kilometers across. This restricts crater dating to large, rather coarse geological provinces. Desire to date smaller features, such as individual lava flows or craters such as Copernicus, has led to interest in counting sub-kilometer-sized craters, which are more abundant. In most planetary regions, a majority of these craters are probably secondary impact craters. Since these cluster around their parent primaries and along rays from primaries, instead of being distributed at random, they are trickier to use for dating. Wilhelms, Hodges, and Pike (1978) find that secondaries as large as 20 km across, which were thrown out of the larger basins, may predominate in the lunar uplands, complicating the situation even more. Furthermore, volcanic processes add uncertain numbers of small craters that may be mistaken for meteorite craters. Clearly, one must be careful when attempting to derive ages by counting small pits. Some investigators have had some success by using morophological criteria to discriminate primaries from secondaries (Neukum and others, 1975).

Table 10-5
Stratigraphic Systems Used for Earth and the Moon

Absolute Time Scale (Gy before present)	Relative Time Scale (Stratigraphic Terminology)			
	Earth		Moon	
	Era	Period (or system)	Period (or system)	Events[a]
0				
	Cenozoic			Mammals Modern continents forming
		Cretaceous		
	Mesozoic	Jurassic		Dinosaurs
		Triassic		Ferns, conifers
		Permian		
		Carboniferous		
Note expansion of scale (0 to 1Gy)		Devonian		Fish
	Paleozoic	Silurian		Early land plants
		Ordovician		
		Cambrian		Earliest well-formed fossils (trilobites, etc.)
		Late	Copernican	
1				
	Precambrian or Proterozoic	Middle		Sporadic large craters forming on Earth and the moon
				Oxygen increasing in Earth's atmosphere
2		Early	?[b]	Soft life forms in Earth's seas; stromatolite fossils forming at seacoasts
			Eratosthenian	
				Decline of mare lava flooding on the moon
3				Crustal and atmospheric evolution on Earth
	Archean	Archean		Early microscopic fossils on Earth (poor record)
				Mare flooding on the moon
			Imbrian	Oldest terrestrial rocks
4				
			Nectarian	"Modern" basins forming on the moon Intense cratering
5			Pre-Nectarian	Magma ocean and earliest features (obliterated by subsequent cratering) Planet formation

[a]Events are on Earth unless specified otherwise.
[b]Date of this division is uncertain.
Note: The Quaternary (0–3) and Tertiary (3–62 My) periods are too brief to show within the Cenozoic Era.

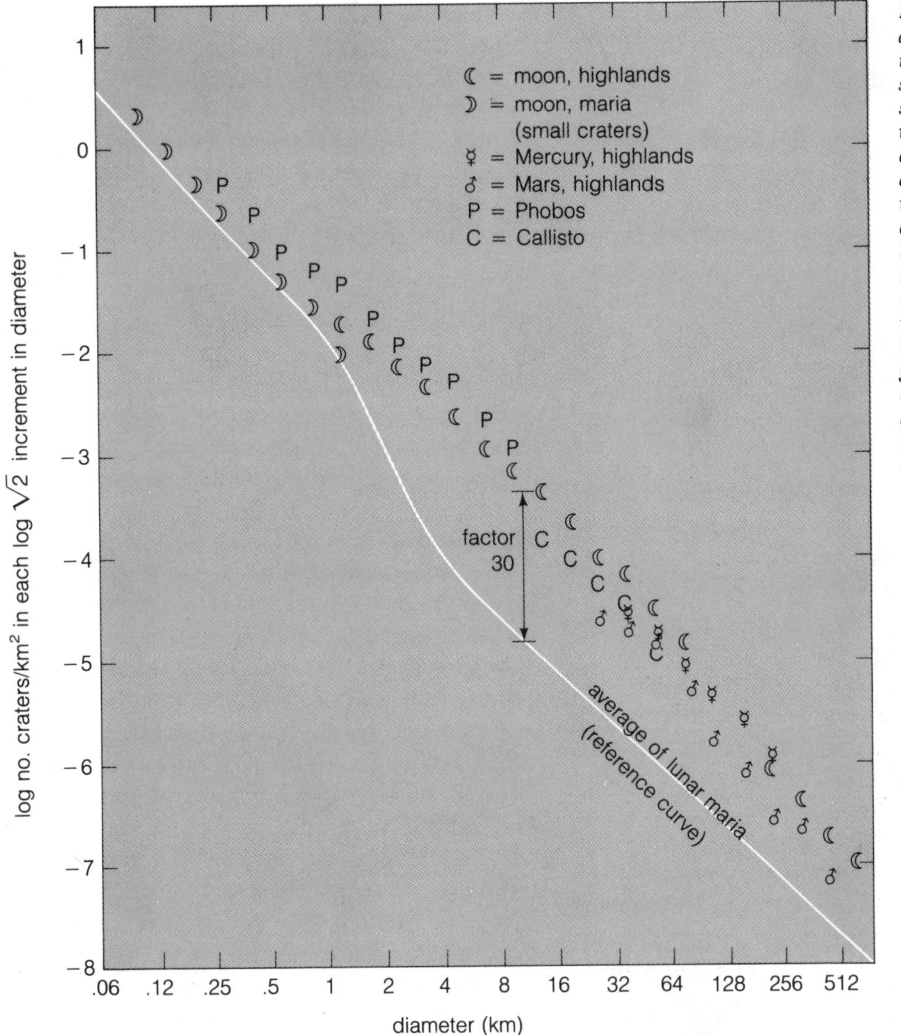

Figure 10-23. Crater counts on some of the most heavily cratered surfaces of the solar system mark a fairly well defined band in the crater-diameter distribution diagram, with about 30 times the crater density on the lunar maria. This band may approach the maximum number of visible craters that can be crowded on a planet, given the size spectrum of impacting meteorites, and it marks surfaces that have developed megaregoliths.

Soderblom and Lebofsky (1972) developed a different technique by noting that if all primaries of a given size began life as bowls of a certain shape, subsequent impacts would have required a fixed amount of time to erode the bowls to a degraded state where their interior walls had a shallow angle of, say, 1°. A parameter called $\boldsymbol{D_L}$ ("***D*** sub ***L***") was defined as the diameter of a crater eroded to this state of degradation. D_L was then used as a measure of the age of the surface. A fair correlation exists between Apollo age data, crater counts, and D_L ages for various surfaces. Since D_L can be found for quite small craters (100 m or so), it is a useful criterion for dating local surfaces.

MICROEFFECTS ON AIRLESS SURFACES

Micrometeorite Effects

We have already discussed cratering and associated effects caused by the impact of *macro*scopic meteorites. Still more abundant are *micro*meteorites. Their integrated smoothing or eroding effect is called **sandblasting**. The coarser plowing effect of somewhat larger (up to centimeter-scale) meteorites is known as **gardening**. **Pitting**, or **microcratering**, is the production of microscopic or subcentimeter-scale craters by impacts of individual micrometeorites on rock surfaces. Most lunar boulders, for example, have an array of microcraters (often called "zap pits"), whose number density depends on the length of time the rock was exposed on the surface. A given rock may be exposed for a while and then turned under by gardening; it may even go through several cycles of exposure.

On various regions of Callisto, Ganymede, and Europa, the most heavily cratered surfaces are generally darkest. This may indicate a process on two-component ice-stone worlds whereby microcratering preferentially sublimes the icy component, while the stony material remains behind. Long exposure to cratering would thus tend to build up a dark, dust-enriched regolith on the surface of these worlds, even if the interior is ice-rich (Hartmann, 1980b). The very slow sublimation of the ice component due to the weak solar heating at 5 AU would help produce the same effect. Saturn's and Uranus' moons show some evidence of the same effect. For example, the highest albedo in the solar system (exceeding 90%) belongs to sparsely cratered Enceladus, which appears to be extensively resurfaced by fresh ice. More cratered Saturnian moons, such as Rhea and Mimas, are darker, with albedos around 60%. Within the Uranian system, the highest albedos belong to the two moons with the most evidence of resurfacing, Miranda and Ariel (though the Uranian system albedos are 20–40%, less than those in the Saturn system, for unknown reasons) (Smith and others, 1986).

Clark (1980) and others have emphasized an interesting related effect. Intimate mixtures of ice and carbonaceous dust tend to show the dark coloration of the dust, even if the dust amounts to only a percent or so of the mixture, because light penetrates through the ice grains until being absorbed by the dust. Thus, dark outer solar system surfaces such as Callisto (albedo 19%) or C-type and D-type asteroids (albedo $\simeq$4%) could contain large percentages of ice in their surface regoliths even though they are very dark. Clark (1980) suggests 30% to 90% ice by weight in Callisto. He finds that a frost layer on such dark materials would produce an obvious ice spectrum only if its thickness exceeds about 0.6 mm.

The Strange, Dark Side of Iapetus

As discovered telescopically, the leading side has an albedo of only 5%, while the trailing side is at least 6 times brighter, with albedo 28 to 60% (see Figures 2-26 and 10-24). The bright trailing side is primarily water ice.

Beginning with Cruikshank and others (1983), various observers first showed that the dark side spectrum is similar to the reddish black D asteroids, and matches a mixture of organic compounds from carbonaceous meteorites, ice, and hydrated silicates (Bell and others, 1985; Clark, 1986). But why is this material all on one face of Iapetus? As first pointed out by Soter (1974), Enceladus is hit by dust knocked off Phoebe, the next moon outward. Phoebe is a retrograde-orbiting black object of spectral class C, probably captured. When meteorites blast dust off its surface, the dust spirals inward toward Saturn, hitting the leading side of prograde Iapetus at unusually high speed. A buildup of a few percent dark dust would darken an icy regolith on Iapetus. The reddish color on Iapetus may come from synthesis of organic compounds utilizing fresh methane-bearing Iapetus ice exposed during the impacts. The irregular outlines of the white/black borders (Figure 10-24) may result from later cratering or endogenic events, such as eruptions of fresh, watery "magma" (Smith and others, 1982).

Glasses

Glass is produced during impacts when molten ejecta is cooled so fast that it does not have time

Figure 10-24. Saturn's moon Iapetus has dark surface material on its leading hemisphere (lower left side in this view from above the north pole) and bright material on its trailing hemisphere (upper right). Boundary of two materials is ragged and well-defined. (NASA, Voyager 2)

to form a crystal lattice structure. Lunar craters up to 10 cm across are commonly lined with glassy material, and larger craters have also splashed glassy blobs or sprayed out fine beads of glass (Figure 6-9) that are mixed with the nearby soil. As a result, most lunar soil has a component of microscopic glassy beads, and the amount increases with the age of the soil. Soils near a young crater at the Apollo 14 site have 10% glass; mare lava soils, 20% to 40%; upland soils, 40% to 85% (Short, 1975).

The glass beads have different colors and brightness, depending on the minerals in the parent rock that melted to produce them. The beads also tend to darken with age as volatile elements are lost during shock events from subsequent gardening. Further, some nonglassy surface soils appear to have picked up a coating of dark iron- and titanium-rich glasses (Short, 1975).

These effects help explain the photometry of the moon and other airless bodies. Generally, the colors are remarkably uniform because gardening has mixed local soils: Mare soils are dark gray, and upland soils lighter gray-tan. The major local variations are associated with young craters (Figures 10-25 and 10-26). Craters produce bright rays, but the rays tend to fade with time as gardening dilutes the bright ray material.

In addition to causing complete melting and the resultant production of glass, small and large impacts send shock waves, with momentary pressures up to several hundred kilobars, through nearby rocks and soils. Resultant **shock damage** can include alterations of the crystal shapes of certain crystals, partly melted fragments, or microfractures.

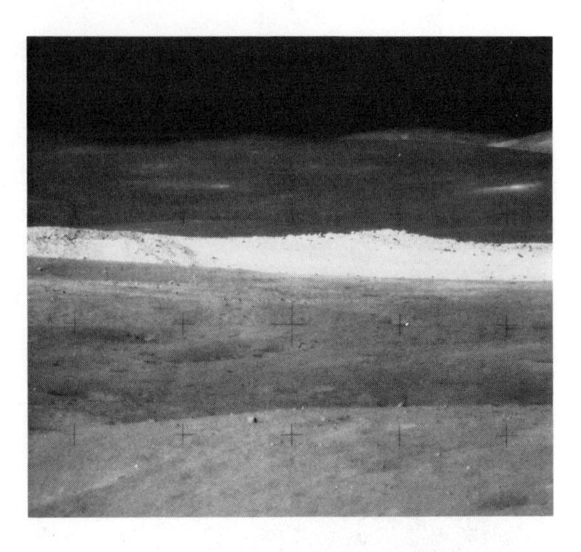

Figure 10-25. A lunar landscape with unusual albedo variation. This is a telephoto view of the rim of the 640-m South Ray crater, the 2-My-old crater approached by Apollo 16 astronauts. The bright rim stands out from the darker rolling highlands. Distant bright patches mark other young craters. (NASA)

Albedo Effects and Sputtering Caused by Irradiation

Atomic particles such as protons and electrons strike airless surfaces. These may come directly from the solar wind or be trapped in the Van Allen belts formed in magnetic fields of planets, where close satellites orbit. Hapke (1965) found experimental evidence that such irradiation might redden and darken rocky soils, accounting for the reddish tinge and low albedo of Mercury and the moon. Astronauts, however, found that the uppermost lunar surface is usually a bit lighter than the soil a few centimeters below. Glasses may play a role, but researchers also suspect an additional effect from radiation bleaching.

Sputtering is a microscopic erosion caused by low-energy ions striking a solid surface. (High-energy particles penetrate farther and produce other effects.) Principal agents in the planetary environment are solar protons and alpha particles in the energy range 1 to 20 keV. Sputtering tends to smooth large irregularities by dislodging atoms on projecting surfaces and removing them to less accessible spots. On the other hand, polished smooth surfaces are microscopically roughened by differential crystalline response to the sputtering process. A cementing of granular material also results.

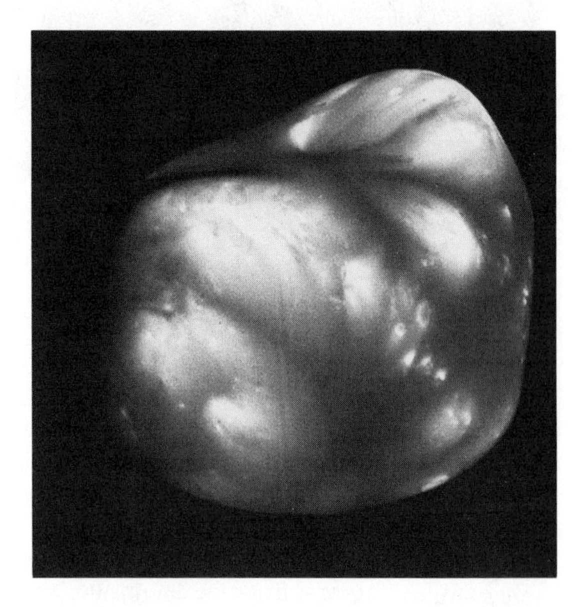

Figure 10-26. Albedo variations associated with the younger craters stand out on the impact-eroded surface of lumpy-surfaced Deimos at nearly full-phase lighting. (NASA, Viking Orbiter)

The moon and other small bodies lose mass to the sputtering process because some rock particles are blasted off into space. KenKnight, Rosenberg, and Wehner (1967) estimated that exposure of the lunar surface for 4.5×10^9 y has resulted in the loss of a layer that is a few centimeters thick.

Compositional Effects

Since impact experiments show that each meteorite dislodges 100 to 1000 times more local material than its own mass, there will be an admixture of only a small fraction by mass of meteoritic material in planetary regolith (Gault, Heitowit, and Moore, 1964). Therefore, the meteorite component gets diluted in planetary soils. Mare soil samples have only about 1.5% to 2% meteoritic (mainly Type 1 carbonaceous chondrite) material, indicating an infall rate of about 4 g/cm^2/Gy on the moon (Ganapathy and others, 1970) in the last few billion years. Additional cratering churns soils to greater depth, so that the highland soils have not accumulated greater concentrations of meteorite material (Taylor, 1975).

Compositional effects also arise on the level of atoms. The solar wind strikes airless surfaces and implants atoms in the soil. The lunar soil has about 0.005% to 0.01% by weight of hydrogen from this source. Since the moon lacks water, this hydrogen may someday be an important resource for astronauts. It is easily removed by mild heating and thereby available for combination with oxygen to convert 0.05% by weight of lunar soil into water (Williams, 1976; Arnold, 1977). Similarly, around 0.01% of the lunar soil is a mixture of rare gases such as helium, neon, and argon, also trapped from the solar wind.

Another change on the atomic level is the production of rare or short-lived isotopes, such as helium 3 or aluminum 26, by **spallation,** the splitting of heavier atoms by impacts with atomic particles. Concentrations of such materials give ways of dating surface exposures of lunar soils. For example, material from the 640-m South Ray crater (Figure 10-25) visited by Apollo 16 astronauts turned out to have been lying on the surface for only 2 My, making this crater one of the youngest sampled. Most randomly sampled lunar soils have roughly 400 My exposure or residence time at the surface, with much variation (Taylor, 1982).

Finally, the Italian researcher Strazzulla has emphasized a compositional effect important in the outer solar system. Cosmic ray or solar wind protons, hitting methane ice on outer solar system surfaces, can break down certain organics or convert them to plain carbon. At the same time, other organic molecules may be built up. Histories of organic materials on surfaces may be complex. Cruikshank and others (1991) note a spectral band at 2.2 μm, probably from C≡N bearing organics, which appears in some D asteriods, some comets, the dark side of Iapetus, and the rings of Uranus. It may mark the first organic material to be found in all four classes of objects—asteroids, comets, moons, and rings.

Asteroid 5145, discovered on a Chironlike orbit in 1992, has the reddest colors yet observed for any interplanetary object; some observers attribute this to a concentration of colored organic compounds of unknown origin (Meuller and others, 1992; Fink and others, 1992).

SUMMARY

Two subjects dominate this chapter—cratering and petrology. Primitive bodies—the ones that have little modification by internal activity—are heavily cratered, possibly saturated by craters during the intense early bombardment that accompanied sweepup of planet-forming debris. Surface petrology divides naturally into silicate surfaces in the inner solar system and carbonaceous/icy mixtures in the outer solar system.

A first-order understanding of silicate surfaces comes from a simple differentiation model in which low-density feldspar minerals accumulate in anorthositic crusts as magma oceans cool. Eruptions of mantle melts and partial melts produce basaltic volcanism on the surface of all terrestrial bodies.

Beyond 2.7 AU most surfaces are dark because of the carbonaceous component, unless thermal activity has produced water eruptions, bright ice flows, or bright frost condensates from escaping vapors.

Water ice and frost are important in the Jupiter-Uranus region, but more volatile ices, including methane (CH_4), nitrogen (N_2), and carbon dioxide (CO_2), are important at still greater distances.

The competition between rock formation and rock pulverization was initially won by cratering, producing deep megaregoliths on the most primitive surfaces. Later eruptions produced lava plains where later catering accumulated to different degrees. Crater counts thus produce useful information about the age, erosional state, and subsurface character of those surfaces.

Of course, not all planets were dominated by primitive, heavily cratered surfaces. On larger, less primitive worlds, lava flows continued to erupt because thermal histories were more protracted. Volcanic edifices built up. Tectonic effects may have faulted or folded the surface, and winds or water may have stripped materials from some regions and deposited them in sedimentary blankets elsewhere. The next chapter will discuss those modifying effects in greater detail.

CONCEPTS

petrology
minerals
rocks
acidic
siliceous
basic
mafic
ultrabasic
ultramafic
feldspars
orthoclase
plagioclase
quartz
silica
pyroxenes
amphiboles
micas
water ice
carbon dioxide ice
ammonia ice
methane ice
olivine
iron oxides
magnetite
maghemite
hematite
goethite
limonite
troilite
graphite
carbonaceous minerals
clay minerals
montmorillonite
nontronite
igneous rock
sedimentary rock
metamorphic rock
lava
intrusive
phaneritic
extrusive
aphanitic
glass
basaltic rock
gabbros
basalts
anorthosite
granitic rock
granite
rhyolite
fractionation
reaction series
sedimentation
weathering
shale
sandstone
breccia
evaporites
duricrust
metasomatism
KREEP basalts
norites
troctolites
Genesis rocks
regolith
megaregolith
prelithospheric chaos
lithospheric evolution era
mare era
postmare era
crater
astrobleme
hypersonic
shock wave
primary impact crater
ejecta blanket
secondary impact crater
ray
bedrock
base surge
simple crater
central peak
complex crater
peak ring
multiring basin
rampart crater
relative age
stratigraphic relationships
absolute age
stratigraphic column
crater retention age
steady state cratering
D_L
sandblasting
gardening
pitting
microcratering
shock damage
sputtering
spallation

PROBLEMS

1. Why did the discovery that the lunar highlands are a relatively uniform mass of anorthositic gabbros and anorthosites imply that a lunar magma ocean had once existed?

2. Why are the lunar highlands heavily cratered and the lunar maria only sparsely cratered?

3. Compare surface rock types that you might expect on an imaginary, 5000-km-diameter planet of bulk composition like Earth given the following conditions:

(a) The planet had never melted.
(b) The planet had a magma ocean with very little convection or vigorous stirring.
(c) The planet had a magma ocean in which there had been vigorous convection and efficient differentiation of siliceous and basic minerals.
(d) The events in (c) had occurred but were followed by extensive volcanism that tapped magmas in the planet's upper mantle.

4. How would discovery of wide regions of granite on Venus imply a history different from that implied by discovery of a wholly basaltic or gabbroic crust?

5. Why is Mars red?

6. Why might evaporite deposits be common on some outer solar system satellites if they had near-surface heat sources such as tidal heating or radiogenic heating? What surface material is most common on these bodies in the absence of such heating?

7. Why are feldspar-rich crusts probably thinner on the larger terrestrial planets than on the moon?

8. If the lunar (or other) magma oceans solidified in the first 100 My or so of planetary history, why are 4.5-Gy-old Genesis rocks so rare on these worlds?

9. Compare the appearance of the surface topography (especially the presence of heavily cratered highlands) on two moonlike worlds (for example, the moon and Mercury) if magma ocean activity or near-surface melting lasted only until 4.4 Gy ago on one but until 3.8 Gy ago on the other.

10. A 1-km-diameter, long-period comet nucleus in retrograde orbit and an equal-sized Apollo asteroid in direct orbit hit the moon. (a) Will they make the same-sized crater? (b) If not, which makes the bigger crater? (c) Roughly what is the diameter of the two craters?

11. You are an astronaut traversing a 10-km-diameter crater on Mercury. The crater sits in a 100-m-deep flow of basaltic lava, overlaying upland material of unknown composition X. You begin on the lava, cross the ejecta blanket, and climb the central peak. Describe some of the rocky or soillike materials you find during this traverse.

12. A volcano on Mars is found to have one-tenth as many craters of all sizes as the average lunar maria. What is its absolute age?

ADVANCED PROBLEMS

13. Consider a typical Apollo asteroid of diameter 2 km approaching the moon on an orbit with aphelion in the outer asteroid belt and perihelion at Earth's orbit.
(a) At what velocity does it approach the Earth-moon system?
(b) At what velocity does it hit the moon?
(c) What energy, in joules, is expended during the collision?
(d) What diameter crater does it make?

14. How do the answers in problem 13 vary if the object falls into Earth instead of the moon?

15. Suppose you were going to dig a 10-km-diameter hemispherical crater in such a way as to throw the material out of the hole (shovelful by shovelful or all at once) about 10 crater diameters upward into the air, allowing the material to disperse outside the rim at random. (a) Calculate the potential energy consumed in this operation. (b) Compare this figure with the kinetic energy indicated to be necessary to create a 10-km-diameter crater and comment on the similarity or difference.

16. Suppose an incoming comet nucleus strikes the moon and ejects 500 times its own mass from the crater that is formed. The fastest 0.2% of this ejecta is moving faster than 2.4 km/s. (a) Does the moon gain or lose net mass during this impact? (b) What would be the answer for the same circumstances during an impact on Mercury?

CHAPTER ELEVEN

PLANETARY SURFACES 2: VOLCANISM AND OTHER PROCESSES OF SURFACE EVOLUTION

A scene on the surface of Mars contains a mixture of volcanic boulders and windblown dune deposits of dust, symbolizing varied processes modifying the surfaces of the more geologically active planets. (NASA, Viking 1 telephoto view, width about 18°)

Planetary surfaces are not all craters and rock-strewn uplands. A glance at the moon reveals not a uniformly cratered surface but rather the prominent features of "the man in the moon"—dark lava plains produced by volcanism that obliterated many cratered areas. Larger worlds are still more geologically active. Mars and Earth, for example, show faulted terrains, vast dune fields, polar ice caps, canyons, and river valleys.

This chapter deals with many processes that have shaped planetary landscapes through geological time. Much of the chapter deals with volcanism and related processes, which have happened to some extent on most worlds. Middle portions of the chapter deal with erosional effects on larger worlds, whose atmospheres and hydrospheres have modified their surfaces. Chemistries and temperatures of surfaces are also discussed. The last part of the chapter attempts to draw all these diverse processes together by summarizing the surface conditions of the planets.

An organizing principle underlies the division between this and the last chapter. Volcanism, tectonics, and other processes discussed here are **endogenic processes**—those that arise from within the planet and its atmosphere. Most processes discussed in the last chapter, such as cratering and solar wind sputtering, are **exogenic processes**—those acting on the planet from the outside.

VOLCANISM AND TECTONICS

Volcanism includes all processes by which material—gas, liquid, or solid—is expelled from a planet's interior. **Tectonics** includes all processes causing movement or distortion of planetary surfaces. Tectonic processes produce folded mountains, faults, rifts, and so on. Volcanism and tectonic activity often occur together but can also occur independently.

Rocks Produced by Volcanism

The last chapter also described the basic rock materials that formed primeval crusts, not so

Figure 11–1. Flow of aa basalt lava. Solidifying crust was broken into rough clinkers by the motion of flow behind it, creating an intricate surface structure. (Pinacate volcanics, Sonora, Mexico; photo by author and Joyce Rehm)

much by conventional volcanic processes as by basic processes of magma-ocean solidification near the surface of the planet. Volcanism can be thought of as eruptions that bring new rock materials up from depth and extrude them onto the surface. The most important rock produced by volcanism is basalt (refer back to Figure 10-1), an extrusive rock rich in plagioclase feldspars but having a composition range that includes minerals such as pyroxenes and olivine.

Basalts are the main rock in the volcanic parks of the American West and Hawaii, the lunar maria (as confirmed by Apollo samples in 1969), and among the boulders of Mars (based on their appearance and Viking soil samples) and among most sites visited on Venus (based on Venera soil chemistry analyses). Geochemical studies of lunar basalts indicate that they erupted from magma sources perhaps 300 km down, probably produced around 4 to 3.2 Gy ago by radiogenic heating.

In the lunar uplands, we find a more complex story. The last chapter showed that the lunar uplands are mostly low-density rocks, primarily anorthosites (nearly pure feldspar), anorthositic gabbros (coarse-grained equivalent of feldspar-rich basalt), and some norites and troctolites (gabbros with special mineralogy). These rocks, called **cumulates**, are formed by mineral assemblages that more or less floated to the top of

Figure 11–2. Tongue of solidified pahoehoe basalt lava lapping onto a road. In the distance the road is cut by an aa flow. (Kilauea, Hawaii; photo by author)

the magma ocean. But erupted, noncumulate rocks are found among them. Most noteworthy are the KREEP basalts (Chapter 10), which are older than mare basalts. Their chemistry suggests that they were erupted from shallower magmas, probably the last magmas in the magma ocean, at depths around 60 km at the base of the anorthositic highland crust (Taylor, 1975).

Basaltic lava flows can take many forms, depending on their rate of flow and viscosity, which in turn depend on the angle of slope, the gas content, temperature, and so on. The two most common forms have Hawaiian names. **Aa** (AH-ah) is rough and clinkery (Figure 11-1) and is formed when solid blocks on the top of the flow are broken and upended by the flow's motion. (A mnemonic device: Think of the pointed top of the *A* as representing the pointy texture). **Pahoehoe** (pa-HOY-hoy) is smooth (Figure 11-2) and sometimes has a glassy surface produced by rapid cooling. Often these varieties are mixed and overlapping.

Other volcanic rock types occur, especially on Earth. Eruptions of rhyolite, a light-colored, siliceous rock, may be even more voluminous on Earth than eruptions of basalt, though they are not as easily recognized, especially after erosion. Many rhyolite deposits were emplaced not in the form of massive lava flows but as clouds of pulverized, red-hot rock fragments suspended in hot gases. These are known as **fluidized eruptions**. There is no flow of liquid, but the cloud of gas and suspended particles is dense enough to flow along the ground like a swirling flash flood or like the base surge in Figure 10-9. The cloud itself is called a **nuée ardente** (glowing cloud). A nuée ardente from Mt. Pelée, in Martinique, raced across the city of St. Pierre at 150 m/s on May 8, 1902, carrying along large boulders, leveling buildings, and killing 29 000 inhabitants. There are terrestrial rhyolitic deposits less than 1 My old and covering 26 000 km^2 (Rittmann, 1962).

A number of other volcanic products are common and of planetological interest. Very fine volcanic fragments—of millimeter scale—are called **ash**. Volcanic fragments about 10 mm in size are called **cinders** (Figure 11-3). **Tuff** is a type of rock formed by consolidation of ash particles, either by deposition into water or by coalescence directly from the air. Tuff is sometimes classified as sedimentary because of its depositional character. **Welded tuff**, or **ignimbrite**, forms when hot, glassy ash particles weld to each other during deposition. Ignimbrites can form vast, blanketing deposits, usually of siliceous composition, as a result of nuée ardente eruptions.

STRUCTURES PRODUCED BY VOLCANISM

If we photograph, fly over, or land upon a planet, we see a great number of surface features. To an untrained observer, these may be bewildering in

Figure 11–3. Ash and cinders from the background cinder cone blanketed and destroyed most of a forest at this site at Kilauea, Hawaii, 20 y prior to this photo. Foreground drainage craters formed when cinders drained into fissures in underlying lava flows. (Photo by author)

their variety. For example, on the planets seen so far, features include ocean coastlines, mountain ranges, craters, lava plains, shifting dune fields, fault scarps, canyon systems, volcanic mountains, and fracture systems. The study of the structure and origin of landforms is a part of geology known as **structural geology**. On Earth, structural geologists deal mostly with folded mountain belts and strata. In planetary science, structural geological problems deal more with volcanic and tectonic landforms. Most planetary geologists believe that a thorough knowledge of all volcanic and tectonic structures of Earth would be the best preparation for interpreting other planetary surface features.

In this field it is important to learn the terminology associated with these structures, which is discussed below. Many of the volcanic landforms are sketched in Figure 11-4, where they may have either positive or negative relief.

Cinder cone (Figures 11-5 and 11-6). A conical hill typically 50 to 300 m in height that is built up around a vent that ejects cinders, ash, and boulders. Smaller cones built as viscous lava spatters and welds into a hill around the vents are called **spatter cones, scoria cones,** and **hornitos**.

Shield volcano (Figures 11-7, 11-8, and 11-9). A gently sloping volcanic mountain built by fluid, basic lava flowing out from a central vent. Shield volcanoes may grow to enormous size. The largest known single mountain of any kind in the solar system is the vast Olympus Mons shield vol-

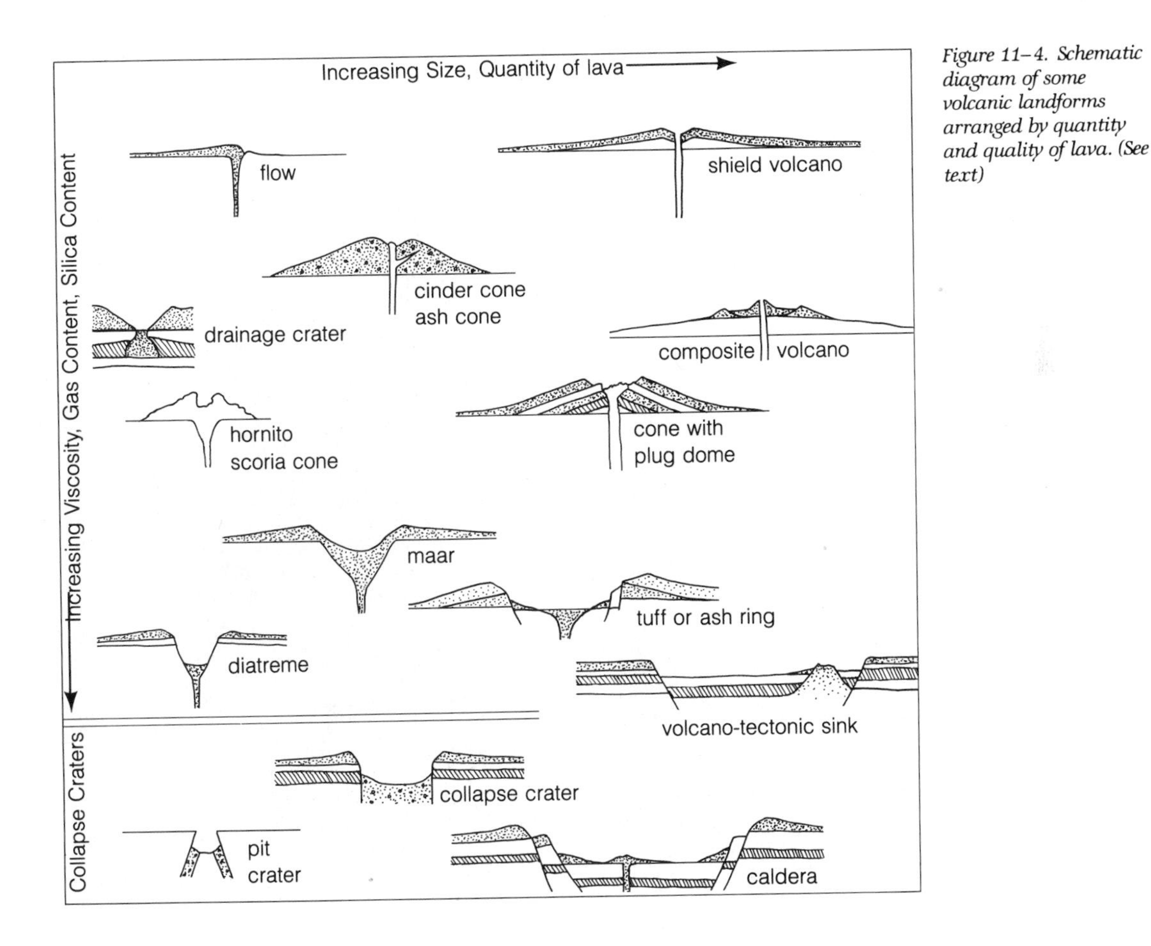

Figure 11–4. Schematic diagram of some volcanic landforms arranged by quantity and quality of lava. (See text)

Figure 11–5. Terrestrial cinder cones and associated lava flow near Flagstaff, Arizona. In middle distance is a diatremelike crater. On the horizon are the San Francisco peaks, an eroded shield volcano dominating this volcanic region. (Infrared aerial photo by author)

Figure 11–6. A possible lunar cinder cone. Several cones and pit craters on the floor of Copernicus suggest that the floor has been filled by volcanic eruptions younger than most lunar maria. Lighting from the right in this vertical view reveals a cone about 1 km across with a summit crater and numerous boulders on its slopes. At right is a C*-shaped remnant of a similar feature. Most lunar eruptions apparently produced lavas that were too fluid to create cones and instead extruded into extensive sheetlike flows. (NASA, Viking Orbiter)*

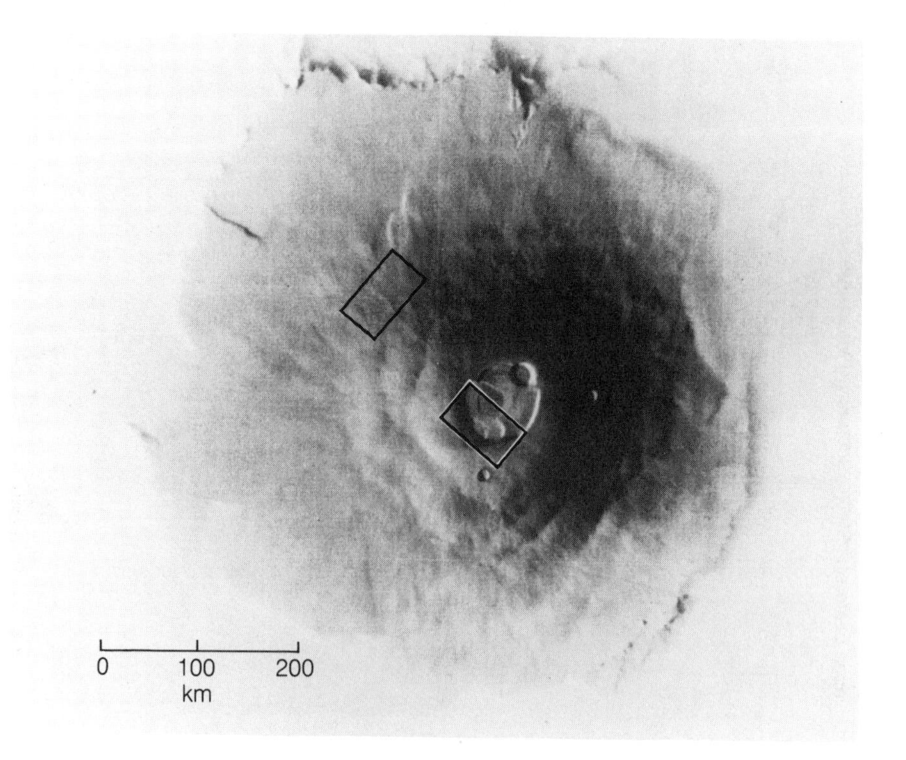

cano on Mars (Figures 11-7 and 11-8). Its base is big enough to cover most of Missouri, and it rises about 24 km above the surrounding lava plains. Among the largest shields on Earth are the Pacific volcanic islands. Mauna Loa, on Hawaii (Figure 11-9) rises about 9 km from its ocean floor base and may be the largest single mountain on Earth. By comparison, Everest rises a bit over 8 km above sea level and less above its own mountainous base.

Composite volcano. A large volcanic mountain with an irregular profile, built by eruptions of viscous and fluid lavas. Often it resembles a shield surmounted by a large cinder cone.

Crater. A term used (unfortunately) for almost any hole in the ground, whether caused by meteorite impact, volcanic explosion, or collapse. The terms listed below specify different types of volcanic craters.

Caldera (Figures 11-10 to 11-13). A volcanic crater, usually 1 km or more across, caused primarily by collapse. It may or may not have a raised rim caused by eruptive activity, and it often surmounts a volcanic mountain. As lava erupts and flows downhill, the overlying rock layers often sag or collapse to make up the vacated volume, producing a summit collapse crater. As shown in Figure 11-10, underground cavities and **lava tubes** up to tens of meters across and several kilometers long are often left when lava drains or erupts, and the roofs of these can also collapse

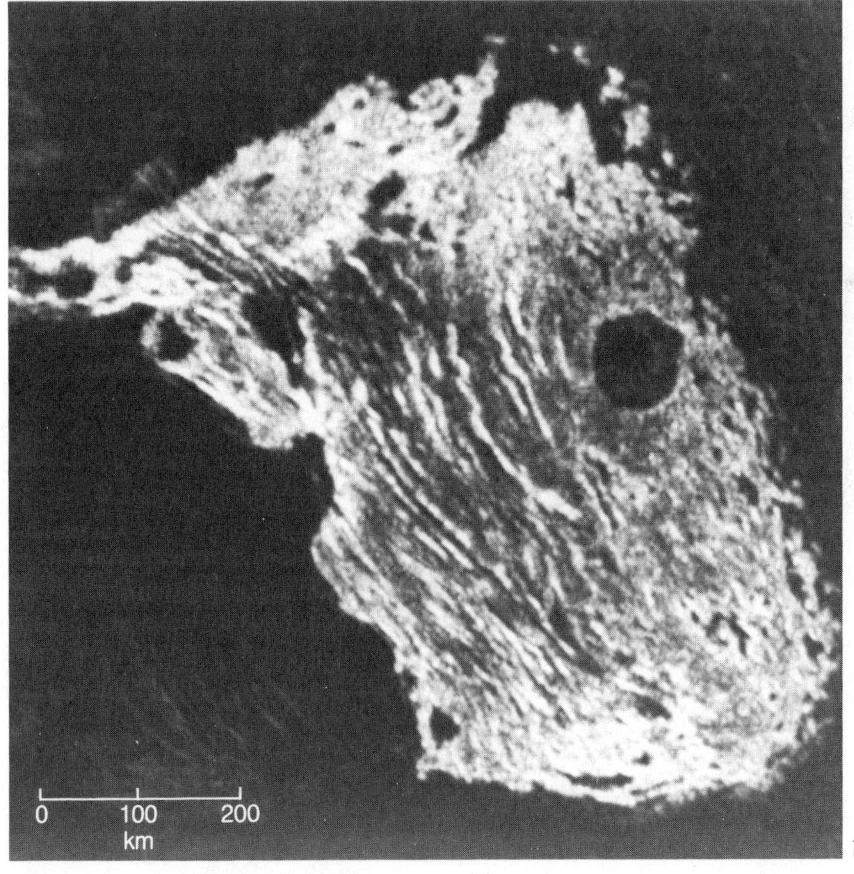

*Figure 11–7. Comparison of Olympus Mons (Mars) to possible shield volcano Maxwell Montes (Venus), at similar scale. (**a**) (opposite page, bottom), Olympus Mons with boxes showing positions of closeups in Figure 11–8. See Chapter 9 opening photo for regional setting. (NASA, Viking Orbiter mosaic) (**b**) Earth-based radar image of Maxwell Montes on Venus. The circular feature resembles a summit caldera. The brightness of the image indicates an extremely rough surface (at the scale of the 13 cm wavelength). Pioneer radar measures indicate a summit elevation of 17 km above the mean planetary radius. Arcs may be rough lava flow or tectonic features. (Courtesy of D. B. Campbell, Arecibo Observatory)*

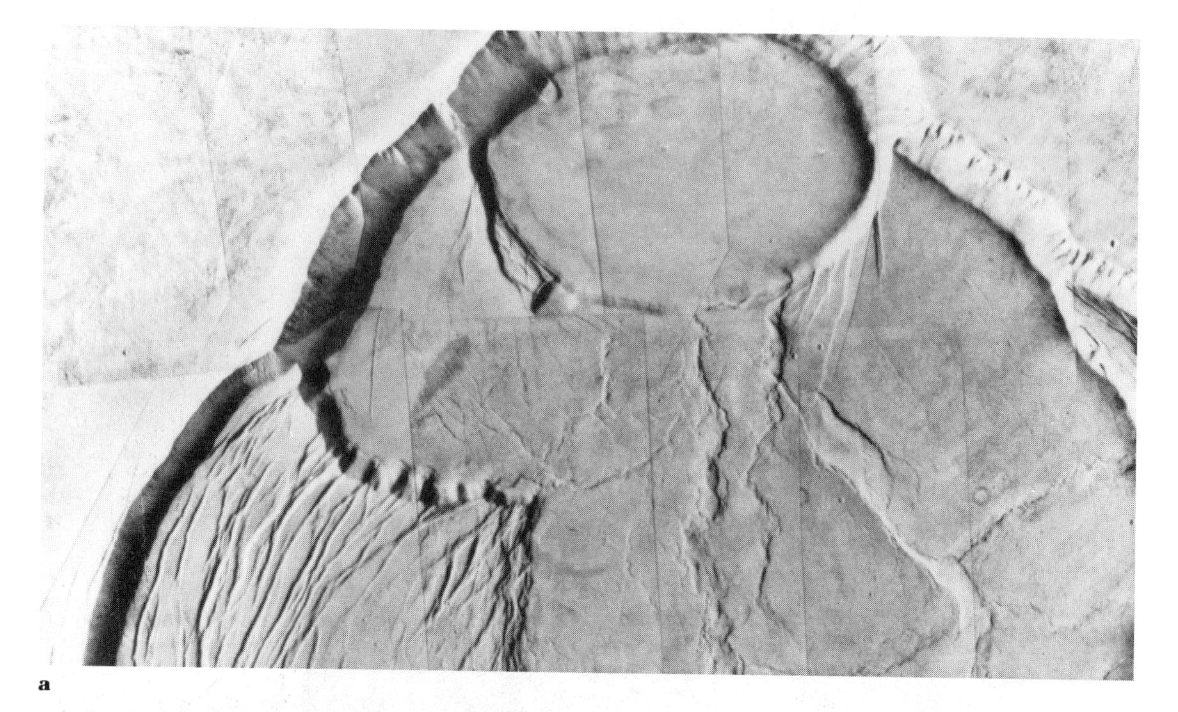

a

b

*Figure 11–8. Features of Olympus Mons (compare Figure 11–7). (**a**) Oblique view of summit calderas. Note the landslide on the rim at top, the slumped and faulted terrain at lower left, and the wrinkle ridges on the lava flows on the central caldera floor. (**b**) Lava flows on the flanks. A prominent fissure follows a ridge crest, resembling the terrestrial feature in Figure 11–9b. At upper center, a fingerlike flow runs downhill toward the upper right, parallel to the fissure. The absence of impact craters testifies to the youth of this surface, which may be only several hundred million years old. Region is about 43 × 55 km. (NASA, Mariner 9)*

to form small craters about 10 to 1000 m across, usually called **pit craters** (Figure 11-12).

Diatreme. A volcanic crater caused primarily by eruptive and explosive activity, often associated with gas-charged eruptions producing ash, that creates a gently raised rim (Figure 11-5, middle distance). A **maar** is a similar crater associated with groundwater that gains access to the magma chamber and causes an explosion. A diatreme often has a lake in its interior. Related to diatremes may be some of the lunar pits called **dark halo craters** (Figure 11-13), which are rounded craters with low rims forming dark aprons of debris that mantle the surrounding topography. Some are located on fissures and may be eruptive sites. They have not been visited by astronauts. Features between diatremes and cinder cones are tuff rings (nearly synonymous terms: ash rings, ash cones, and tuff cones), which have fairly high rims and floors that may be barely depressed. Diamond Head and many other Hawaiian craters (Figure 11-14) are examples.

a

b

Figure 11–9. Features of the terrestrial shield volcano Mauna Loa, in Hawaii, for comparison with Figures 11-7 and 11-8. (**a**) *Collapse calderas mark snow-covered summit, with a profile of the Mauna Kea shield volcano in background. Pits are smaller but similar in form to the summit calderas of Olympus Mons. Striated slopes resemble the texture shown in Figure 11-8b. (U.S. Geological Survey, D. W. Peterson)* (**b**) *Downhill view of a fissure on a ridge built by eruptions along a rift zone radial to the summit. The fissure has been widened by posteruption collapses along its edges. The feature resembles that in Figure 11-8b. (Photo by author)*

Figure 11–10. Road cut through basalt flows reveals subsurface cavity typical of basalt flows. (Kilauea, Hawaii; photo by author)

Figure 11–12. Inner rim of a pit crater is a vertical wall with a succession of basalt flows that built the surrounding countryside. Wall reveals a feeder dike (left of center) along which some lava may have erupted. The flat rim indicates little eruptive activity from the crater itself. (Mauna Loa summit, Hawaii; photo by author)

Figure 11–11. Caldera about 1.6 km across, with eroded rim of tuff beds. Floor is filled with eroded tuff and windblown dust, reminiscent of Martian craters. (MacDougal Crater, Pinacate volcanics, Sonora, Mexico; infrared aerial photo by author)

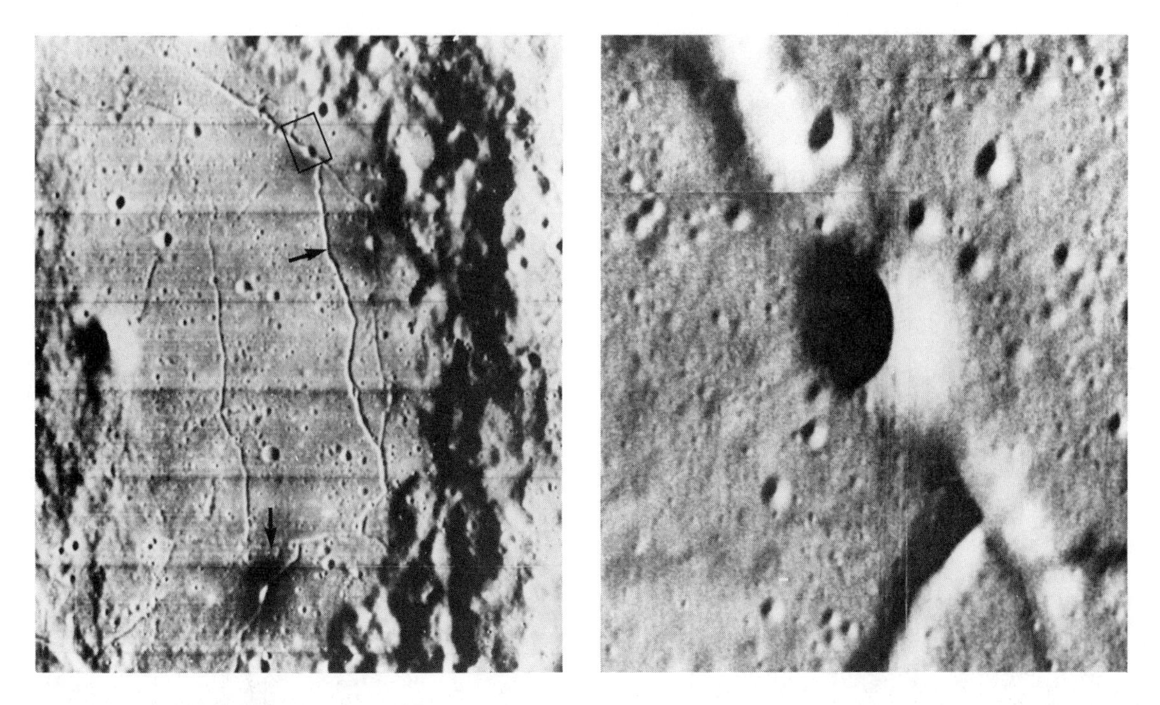

Figure 11–13. Medium- and high-resolution views of dark halo craters (arrows) on the floor of the lunar crater Alphonsus. Boxed example is shown in more detail at right. (NASA, Ranger 9)

Volcano-tectonic sink. This is a depression larger and more irregular than a caldera, sometimes exceeding 30 km in diameter.

WHAT CAUSES VOLCANISM?

How can we explain the rise of dense, molten rock from the interiors of Earth, Io, Mars, the moon, and other planets? What makes the planet spew forth with dazzling fire fountains, sluggish flows, and other forms? The basic cause appears to be that magma is slightly less dense than most of the surrounding solid rock of the upper mantle (Eaton and Murata, 1960). This means that the magma tends to rise just as a cork floats in water. At the magma source, many kilometers below the surface, the weight of the overlying rock layers pressing on the magma exceeds the weight of a (less dense) magma column just reaching to the surface. Therefore, if any cracks permit access to the surface, the magma gets pushed out. Combining this scenario with measurements of Hawaiian lava densities, Eaton and Murata (1960) calculated a source region for Hawaiian magma 60 km below the surface. This figure agrees with seismic evidence that magma originates at that level.

Of course, other factors may come into play. If tectonic forces squeeze the magma source region, they add to the pressure driving the magma upward. An analogue for this would be a plastic bottle filled with water and buried in sand with the bottleneck flush with the surface. Pressure or weight on the sand would "erupt" a flow of water. Another factor is volatile content. As magma rises close to the surface, the pressure is less, and dissolved gases may begin to form bubbles. The bubbling gas may froth up the magma and create enough pressure near the vent to shoot out fountains of glowing material. An analogue for this would be agitated soda pop, rather than water, in the plastic bottle. Sudden release of the bottle cap (representing the sudden access of

Figure 11–14. Hawaiian tuff ring, eroded and breached at one end by the sea. For scale, note swimmers and beach at right. (Hanauma Bay, Honolulu; wide angle photo by author)

magma to the surface) would produce fountaining. If the amount of gas is modest, it may produce vesicular lava, but if the amount is large, it may disrupt the magma into cinders and ash.

Even the geometry of the event may affect the type of eruption (Wilson and Head, 1981). For example, a narrow constriction could produce high pressure and a high fountain, while a large opening or long rift might produce a gentler flow. The shape of the vent often changes during the eruption, owing to erosion of the side walls by the lava, plus tectonic motions that may open new vents. As indicated by Figure 11-15, actual eruptions may be very complex, since the magma conduits may be linked by many interconnected passages, with eruptions breaking out and ending in various places. Early opening of a vent may yield a gentle eruption, while the buildup of pressure could produce an explosive eruption that blows away overlying crust.

STRUCTURES PRODUCED BY TECTONIC ACTIVITY

Volcanic and other areas may be subjected to **tectonism**, or formation of landforms by compressive or stretching forces. Figure 11-16 reviews some types of **faults**, or fractures along which offset motion has occurred.

Thrust faults and **strike-slip faults** predominate where compressive forces have squeezed an area.

Normal faults and **grabens** (a German-derived term; sometimes *graben* is also used as the plural form) are produced where an area is stretched, as in rift zones. Grabens are prominent in Figures 9-17, 11-8a (lower left), 11-16, and 11-17.

a

b

c

Figure 11–15. Three different styles of eruptions. (**a**) *Fire fountain eruption of Kilauea Iki, Hawaii, in 1959. The spray of hot ash and cinders reached as high as 600 m, one of the highest such fountains ever recorded. Note debris on the road. Posteruption view of the devastated forest is shown in Figure 11-3. (U.S. Geological Survey, J. P. Eaton)* (**b**) *Molten lava splashing a few meters into the air, with twisted pahoehoe in foreground. (Boone Morrison)* (**c**) *A nonmagmatic eruption sprays hot water containing dissolved minerals. The rim is built around the vent by the deposition of evaporite minerals left by the eruptions. (Castle Geyser, Yellowstone National Park; photo by author)*

Lineament is a more general term for any linear structural feature, typically hundreds of meters to kilometers in length. It may be a fault, a graben, or a linear array of pits or cinder cones lying on some hidden underground fracture. Many lineaments are associated with volcanism (Figure 11-17), but others are not. The latter include radial structures around multiring impact basins (as were shown in figure 10–18), probably involving impact-generated fractures.

VOLCANIC AND TECTONIC LANDFORMS OF THE MOON AND PLANETS

The terminology of terrestrial volcanic landforms developed through years of geological fieldwork, and many terms mentioned above have been extended to other worlds. At the same time, an additional terminology of lunar landforms developed independently through telescopic observations. Features were given picturesque names simply to describe their appearance, not their origin. Many of these names have been retained and transferred to features on other planets, giving us a set of terms that are noncommittal about mechanisms of formation. Most of these features have not been examined at close range by astronauts or spacecraft, but most are believed to be volcanic or tectonic in origin.

MATHEMATICAL NOTES ON VOLCANIC ERUPTIONS

Consider the hydrostatic pressure on a magma chamber below rock layers of density ρ_R at depth z. Suppose this is just equal to the pressure exerted by the weight of the magma of lower density ρ_M extending up to the top of a volcano of height h. This would be the greatest height that such a volcano could reach, since no pressure is available to push the magma any higher.

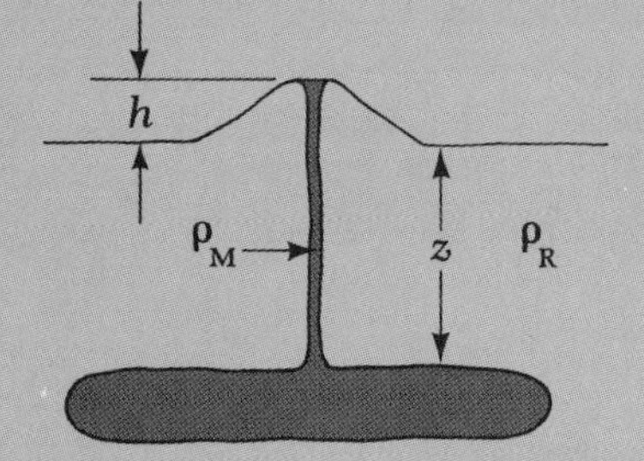

Thus, the pressure P at the depth of the magma chamber is

$$P = \rho_M g(z + h) = \rho_R g z$$

where g = gravitational acceleration.
Then,

$$h = \frac{\rho_R - \rho_M}{\rho_M} z \simeq \frac{\Delta\rho}{\rho_M} z$$

where

$\Delta\rho$ = difference in density between magma and surrounding rock.

■ *Example*

If lunar magma has few volatiles and is thus only (say) 40 kg/m^3 less dense than lunar rock, then the maximum height h of a volcano originating from a 150-km-deep magma zone would be only about (40/3000)150 = 2 km, which is considerably less than the relief of large shield volcanoes on Earth (5 to 10 km) or Mars (up to 24 km). This effect might account for the paucity of tall lunar volcanic structures.

■ *Example*

Eaton and Murata (1960) analyzed volcanism at Kilauea, Hawaii, as shown, using a measured magma density of 2770 kg/m^3. Thus, adding the pressures contributed by each layer,

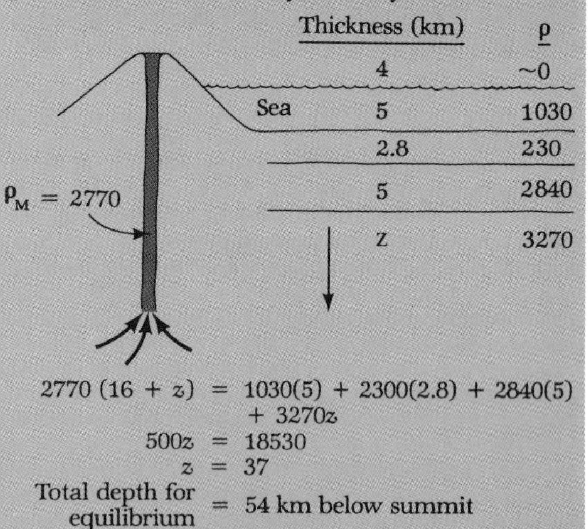

$$\begin{aligned} 2770(16 + z) &= 1030(5) + 2300(2.8) + 2840(5) \\ &\quad + 3270z \\ 500z &= 18530 \\ z &= 37 \end{aligned}$$

Total depth for equilibrium = 54 km below summit

They concluded that this is, in fact, the source depth of Kilauea magmas, since "great swarms of tiny to moderate, sharp earthquakes, totaling several thousand during the few days they last, emanate from depths between 45 and 60 km beneath the summit of Kilauea. These are the deepest quakes . . . in Hawaii, . . . 4 times deeper than the Mohorovičić discontinuity under Kilauea."

Mare. As noted in Chapter 2, this term originally applied to any dark area on a planet (once thought to be a sea). Though Martian maria turned out to be associated with windblown dust deposits, the term is now generally applied to lava-covered plains (see Figure 11-18a, right foreground) that are dark because of the dark, basic rock. As correctly argued as early as 1949 by astrophysicist

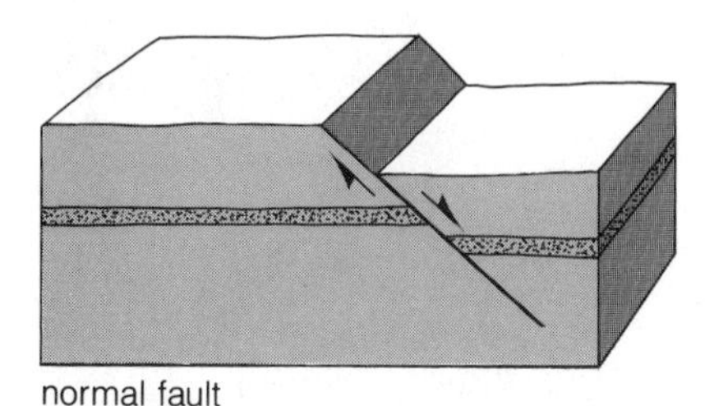

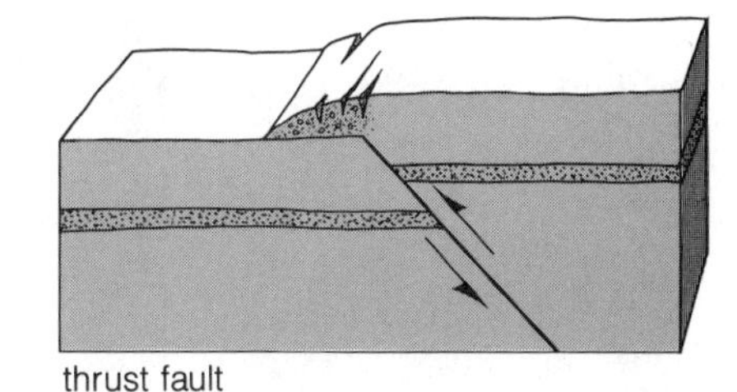

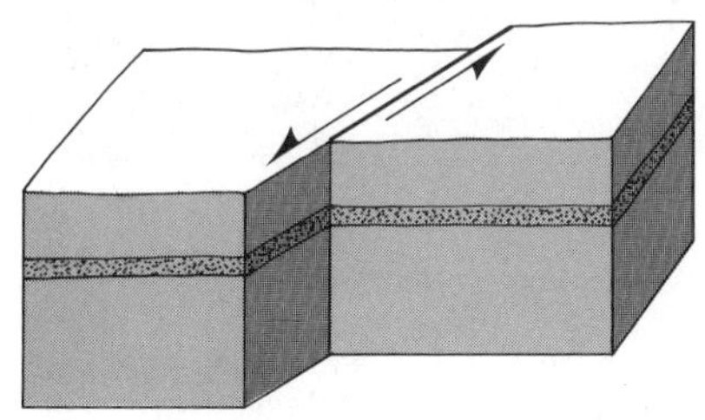

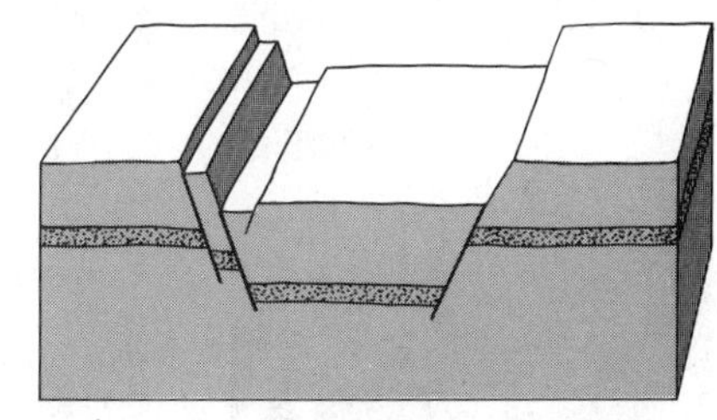

Figure 11–16. Examples of fault structures.

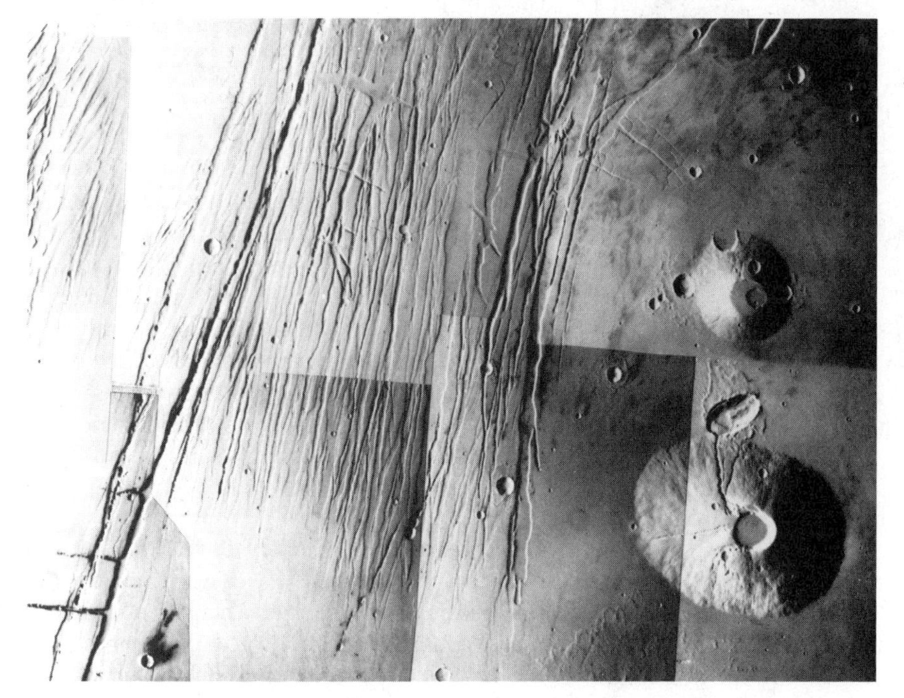

Figure 11–17. A lineament family of faults and grabens beside two volcanoes on the north flank of Mars' Tharsis volcanic field. These and other lineaments run roughly radial to the Tharsis complex and may have been produced by updoming (see Wise, Golombek, and McGill, 1979). Photo width about 600 km. (NASA, Viking Orbiter 1 mosaic)

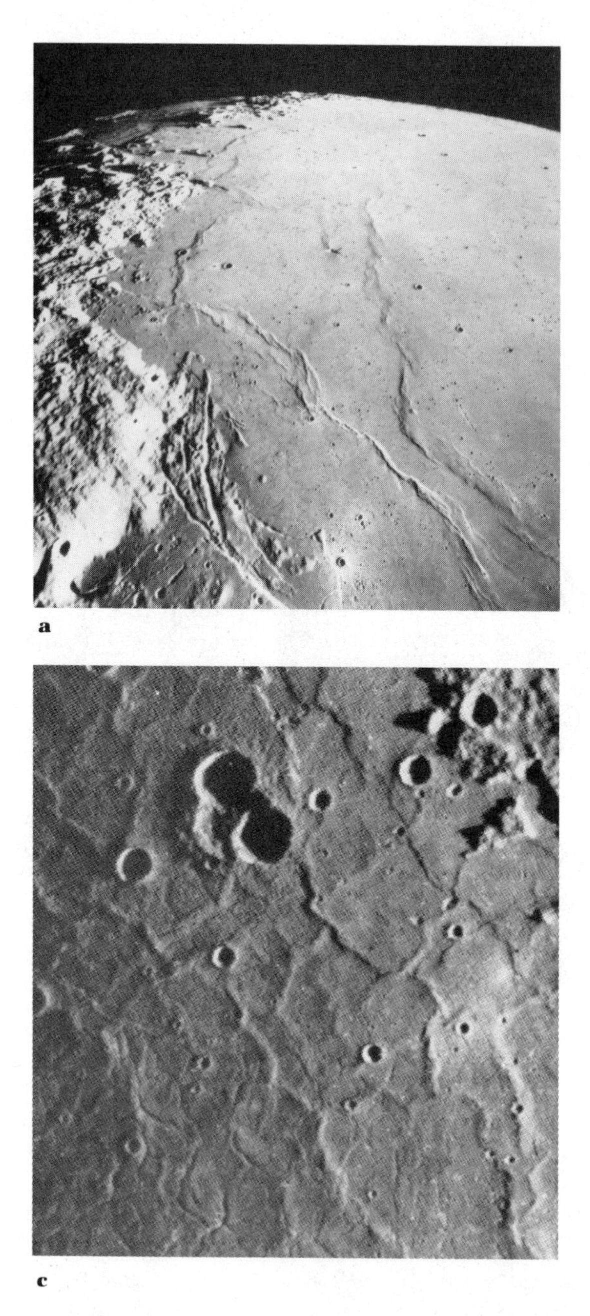

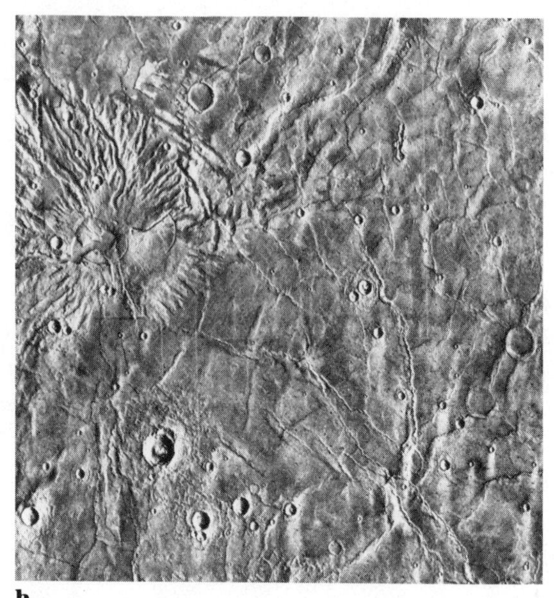

Figure 11–18. Wrinkle ridges on three planets. (**a**) *Southwestern part of Mare Serenitatis on the moon. Some of the ridges and other features lie radial to the Imbrium basin (upper left on horizon). Rilles cut the older terrain on the mare "shelf," lower left. Foreground width 150 km.* (**b**) *Tyrrhena Patera volcano (upper left, cut by rilles) and nearby volcanic plains, laced by wrinkle ridges. The ring features, as along right margin, have an uncertain origin; they might be partly buried impact craters or volcanic features.* (**c**) *Marelike surface in the Caloris impact basin. Same scale as in part* **b**; *width about 320 km. (NASA)*

Ralph Baldwin, lunar maria were formed by very fluid lavas. Apollo rocks indicate unusually high eruption temperatures of 1400 to 1600 K, increasing their fluidity (Epstein and Taylor, 1970). The flatness of mare lava fields and the lack of large volcanic mountains is apparently due to the high fluidity of the lavas.

Wrinkle ridge (Figure 11-18). A crinkly type of ridge, often a few kilometers across and 100 km or more in length and usually found in mare lava plains. These ridges often arc around edges of mare lavas in circular basins and may be compressional ridges formed during the final stages of lava filling and sagging in these basins. Some may be sources of lava flows. They are not recognized on Earth, possibly being disguised by erosion and deposition.

Dome. A roughly circular hill with a conical or rounded profile and sometimes a summit pit. Domes may be volcanoes. They often appear in mare areas. Particularly rough ones can be seen in Figure 11-19a; the lunar cone in Figure 11-6 might also be called a dome.

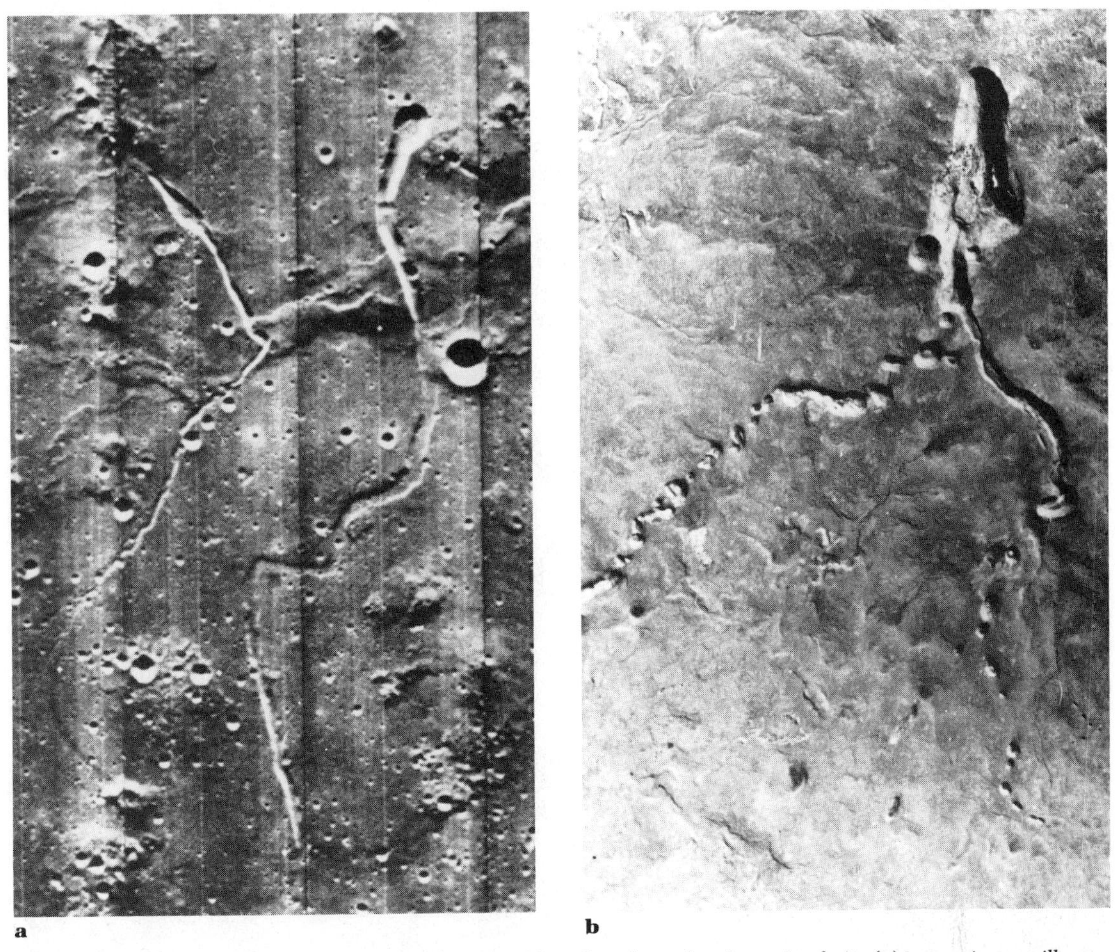

Figure 11–19. Comparision of lunar rilles with a terrestrial lava flow channel and associated pits. (**a**) *Lunar sinuous rilles among Marius Hills dome field.* (**b**) *Icelandic flow channel and collapse pits. Such comparisions are instructive, though one must be careful of scale differences. As in this case, lunar examples are often larger in scale. (NASA)*

Rilles (Figures 11-19, 11-20, and 11-21). Long valleys of two main types. A **sinuous rille** is a winding valley superficially resembling a channel cut by a river or lava flow. A **linear rille** is straight sided and more like a graben. These types, which can grade into each other, may involve four origin mechanisms, operating singly or jointly: (1) collapse of lava tubes; (2) buildup of levees of spattered lava along the sides of the still-molten central part of a flow, thus defining a channel for the flow; (3) draining of the central molten part of a flow away from its borders or levees, thus leaving a depressed channel (terrestrial flows do not seem efficient at eroding channels out of underlying rock); and (4) graben formation. The Apollo 15 astronauts' brief visit to the sinous Hadley Rille (Figure 11-21) did not allow enough fieldwork to establish its origin.

Crater chain (Figure 11-22). An alignment of craters, usually not more than a few kilometers in diameter, but often many tens of kilometers in length. Some crater chains are associated with rilles, as in Figures 11-18 and 11-22. Flat rims

Figure 11–20. Terrestrial lava flow channel possibly analogous to a sinuous rille. Channel was opened partly by a lava tube collapse (especially around the "head," right middle distance); lavas associated with the background cinder cone also flowed into and down the channel. (Mauna Loa, Hawaii; photo by author)

suggest collapse pits, while raised rims suggest eruptive vents along fissures.

LARGE-SCALE AND GLOBAL LINEAMENT SYSTEMS

Many lineaments, such as rilles, faults, and crater chains, appear not just as local features, but also in vast parallel or radiating swarms. Some radiate from large impact basins, but others involve uncertain global forces.

Tectonic Patterns Associated with Impact Basins

Because we have treated impact structures and volcanic structures separately, we have only hinted

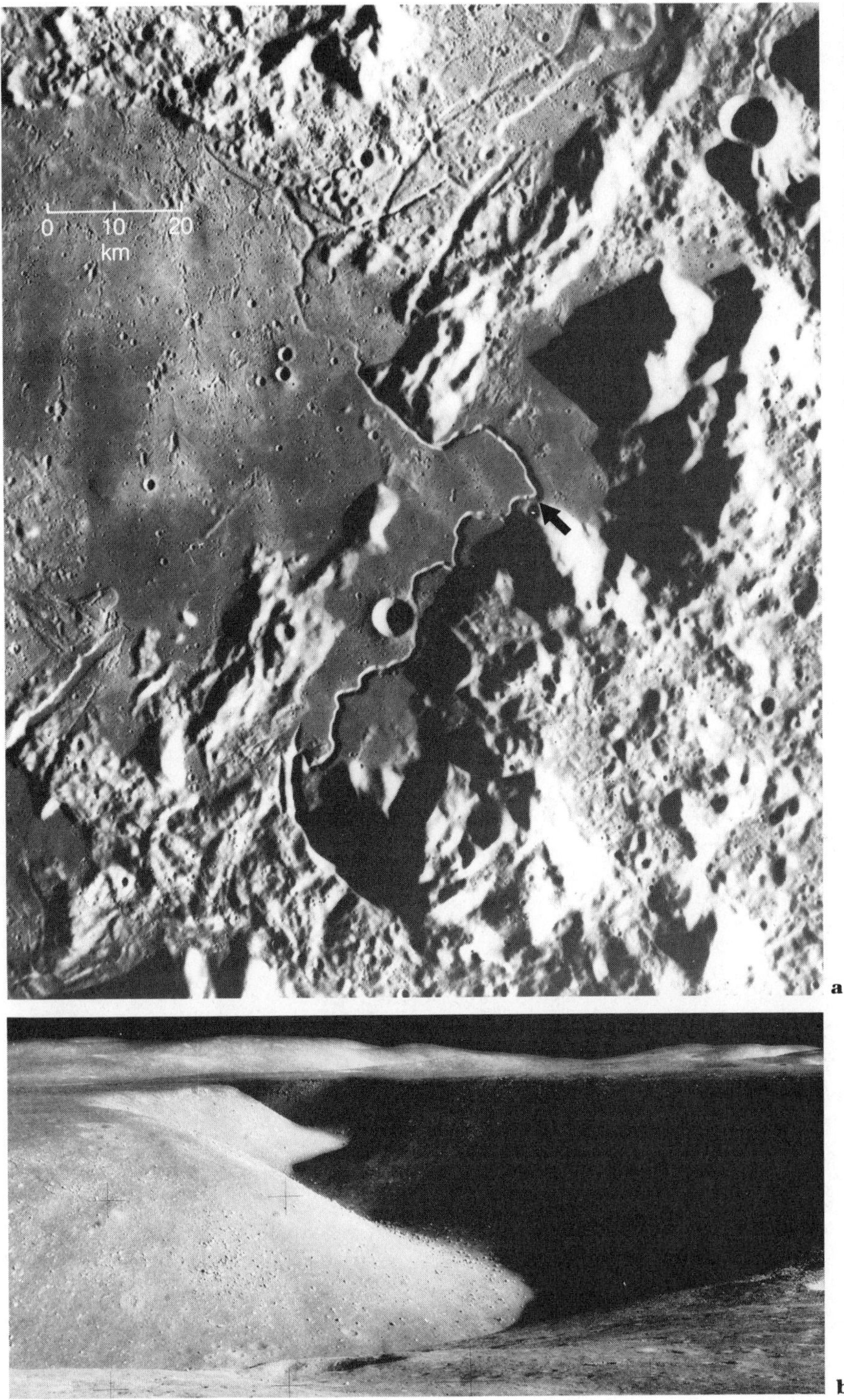

Figure 11–21. Two views of Hadley Rille, a lunar sinuous rille. (**a**) *Vertical view shows Imbrium basin's Apennine Mountain rim (right) and the Mare Imbrium edge (left). Rille originates in the depression in the mountain shadows at bottom. Arrow shows position and direction of view in* **b**. (**b**) *Ground view, looking north from the sharpest bend at the base of the prominent Apennine peak. See also the close-up of the rille wall, Figure 10-4b. Rille width about 1.3 km. (NASA, Apollo 15)*

Figure 11–22. Davey crater chain on the moon, photographed from orbit with a camera hand-held by Apollo 14 astronauts. Alignment of such craters suggests an internal eruptive mechanism along a subsurface fissure. (NASA)

at the complex landforms that develop as volcanism and tectonic activity modify the primeval, impact-dominated landscapes of planets. The large impacts that formed multiring basins 4.6 to 3.9 Gy ago probably created patterns of concentric and radial fractures in the lithospheres around the basins. These were immediately masked by radially striated ejecta blankets; later the underlying fractures helped control the distribution of volcanic eruptions. Magma reached the surface along some concentric fractures. Portions of premare craters and basin rims broke

*Figure 11–23. Latticelike grid systems of lineaments, apparently formed by tectonic development and lava flooding along radial and concentric fractures around impact basins. (**a**) Ridges primarily radial to the Imbrium basin. These form the Haemus Mountains, which are the rim of the Serenitatis basin (left edge), which predates Imbrium. Width about 200 km. (Earth-based telescopic photo; Hale Observatories) (**b**) Radial and concentric ridge pattern in the rim of Caloris basin, Mercury. Basin floor is at bottom. Width about 450 km. (NASA, Mariner 10)*

a

b

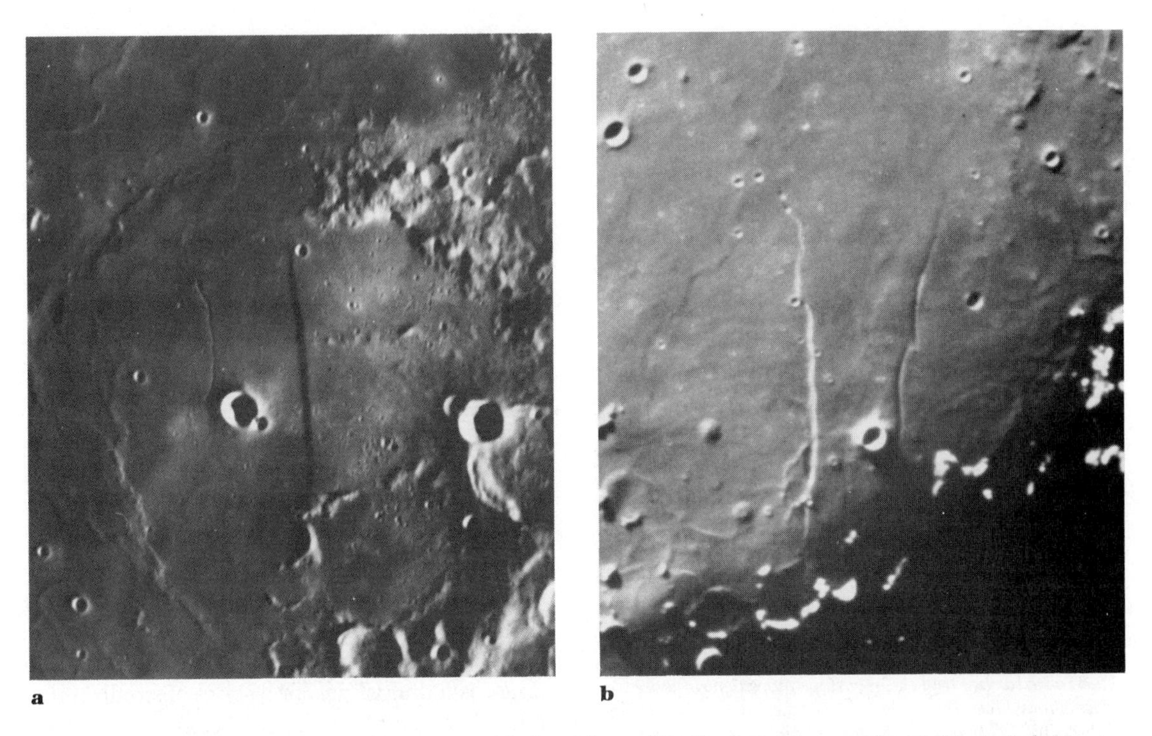

Figure 11–24. Two similar examples of lunar normal faults with parallel rilles, both lying roughly radial to the Imbrium basin and both with downthrown side (left) toward the interior of a mare-filled basin. They may represent basin sagging along premare Imbrium faults. (**a**) *The Straight Wall in Mare Nubium. (Earth-based photo; University of Arizona)* (**b**) *The Cauchy Fault in Mare Tranquillitatis. Note adjacent domes, one with a summit pit. (Earth-based photo; Yerkes Observatory)*

apart along radial and concentric fractures, producing the latticelike landforms of Figure 11-23, called **grid systems**. Arc-shaped maria also formed along concentric basin rings, showing that lava extruded along these faults, as seen in Figure 10-16a (center right). Even postmare faulting in the lava plains was probably affected by the basin-created fractures in the submare lithosphere, as suggested by Figure 11-24.

Figure 11-25 shows the result, a giant impact

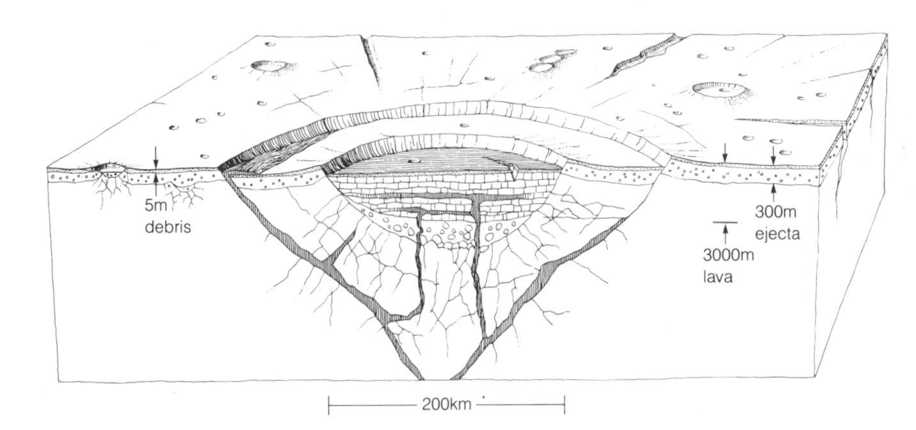

Figure 11–25. Schematic cross section of a multiring basin system, showing fracture patterns and later flooding by lavas.

Figure 11–26. Contact between an upland hill (lower left) and mare surface (right). Downslope movement of loose debris may have caused the convex "toe" at the base of the hill, as well as boulder outcrops, texture, and relative paucity of craters on the hill surface. Picture width 4.3 km. (Flamsteed hills; NASA, Orbiter 3)

scar, modified by lava flooding and tectonic activity along its concentric and radial fractures. Sagging of the central basin as lava extrudes and weighs down the surface creates additional tectonic modification, including concentric rilles (grabens) and wrinkle ridges in the mare surface (compare with Figure 10-16).

Global Lineament Systems

On the moon, the grid system is strongly developed along the northwest-southeast and northeast-southwest directions (Fielder, 1961; Strom, 1964). In places, the surface looks predisposed to break in these directions during the production of otherwise symmetrical basin radial systems. Mars shows evidence of a similarly oriented preferential direction among its various lineaments (Binder, 1966). These systems suggest early, global stresses in the surface layers of these planets. Possibly, relatively rapid changes in the rotation rate or rotation axis, due to large impacts or tides, created such stresses or fractures in the lithospheres. Subsequent stresses could have been relieved by fractures in the preferred directions.

MASS MOVEMENTS

Even if a planetary landscape had negligible impacts, no volcanism, no major fracturing, and no atmosphere to transport debris, it would change. Regolith or soil can move in response to earthquakes or gravity, even under seemingly insignificant stress. Various kinds of mass motions of this type are described in standard textbooks on physical geology. Of special interest are motions such as **slumping**, whereby material on gentle slopes (considerably less than the angle of repose) migrates downhill. The motion can be discontinuous, as in landslides, or a series of small slumping movements, or virtually continuous **creep**. On Earth, this process is aided by water, not acting as a lubricant under the flow, but as a destroyer of the surface tension created by the normally small amounts of moisture between grains of the soil.

On the moon, an interesting difference between the hilly uplands and mare plains may be due to creep (Milton, 1967). As shown in Figure 11-26, upland hills, where they contact the maria, often have a rounded "toe" at their bases; and the slopes have fewer craters than do the maria, even though the upland surfaces are believed to be older. Both these effects may be due to a slow creep of the regolith downslope, perhaps during seismic tremors caused by impacts or moonquakes. Boulders on the slope are exposed by this process, and craters at the bottom of the slope are sometimes invaded and filled in. Tracks in the dust on lunar hillsides show the paths followed by some of these boulders that have rolled downhill. Figure 11-27 shows a more dramatic downslope movement on Mars—the collapse of the wall of Valles Marineris in enormous landslides.

Figure 11–27. Landslides in Valles Marineris, Mars. Striated fans are flows of material spreading from concave collapse sites in far and near (lower left) walls of the canyon. Note hummocky collapsed terrain (upper center). Canyon is about 2 km deep. Picture width 63 km. (NASA, Viking Orbiter 1 mosaic)

ATMOSPHERIC EFFECTS ON PLANETARY SURFACES

The last chapter showed that a planet with no atmosphere acquires a regolith as a result of meteorite and micrometeorite cratering. We can estimate that most airless planets acquire a few millimeters of regolith every million years. Thus, even geologically recent lava flows would have fine dust on their surfaces without atmospheric protection. Therefore, if we add even a thin atmosphere to such a planet, the highly mobile dust will be moved by the wind.

Windblown Deposits on Mars

Mars and arid parts of Earth have developed astonishingly similar landforms as the result of wind action, as shown in Figure 11-28. In fact, research on Martian surface features has emphasized that traditional geology texts underrate the importance of wind action on Earth. The study of geology evolved in European climes, and geological study of arid regions suffered because they were economically unimportant (until the oil discoveries of this century). Yet many of Earth's equatorial regions, like Mars, are arid and dominated by wind-formed landscapes.

Even before the detailed mapping by Mariner 9 and landings by Vikings, researchers recognized that the thin Martian winds could carry Martian dust grains smaller than about 0.3 mm, while the wind would pick up larger grains, enabling them to hop erratically over the surface—a process called **saltation**. This phenomenon is familiar to anyone who has lain on a windy, sandy

a

b

c

Figure 11–28. The closest analogs of Martian landscapes on Earth occur in the most arid regions. (**a**) *In this coastal desert region of Peru, rainfall is so rare that landforms are controlled mostly by wind, which has stripped fine material and left a "pavement" of various-sized boulders and gravels. Rare (every few decades?) storms cause runoff that erodes channels, which resemble those on Mars. (Photo by author)* (**b**) *Scattered basalt boulders and in a once-glaciated area on the arid summit of Mauna Kea volcano, Hawaii. (Photo by author)* (**c**) *Scattered basaltic boulders and dust at Viking 1 landing site in Chryse Planitia, Mars. (NASA)*

beach (see Sagan and Pollack, 1969). From Earth, large-scale Martian dust storms have been seen, proving the existence of Martian dust transport. Saltation can play a striking role in erosion, as shown in Figure 11-29. Since the saltating grains stay within a meter or so of the ground, they undercut rock outcrops, wearing fresh supplies of dust from rock surfaces while old debris weathers into fine, iron-oxide red dust.

Most patchy markings seen on Mars through Earth-based telescopes are apparently due to thin dust deposits, either lighter or darker than the background (Veverka, Thomas, and Greeley, 1977). Because the Martian winds are seasonal, the markings often change seasonally and annually, as seen in Figure 11-30. For example, new dust storms often start in the summer in the southern hemisphere, sweeping new masses of dust to other regions of the planet. This explains the seasonal variations in the dark markings, interpreted incorrectly by Lowell and other early observers as evidence of growing vegetation.

One form taken by the thin dust deposits is a pattern of long **wind streaks** downwind from craters and other obstacles, as seen in Figure 11-31. As the wind sweeps across the Martian plains, dust is often deposited on the leeward sides of craters. It may be darker or lighter than the background. In some regions streaks from recent dust storms overlap earlier, fainter streaks in other

Figure 11–29. "Mushroom Rock," a wind-eroded outcrop of basalt in the dusty wastes of Death Valley. Saltating sand grains have cut into the lower parts of the rock and will eventually sever the upper part. (Photo by author)

directions, created by earlier storms in other seasons when prevailing wind directions were different. As shown in Figures 11-31b and 11-32, similar effects can be found on Earth. Wind tunnel studies with simulated Martian conditions produce similar streaks on the leeward side of scale model craters (Greeley and others, 1974). Spacecraft monitoring of streaks and associated dust storms shows that the global Martian wind patterns roughly repeat from year to year.

Localized "drifts" around the Viking 1 lander were found to lie in the same orientation as the larger-scale streaks in that region observed from orbit (Greeley, Papson, and Veverka, 1978).

Dune fields, or large masses of wind-sculpted dunes, are common on both Mars and Earth (Breed, 1978). Winds pile fine materials into ripple patterns on a vast range of scales, from centimeter-sized waves to graceful ridges kilometers apart (Figures 11-33 and 11-34). When Martian dunes were discovered during the Mariner 9 orbital mapping program in 1971–1972, many planetary geologists turned to the classic work by Bagnold (1941), who made the first serious studies of dunes while stationed in the African desert in the 1930s. On Mars, since crater rims disturb the wind flow, many of the most remarkable dune fields are formed in crater bottoms. The most spectacular dune field in the solar system rings the north polar cap of Mars (Cutts and others,

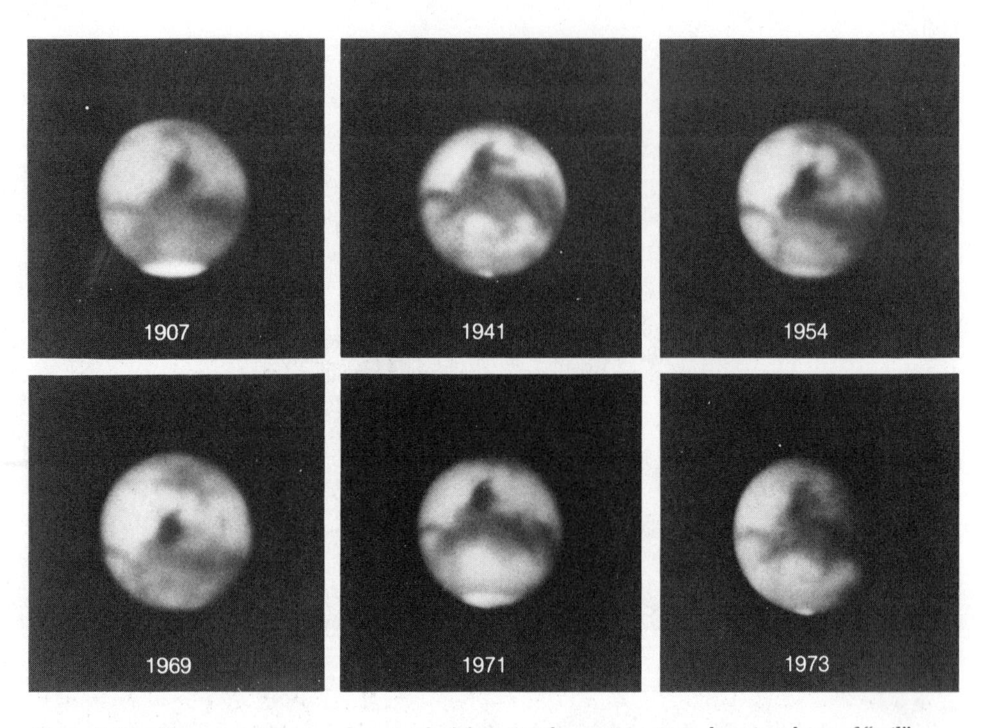

Figure 11-30. Changes on Mars as photographed from Earth over 66 y. Note changing shape of "tail" extending east from Syrtis Major, especially between 1941 and 1954, and brightening of Hellas bright region (bottom) between 1969 and 1971. Changes probably involve variable dust deposits. (Lowell Observatory)

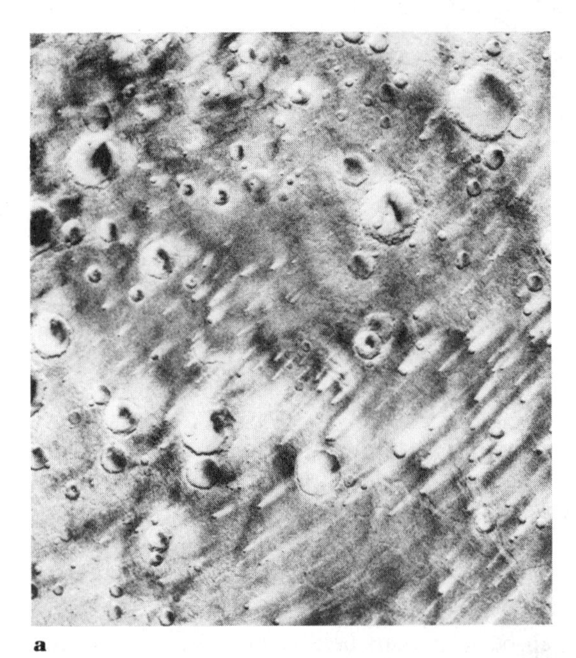

a

b

*Figure 11-31. Wind streaks on Mars and Earth. (**a**) (left) Intermediate-sized craters in this region have downstream streaks; dark deposits in larger craters are probably dune fields. (NASA) (**b**) (above) Light-toned streak left in the wake of a bush on darker background material, Williams Wash, western Arizona. Similar streaks appear in the background. (Photo by author)*

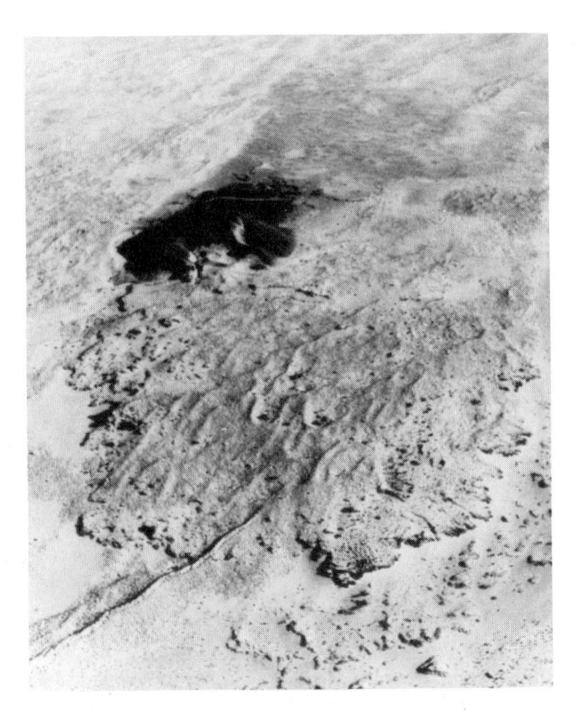

Figure 11-32. Cinder cone and associated basaltic lava flow (extending toward camera), covered by a thin deposit of light, windblown dust. The flow structure can be seen, but the albedo is lightened. Dark dust derived from a cinder cone is deposited beyond the cone in a wind streak pattern. (Pinacate volcanics, Sonora, Mexico; aerial photo by author)

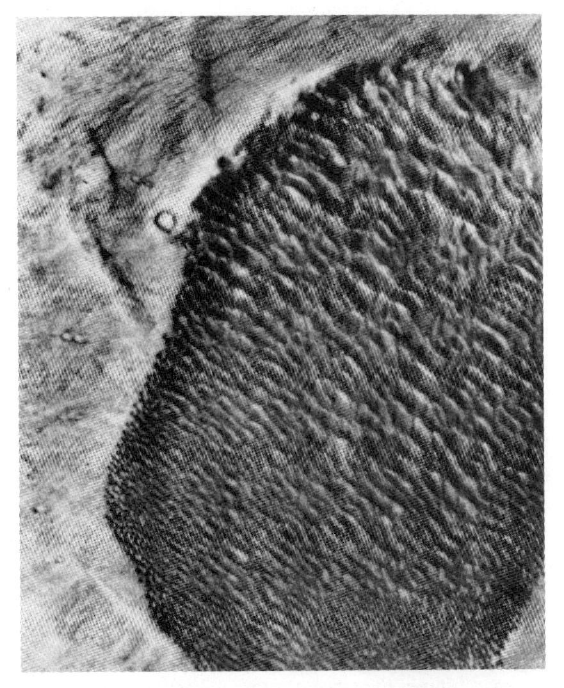

Figure 11-33. Dark dune field forms a 35 × 55 km patch on the floor of a Martian crater. (NASA, Mariner 9)

1976). Its area of 5×10^6 km^2 exceeds that of the Sahara and Arabian deserts combined.

In the polar regions of Mars, thick deposits of sediments have accumulated. These are revealed in orbital photos by the contourlike edges of the stratified beds, as seen in Figure 11-35. Researchers believe that the annual condensation of carbon dioxide and water frost deposits in the polar regions (see next chapter) plays a role in the dust deposition (Squyres, 1979; Cutts, Blasius, and Roberts, 1979). Probably the volatiles condense on dust particles and then carry them to the surface. Annual deposits of dust and ice may form, with the dust left behind during the spring "thaw" (or more properly, sublimation). Variations in dust storm intensity from year to year and in the polar climate from one era to another may aid in creating layered deposits of different thicknesses and strengths. Crater counts in the Martian polar sediments are very low (Figure 10-22e), suggesting that resurfacing occurs at the 1-km scale in 0.1 Gy or less. Deposition, stratification, and erosion are probably still occurring, creating vivid polar landscapes, as suggested by Figure 11-35b.*

HYDROSPHERIC EFFECTS ON PLANETARY SURFACES

Any near-surface layer of a planet in which water is an important agent is called a **hydrosphere**. Earth is currently the only known planet with an active hydrosphere.

---

*Note in passing that our manned and unmanned landers on various planets have mostly been targeted at "smooth" landing sites and thus have not yet shown us the more spectacular planetary landscapes.

Figure 11–34. Changes in fine-scale dune structure at the foot of a 1-m boulder on Mars occurred sometime during the 112-d interval between these two pictures. A 30-cm tongue of dust slipped from the dune face, right center. Boulder is mantled by dust settled out of the air. (NASA, Viking 1)

Figure 11–35. (**a**) *Orbital view of stratified terrain about 15° from Mars' south pole. Width 40 km. Individual strata are about 20 to 50 m thick. Lack of craters suggests that the deposits are very young, perhaps still forming. Striations (top) may be due to wind erosion. (NASA, Mariner 9)* (**b**) *A landscape, page 371, in the Martian polar regions showing stratified deposits similar to those in the orbital view shown in part* **a.** *View includes other features of the polar terrain, such as eroded landforms, frost deposits on the slopes, and the dune field. (Painting by Pamela Lee)*

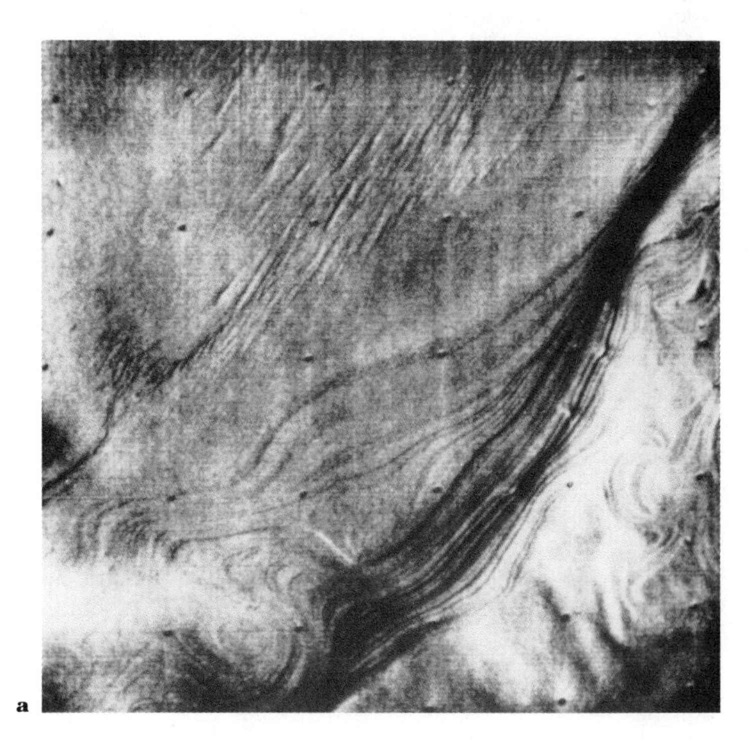

a

Water on Present-Day Mars

The few percent water in the surface layers of Mars is found in three forms: chemically bonded in minerals in the soil, frozen in ice deposits at the two poles, and frozen in permafrost beneath the surface. The first two forms were discovered spectroscopically from Earth and confirmed by Viking. Viking lander soil samples released 0.1% to 1% by weight of water after being heated to 620–770 K. The subsurface permafrost layers are more conjectural; they are supported by topographic features resembling collapses when ice melts and water drains out of an area, as seen on Earth (Carr and Schaber, 1977). Also, the peaking of atmospheric water vapor in early summer as local soil is warmed may indicate a permafrost reservoir buried at depths of 10 cm to 1 m in both polar regions above 40° latitude (Farmer and Doms, 1979).

The Channels of Mars

Planetary scientists, knowing that the amount of water on Mars is very small, were astonished in 1972 when photographs from Mariner 9 revealed numerous features that appear to be well-preserved dry riverbeds. These came to be called **channels**—not to be confused with the largely imaginary "canals" popularized by Lowell around 1900 (see Chapter 2). Examples are shown in Figures 2-16, 11-36, and 11-37.

It is hard to believe that these channels could have been produced by liquid water under present Martian conditions. On much of the Martian surface, the atmospheric pressure is less than the 6.1 mbar needed for water to remain liquid; water would spontaneously boil away into the gas phase. On the rest of Mars, the temperature is usually below freezing, and any liquid water that did not evaporate into the gas phase would freeze.

b

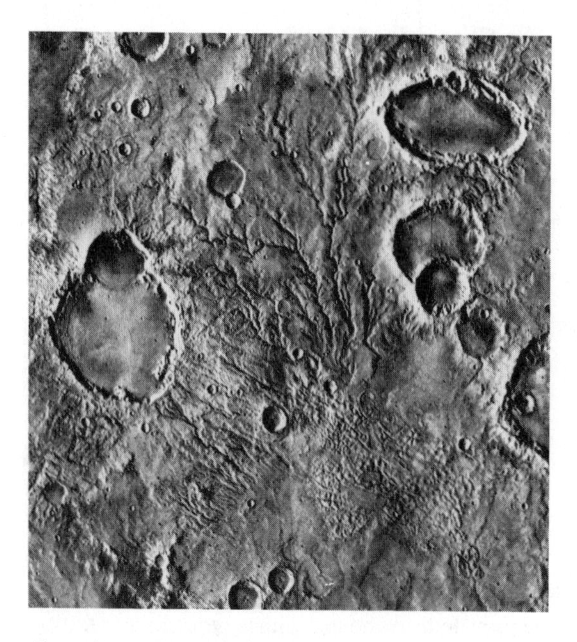

Figure 11–36. Martian channels and tributaries in the cratered uplands northeast of Argyre basin. Width of picture 310 km; main channels about 2 km wide. (NASA, Viking)

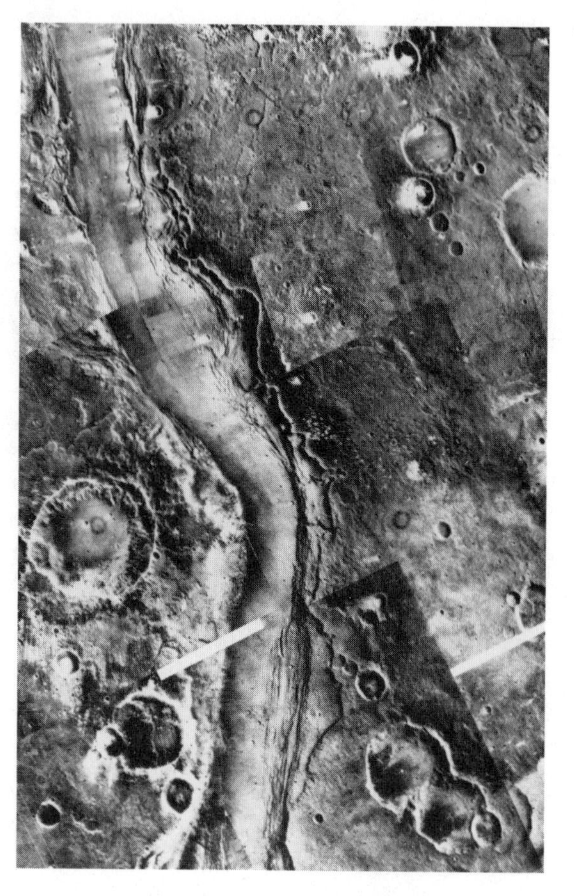

Figure 11–37. Martian channel Ares Vallis, 20 km wide and about 1 km deep. Note terraces and the scarcity of superimposed craters. Flow is toward the top, where a channel cuts off the mouth of an older channel (upper left). (NASA, Viking mosaic)

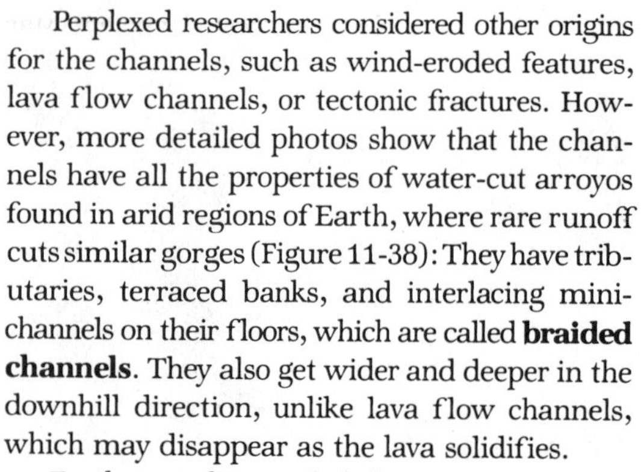

Perplexed researchers considered other origins for the channels, such as wind-eroded features, lava flow channels, or tectonic fractures. However, more detailed photos show that the channels have all the properties of water-cut arroyos found in arid regions of Earth, where rare runoff cuts similar gorges (Figure 11-38): They have tributaries, terraced banks, and interlacing minichannels on their floors, which are called **braided channels**. They also get wider and deeper in the downhill direction, unlike lava flow channels, which may disappear as the lava solidifies.

Further study revealed three types of channels: (1) Some are large and have tributary systems that fan out into the Martian desert. They originate over a large area (Figures 2-16 and 11-36). (2) Other large channels originate in rugged, depressed regions, such as that shown in Figure 11-39. These regions are called **chaotic terrain**, and they resemble collapsed terrain on Earth produced by the melting of permafrost and the outflow of water. (3) A third type of channel is more abundant but smaller. It occurs as a channel network and is common in the dark, cratered uplands. All channels are most common in the Martian equatorial (warmer) regions and less

Figure 11–38. Terrestrial arroyo in Death Valley. Note terraces and braided deposits on the floor, as found in many Martian channels. (Photo by author)

common in higher latitudes and polar strata (Pieri, 1976).

Some of the larger channels empty into broad plains, such as Chryse Planitia, the plain where Viking 1 landed. Here, water seems to have swept across the whole area, eroding craters and cutting through wrinkle ridges, as seen in Figure 11-40.

A few areas, such as shown in Figures 2-16 and 11-36, exhibit **dendritic** channel patterns. These are treelike branching patterns of tributaries. The water that probably carved these cannot have originated in a single collapse feature. Some researchers have suggested ancient Martian rainfall to account for the dispersed water source (Masursky and others, 1977)—a radical suggestion for a planet whose present atmosphere and surface cannot sustain much liquid water!

Obviously, the Martian channels force us to suggest that the ancient Martian surface environment was wetter than today. How old are the channels? Lacking rock samples, we can only judge from crater counts, which are low. The numbers of craters are in that unfortunate middle range to which assigning an absolute age is difficult.

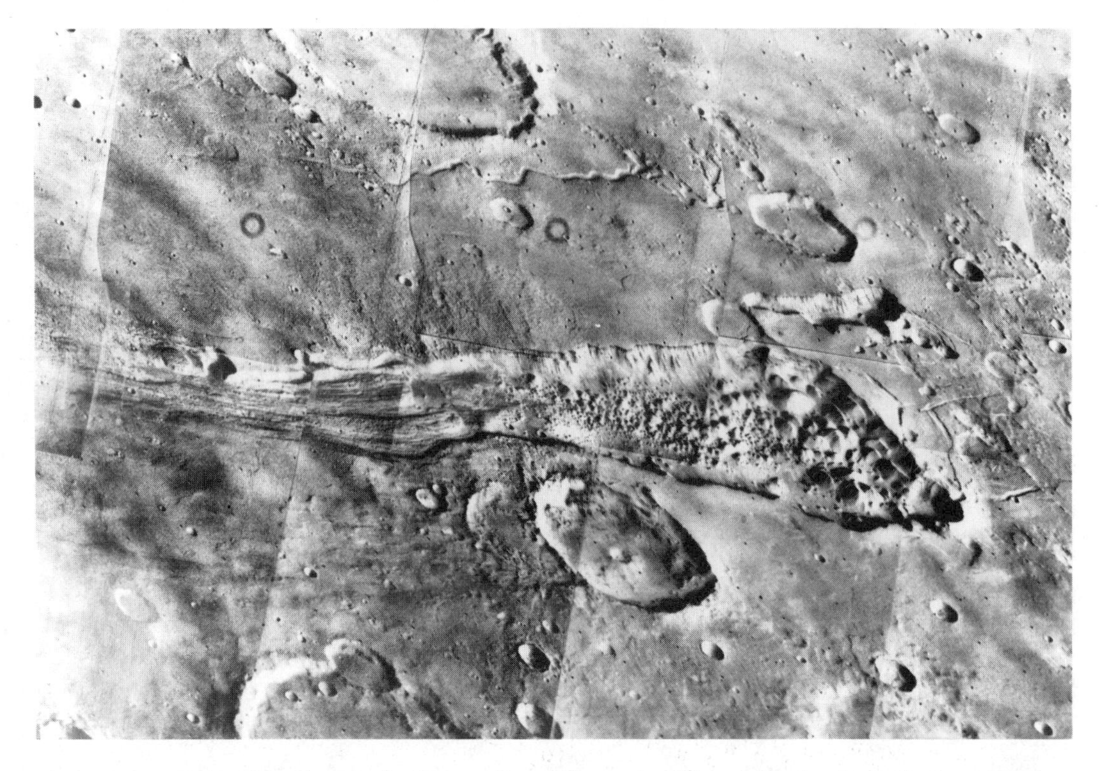

Figure 11–39. Martian channel emanating from a collapsed "box canyon" containing chaotic terrain. Note striated flow deposits on channel floor. Restricted area of source indicates that the origin of the flowing liquid was associated with the formation of the chaotic terrain. (NASA, Viking mosaic)

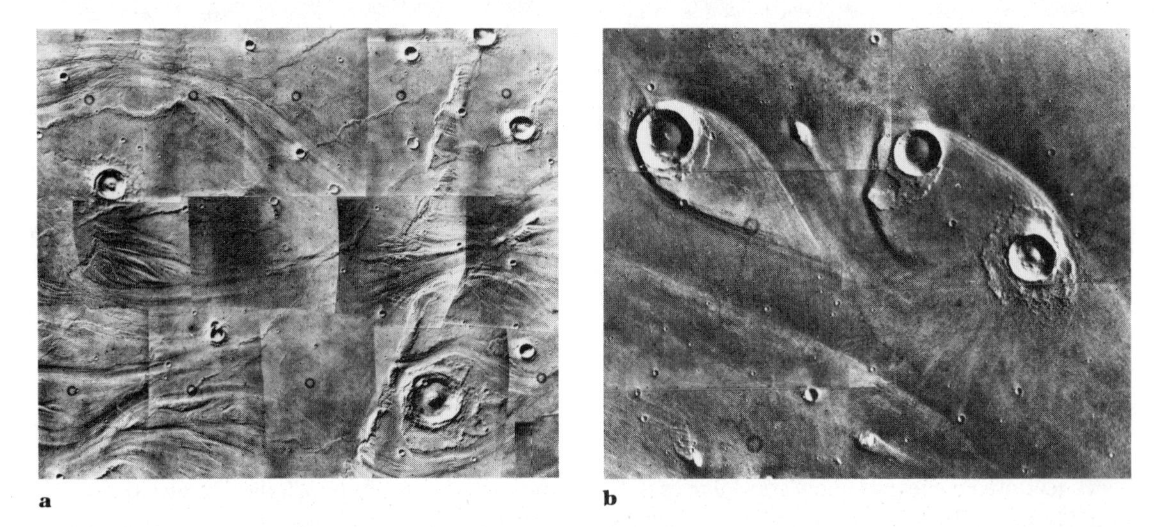

Figure 11–40. Portions of Mars' Chryse Planitia modified by a presumed flow of water. **(a)** *Flow features appear throughout this 200 × 250 km lava plain. Note wrinkle ridges cut by flow. Flow was from west (left), where channels empty into the plain from the highlands.* **(b)** *Streamlined "islands" left in the wake of preflow craters. The tear-shaped island is 15 × 41 km. (NASA, Viking mosaics)*

An extended episode or episodes of channel formation (especially the fine dendritic systems) has been placed somewhere between 0.4 and 3.5 Gy ago by several investigators (Masursky and others, 1977; Hartmann, 1978).

The degree of difference between modern and ancient Martian climates needed to explain the channels is debatable. The channels emanating from chaotic terrain require the least climate change. They have been pictured as catastrophic floods caused by sudden, massive melting of underground permafrost, perhaps by geothermal activity in localized areas. Volumes of collapsed terrain and channels imply water flow rates of a few hundred million cubic meters per second (Masursky and others, 1977).

This figure compares with 2×10^4 m^3/s for the Mississippi River, and about 2×10^7 m^3/s for the prehistoric Lake Missoula flood. The latter was a catastrophic flooding of eastern Washington State, caused by the sudden breakup of a glacial dam across a prehistoric lake. The terrain caused by this flood is called the Channeled Scablands of eastern Washington and bears a remarkable resemblance to some Martian channels (Komar, 1979). Such volumes of water are so great that flows could have traveled a long distance on Mars before evaporating or freezing. Wallace and Sagan (1979) calculate that 20-m-thick layers of ice would have formed on the surfaces of Martian rivers and retarded evaporation, even under present conditions. As early as the days of Lowell, astronomers pointed out that solution of salts or other minerals in Martian water would lower the freezing point significantly, giving additional opportunity for flow.

On the other hand, the channels with dispersed sources seem to require an ancient Martian climate that was warmer and had a higher atmospheric pressure, allowing dispersed liquid water. One possibility is that the channels are very old and date from an early era (3 to 4.5 Gy ago?) when Mars had a denser, primordial atmosphere that later leaked off into space. Another possibility is that greater amounts of sunlight on the ancient Martian poles sublimed more of the polar CO_2 and H_2O ice into the air. These mechanisms for Martian climate variations will be discussed in the next chapter.

Interestingly, while empirical relations have been derived among terrestrial river parameters such as sinuousity, rate of flow, and slope, planetary scientists have had problems in applying these relations to a theoretical understanding of the features of Martian channels, which are subject to gravity a third of that on Earth. Research in this area is still under way.

Patterned Ground on Mars

Patterned ground is terrain with an interconnecting system of polygon- or lattice-shaped furrows and hummocks. On Earth, it is typically produced by expansion and contraction of water-rich underground soil during freeze-thaw cycles of permafrost, and the polygons may exceed 100 m in dimension.

On Mars, several features related to patterned ground have been seen. As shown in Figure 10-19, polygonal patterned ground has been found on many Martian plains, but the polygonal patterns are bigger than on Earth, typically 5 to 20 km. Conceivably, the scale could be enlarged by deeper or thicker volumes of permafrost than on Earth. The patterns have thus been interpreted as possible evidence for Martian permafrost, though they may instead relate to solidification of the lavas that are believed to have formed these plains (Carr and Schaber, 1977).

At the Viking 2 landing site in the Utopia plain, patterns of shallow "gullies" run near the lander, as seen in Figure 11-41. These are nearer to the scale of furrows in terrestrial patterned ground and may also indicate underground Martian permafrost.

GEOCHEMICAL CYCLES

Terrestrial geologists have defined the **geochemical cycle** as a schematic description of the

Figure 11–41. Furrow across middle foreground (left center to lower right) of this Martian landscape in Utopia Planitia may be part of a larger patterned ground system associated with underground permafrost. Flaky texture of the duricrust is visible between rocks in the lower right corner where sampling was conducted. (NASA, Viking 2)

complete history of rock material after it is injected into the planetary surface environment from deep magma sources. The concept can usefully be applied to other planets.

Each planet has its own geochemical cycle. On a small, airless, primitive world, the possible histories of rock material are limited. Lava on the surface may be shock damaged, pulverized, reburied, and/or welded into coherent breccia before remelting occurs. The cycle is slow, because of general geological inactivity. On planets with atmospheres and active interiors, such as Earth, Venus, and Mars, the cycle may be faster and more complex.

Figure 11-42 shows the cycle known in most detail—that for Earth. Magmas may produce atmospheric, hydrospheric, and rock products. The rock materials may break down into sediments, be metamorphosed, or be remelted. Most of the continental materials have probably been through the cycle several times. The cycles for Venus and Mars are less understood, but several important elements may be sketched.

Carbon Dioxide, Carbonates, and the Urey Reaction

Harold Urey, in his pioneering 1952 book on the geochemistry of the planets, pointed out the importance of the reaction between CO_2 gas and the silicate minerals to form carbonate minerals and quartz. An example is

$$\underset{\text{enstatite}}{MgSiO_3} + \underset{\text{carbon dioxide}}{CO_2} \rightleftharpoons \underset{\text{magnesite}}{MgCO_3} + \underset{\text{quartz}}{SiO_2}$$

High temperatures drive this reaction to the left, while the presence of water drives it to the right. This and similar reactions are sometimes called the **Urey reaction**.

The great importance of this type of reaction is that it allows us to appreciate the remarkable difference between the surface environments of the two similar-sized planets Earth and Venus. Urey pointed out that volcanism would produce carbon dioxide gas on both planets (see next chapter); but on Earth, oceans dissolve most of this gas, and the weak carbonic acid thus formed reacts with the silicate rocks to form carbonate

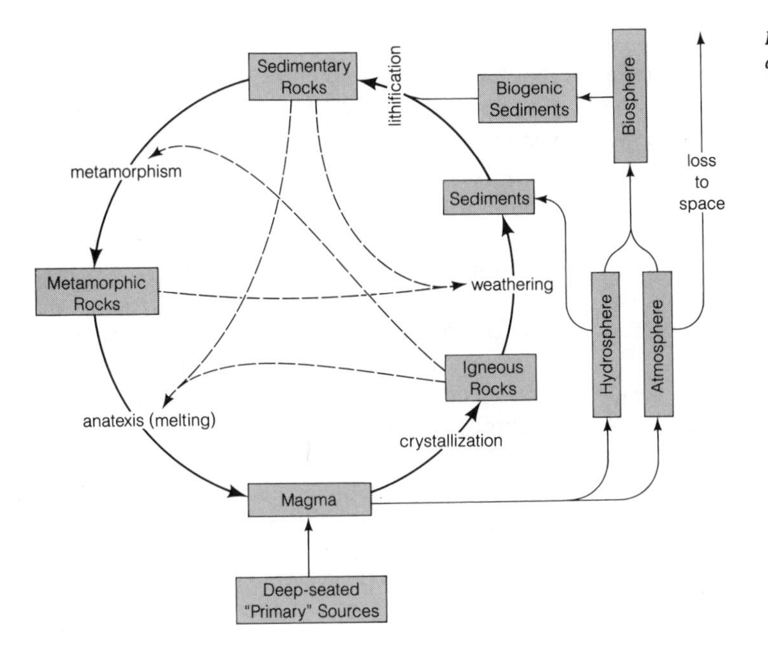

Figure 11–42. The terrestrial geochemical cycle.

rocks. Thus, most of the CO_2 on Earth is tied up in carbonate rocks and in the oceans. Urey (1952) hypothesized that Venus has long been deficient in H_2O, so the CO_2 remains gaseous. Hence the reaction on Venus is driven to the left, and most of Venus' CO_2 is in its massive atmosphere instead of in rocks.

The Chemistry of Venus

Recent studies of Venus complicate the picture, though the basic idea remains the same. In 1967, Connes and others made the surprising discovery that hydrogen chloride (HCl) and hydrogen fluoride (HF) are trace constituents of the carbon-dioxide-dominated Venus atmosphere. Later, Sill (1973) and Young (1973) showed that the droplets forming the Venus clouds are sulfuric acid. Clearly, Venus has an unearthly chemistry!

Chemical analyses soon showed that the hot (750 K!) atmosphere of Venus would react with the rocky surface to affect the surface chemistry. Lewis (1968) attempted to explain the atmospheric chemistry in this way and suggested that even mercury and lead could be driven out of the crustal rocks and into the atmosphere, with lead possibly condensing out at the poles. The Russian geochemists Khodakovsky and others (1979) made more detailed analyses of the chemistry of rocks containing Si, Ti, Al, Fe, Mn, Mg, Ca, and other elements in the presence of CO_2, CO, and H_2O at several altitudes and temperatures on Venus. Consistent with the above discussion, they found that the CO_2 would remain in the atmosphere because of a lack of water to dissolve it. Any water may have been driven into the rocks by oxidizing iron minerals:

$$2FeO + H_2O \rightarrow Fe_2O_3 + H_2$$

the same reaction that may have used up some of the water on Mars and produced the rust-red, oxidized iron minerals there.

Hunten (1973) considered a different reaction that would use up water:

$$CH_2 + 2H_2O \rightarrow CO_2 + 3H_2$$

The idea here is that Venus may have accreted from hydrocarbon-rich planetesimals, with about 25% as much water as Earth's planetesimals. The available water was then consumed in oxidizing the hydrocarbons (represented as CH_2 in the equation) and perhaps iron minerals, as in the earlier equation. The hydrogen from these reactions, being very light, escaped readily into space and left the carbon-dioxide-rich atmosphere behind. On Earth, water was left over after the hydrocarbons were oxidized.*

To be more realistic, such discussions have to include reactions among all the important mineral components. However, we probably do not know all of them.

The Chemistry of Mars

One important process on Mars has been the oxidation of iron minerals, as mentioned above. Interestingly, this may tie in with the probable presence of flowing water in the past. Everyone is familiar with the expanses of red-colored deposits in the deserts of the U.S. Southwest, and geologists have found that similar deposits of sediments, called **red beds**, are common in many parts of Earth's stratigraphic sequence. Studies show that fresh terrestrial sediments derived in humid regions tend to be brownish and blackish and involve hydrated iron oxides, often in amorphous or poorly crystallized states. Arid regions with intermittent supplies of moisture produce the red iron oxide minerals. A fluctuating water table, alternately wetting and drying the sediments, especially encourages production of red beds (Van Houten, 1973). Thus, the postulated episodes of a moist Martian climate would be ideal for producing the observed red Martian dust.

If certain regions of Mars were once or episodically exposed to water (either by flow or upward percolation of melted permafrost), then another chemically important phenomenon there may have been the creation of evaporite deposits. As shown in Figure 11-43, these can produce spectacularly differentiated salt deposits on Earth. The term *salt*, in this case, includes not only NaCl, but also other minerals, especially sulfates, such as anhydrite ($CaSO_4$) and gypsum ($CaSO_4 \cdot 2H_2O$).

As mentioned in the last chapter, production of such evaporites tends to bond surface dust into crumbly layers called **duricrust**. Supporting the idea that water has played a role in the Martian mineral chemistry, duricrust was found on Mars (Figure 11-41) and found to be 20% to 50% richer in sulfur than the loose dust (Mutch and others, 1977). Further, even the loose dust has about 100 times the sulfur content of average materials in Earth's crust (Clark and others, 1977). These results suggest that Mars' soil is rich in sulfate evaporites.

TEMPERATURES OF PLANETARY SURFACES

If planetary surfaces consisted only of solid rock and had no atmospheres, their temperatures would be fairly simply determined by the amount of solar radiation striking them. At dawn, the surface would warm rapidly until it radiated energy at the same rate as it absorbed energy. The temperature would peak soon after noon and then drop at night as the surface radiated its energy into space.

In reality, various factors complicate this. One is that most planets have a dusty regolith that acts as a good insulator, preventing much heat from being transmitted more than a few centimeters into the ground during the daylight period. The subsurface thus shows much less temperature variation during a day than the surface does. Further, as we will see in the next chapter, atmospheres greatly alter the temperature by absorbing sunlight or the reradiated planetary thermal radiation at some wavelengths and not at others. Another complication is albedo: Light-colored rocks absorb much less sunlight than dark rocks.

*The initial water budget of Venus is discussed further in the next chapter.

Figure 11–43. Evaporite beds: "The Devil's Golf Course," Death Valley. The deposits are 95% pure salt plus silt. They are formed by evaporation of water draining onto the plain, which is the floor of a large graben. (Photo by author)

Table 11-1 summarizes some observational and theoretical data on planetary temperatures. The measured day and night temperatures in the table are either direct measurements or estimates closely based on measurements of thermal properties. The calculated average temperatures are based on the simplified theoretical model given in the accompanying mathematical box. Scanning the table, one sees that the dominant effect in determining temperatures is simply the distance from the sun. The considerable albedo differences among Saturn's moons, for instance, make only about a 30 K difference in temperature. Outstanding anomalies are the 730 K (855°F) surface temperatures of Venus, caused by its special atmospheric properties (see the next chapter), and the erupting hot spots of Io, ultimately produced by Jupiter's tidal heating effects.

SYNTHESIZING AND SUMMARIZING PLANETARY SURFACE PROCESSES

To summarize this and the preceding chapter, we will review surface environments planet by planet. It is an exciting challenge to attempt to draw together all this information so that we can produce comprehensive pictures of the surfaces of other planets. We ask not simply for predictions of numerical measurements to be made by instruments, but for an idea of the landscape, history, and "feel" of each planet as it may be experienced by the first human explorers. Table 11-2 summarizes some of this information, with the planets entered in order of size so that the trend can be seen from smaller bodies with primitive materials to larger bodies with more complex, differentiated surfaces—a trend made more complicated by the additional changes from silicate mineralogies in the inner solar system to carbonaceous and ice mineralogies in the outer solar system.

PHOBOS, DEIMOS, AMALTHEA, AND OTHER SMALL BODIES

Phobos and Deimos have similar colors and albedos (Veverka, Duxbury, and Thomas, 1978). The material has a low albedo of only 5% as seen on pages 146 and 471 (Pang and others, 1980). Observers disagree on whether the spectra are neutral (like C-class asteroids) or reddish (like Ds). The densities (about 1950 kg/m^3) measured by Viking and PHOBOS-2 are also consistent with carbonaceous composition. Close-up photos of Phobos and Deimos (Figures 2-18, 5-7a, 7-5a, 7-9, 10-26, and page 118) and mid-range photos of Amalthea, Hyperion (Figure 2-27), S11 (Figure 7-5c) and other small Saturn moons reveal heavy (saturated?) cratering on all these bodies.

In spite of some similarities, the surfaces of small moons differ. The textures of even Phobos and Deimos differ at a scale of 10 to 100 m. The terrain on Deimos seems blanketed, with boulders protruding here and there; the craters on Phobos seem more rugged, and the terrain is cut by the striking grooves best seen in Figure 5-7a.

Soter (1971) pointed out that while meteorites easily knock debris off Phobos and Deimos, this dust cannot easily escape from Mars' gravitational field. Instead, it goes into Martian orbit,

Table 11-1
Planetary Temperatures, Measured and Calculated

Planet/Layer	Measured temperature (K) Day	Measured temperature (K) Night	Calculated mean (or typical) temp. (K)	Bond albedo (assumed or measured)	Effective emissivity (assumed)	References and notes
Mercury						
Soil surface	700 max.	100	452	0.10	0.95	Mariner 10; Strom (1979).
Venus						
Cloud tops, 65-km altitude	215–240	215–240	261	0.76	0.65	Pioneer Venus IR; Taylor (1979).
Surface air	721–731	732	—	—	—	Pioneer Venus 4 landers; Russian landers; Seiff and others (1979). Strong greenhouse effect.
Earth						
Surface air	277–310	260–283	281	0.39	0.65	
Moon						
Soil surface	380 max.	100 min.	280	0.11	0.95	Taylor (1975). Subsurface, 210 K; soil = excellent insulator.
Mars						
Surface air	240 max.	190 min.	—	—	—	Viking landers; Hess and others (1977).
Soil surface	250–280	150–180	—	—	—	Viking orbiter IR; Kieffer (1977).

Vesta Soil surface	—	—	175	0.24	0.95	—
Ceres Soil surface	—	—	171	0.05	0.95	—
Jupiter Cloud tops or haze	120–150	—	120	0.44	0.65	Voyagers, Hanel and others (1979a, 1979b). $P = 150-800$ mbar.
Io Surface	125–135	—	106	0.56	0.8	Voyagers, Hanel and others (1979a, 1979b).
Warm spots	270–310?	—	—	—	—	Voyagers; Hanel and others (1979a, 1979b). 180-K spots cover 5%.
Eruptions	385–600?	—	—	—	—	Voyagers; Hanel and others (1979a, 1979b); Witteborn, Bregman, and Pollak (1979). Molten sulfur at 385 K.
Europa Surface	125 max.	85	103	0.58	0.9	Voyagers; Hanel and others (1979a, 1979b).
Ganymede Surface	145	85	107	0.48	0.95	Voyagers; Hanel and others (1979a, 1979b).
Callisto Surface	153 max.	79 min.	122	0.13	0.95	Voyagers; Hanel and others (1979a, 1979b).
Small moons Surface	140–170	—	126	0.03	0.95	Amalthea, quoted by Morrison (1977, p. 283).
Saturn 1-bar level	120–160	—	88	0.46	0.65	Orton (1979).
Icy moons Surface	67–93	—	70	0.7	0.9	Hanel and others (1982).
Titan Cloud tops	90	—	97	0.21	0.65	—
Surface	82–136	—	—	—	—	Morrison (1977, p. 281).
Iapetus Trailing	—	—	85	0.35	0.9	—
Iapetus and Phoebe Leading	⩾110	—	94	0.05	0.9	Hanel and others (1982).
Uranus 1-bar level	78	—	59	0.56	0.65	Hanel and others (1986).
Icy moons Surface	84–86	—	62	0.27	0.9	Hanel and others (1986).
Neptune 1-bar level	69	—	48	0.51	0.65	Conrath and others (1989).
Icy moons Surface	38	—	33	0.85	0.9	Conrath and others (1989).
Rocky moons Surface	—	—	52	0.1	0.9	—
Pluto and Charon Surface	—	—	37	0.6	0.9	Cruikshank and Silvaggio (1980).

spreading along the orbits of Phobos and Deimos. The velocity required to knock debris off these satellites is only around 10 m/s, but the velocity to make the debris escape Mars entirely is about 960 m/s for Phobos and 550 m/s for Deimos. Thus, impacts on these satellites, especially Phobos, would eject debris, forming "dust belts" around Mars. These would eventually be

MATHEMATICAL NOTES ON PLANETARY SURFACE TEMPERATURES

Any surface heated by radiation approaches an equilibrium temperature where the amount of radiation received equals the amount radiated for any given area over any given time. Therefore, the equilibrium temperature of a planetary surface can be calculated by equating the flux of incoming solar radiation with the flux of outgoing thermal radiation. The Stefan-Boltzmann radiation law states that for a blackbody:

$$E = \sigma T^4$$

where

E = energy flux ($J/m^2 \cdot s$) radiated by body of temperature T

The Stefan-Boltzmann constant σ was defined in Table 2-2. We also defined the solar constant $F_\odot$ there, which is the flux of energy ($J/m^2 \cdot s$) received by a surface 1 AU from the sun. Since the flux of sunlight declines as the inverse square of the distance from the sun a, the flux of sunlight striking any planet is $F_\odot/a^2$, where a is given in astronomical units. Thus, for any blackbody, we would have the equilibrium condition, where incoming flux equals outgoing flux:

$$\frac{F_\odot}{a^2} = \sigma T^4$$

But a planet isn't a blackbody. The amount of incoming radiation actually absorbed is $(1 - A)$ times the total, where A is the Bond albedo, defined in Chapter 2. Further, for a spherical planet with zero obliquity, only the equator receives full illumination; a factor $\cos \phi$ needs to be inserted to correct for other latitudes. (For planets with high obliquity, the situation is still more complex.) Finally, if the planet rotates with moderate speed, the surface doesn't come into full thermal equilibrium; it heats up during the day and cools at night. A given zone of latitude absorbs sunlight across its diameter (as seen from the sun) but reradiates around its full circumference, so another correction factor of diameter/circumference, or $1/\pi$, is added.

Similarly, the right side of the equation needs a correction for a nonblackbody. A gray body radiates e times as much flux as a blackbody of the same temperature, where e is a property of the given material, a fraction called the emissivity. The emissivity for brick, glass, and rough dull metal is around 0.8–0.9, but for shiny metals it can be as low as 0.02–0.1.

With these corrections, our estimate of temperature T at latitude ϕ on a rotating planet is given by the equation

$$\frac{1 - A}{\pi} \cos \phi \frac{F}{a^2} = e\sigma T^4$$

If we insert numerical values and assume we are interested in a "typical" latitude of 30°, this equation reduces to

$$T = \sqrt[4]{\frac{1 - A}{ea^2}}\, 285\ \text{K}$$

where the Bond albedo A, emissivity e, and solar distance a are known or can be estimated. Often only the geometric albedo is measured, but the Bond albedo is usually not very different (see Chapter 2 for clarification of these terms).

swept up by the satellites themselves. Each new crater thus provides a source of sandblasting that erodes the satellite and yet ultimately adds to the regolith. Such a regolith, at least 1 mm deep, has been found on Phobos by photometric and infrared thermal observations (Pollack, 1977). Differences in the time since the last major impact(s) might thus account for the different surface textures of these satellites and the high visibility of the Phobos grooves, which appear to be surface expressions of fractures from a crater almost large enough to fragment the satellite.

Table 11-2
Dominant Surface Material on Selected Bodies

Object	Approx. diameter (km)	Surface material
Saturn ring bodies	<0.1	H_2O ice
Deimos	12	Dark, carbonaceous silicates
Phobos	22	Dark, carbonaceous silicates
J7 Elara	80	Dark, carbonaceous silicates (+ H_2O ice?)
J6 Himalia	180	Dark, carbonaceous silicates (+ H_2O ice?)
S9 Phoebe	220	Dark, carbonaceous silicates (+ H_2O ice?)
J5 Amalthea	270 × 155	Dark, reddish (carbonaceous and sulfur?) soil
S2 Enceladus	500	H_2O ice (virtually pure)
1 Ceres	914	Dark carbonaceous hydrated silicates (+ NH_3-bearing silicates or H_2O ice?)
S3 Tethys	1048	H_2O ice
S4 Dione	1120	H_2O ice
U1 Ariel	1160	H_2O ice + dust?
P1 Charon	1190	H_2O ice
S8 Iapetus	1440	Dark, carbonaceous silicates (leading side); H_2O ice (trailing side)
U4 Oberon	1520	H_2O ice + dust?
S5 Rhea	1530	H_2O ice
U3 Titania	1580	H_2O ice + dust?
Pluto	2300	CH_4 ice
N1 Triton	2700	N_2 ice + Ch_4 ice
J2 Europa	3130	H_2O ice
Moon	3476	Basaltic soil and rock
J1 Io	3630	Sulfur compounds
J4 Callisto	4840	Dark (carbonaceous?) soil + H_2O ice
Mercury	4878	Basaltic but not ultrabasic soil (based on spectra)
S6 Titan	5150	? (obscured by clouds, differentiated?)
J3 Ganymede	5280	H_2O ice and (carbonaceous?) soil
Mars	6787	Basaltic, chemically weathered soil and rock
Venus	12,104	Basaltic and some granitic soil and rocks
Earth	12,756	H_2O liquid; basaltic and granitic soil; rock
Neptune	49,528	H_2 liquid?
Uranus	51,118	H_2 liquid?
Saturn	120,660	H_2 liquid?
Jupiter	142,800	H_2 liquid?

Source: Data based on Cruikshank (1979; private communication, 1981); Degewij (private communication, 1980); Strom (1979); Veverka, Thomas, and Duxbury (1978); Burns (1977); King and four others (1992).
Note: Table lists bodies for which observations or inferences are available; see also Tables 7-1 and 7-2 for related asteroidal data.

Craters lying along the Phobos grooves are a real puzzle. They are too well aligned to mark secondary impacts, and they can't mark simple drainage of regolith into cracks, because they have raised rims. Hartmann (1980) suggested that Phobos originally contained volatiles that vented through the fracture-grooves, piling up regolith around the vents. Fanale and Salvail (1990) calculated that Phobos could still contain ices, and might be leaking H_2O molecules. Independently, Soviet investigators on the PHOBOS-2 mission reported detection of H_2O molecules at about the flux predicted by Fanale and Salvail. Phobos may thus be one of many volatile-rich, C-type asteroids scattered by Jupiter from the outermost belt, and captured by Mars (Hartmann, 1990).

Another small moon, Jupiter's Amalthea (155 × 270 km), is unusually red and may be coated with sulfurous material that was blown off Io and spiraled inward toward Jupiter, similar to the contamination of Iapetus by Phoebe dust, described in the last chapter.

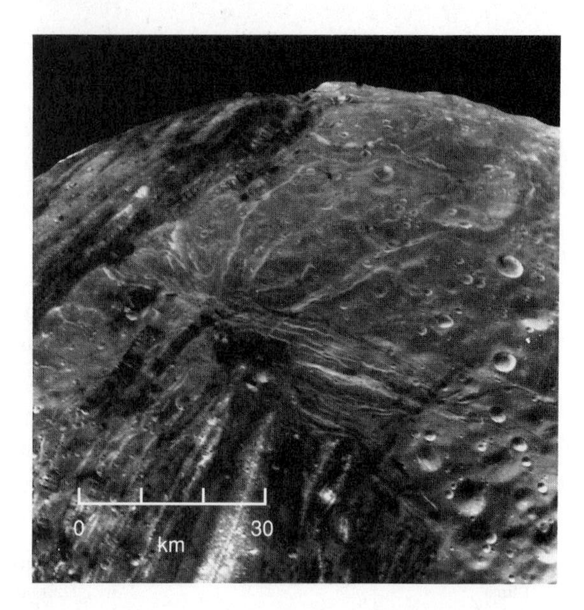

Figure 11–44. Closeup of Miranda shows old cratered terrain (right) cut by swaths of bright and dark grabens. Note profile of graben crossing the limb, notable as a 4-km-deep valley (top center). Contrary to some depictions, the fault surfaces are not vertical, but slope at about 45°. (NASA Voyager 2)

MIRANDA

Stepping up to a larger size, one of the most unexpectedly complex surfaces in the solar system is that of Miranda (Figure 1-1, page 4). Smallest of Uranus' five major moons, 470-km Miranda was expected to be the least geologically evolved. However, Voyager photos showed a unique, incredibly complex surface. Swaths of ridges, grooves, and 4-km-high faulted cliffs form peculiar angular patterns, interspersed with ordinary, heavily cratered surfaces (Figure 11-44).

Why should such a small moon have so much deformation? The first theory, proposed by the Voyager team, noted that moons close to giant planets have a much higher impact rate than more distant moons, because the planet attracts asteroids and comets toward itself. Miranda might have been fragmented and reassembled one or more times, leading to a chunky, irregular structure. Heat released as gravitational contraction led to a more spherical shape and could have caused geologic activity (Smith and others, 1986).

But what about tidal heating, of the sort that powered Io's volcanoes? Originally, it was thought too weak to do the job. However, Dermott and others (1987) and Marcialis and Greenberg (1987) analyzed Miranda's dynamics and concluded it may have been temporarily trapped in resonances in the past, and forced into higher eccentricity orbits and/or chaotic rotation that may have led to heating. Thus, the heat sources remain uncertain, but tidal heating now seems likely.

THE INTERMEDIATE-SIZED SATELLITES OF SATURN

Mimas (390-km diameter), Enceladus (500), Tethys (1048), Dione (1120), and Rhea (1530) have cratered surfaces of clean, bright water ice, with albedos 60% to 95% (Cruikshank, 1979; Smith and others, 1982). Though mostly larger than Miranda, they show some related features. Several have fracture systems.

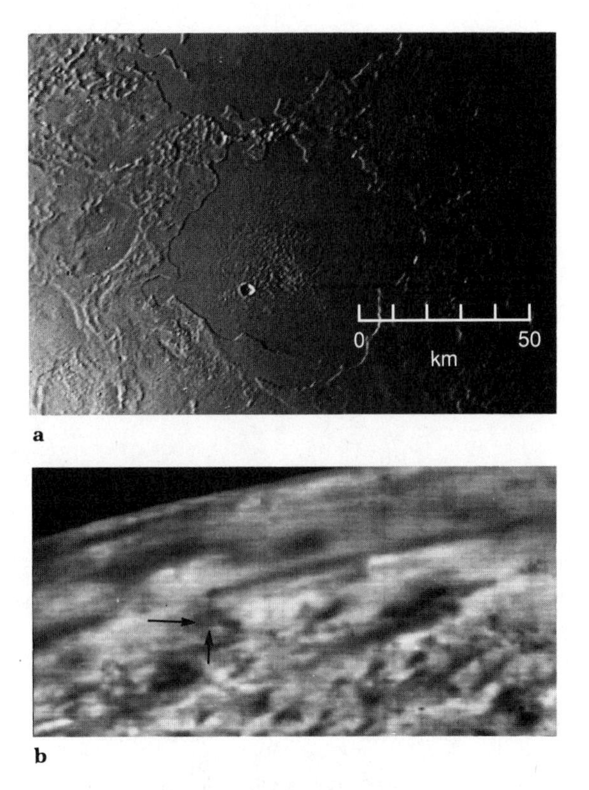

Figure 11–45. Features of Triton. (**a**) *"Ice lake" of uncertain origin may mark an ancient, degraded impact or tectonic feature. Deposits around it seem clearly layered (see overlapping layers at bottom center).* (**b**) *Dark volcanic smoke plume is the long dark streak drifting from left to right across image. Plume rises vertically from a vent on the surface (arrows). Vent details were not resolved, but elevation of plume was confirmed by stereoscopic pairs of images. Horizon at top. (NASA Voyager 2)*

As discussed on pages 43 and 291, Enceladus is especially important because it appears to offer a "missing link" between heavily cratered old surfaces (Rhea, Dione, Callisto) and smoother or fractured young surfaces (Europa, Ganymede, Miranda). The ages of the young plains have been estimated at less than 1 Gy by crater counts (Smith and others, 1982). Whether tidal heating could have produced volcanic eruptions of watery "magma" is controversial (see page 290), but some heating and eruptive mechanism seems to have been at work. The E ring of Saturn—a weak ring whose density peaks near Enceladus' orbit—suggests that eruptions have thrown material off Enceladus into orbit around Saturn.

As with Miranda (facing page), Voyager scientists estimated the cratering rate in the Saturn system to be so high that the inner moons, especially Mimas, may have been fragmented and reassembled several times. Mimas' and Tethys' heavy cratering includes one crater on each body almost large enough to have shattered the body. Similar satellite breakups might have contributed to the plethora of small inner moonlets and ring material.

TRITON AND PLUTO

Even before Voyager got to Neptune, scientists expected a strange surface on its moon Triton. Cruikshank and Silvaggio discovered atmospheric methane gas and surface methane ice in 1979. Color changes suggested atmospheric hazes; Cruikshank and others (1989) suggested nitrogen frost or even seas of liquid nitrogen. Voyager found no seas but an extraordinary world (Figure 2-36). The tenuous atmosphere consists mainly of nitrogen with traces of methane, and a total surface pressure of only 0.01 mbar (compare with Mars' 6 mbar). Alternate illumination of the two poles by the sun (82 years each due to Triton's high orbit inclination) leads to seasonal transfer of frost deposits from one hemisphere to the other. The south cap had been illuminated for 30 years when Voyager arrived, but still had a frost layer of N_2 ice, with albedo $> 80\%$. The very young surface is mostly N_2 and CH_4 ice with some carbonaceous dirt; there is no spectral evidence of H_2O ice. Unfamiliar geologic units include layered, smooth "ice lakes" (Figure 11-45a) and cantaloupe-textured terrain (Figure 10-21).

Most surprising of all were at least two active vents, with plumes of dark smoke rising some 8 km vertically, then streaking off parallel to the surface in upper-atmosphere winds (Figure 11-45b). Numerous parallel dark surface streaks mark windblown deposits of debris from such

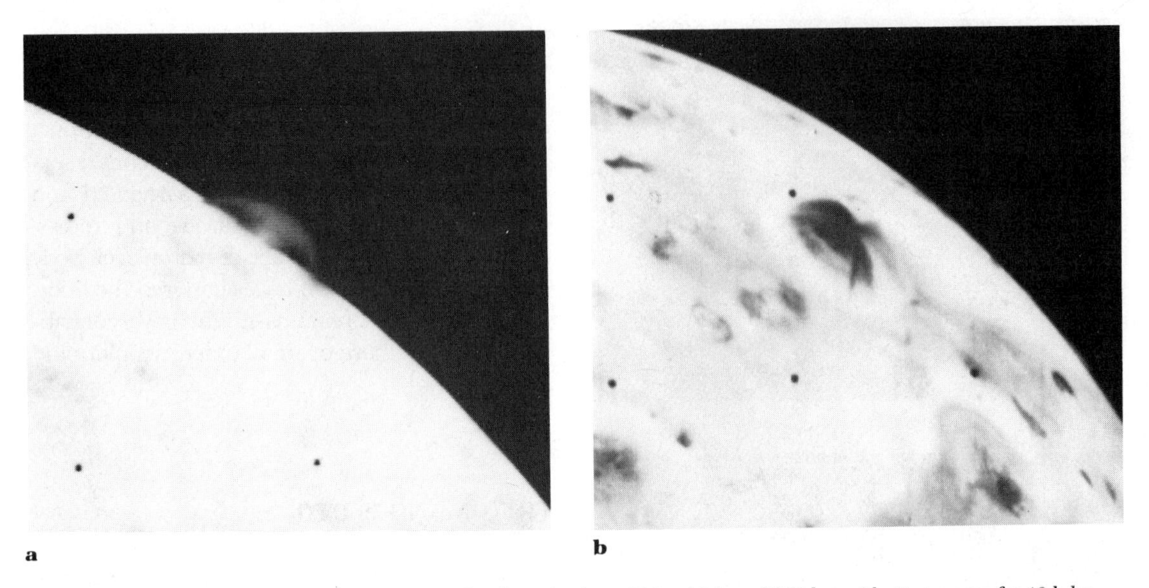

Figure 11–46. An erupting volcanic plume on Io. The plume is about 70 km high and 250 km wide. It was seen for 18 h by Voyager 1 and was still active 4 mo later, as revealed by Voyager 2. (**a**) *Plume silhouetted on horizon.* (**b**) *Plume seen obliquely against Io's disk. Note distinct arcs of ejecta, indicating complex jetting at vent. (NASA, Voyager 1)*

vents. Researchers presume that the vents are driven by gas, probably N_2 jetting from below the surface (Smith and others, 1989). Some of the required heat may come from its being kept molten for about 1 Gy by tidal forces after capture by Neptune (Goldreich and others, 1989). Such a capture event would also explain Triton's odd, inclined, retrograde, circular orbit.

Numerous researchers have commented on the similarity of Triton and Pluto, whose diameters are 2700 and 2300 km, respectively. They both have densities close to 2000–2100 kg/m^3, tenuous atmospheres of comparable surface pressure, and atmospheres and surface frosts involving CH_4 and N_2. Pluto's strong CH_4 surface frost spectral signature may involve a peculiar transfer of volatiles from its moon Charon.

As remarked above, Triton may have been an interplanetary body that was captured by Neptune. Thus, instead of viewing Pluto as a planet and Triton as a moon, it may be more fruitful to visualize both as only among the largest bodies of the inner Oort cloud. Who knows what revelations will come when the first close-up images of Pluto and Charon's surface geology are made?

IO

At still larger size, the most volcanically active world in the solar system is 3630-km Io, with its dark volcanic vents continuously spewing elegant, umbrella-shaped plumes as much as 280 km into space (Figure 11-46). During 6½ days of monitoring in 1979, Voyager 1 found eight erupting volcanoes. Four months later, Voyager 2 found at least six still erupting and one or two possible new eruptions. Eruptions have subsequently been detected with Earth-based telescopes at Mauna Kea Observatory (Johnson and others, 1988).

Based on their height, the plumes were ejecting material at 0.5 to 1.0 km/s. Exact temperatures were not measured, but models suggested 270 to 310 K for material in the plumes, with the material at the vent perhaps much hotter (Hanel and others, 1979a, 1979b). Since much of Io's surface is believed to consist of sulfur compounds, molten sulfur at 385 K probably exits. The Mauna Kea observations gave 900 K, nearly 200 K above sulfur's boiling point—indicating silicate lavas, not sulfur, may have been involved.

Io volcanoes usually display a blackish cen-

Figure 11–47. Volcanic calderas and flows on Io. Close-up of irregular caldera about 50 km across, surrounded by radiating flows. Photo shows some evidence of stratified terrain; smallest details about 0.5 km across. (NASA, Voyager 1)

tral caldera about 100 km across. This is often surrounded by a lighter patchy region, sometimes crossed by radial, red flows that presumably run downhill from the caldera, as seen in Figure 11-47. Relief is very slight. This region is often surrounded by orange deposits stretching hundreds of kilometers, often accompanied by outlying whitish deposits.

These properties may relate to those of sulfur lavas (Sagan, 1979). At around 500 K, sulfur is black and has the fluidity of basaltic lavas. As it cools to around 470 K, it becomes red and remains lavalike. This would account for black calderas with red, lavalike flows. However, sulfur becomes even more fluid—almost watery—at temperatures around 400 K, at which point the color is yellow-orange. This could account for the wider deposits of yellowish materials beyond immediate caldera rims. Of course, varying compositions of sulfur compounds may also play a role in the colors of Io, but it is interesting that simple sulfur chemistry may explain the gross color patterns of Io.

Io volcanic eruptions have also been studied by extraordinary observations from Earth. About a year before the Voyager flybys, several observers reported a dramatic flare-up of the 3- to 5-μm thermal radiation from Io, which lasted a few hours (Witteborn, Bregman, and Pollack, 1979). The explanation was unknown, though a 600-K hot spot roughly 50 km in diameter was proposed.

A similar event was observed from Earth between the two Voyager flybys (Sinton, 1979). This time, the geometry permitted approximate location of the source on Io, and Voyager 2 confirmed the formation of a new caldera surrounded by new deposits at this site, which were not present four months earlier. Sinton suggested that these events are explosions, possibly caused when a fracture leads from the surface to underground molten sulfur. The pressure release leads to sudden production of vapor in the 1200-K to 1300-K sulfur, which would produce enough energy to blow off part of the crust and expose violently boiling sulfur. This sulfur would then form a crust and cool. This scenario checks with the 23% drop in thermal flux reported during the first hour of Sinton's observation and with a temperature of 650 to 700 K reported by Voyager 1 for one 3- to 5-km hot spot (quoted by Sinton, 1981). Sinton's further observations led to a model with two classes of hot spots at about 300 and 600 K.

The Earth-based observations of spectacular, brief flare-ups, probably during caldera formation, suggest that even more violent volcanism occurs on Io than was seen by Voyager.

While most of the Io volcanoes do not involve high, conical peaks, some impressive mountains have been found. Some calderas and other features have gently sloping rims, and the flows radiating away from some vents testify to volcanic mountains (Carr and others, 1979). Other raised features include plateaulike flatlands (Figure 11-48), which are sometimes cut by cliffs. Some of these appear to be tectonic, uplifted features bounded by faults.

More puzzling are layers of stratified features with eroded-looking cliffs and valleys, found especially in the south polar areas. The polar regions are some tens of Kelvin colder than Io's equatorial regions, and Voyager scientists have suggested that the polar strata of Io are deposits involving SO_2 emitted by volcanoes (Voyager 1 found an SO_2 cloud with partial pressure of 0.0001 mbar near a volcanic plume), which condenses in the colder polar regions (McCauley, Smith, and Soderblom, 1979). Just as Mars has polar strata involving CO_2 and dust precipitation, Io may have polar strata involving SO_2. These strata suggest that SO_2 below 1-km depth has a liquid form and, perhaps with molten sulfur, works its way toward the surface. There the pressure is low enough to allow some of the liquid to change to gas; it expands 5000 times, erupting in the form of SO_2 snow fountains and causing collapse and erosion of the fault faces, which create the eroded scarps seen in the photos.

Figure 11-49 characterizes some of the phenomena of Io as they might appear to a surface visitor.

EUROPA, GANYMEDE, CALLISTO, AND OTHER LARGE SATELLITES

Jupiter's satellites, and presumably others, were clearly exposed to heavy cratering between their formative era and today, since Callisto and parts of Ganymede are nearly saturated with craters. The less-cratered regions appear to have been resurfaced too recently to have accumulated the full complement of craters. The question is, how were they resurfaced? Figure 11-50 suggests the interesting possibilities. Among these bodies, the darkest surfaces are most cratered and the light surfaces least, with the bright surface of Europa being relatively pure H_2O ice according to spectral data. These facts suggest that surface melting or eruption of a watery "lava" allowed the silicate dirt to sink, with subsequent freezing creating a clean ice surface. What, then, was the heat source?

Peale, Cassen, and Reynolds' (1979) original study of Io's tidal heating, which predicted Io's volcanism, found about 5% as much tidal energy generation in Europa as in Io under present conditions. Perhaps tidal and radiogenic energy sources heated Europa enough after the end of the early intense cratering to melt the surface layers, causing eruption of water (Europa's "magma") and resurfacing with ice. Apparently heating was insufficient to drive off all the water, as happened on Io. Ganymede seems to have heated enough to crack the surface in places, causing limited resurfacing in the light-colored, grooved terrain. Some old craters were flooded with lighter ice, as shown in Figure 11-51, but much of the old, dark, cratered terrain, which resembles that of Callisto, was left. Bright, smooth, circular patches are presumably icy scars of old, filled-in craters. They are a new type of planetary feature and were given the unusual name **palimpsest** by the Voyager team; the term refers to a reused writing surface on which the original writing has been erased. Figure 11-51 also shows one of the systems of parallel arcuate rings, which may be remains of multiring basin systems flattened by isostatic filling of depressions by the

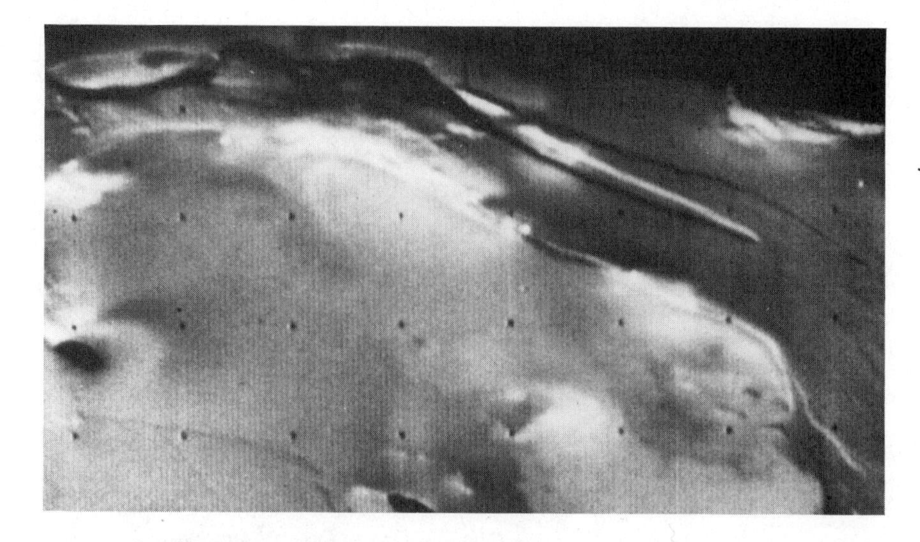

Figure 11–48. Oblique view of structure brought out by low solar illumination along Io's terminator. The finger-shaped valley is about 300 km long and 50 km wide. These and similar structures appear to involve stratified deposits. (NASA, Voyager 1)

Figure 11–49. Imaginary wide-angle view of an erupting volcano on Io, with sulfur flows in foreground. Jupiter subtends an angle of nearly 20°; picture width about 110°. (Painting by author)

Figure 11–50. The sparsely cratered ice plains of Europa and several other satellites in the outer solar system indicate that watery magmas erupted in the past and flooded large regions. In this imaginary reconstruction, a geyser (background) accompanies massive water eruptions on Europa, creating temporary surf and a thin transient atmosphere. Details of such hypothetical events remain to be studied. (Painting by Michael Carroll)

viscous ice lithosphere. As discussed on p. 36 this process may have flattened or obliterated many old, large craters on Ganymede, Callisto, and other icy worlds, while leaving smaller surface features intact.

In a pre-Voyager attempt to explain the sodium glow around Io (see Chapter 2) and other Io peculiarities, Fanale, Johnson, and Matson (1974) raised an interesting point about the surfaces produced by heating and partially melting ice-rich or carbonaceouslike worlds. They pointed out that modest heating of a carbonaceous-chondrite-like body would produce water containing dissolved minerals; this would percolate to the surface, where the water could evaporate, leaving behind evaporite deposits rich in salt (sodium chloride) and sulfur compounds such as those in Figure 11-43. In support of their model, Fanale, Johnson, and Matson noted that gypsum ($CaSO_4 \cdot 2H_2O$), epsomite ($MgSO_4 \cdot xH_2O$, or magnesium sulfate hydrated with variable amounts of water), and bloedite ($MgSO_4 \cdot Na_2SO_4 \cdot xH_2O$) have been identified in carbonaceous chondrites. (Recall that sulfur was also found to be concentrated in the proposed evaporite deposits near the Viking 2 lander.)

Why are evaporites not more common among satellites of the outer solar system? Apparently most water reaching their surfaces froze too rapidly to leave evaporite deposits behind. Only on Io was there such repeated and intense heating that virtually all surface water was eventually driven away, leaving a sulfurous crust. Nonetheless, some regions of some bodies might have been reheated enough to produce concentrations of evaporite minerals.

At the opposite extreme, bodies that never melted significantly never experienced differentiation of the H_2O and the dark carbonaceous dust and hence show the dark coloration and spectrum of the dust. Because modest differences in heating produce gross differences in geochemical differentiation of such materials, the surface chemistries of the outer solar system worlds may be very indicative of their past histories.

Of other large moons, such as Titan and Triton, we have little observational data. In Triton's low-temperature part of the solar nebula, methane (CH_4) is anticipated as a major ice condensate. Spectroscopic work indicates that CH_4 ice is probably present on or near Triton's surface (see Chapter 12 for observational details, whose interpretation is controversial). Spectra imply definite differences between Triton's surface and the types of icy surfaces of the Jupiter and Saturn satellites; candidate surfaces include CH_4 ice, unusual structures of H_2O ice, or clathrates. In the case of Titan, the known properties of the atmosphere guarantee an exotic, if still unknown, surface (see p. 444).

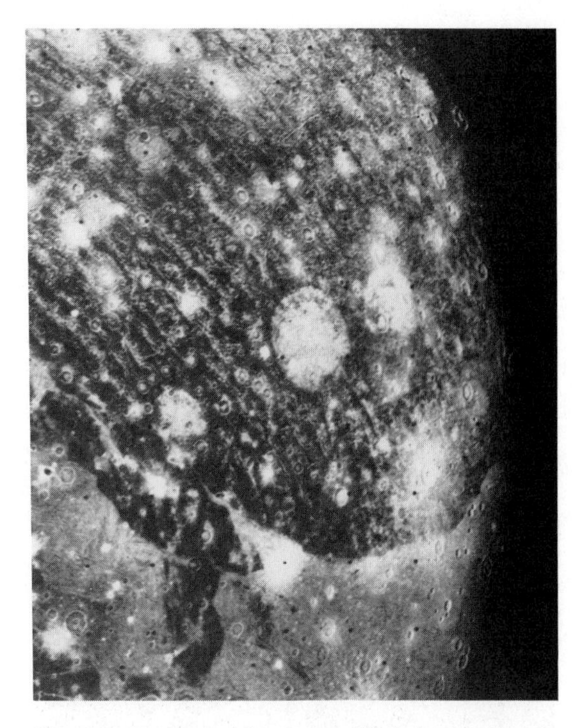

Figure 11–51. Contrast between an old, dark, cratered surface and younger, grooved terrain on Ganymede. The old surface preserved arcs that are probably part of a multiring basin system similar to those seen on Callisto (compare Figure 10-17). Bright ovals, or palimpsests, are probably old craters filled in by ice. (NASA, Voyager 2)

THE MOON

The planetary surface whose history is best known is probably the moon. In the beginning, there was intense cratering (Figure 6-5), repeatedly pulverizing the thin lithosphere that was trying to form on a magma ocean surface. After a few hundred million years, a heavily cratered surface emerged, underlain by a thickening lithosphere (Figure 10-5). Figure 11-52 shows the evolution of this surface seen from an imaginary point in space. In Figure 11-52a, we see the heavily cratered pre-Imbrium uplands, about 4.3 Gy ago, in the region that was to become Mare Imbrium. Volcanism during this period produced lava flows of KREEP basalt, found by astronauts in the lunar highlands. But surfaces of those flows were rapidly brecciated by the intense bombardment. Figure 11-52b shows the same scene about 3.8 Gy ago, soon after the enormous impact that formed the huge Imbrium basin. The Imbrium ring system and ejecta are prominent. Already some lava flows from the partially molten zone below the lithosphere have erupted through the ring fractures created by the impact, producing an appearance like that of the Orientale basin today. In Figure 11-52c, about 3.4 Gy ago, the mare-forming eruptions are continuing. Mare Imbrium and Oceanus Procellarum nearly have their present configurations (see Figure 2-10). In Figure 11-52d, mare flooding has ceased, and impacts have created a few more craters, such as Copernicus, with its impressive ray system. The scene shows the present appearance.

Figure 11-53 gives some loosely corresponding surface views. Figure 11-53a is an imaginary

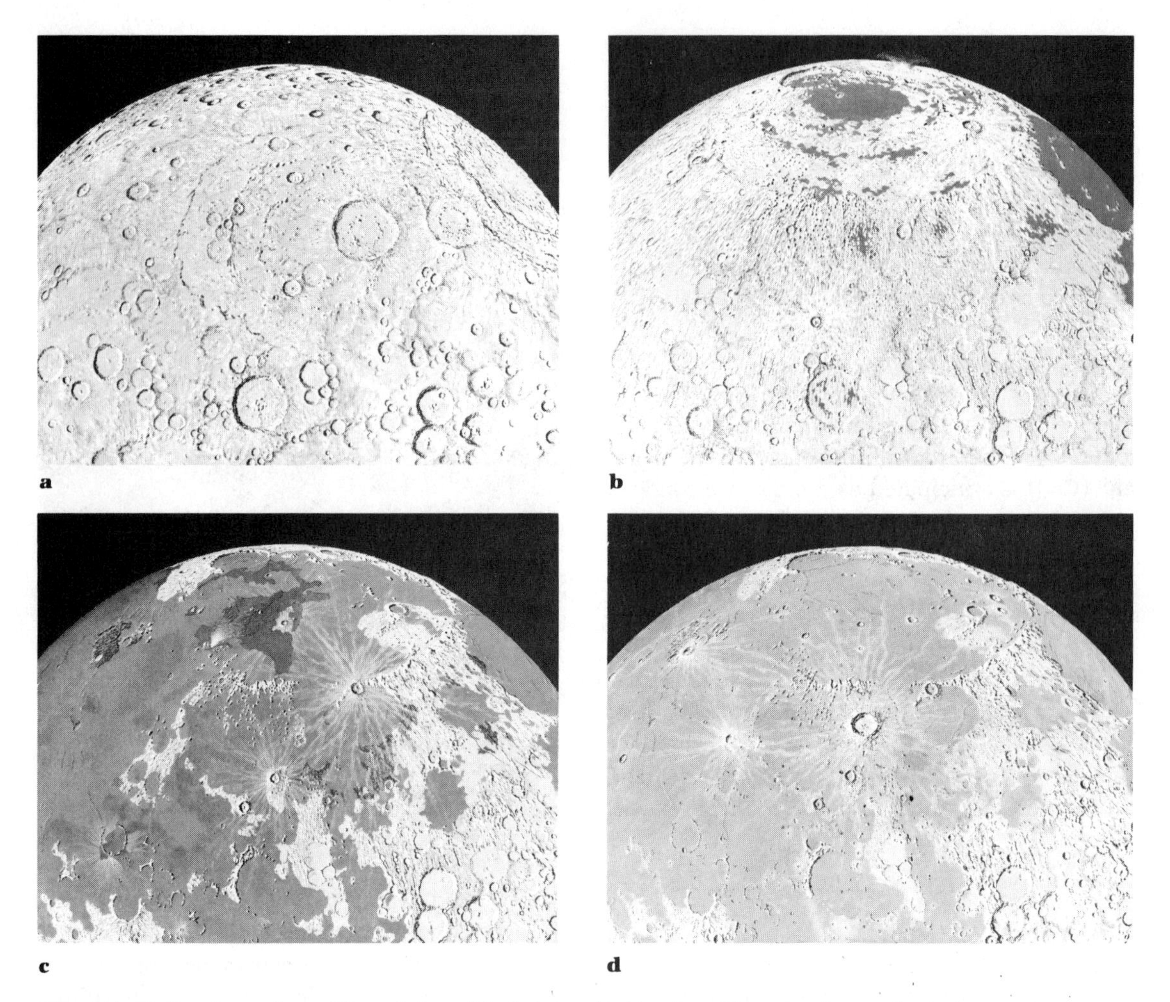

Figure 11–52. Imaginary views of the evolution of the Imbrium basin region of the moon, from prebasin to present configuration. See text for further explanation. (Paintings by Don Davis, U.S. Geological Survey, courtesy D. E. Wilhelms)

view of a fresh mare lava flow of fluid, pahoehoelike basalt, based on the terrestrial appearance of such flows. This is a view not encountered on the present-day moon, because cratering has beaten the surfaces of these flows into rubble. For example, in Figure 11-53b, an impact has punctured a thin regolith cover and exhumed boulders of intact lava. More typical of today is Figure 11-53c, which shows a vista of a regolith-covered mare with upland mountains in the background and astronauts' tracks in the powdery soil.

A major unresolved controversy is the nature of certain light-colored, moderately cratered plains in the lunar uplands. These show some indication of being ancient lava flows partly filling the spaces between crater rims. They have come to be called the **Cayley formation**. Apollo scientists targeted the Apollo 16 mission to one of these areas, hoping to collect ancient upland lava samples. Instead, the astronauts found highly brecciated material with ages typically 4.0 to 4.4 Gy. Many lunar scientists then interpreted these areas as sheets of ejecta that swept across the

a

b

*Figure 11–53. Different stages of lunar surface evolution. (**a**) Imaginary view of fresh pahoehoelike lava flow prior to pulverization by impact gardening. (Painting by Hiroki Morinoue) (**b**) Boulders excavated by cratering near the edge of Mare Serenitatis. (NASA, Apollo 17) (**c**) Powdery regolith soil at Apollo 15 landing site, on the edge of Mare Imbrium. (NASA, Apollo 15)*

c

uplands from the ancient basin-forming impacts, perhaps in the form of immense base surges (see Taylor, 1975, for a summary).

However, many small, dark halo impact craters in Cayley formation units uplands appear to penetrate a light-colored veneer and throw out darker ejecta (Schultz and Spudis, 1979). The ejecta has been shown spectrally to be basaltic (Hawke and others, 1985). Similarly, the Galileo spacecraft, flying past the moon's far side, discovered that darker Cayley-like plains generally have a more basaltic spectrum than brighter uplands. Probably many upland plains mark ancient lava flows, possibly the KREEP basalts. Lava plains that formed before a certain date, around 4.0 Gy ago, received a substantial dusting of bright ejecta, hiding their dark color and making them look, albedowise, like uplands.

Thus, while the moon looks sharply divided into two surface units—bright uplands and dark mare patches making the "man in the moon"—the physical evolution of these units may not be so distinct. The "true uplands" are the light-colored anorthositic rock that formed the early lithosphere about 4.4 Gy ago. Dark basaltic lavas that erupted on that surface prior to about 4.2 Gy ago were more or less chewed up by the intense early meteoritic bombardment (Figure 6-5) and converted back into rugged, light-colored uplands. Thick lava flows that formed about 3.9 to 3.8 Gy ago caught the tail end of the bombardment and received perhaps 3 to 20 times the amount of cratering as the dark lunar maria; they were thus masked by a veneer of overlapping rays and ejecta sheets. Only the lava flows formed after about 3.8 Gy retained their familiar dark coloration. Because most of the moon's surface formed either before 3.9 or after 3.8 Gy ago, most of the surface is either bright upland or dark mare, with few intermediate areas.

Does Lunar Volcanism Occur Today?

For a long time the moon was regarded as a dead world. Indeed, this accounted for a certain lack of interest in the moon by astronomers until the middle 1900s. A few amateur astronomers had reported changes in certain lunar structures, but none of the reports were documented well enough to be taken seriously by most professional astronomers. However, on November 19, 1958, the Soviet astronomer Kozyrev (1959) obtained a photographic spectrum indicating an emission of gas in the lunar crater Alphonsus. Next, on two occasions in 1963, a number of observers at Lowell Observatory saw red glowing spots in and near the crater Aristarchus (Greenacre, 1963). Such events came to be known as **lunar transient phenomena**. Earlier historical examples, less well verified, are also known.

A statistical study suggests that lunar transient phenomena are most frequently reported when the moon passes through its perigee point, when lunar tidal forces are largest (Middlehurst and Chapman, 1968). Apollo seismic observations showed that a particular type of moonquake concentrates at this time. Observations by orbiting astronauts and their instruments turned up no evidence for active eruptions, but measurements of radioactive radon-222 gas, produced by uranium decay, showed peaks over Aristarchus, Grimaldi, and the edges of maria—possibly indicating gas leakage through fractures in these areas (Taylor, 1975). (Radon emissions were correlated with a major Japanese earthquake in 1978; Wakita and others, 1978). In short, significant lava eruptions seem unlikely at the present time, but gas emissions or perhaps even small ash ventings may occur occasionally.

The lunar surface is far from fully known. Because Apollo landings were targeted on rather flat places for safety reasons, we have not yet seen surface vistas at the rims of large craters, along major fault scarps such as Straight Wall, around possible eruption sites, or among domes. A few places are young enough not to have been smoothed by regolith formation. Figure 11-54 shows an example of a strikingly rugged terrain that must have a spectacular appearance from the ground.

Figure 11–54. Interior of the 90-km crater Tycho, one of the youngest and roughest extensive regions on the moon. Lunar samples and crater counts suggest ages of 100 to 270 My. (**a**) *Tycho and surroundings.* (**b**) *Close-up of rough floor northeast of the central peak along the wall shadow. Width 6.5 km; smallest features about 10 m. (NASA, Orbiter 5)*

MERCURY

Three flybys by Mariner 10 in 1974 and 1975 showed that Mercury is similar to the moon in having cratered uplands, multiring basins, and plains with the morphological features of lunar mare lavas. The Mariner findings have been reviewed by Strom (1979).

The major difference between the surfaces of Mercury and the moon is the presence of more intercrater plains on Mercury, somewhat resembling the lunar Cayley formation. These can be seen outside the Caloris basin by comparing Figures 10-16a and 10-16b. A possible interpretation is that since Mercury is the larger planet, its cooling took longer and its asthenospheric source for volcanic eruptions lasted longer, as seen in Figures 9-24 and 9-25. Also, as seen in Table 6-1, Mercury's cratering rate probably averaged about twice the moon's. Thus, while a Mercurian lava plain that formed 3.5 Gy ago would have escaped the early intense bombardment and would not have been converted into cratered uplands, it would have accumulated more cratering than a 3.5-Gy-old lunar mare. At present it would be masked by a veneer of rays and secondary ejecta, thus creating the upland plain appearance seen on the photos.

Another moon-Mercury difference is that Mercury apparently contracted sufficiently to create very large thrust faults (Figure 9-26), as discussed in Chapter 9. Nevertheless, an astronaut plunked down at a random spot on Mercury might have a hard time telling from the landscape alone (without looking at the 2.5 times larger and 6.5 times brighter sun) that he or she was not on the moon.

MARS

Mars, being still more massive than Mercury, had still more prolonged volcanism, which resurfaced many of the early cratered surfaces. More importantly, the atmosphere and its winds moved materials around, creating transient bright and dark markings, dune fields, stratified deposits, wind streaks, and deposits in crater floors. One interesting problem is whether the large-scale dark markings seen from Earth are *all* merely products of wind deposition. Alternatively, for example, many dark areas might be regions with more rock outcrops, producing darker brown surfaces that weather to the lighter, reddish dust characteristic of Martian deserts. Both of the boulder-strewn plains photographed by the two Viking landers were, incidentally, in bright regions of Mars.

Continuing volcanism on Mars produced basaltic flows and built up the largest known volcanic edifices in the solar system, such as the towering, youthful-looking Olympus Mons and neighboring volcanoes of the Tharsis region. They are not known to be active, but neither are they known to be extinct!

Wind circulation particularly has built up dunes and strata in the polar regions. Under present climatic conditions, water ice appears almost permanently frozen at the poles and in permafrost. In the winter, carbon dioxide condenses at the poles, much of it on dust particles in the atmosphere, which then precipitate onto the ground. This dusty carbon dioxide snow has two effects: It coats the ground with a broad, transient, winter carbon dioxide polar cap, and it adds new dust to the sedimentary strata at the poles. Erosion has cut into the strata in many places; there are cliffs eroding back and partially exposing ancient impact craters that were once buried by strata.

The major mystery of Mars is the evidence that the climate was much different long ago. This evidence includes (1) the channels; (2) crater numbers and morphology indicating that craters were eroded and filled faster at some time in the past (1 to 3 Gy ago?) than today; and (3) theoretical calculations that the axial tilt of Mars varied in the past, allowing more or less sunlight on the polar caps. The subject of climatic change will be considered in more detail in the next chapter. At this point, simply remember that in spite of the cold, arid conditions on Mars today, there were probably episodes of permafrost melting and catastrophic flooding in certain regions in the past, and the whole environment may have been moister and warmer at times.

Another mystery of Mars has been the question of life, reviewed historically in Chapter 2. Although the Viking landers found neither life nor even any organic molecules to a level of a few parts per billion, the evidence for a more benign climate in the past raises questions about ancient biochemical processes on the red planet. These will be considered in the last chapter.

Of all extraterrestrial planetary surfaces, that of Mars comes closest to matching conditions on Earth. In terms of resources of air and water, it might offer the most attractive planetary habitat for future astronaut-scientists and colonists.

VENUS

The surface of Venus is enigmatic. We know that the atmosphere produces an extraordinary environment with temperatures around 750 K (891°F) day and night and with pressure 90 times that on Earth's surface. The surface air is "heavy," having a density about 50 times that of air on Earth, but still only 5% that of water. Though Venus is completely overcast by clouds, the bottom of the dense cloud layer is normally about 44 km above the surface, and the lower atmosphere is surprisingly bright and clear, something like a very cloudy day on Earth.

Wind velocities measured at or near the surface by three Soviet probes (Veneras 8, 9, and 10) and two American probes (Pioneer "Day" and

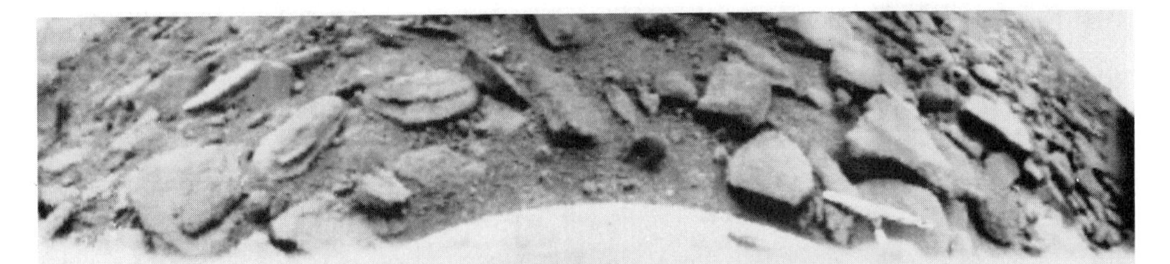

a *and* **b**

c *and* **d**

Figure 11–55. Panoramas on the surface of Venus, photographed by Russian landers, partly visible at bottom center. (**a**) *Boulders and gravel at Venera 9 landing site. T-shaped probes are about 40 cm wide.* (**b**) *Sheet-like rock outcrops and loose (darker) soil at Venera 10 site.* (**c** *and* **d**) *Views in two directions showing gravel and platy rock texture at Venera 14 site. The platy, sheet-like outcrops are believed by some analysts to result from thin flows of very fluid basaltic lavas. Different rock forms indicate active erosion process.* (**a** *and* **b** *courtesy of C. P. Florensky, Vernadsky Institute, Moscow; computer reconstruction by Institute of Problems of Information Translation, U.S.S.R. Academy of Sciences.* **c** and **d** courtesy NASA, provided by U.S.S.R.)

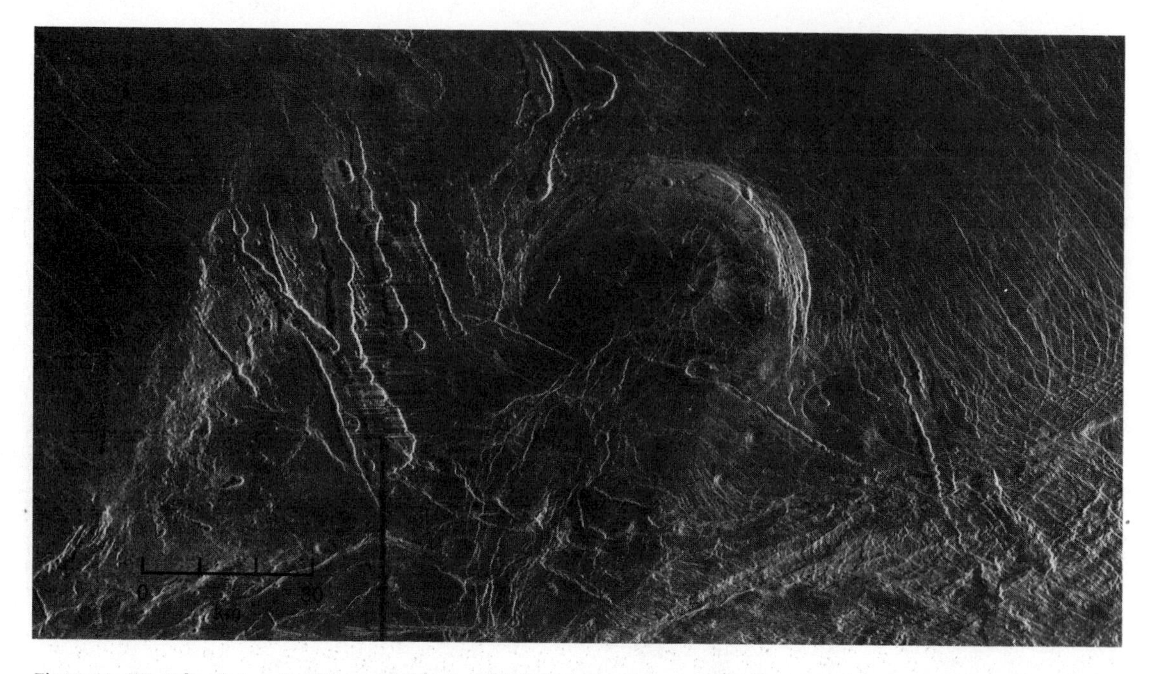

Figure 11–56. Volcanic terrain of Venus at the edge of Lakshmi Planum, an elevated smooth plain. Partly formed circular feature is probably a volcanic collapsed caldera, named Siddons. To its left are lava flow channels and possible collapsed lava tubes (compare Figure 11-19, page 359). At lower right is intensely deformed terrain. (NASA Magellan radar image)

"North") were very low, about 1 ± 1 m/s. This accounts for the low dust content found in the surface atmosphere by the Russian probes (Keldysh, 1977). White (1981) calculated little dust transport and found rates of wind saltation of dust grains on Venus:Earth:Mars to be 1:10:250. Nonetheless, Magellan radar mapping surprised observers with images of scattered dunes and windblown streaks of ejecta from impact craters (Greeley and ten others, 1991).

Photos from Venus' surface made by Venera landers (Figure 11-55) show different kinds of rocky landscapes. Venera 9 landed on a hillside (judging from the camera tilt) and showed gravel and loose angular boulders up to about 0.5 m across. Venera 10 landed on a smoother surface, apparently soil and sheetlike outcrops of rock.

Veneras 13 and 14 landed in March, 1982 with refined equipment. Venera 13 photos showed soil and flat rock outcrops like those of Venera 10, with a few slab-like rocks. Venera 14 photos (Figure 11-55c, d) showed a remarkable slabby or platy texture in flat-lying rock terrain.

Soviet VEGA 1 and 2 missions, on their way to Halley's comet, also dropped landers that carried no cameras but measured compositions.

All seven Soviet landers that gave composition data (Veneras 8–10, 13–14, and VEGA 1–2) landed in plains near the equator, some on the fringes of uplands. Two types of composition measurement were used. Gamma-ray spectroscopy (Venera 8, 9, 10, Vega 1, 2) measured the amount of radioactive K, U, and Th in surface materials, testing whether the rock is basaltic or granitic. X-ray fluorescence (Venera 13, 12, Vega 2) measured the content of Si, Ti, Al, Fe, Mn, Mg, Ca, K, S, and Cl in 1 cm^3 drilled from beneath the lander. At five sites, the composition resembled tholeitic basalt—the type of basalts found in Earth's sea-floor crust, extruded from the upper

Figure 11–57. The most representative surface on Earth. (Photo by author)

mantle. The Venera 13 site also gave a basaltic composition of a high-K type known as leucitic or subalkaline. The Venera 8 site was the most provocative. Situated about 5000 km east of Phoebe Regio, probably on a somewhat raised area that seems in the midst of volcanic flows in the Magellan imagery, it has a granitic composition, possibly a syenite type related to the Earth's continents (Saunders and others, 1991). If this is right, Venus may be only a little less differentiated than Earth in terms of measured rock chemistry, as we would expect on basis of size. In general, the dominance of basalts is consistent with Chapter 9's picture of a planet where lack of plate-active tectonics has inhibited crustal recycling and granitic continent buildup. After examining the evidence, Magellan researchers affirmed a general law of the terrestrial planets: "Basaltic volcanism is a dominant process on terrestrial planet surfaces" (Saunders and others, 1991).

How old are Venus' volcanic flows? There has been a persistent suspicion that volcanism is active on Venus today, but no one has proved it. Among the evidence: rapid fluctuations in SO_2 gas content that might come from volcanic exhalations (Esposito, 1984), and (controversial) indications of concentrations of lightning around volcanic peaks (measured from radio signals). From counts of craters, divided by the estimated cratering rate, Magellan researchers (Phillips and others, 1991) concluded that various units on Venus span ages ranging from 800 My to less than 100 My, and that Venus is volcanically active today.

EARTH

Figure 11-57 reminds us that our most common "landscape" is unique in the solar system—and is not the one on which we spend most of our time. Even the great range of environments on Earth hardly begins to match the variety found among other worlds, small and large. The astonishing individuality found among places such as Phobos, Amalthea, Enceladus, Dione, Iapetus, Europa, Io, Titan, Ganymede, Mars, Venus, and Earth emphasizes that each planetary body is its own special world—to a greater extent than we would have guessed a decade or two ago.

CONCEPTS

volcanism
tectonics
endogenic processes
exogenic processes
cumulates
Aa
pahoehoe
fluidized eruption
nuée ardente
ash
cinders
tuff
welded tuff
ignimbrite
structural geology
cinder cone
spatter cone
scoria cone
hornito
shield volcano
composite volcano
crater
caldera
lava tube
pit crater
diatreme
maar
dark halo crater
volcano–tectonic sink
tectonism
fault
thrust fault
strike-slip fault
normal fault
graben
lineament
mare
wrinkle ridge
dome
rille
sinuous rille
linear rille
crater chain
grid system
slumping
creep
saltation
wind streaks
dune field
hydrosphere
channel
braided channel
chaotic terrain
dendritic
patterned ground
geochemical cycle
Urey reaction
red bed
duricrust
palimpsest
Cayley formation
lunar transient phenomenon

PROBLEMS

1. (a) Why do terrestrial lava flows have different textures, such as aa and pahoehoe? (b) Why do lava flows found on the moon today lack these textures?

2. (a) What does the absence of large lunar volcanic mountains and the paucity of lunar cinder cones imply about the fluidity of lunar lavas? (b) Why are maars unlikely on the moon?

3. (a) What causes volcanic eruptions? (b) Do volcanic areas necessarily have totally molten zones (with the temperature exceeding the melting point of all rock minerals) below them within 100 km of the surface?

4. Why were lava eruptions on the moon (or on other planets) 3.5 Gy ago most likely to occur in or near recently formed, large impact basins?

5. (a) Why are Martian channels unlikely to form under present Martian conditions? (b) What past conditions might have produced them? (c) What do Martian polar strata imply about the Martian climate's history?

6. In the context of past and future evolution of life on Earth or on other planets, what is the philosophical importance of evidence about the existence or nonexistence of dramatic global climate fluctuations on Mars or other planets?

7. Describe the history and current location of H_2O, if any, on (a) Mars, (b) Venus, and (c) Io.

8. Why would Figure 11-35b be internally inconsistent if the sunlight were coming from overhead?

9. How are the surfaces and histories of Phobos, Deimos, J6 Himalia, J7 Elara, S9 Phoebe, belt asteroid 10 Hygiea, and Trojan 624 Hektor and certain other asteroids possibly related?

10. Name at least one planet besides Earth and at least one satellite that could have once had large oceans of liquid water. Cite evidence.

11. Suppose future astronauts colonize planetary bodies. Discuss how they might obtain building stone, H_2O, iron, sulfur, molecular hydrogens, and molecular oxygen on (a) the moon, (b) asteroid 1580 Betulia, (c) asteroid 16 Psyche, (d) Mars, (e) Ganymede, and (f) Io.

ADVANCED PROBLEMS

12. Smith and others (1979) discuss a 20-km-thick crust of sulfur and silicates on Io, with perhaps a 1-km-deep layer of liquid sulfur involved in the eruptions. Johnson and others (1979) used the geometry of the erupting plumes to calculate an average resurfacing rate all over Io of 3×10^{-4} to 0.1 cm/y. The radius of Io is 1820 km. If a layer roughly 10 km deep is assumed to cycle through the eruptions, how long does it take for all this material to go through the cycle of eruption, burial, remelting, and reeruption? Has most of the material been through the cycle less than once, once, or more than once?

13. Volcanologist A. Rittmann (1962) discussed nine terrestrial eruptions with eruptive rates of about 0.03 to 3000 km^3/y (for individual volcanoes). Taking these as lower and upper limits on the eruption rates of Io volcanoes and assuming that there are eight eruptions on the average at all times, calculate the cycle time for the upper 10 km of Io's crust. Compare this with the result in problem 12.

14. (a) Mars is at perihelion. A polar ice field with dirty ice having an albedo of 40% is located so that the sun is 10° above the horizon. The emissivity of the ice-soil mixture is 0.5, and half the sunlight gets through the Martian atmosphere at the slant angle mentioned. Calculate the equilibrium temperature of the Martian surface. (b) A landslide occurs, exposing a cliff face with an 80° slope, which is fully exposed to the sun. Calculate the equilibrium temperature of the ice-soil mixture in the cliff face. (c) Comment on the future erosion of the cliff area.

CHAPTER TWELVE

PLANETARY ATMOSPHERES

When studying planetary atmospheres, we want to learn their origin and evolution, the principles that determine their structure, and their meteorological conditions. Atmospheric physicists range from those concerned with present surface conditions to those concerned with long-term climatic changes. One vigorous area of research concerns interactions between atoms and magnetic fields at the fringe of the atmosphere, where the solar wind interacts with planetary gas. This study, sometimes called simply **particles and fields**, has expanded greatly with spacecraft measures near all the planets out to Saturn, but its analyses are somewhat beyond the scope of this book. Another growing area is research on climatic effects of natural or artificial atmospheric changes, such as volcanic ash in the stratosphere, industrial CO_2 increases, or chemical damage to the ozone layer.

Swirling clouds covering Jupiter demonstrate complex atmospheric processes. Various chemical compounds give the clouds different hues, and atmospheric turbulence affects the clouds' motions. (NASA, Voyager)

ORIGIN OF PLANETARY ATMOSPHERES

Primitive Atmospheres

As we discussed in Chapter 5, the planets were immersed in a gaseous medium when they formed. This medium was rich in hydrogen and had a density probably not more than (10^{-3} kg/m^3, and a temperature not more than a few hundred Kelvin near the surfaces of the accumulating planetesimals. The corresponding pressure in the nebular gas was less than 2% of Earth's surface atmospheric pressure today.

Such gas must have concentrated in the gravitational fields of the primeval planets to form **primitive atmospheres**. These would have had nearly solar composition except that the light gases, such as H, would have tended to float to the top of the atmosphere and escape into the solar nebula (see the later section on atmospheric escape), while heavier gases—such as argon (Ar), neon (Ne), and ammonia (NH_3), would have been more concentrated toward the surface.

Inert Gases as Tracers of Early Conditions

Could our present atmosphere simply be a remnant of such a primitive atmosphere? No. Consider the amount of neon (Ne) in our present atmosphere. Neon is a heavy inert gas, and none is produced by radioactive decay. Because it is heavy, virtually none has escaped into space, and because it is inert, it has not chemically bonded into the crustal rocks or oceans. Thus, the Ne in our present atmosphere is a good tracer of the primitive atmosphere. By adding the proportionate amounts of other solar nebula gases to the existing Ne, we can calculate the corresponding mass of the primitive atmosphere, which turns out to be only 0.9% of the mass of the present atmosphere (Walker, 1977, p. 182). The conclusion is that the present atmosphere is much too extensive to be a remnant of a massive primitive atmosphere.

The above calculation suggests two possibilities: (1) The primitive atmospheres of terrestrial planets were small, only a fraction of the present atmospheres (except for Mercury); or (2) the primitive atmospheres may have been much more massive but were swept away from the planets as the nebula was cleared by violent T-Tauri-phase solar winds. The second idea has been invoked to help explain chemical properties of the planets and the presence of captured satellites.

Measurements of argon in Venus' atmosphere by Pioneer and Venera probes created a stir when they were reported in 1979. Argon has several isotopes. Argon 40 is a radioactive decay product, and thus its concentration measures not a primitive atmosphere, but rather the amount of argon created inside Venus by decay of potassium and then outgassed during volcanism. Argon 36 and argon 38 are not radiogenic and thus are tracers of the primitive atmosphere in the same way as neon, discussed above. Prior to the Venus measurement, some researchers had theorized that planetesimals in Venus' vicinity were so much hotter than those of Earth that less argon was adsorbed on their surfaces. Venus was thus predicted to have less argon 36 and argon 38 than Earth. The Venus measurements, combined with Viking Mars measurements, established just the reverse trend: concentrations (in units of 10^{-12} of argon 36 and argon 38 per gram of planet mass) on Venus, Earth, and Mars of about 5000, 46, and 0.2, respectively (Pollack and Black, 1979). A possible explanation is that the temperature in the nebula did not rise as fast as thought toward the sun, and so the nebular gas pressure increased sharply enough toward the sun to drive these amounts of argon into the planetesimals. Or, perhaps, the closer the planetesimals to the sun, the more solar wind argon atoms were trapped in the planetesimal materials. The matter remains controversial.

Secondary Atmospheres of Earth, Mars, and Venus

Secondary atmospheres are atmospheres that have been produced or significantly altered by gases exhausted from the planetary interior. Although the giant planets have gravitational fields strong enough to retain their primitive atmospheres, smaller planets lost much of their primitive atmospheres. As shown in Figure 12-1, these were replaced by gases from the interiors of the planets. In the case of Earth, new gases were also added by biological processes.

What kinds of gases were added by volcanic processes? In a classic paper, Rubey (1951) showed that these included many of the gases now in Earth's atmosphere. Rubey made an inventory of existing volatiles at Earth's surface and, with some assumptions about the primitive atmosphere, calculated the distribution of gases that must have been added from some source in the last few billion years to explain today's atmospheric and surface volatile inventory. Table 12-1 shows the rough agreement between gases coming out of volcanoes and the gases required by Rubey's analysis. Thus, volcanic gases appear to account for much of Earth's present secondary atmosphere.

Some controversy remains about the nature of the original primitive atmosphere. Was it strongly reducing (rich in hydrogen and hydrogen compounds) like the nebular gas and Jupiter's atmosphere? Or was it more oxidizing like

Figure 12–1. Evolution of a secondary atmosphere occurs as volcanic gases, mostly carbon dioxide and water, are emitted from planetary interiors. This occurs during both large-scale eruptions and small-scale venting. **(a)** *The plume from a volcanic eruption at Mauna Ulu volcano, Hawaii. (Photo by Boone Morrison)* **(b)** *Local emission from a volcanic fissure. The light-colored deposits are sulfur condensates. (Kilauea volcano, Hawaii; photo by author)*

today's atmosphere, or perhaps more CO_2-rich like the atmospheres of Mars and Venus? An important extension of Rubey's work was a study by Holland (1962), who showed that metallic iron in the early, undifferentiated Earth would have combined with and removed oxygen from the first volcanic gases as they ascended to the surface, making them rich in H_2, H_2O, CO, and H_2S. As a result of this study, researchers for many years assumed that the original primitive atmosphere and even the early secondary atmosphere were strongly reducing and rich in hydrogen compounds. Today, researchers are more inclined to believe that reducing conditions were very short-lived and soon replaced by a secondary atmosphere neither strongly reducing nor strongly oxidizing, but perhaps rich in CO_2 and N_2.

In any case, it is clear that today's oxygen-rich conditions evolved slowly because of two processes. First, dissociation of H_2O molecules into H and O occurred in the upper atmosphere when H_2O molecules were struck and split by energetic ultraviolet sunlight. Second, photosynthesis began after green plants evolved. Analysis of ancient sediments clearly show that the sediments deposited before about 2.5 Gy ago formed in oxygen-poor environments and that the O_2 began to increase rapidly after about 2.0 Gy ago. Fossils from this time show the emergence of photosynthetic plants that gave off O_2 (Walker, 1977).

Table 12-1
Gases Added to the Atmosphere by Volcanic Outgassing of Earth (percent composition by weight)

Gas	Observed from eruptions that may tap the mantle (Hawaiian volcanoes)[a]	Observed from conti-nental fumaroles, and geysers	Volatiles outgassed according to calculation by Rubey (1951)
H_2O	57.8	99.4	92.8
CO_2[b]	23.5	0.33	5.1
Cl_2	0.1	0.12	1.7
N_2	5.7	0.05	0.24
S_2	12.6	0.03	0.13
Others	<1	<1	<1
Total	100	100	100

[a]Figures from Hawaiian volcanoes. This source, which lies above a thin oceanic crust, is the most primitive, deep-seated magma available. (Continental fumaroles, on the other hand, represent material that has been chemically reprocessed and contaminated with groundwater, etc.)
[b]Plus small amounts of carbon monoxide (CO).

These ideas of atmosphere evolution help explain relationships among the atmospheres of Venus, Earth, and Mars. At first glance, they appear entirely different: Venus has a massive CO_2 atmosphere; Earth has N_2–O_2; and Mars has a thin CO_2 atmosphere. But Table 12-2 shows that, contrary to our first impression, Venus, Earth, and Mars have similar total inventories of volatiles, once we count the volatiles bound up in rocks and permafrost. In fact, Venus and Earth have nearly identical total CO_2 inventories! Table 12-1 suggests that secondary atmospheres outgassed onto these planets by volcanism should be mostly H_2O and CO_2 with traces of other materials such as N_2. In support of this, all three planets have very abundant CO_2, which probably came from volcanic outgassing—an assumption consistent with Venusian and Martian evidence for volcanism.

The amounts of H_2O and N_2 may have depended strongly on initial bulk compositions. For example, Venus may have contained less H_2O than Earth at the outset because of its closer position to the sun. Also, as discussed later in this chapter, the higher temperatures of Venus probably contributed to H_2O dissociation and loss from the atmosphere.

Recent measurements of the D/H (deuterium/hydrogen) ratio by a Pioneer probe in Venus' atmosphere show that Venus outgassed at least 0.3% of the water in Earth's oceans, and possibly a full ocean's worth (Donahue and others, 1982). To quote the title used by these authors, "Venus was wet." D/H gives a measure of past hydrogen content (hence H_2O, since most H would have united with O) because deuterium is twice as heavy as H, and hence escapes more slowly into space. Given a present-day D/H measure, we can calculate backward to past values and estimate the total past H that passed through venus' atmosphere.

The richness of N_2 in Earth's atmosphere is something of a problem; it may be strongly affected by biological processes. Mars, with the smallest mass and internal heating, may have produced the least gas per kilogram, but the proportions may still have matched those in Table 12-1. Mars' volatiles are difficult to inventory for two reasons. First, the amounts of carbonate rocks and H_2O permafrost below the surface are uncertain. If water was more abundant in the past, much CO_2 could have been tied up in carbonate rocks as on Earth (Anders and Owen, 1977; Pollack, 1979). Second, Viking gave evidence that Mars once had a more massive atmosphere, much of which leaked off into space because of the low gravity (McElroy, Yung, and Nier, 1976).

STRUCTURE AND CONDENSATES OF PLANETARY ATMOSPHERES

The same principles discussed in the beginning of Chapter 9 can be used to compute the structure of a planetary atmosphere. The hydrostatic equation $dP = -\rho g\, dz$ applies to both planetary atmospheres and interiors because both behave like fluid systems. (The minus sign appears here since z is chosen to increase upward as P decreases.)

Temperature Structure

In Chapter 9, when we tried to compute the internal structure of a planet, we saw that the

Table 12-2
Inventories of Selected Outgassed Volatiles on Terrestrial Planets (10^{-9} kg/kg planetary mass)

Volatile	Venus	Earth	Mars
H_2O			
Atmosphere	60[a]	3[b]	0.02
Oceans and polar caps	—	250 000[a,b]	5 000?[a,c]
Crust	160 000?[c]	30 000[b]	10 000?[a,c]
Total	160 000?	280 000	15 000?
CO_2			
Atmosphere	100 000[a,c]	0.4[b]	50[a,d]
Polar caps	—	—	10[d]
Crust	—	100 000[a,b]	>900?[e]
Total	100 000	100 000	>1000?
N_2			
Atmosphere	2 000[a]	2 000[a]	300 xa
^{40}Ar			
Atmosphere	4[a]	11[a]	0.5[a]

Note: Entries with two references are averages.
[a]Pollack and Black (1979).
[b]Walker (1977).
[c]Khodakovsky and others (1979).
[d]Hess, Henry, and Tillman (1979).
[e]Fanale and Cannon (1979) estimate 6×10^{17} kg of CO_2 adsorbed in nontronite clays in Martian polar layered terrain and associated regolith, substantially exceeding the CO_2 in the atmosphere or polar caps.

equation of state became important. The same applies to atmospheres except that in this case the equation of state is much simpler, namely the **ideal gas law**, which applies to any gas. The ideal gas law relates the pressure P to the density ρ, the temperature T, and the composition, which in the case of gases can be indicated simply by the mean molecular weight μ. The ideal gas law reads

$$P = \frac{\rho}{\mu M_H} kT$$

where M_H is the mass of a hydrogen atom and k the Boltzmann constant (see Table 2-2).

Suppose that we now try to compute a crude model of the atmosphere by using the hydrostatic equation and measurements of the surface P, ρ, and T. Following the same scheme as in Chapter 9, we could divide the atmosphere into (say) 100 layers, each 1 km thick, and try to compute the pressure and other conditions at each layer. The atmosphere is so thin that we can treat g as a constant, to a first approximation.

Using the hydrostatic equation, we insert the known surface values of ρ, g, and $dz = 1$ km. We compute dP, the change in pressure between the ground and the top of the first layer. We now have P_1, the pressure 1 km above the ground.

Now we are stuck, because we have no way of getting the density at that level to use in our next calculation for the second kilometer. If we had measures of ρ at all levels we could proceed. Or if we had measures of T at all levels, we could proceed by using the ideal gas law to compute ρ, since we would then know P, T, and the composition (assumed to be uniform).

In other words, to compute a reliable model of an atmosphere, given the conditions at any one level, we have two alternatives: We must either (1) have the values of P, ρ, or T or (2) make some simplifying assumption about the behavior of one of these.

One of the simplest and conceptually most useful assumptions is that the temperature is constant at all levels (that is, that we have an **isothermal atmosphere**). At first sight this assumption seems a gross mistake, since we know that the air gets noticeably cooler as we ascend even a modest mountain. But the changes in atmospheric temperature are only some tens of Kelvin out of about 300 K (that is, a change of only a few percent rather than a substantial change). Model atmospheres assumed to be isothermal are among the simplest to calculate (see the accompanying mathematical notes).

One useful concept that comes from such considerations is the **scale height**, or vertical distance over which the density decreases by a specified factor, usually 0.368 (which is $1/e$, a value that comes out of the mathematical theory). Sometimes the decimal scale height is used instead (the height over which the density decreases by a factor of 1/10). At Earth's surface the scale height is about 6 km, on Venus about 15 km, on Mars about 18 km, and on Titan roughly 140 km. Generally, lower gravity results in a much more extended atmosphere with a larger scale height and a lower mean density. Such figures emphasize that we live within walking distance of interplanetary space.

More sophisticated models of atmospheres quickly make us deal with the fact that real atmospheres are not isothermal. In fact, if we could create an isothermal atmosphere and then let nature take its course, the atmosphere would soon develop its own temperature structure. The two major sources of heat energy causing this new distribution of temperature are (1) diurnally (day–night) varying inputs of radiation from the sun (visible radiation) and from the planet (infrared radiation) and (2) condensation or evaporation of certain atmospheric constituents (heat input governed by the latent heat of condensation). In order to see how these sources affect atmospheric structure, we must first discuss how energy is distributed through the atmosphere.

Rayleigh Scattering, Blue Skies, and Pink Skies

In 1868, English physicist John Tyndall noticed the blue glow of a light beam as it passed through a liquid suspension of tiny colloidal particles whose diameters were less than the 0.5-μm wavelength of visual light. He concluded that the blue color of the sky must be due to sunlight interacting with similar particles in the air, called **aerosols**. By 1871, another English physicist, John Strutt, also known as Lord Rayleigh, showed that light interacts with these small particles in a way called **scattering**. If light encounters particles smaller than its own wavelength, it does not reflect off them in the normal way, but rather is scattered in various directions in such a way that the intensity of the scattered light increases dramatically with shorter wavelengths.* That is, the scattered light looks much bluer than the original light. This type of scattering is called **Rayleigh scattering**. For a few years, physicists thought the sky was blue only because of the presence of tiny water droplets, salt crystals, or other aerosols (Young, 1982). But in 1899, Lord Rayleigh discovered that even air molecules cause Rayleigh scattering. Even a sky without aerosol particles would be blue.

If we look at the clear daytime sky in a direction away from the sun, we see blue light scattered off gas molecules and tiny aerosol particles. If we look at the sun, we see the unscattered portion of the light, from which some of the blue has been removed. If we look at the setting sun, the sunlight has come along such a long path through the atmosphere that most of the blue light has been removed, and the setting sun looks vivid red in color. If we look at the sky on a foggy

*The intensity of the scattered light is proportional to (wavelength)$^{-4}$.

MATHEMATICAL NOTES ON ATMOSPHERIC STRUCTURE

A great deal can be learned about atmospheres by simply combining the hydrostatic equation and the ideal gas law, under the assumption that the atmosphere is isothermal. By substituting the ideal gas law's expression for ρ into the hydrostatic equation and by assuming T is constant, we have

$$dP = \frac{-\mu M_H P g}{kT} dz$$

This differential equation is readily solved. Dividing by P, we have on the left dP/P, which coincidentally is identical to $d \ln P$. Integrating this logarithmic form from the surface ($z = 0$) to any arbitrary height, we find that

$$P = P_0 \exp\left(\frac{-\mu M_H g}{kT} z\right)$$

where P_0 is the pressure at the ground level. We thus have solved for the pressure structure P as a function of z.

Two concepts are immediately evident. First, the density structure in an isothermal atmosphere is just the same as the pressure structure, since, by the ideal gas law with T constant, we have $P/P_0 = \rho/\rho_0$. The second concept is the concept of scale height. If we define

$$\text{scale height} = H = \frac{kT}{\mu M_H g}$$

This expression was used to calculate the scale heights quoted in the text. Note that scale height depends on T and hence changes with the height in real (nonisothermal) atmospheres.

We thus have the simple expression

$$P = P_0 e^{-z/H}$$

We can see that P falls by $1/e$ for each increase in height by H. The expression gives the pressure and hence density at all heights in any reasonably isothermal part of an atmosphere.

day, the air is full of moderate-sized water droplets—particles larger than a wavelength. Hence Rayleigh scattering does not occur and we see the neutral whitish color of sunlight scattered and reflected throughout the sky.

The nature of gas molecules and aerosol particles in other planetary atmospheres is very important to understanding how sunlight interacts with these atmospheres and how they look from above or from the surface.

On Mars the air is thin so that there is less Rayleigh scattering by gas molecules. This would make the sky dark except for the fact that much fine (μm-scale) reddish dust from the surface is suspended in the air. The scattered light from these larger-than-a-wavelength particles acquires their color, giving the sky a pinkish-tan cast.* The Martian sky in the direction of sunrise and sunset, as shown in Figure 12-2, shows a darker band near the horizon because of the greater concentration of dust near the surface.

Radiative Transfer of Heat Energy

The theoretical treatment of the radiative transfer of heat is exceedingly complex and is the basis of much of astrophysics. We can only touch on it here. (It was mentioned briefly in Chapter 9 with regard to heat transfer in planetary interiors.)

The keys to the theory of radiative transfer are the concepts of opacity and optical depth. **Opacity** is a measure of how much radiation is absorbed** over a given distance through the atmosphere. In a dense fog on Earth, the opacity is very high, and we can see perhaps 10 m. On a clear day we can see 100 km. The **optical depth** helps to express the effect of opacity. The optical depth is a dimensionless number that expresses the amount of radiation lost from a beam of light. The optical depth is defined so that at optical depth 1, most of the light has been lost out of the original beam. Optical depth 10 implies that virtually no light is getting through. Consider a beam of light approaching an atmosphere from space. At the top, the optical depth is zero. If the atmosphere is cloudy, like that of Venus, optical depth 1 may occur near the cloud tops. But if the atmosphere is clear, like much of Earth's, then we may have an optical depth of only 0.5 or some other fraction at the ground.

The subtle point about opacity and optical depth is that they vary with the wavelength (color) of the radiation. If we happen to pick a wavelength that some atmospheric constituent, say carbon dioxide, strongly absorbs, the opacity will be very high at that precise wavelength even though it may be nearly zero at nearby wavelengths. In other words, if we looked through a filter of precisely that color, the atmosphere would look entirely opaque, but if we looked through a filter of a slightly different color, the atmosphere would appear transparent. To say it still another way, optical depth 1 would occur high in the atmosphere at the absorbing wavelength but low in the atmosphere (or not at all) at a nearby wavelength.

To give another example, an atmosphere could be quite transparent in the visible wavelengths but quite opaque at infrared wavelengths. This is just the case on Earth, as shown by Figure 12-3. The atmosphere is moderately clear at visual wavelengths, but water vapor absorbs much of the infrared. Earth-based astronomers trying to detect diagnostic infrared spectral absorptions by certain minerals or ices on other planets must therefore observe from very high mountains or spacecraft above most of Earth's water vapor.

Another example of absorption of certain wavelengths, this time by methane, is dramatically seen in Figure 12-4.

---

*Before the Viking landings, scientists and artists had pictured a dark blue Martian sky, like that at very high altitude over Earth. The first Viking photos showed a much brighter sky than expected; controllers applied color controls that gave a blue sky tint to the first photos handed out to the press! Only after a day of calibration did the Viking scientists discover that reddish light dominates in the delicate balance of sky colors; properly processed photos show a pink Martian sky.

**Or scattered. Scattering and absorption are two distinct mechanisms for removing energy from a beam of light. For simplicity we shall minimize the difference between these mechanisms, although the difference is crucial in advanced radiative transfer theory.

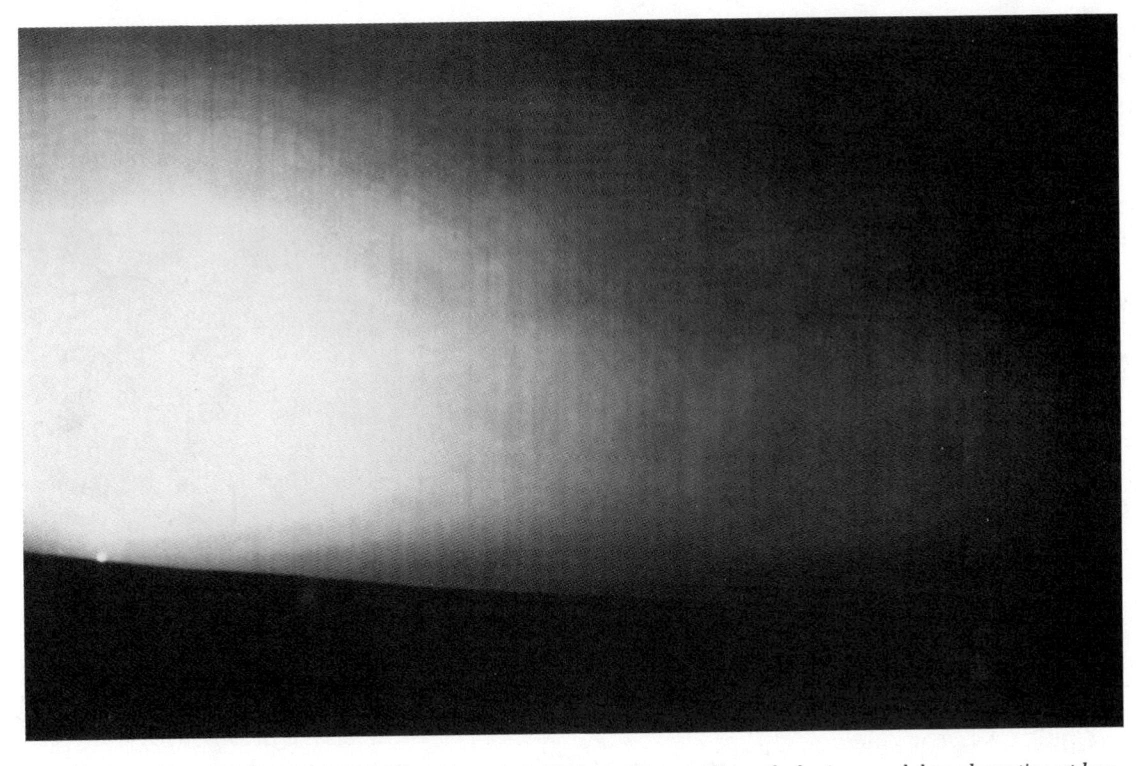

Figure 12–2. Viking 2's 631st sunrise on Mars, June 14, 1978. The sun is poised over the horizon, and dust absorption at low elevation causes darkening along the horizon. (NASA)

To see why an atmosphere develops an irregular temperature profile, suppose that we could somehow create a perfectly isothermal atmosphere. Let us examine the consequences with respect to radiative transfer theory. Consider two cases. In the first case, suppose that the atmosphere was absolutely transparent at all wavelengths. Then the radiation from the sun would come through the atmosphere without any absorption. It would strike and heat the ground, causing the ground to radiate in the infrared, as in Figure 12-5a. The infrared radiation would also escape into space without affecting the atmosphere, since the atmosphere has no opacity at infrared wavelengths. Thus there would be no radiative influence of the sunlight on the atmosphere. No radiative energy would be absorbed at any wavelength or at any level in the atmosphere, and so the atmosphere would remain cold and isothermal (neglecting conduction effects, where the atmosphere contacts the heated ground). This is like the situation on mountain ski slopes, where the air is dry and cold, but the sunlight strong and warming.

For our second case, suppose that there is a small amount of opacity at various wavelengths. Some incoming sunlight and some outgoing infrared radiation would be absorbed. Thus the sunlight would heat the atmosphere. Sunlight can cause heating only if there is some opacity causing radiative energy to be absorbed (Figure 12-5b). This is like the situation at the seacoast, where the air is humid and warm.

To see one reason why the temperature structure of an atmosphere can be quite irregular, let us consider a still more realistic example. Suppose that the various absorbing constituents are arranged in layers. Take Earth as an example.

*Figure 12–3. Different opacities of Earth's atmosphere at visual wavelengths and infrared wavelengths are vividly shown in these two pictures. (**a**) Image at visual wavelengths (0.1–44 μm). Land masses of Africa and Europe, with scattered opaque clouds. (**b**) Image at infrared wavelengths (5.7–7.1 μm). The water vapor absorption bands show diffuse, moist cloud masses; the image does not penetrate to the ground. (European Meteorological Satellite images; courtesy C. R. Chapman)*

a

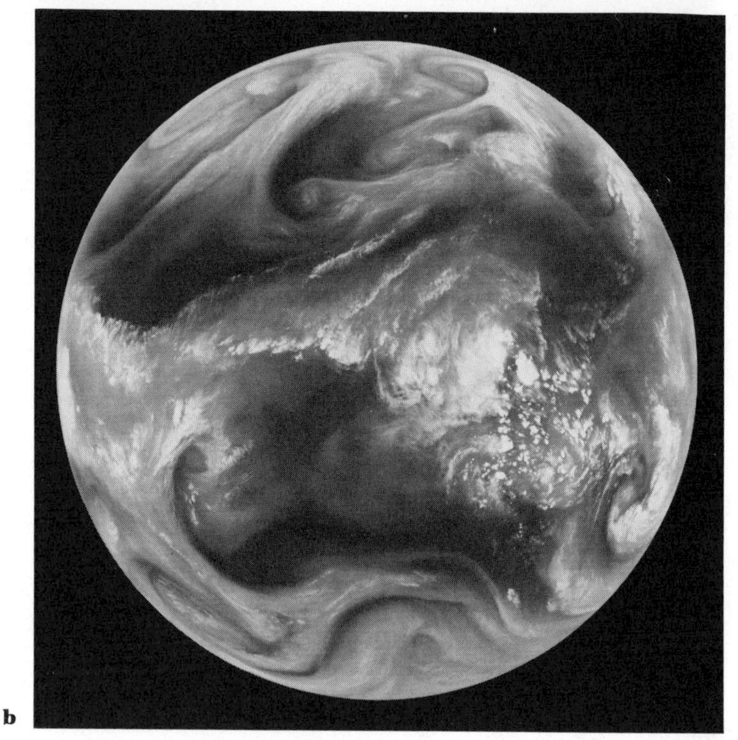

b

Figure 12–4. Photographs of Saturn made with films and filters sensitive to different colors of light. Left = blue (450 nm), the most common view (see Figure 2-25). Right = wavelength absorbed by methane gas (897.5 nm); abundant methane in the atmosphere absorbs most of the light at this wavelength making the globe dark. The rings remain bright, being covered with ice that reflects sunlight at this wavelength. (Lunar and Planetary Laboratory, University of Arizona)

Ozone (O_3), which strongly absorbs ultraviolet sunlight, is created and concentrated by photochemical reactions only in a layer at high altitudes. Water vapor, which is a very strong absorber of infrared, is created primarily at low levels by the evaporation of surface water. If these two constituents are the main absorbers, some of the incoming sunlight will be absorbed at high altitudes, heating the upper atmosphere, and most of the outgoing planetary infrared radiation will be absorbed at low levels, causing the air temperature near the surface to be higher than that at intermediate altitudes. This in fact is the situation on Earth.

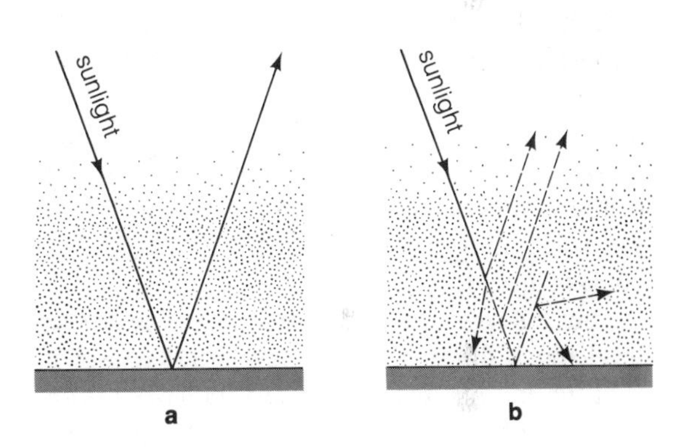

*Figure 12–5. Transmission of sunlight through (**a**) a hypothetical transparent atmosphere and (**b**) a more realistic atmosphere with absorption and reemission of radiation.*

The Greenhouse Effect

The **greenhouse effect** is a strong heating of the lower atmosphere of a planet due to selective absorptions in the infrared. We can easily understand it by reviewing the above model. Suppose, as is the case on Earth, that the atmosphere is *relatively* transparent in the visible wavelengths but has some strong absorbing medium in the infrared, such as water vapor or carbon dioxide. The surface of the planet is heated to a few hundred Kelvin by whatever solar radiation gets through. Any surface heated to such temperatures will radiate in the infrared, and this outgoing infrared radiation will get absorbed by the atmosphere. What happens to that energy? The atmosphere cannot keep absorbing infrared energy

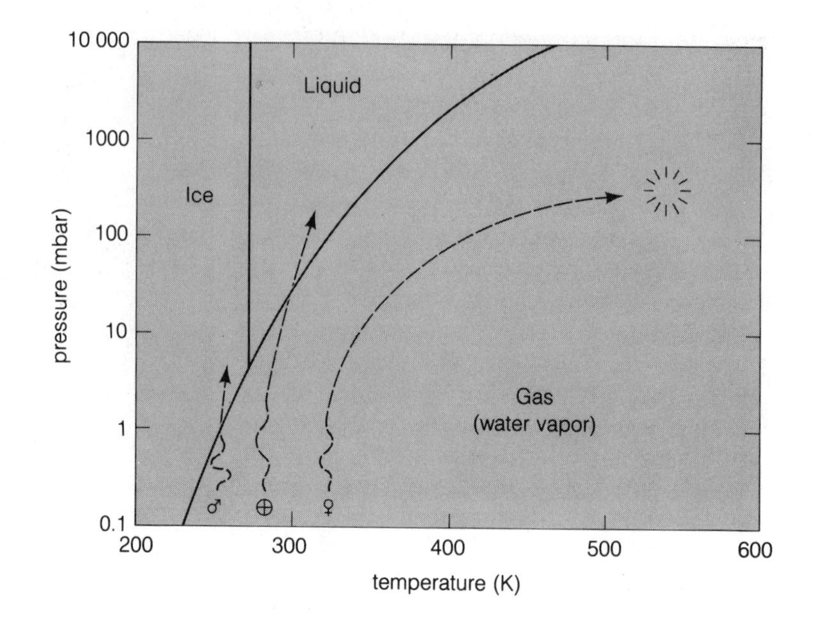

Figure 12–6. Phase diagram for water, showing phases existing at various temperature/pressure combinations. Dashed lines are schematic evolutionary tracks for H_2O on Mars, Earth, and Venus. Wiggles represent hypothetical climatic oscillations. The H_2O-caused greenhouse warming sets in at about 1 mbar of water vapor pressure.

indefinitely without getting warmer, so the atmosphere heats up and itself radiates in all directions. The lower atmosphere and the ground heat up, radiating still more infrared, until the amount of infrared energy escaping from the top of the atmosphere is equal to the amount of visible solar energy coming in. Only then is equilibrium achieved.

But the new equilibrium may leave the ground level much hotter than the few hundred Kelvin that we initially expected from the solar input. This greenhouse effect is named after the same phenomenon that occurs in a greenhouse, where the glass panes are analogous to our infrared absorber. Glass lets in sunlight, but it blocks the outgoing infrared and traps the warm air, causing the greenhouse to get warmer until enough infrared escapes to reach equilibrium.

The greenhouse effect is familiar in nature on Earth. For example, during a cloudy night, when much water vapor is in the air, the lower atmosphere absorbs heat and the air is likely to stay warm; during a sparklingly clear, dry night, the temperature may drop rapidly because the ground can radiate in the infrared the energy it gained from the sun in the daytime. For this reason cool nights are common in dry deserts.

As deduced by Carl Sagan as early as 1962, the greenhouse effect probably accounts for the 750-K surface temperature of Venus. The 90 bar of CO_2 on Venus creates a greenhouse far exceeding that of the 1 bar of O_2 and N_2 on Earth. The greenhouse effect is probably crucial to the different histories of Venus, Earth, and perhaps even Mars. Study the **phase diagram** in Figure 12-6, which shows the pressures and temperatures producing stability for different phases of H_2O. The temperature is low enough on Mars that, as the secondary atmosphere built up there, volcanically emitted H_2O probably condensed directly as ice, preventing an H_2O greenhouse effect, which appears when about 1 mbar of H_2O is available. Earth's initially warm temperature allowed H_2O to condense as liquid. On Venus, the strong CO_2 greenhouse, plus the increasing H_2O greenhouse and Venus' initially warmer condition, raised the temperature so high that the evolutionary path missed the liquid regime, and the gaseous water molecules eventually dissociated and/or reacted with other materials.

Figure 12-6 emphasizes that the creation of an Earth-like planet is a somewhat delicate matter. According to this view, if Earth had been only about 5% closer to the sun, it, too, would have suffered Venus' fate, and life would not have evolved (Rasool and de Bergh, 1970). Similarly, a slight heat input to Mars might trigger a substantial greenhouse effect, causing dramatic warming.

Condensable Substances—Moist Atmospheres

Meteorologists consider an atmosphere **moist** when it contains substances that can condense into liquid or solid form. On Earth, water vapor is the familiar example, condensing to form cloud layers composed of water droplets or ice crystals. Such cloud layers may affect planetary atmospheric temperatures in two ways: (1) Clouds can interrupt the flow of radiation with an abrupt layer of high opacity and (2) enough heat can be released when some substances condense to affect the atmospheric structure. This is the case with water vapor on Earth; indeed, water vapor is so important that a model of the lower Earth atmosphere will be substantially wrong without taking it into account, as follows.

Suppose an imaginary parcel of air convects upward in the atmosphere (review the description of convection in Chapter 9). If there were no water vapor, it would tend toward a certain equilibrium temperature and pressure. If the parcel were moist to start with, the cooling might make water vapor condense, creating a puffy **cumulus cloud**, liberating heat, and causing the parcel to tend toward a different equilibrium temperature than if it were dry. Thus condensable substances change the temperature structure of atmospheres.

The temperature structure is often expressed in terms of the **temperature gradient**, which is the change in temperature per kilometer as we ascend into the atmosphere. Since this quantity is negative near Earth's surface, meteorologists often use for convenience the **lapse rate**, which is simply the negative of the temperature gradient. Thus on Earth, the typical near-surface temperature gradient is -6.5 K/km, while the lapse rate is 6.5 K/km. On Venus the lapse rate near the surface is about 8 K/km (Seiff and others, 1979), and on Mars about 3 K/km (Seiff and Kirk, 1977), according to measurements by landers.

The Sulfuric Acid Condensate Clouds of Venus

An extraterrestrial example of condensate clouds occurs on Venus, where the clouds are made from droplets of sulfuric acid (H_2SO_4, commonly used as battery acid) rather than water. Pioneer Venus' 1978 parachute probes confirmed early evidence that Venus' clouds are H_2SO_4, and showed that the clouds are concentrated in a layer from 48 to 58 km above the surface. The H_2SO_4 droplets in this region have diameters of about 1 to 10 μm. Above this cloud deck, a haze of 1- to 3-μm droplets extends upward to at least 68 km (Figure 12-7). Below the cloud deck a haze of 1- to 2-μm droplets extends downward to a sharp cutoff at 31 km. Below that the atmosphere is dense but clear (Knollenberg and Hunten, 1979a, 1979b).

Theoretical work combined with other Pioneer Venus observations have revealed how these clouds were probably formed. At altitudes of around 80 km, under the influence of solar ultraviolet, chemical reactions among minor atmospheric constituents produce small droplets of H_2SO_4, which then begin to fall slowly. Between 48 and 58 km is a region with convective updrafts and downdrafts; thus the particles are likely to be caught there, growing slowly during a lifetime of a few months. This is the main cloud deck (Toon, 1979; Knollenberg and Hunten, 1979a, 1979b). Droplets that get too big eventually drop out of this layer into the warmer region below.

While temperatures range between 288 and 364 K (59°F to 196°F) in the cloud deck, they reach 493 K (428°F) about 20 km below the clouds, and so the droplets rapidly evaporate as they fall out of the cloud. This explains both the smaller particle size observed in the sulfuric acid "driz-

Figure 12–7. When Venus passes between Earth and the sun, sunlight backlights the atmosphere. A thin, high layer of haze scatters sunlight so that the atmosphere is visible all around the disk. This was the first evidence for an atmosphere on Venus, discovered in 1761. The sun is far out of the picture to the upper left. (New Mexico State University Observatory; courtesy B. A. Smith)

zle" falling out of the cloud and the termination of the drizzle below 31 km. Condensation and movement of the cloud particles involve a heat flux about one-fourth of the solar flux at that altitude. The clouds are believed to be stratuslike forms, not puffy cumulus clouds (Knollenberg and Hunten, 1979b).

Water and Carbon Dioxide Condensates of Mars

Mars has an important cycle of condensation involving the abundant CO_2 and sparse H_2O in its atmosphere. Many of the phenomena can be understood with the phase diagram shown in Figure 12-8. The solid curves show the conditions for condensation of CO_2 and H_2O. The **triple point for H_2O**, the only conditions under which ice, liquid water, and water vapor can coexist, is marked at $P = 6.1$ mbar and $T = 273$ K. Curve *a* marks the daily range of conditions for a representative high-altitude region on Mars, such as the Tharsis volcanoes. Usually, H_2O is frozen and CO_2 is gaseous. Since $P < 6.1$ mbar, the warmest days do not produce stable liquid water; instead, they cause ice to sublime into water vapor. Cold nights or winter days may cause condensation of CO_2 gas into CO_2 frost (solid CO_2 is the familiar "**dry ice**"). Curve *b* shows conditions in a low-altitude region, such as the floor of Valles Marineris. Here, $P > 6.1$ mbar, so that a balmy summer Martian afternoon at 35°F might produce small amounts of liquid water in the soil. Since the ambient air pressure is not provided by water vapor but by quite dry CO_2 gas, this water will evaporate rapidly into the dry air.

Figure 12-9 shows a consequence of Figure 12-8: the dramatic result of winter's onset at the modest latitude of 44° N on Mars. Image a is a summer view of the rocky plain as it appeared soon after Viking 2's landing, but image b shows the scene during a 100-d winter period when frost remained on the surface. The complex mechanism of frost deposition deduced by Viking scientists involves dust particles blown into the atmosphere by dust storms. These act as condensation nuclei for H_2O ice, but the particles still do not grow big enough to fall rapidly out of the atmosphere. However, as winter nights grow colder, the CO_2 ice regime on Figure 12-8 is reached, causing CO_2 ice to condense onto the dust/water particles, giving them a heavy enough coating to precipitate onto the ground (Pollack, 1979). The daytime sun, however, warms the CO_2 frost and sublimes it back into the atmosphere, leaving "snow," or, more accurately, a frost of water-ice-coated dust grains, as seen in Figure 12-9b.

In Figure 12-8, the conditions on Earth are shown in the upper right, where we see transitions from the solid to liquid H_2O regimes. Again, since Earth's atmosphere is not pure H_2O, the liquid water is not entirely stable. It evaporates into the air. If enough liquid water is available, an equilibrium water vapor pressure ("100% humidity") is reached.

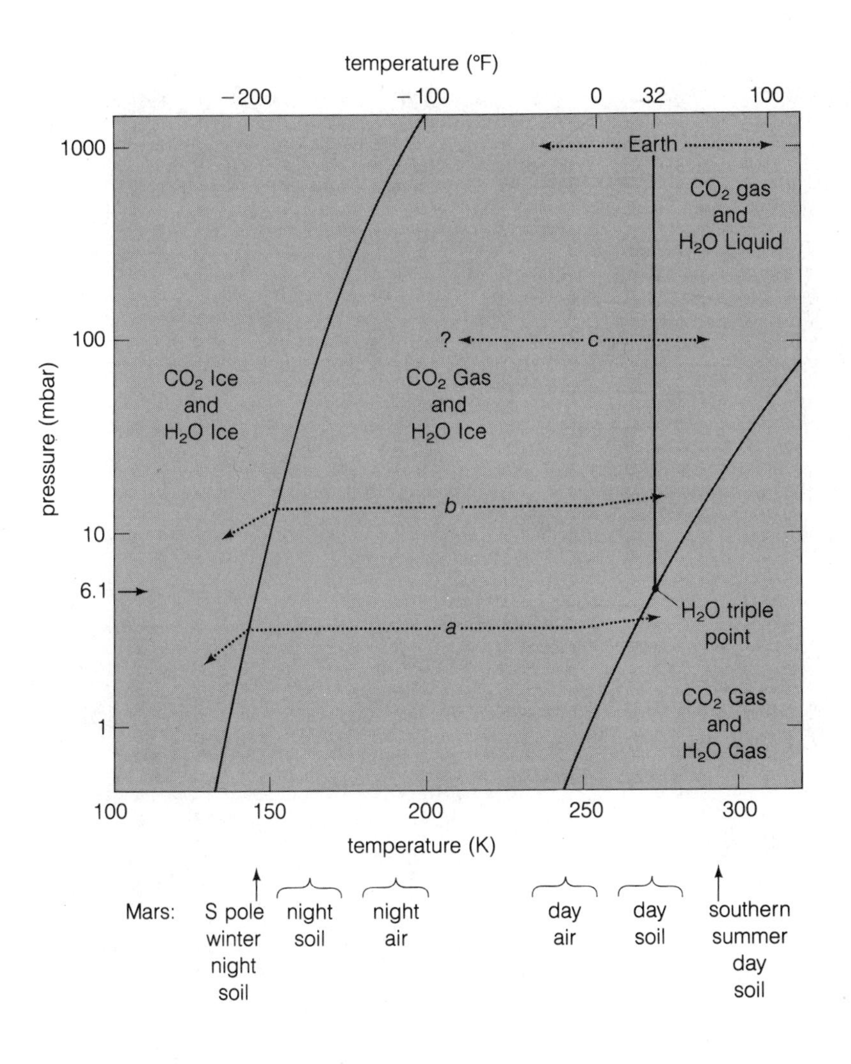

Figure 12–8. Phase diagram for a planetary environment with carbon dioxide and water, accounting for various Martian phenomena. Curves a *and* b *show conditions at high and low surface elevations on Mars, respectively. See the text for further discussion. (After a diagram by Mutch and others, 1976)*

Curve *c* shows a hypothetical condition for ancient Mars if the air pressure then was higher than it is today. In this case, a greenhouse effect might have warmed the air, liquid water would have been stable on warmer days, and, if enough water was available, it might have produced the channels by erosion.

Martian Clouds

On present-day Mars, areas warmed by the morning sun emit water vapor to the air. About half an hour after dawn, this freezes to form clouds or ground fogs of water ice crystals in some regions. Ground fog clinging to the floors of channels and canyons is common. Higher clouds form, as on Earth, where air gets lifted over elevations, such as the slopes of volcanoes in the Tharsis region. Poleward of latitudes about 65°, winter conditions lead to freezing of even the CO_2 at about 148 K and a winter **polar hood** of CO_2 clouds and haze (CO_2 ice particles) hangs over the polar regions.

Some of these features were recorded decades ago by Earth-based telescopic observers who classified Martian clouds into three groups. **White**

Figure 12–9. Summer and winter on Mars. (**a**) *Viking 2 view of rocky Utopia plain as it appears during most of the 687-d Martian year.* (**b**) *For about 100 d during the Martian winter at the same site, a partial layer of water-type frost covers the ground. See the text for details. (NASA)*

clouds are probably mostly H_2O ice crystals and may include crystal hazes of CO_2 ice around the poles and at altitudes of 50 km, where CO_2 ice can condense (Briggs and others, 1977). **Blue clouds** or blue hazes are clouds prominent on blue-sensitive photos. They, too, are probably composed of H_2O and CO_2 ice crystals, but they may involve smaller particles. Kuiper (1964) simulated blue haze by tiny ice particles. Both kinds of clouds are common at dawn and dusk on Mars. **Yellow clouds**, which are prominent in red and yellow photos, are dust storms composed of dust particles a few micrometers across.

Martian Polar Caps

The coldest Martian regions are the winter poles, tipped away from the sun by 25° and shrouded in darkness for half a Martian year, or 11 Earth mo. The CO_2 that freezes out of the atmosphere at 148 K forms a frost layer on the ground, either by forming there or precipitating from the air.

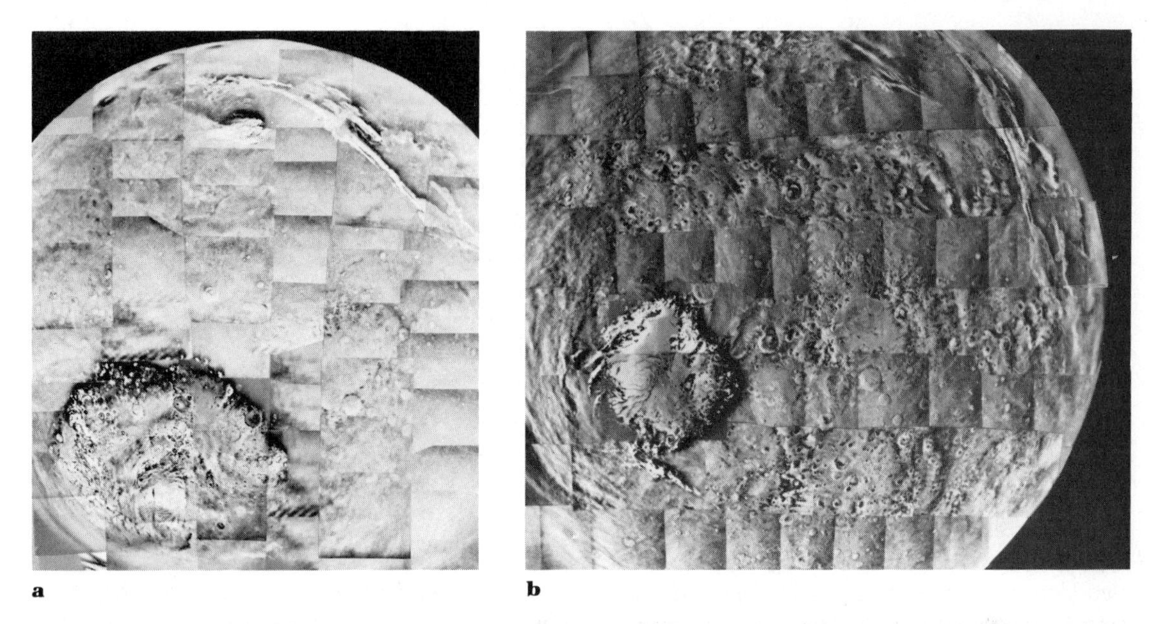

Figure 12–10. Two views of the Martian southern hemisphere, showing polar cap evolution. (**a**) *Mid-spring view shows polar cap with radius of 25°. Photo was made 5 wk after the start of a major global dust storm. Valles Marineris canyons (upper right) are filled with bright dust clouds, and a 600-km-wide dust cloud, reaching as high as 12 km, is seen to their left.* (**b**) *Five days before the start of summer, the polar cap has shrunk to 12° radius, almost the configuration of the permanent water ice cap. The dust storm has mostly dissipated, though Valles Marineris still appears dusty. (NASA, Viking; courtesy Geoffrey A. Briggs)*

So much CO_2 freezes out of the atmosphere that the total air pressure varies by 26% seasonally, due to the annual shuttling of CO_2 back and forth from one winter polar cap to the other (Hess, Henry, and Tillman, 1979).

The cycle of polar events is complex and interesting. It is partly shown in Figure 12-10. The polar temperature is almost always below the freezing point for water, even in full sunlight, so that a small, permanent polar ice cap of frozen H_2O exists, slightly off center at each pole and reaching about 10° from the north pole and about 6° from the south (where summer corresponds with the perihelion passage and is thus warmer). Viking orbital observations confirmed that the north permanent cap is H_2O ice (Kieffer and others, 1976). The south permanent cap is also believed to be H_2O or H_2O–CO_2 **clathrate** (mixture in which small carbon dioxide molecules are trapped in the more open crystal structure of water), although CO_2 dominance has not been ruled out (James and others, 1979). As local autumn temperatures drop, the polar hood forms; and the transient CO_2 ice cap forms, reaching 33° radius from the south pole and 25° from the north pole (where winter, coinciding with perihelion passage, is warmer and shorter). The depth of these CO_2 and H_2O deposits is unknown, though the amount of CO_2 transferred rules out a thickness more than a few tens of meters; the CO_2 transient cap has been estimated to be 50 cm thick (Hess, Henry, and Tillman, 1979).

CLIMATE EVOLUTION: WAS THE ANCIENT MARTIAN CLIMATE RADICALLY DIFFERENT?

A possible interpretation of the atmospheric facts presented so far is that Mars, during its total history, degassed more CO_2, H_2O, and N_2 than are present in its atmosphere today, but that these gases dissipated so rapidly that the atmosphere was never substantially denser and never had a dramatically different geological or biological effect on the planet. Even the Martian channels (see the previous chapter) could be explained by geothermal or other events that produced large, localized flows of water not involving global environmental effects.

Nonetheless, from the time of the channels' discovery in 1972 by Mariner 9 through the period of Viking measurements in the late 1970s, a suspicion has grown that Mars has experienced extraordinary environmental changes during its evolution. A wide-ranging list of observations and inferences is relevant:

1. The distribution of the channels suggests widespread water flow, and the forms of the dendritic channels suggest finely dispersed water sources, possibly rainfall (Masursky and others, 1977; see also Chapter 11).

2. The number of craters superimposed on the channels suggests that even the most recent major fluid flows were more than 100 My ago. Most presently visible channels probably formed 0.4 to 3.5 Gy ago (Malin, 1976; Hartmann, 1978; Masursky and others, 1977; see also Chapter 11).

3. Layered polar terrain suggests erosion-deposition cycles. The scarcity of craters in this terrain suggests that these cycles are recent, less than about 400 My in age, and possibly seasonal. But **unconformities** (discontinuous contacts between strata with different tilts) indicate major breaks in the depositional cycles over longer times and possible episodicity of erosional conditions (Cutts and others, 1976).

4. Crater size distributions in the older regions of Mars and the crater morphologies suggest obliteration of many old, small craters due to enhanced erosion and deposition all over Mars; this erosion was possibly episodic, and ended 1 to 4 Gy ago (Hartmann, 1973; Chapman and Jones, 1977). This evidence suggests a denser atmosphere during that period. Erosion of craters has been distinctly less in the last 1 Gy.

5. Major climatic changes have occurred on Earth. These include not only the short-term ice ages (10^5-y time scale; North America as far south as Chicago and New York were buried under a nearly continuous ice sheet 18 000 y ago), but also rather abrupt changes on a longer time scale. Widespread extinctions of most land *and* sea species in intervals of a few million years occurred 63 and 230 My ago (see the next chapter). The ice ages are probably controlled mostly by periodic changes in precession (21 000-y dominant periodicity), and obliquity (41 000-y dominant periodicity). These astronomical effects change the amount and duration of sunlight striking the polar caps each year and thus affect the polar ice thickness, the ocean level, and the world climate (Hays, Imbrie, and Shackleton, 1976).

6. Mars undergoes similar changes, which have been well analyzed. These include 0.12- and 1.2-My cycles of changes in obliquity, which ranges from about 17° to 34°, thus tipping the poles more or less toward the sun. This in turn causes more than a 60% variation in the average annual **insolation** (total solar flux striking the surface) at the pole (Ward, 1974). Also, changes in eccentricity with 0.1- and 2-My cycles cause a variation of about 23% in polar insolation (Murray, Ward, and Yeung, 1973). Theoretical calculations of model atmospheres (reviewed by Pollack, 1979) show that such changes in polar insolation could produce dramatic changes in atmospheric conditions by subliming polar ice. This could raise Martian atmospheric pressure from 6 mbar to as much as 1 bar if enough CO_2 were available to be sublimed off the poles. Probably the presently available CO_2 is not this great. The time scale for these changes is almost certainly too short for them to account for the channels (Hartmann, 1978), but they do allow for cyclic variation of atmospheric pressure by tens of millibars or more every few million years. These cycles may have involved enhanced dust transport, causing the polar sedimentary layers.

7. Remarkable related effects were found by Ward, Burns, and Toon (1979), who showed that the present mean obliquity is influenced by the Tharsis volcanics, which form a massive bulge on one side of the planet, affecting the moment of inertia. Subtracting this bulge so as to model a pre-Tharsis Mars, these researchers found an obliquity range from about 25° to 43°. The 43° value tipped the polar ice cap much more toward the sun than is possible in modern times. Whenever Mars' obliquity approached the higher values, sunlight could have fallen on one pole strongly enough to melt the H_2O deposits (which in modern times form the "permanent" residual ice cap). This means that pre-Tharsis Martian history should have been dotted by episodes when all the summer pole's H_2O evaporated and moved through the atmosphere toward the winter pole. Conceivably this could allow rainfall on the way if the atmosphere were warm and dense enough.

8. Mapping and crater counts show that the channels generally predate the formation of the Tharsis volcanics, a fact consistent with the hypothesis of a wetter climate before the Tharsis bulge formed.

9. Viking observations mentioned in connection with Table 12-2 indicate that much more gas erupted during Martian history than is present in the atmosphere now, presumably during volcanism associated with the Tharsis lavas. This includes an estimated 140 mbar or more of CO_2, "more than enough to permit the existence of liquid water," and an amount of water "equivalent to a 9-m layer of liquid over the entire planet" (Anders and Owen, 1977). An independent Viking study of nitrogen isotopes suggests enough nitrogen outgassing to correspond to a total atmospheric pressure of about 150 mbar to more than 1000 mbar. Of course, we don't know that all this gas was ever in the atmosphere at once, but the results suggest that the mean atmospheric pressure during the major Tharsis eruptions was much greater than it is today.

10. If Mars ever had a CO_2 or CO_2–H_2O atmosphere with a surface pressure higher than about 300 mbar, the surface climate could have been strikingly warm because of the same greenhouse effect that warms Venus' atmosphere. For example, in a 1000-mbar atmosphere of CO_2 with 77% H_2O humidity, Cess, Ramanathan, and Owen (1980) find a global mean temperature of 294 K (70°F). Such an environment could produce much fluvial activity and erosion.

Thus, a plausible scenario for Mars is that its atmosphere during early Tharsis activity, perhaps more than 2 or 3 Gy ago, was denser and warmer and became moist intermittently due to polar ice melting; rivers may then have flowed on the surface, as suggested in Figure 12-11. The pressure of the atmosphere slowly declined. After Tharsis eruptions built up a bulge on one side of the planet, obliquity conditions permitted melting of the polar H_2O ice became rare. Most water became permanently frozen, leaving an arid planet dominated by blowing dust. Only new data from Mars will clarify if this scenario is correct.

The importance of these ideas is threefold. First, they have spurred a new look at the climatic history of Earth. Second, the question of liquid water on Mars is critical to understanding possible Martian biochemical evolution (see the next chapter). Third, the demonstration of frequent, radical oscillations between habitable and nonhabitable conditions on any planet would strongly affect our ideas of the evolution of intelligent life outside the solar system.

DYNAMICS OF PLANETARY ATMOSPHERES

Having considered the evolution of certain atmospheres, we now consider phenomena of present-day atmospheres. Why do winds blow? Why shouldn't winds blow themselves out and remain calm forever? In this section we will show that winds are a consequence of solar radiation input and that wind and cloud patterns are associated with planetary rotation.

a

b

*Figure 12–11. Possible effects of Martian climatic change: before and after views of a Martian channel. (**a**) Present-day appearance of a channel. (Painting by author) (**b**) Reconstruction of the same scene under ancient climatic conditions during sudden flooding. A denser atmosphere may have produced occasional storm conditions. (Painting by Ron Miller)*

Vertical Mixing by Convection

Convection offers one way to drive atmospheric circulation, both on local and regional scales. Recall from Chapter 9's discussion of convection that convective flow will start in a medium if the temperature gradient becomes **superadiabatic**, that is, steeper than that experienced by an adiabatically rising parcel.

Thus we see one way that winds could start. Suppose we magically calmed all the winds on a planet. Sunlight strikes the planet. Perhaps a certain patch of dark ground gets quite hot and heats the parcel of air above it by conduction. That parcel expands and rises, causing new air to rush in horizontally to fill the potential vacuum. Already this is a local wind. If the temperature gradient happens to be superadiabatic, the process is self-regenerating and a substantial convective uplift may begin. Air passengers flying by might experience substantial choppiness as their plane flies through the ascending and descending currents.

Martian Dust Storms

This process of local heating not only causes winds but is probably a key to the summer onset of global dust storms in Mars' southern hemisphere. Mars passes through perihelion about a month before the beginning of summer in the south, so that maximum solar insolation occurs south of the equator in late spring. Mars' high eccentricity causes it to be 17% closer to the sun at perihelion than at aphelion, so the insolation difference between northern and southern summers is substantial.

Summer heating of the ground and adjacent air, primarily in the southern hemisphere, is strong enough to drive upward-swirling air masses into the atmosphere, possibly resembling the Southwest's dust devils. The dust itself absorbs sunlight and makes the air still warmer, adding to convective turbulence and causing a positive feedback effect. Thus, the storms, once started, spread in a day or so into massive dust clouds (Figures 12-10a and 12-12) that eventually carry dust across much of the globe (Golitsyn, 1973; Gierasch and Goody, 1973).

The dust particles lifted in Martian storms are very small, about 1 μm across. Calculations suggest that high winds of at least 50 m/s (110 mph) are needed to raise dust in the air of Mars, contrasting with only about 8 m/s needed on Earth (Mutch and others, 1976, p. 242). Viking Landers 1 and 2, at 22° and 48° N latitude, respectively, measured arrival of dust at their positions from storms that started in the southern hemisphere. The arrival at the Viking 1 site was accompanied by increased winds with gusts up to 26 m/s (58 mph). As seen in Figure 12-13, the Viking 1 lander also photographed extensive dune fields of dust sculpted by the Martian winds.

Global Circulation of the Planetary Atmospheres

Let us now consider the whole globe instead of concentrating on a local area. Suppose again that we could magically stop all winds. Stagnant air at the equator would then be heated and tend to rise, while cold air at the poles would tend to sink. This would set up a circulation pattern with air moving down at the poles toward the equator and then upward in the equatorial regions. This very simple model does not accurately predict planetary circulation patterns (although southward-moving weather patterns are familiar in the United States, especially during winter), but it does show that the continuous input of solar energy is a continuous driving force that stirs the atmosphere.

In the real world, several factors complicate the flow of air:

1. The **Coriolis force** is an apparent force due to Earth's rotation. The equator of Earth is moving eastward at about 464 m/s (1000 mph). Therefore, air masses moving away from the slow-moving poles appear to lag behind and shift

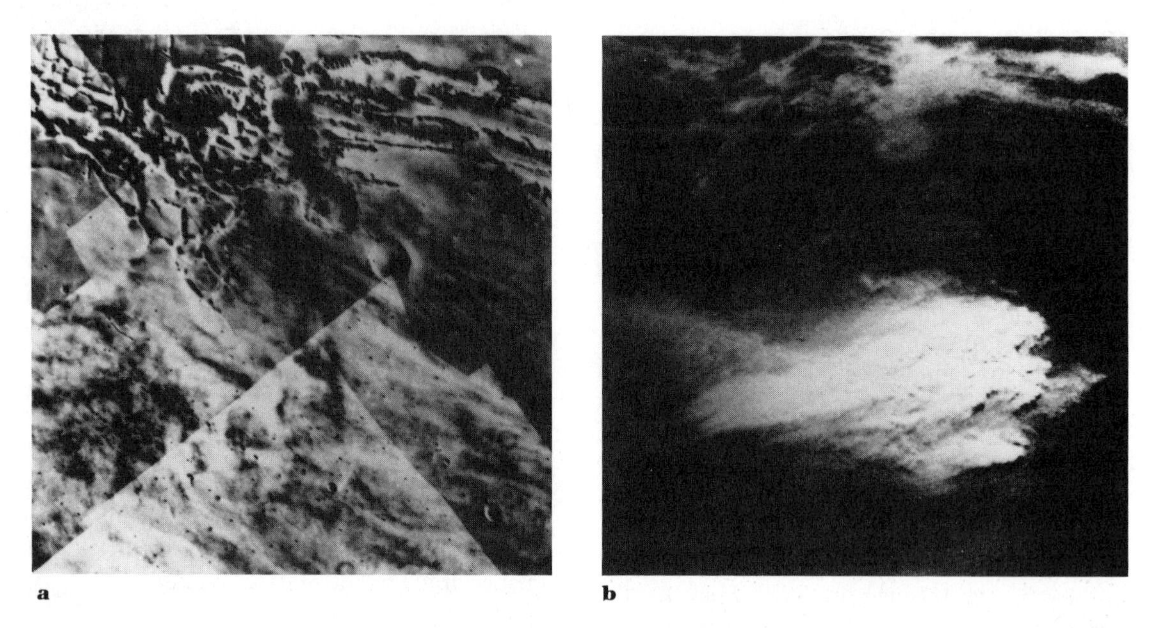

*Figure 12–12. Development of a Martian dust storm. (**a**) Typical view of high plains south of Valles Marineris canyons (top) shows scattered water ice clouds in early winter. (**b**) Mid-spring view (same as in Figure 12-9a) shows a turbulent dust cloud 12 km high and 600 km wide (the width of Colorado) covering the same area. The canyons are filled with low dust clouds. The date is about 35 d before perihelion passage, when strong solar heating warms the ground at southern latitudes and promotes convection, raising dust. (NASA, Viking)*

*Figure 12–13. The dune fields of mars, photographed under back-lighting (**a**) and front lighting (**b**). Rocks in the foreground range up to about a foot in length. Viking investigators left open the question of whether these were actively moving dunes or stabilized remnants of eroded sediment cover. (NASA, Viking 1 lander)*

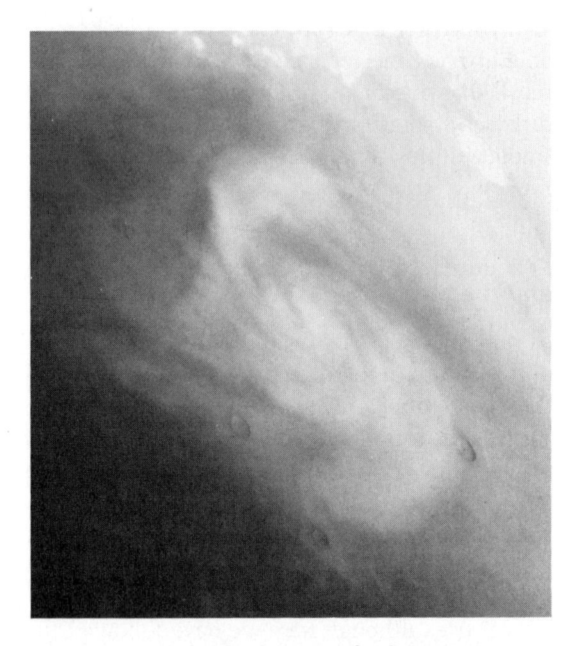

Figure 12–14. A cyclonic storm at 65° N latitude on Mars, associated with the boundary between cold air over the polar ice cap (outlying white frost patches, upper right) and warmer air from the mid-latitudes. The season is mid-summer, and the clouds are water ice crystals condensed from water vapor that evaporated from warm soil during the summer. Winds near the storm edge are estimated at 26 m/s (58 mph). The frost-filled crater Korolev, with a 90-km diameter, is in the upper right. (NASA, Viking 1)

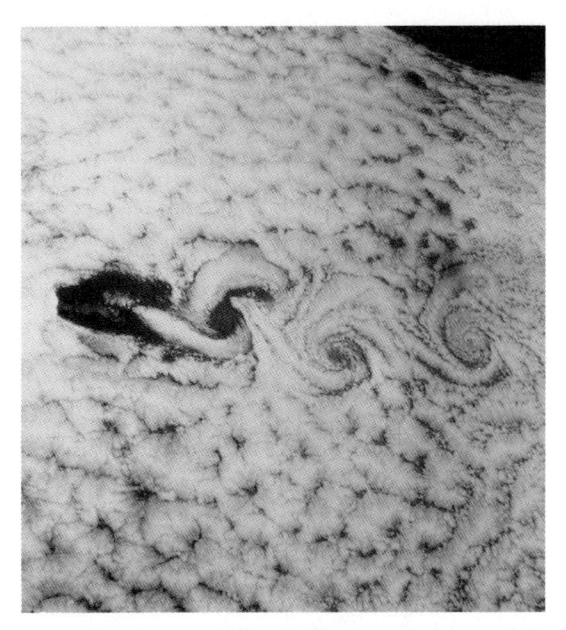

Figure 12–15. Series of cyclonic eddies produced as airflow is disturbed by Guadalupe Island, off Baja California. (NASA, Skylab 3)

westward, while poleward-moving air pulls ahead to the east. These Coriolis deflections produce the spiral motions (with opposite directions in the north and south hemispheres) associated with cyclonic storms, where air tries to rush radially into low-pressure regions. Several examples are prominent in Figure 12-14b; Figure 12-15 shows that the same effect occurs on Mars.

2. Land masses and their relief affect airflow, both on a small scale (Figure 12-14) and on a continental scale.

3. Fluid mechanics systems, depending on their dimensions and other properties, tend to set up turbulence and cellular patterns. For example, if you try to push water around in a bathtub with your hand, much of the motion you impart to the water ends up in turbulent eddies instead of a smooth flow. It is extremely difficult to predict the details of the water motion. In the same way, no simple model of a global atmosphere can predict the details of atmospheric motions visible on a planet from day to day. (Recall the discussion of turbulence in Chapter 3.)

ATMOSPHERIC LEVELS AND UPPER ATMOSPHERES

The principles discussed above can be applied to any part of any planetary atmosphere. However, certain additional phenomena enter the picture, especially in the upper atmospheres of the planets, because the upper atmospheres interact directly with the cosmic environment (for example, with the solar wind). The terminology used to discuss the general structure of planetary atmospheres at different levels derives from Earth's atmosphere and is based primarily on temperature effects. Let us begin with the lowest levels.

Troposphere, Tropopause, and Stratosphere

In a general planetary atmosphere (but particularly in the case of Earth), one might expect the lapse rate to be greatest near the ground, because infrared absorbers such as water vapor are concentrated here and also because this is where the atmosphere can be heated by contact with the ground, often producing superadiabatic lapse rates. At higher altitudes, the air is thinner and more transparent to radiation, and little energy is absorbed in each kilometer of height. At such altitudes, therefore, there would be a much lower lapse rate, as was the case in our nonabsorbing isothermal atmosphere.

This intuitive reasoning is correct in that the planetary atmospheres generally do have low-altitude regions of high lapse rates, turbulent motion, and cloud-forming activity and higher regions of nearly uniform temperature, smooth airflow, and lack of turbulence. This structure is shown for Earth in Figure 12-16.

The region in which the temperature gradient approaches zero is known as the **tropopause**; it is quite abrupt in the terrestrial atmosphere. It lies at a height of about 18 km at the equator and as low as 7 km at higher latitudes. The region below the tropopause is the **troposphere** (from a Greek root referring to change or mixing). It is characterized by greater lapse rates, more turbulent motions, and chemical mixing. The region above is the **stratosphere**, characterized by a more nearly isothermal temperature profile and smoother motions. Winds near the tropopause are often laminar, or uniform, and of high speed. These winds are known on Earth as the **jet stream**. Air travelers will recall that once a jet airliner gets well above the turbulent cloud layer, the flight becomes much smoother. This is because the aircraft has entered the lower stratosphere.

Temperature Irregularities and the Mesosphere

In general, the region from the tropopause to the higher realm of increasing temperature is called the **mesosphere**. There is some confusion between this term and the term *stratosphere* when applied in general planetology, because the atmospheres of different planets may have different patterns of temperature irregularities. Chamberlain (1962) proposed that the term *stratosphere* be abolished altogether in planetological use, although here we have followed the terrestrial use: The stratosphere is the nearly isothermal part of the mesosphere bordering and immediately above the troposphere, and it may in principle constitute the entire mesosphere.

The best example of complex structure in a mesosphere is the terrestrial **ozone layer**. As the solar ultraviolet comes down through the atmosphere, it is absorbed by oxygen molecules (O_2), dissociating them into pairs of single oxygen atoms. The oxygen atoms may combine with other oxygen molecules to form ozone molecules (O_3). The ozone itself absorbs ultraviolet sunlight, especially in certain color regions known as the **Hartley-Huggins bands**. If this high-energy ultraviolet radiation were not stopped, it would be lethal to life on Earth as we know it. In the region far enough down for oxygen molecules to be sufficiently dense, these energy absorptions heat the air significantly. The heating effect is maximized at about 50 km up, where there is a temperature maximum about 100 K greater than the tropopause temperature. The compositions and radiative properties of other planetary atmospheres may cause many variations in mesosphere structure (see the end of this chapter).

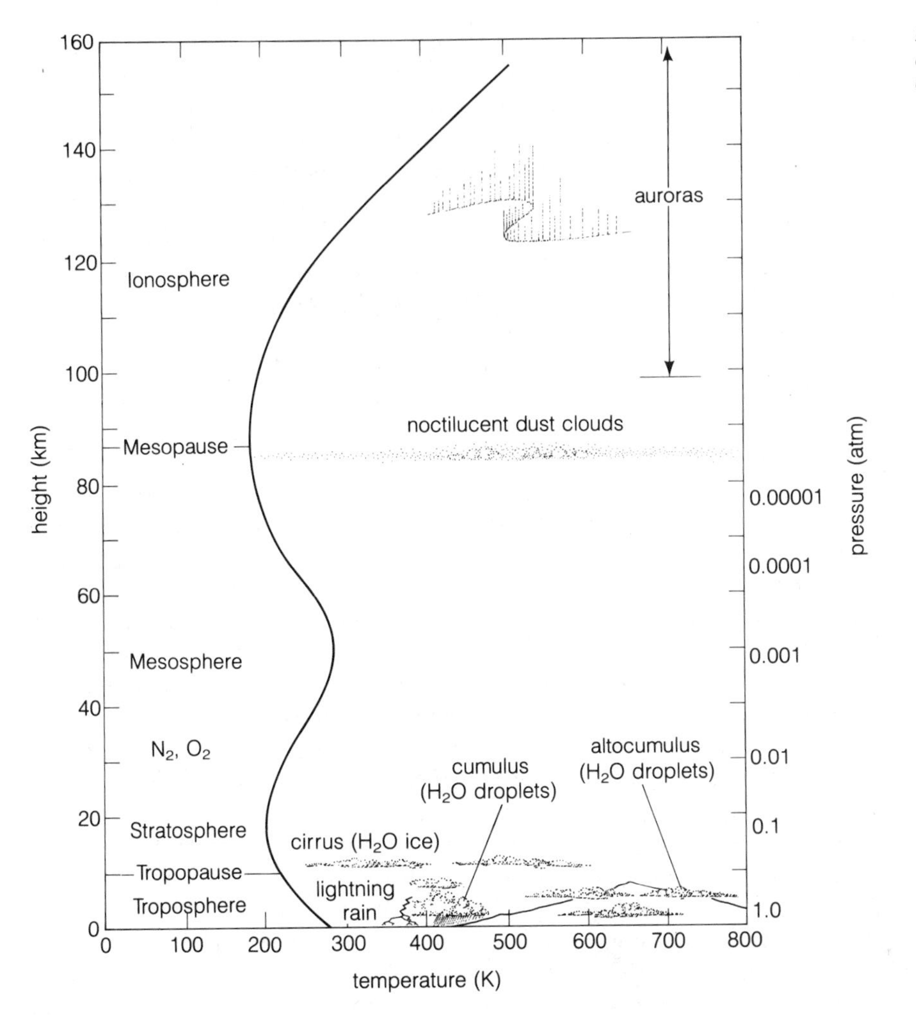

Figure 12-16. Structure and features of Earth's atmosphere.

Thermospheres, Ionospheres, and Exospheres

In their outer reaches, the planetary atmospheres grade into the interplanetary gas with its high-energy particles. The **thermosphere** is defined as the outer region where the temperature increases with height. An important characteristic of the thermosphere is that in the rarefied gas, molecules can move substantial distances before colliding with other molecules. Thus the **mean free path**, or typical distance between collisions of molecules, becomes much longer than in the denser lower atmosphere; separation of gases by diffusion becomes important, whereas in the lower atmospheres, high density and turbulent mass motions ensure mixing. Molecules of the lighter gases, especially hydrogen and helium, have higher velocities and thus can rise to great heights on nearly ballistic trajectories. This produces a degree of stratification; the lightest gases concentrate in the highest layers.

The planets are cold objects moving through the hot interplanetary gas. The solar wind sweeps out past the planets with an average velocity of 600 km/s near Earth. As this corresponds to a kinetic temperature of millions of Kelvin,* it is not surprising that the atmospheric boundaries are heated somewhat. The presence or absence of a planetary magnetic field, which deflects this onrush of ionized gas, thus affects thermosphere temperatures.

Rocket and satellite observations have shown that the temperature of the terrestrial thermosphere increases from about 180 K at 80 km to a high, variable value, ranging from 700 to 1800 K above 200 km, depending on solar activity (that is, the sunspot cycle), the time of day, and other factors. The primary cause of such high temperatures is absorption of solar ultraviolet radiation at altitudes between 80 and 200 km. High-energy particles from the solar wind may penetrate the planetary magnetic field and interact with the high atmosphere, generating additional heating.

The **ionosphere** is a portion of the thermosphere in which charged particles, or ions, are abundant. These ions are negatively charged electrons and positively charged atoms or molecules from which the electrons have been knocked off by photons of solar ultraviolet. These photons may impact so much energy that the electrons are knocked off with much higher velocities than the local equilibrium thermal velocities characterizing the neutral molecules. Indeed, measurements show that equilibrium does not exist between the radiation, the ions, and the neutral particles. Observed **electron temperatures** are often as much as twice the gas temperatures defined by the neutral particles.

Ions are so numerous that the ionosphere is an electrically conducting layer. Hence it drastically affects the passage of radio waves. Prior to 1901 it was confidently predicted that radio would be useful only over short ranges, since electromagnetic radiation travels in straight lines. Earth's ionosphere was discovered in 1901 when Marconi astonished the world by beaming radio signals across the Atlantic. Two physicists, Kennelly and Heaviside, independently but simultaneously explained Marconi's discovery by postulating a conducting, ionized layer high in Earth's atmosphere. Once called the Kennelly–Heaviside layer, the ionosphere absorbs and reflects radio waves, allowing "bounce beaming" over long distances.

Earth's ionosphere ranges from about 80 to 300 km in altitude, and it contains two main maxima in electron density, the lower E region (90 to 120 km) and the upper F region (150 to 300 km). During the daytime the latter often has two maxima, F_1 and F_2. All the layers, especially the uppermost F_2 region, are highly variable and are affected by the time of day, the 11-y sunspot cycle, magnetic storms, and so on. Electron densities in the F_2 region reach as much as 10^6 electrons/cm^3.

Similar ionosphere structures were found by Pioneer and Viking probes on Venus and Mars, with ion densities reaching 10^5 ions/cm^3 at altitudes of 140 and 130 km, respectively (von Zahn and others, 1979).

The outer edge of the thermosphere is the **exosphere**, a level above which at least half of the upward-moving molecules do not hit another molecule, instead following a long trajectory into space and possibly escaping. The exosphere temperature is very important to the evolution of the atmosphere. If the exosphere temperature is high, exosphere molecules will have high velocity and can escape into space. Certain molecules that are efficient radiators in the infrared may serve as thermostats that keep the thermosphere temperature down. "Thermostat molecules" are thus crucial to the atmosphere's history. The absence of such molecules forces the temperature up until outward radiation offsets the incoming solar radiation.

*The reader may wonder how the planets can be embedded in a multi-million-Kelvin gas without being vaporized. The important point is that temperature is defined in kinetic theory by the energy (or velocity) of the atoms and molecules. (Physics students should recall the equality of thermal and kinetic energy of an atom: $\frac{3}{2}kT = \frac{1}{2}\mathbf{mv}^2$.) But for a hot gas to heat a solid mass that it contacts, the gas must be fairly dense. If a gas is as tenuous as the interplanetary medium, the solid mass simply will not be hit by enough atoms to absorb enough energy to maintain a high temperature. If the solid mass were somehow heated to as high a temperature as the tenuous gas, it would at once radiate at a furious rate until it cooled.

An example of the effect of thermostat molecules occurs in Mars' exosphere. Without a good infrared radiator, its temperature would be about 2000 K. Chamberlain (1962) pointed out that CO, produced from the abundant CO_2 on the planet, would be a good radiator. Chamberlain's original calculations suggested an exosphere temperature as low as 1100 K, and subsequent theorists suggested still lower values of only a few hundred Kelvin. Viking entry probes measured an exosphere temperature of 210 K near an altitude of 175 km (Hanson and others, 1977). Similarly, Pioneer entry probes on Venus found an exosphere temperature of 285 K at an altitude of about 165 km (von Zahn and others, 1979). Because exosphere temperatures control escape rates into space, even short-term changes in exosphere temperatures, perhaps due to changes in the quantity of exosphere thermostat molecules, could create important episodes of enhanced atmosphere escape rates.

ESCAPE OF PLANETARY ATMOSPHERES

The theory of gravitational escape of gases from planets was developed by Jeans (1916). If one of the upward-moving molecules in the exosphere does not hit another molecule, it will follow a ballistic path—an elliptical orbit arcing above the atmosphere and eventually returning. If the molecule should happen to be one of the fastest-moving molecules and have a velocity greater than the escape velocity for that particular altitude, it will follow a parabolic or hyperbolic orbit and escape from the planet (see Chapter 3). Every time this happens, the planet's atmosphere permanently loses one molecule.

Jeans, and later Spitzer, calculated the rate at which this happens. Each molecular species (H_2, N_2, O_2, CO_2, and so on) may be treated separately. The principle known as the **equipartition of energy** states that, on the average, each kind of molecule has the same energy as any other kind. In order to have the same energy, the small ones must move fast, and the big ones must move slowly. Therefore, on the average, the light hydrogen molecules will move faster than the heavier helium molecules, which will move faster than the still heavier oxygen molecules, and so on.

The fastest-moving molecular species, hydrogen, will have the most molecules moving faster than the escape velocity and will be the first type of gas to be depleted in the atmosphere of any planet. Because of its strong gravitational field, Jupiter has apparently retained much of its original hydrogen, while Earth—with its smaller gravitational attraction and lower escape velocity—has lost its hydrogen but retained most heavier gases.

Four phenomena complicate this picture:

1. In any molecular species there is a distribution of speeds. For example, in the Martian exosphere, the *mean* hydrogen velocity would be about 2.3 km/s, about half of Mars' 5-km/s escape velocity. Nonetheless, a certain fraction of H atoms would move faster than 5 km/s, facilitating a slow but steady escape.

2. Solar ultraviolet radiation may break up certain molecules, a process called **photodissociation**. For instance, H_2O (molecular weight 18) may be too heavy to escape, but if it breaks into H and OH in the exosphere, the H may rapidly escape from the planet. Even the OH may break apart. H_2O may thus be rapidly depleted as H leaks off and O oxidizes other materials.

3. Atoms and molecules of the solar wind stream by planets at hundreds of kilometers per second. If these collide with exosphere molecules, they may kick the target molecules out of the atmosphere or at least dissociate them. Today's solar wind thus helps deplete atmospheres, and a primeval super solar wind might have been important in stripping primitive atmospheres off planets.

4. Heavy molecules tend to concentrate in the lower atmosphere; thus their escape rate from the exosphere is limited by the rate at which they can diffuse upward.

These ideas apply to Venus and its missing H_2O. Any original H_2O was too heavy to escape in the age of Venus. But if H_2O broke into H and O atoms, the H could escape more easily. However, the low 285 K exosphere temperature found by Pioneer entry probes implies 20 Gy even for hydrogen to escape! Perhaps the exosphere was hotter in the past. At 1000 K, the escape time could be much less than the age of Venus. Some evidence favors this. The ratio of deuterium, or the heavy isotope of hydrogen, to ordinary hydrogen, usually written D/H, is much higher on Venus than on Earth. The usual interpretation is that when H_2O broke into H and O, the lighter H

MATHEMATICAL NOTES ON ATMOSPHERIC ESCAPE

The expression for the lifetime of a planetary atmosphere is too complex to derive here. However, a rule of thumb emerges from the fact that the escape time depends rather critically on the escape velocity of the planet, for which an expression is derived in Chapter 3. This rule of thumb can be expressed as follows: An atmosphere will escape from a planet in a time less than the age of the solar system if

$$V \geqslant f v_{esc}$$

where

V = root-mean-square molecular velocity $= \sqrt{3kT/m}$

f = constant, about 0.25

v_{esc} = escape velocity from planet

m = molecular mass for given molecular species

k = Boltzmann constant

Sharonov (1958) writes a more general rule of thumb to take into account that the exosphere temperature will tend to decline roughly with the square root of the distance from the sun, to the first order (according to simple laws of radiative heating). Combining this with the equation and definitions above, Sharonov concludes that the ability to retain an atmosphere might be called the relative atmosphere retentivity, $\mathcal{R}$, which is defined as the ratio

$$\mathcal{R} = \frac{v_{esc}}{V} \propto v_{esc} a^{1/4}$$

where a is the distance of the planet or satellite from the sun. Planets are listed by relative retentivity $\mathcal{R}$ in Table 12-3.

Spitzer (1952) derived the time t to reduce atmospheric density of $1/e$ of its original value by thermal escape (neglecting nonthermal effects such as photodissociation and solar wind):

$$t = \frac{\sqrt{6\pi}}{3g} V \frac{e^Y}{Y}$$

where

g = surface gravity of planet $= \dfrac{GM}{r^2}$

$$Y = \frac{GMm}{kT_x R}$$

G = Newton's gravitational constant

M = planetary mass

R = planetary radius

T_x = exosphere temperature

m = molecular mass = molecular weight × mass of H atom

This expression applies to an idealized isothermal atmosphere. However, with a more realistic temperature structure, escape can be as much as 10^5–10^6 times faster. This discrepancy shows the importance of understanding exosphere temperature structure.

atoms escaped much faster than the heavier D atoms, leaving a concentration of D behind. Grinspoon (1987) suggests an alternative model, in which Venus was initially dry, and the current D/H was built up by water provided by comet impacts. Certainly the question of the original water contents and sources on Venus and Earth, and the relation to Mars, is a major issue in planetary science (see review by Pepin, 1991).

SUMMARY: ATMOSPHERES OF SELECTED WORLDS

Table 12-3 summarizes planetary atmospheres with the planets listed in order of ability to retain atmospheric gases. A dramatic transition from air-rich to airless worlds occurs around the size range of Triton, Ganymede, and Titan.

A note is in order about the units used to report atmospheric compositions. Table 12-3 lists fractional numbers of molecules of each species, or **mixing ratios**. Meteorologists commonly specify composition by mixing ratios rather than by, say, percentage by weight (the units usually used in discussing planetary interiors). Telescopic observers usually report compositions in terms of the equivalent amount of gas traversed by the light beam on its way to the telescope. This is the gas traversed either above the surface or above some opaque cloud layer. The unit often used is the **kilometer-atmosphere**, the amount of gas traversed in a 1-km path under standard

Table 12-3
Planets Listed in Order of Atmospheric Retentivity

Planet	Relative retentivity R[a]	Surface pressure[b]	Constituents (% by no. molecules or volume)
Jupiter	250	High	90 H_2, 10 He, 0.07 CH_4
Saturn	170	High	97 H_2, 3 He, 0.05 CH_4
Neptune	147	High	>74 H_2, <25 He, >1 CH_4 at depth
Uranus	120	High	83 H_2, 15 He, 2 CH_4 at depth
Earth	30	1000	78 N_2, 21 O_2, 1 Ar
Venus	25	90 000	96 CO_2, 3 N_2, 0.1? H_2O
Mars	15	10	95 CO_2, 3 N_2, 1.6 Ar
Titan	12	1500	>90 N_2, 1–10 CH_4, ? Ar
Ganymede	11	$<10^{-8}$	
Triton	10?	0.016	N_2, CH_4
Io	10	$\sim 10^{-7}$?	? SO_2
Mercury	9	<0.3	
Europa	8	—	
Pluto	7	0.01?[c]	? CH_4, ? N_2
Callisto	6	—	
Moon	6	—	
Rhea	3	—	
Iapetus	3	—	
Dione	2	—	

Sources: Orton (1978); Hanel and others (1979); Oyama and others (1979); Owen and others (1977); Cruikshank and Silvaggio (1979, 1980); Owen (1982); Broadfoot and others (1979); Kumar (1979); Belton, Hunten, and McElroy (1967); Ingersoll (1981); Barbato and Ayer (1981); Stone and Miner (1986, 1989); Ingersoll (1990).
[a]A measure based on escape velocity and exosphere temperature, discussed in the accompanying mathematical notes.
[b]Figures in millibars.
[c]At perihelion.

atmospheric conditions, usually taken as 20°C and 1 atm (1 bar) of pressure. Frequently this same unit is defined as the **kilometer-amagat**, where *amagat* refers to a slightly more rigorously defined density equal to 0.0446 g·mole/liter. A gram-mole is the number of grams equal to the molecular weight of the gas species.

The following sections summarize historical and modern studies for each planet.

Venus

The atmosphere of Venus was discovered by a Russian scientist, Lomonosov, during a solar transit of Venus in 1761. Lomonosov observed sunlight backlighting this atmosphere, as seen in Figure 12-6. The composition remained a matter of controversy for more than two centuries. When the earliest observers realized that Venus was covered by clouds, they assumed that they were seeing dense water vapor clouds like those of Earth. Science fiction writers depicted Venus as a rainy, steamy swamp populated by dinosaurs or some fictional equivalent. However, careful observations during the 1950s and early 1960s failed to give evidence of substantial amounts of water vapor. A variety of (sometimes exotic) theoretical models of the atmosphere followed: models with planet-wide oil fields; models in which all the water was tied up in low-level clouds, ice crystals, and so on; and models with tremendous winds and blowing dust that would generate the high surface temperatures by friction. After some controversy, most investigators accepted that the clouds and general Venus meteorology are not dominated by effects of water, since the observed amount of water vapor above the clouds is very small.

In 1928, American astronomer Frank Ross announced the discovery of hazy cloud patterns visible on ultraviolet photos but virtually invisible to the eye. Chapter 2 detailed unsuccessful attempts to determine Venus' rotation period by tracking these patterns from Earth. These attempts led only to discovery of a west-to-east, 4-d cloud circulation but did not reveal the planet's rotation.

Mariner 2's flight past Venus in 1962, together with nearly simultaneous Earth-based observations, led to the discovery that Venus is very hot, with surface temperatures initially estimated at 400 to 700 K (Sagan, 1962), which Sagan correctly attributed to a greenhouse effect. Chapter 2 detailed the discoveries of the ensuing series of Soviet probes that were parachuted into the atmosphere, culminating with the first successful surface measurements by Venera 7 in 1970. This probe transmitted 23 minutes of data from the surface before failing and reported a temperature of 748 K and a pressure of 90 bar!

A further breakthrough came in 1972–1973 as several scientists independently identified the Venus clouds as being made of 2-μm droplets of 78% to 90% sulfuric acid.

The Pioneer Venus mission parachuted four probes through Venus' atmosphere on December 19, 1978, adding numerous details to the above results. These included detection of minor constituents such as 180 ppm of SO_2, 67 ppm of Ar, and 20 ppm of CO, in addition to the gases listed in Table 12-3 (Oyama and others, 1979). The atmospheric structure derived from Pioneer and other measurements is shown in Figure 12-17.

Daily monitoring of the Venusian clouds by space vehicles in the 1970s revealed strong westward cloud circulation at about 100 m/s, as shown in Figure 12-18. Interestingly, such photos confirmed controversial features that had been dimly glimpsed and sketched by visual observers of previous decades: a semipermanent "cusp cap," or polar cap of bright haze, often bordered by a darker band (Figure 12-19). Figure 12-20 shows the polar cloud cap as it would appear from above. Such features are prominent only in ultraviolet light. Spacecraft studies showed that the bright cap and bright bands are high hazes composed of 0.25-μm particles, smaller than the 1- to 10-μm particles in the main cloud deck described earlier in this chapter.

A curious feature of Venus that has not been

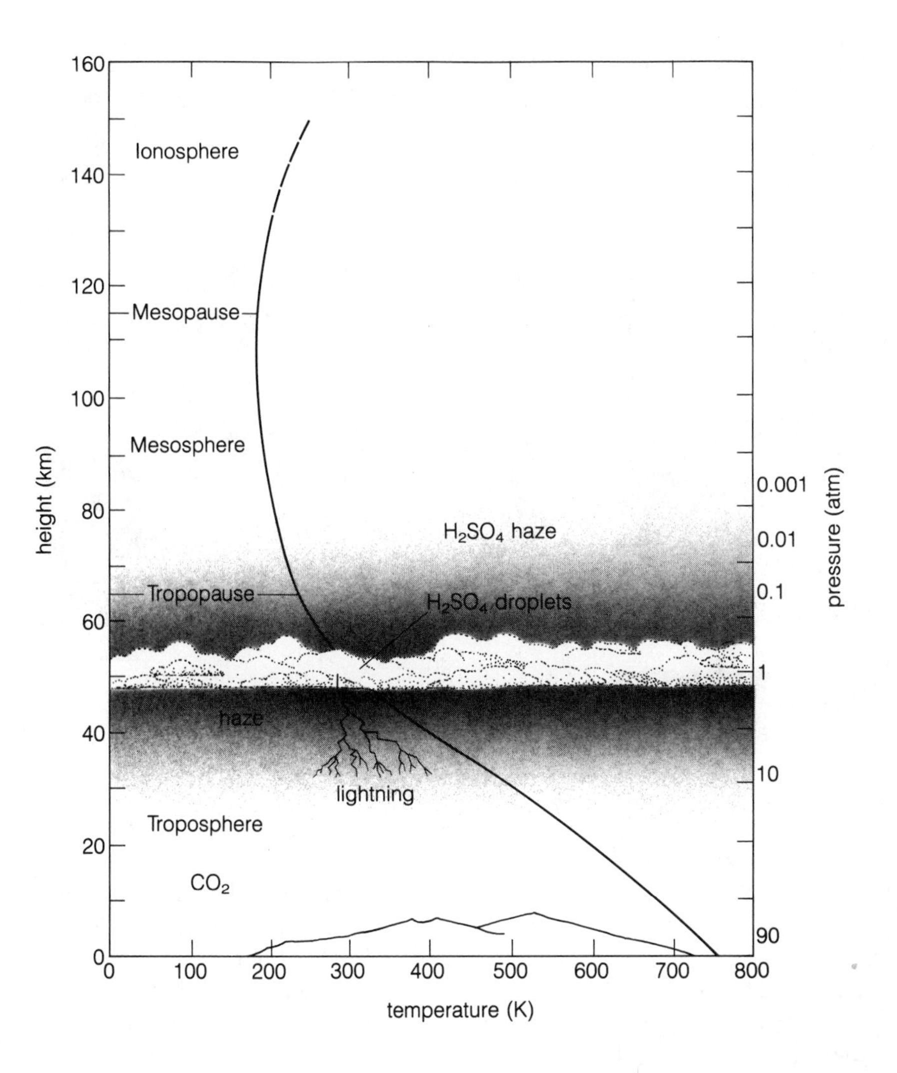

Figure 12–17. Structure and features of Venus' atmosphere, based on Pioneer and other measures.

accounted for is the **ashen light**, which is a faint glow of the dark hemisphere of Venus. It has been seen by a number of visual observers, particularly near inferior conjunction, when the dark side of the planet is directed toward Earth. Theoreticians have been unable to account for it, in spite of the fact that it has been reported even in telescopes with apertures as small as 30 cm. Usually described as tinted with a coppery color, it is probably a high-altitude phenomenon similar to the terrestrial aurora or airglow (Levine, 1969).

Airglow in the form of certain spectral emission lines has been found on the dark side of Venus by Soviet and American spacecraft, but the glow is too faint to account for ashen light. Venera 11 and 12 probes in 1978 and the subsequent analysis of Pioneer data obtained at about the same time revealed strong evidence for lightning discharges on Venus (Figure 11-55), leading to suggestions that the lightning might be abundant enough to account for the ashen light (Ksanfomaliti, 1980). The past history of water and other volatiles in Venus' atmosphere is an area of active current research.

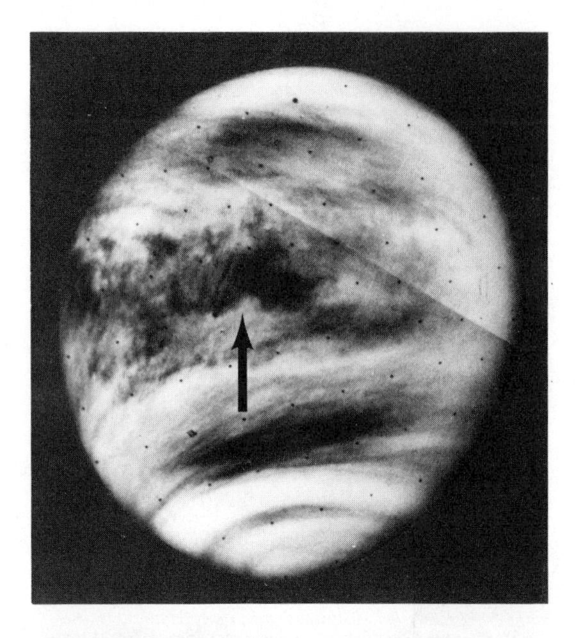

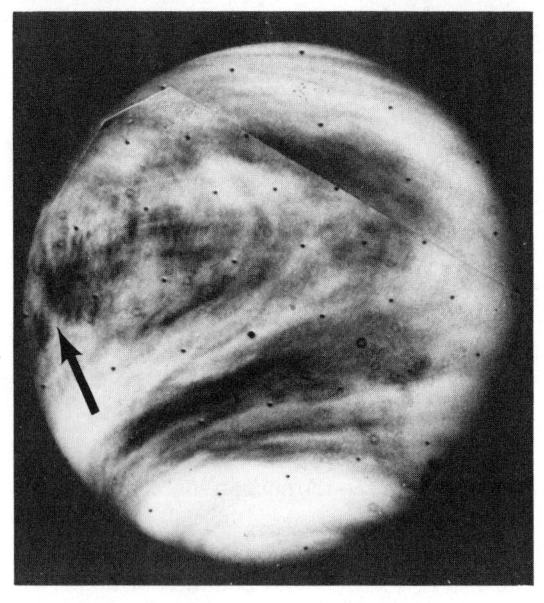

Figure 12-18. Three ultraviolet images of Venus, with arrows indicating a westward drift of cloud features during a 14-h period on February 7, 1974. (NASA, Mariner 10)

Earth

Figure 12-16 summarized our atmosphere's structure. A principal point emphasized in this chapter is the evolution of our atmosphere from a possible primitive atmosphere rich in hydrogen compounds, through an early secondary atmosphere poor in oxygen, to the modern oxygen-rich atmosphere. Water vapor and carbon dioxide were the most important gases added by volcanic degassing to form the secondary atmosphere. Most of the oxygen was added by biological processes, considered in more detail in the next chapter.

Mars

Chapter 2 chronicled the drastic decline in estimates of Martian atmospheric density, from Herschel's estimate of an Earth-like environment to modern measurements of 6 mbar surface pressure. Among the significant meteorological events in this thin atmosphere (Figure 12-21) are the formation of whitish and bluish H_2O and CO_2 clouds, the seasonal evolution of global or regional yellowish dust storms, and the seasonal changes in pressure as CO_2 condenses into frost at the polar caps.

The discussion in this chapter, together with the existence of the mysterious channel systems (Chapter 11), suggests that today's dry, dusty, frozen Mars is not the Mars of the past. The total of volatiles outgassed from Mars was almost certainly more than is now visible. Rossabacher and Judson (1981) assert that storage in the polar caps, atmosphere storage, and loss into space account for only about 10% of all H_2O degassed onto Mars and that the rest is stored in a "cryosphere" of permafrost. McElroy, Yung, and Nier (1976) conclude from present nitrogen isotope ratios that as much as 30 mbar of N_2 was degassed in the past. Model atmospheres with 500 to 1000 mbar surface pressure of CO_2 allow liquid water to be stable over much of the surface partly because of greenhouse warming (Hoffert and others, 1981; Toon and others, 1980). Such atmospheres are plausible products of early Martian volcanism. These are just examples of data supporting the belief that the first 1 to 3 Gy of Martian history witnessed a denser, warmer atmosphere with frequent liquid water activity (Toon and others, 1980; Carr, 1981). The atmosphere dwindled, greenhouse warming decreased, and the production of the Tharsis volcanics helped produce a situation where much water now remains permanently frozen at the poles, as well as in widespread permafrost layers.

Figure 12-19. A view of Venus from above the southern hemisphere (lat. −30°) shows the broad, bright south polar cloud cap, bordered by a concentration of dark bands (for example at lower left). (NASA, Pioneer orbiter photo, Feb. 11, 1979)

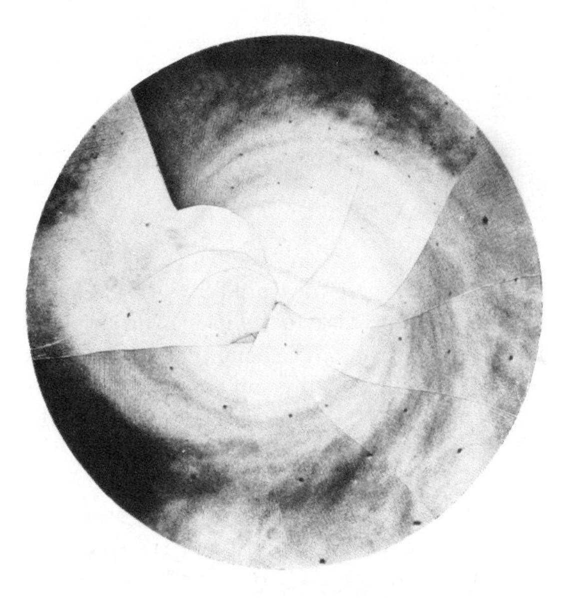

Figure 12-20. Photo-mosaic indicating the spiral structure of Venus' south polar cloud cap and its darker border, as seen from above the pole. (NASA, Pioneer images processed by Sanjay Limaye, Space Science and Engineering Center, University of Wisconsin)

Jupiter and the Giant Planets

Earth-based spectra of Jupiter and Saturn long ago indicated absorptions by methane and ammonia (Kuiper, 1952), but scientists suspected that hydrogen and helium were also retained by the strong gravities of these planets. Molecular hydrogen was eventually detected by a series of Earth-based observations (reviewed by Owen, 1970). On Jupiter and Saturn, respec-

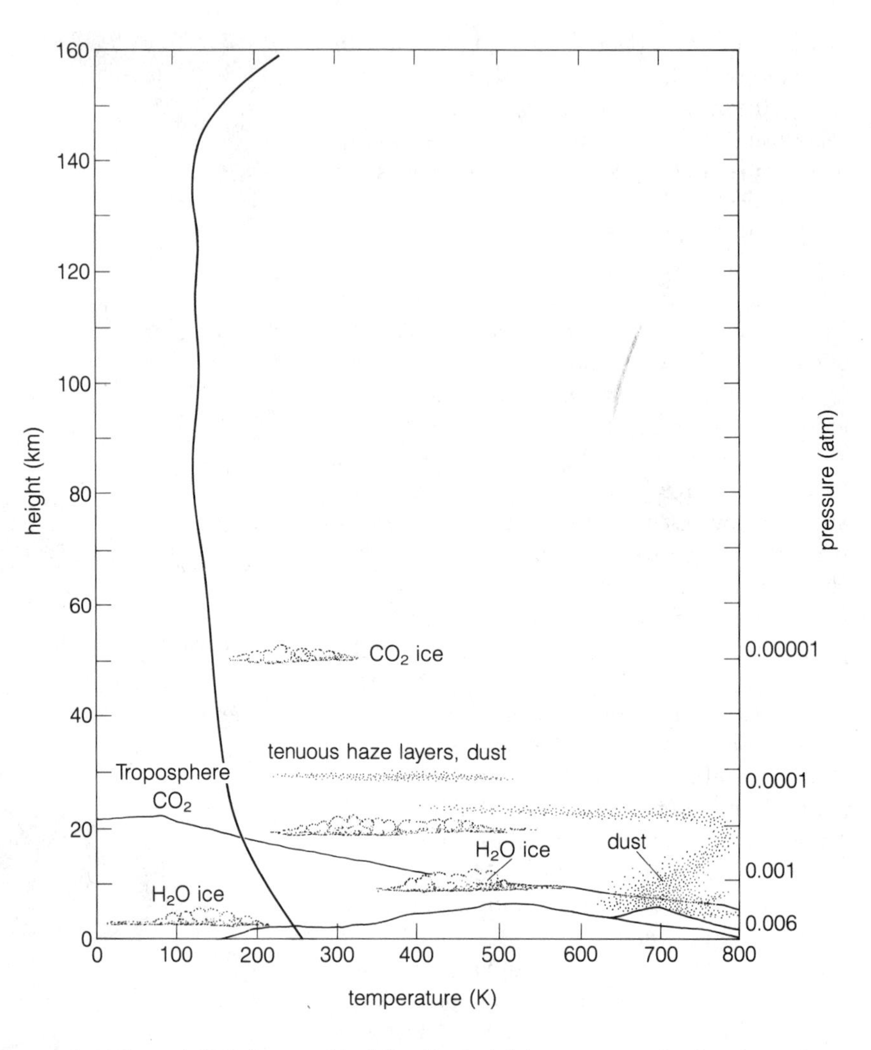

Figure 12–21. Structure and features of Mars' atmosphere, based on Viking and other measures.

tively, the Voyagers measured 81% and 88% by mass of H_2, very close to the solar hydrogen content of 79% by mass. The helium percentages for Jupiter, Saturn, and the sun were measured at 18%, 11%, and 20%, respectively (Ingersoll, 1981). On both Jupiter and Saturn, the Voyagers also identified many trace constituents composing less than 1% of the mass. These included ammonia (NH_3), methane (CH_4), phosphine (PH_3), ethane (C_2H_6), and acetylene (C_2H_2) (Hanel and others, 1979; 1981).

The ever-changing, colorful cloud patterns of Jupiter and Saturn have remarkably intricate structure, as shown by Figures 12-22 and 12-23. Dark **belts**, bright **zones**, and a few prominent features such as Jupiter's Great Red Spot are relatively stable, but smaller features change from day to day. Whole belts and zones change intensity from year to year, as can be monitored with a backyard telescope. As sketched in Figure 12-24, Voyager temperature maps revealed that the bright zones are high, upwelling clouds in the cold, upper atmosphere. The darker, more colorful belts are regions where we see deeper, warmer levels. Saturn's clouds are generally less colorful than Jupiter's and less broken by prominent oval disturbances. However, the strong eastward and westward jet streams, moving at

Figure 12–22. Convoluted cloud patterns of belts and zones on Jupiter, interrupted by the Great Red Spot, center. Note Jupiter's limb, lower right (NASA, Voyager 1)

hundreds of meters per second relative to each other at different latitudes and altitudes, cause shearing of clouds on both planets, as described in Chapter 2. The Red Spot is one such turbulent eddy, circulating in a counterclockwise direction. Low temperatures in the Red Spot show that it lies higher than surrounding clouds. White oval disturbances occur on both Jupiter and Saturn; Voyager 2 observed a Saturnian example measuring 5000 × 7000 km and circulating with 100 m/s winds.

The rapid changes in the cloud markings are the first clue to the structure of giant planet atmospheres. The changes suggest convection

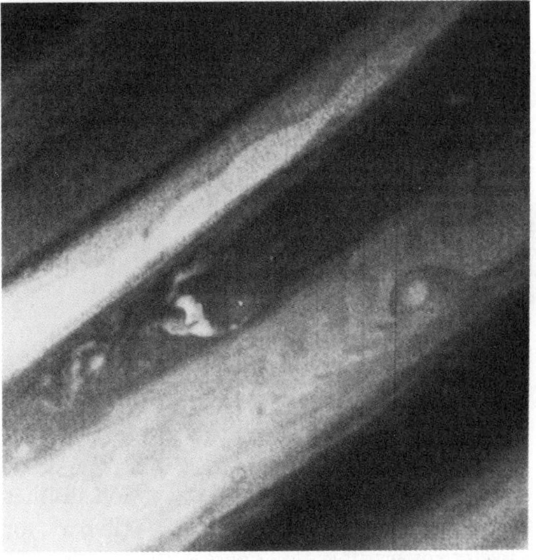

Figure 12–23. Small-scale cloud details in the atmosphere of Saturn. Bright oval (right) and spots in the dark belt are convective clouds. A wavelike longitudinal cloud stretches along the bright zone at the top. Smallest cloud features are about 175 km across. North is toward upper left. (NASA, Voyager 2)

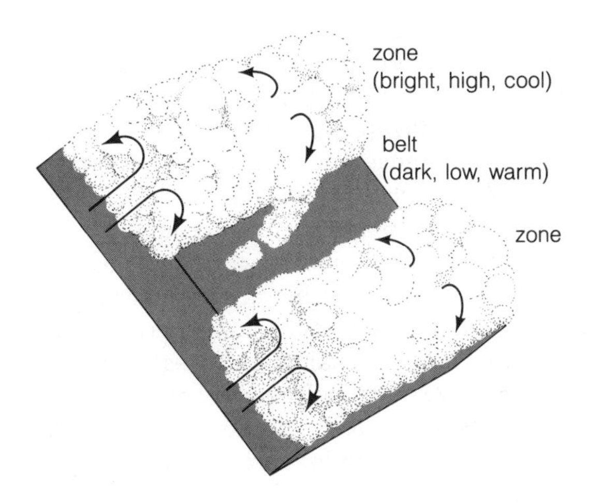

Figure 12–24. (left) Convective structure of belts and zones deduced from spacecraft measures and cloud motions. Zones are bright upwelling clouds; belts are descending areas exposing lower, warmer regions. Patterns acquire greater complexity due to shearing motions caused by Coriolis forces.

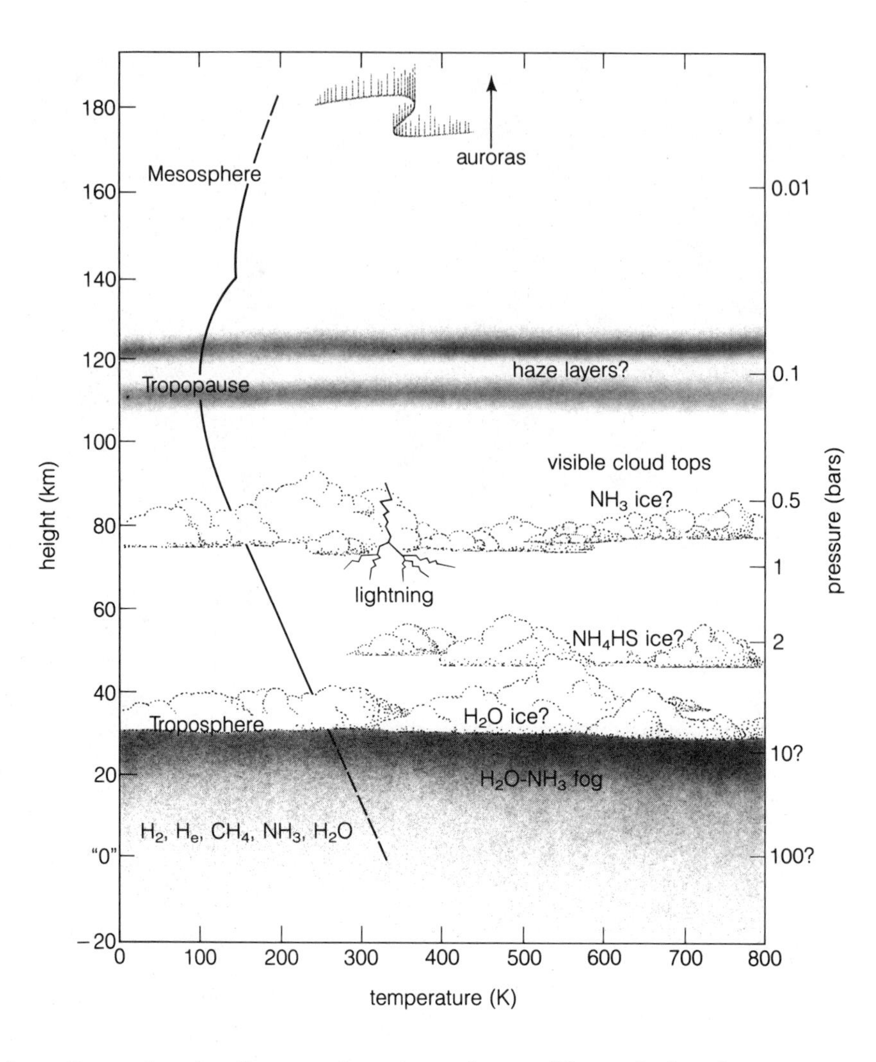

Figure 12–25. Structure and features of Jupiter's atmosphere, based on Voyager and other measures.

driving ascent from below. Convection implies that the temperature gradient is adiabatic or slightly superadiabatic. These ideas led to early models of the temperature structure of Jupiter's atmosphere, based on Kuiper's (1952) calculation of an adiabatic temperature gradient of -3 K/km below the clouds. Voyager data (Figure 12-25) confirmed this reasoning, measuring a temperature gradient closer to -2 K/km. Figure 12-26 shows a similar atmospheric structure diagram for Saturn.

The source of cloud colors on Jupiter and Saturn is uncertain. Prinn and Owen (1976) showed that reactions among the major and minor constituents, such as sulfur and phosphorous, can produce colored compounds such as ammonium polysulfides ($[NH_4]_xS_y$), hydrogen polysulfides ($H_x S_y$), and phosphorous crystals (P_4). Their tones of yellow, orange, brown, and red may explain the colors. Other compounds, such as organic molecules, have also been proposed.

Voyager discovered remarkable visual phenomena in Jupiter's nighttime atmosphere: aurora and lightning. Both of these were recorded photographically in visual wavelengths as shown in Figure 12-27. Ultraviolet auroral emissions near the poles were also detected. Voyager instruments recorded similar ultraviolet aurora within

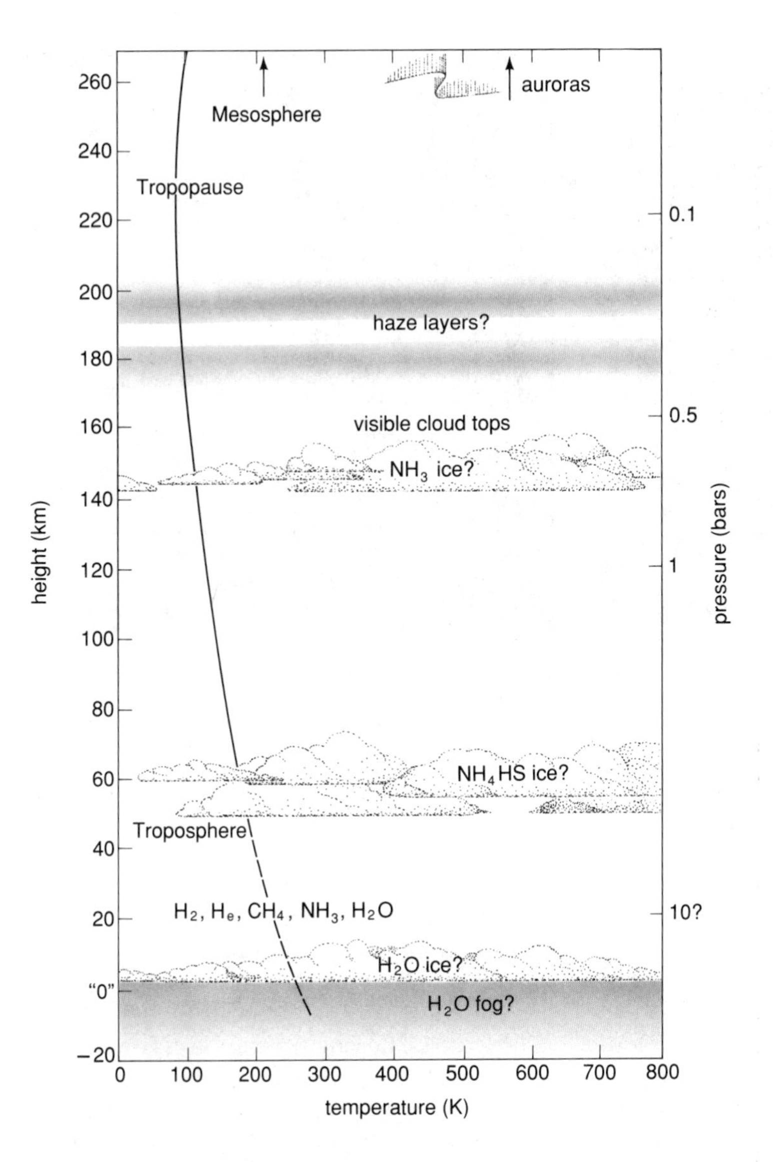

Figure 12–26. Structure and features of Saturn's atmosphere, based on Voyager and other measures.

12° of the poles of Saturn. Interestingly, "frequent and powerful thunderstorms in the Jovian atmosphere," causing lightning at 10 000 times the terrestrial rate had been predicted as early as 1975 by Bar-Nun, based on his hypothesis that observed Jovian acetylene was synthesized by lightning. Ironically, Lewis (1980) later concluded that the acetylene is not efficiently produced by the lightning, so that the prediction, though correct, has questionable status.

The atmospheres of Uranus and Neptune show some similarities and some differences when compared with those of Jupiter and Saturn. Most pronounced is the blue color. This stems from a combination of two effects. The clouds are overlain by a thick layer of gaseous haze in which Rayleigh scattering produces a sky-blue color. Perhaps more importantly, methane gas absorbs red and orange light, producing an overall blue tone with a faint greenish tinge.

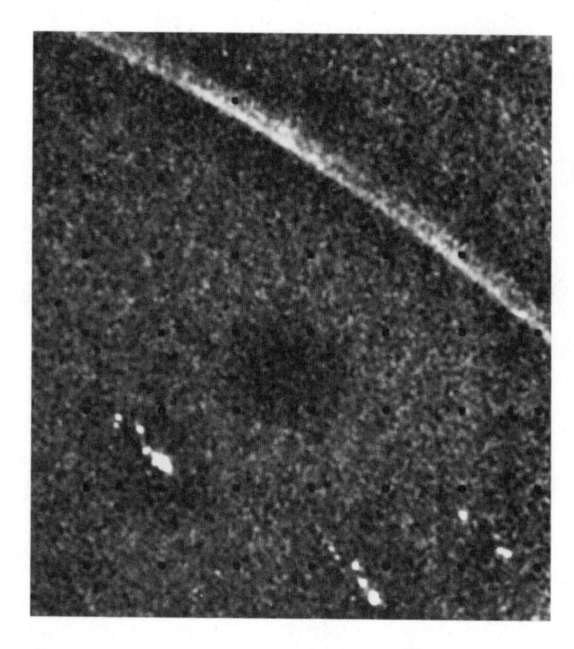

Figure 12–27. Lights in Jupiter's night sky, seen from above. This three-minute Voyager orbiter exposure shows aurora over north pole (bright limb, across top) and bright lightning flashes among the clouds. The aurora is brighter than Earth's; the lightning is comparable to superbolts near the tops of terrestrial tropical thunderstorms. (NASA, Voyager 1)

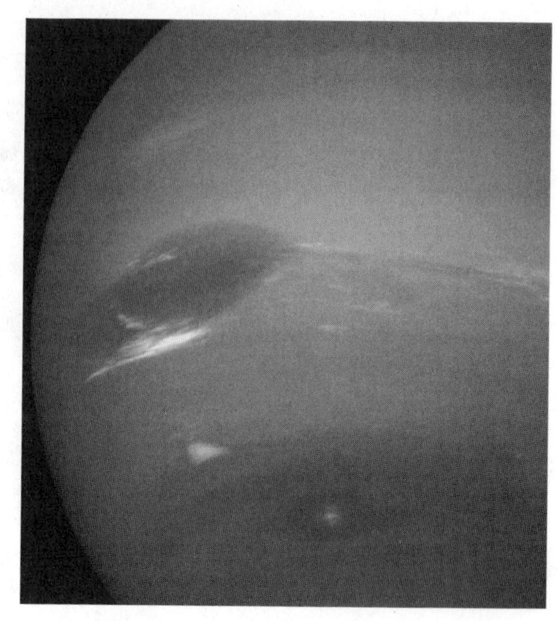

Figure 12–28. Close-up of a portion of Neptune surprised Voyager scientists by showing prominent cloud patterns in its atmosphere. Dark oval clouds, such as the prominent "Great Dark Spot," mark circulating storm systems, capped by white condensed clouds higher in the atmosphere. (NASA Voyager 2 photo)

Voyager showed Uranus to have a nearly featureless blue globe, with only the palest features visible through the haze. Recall that Uranus has little excess internal radiation, compared to its absorbed sunlight (page 267), so that there is little energy input from below to drive big storm systems. Neptune, in contrast, turned out to have some pronounced dark belts, a few bright white high-level clouds, several small dark ovals, and a Great Dark Spot 30 000 km across, 2½ times bigger than Earth. It is similar in behavior to Jupiter's Great Red Spot; both have counterclockwise (anticyclonic) rotation. Various wind speeds were measured on various cloud features. Unlike the atmospheres of Jupiter and Saturn, Uranus' and Neptune's atmospheres circulate slower at equatorial latitudes than the rotation of the planet itself; winds oppose the rotation direction.

Although they have less cloud variety than Jupiter and Saturn, the poles and equators of Uranus and Neptune share similar temperatures, implying that considerable global circulation must occur to equalize the temperatures.

All four giants have atmospheres that are very roughly 3/4 H_2 by mass and 1/4 He, with traces of other gases. Declining temperatures cause more ammonia (NH_3) to freeze out in the form of cloud crystals as we move from Jupiter to Neptune, and the methane (CH_4) absorption becomes more pronounced.

Io

Many of the remarkable phenomena involving Io's thin atmosphere are described in Chapter 2. Pioneer 10 detected a thin ionosphere around Io, confirming the presence of gas, which had been suspected from ground-based observations. Then ground-based observers discovered an extended cloud of sodium (Na), sulfur (S), and

Figure 12–29. Part of the glowing ring of gas atoms blown off Io and extending around Jupiter (right) near Io's orbit. The ring is tipped, relative to Jupiter's equator, because the atoms spread out by moving with Jupiter's rotating magnetic field, which is tipped relative to Jupiter itself. The diamond ring effect is caused by looking tangentially along ribbon-like sheets that extend north and south along magnetic field lines. Gas temperature in the toroidal ring range 20 000 to 400 000 K. The image is a 10-minute exposure made in the light of glowing, ionized sulfur (SII) atoms; the image of Jupiter was a separate exposure inserted to show scale and orientation. (Jet Propulsion Laboratory, courtesy John Trauger)

other atoms that extends many Io diameters from Io and stretches in both directions around the orbit. These atoms emit a remarkable auroral glow in this cloud, as seen in Figure 12-29. The emission in the yellow D line of the Na spectrum is especially strong. This glow is brightest when Io is moving toward or away from the sun at the two points of elongation (as seen from the inner solar system). The reason is that the glow is stimulated by solar photons that excite the Na atoms. The solar spectrum lacks photons of the D wavelength because of a deep absorption of the Na D line by the solar atmosphere. Only when Io moves toward or away from the sun is the incoming solar light Doppler shifted enough so that the sodium atoms are illuminated by solar light with the D line wavelength. The yellow sodium glow has been reported to have a brightness several times that of the faintest auroras visible to the eye on Earth. Thus it should be visible to an observer on Io when Jupiter is half lit by the sun.

Another phenomenon involving volatiles on Io is the probable deposition of sulfur dioxide (SO_2) frost from gases produced by the volcanic eruptions. Whitish SO_2 frost deposits appear likely to form under relatively cold conditions in Io and may be the explanation of the brightening *sometimes* observed for 10 to 15 minutes after Io comes out of eclipse (see Chapter 2). If eruptions are providing abundant SO_2 at the time of the eclipse, the effect might be strong; if not, it might not occur. Veverka and others (1981) monitored three Io eclipses from Voyager spacecraft and saw only a 3% brightening of the south polar region for 3 minutes after the eclipse ended. The phenomenon remains intriguing.

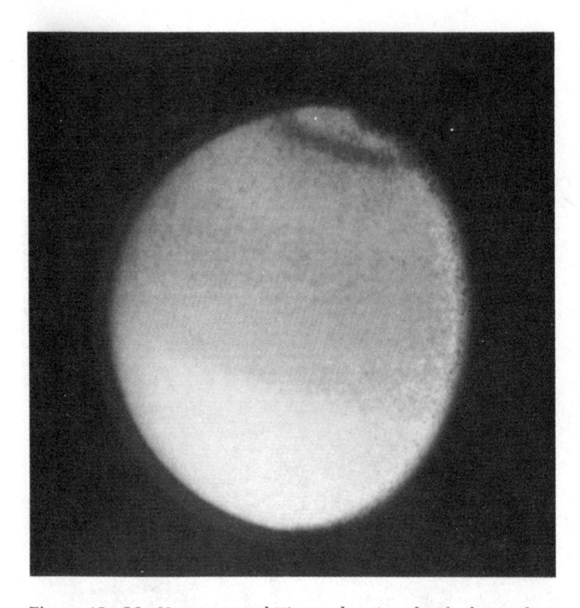

Figure 12–30. Haze-covered Titan, showing the dark north polar band at 60° N latitude, and the brighter tone of the south hemisphere relative to the north. Voyager scientists speculate that the cloud features may change with the seasons. A band of sub-micron-size haze extends above the daylit side. (NASA, Voyager 2)

Titan

Titan is one of the most fascinating of the air-rich worlds, partly because its N_2-rich atmosphere resembles Earth's, and partly because its haze cloaks the surface in mystery (Figure 12-30). The atmosphere was discovered in 1944 when Kuiper detected absorption bands of CH_4 in the spectrum. This marked the first discovery of an atmosphere on a satellite, and Titan remains the only satellite with a dense atmosphere.

Geochemical theory prior to Voyager 1's 1980 flyby suggested that large amounts of N_2 might have been produced by volcanic activity and resultant dissociation of NH_3 gas (Atreya, Donahue, and Kuhn, 1978), but N_2 could not be confirmed from Earth because it lacks absorptions in accessible parts of the spectrum. Voyager data showed that the atmosphere is at least 90% N_2 by volume, and that the CH_4 constitutes only a few percent. Voyager data also revealed the striking fact that the surface atmospheric pressure on Titan is 1.50 ± 0.02 bars—closer to the terrestrial value than found on any other world (Stone and Miner, 1981; Morrison and others, 1986)!

Polarization and photometric studies led to the 1973 discovery of the haze layer obscuring Titan's surface (Veverka, 1973; Zellner, 1973). The haze is a photochemical smog produced when sunlight interacts with the CH_4 and other compounds in the atmosphere. Podolak, Noy, and Bar-Nun (1979) theorized that the haze consists of particles of polyethylene, a polymer of the gas ethylene (C_2H_4), which has been detected in Titan. Voyager 1 detected about 1% of methane (CH_4) by volume in the mid-atmosphere, and smaller traces of ethane (C_2H_6), acetylene (C_2H_2), ethylene (C_2H_4), hydrogen cyanide (HCN), and possibly methylacetylene (C_3H_4) and propane (C_3H_8) (Hanel and others, 1981). These compounds support the idea that the haze layer is a colored photochemical smog.

The measured surface temperature of Titan is a frigid 94 K (−290°F). This is strikingly near the triple point temperature for methane, 90.7 K (methane's triple point pressure is 117 mbar, reached some tens of km above the surface). This means that CH_4 might exist as ice, liquid, or gas in Titan's lower atmosphere, just as water does in Earth's lower atmosphere, where the environment has a similar relation to the triple point for H_2O (273.01 K; 6.1 mbar). The conditions in Titan's lower atmosphere ensure formation of CH_4 condensate clouds, and many researchers have come to expect CH_4 rain, CH_4 snow, and CH_4 rivers, lakes, or oceans on the surface. Figure 12-31 indicates the atmospheric structure.

During early post-Voyager press conferences scientists characterized Titan in colorful terms. Voyager scientist Donald Hunten commented that octane would be synthesized in the atmosphere and that he visualized "it raining frozen gasoline on Titan." Voyager scientist Von Eshelman characterized the surface as "a bizarre, murky swamp (with) hydrocarbon muck." Owen (1982) pictures a methane ocean with good surface visi-

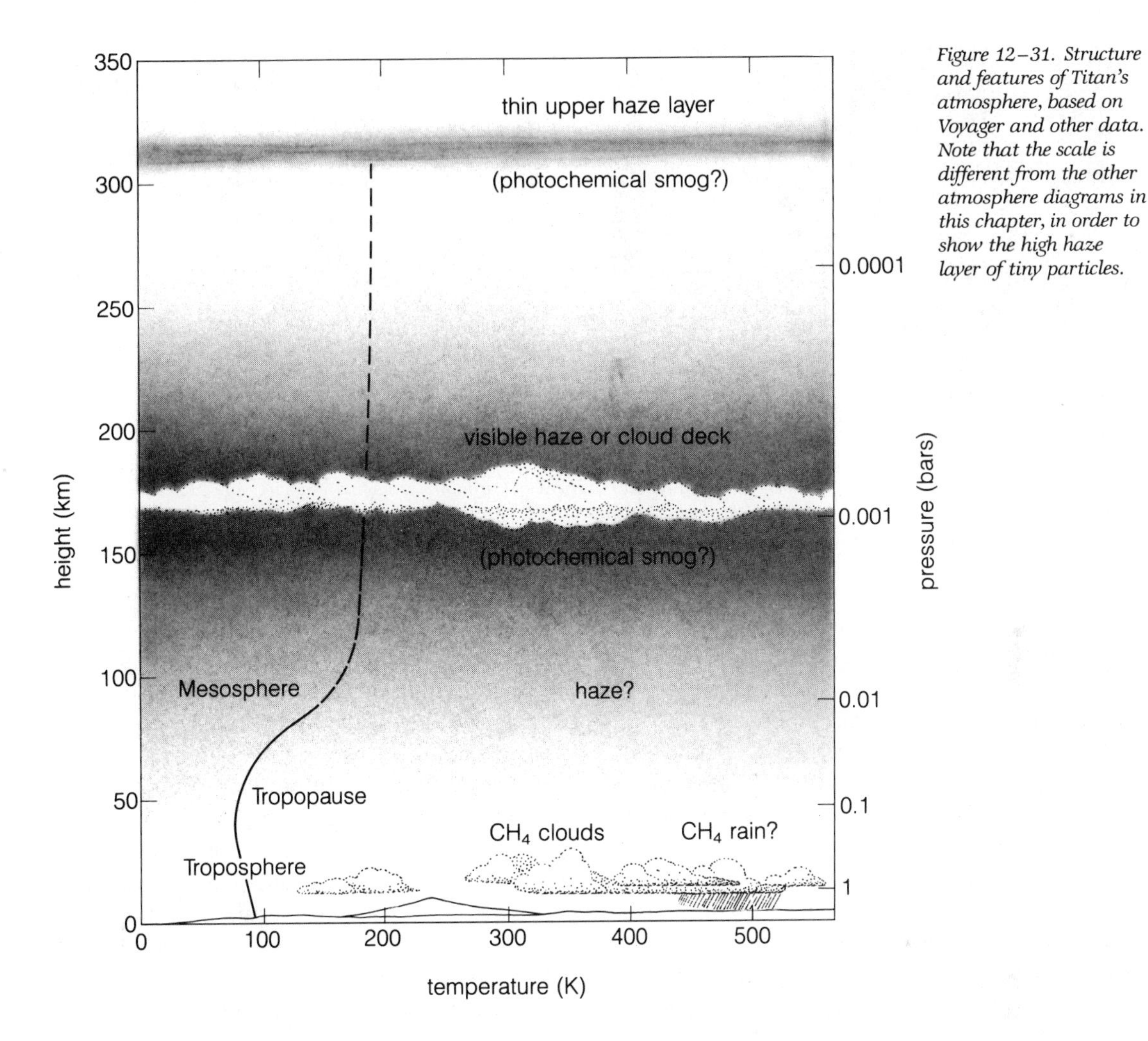

Figure 12–31. Structure and features of Titan's atmosphere, based on Voyager and other data. Note that the scale is different from the other atmosphere diagrams in this chapter, in order to show the high haze layer of tiny particles.

bility (occasionally marred by methane rainstorms) but a low midday light level resembling that of a full-moonlit night on Earth. He suggests minimal winds. He also remarked that because of the virtual invisibility of celestial objects (obscured by the low methane clouds and mid-level smog) and the absence of a magnetic field, navigating a boat on Titan's oceans would be difficult!

Other Small Bodies: Thin Atmospheres?

Probably no other bodies in the solar system have substantial atmospheres, but searches are under way using sensitive spectroscopic equipment to seek subtle absorptions produced by gases such as CH_4. Some of the results are controversial.

Cruikshank and Silvaggio (1980) found absorptions in Pluto's spectrum, which they interpreted as evidence of CH_4 ice on Pluto's surface. They calculated that this should sublime to maintain a thin atmosphere whose surface pressure would change markedly as Pluto moved from aphelion to perihelion. Current data suggest that the surface pressure could range from 0.0006 to as high as 0.01 mbar during Pluto's 248-year "year."

As remarked in the last chapter, Triton and Pluto may be linked in having similar origins as

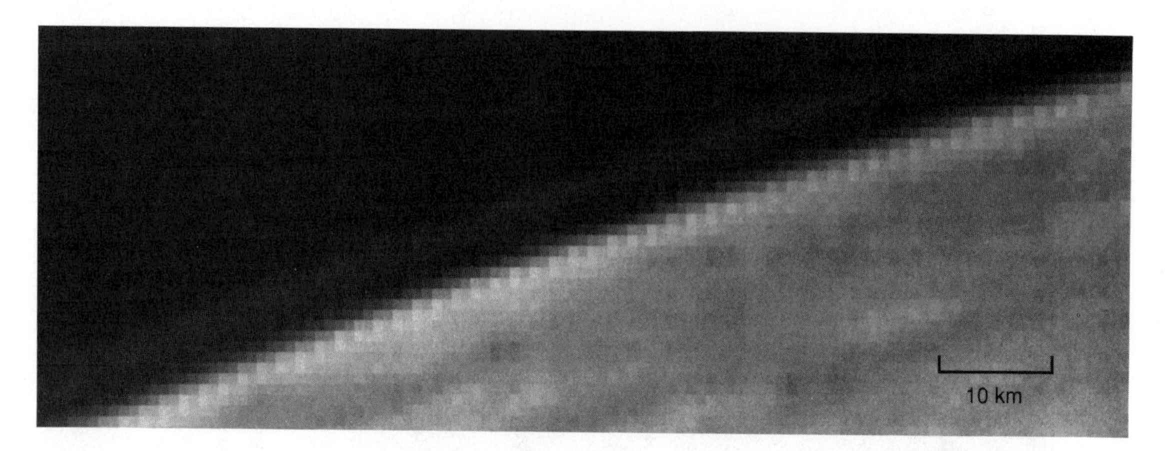

Figure 12–32. Triton's surprising haze layer. Although the atmosphere of Triton has a surface pressure only 0.02 mbar, Voyager photos revealed a haze layer about 3 km thick, floating about 3 km above the surface. The haze may be either condensates or smoke from the eruptive vents on the surface. (NASA Voyager 2 photo)

interplanetary bodies. Indeed, as we consider atmospheric processes on small icy bodies, we can see more clearly than ever that modern studies are blurring the classic distinctions between planets, moons, and comets.

Consider three bodies, seemingly radically dissimilar: Pluto (or rather the Pluto/Charon system), Triton, and Chiron. Each has its own unique properties, but we can now see links in the behavior of ices and gases. All have very volatile ices whose behavior is important in forming the total "personality" of each body.

In Pluto's case, a strange atmospheric transfer has transpired between Pluto and its moon Charon. During Pluto's "summer" at perihelion, methane ice sublimes into a gas on both bodies, but Charon is so small that the gas molecules can escape—essentially turning Charon into a comet. Nearby Pluto has stronger gravity, and so it not only retains its own methane, but captures some of that leaking off Charon. As a result, Charon's surface is nearly denuded of methane. Spectra reveal that Pluto has a methane-ice surface, but Charon has a water-ice surface.

In the case of Triton, the high orbit inclination leaves first one pole in sunlight, then the other—reminiscent of Uranus. As a result, nitrogen and perhaps methane ices sublime from the sunny pole into the atmosphere, only to recondense at the cold opposite cap. The southern cap of Triton has been exposed to the sun for several decades, but Voyager photographed a bright cap covering the polar region. Darker surface exists at lower latitudes, but a faint frost is seen at the terminator, where new condensation seems to be occurring (Smith and others, 1989). Thin clouds and haze layers were also seen, as shown in Figure 12-32. The seasonal production of hazes and transfer of material from one pole to the other, together with the active eruptions, probably explain some spectral changes that have been observed on Triton.

Chiron is another distant interplanetary body where volatilization of ices into gases changes everything. As remarked in Chapter 2, it was cataloged as asteroid 2060 but is now acting as a comet. Meech and Belton (1990) have outlined a unique atmospheric effect that makes Chiron produce a "dust atmosphere" in its coma, different from other comets. Because Chiron is so much bigger than other comets, its gravity is stronger. In addition, the ice that is subliming (CO_2?) probably has a much larger molecular weight than the H_2O that powers more familiar comets; therefore the molecular velocities are slower. The combination of high gravity and low gas velocity means that many dislodged dust particles move too slowly to escape and be blown

Figure 12–33. Ganymede has polar frost deposits. The north pole is at the top of this image, and the edge of the polar cap is prominent where it crosses the large, circular patch of ancient dark crust. (NASA, Voyager 2)

into a comet tail by the solar wind. Thus, each comet outburst produces a long-lived "atmosphere" of dust around Chiron itself, which only gradually settles back as the gas production weakens.

These examples show the blurry distinction between cometary and atmospheric phenomena among small icy bodies. Volatile frosts also produce interesting and transient phenomena on larger bodies. An interesting example is Ganymede, where Voyager photos show polar caps that may be thin deposits of H_2O ice, as seen in Figure 12-33. The deposit must be thin, since underlying surface structure and albedo variations are seen and since the cap brightness does not change much from the pole to the cap edge at about 40° to 45° latitude. This is close to the calculated limiting latitude below which solar warming would sublime an H_2O frost cap (Smith and others, 1979). Of course, once volatiles migrate to a polar region where they freeze out, they may be permanently trapped, like the polar H_2O ice cap of Mars. Thus, many of the outer satellites, if volatiles were once emitted on their surfaces, may have polar deposits and overlapping polar caps of CH_4, H_2O, and other frozen materials. Slow (seasonal?) sublimation of such materials could provide very thin atmospheres whose pressures would equal the vapor pressures of these frosts under the ambient temperature conditions at the frost deposit sites.

CONCEPTS

particles and fields
primitive atmosphere
secondary atmosphere
ideal gas law
isothermal atmosphere
scale height
aerosols
scattering
Rayleigh scattering
opacity
optical depth
greenhouse effect
phase diagram
moist (atmosphere)
cumulus cloud
temperature gradient
lapse rate
triple point for H_2O
dry ice
polar hood
white clouds (on Mars)
blue clouds (on Mars)
yellow clouds (on Mars)
clathrate
unconformities
insolation
superadiabatic
Coriolis force
tropopause
troposphere
stratosphere
jet stream
mesosphere
ozone layer
Hartley-Huggins bands
thermosphere
mean free path
ionosphere
electron temperature
exosphere
equipartition of energy
photodissociation
mixing ratio
kilometer-atmosphere
kilometer-amagat
ashen light
belt
zone

PROBLEMS

1. Describe how compositions of atmospheres evolve with time. Start with the primitive composition and then discuss the production of secondary atmospheres and thermal escape.

2. Why is a cloudy night likely to be warmer than a clear night if other conditions are the same? Why is a night in a clear, high desert colder than a night on the beach if the noontime temperatures are the same? Relate these events to the greenhouse effect.

3. Why does Titan, which resembles a terrestrial planet, have a nitrogen-methane atmosphere when terrestrial planets have nitrogen-carbon dioxide–water vapor atmospheres (or volatile inventories)?

4. Describe the surface climate on an imaginary planet in Earth's orbit with an atmosphere that has white clouds of optical depth 10 in the visual region of the spectrum but optical depth 0.1 in most of the infrared.

5. Why might one expect Mars' atmosphere to have changed more in the past than Venus' or Earth's? Cite evidence that this happened.

6. Why do low-pressure storm systems develop cyclonic flow patterns, and why are the patterns in the two hemispheres mirror images of each other?

7. Describe the annual cycle of events at one of Mars' poles.

8. Why is exosphere temperature critically important to an atmosphere's evolution?

ADVANCED PROBLEMS

9. How much does 1 m^3 of the air around you weigh (as expressed in kg)?

10. If the nominal surface pressure on Mars is 7 mbar, calculate the pressure in a canyon 4 km below this level and at the top of a volcano 20 km above this level.

11. Which planet is better screened against meteorites, a Titan-sized planet with 1-bar surface pressure or an Earth-sized planet with 1-bar surface pressure?

12. (a) Calculate an atmospheric structure for an ancient Mars with an opaque cloud deck at 4 km altitude with albedo 40%. Assume dry CO_2 composition and ignore greenhouse effects. Calculation should show temperature trends above and below the clouds. (b) Describe how results would change with a greenhouse effect caused by CO_2 and large amounts of water vapor.

13. (a) Calculate thermal escape times for H, He, and H_2O on Mars with exosphere temperatures of 210 K and 1000 K. Comment on the volatile history on Mars if transient events ever heated Mars' exosphere to 1000 K for 1 My. (b) Why is the calculation inadequate to fully explain the escape of Martian volatiles?

CHAPTER THIRTEEN

LIFE: ITS HISTORY AND OCCURRENCE

The sun and the sea are two components believed necessary to create an environment in which biochemical reactions can lead to the formation of life. Sunlight or some other source of energy is required to encourage the necessary chemical reactions, and liquid water to provide a medium in which complex organisms can evolve. (Photo by author)

One of the most intriguing questions in astronomy is whether planets elsewhere in the universe harbor what we are pleased to call "intelligent life." Extraterrestrial life must either exist or not exist. Either case has striking consequences. The nineteenth-century Scottish writer Thomas Carlyle sardonically said that other worlds offer "a sad spectacle. If they be inhabited, what a scope for misery and folly. If they be not inhabited, what a waste of space."

What can planetary science say about these two possibilities? Meteorites have supplied some evidence of organic (carbon-based) chemistry in space; the **organic molecules** (complex carbon-based molecules) recently found in interstellar gas are provocative; and Mars' absence of life limits the chemical possibilities on other planets.

In this chapter, we will consider the issue in several steps. First, we will discuss what we mean by *life* and what conditions life would require to exist on planets. Second, we will review the long process that led to the evolution of life on Earth, as best we understand it. Third, we will ask whether there are planets that could support a similar evolution of life elsewhere in the universe. Finally, we will try to estimate the probability of intelligent life actually existing elsewhere and whether we might contact it, or it, us.

THE NATURE OF LIFE

What do we mean by "life"? **Life** is not a status but a process—a series of chemical reactions using carbon-based molecules, by which matter is taken into a system and used to assist the system's growth and reproduction, with waste products being expelled. The system in which these processes occur is the cell. All known living things are composed of one or more cells.* A **cell** is in essence

*Viruses might be an exception. They are simpler than cells, yet can reproduce themselves using materials from host cells. Biologists disagree on whether viruses should be considered a form of life.

a container filled with an intricate array of organic and inorganic molecules (protoplasm). Codes for cellular processes are contained in very complex molecules (such as the famous DNA) located in a central body called the nucleus.

The elements most prominent in the organic molecules are carbon, hydrogen, oxygen, and nitrogen—all common in planetary matter. Phosphorus, also important to life (in small amounts), is also widely available. Carbon is especially critical because it can combine to make long chains of atoms—large molecules that encourage the complicated chemistry of genetics, reproduction, and so on. That is why the term **organic chemistry** (as well as *organic molecules*) refers not specifically to life forms but more generally to all chemistry (and molecules) involving carbon.

We often make the mistake of thinking of ourselves as static beings instead of dynamic systems. We casually assume that we are constant entities, as if our identities solely depended on the form of our bodies. But our bodies today are not the same ones we had seven years ago. Hardly a cell is still alive that was part of that body. This dynamic conception of life is a far cry from the conception only a few generations back, when bodies were thought of as semipermanent machines whose parts gradually wore out. Even our seemingly inert skeletons are living and changing, always replacing their cells. We *must* keep changing—the cells must keep processing new materials to stay alive. When the processing stops, we call the condition death.

The nature of living beings is illustrated in an analogy made by the Russian biochemist A. I. Oparin (1962). Consider a bucket that has water pouring in at the top from a tap and flowing out at the same rate through a tap in the bottom. The water level in the bucket stays constant, and a casual observer would call it a "bucket of water." But it is not like an ordinary bucket standing full of water. The water at any instant is not the same water as at any other instant, yet the outward appearance is constant. We are like buckets with water and nutrients and air flowing through us, but with other, much more complex attributes, such as the ability to reproduce and to be affected by genetic changes that let us evolve from generation to generation.

This metaphor gives us some clue to the kinds of processes involved in the origin of life. We are looking for a process in which complex carbon-based molecules can enter cell systems that (1) draw material from the incoming molecules to create new molecules, (2) incorporate the new molecules into new structures, (3) eject unused material, and (4) reproduce themselves.

Scientists therefore usually choose to define life by these specific carbon-based processes. Often at this point people ask, "What about some unknown form of consciousness?* Or what about some unknown chemistry based on other elements, such as silicon, that can form big molecules?" The answer is that we have never observed or experimented with such life forms, so we can say nothing substantive about them. If we admit they are plausible, we simply increase the probability of life or consciousness in the universe. But researchers usually restrict their discussion to forms on which chemical and behavioral data are available.

Whatever other concepts we invent—civilization, religion, technology, art, war, love, communication—it is the chemical processes of life that define us, just as they define the spiders, sea urchins, elephants, moths, amoebas, redwoods, and all the other incredibly varied living creatures around us. To judge whether life may exist on other planets—whether other planets are already "taken"—we must find out how those processes got started on Earth.

THE ORIGIN OF LIFE ON EARTH

In addition to the life-forming elements—especially carbon (C), hydrogen (H), oxygen (O), and

*For example, in *The Black Cloud*, astronomer Fred Hoyle imagined an interstellar nebular cloud with matter and electromagnetic fields organized in such a way that it developed a consciousness or will of its own.

nitrogen (N)—water was crucial to the development of life on Earth. Several facts indicate this:

1. Most of our body weight is made up of water. The percentage is even higher for plants.

2. Body fluids, like tears, have a saltiness similar to that of the oceans. The oceans are our heritage. As embryos, we at first are immersed in fluid and develop bodies more like fish than like mammals—for example, we have gills.

3. Organisms deprived of liquid water quickly die. And dead organisms are shriveled and dried.

4. Taking a more theoretical view, water provides a medium in which organic molecules can be suspended and can interact—a likely place for life to begin.

So a good starting question is, what was the early history of C, H, O, N, and H_2O on Earth? The nebula from which the planets formed must have been rich in compounds that would form from cosmic gases. As is clear from Chapter 5, these compounds included hydrogen molecules, ammonia, methane, water, and perhaps some nitrogen molecules (respectively, H_2, NH_3, CH_4, H_2O, and N_2). Note which atoms were involved: C, H, O, and N—the elements of life!

Probably within a hundred million years after Earth formed, heat from internal radioactivity and impacts melted parts of Earth's interior, unleashing volcanic activity. Volcanic activity released gases, especially water vapor (H_2O) and carbon dioxide (CO_2). Thus, even if there were no oceans at the beginning, by about 4.5 to 3.5 Gy ago there were probably bodies of surface water under an atmosphere that was rich in hydrogen, ammonia, methane, water, and nitrogen and that was becoming richer in oxygen.

Chemical reactions occurring naturally in such an environment could have produced building blocks of life, as has been proved by laboratory experiments. In the 1950s, chemist S. L. Miller put a gaseous mixture of hydrogen, ammonia, methane, and water vapor (to represent the primitive atmosphere) over a pool of liquid water (primitive seas) and passed electric sparks through it (simulating energy sources such as lightning). After several days, the pool began to darken. The water now contained a solution of **amino acids**, the class of molecules that join to form **proteins**, the huge molecules in cells. This so-called **Miller experiment** proved that the complex organic chemicals necessary for life could be built in a natural environment (Miller, 1955). An example of a reaction in the Miller experiment is the production of the amino acid glycine:

$$\underset{\text{ammonia}}{NH_3} + \underset{\text{methane}}{2CH_4} + \underset{\text{water}}{2H_2O} \xrightarrow{\text{energy}} \underset{\text{glycine}}{C_2H_5O_2N} + \underset{\text{hydrogen}}{5H_2}$$

Further geochemical studies have suggested that Earth's early atmosphere may not have been so primitive—that is, so much like the nebular gas—as first thought. It may have been less rich in hydrogen compounds and more rich in volcanic gases such as CO_2, CO, and N_2. Solar ultraviolet light striking such an atmosphere tends to produce water and hydrogen cyanide (HCN). Further experiments showed that a Miller-like reaction produces amino acids even under these conditions (Goldsmith and Owen, 1979):

$$\underset{\text{hydrogen cyanide}}{3HCN} + \underset{\text{water}}{2H_2O} \rightarrow \underset{\text{glycine}}{C_2H_5O_2N} + \underset{\text{cyanamide}}{CN_2H_2}$$

Other experiments showed that many kinds of energy sources—including ultraviolet light from the sun, volcanic activity, and even meteorite impacts—could produce similar reactions. Thus we can conclude that building blocks of life *would* have formed on early Earth and in similar environments elsewhere. The conclusion has been confirmed by the discovery of extraterrestrial amino acids (but not fossils or life forms) in several carbonaceous chondrite meteorites (Kvenvolden and others, 1970; Cronin, Pizzarello, and Moore, 1979). Amino acids can form in either a "right-handed" or "left-handed" symmetric

Figure 13–1. Broths rich in amino acids and complex organic molecules formed in long-lived tidewater pools on ancient Earth. Evaporation of water could have concentrated the remaining organic materials, allowing complex reactions. The resulting products could have "fertilized" oceans with living organisms or protoliving materials. (Photo by author)

molecular structure, but on Earth, some biological or prebiological process led to strong preponderance of only one of the two structural options (called a *nonracemic mixture*). The meteoritic amino acids were proved extraterrestrial because their molecular structures had equal proportions of the two structures (called a *racemic mixture*).

From such results, researchers have concluded that molecular organic materials existed in Earth's primitive oceans, probably less than 0.5 Gy after Earth formed. Such material may have accumulated in isolated tidewater ponds (Figure 13-1) and concentrated as water evaporated, leaving heavy molecules behind to interact and form complex substances.

Related discoveries have been made by Viking landers on Mars. Many scientists were surprised when Viking revealed not only no life forms on the Martian surface, but also no organic molecules in the soil, to a level of a few parts per billion! In retrospect, we see that Mars' surface is exposed to solar ultraviolet rays, unlike Earth's surface, which is largely screened by the ozone layer. These rays can not only damage cells in our skin, causing sunburn, but also break apart organic molecules. The Martian dust, frequently moved by the wind, may be sufficiently irradiated to prevent any buildup of organic molecules on Mars or any preservation of organic molecules that might have formed in the past. The same ultraviolet light produces unfamiliar reactive chemical states in Martian minerals. One of the Viking tests for life on Mars was to add chemical nutrients to the soil and see if they were metabolized by any microscopic creatures. Indeed, the nutrients did lead to the production of organic molecules and the release of CO_2 gas. This result is now attributed to the reactive states produced in the soil by ultraviolet radiation. Indeed, laboratory experiments in 1977 duplicated many Viking results using simulated Martian soil (iron oxide minerals exposed to ultraviolet light and CO_2). These results may indicate ways in which chemical reactions on a lifeless planet with no organic molecules can synthesize organic molecules by utilizing atmospheric CO_2 under certain conditions (Horowitz, 1977).

The next step is less certain. Florida biologist S. W. Fox has shown that simple heating of dry amino acids (as might happen on a dry planet such as present-day Mars) can create protein molecules. When water is added, these proteins assume the shape of round, cell-like objects called **proteinoids**, which take in material from the surrounding liquid, grow by attaching to each other, and divide. Though they are not considered living, they resemble bacteria so much that experts have trouble telling the difference in microscope views.

Engel and Nagy (1982) reported finding nonracemic mixtures of amino acids, like those in terrestrial proteins, inside an uncontaminated piece of the Murchison carbonaceous chondrite. This raises the possibility that some proto-biological processes went further than once thought, producing protein-related amino acids inside moist carbonaceous chondrite parent bodies. This would be an exciting discovery about the origins of life, but the findings are still very controversial.

Possibly related to proteinoids are objects discovered in the 1930s by Dutch chemist H. G. Bungenberg de Jong. When proteins are mixed in solution with other complex molecules, both sets of substances spontaneously accumulate into cell-sized clusters called **coacervates**. The remaining fluid is almost entirely free of complex organic molecules (Oparin, 1962).

The next step toward recognizable life forms is still more uncertain, but many biologists theorize that cell-like proteinoids and coacervates in primeval pools of "organic broth" began reacting with fluids in the pools and with each other, accumulating more molecules and growing more complex, as suggested by Figure 13-2. Eventually these could have evolved into biochemical systems capable of reproducing.

Whatever the processes, microscopic cellular life must have arisen between 4.6 and 3.5 Gy ago, because life's fossil remains have been discovered in rocks after that period. In a sequence of rocks from Greenland, 3.7-Gy-old rocks do not show traces of ancient life, but rocks younger than about 3.1 Gy do. Also, 3.4-Gy-old rocks found in South Africa in 1977 contained fossil methane-producing bacteria (Gould, 1978).

Among the best evidences of early life are fossil remains of **stromatolites**, colonies of blue-green algae with a cabbagelike structure. A good example was reported from Australia and dated at 3.5 Gy (Walter, Buick, and Dunlop, 1980). Nisbet (1980) lists other 2.7- to 3.5-Gy-old stromatolites from Canada, Zimbabwe, and Australia, with possible evidence of 3.7-Gy-old rocks with biological carbon isotope chemistry from western Greenland.

An interesting factor is that life probably could not have evolved very far before about 4.2 Gy ago because of the intense early meteorite bombardment, though biochemical evolution must have begun in that period. Thus, life forms such as bacteria and blue-green algae apparently evolved rapidly around 3.4 to 3.7 Gy ago. Life arose "as soon as it could; perhaps it was as inevitable as quartz or feldspar" (Gould, 1978).

Earth was not a passive backdrop. Craters formed and filled. Ocean and land configurations changed. The atmosphere evolved from **reducing conditions** (dominated by hydrogen compounds) toward **oxidizing conditions**.

Oxygen content in the initial atmosphere was small, probably only a few percent. However, the first plant organisms produced molecular oxygen (O_2), and the oxygen content began to rise. Empirical evidence for this is that rocks formed and buried more than about 2.5 Gy ago have been found to be suboxidized, or formed under more oxygen-poor conditions than exist today (Goldsmith and Owen, 1979). Oxidized red beds are rare before 2 Gy ago and common afterward (Walker, 1977).

As soon as some oxygen became available, solar ultraviolet light broke some O_2 molecules, and the free pair of O's joined with other O_2 mol-

Figure 13–2. The importance of fluid medium for primitive evolution is suggested by these photos of single-celled organisms engulfing food from surrounding fluid. (**a**) *An amoeba flows to surround a nearby food particle. (Optical microscope photo, S. L. Wolfe)* (**b**) *The protozoan* Woodruffia *ingests a paramecium. (Electron microscope photo; T. K. Golder)*

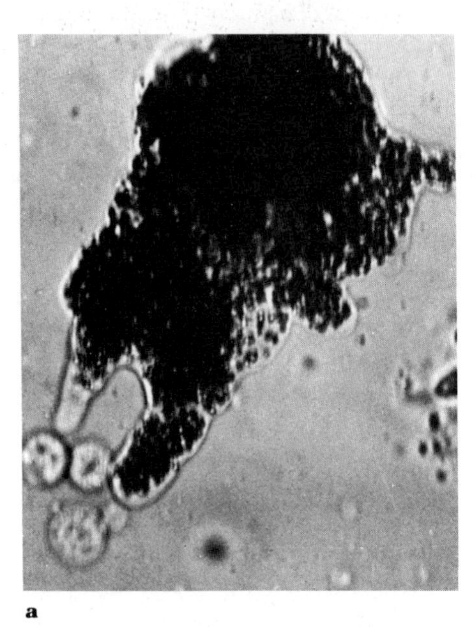

a

ecules to make **ozone** (O_3). This led to production of an **ozone layer** high in Earth's atmosphere, which absorbs nearly all the incoming ultraviolet. This was an important step, because too much ultraviolet breaks up complex molecules and thus could prevent life—the apparent fate suffered by biochemistry on Mars. Thus, paradoxically, the early presence of solar ultraviolet may have helped life start by providing energy to break up old molecules and form new ones, but its absence later allowed life to evolve safely on the surface.

Living things could hardly be unaffected by all these changes. Whereas the earliest life forms developed and existed without oxygen, life now had to adjust to oxygen. Such environmental changes favored evolution—when a mutation trait appeared that happened to be well adapted to the new environment, it was likely to have more offspring, promoting retention of the new trait and thus evolution of new species.

Yet by 2 Gy ago, we still would have scarcely recognized our Earth. The land was still barren. Some areas must have looked like today's deserts or like Mars. Some areas were moist and washed

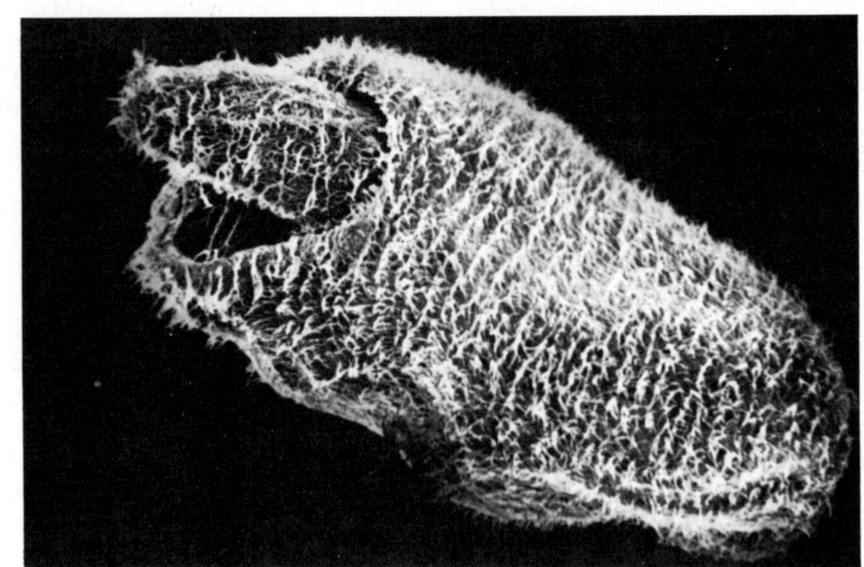

b

by rains, but instead of luxurious forests there were only bare dirt, eroded gullies, and grand canyons. Brown vistas stretched to the sea.

Most life was still in the oceans, soft bodied and rarely producing fossils. Stromatolites increased in abundance 2.5 to 2.0 Gy ago, living at the boundary of water and rock along seacoasts. Today they are found only in rare, oxygen-poor salt marshes on some seacoasts. They produce oxygen by photosynthesis, and their abundance 2.5 to 2.0 Gy ago is more evidence for the upswing in O_2 production after that time (Walker, 1977; Goldsmith and Owen, 1979). It is strange to realize, as Figure 13-3 emphasizes, that during most of Earth's history, life would have been hard to find on representative *land*scapes on Earth.

One sign of the adaptability of life forms once they evolved is the rapid proliferation of advanced species, as indicated in the geological time scale shown in Table 10-5. While it took about half the available time to go from complex molecules to algae and bacteria, it took only the last 12% of Earth history to go from the first hard-bodied sea creatures to humans. Biologists attribute evolution in general and this rapid proliferation in particular to **natural selection**, the greater production of offspring by those individuals best adapted to the ever-changing environment. Details of the natural selection process are uncertain. Prototypes of new species may have evolved in obscure ecological niches and then emerged rapidly when sudden environmental changes (such as ice ages) caused extinction of earlier species better suited for earlier conditions.

Whatever the specific mechanism, we can say from the fossil record that Earth experienced evolution from nonliving organic chemicals to small organisms in a few billion years, and that these organisms evolved in less than a billion years to species with self-conscious intelligence.

PLANETS OUTSIDE THE SOLAR SYSTEM

From all we have just said, we conclude that if planetary surfaces with the necessary conditions—liquid water and C–H–O–N chemicals—exist anywhere, life is likely to evolve on them. And advanced species will probably appear eventually. But are there any such planets?

Figure 13–3. First signs of life encroaching on land did not occur until about halfway through Earth's history. This scene about 2 Gy ago shows two such signs: a colony of stromatolites (left shoreline) and some lichens or moss on the rocks (right foreground). The rest of the land (left distance) was probably barren and eroded. Earth had few surface signs of life during the first 90% of its history. (Painting by Ron Miller)

Chapters 4 and 5 showed that the planet-forming process is likely to be a normal evolutionary development in a cocoon nebula around a single star. Since such planets are less likely to form in a double or multiple star system, an informed guess would be that between 1% and 30% of all stars have planetary companions.

However, even if planets exist near some other stars, there is no guarantee that they are habitable. Astronomers have proposed several conditions for a habitable planet:

1. The central star should not be more than about 1.5 $M_\odot$, so that it will last long enough for substantial life to evolve (at least 2 Gy), and so that it will not kill evolving life with too much ultraviolet radiation.

2. The central star should be at least 0.3 $M_\odot$ to be warm enough to create a reasonably large orbital zone in which a planet could retain liquid water.

3. The planet must orbit at the right distance from the star, so that liquid water will neither evaporate nor permanently freeze.

4. The planet's orbit must be almost circular to keep it at the proper distance and prevent too drastic seasonal changes.

5. The planet's gravity must be strong enough to hold a substantial atmosphere.

Dole (1964) reviewed these criteria and concluded that roughly 6% of the stars, mostly from 0.9 to 1.0 $M_\odot$, have habitable planets. However, the figure remains highly uncertain.

HAS LIFE EVOLVED ELSEWHERE?

If habitable planets exist and if life evolves readily under habitable conditions, is life abundant throughout the universe? There are additional factors to consider. For one thing, planetary environments change with time, so that today's habitable planet may not be habitable tomorrow. Can life survive such changes?

Effects of Planetary and Astronomical Processes on Biological Evolution

Basic planetary or stellar processes may be involved in encouraging or hindering biological evolution. For example, Earth's fossil records indicate an episode called the "great dying" at the end of the Permian Period about 250 My ago. As many as 96% of all land and sea species became extinct in a few million years or less (Raup, 1979). This is only the biggest of the so-called mass extinction events. Raup and Sepkoski's (1986) study of 11,800 genera of marine animals during the last 270 My shows: 61% of existing genera disappeared during the Permian great dying 250 My ago; 49% of existing genera, including dinosaurs, became extinct at the end of the Cretaceous Period 65 My ago; 43% died out at the end of the Triassic Period 210 My ago; 30 percent at the end of the Jurassic about 140 My ago; plus smaller extinctions grading down into the noise level of evolution. How can these sudden mass extinctions be explained? Among possible answers:

1. Interior convection could have caused plate-tectonic crustal splitting, hence changing sea levels, ocean currents, wind patterns, and seasonal extremes (for example, Gartner and McGuirk, 1980). This process has been responsible for isolating land masses such as Australia and allowing different species to evolve there.

2. Volcanic eruptions may have spewed enough dust into the high atmosphere to reduce the sunlight reaching the surface. For example, widespread sunlight decreases of up to 25% occurred after the 1883 Krakatua (Sumatra), 1912 Katmai (Alaska), and 1982 El Chichon (Mexico) eruptions. Larger, rarer eruptions might have cut sunlight for some years, causing the decline of some species and the ascendancy of new ones.

3. Slight changes in the tip of the planetary axis to the plane of the ecliptic, caused by gravitational forces, are believed to have caused major climatic changes, such as the ice ages (for example, Imbrie and Imbrie, 1980).

4. Slight changes in the sun's radiation may have changed climates.

5. Irradiation from a nearby supernova could have affected the climate or directly damaged organisms.

6. A radical breakthrough on the whole question of mass extinctions came in the 1980s. The second-biggest extinction, defining the end of the Cretaceous and beginning of the Tertiary Period, was virtually proven to be the result of an asteroid or comet impact. This event is known as the K-T extinction event, a name derived from geologic abbreviations for the Cretaceous and Tertiary. Alvarez and others (1980) first presented evidence for the asteroid impact idea after measuring excesses in the iridium (Ir) content of 65-My-old limestones; these iridium impacts turned out to be 20 to 160 times that of limestones of other ages. Because Ir is enriched in meteorites relative to Earth, they suggested that a 6- to 14-km-diameter asteroid struck Earth and ejected massive amounts of meteoroid-enriched dust into the atmosphere. This hypothesis has had a tremendous impact on the paleontological community and caused a burst of interesting research. Within months, other investigators confirmed the Ir excess in 65-My-old sediments from many parts of the world. The effects of the impact are hard to contemplate. We know that some 75% of species that had evolved (notably dinosaurs) became extinct within a few million years, and new species (notably mammals) became dominant (Russell, 1982). However, detailed studies of the fossil record show that this did not happen all at once, and various species were affected differently. No land animals larger than 25 kg survived, but freshwater animals and plants were much less affected (Russell, 1982).

At first, paleontologists were loathe to accept a catastrophic impact. But the evidence has broadened. The K-T boundary layer was found to contain shocked quartz grains—evidence of shock pressures created during meteorite impact but not during volcanism. Also, the layer contains soot corresponding to burning down most of the world's forests at the end of the Cretaceous.

Researchers have hypothesized impact consequences to explain these observations. First, the dust thrown into the stratosphere absorbed sunlight and is calculated to have caused 3 to 6 mo of daytime darkness, consistent with a reported extinction of some 49% of genera of light-dependent floating marine organisms, but only 20% of (less light-dependent) land genera (Kerr, 1981; Ahrens and O'Keefe, 1982). Asteroid or comet impact may damage the ozone (O_3) layer. Turco and others (1981) found that the Tunguska explosion of 1908 generated as much as 30 million T of nitrogen oxide (NO) in the stratosphere, reacting to deplete up to 45% of the O_3 in the northern hemisphere, consistent with Smithsonian measures of 1909–1911. If the impact occurred in the ocean (as is probable since water covers 6/7 of Earth), giant tsunamis would devastate coastal areas. Still more gruesome, cratering expert Jay Melosh calculated that debris falling back into the atmosphere would create a worldwide radiant heat pulse from the sky, broiling exposed life forms and possibly explaining the forest fires. Creatures in burrows, underwater, or in rainstorms and snowstorms might survive this heat, explaining complexities in the extinction patterns.

Where is the crater, which should have been on the order of 100 to 200 km diameter? Of course, it might have been on a seafloor that has since been destroyed by subduction. Arizona researcher Alan Hildebrand and others noted that K-T boundary deposits around the Caribbean had tsunami deposits, and they have identified a possible K-T impact crater, about 180 km across, buried under sediments straddling the Yucatan coast (Hildebrand and Boynton, 1991).

Interestingly, the next largest extinction event, 210 My ago, coincides with the creation of the 100-km Manicougan crater, in Canada. Craters smaller than, say, 50 km may cause effects that are lost in the noise of biological evolution. Craters larger than the 100–200 diameter range can be expected to form every 70 to 250 My (Hartmann and Miller, 1991).

Strangely, the largest extinction event, the "great dying" at the end of the Permian Period, 250 My ago, shows no evidence of an impact, and its cause remains unknown.

If Earth (or any other planet) suffers catastrophic impacts that wipe out species every few hundred My, this is a radical change in classic Darwinism, which pictures evolution as spurred only by competition among species within environments that change slowly due to terrestrial forces. Such disasters might seem to hinder evolution of life. But while wiping out some species, they may promote the emergence of other species. For example, if some aspects of intelligence had first appeared in a benign, constant environment where food was plentiful, these traits would have had little value. But if ice ages destroyed the mild environment, then the cleverer groups might have emerged from their previous obscurity.

One of our problems is that we just don't know whether Earth has experienced rare special effects that have promoted life and intelligence. Goldsmith and Owen (1979) suggest that the mere presence of our unusually large moon is such an effect: Its tides helped produce tidal pools, where organic molecules could concentrate, and its tidal forces have helped hold Earth's axis in a fixed position, preventing axis excursions (such as probably happened on Mars) that could cause drastic climatological oscillations.

Perhaps it is no coincidence, then, that the time scale of biological change is comparable to the time scale of geological change. Life's response to changing environments is to change itself, on time scales as short as a million years. Old species die and new species prosper under the new conditions. We must not underestimate the possibility, however, that environmental changes on still shorter time scales prevented life from evolving on many or most planets.

Adaptability and Diversity of Life

The great variety of ancient and modern species on Earth and the variety of environments in which they have thrived suggest that, given time, life could also evolve to fit a wide range of conditions on other planets. Even humans have a remarkable adaptability. We thrive from equatorial wet jungles to dry deserts to arctic plains to Andean summits, where air pressure is barely half that at sea level. During the last ice age, we survived by migrating.

But the limits for survival of simpler life forms are even wider. Microflora are known in supercooled Antarctic ponds that remain liquid at 228 K (−49°F) because of dissolved calcium salts. And bacteria are known in Yellowstone Hot Springs at a temperature of 363 K (194°F)—a 59% variation in temperature (Sagan, 1970). In laboratory experiments, common bacteria have survived in liquid cultures at least 24 h in CO_2 atmospheres at 433 K (320°F), extending the range to a 90% variation in temperature.

Habitable pressure regimes show an even greater range. Bacteria exist at altitudes where the atmospheric pressure is only 0.2 bar, and more advanced organisms live in ocean depths with pressures of hundreds of bars (Sagan, 1970). This pressure range exceeds a factor of a thousand.

Experiments with terrestrial life forms in nonterrestrial environments have shown grass seeds germinating in atmospheres of C, H, O, and N compounds, insects with normal behavior at pressures as low as 100 to 160 mbar, insects surviving brief exposure to Martian surface pressures, and bacteria surviving in CO_2 at conditions between those of Earth and Venus (Siegel, 1970).

Conversely, alien organisms might survive terrestrial conditions and have devastating effects on Earth. There are historical examples of similar events. The plague caused by Asian bacteria introduced into Europe in the 1300s killed about a quarter of all Europeans, and as many as three-quarters of the inhabitants in some areas. Diseases introduced into Hawaii after the first European contact in 1778 killed about half of all Hawaiians within 50 years. Some 95% of the natives of Guam were wiped out by disease within a century of continued European contact (Underwood, 1975). For these reasons, early Apollo astronauts were quarantined until it was clear that they carried no lunar organisms. And future samples from Mars may be analyzed only in space labs, well above Earth's atmosphere.

Thus, change and evolution in life populations are caused not only by life's adaptations to new environments but also by the invasion and destruction of some populations by others. With these facts as well as cultural competition in mind, anthropologist D. K. Stern (1975) has remarked, "It is likely that the meeting of two alien civilizations will lead to the subordination of one by the other."

Appearance of Alien Life

If alien life did evolve, could we expect it to evolve through any stages similar to our own? Opinion is divided on this point. MIT physicist Philip Morrison (1973) has emphasized the **convergence** effect, whereby species with similar capabilities in similar habitats evolve to look alike. For example, three different species of large sea creatures are "designed" for fast swimming in the ocean and look alike: the extinct reptile Ichthyosaur; the shark, which is a fish; and the mammal that returned to the sea, the dolphin. Similarly, aliens living on planetary surfaces in gaseous atmospheres and using "intelligence" to manipulate their surroundings with tools might well have bilateral symmetry, appendages used as hands, a pair of eyes designed (like ours but unlike most animals') for stereo vision, and so on.

On the other hand, the famous paleontologist, George Gaylord Simpson (1964, 1973) argues that although life is likely to start, the long chain of environmental changes and evolutionary steps that produced humans is unlikely even to be approximated elsewhere, so that there is likely to be a "nonprevalence of humanoids." Simpson labels the whole attempt to estimate probabilities of alien life as nearly meaningless because of our lack of experiments or examples.

Physicist W. G. Pollard (1979) counters that we do have examples. He notes that about 180 My ago, Australia broke off Gondwanaland and can be thought of as an Earth-like planet, "A," where evolution continued independently from a primarily reptilian stock. Similarly, South America broke off Africa 130 My ago and can be viewed as an independent planet "S." Independent evolution also continued on planet "E" (Earth, made up of Africa and adjoining land). In the same period of 130 to 180 My ago, independent evolution diverged, rather than converging on humans. Humans appeared only on E, primates developed on S, and marsupials on A. Parker (1977) argues that humans were unlikely to appear by continued evolution on A. Humanlike creatures, appearing on E only about 4 My ago, have existed for only 0.1% of Earth's history.

In summary, natural selection seems to produce species capable of occupying any habitable environment. Thus, we should not be surprised if life has evolved on another planet. But this life may look very strange to us, even if it displays recognizable intelligence. After all, if mushrooms and corals and woolly mammoths and Venus' flytraps all evolved on one planet, how much greater may be the differences between life forms on two different planets? Feathers and fur and sex and seeds and symphonies may be products of Earth only.

Effects of Technological Evolution on Biological Evolution

If life will evolve when the right conditions exist, and if those conditions probably do exist elsewhere, then what can we predict about life elsewhere? Should we predict intelligence and civilizations? What do these terms mean? Should we assume that other civilizations will achieve space flight or might visit us some day? This raises the question of technology and its role in the evolution of life. **Exobiologists** (biologists concerned with possible life on other worlds) have only one inhabited world to study as an example, so their answers to these questions are uncertain. Many exobiologists have assumed that intelligence involves, among other things, use of tools to modify the environment; hence, technology.

But as we have seen, while limited environmental change can be helpful, environmental change that is too much or too fast can be fatal! As Pulitzer Prize–winning naturalist René Dubos points out, we are umbilical to Earth, and if we alter our planet too much before acquiring an ability to leave it, we are finished. For this reason, consideration of our own case leads to the conclusion that the development of technology may actually end civilization on some worlds. This is hardly wild speculation, since we see a few nominally moral, intelligent technologists on our own world spending entire careers devising weapons solely to deal death to our own species.

In the past, war did not threaten our whole species, because conflicts involved only a small percentage of the world. But today, nuclear, biological, and other types of weapons could involve the whole world. For example, the radioactive strontium 90 produced by nuclear explosions has a half-life of 28 y. It was blasted freely into the atmosphere before the nuclear test-ban treaty of 1963 (the Moscow Agreement). A year after the first H-bomb tests by the United States in the Pacific, strontium 90 deposits in American soil increased soil radioactivity by 0.5%. A few weeks after a 1976 Chinese nuclear test, airborne radioactive debris fell onto the United States, increasing radiation levels. Obviously, a sufficiently massive nuclear exchange could devastate not only civilization but also future forms of life, whose genetic pool would be exposed to high radiation levels for decades. Humanity has thus proved that a planetary culture could wipe itself out by conscious design of weapons, as irrational as that may seem.

We have also proved that this disaster could happen by mistake. As our technology assumes a planetary scale, so do our accidents. Problems as diverse as nuclear power and aerosol spray cans illustrate the issue. Many currently planned nuclear power plants will produce radioactive plutonium wastes, among the most toxic of known materials. Although safe when sealed, some kilograms of plutonium dust accidentally spilled into the air could devastate life in whole states. Yet many kilograms are already being processed, and several governments promote greater dependence on nuclear power plants until other energy forms become available.

Accidents are not the only danger: We have seen our society spawn terrorists and madmen. With some grams of stolen plutonium 238, such characters could threaten whole cities. Radioactivity presents a known technological danger that our society has allowed to develop in spite of knowing about it ahead of time.

The seemingly harmless aerosol can is a different story. In 1974, several scientists realized from theoretical calculations that when Freon (the propellant gas used in such cans) is released into the air, it reaches Earth's protective ozone layer high in the stratosphere. Here its breakdown products act to destroy the ozone layer. Depletion of ozone would let more solar ultraviolet radiation reach the surface, increasing hazards of skin cancer. Even as more Freon was being sold, planetary astronomers and chemists published refined calculations showing that to deplete the Freon already in the atmosphere could take as long as a century. Thus, our generation's consumer gimmicks could be creating future health hazards. After a 1975 report by the National Academy of Sciences, confirming the ozone danger, moves were made in 1976 and 1977 by American regulatory agencies to phase out Freon aerosol propellants by 1979—an example of a technological danger that almost escaped our notice.*

If humanity is any example, long-term survival of a planetary culture is not assured. Although we have been around less than a hundredth of a percent of the age of the universe, we are already having close brushes with global disaster. Thus, we can speculate that if evolution produces intelligent societies that remain tied to one planet, many of them may last only a fraction of a percent of the age of the universe—in which case there is little chance that a given culture will be around at the same time we are.

*But we should be haunted by the words of Harvard planetary physicist Michael McElroy, who performed some of the earliest Freon calculations: "What the hell else has slipped by?"

More optimistically, we have succeeded in identifying the **cultural hurdle** that we (and perhaps intelligent species on any planet) must surmount: the transition from scattered competing nation-states with the capability to damage the planet to a stable planetary or interplanetary technological society of intelligence and imagination. Perhaps we and some other cultures will cross this hurdle. After all, we have identified some perils in time. Perhaps some cultures have completed the transition from being planetary cultures to being interplanetary, thus ensuring their survival against ecological disaster on any one planet. Such cultures might last and be detectable for billions of years instead of thousands.

Our conclusion so far is that a fraction of the stars have planets, a fraction of the planets ought to be inhabited, and a fraction of the inhabited planets ought to have either civilizations or relics of destroyed civilizations. The next question is whether any actual evidence for extraterrestrial carbon-based life as we know it exists today.

Alien Life in the Solar System?

According to the ground rules, a habitable planet has (or has had) liquid water and an atmosphere in which complex carbon-based organic chemistry can proceed—and these conditions must have lasted long enough for a specific degree of complexity to have evolved. If the specified degree is to be intelligence, or meter-scale animals made mostly out of liquid water (like ourselves), then we can almost certainly rule out every body in the solar system but Earth, since conditions elsewhere do not permit long-term bodies of liquid water.

But if we ask about primitive life forms in the solar system, rather than intelligence, the situation is a little more promising. As noted, meteorites indicate that amino acids have been synthesized beyond Earth. On Mars, they may have been synthesized in the past. As described on p. 452, ultraviolet light destroyed organic material on the surface to a level of less than a few parts per billion. Yet even today, native Martian soil has the ability to synthesize organic molecules from nutrients. Levin and Stratt (1981) assert that a Viking-class instrument failed to detect organic material in a certain Antarctic soil sample in which another Viking-class instrument did detect microorganisms! They argue that Viking may possibly have detected primitive life-forms when its experiments detected synthesizing of organic molecules from nutrients. Most analysts doubt this view, and suspect that Mars is probably now sterile. Perhaps Mars never had sufficiently long-lived bodies of liquid water to allow living things to evolve. However, some Viking scientists believe that Martian microbes could exist below the surface, perhaps in geothermal regions where water often melts and solar ultraviolet does not penetrate. Geothermal areas have also been hypothesized on Titan and other icy satellites.

In this regard, the recent discovery of dense animal populations around geothermal vents at a depth of 2550 m (at a pressure of 260 bar) in the Pacific is intriguing. The organisms that form the base of the local food chain draw their energy not from sunlight but from geothermal sources, overturning the notion sunlight is always life's main energy source (Karl, Wirsen, and Jannasch, 1980). A discovery of similar interest by Florida State University biologist E. I. Friedmann was reported in 1978: In apparently lifeless, dry valleys in Antarctica, microorganisms such as bacteria, algae, and fungi thrive in pore spaces *inside* rocks. Though the rock may be in a frozen area, sunlight heats the rock itself and creates a microscopic niche suitable for life. Life, seemingly inexhaustibly accommodating, fills the niche.

In summary, there is almost certainly no extraterrestrial intelligent life in the solar system today, and probably no other life forms big enough to be visible to the naked eye. There remains a low chance of microscopic, primitive life forms elsewhere in the solar system. However, complex but nonliving organic materials have been created in other parts of the solar system. Evidence from our solar system favors the argument that life might have evolved on planets near other stars if those planets had enduring, temperate, moist environments.

Alien Life Among the Stars?

The next step is to consider alien life on planets near other stars. There is no direct evidence, but American, Soviet, and other scientists in international meetings have put together a method for considering the possibilities (Sagan, 1973). The logic is to try to estimate the various fractions—of stars having planets, of those planets that are habitable, of those habitable planets where conditions remain favorable long enough for life to evolve, of those planets where life does evolve, of those where intelligence evolves, and of the planet's life during which intelligence lasts. The formalism is sometimes called the **Drake equation**, after American astronomer Frank Drake, who developed the fractional scheme.

Table 13-1 shows optimistic (upper limit) and pessimistic (lower limit) estimates of the various fractions, and the consequent estimates of the upper and lower limits on the fraction of stars that might have civilizations today. The optimistic figure is a few percent; the pessimistic case is only one star in 10^{19}. Note the 17 order of magnitude uncertainty!

Given these figures, how far away would the civilizations be? Figure 13-4 shows the answer by plotting the radial distance required to include a given number of stars.

In the first case, the nearest civilization might be only 15 ly away—amazingly close. At the speed of light that civilization might be reached in only one generation. In the pessimistic case, the closest civilization would probably not be in our own galaxy, but roughly 300 million light-years away in a distant galaxy. Nonetheless, given the innumerable galaxies, it is hard to avoid the conclusion—even with the most pessimistic view—that millions or billions of technological civilizations exist outside the solar system. If our reasoning is right, then at this moment intelligent creatures may be pursuing their own ends in unknown places under unknown suns.

Table 13-1
Estimated Fraction of Stars with Planets That Have Intelligent Life

Criterion	Fraction: Lower limit (?)	Fraction: Upper limit (?)
1. Stars having planets	10^{-2}	0.3
2. Criterion 1 stars ever having habitable conditions on at least one planet	10^{-1}	0.7
3. Criterion 2 planets on which habitable conditions last long enough for life to evolve	10^{-1}	1
4. Criterion 3 planets on which life evolves	10^{-1}	1
5. Criterion 4 planets on which habitable conditions last long enough for intelligence to evolve	10^{-3}	0.9
6. Criterion 5 planets on which intelligence evolves	10^{-1}	1
7. Criterion 6 planets on which intelligent life endures	10^{-7}	10^{-1}
8. Fraction of duration of intelligent life during which it retains an interest in contact with Earth-like civilization	10^{-3}	1
Fraction of all stars with planets that bear intelligent life	10^{-19}	2×10^{-2}
Implication: Distance to nearest civilization	3×10^{8} light years	15 light years

Where Are They?

American and Russian radio astronomers have listened for radio messages from alien civilizations and have found none. Nor do our skies seem to be overrun with alien visitors trying to contact us. Yet, if the upper limit estimate in Table 13-1 is anywhere near correct, aliens would be nearby. Why haven't we heard from them?

One answer might be that we *are* being visited—witness the flying saucer reports. But although some UFO reports are intriguing, verifiable evidence for alien spaceships is abysmally poor. The largest scientific study of the UFO phenomenon (Condon and others, 1969), on which I was a field investigator, showed, in my opinion, no evidence that any UFOs are extraterrestrial. Some interesting phenomena, possibly atmospheric, were recorded, but many other seemingly sensational cases, especially photographic ones, were proved to be frauds or honest mistakes. The fact that the layperson hears most of these cases through the conduit of the popular press proved to be one of the main causes of misinformation. Many reports of seemingly strange events are given front page treatment on the first day; a few days later many are explained, but this is scarcely news and is relegated to a back page, if printed at all. The same reports, including photographs established as frauds by various investigators, may be picked up and run again by tabloids, who merchandize pseudo-science at supermarket checkout stands. For these reasons, most people think that the UFO issue is much more mysterious than it really is.

Nevertheless, no one has established that alien visits are *not* occurring today, but if they are, they are sufficiently rare that we have no evidence of them. The verifiable fact that no obvious historical visits have occurred at least shows that space is not chock-full of alien spacecraft trying to make contact with us.

Another answer to the question may be that the pessimistic figures are right, and that the nearest civilizations are in distant galaxies. Even their radio messages, if any, could be 10 My old

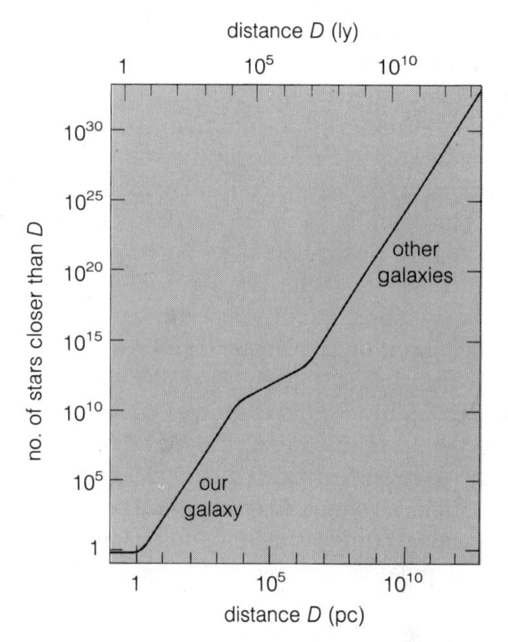

Figure 13–4. Distance in parsecs (bottom) and light-years (top) required to encounter the number of stars indicated at left. If one star in 10^9 has life-bearing planets, the nearest one might be within a few thousand parsecs.

by the time we receive them, and their spaceships would be unlikely to reach Earth if limited to speeds less than that of light, as current physics requires. But most investigators of the problem place the probability well above our lower limit. At a 1971 Soviet-American conference on this problem, the favored estimate came out to be one civilization per 10^5 stars. This would put a million civilizations in our galaxy and the nearest civilizations only a few hundred light years away. Why, then, are there no frequent visits to Earth?

Another answer is that biological evolution need not produce creatures who have a desire to build civilizations or travel through space. Not even all human societies evolve toward technology. Are humans fated to be explorers, bridgebuilders, and businesspeople rather than artists, athletes, or daydreamers? Is the stereotyped aggressive Westerner more representative of the essence of humanity than the stereotyped contemplative Easterner? May not our aggressive technocracy be just one type of *cultural* activity rather than a universally achieved stage of *biological* evolution? Historically, many patterns we once assumed to be biologically imposed have turned out to be merely culturally imposed.

If humanity is not predestined to develop a technological civilization, how much less certain is the course of social development on other worlds. It is absurdly anthropocentric to suppose that beings on other planets would resemble us psychologically, or socially. Consider again the variety of highly evolved life forms on our planet alone. Ants live in ordered societies that do not appear to regard individual survival as important. Dolphins communicate and have brains that seem almost comparable to ours, but they have no manipulative organs and hence no technology.

Perhaps we cannot expect aliens to be motivated by emotions that mean much to us. We certainly cannot expect, as always happens in grade-B science fiction movies, that humanlike aliens will walk out of saucers and invite us to join their democratically constituted United Planets, a galactic organization structured by documents that are curiously reminiscent of the U.S. Constitution. So why assume that other civilizations might even care to try to visit or contact us?*

Evolutionary Clocks and the Explorative Interval

This brings us to the fourth and perhaps most significant answer for why we have no evidence of aliens: We may be farther from aliens in evolutionary time than in physical space. Biological evolution is so persistently experimental that even if another planet started evolving at exactly the same time as ours, and even if its biochemistry produced creatures like us, those creatures are not likely to be in a phase of evolution similar to ours. If the evolutionary "clocks" on the two planets got only 0.02% out of synchronization, they would be 1 My ahead of us or behind us. Thus, even in the unlikely event that other planets produce civilizations recognizable to us, we would have to contact one of those civilizations in a very narrow time interval in order to see any recognizable common interests.

Evolution may pass through only a brief **explorative interval**, during which societies on one planet would care to reach other planets; beyond that stage communication or space exploration might be no more attractive than a national program on our part to communicate with chimpanzees, ants, or dolphins. To be sure, a few of our scholars try this, but they "contact" an infinitesimal fraction of these creatures. What fraction of the anthills or dolphin schools have we humans tried to contact? By the same token, our solar system might be ignored by advanced aliens. Aliens a million years ahead of us might be no more interested in us than we are in ants.

How long might an explorative interval last? We have used tools for about 2 My, and it appears safe to assume that we will have progressed far beyond current technology in another million years, if we survive. Our explorative interval might be a few million years, then, or less than 0.1% of the history of the planet. This is the basis for the lower factor listed in criterion 8 in Table 13-1.

If civilizations are 500 ly apart, then interstellar voyages and messages would take a millennium or more. There might be little incentive for the effort. Large spaceships in which many generations could live and die during interstellar voyages—self-contained "planets" that are a staple of science fiction stories—have also been hypothesized in published models of interstellar colonization by other species (Kuiper and Morris, 1977). In any case, several authors have con-

*Woody Allen (1980) has summed up much of this more succinctly: "If saucers come from outer space, why have their pilots not attempted to make contact with us, instead of hovering mysteriously over deserted areas? My own theory is that for creatures from another solar system, 'hovering' may be a socially acceptable mode of relating. . . . It should also be recalled that when we talk of 'life' on other planets we are frequently referring to amino acids, which are never very gregarious, even at parties."

Figure 13–5. Even if Earth has been visited on 10 000 occasions by alien expeditions, the last visit would probably have been too long ago to have left any record likely to be identified by historical or archaeological records, unless large, prominent artifacts were deliberately left. (Painting by Jim Nichols)

cluded that, given the plausible number of ships in transit at any one time from all civilizations, a visit to a specific random planet, such as Earth, would be rare.

This brings us to a fifth possible answer to the question of alien visits: They may have happened in the remote past. This is the "ancient astronaut" hypothesis, popularized in several pseudo-science books. Figure 13-5 illustrates a problem with this idea. For example, even if 100 visits occurred at random intervals in only the last 1% of Earth's history, they would still average 460 000 y apart! Given the rate of terrestrial erosion by water and glaciation, the prospects for finding evidence of such ancient visits are poor! As for historical visits, there is no good evidence. The Soviet and American collaborators I. S. Shklovskii and C. Sagan (1966) surveyed archaeological and mythological literature even before the hypothesis was a popular fad and found no compelling evidence for ancient astronauts. Neither Earth, the moon, Mars, nor Venus is littered with ancient alien artifacts, and not a single mysterious artifact has been advanced as physical evidence of ancient astronauts. As illustrated by Figure 13-6 popular pseudo-science books have greatly exaggerated the mystery surrounding many archaeological structures and artifacts (Story, 1976).

Radio Communication

Possibly our period of isolation may be nearing an end. For half a century we have been broadcasting radio communications among ourselves. Already our unintentional but weak alert is more than 50 ly out from Earth. Aliens may one day pick up our signals and send radio messages (or an expedition?) in return. Radio astronomers in both the United States and the Soviet Union are therefore still conducting modest programs to listen for such messages with large radio telescopes.

In one such search, for example, radio astronomers listened for broadcasts near 1420 MHz (the 21-cm-wavelength that marks an astronomically important radiation from neutral hydrogen atoms

Figure 13–6. Lines as long as a few miles in the Peruvian desert were claimed as alien landing sites requiring stupendous technologies beyond our means. Ground studies show that these lines were made by Indians around 1800 y ago and required technology no more stupendous than a broom to sweep aside dark stones from desert pavement. The lines are currently threatened by off-road vehicles, whose tracks show scale. (Photo by author)

in space), targeting all solar-type stars not known to be members of multiple star systems, within about 82 ly (see Table 4-2); this included 185 stars. No artificial broadcasts were identified (Horowitz, 1978). Goldsmith and Owen (1979) summarize other searches involving more than 600 stars. All were negative.*

Larger listening instruments have been proposed in both countries, as shown in Figure 13-7.

A message from 1000 ly away would come from a civilization 1000 y in the past, and no answers to our questions could come back for 2000 y. Such communication would be unlike dialogue, but, as physicist P. Morrison has pointed out, more like our receipt of "messages" (books, letters, plays, and artworks) from ancient civilizations such as Greece.

Meanwhile, we have sent a few messages of our own. The Pioneer 10 spacecraft, which flew by Jupiter and left the solar system in 1973, carried a plaque designed to convey our appearance and location to possible alien discoverers of the derelict spacecraft. The first radio message was a test message sent from the large radio telescope at Arecibo, Puerto Rico, in 1974, beamed toward globular star cluster M 13, 27 000 ly away.*

---

*The NASA search for extraterrestrial intelligence by radio listening was halted in late 1981 as the result of an amendment attached to the 1982 Federal budget by Senator Proxmire, explicitly forbidding expenditures for this purpose, on the grounds that it is a "ridiculous waste." I hope that current efforts to reinstate the U.S. program will be successful, as such passive listening puts either positive or negative boundary conditions on one of the most fundamental questions about our cosmic surroundings: are technological civilizations like ours filling space with radio transmissions?

*Astronomers in these projects have been deluged with letters ranging from support to complaints about the nudity of the human figures on the Pioneer 10 plaque. The radio astronomers promptly received a telegram: "Message received. Help is on the way—M 13." Its authenticity might be questioned, because the round trip radio signal travel time to M 13 is 54 000 y.

Figure 13–7. A proposed facility for interstellar communication, "Project Cyclops," envisions an array of radio telescope antennae, each larger in diameter than a football field. The scale is indicated by the multistory control building, right. Such a facility would listen for signals from alien civilizations. American and Soviet scientists have proposed similar facilities. (Artist's conception, NASA)

SUMMARY

The discovery of firm evidence for alien civilizations, or even alien life forms, could be a pivotal development in human history, as suggested in Figure 13-8.

Yet questions of exobiology, and especially of intelligent life on other worlds, leave us with a mystery. Experimental evidence indicates rather strongly that life should start on other planets if liquid water, energy, and the right chemicals are present. Astronomical evidence suggests, but does not prove, that habitable planets ought to exist elsewhere in the universe. Biological evidence shows that life is adaptable and that species can evolve to fit different environments, from ocean depths to low-pressure atmospheres of different compositions.

While the limited evidence indicates that other life forms should exist, there is no evidence that they do or that they have tried to communicate with us. We can only speculate about the reasons. Perhaps they are too far away. Perhaps most civilizations destroy themselves before successfully exploring the universe. Perhaps evolution carries them beyond a stage where they would care to communicate with us. Perhaps they are unrecognizable.

Arthur C. Clarke has remarked that any technology much advanced beyond your own looks like magic to you. Perhaps we are too limited by our own concept of civilization. After all, one creature's civilization may be another's chaos, as shown by Mohandas Gandhi's remark when asked what he thought of Western civilization: He said it wouldn't be such a bad idea. It seems likely that our first contact with aliens (if they exist) might be as incomprehensible as the dramatized contact in Clarke's novel and the Kubrick-Clarke film, *2001: A Space Odyssey*.

Clearly we have been reduced to speculation by a lack of facts. Indeed, the whole field of exobiology has been criticized as a science without any subject matter. Exobiology recalls Mark

a

b

Figure 13–8. Discovery of positive evidence for extraterrestrial civilizations could be a dramatic and pivotal event in human history. This could occur by (**a**) *us going out and finding such evidence (discovery of alien obelisk on the moon, as shown in the MGM release* 2001—A Space Odyssey, *© 1968 Metro-Goldwyn-Mayer Inc.) or* (**b**) *such evidence arriving on Earth (arrival of the "mother ship" as shown in* Close Encounters of the Third Kind, *© 1977 Columbia Pictures Industries, Inc.)*

Twain's comment "There is something fascinating about science. One gets such wholesale returns of conjecture from such a trifling investment of fact." In the same vein, exobiology spokesman Philip Morrison (as quoted by Simpson, 1973) has admitted of exobiology, "Here is a body of literature whose ratio of results/papers is lower than any other." The only way to reduce the conjecture and increase the proportion of fact is to pursue research in many related fields—physics, chemistry, geology, meteorology, biology—and listen to the skies with radio telescopes. There may be surprises waiting out there.

CONCEPTS

organic molecules
life
cell
organic chemistry
amino acid
protein
Miller experiment
nonracemic mixture
racemic mixture
proteinoid
coacervate
stromatolite
reducing conditions
oxidizing conditions
ozone
ozone layer
natural selection
convergence
exobiologist
cultural hurdle
Drake equation
explorative interval

PROBLEMS

1. Compare Alpha Centauri, Barnard's star, and Sirius in terms of the probability of detecting radio broadcasts from intelligent creatures. (See Table 4-2.)

2. Describe several ways in which Earth's internal evolution has affected the evolution of life on Earth.

3. Describe several ways in which extraterrestrial events, such as solar or stellar evolution, might have affected evolution of life on Earth or other planets.

4. Describe ways in which technology could affect the survival of intelligent life on Earth or on other planets. Construct scenarios of (a) the possible destruction of life and (b) guaranteeing the survival of life. (*Hint:* In part [b] consider the impact of space travel.)

5. In your own opinion, what would be the long-range consequences of:

(a) arrival of an alien spacecraft and visitors in a prominent place, such as the United Nations building;
(b) discovery of radio signals arriving from a planet about 10 ly distant and asking for two-way communication; and
(c) proof (by some unspecified means) that life existed *nowhere* else in the observable universe.

6. Given the assumption that our technology has the potential for creating planet-wide changes in environment, defend the proposition that *if* life on other worlds produces technologies like ours, then that life is likely either to have become extinct or to be widely dispersed among many planets by means of space travel.

7. In view of the devastation wrought on many terrestrial cultures by contact with more technologically advanced cultures, which would you say is the safest course: (1) aggressive broadcasting of radio signals to show where we are in hopes of attracting friendly contacts, (2) careful listening with large radio receivers to see if there are any signs of intelligent life in space, or (3) neither broadcasting nor listening, but just waiting to see what happens?

(a) How would it affect the results of our listening program if other intelligent species had reached the first, second, or third conclusion?
(b) If we listen at many frequencies and pick up no artificial signals, does that prove that life has not evolved elsewhere in the universe?

ADVANCED PROBLEMS

8. If an Earth-sized planet (diameter 12 000 km) was circling a star 1 pc away (3×10^{16} m):

(a) What would be its angular diameter when seen from Earth?

(b) Could this be resolved by existing telescopes?

(c) If the planet orbited 1 AU (1.5×10^{11} cm) from its stars, what would be its maximum angular separation from the star as seen from Earth?

(d) Could this angle be resolved?

9. Derive expressions for the inner and outer radii of a toroidal "zone of habitability" around any star of luminosity L (in units of solar luminosity) if an inhabited planet orbiting the star has an effective Bond albedo A and emissivity e, a circular orbit, and liquid water somewhere on its surface in order to support life.

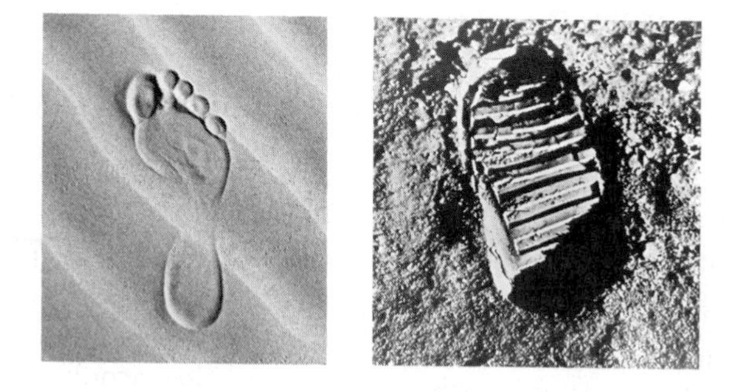

APPENDIX

PLANETARY DATA TABLE

(next six pages)

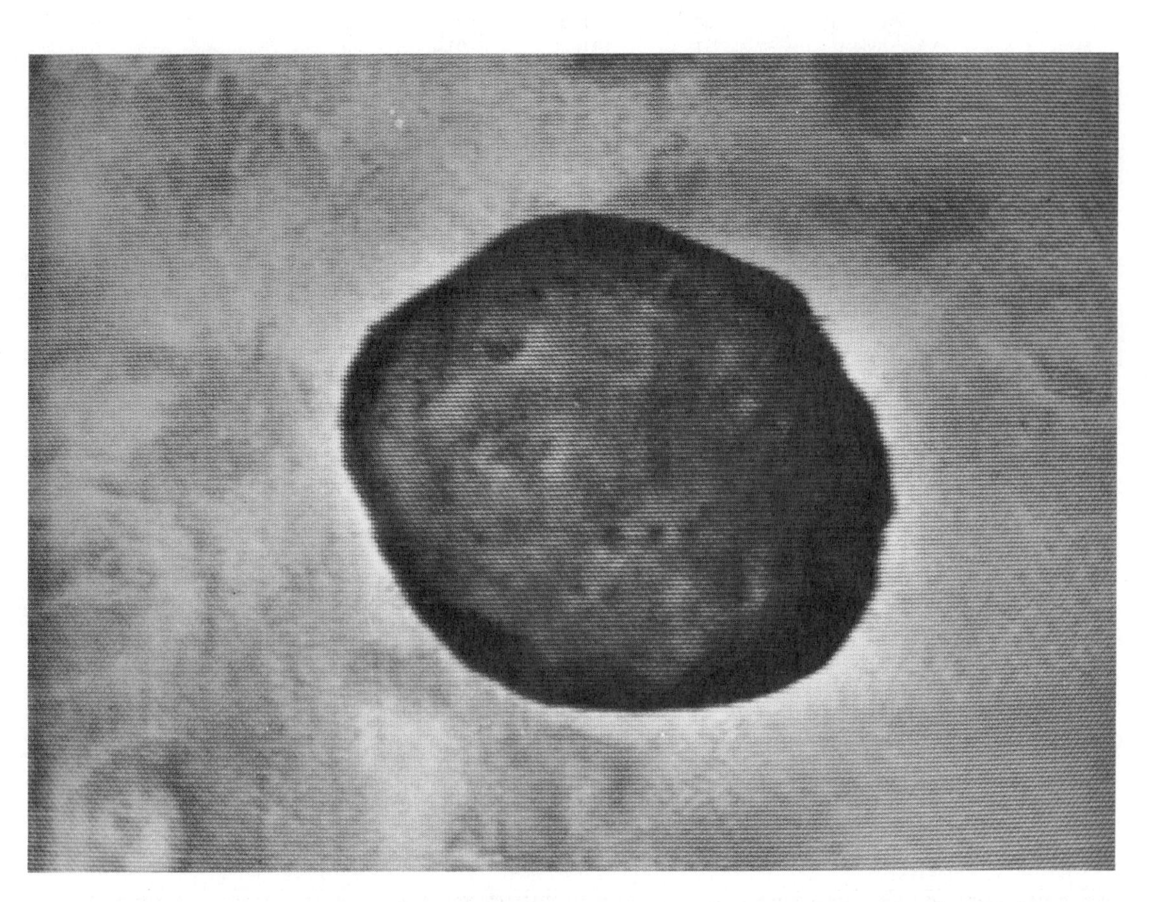

Phobos hanging in front of craters and dusky markings of Mars, as seen from the Russian PHOBOS-2 probe, 1989. (Courtesy B. Zhukov, IKI [Institute for Cosmic Investigations], Moscow)

Data on planets, moons, and

Object[b]	Semi-major axis (AU for planets & asteroids; 10^3 km for satellites)	Revolution period (days unless marked otherwise)	Orbit inclination (degrees; with respect to ecliptic for planets & asteroids; w.r.t. planet equator for satellites)	Orbit eccen-tricity	Obliquity (degrees)	Rotation period (days; R = retrograde)	Diameter (km)
Sun	—	—	—	—	7.25	25.4	1,391,400
Mercury	0.387 AU	87.97	7.00	0.206	0	58.65	4878
Venus	0.723 AU	224.70	3.39	0.007	178	243.0 R	12,104
Earth	1.000 AU	365.26	0.00	0.017	23.4	1.00	12,756
*Moon	384.4	27.32	18 to 29	0.055	6.7	27.32	3476
Mars	1.524 AU	686.98	1.85	0.093	25.0	1.026	6787
*Phobos	9.38	0.319	1.0	0.018	~0	0.319	27 × 19
*Deimos	23.50	1.262	2.8	0.002	~0	1.262	15 × 11
Largest Belt Asteroids							
1 Ceres	2.768 AU	4.61y	10.60	0.077	?	0.378	914
4 Vesta	2.362 AU	3.63y	7.14	0.090	?	0.223	500
2 Pallas	2.773 AU	4.62y	34.80	0.233	?	0.328	522
10 Hygiea	3.138 AU	5.56y	3.84	0.118	?	~0.75	443
511 Davida	3.181 AU	5.67y	15.90	0.172	?	0.363	336
Jupiter	5.203 AU	11.86y	1.30	0.048	3.08	0.410	142,800
J16 Metis	128	0.295	0.0	0.0	~0?	0.30	40
J15 Adrastea	129	0.298	0.0	0.0	~0?	0.29	24 × 16
J5 Amalthea	181.5	0.498	0.45	0.003	~0	0.498	270 × 155
J14 Thebe	222	0.674	0.8?	0.01	~0?	0.674	100
*J1 Io	422	1.769	0.03	0.000	~0	1.769	3630
*J2 Europa	671	3.551	0.46	0.000	~0	3.551	3130
*J3 Ganymede	1071	7.155	0.18	0.002	~0	7.155	5280
*J4 Callisto	1884	16.689	0.25	0.008	~0	16.689	4840
J13 Leda	11,110	240	26.7	0.146	?	?	~16
J6 Himalia	11,470	250.6	27.6	0.158	?	?	~180
J10 Lysithea	11,710	260	29.0	0.12	?	?	~40
J7 Elara	11,740	260.1	24.8	0.207	?	?	~80

[a]Data come from various sources including reports on Jupiter and Saturn systems by Voyager team members (*Science* 1 June 1979; 23 Nov. 1979; 10 April 1981; 29 January 1982); tabulations in *Proc. Lun. Plan. Sci. Conf.* 1979, *9:* inside covers; S. R. Taylor, *Planetary Science: A Lunar Perspective*, Houston: Lunar and Planetary Science Institute (1982); J. K. Beatty, et al. (editors), *The New Solar System*, Cambridge: Sky Publishing Corp. (1990); R. Binzel, T. Gehrels, M. Matthews (editors), *Asteroids II*, Tucson: University of Arizona Press (1989); and additional scientific reports.

other objects in the solar system.[a]

Radius (km)	Mass (kg)	Mean density (kg/m³)	Visual geometric albedo	Escape velocity at surface (m/s)	Surface materials (known or probable)	Atmosphere (Main constituents)
695,700	1.99 (30)	1410	—	617,800	Ionized H and He gas	H + He
2439	3.30 (23)	5420	0.12	4300	Igneous rock (basaltic?)	None
6052	4.87 (24)	5250	0.59	10,400	Basaltic (+ granitic?) rock	CO_2
6378	5.98 (24)	5520	0.39	11,200	Water; basaltic + granitic rock	$N_2 + O_2$
1738	7.35 (22)	3340	0.11	2380	Anorthositic + basaltic rock + dust	None
3398	6.42 (23)	3940	0.15	5000	Basaltic rock + dust	CO_2
14 × 10	9.6 (15)	~1900	0.05	11	Carbonaceous rock + dust	None
8 × 6	2.0 (15)	~2100	0.05	6	Carbonaceous rock + dust	None
510	?	?	0.06	~650	Carbonaceous rock or dust	None
274	~2.5 (20)	~2900	0.24	~350	Basaltic rock or dust	None
269	?	?	0.08	~340	Silicate rock	None
222	?	?	0.05	~280	Carbonaceous rock or dust	None
171	?	?	0.04	~220	Carbonaceous rock or dust	None
71,400	1.90 (27)	1314	0.44	59,500	?	H_2 + He
20	?	?	low	~20	Rock	None
12 × 8	?	?	0.04	~20	Rock	None
135 × 78	?	?	0.05	~120	Rock + sulfurous coating?	None
50	?	?	~0.10	~50	Rock	None
1820	8.89 (22)	3530	0.63	2560	Sulfur compounds	Thin SO_2
1565	4.79 (22)	3030	0.64	2040	H_2O ice	None
2640	1.48 (23)	1930	0.43	2740	H_2O ice + carbonaceous dirt	None
2420	1.08 (23)	1790	0.17	2420	H_2O ice + carbonaceous dirt	None
~8	?	?	?	~4	Carbonaceous dirt + H_2O ice?	None
~90	?	?	0.03	~90	Carbonaceous dirt + H_2O ice?	None
~20	?	?	?	~10	Carbonaceous dirt + H_2O ice?	None
~40	?	?	0.03	~30	Carbonaceous dirt + H_2O ice?	None

(continued)

[b]A traditional letter/number designation scheme is indicated for some of the satellites. Generally, the larger ones discovered before the Voyager flights were given the first, lowest numbers, and others were numbered in order of distance from the planet.

*These are the most-discussed satellites and most imporant in terms of their unique features. Students should know some of the interesting characteristics of each of these moons.

Data on planets, moons, and

Object[b]	Semi-major axis (AU for planets & asteroids; 10^3 km for satellites)	Revolution period (days unless marked otherwise)	Orbit inclination (degrees; with respect to ecliptic for planets & asteroids; w.r.t. planet equator for satellites)	Orbit eccen-tricity	Obliquity (degrees)	Rotation period (days; R = retrograde)	Diameter (km)
Jupiter *(continued)*							
J12 Ananke	20,700	617	147	0.169	?	?	~30
J11 Carme	22,350	692	163	0.207	?	?	~44
J8 Pasiphae	23,300	735	147	0.40	?	?	~35
J9 Sinope	23,700	758	156	0.275	?	?	~20
Trojan Asteroids							
624 Hektor	5.12	11.58y	18	0.02	~80	0.29	~300 × 150
911 Agamemnon	5.15	11.69y	22	0.07	?	0.33	~148
Saturn	9.539	29.46y	2.49	0.056	26.7	0.426	120,660
S18 Pan	133.6	0.576	0	0	?	?	~20
S15 Atlas	137.6	0.601	0.3	0.002	?	?	50 × 20
S16 Prometheus	139.4	0.613	0.0	0.003	?	?	140 × 80
S17 Pandora	141.7	0.629	0.05	0.004	?	?	110 × 70
*S11 Epimetheus	151.4	0.694	0.34	0.009	?	?	140 × 100
*S10 Janus	151.4	0.695	0.14	0.007	?	?	220 × 160
*S1 Mimas	185.54	0.942	1.5	0.020	~0	0.942	390
*S2 Enceladus	238.04	1.370	0.0	0.004	~0	1.370	500
*S3 Tethys	294.67	1.888	1.1	0.000	~0	1.888	1048
S13 Calypso[c]	294.67 (Trail)	1.888	~1?	0.0	?	?	34 × 22
S14 Telesto[c]	294.67 (Lead)	1.888	~1?	0.0	?	?	34 × 26
*S4 Dione	377	2.737	0.0	0.002	~0	2.74	1120
S12 Helene	377	2.74	0.15	0.005	?	?	36 × 30
*S5 Rhea	527	4.518	0.4	0.001	~0	4.52	1530
*S6 Titan	1222	15.94	0.3	0.029	~0?	15.9?	5150
*S7 Hyperion	1484	21.28	~0.5	0.104	?	?	350 × 200
*S8 Iapetus	3562	79.33	14.72	0.028	~0	79.33	1440
*S9 Phoebe	12,930	550.4	150	0.163	?	0.4?	220
2060 Chiron	13.70	50.7y	6.92	0.379	?	0.25	200

[a]Data come from various sources including reports on Jupiter and Saturn systems by Voyager team members (*Science* 1 June 1979; 23 Nov. 1979; 10 April 1981; 29 January 1982); tabulations in *Proc. Lun. Plan. Sci. Conf.* 1979, *9:* inside covers; S. R. Taylor, *Planetary Science: A Lunar Perspective,* Houston: Lunar and Planetary Science Institute (1982); J. K. Beatty, et al. (editors), *The New Solar System,* Cambridge: Sky Publishing Corp. (1990); R. Binzel, T. Gehrels, M. Matthews (editors), *Asteroids II,* Tucson: University of Arizona Press (1989); and additional scientific reports.

[b]A traditional letter/number designation scheme is indicated for some of the satellites. Generally, the larger ones discovered before the Voyager flights were given the first, lowest numbers, and others were numbered in order of distance from the planet.

other objects in the solar system.[a] *(continued)*

Radius (km)	Mass (kg)	Mean density (kg/m^3)	Visual geometric albedo	Escape velocity at surface (m/s)	Surface materials (known or probable)	Atmosphere (Main constituents)
~15	?	?	?	~8	Carbonaceous dirt + H_2O ice?	None
~22	?	?	?	~12	Carbonaceous dirt + H_2O ice?	None
~35	?	?	?	~14	Carbonaceous dirt + H_2O ice?	None
~20	?	?	?	~10	Carbonaceous dirt + H_2O ice?	None
~150 × 50	?	?	0.02	~120	Carbonaceous dirt + H_2O ice?	None
~74	?	?	0.04	~80	Carbonaceous dirt + H_2O ice?	None
60,330	5.69 (26)	690	0.46	35,600	?	H_2 + He
~10	?	?	?	~10	H_2O ice?	None
25 × 10	?	?	0.4	15	Dirty H_2O ice?	None
70 × 40	?	?	0.6	50	H_2O ice?	None
55 × 35	?	?	0.6	35	H_2O ice?	None
70 × 50	?	?	0.5	45	Dirty H_2O ice?	None
110 × 80	?	?	0.5	75	Dirty H_2O ice?	None
196	4.5 (19)	1200	0.6	170	H_2O ice	None
250	8.4 (19)	1200	0.95	205	H_2O ice	None?
530	7.5 (20)	1210	0.7	435	H_2O ice	None
17 × 11	?	?	0.8	11	H_2O ice?	None
17 × 13	?	?	0.6	12	H_2O ice?	None
560	1.05 (21)	1430	0.5	500	Dirty H_2O ice	None
18 × 15	?	?	0.5	14	Dirty H_2O ice?	None
765	2.49 (21)	1330	0.6	660	H_2O ice	None
2575	1.35 (23)	1880	0.2	2640	Liquid CH_4 or ices?	N_2
205 × 110	?	?	0.3	160	Dirty ice?	None
730	1.88 (21)	1160	0.05 (L), 0.5 (T)[d]	590	H_2O-icy dirt(L); H_2O ice(T)[d]	None
110	?	?	0.06	90	Ice + carbonaceous dirt?	None
100	?	2000?	~0.04		Carbonaceous dirt + ice	Coma

(continued)

[c]Calypso and Telesto are in L_5 and L_4 Lagrangian point positions, respectively, relative to Tethys.

[d]L = Leading hemisphere. T = Trailing hemisphere.

*These are the most-discussed satellites and most imporant in terms of their unique features. Students should know some of the interesting characteristics of each of these moons.

Data on planets, moons, and

Object[b]	Semi-major axis (AU for planets & asteroids; 10^3 km for satellites)	Revolution period (days unless marked otherwise)	Orbit inclination (degrees; with respect to ecliptic for planets & asteroids; w.r.t. planet equator for satellites)	Orbit eccen- tricity	Obliquity (degrees)	Rotation period (days; R = retrograde)	Diameter (km)
Uranus	19.18	84.01y	0.77	0.047	97.9	0.75	51,118
U6 Cordelia	49.7	0.335	~0.14	~0	0?	?	~30
U7 Ophelia	53.8	0.376	~0.09	~0.01	0?	?	~30
U8 Bianca	59.2	0.435	~0.16	~0	0?	?	~40
U9 Cressida	61.8	0.464	~0.04	~0	0?	?	~70
U10 Desdemona	62.7	0.474	~0.16	~0	0?	?	~60
U11 Juliet	64.4	0.493	~0.06	~0	0?	?	~80
U12 Portia	66.1	0.513	~0.09	~0	0?	?	~110
U13 Rosalind	69.9	0.558	~0.28	~0	0?	?	~60
U14 Belinda	75.3	0.624	~0.03	~0	0?	?	~70
U15 Puck	86.0	0.762	~0.31	~0	0?	?	150
*U5 Miranda	129.8	1.414	3.40	0.00	0	1.414	470
*U1 Ariel	191.2	2.520	0.00	0.00	0	2.520	1160
*U2 Umbriel	266.0	4.144	0.00	0.00	0	4.144	1170
*U3 Titania	435.8	8.706	0.00	0.00	0	8.706	1580
*U4 Oberon	582.6	13.463	0.00	0.00	0	13.463	1520
Neptune	30.06	164.8y	1.77	0.009	29.6	0.80	49,528
N3 Naiad	48.0	0.296	~0.0	~0	0?	?	~50
N4 Thalassa	50.0	0.312	~4.5	~0	0?	?	~80
N5 Despina	52.2	0.333	~0.0	~0	0?	?	~180
N6 Galatea	62.0	0.429	~0.0	~0	0?	?	~150
N7 Larissa	73.6	0.554	~0.0	~0	0?	?	~190
*N8 Proteus	117.6	1.121	~0.0	~0	0?	?	~400
*N1 Triton	354.8	5.877	157	0.00	0	5.877	2700
*N2 Nereid	5513.4	360.16	29	0.75	?	?	~340
Pluto	39.53	247.7y	17.15	0.248	122.5	6.39	2300
*P1 Charon	19.6	6.39	98.8		0	6.39	1190

[a]Data come from various sources including reports on Jupiter and Saturn systems by Voyager team members (*Science* 1 June 1979; 23 Nov. 1979; 10 April 1981; 29 January 1982); tabulations in *Proc. Lun. Plan. Sci. Conf.* 1979, *9:* inside covers; S. R. Taylor, *Planetary Science: A Lunar Perspective,* Houston: Lunar and Planetary Science Institute (1982); J. K. Beatty, et al. (editors), *The New Solar System,* Cambridge: Sky Publishing Corp. (1990); R. Binzel, T. Gehrels, M. Matthews (editors), *Asteroids II,* Tucson: University of Arizona Press (1989); and additional scientific reports.

[b]A traditional letter/number designation scheme is indicated for some of the satellites. Generally, the larger ones discovered before the Voyager flights were given the first, lowest numbers, and others were numbered in order of distance from the planet.

other objects in the solar system.[a] *(continued)*

Radius (km)	Mass (kg)	Mean density (kg/m^3)	Visual geometric albedo	Escape velocity at surface (m/s)	Surface materials (known or probable)	Atmosphere (Main constituents)
25,559	8.68 (25)	1290	0.56	21,300	?	H_2 + He
~15	?	?	<0.1	~10	Carbonaceous dirt + ice (?)	None
~15	?	?	<0.1	~10	Carbonaceous dirt + ice (?)	None
~20	?	?	<0.1	~20	Carbonaceous dirt + ice (?)	None
~35	?	?	<0.1	~30	Carbonaceous dirt + ice (?)	None
~30	?	?	<0.1	~30	Carbonaceous dirt + ice (?)	None
~40	?	?	<0.1	~40	Carbonaceous dirt + ice (?)	None
~55	?	?	<0.1	~50	Carbonaceous dirt + ice (?)	None
~30	?	?	<0.1	~30	Carbonaceous dirt + ice (?)	None
~35	?	?	<0.1	~30	Carbonaceous dirt + ice (?)	None
75	?	?	0.07	~70	Carbonaceous dirt + ice (?)	None
235	6.89 (19)	1350	0.34	200	H_2O ice	None
580	1.26 (21)	1660	0.40	560	H_2O ice	None
585	1.33 (21)	1510	0.19	537	H_2O ice	None
790	3.48 (21)	1680	0.28	770	H_2O ice	None
760	3.03 (24)	1580	0.24	710	H_2O ice	None
24,764	1.02 (26)	1640	0.51	23,300	?	H_2 + He
~25	?	?	~0.06	~25	Carbonaceous dirt + ice (?)	None
~40	?	?	~0.06	~40	Carbonaceous dirt + ice (?)	None
~90	?	?	~0.06	~90	Carbonaceous dirt + ice (?)	None
~75	?	?	0.054	~75	Carbonaceous dirt + ice (?)	None
~95	?	?	0.056	~95	Carbonaceous dirt + ice (?)	None
~200	?	?	0.060	~200	Carbonaceous dirt + ice (?)	None
1350	2.14 (22)	2070	~0.75	1452	N_2 ice, CH_4 ice	N_2, CH_4
~170	?	?	0.14	~170	Carbonaceous dirt + ice (?)	None
1150	1.29 (22)	2030	~0.5	1100	CH_4 ice, (N_2 ice?)	Thin CH_4
595	1.77 (21)	~2000	~0.4	~590	H_2O ice	None

[a]These are the most-discussed satellites and most imporant in terms of their unique features. Students should know some of the interesting characteristics of each of these moons.

REFERENCES

CHAPTER ONE

Bowen, C. D. 1963. *Francis Bacon.* Boston: Little, Brown.

Criswell, D. R., ed. 1977. *New Moons.* Houston: Lunar and Planetary Institute.

Dingle, H. 1960. "Philosophical Aspects of Cosmology." *Vistas in Astronomy 1:* 162.

Dyson, F. 1969. "Human Consequences of the Exploration of Space." *Bull. Atomic Scientists,* September, p. 8.

Gaffey, M. J., and T. McCord. 1977. "Mining the Asteroids." *Mercury 6:1* (No. 6, Nov.–Dec.)

Hartmann, W. K., R. Miller, and Pamela Lee. 1984. *Out of the Cradle.* New York: Workman Publishing Co.

Kuhn, T. S. 1962. *The Structure of Scientific Revolutions.* Chicago: University of Chicago Press.

Nicks, O. W., ed. 1970. *This Island Earth.* Washington, D.C.: *NASA SP-250.*

Novak, B. 1980. *Nature and Culture.* New York: Oxford University Press.

Udall, M. K. 1969. "The Environment—What You Can Do." *Congressman's Report 9,* May 20.

Wetherill, G. 1979. "Apollo Objects." *Sci. Am.,* March, p. 54.

CHAPTER TWO

Alexander, A. F. O. 1962. *The Planet Saturn.* New York: Macmillan.

Andersson, L., and J. Fix. 1973. "Pluto: New Photometry and a Determination of the Axis of Rotation." *Icarus 20:* 279.

Beatty, J. K., and others, eds. 1981. *The New Solar System.* Cambridge, Mass.: Sky Publishing.

Betz, A. L., E. C. Sutton, R. McLaren, and C. McAlavy. 1977. "Laser Heterodyne Spectroscopy." In *Proceedings of the Symposium on Planetary Atmospheres,* ed. A. V. Jones. Ottawa: Royal Society of Canada.

Bigg, E. K. 1964. "Influence of the Satellite Io on Jupiter's Decametric Emission." *Nature 203:* 1008.

Binder, A. B., and D. P. Cruikshank. 1964. "Evidence for an Atmosphere on Io." *Icarus 3:* 299

Binzel, R. P. 1990. "Pluto." *Sci. Am. 262:* 50.

Brown, R. A. 1973. "Optical Line Emission from Io." In *Exploration of the Planetary System,* ed. A. Woszczyk and C. Iwaniszewska. Boston: Reidel.

Carusi, A., G. Valecchi, and L. Kresak. 1981. "Orbital Patterns at Close Encounters." Paper presented at International Astronomical Union Colloquium 61 on Comets, Tucson.

Chapman, C. R. 1968. "The Discovery of Jupiter's Red Spot." *Sky and Telescope 35:* 276.

———. 1977. *The Inner Planets.* New York: Scribner's.

Christy, J. W., and R. Harrington. 1978. "The Satellite of Pluto." *Astron. J. 83:* 1005.

Clark, R. N. 1980. "Ganymede, Europa, Callisto, and Saturn's Rings: Compositional Analysis from Reflectance Spectroscopy." *Icarus 44:* 388.

Cruikshank, D. P., and C. R. Chapman. 1967. "Mercury's Rotation and Visual Observations." *Sky and Telescope 34:* 24.

Cruikshank, D. P., C. Pilcher, and D. Morrison. 1976. "Pluto: Evidence for Methane Frost." *Science 194:* 835.

Cuzzi, J. N., and J. B. Pollack. 1978. "Saturn's Rings: Partial Composition and Size Distribution as Constrained by Microwave Observations." *Icarus 33:* 233.

Dyce, R. B., G. H. Pettengill, and I. I. Shapiro. 1967. "Radar Determinations of the Rotations of Venus and Mercury." *Astron. J. 72:* 351.

Elliot, J. L., E. Dunham, and D. Mink. 1977. "The Rings of Uranus." *Nature 267:* 328.

Fanale, F., T. Johnson, and D. Matson. 1974. "Io: A Surface Evaporite Deposit?" *Science 186:* 922.

Florensky, C. P., A. Basilevsky, and A. Pronin. 1977. "The First Panoramas of the Venusian Surface: Geological-Morphological Analysis of Pictures." *COSPAR Space Research 17:* 645.

Focas, J. H., and A. Dollfus. 1969. "Optical Characteristics and Thickness of Saturn's Rings Observed on the Ring Plane in 1966." *Astron. Astrophys. 2:* 251.

Goldreich, P., and S. J. Peale. 1967. "Spin-orbit Coupling in the Solar System II: The Resonant Rotation of Venus." *Astron. J. 72:* 662.

Goldreich, P., and S. Soter. 1966. "Q in the Solar System." *Icarus 5:* 375.

Goldreich, P., N. Murray, P. Longaretti, and D. Banfield. 1989. "Neptune's Story." *Science 245:* 500.

Greenberg, R., D. Davis, W. Hartmann, and C. Chapman. 1977. "Size Distribution of Particles in Planetary Rings." *Icarus 30:* 769.

Guerin, P. 1970. "The New Ring of Saturn." *Sky and Telescope 40:* 88.

Hartmann, W. K. 1980. "Surface Evolution of Two-Component Stone/Ice Bodies." *Icarus 44:* 441.

Hartmann, W. K., and O. Raper. 1974. *The New Mars.* Washington, D.C.: NASA SP-337.

Hayes, S. H., and M. Belton. 1977. "The Rotational Periods of Uranus and Neptune." *Icarus 32:* 383.

Hoyt, W. G. 1976. *Lowell and Mars.* Tucson: University of Arizona Press.

Ingersoll, Andrew P. 1990. "Atmospheres of the Giant Planets." In *The New Solar System.* Cambridge, Mass.: Sky Publishing Corp.

Jewitt, D. 1979. "Discovery of a New Jupiter Satellite." *Science 206:* 951.

Keeler, J. E. 1895. "A Spectroscopic Proof of the Meteoric Constitution of Saturn's Rings." *Astrophys. J. 1:* 416.

Keldysh, M. V. 1977. "Venus Exploration with the Venera 9 and Venera 10 Spacecraft." *Icarus 30:* 605.

Kowal, C. T. 1978. "Letter to the editor." *Sky and Telescope 55:* 195.

Kuiper, G. P. 1944. "Titan, a Satellite with an Atmosphere." *Astrophys. J. 100:* 378.

———. 1956. "The Formation of the Planets, I, II, III." *J. Roy. Astron. Soc. Can. 50, 57:* 105, 158.

———. 1957. "Further Studies on the Origin of Pluto." *Astrophys. J. 125:* 287.

Kuiper, G. P., D. P. Cruikshank, and U. Fink. 1970. "Letter to the editor." *Sky and Telescope 39:* 80.

Larson, S. 1977. "Faint Inner Satellites of Saturn." Presented at the 8th annual meeting of the Division for Planetary Sciences, American Astronomical Society, Honolulu, Jan.

Lebofsky, L. A. 1975. "Stability of Frosts in the Solar System." *Icarus 25:* 205.

Lebofsky, L. A., T. V. Johnson, and T. B. McCord. 1970. "Saturn's Rings: Spectral Reflectivity and Compositional Implications." *Icarus 13:* 226.

Luu, Jane X., and D. Jewitt. 1988. "A Two-part Search for Slow-moving Objects." *Astron. J. 95:* 1256.

Lyttleton, R. A. 1936. "On the Possible Results of an Encounter of Pluto with the Neptunian System." *Monthly Notices Roy. Astron. Soc. 97:* 108.

———. 1968. *Mysteries of the Solar System.* New York: Oxford University Press.

Macy, W., and W. Sinton. 1977. "Detection of Methane and Ethane Emission on Neptune but not on Uranus." *Astrophys. J. 218:* L74.

McCord, T. B. 1969. "Comparison of the Reflectivity and Color of Bright and Dark Regions on the Surface of Mars." *Astrophys. J. 156:* 79.

McKinnon, W. B., and H. Melosh. 1980. "Evolution of Planetary Lithospheres: Evidence from Multiringed Structures on Callisto and Ganymede." *Icarus 44:* 454.

Minton, R. B. 1973. "The Red Polar Caps of Io." *Comm. Lunar Planet. Lab. 10:* 35.

Murray, B. C., M. Belton, G. Danielson, M. Davies, D. Gault, B. Hapke, B. O'Leary, R. Strom, V. Suomi, and N. Trask. 1974. "Venus: Atmospheric Motion and Structure from Mariner 10 Pictures." *Science 183:* 1307.

Owen, T. C., and G. P. Kuiper. 1964. "A Determination of the Composition and Surface Pressure of the Martian Atmosphere." *Comm. Lunar Planet. Lab. 2:* 113.

Peek, B. M. 1958. *The Planet Jupiter.* New York: Macmillan.

Pettengill, G. H., and R. B. Dyce. 1965. "A Radar Determination of the Rotation of the Planet Mercury." *Nature 206:* 1240.

Phillips, R., R. Grimm, and M. Malin. 1991. "Hotspot Evolution and the Global Tectonics of Venus." *Science 252:* 651.

Pilcher, C. P., C. R. Chapman, L. A. Lebofsky, and H. H. Kieffer. 1970. "Saturn's Rings: Identification of Water Frost." *Science 167:* 1372.

Pollack, J. B. 1973. "Mariner 9 Television Observations of Phobos and Deimos, 2." *J. Geophys. Res. 78:* 4313.

———. 1975. "The Rings of Saturn." *Space Sci. Rev. 18:* 3.

———. 1978. "Near-Infrared Spectra of the Galilean Satellites: Observations and Compositional Implications." *Icarus 36:* 271.

Pollack, J. B., D. Strecker, F. Witteborn, E. Erickson, and B. Baldwin. 1978. "Properties of the Clouds of Venus, as Inferred from Airborne Observations of Its Near-IR Reflectivity Spectrum." *Icarus 34:* 28.

Prinn, R. G., and T. Owen. 1976. "Chemistry and Spectroscopy of the Jovian Atmosphere." In *Jupiter,* ed. T. Gehrels and M. Matthews. Tucson: University of Arizona Press.

Reese, E. J., and B. A. Smith. 1968. "Evidence of Vorticity in the Great Red Spot of Jupiter." *Icarus 9:* 474.

Saunders, R. S., and G. H. Pettengill. 1991. "Magellan: Mission Summary." *Science 252:* 247. (See also succeeding articles.)

Smith, B. A. 1967. "Rotation of Venus: Continuing Contradictions." *Science 158:* 114.

———. 1990. "The Voyager Encounters." In *The New Solar System,* ed. J. K. Beatty and A. Chaikin. Cambridge, Mass.: Sky Publishing Corp.

Smith, B. A., and Voyager Imaging Team. 1979. "The Jupiter System through the Eyes of Voyager 1." *Science 204:* 951.

Smith, B. A., and others. 1981. "Encounter with Saturn: Voyager 1 Imaging Science Results." *Science 212:* 163.

Smith, B. A., and 64 others. 1989. "Voyager 2 at Neptune: Imaging Science Results." *Science 246:* 1422.

Soderblom, L. A. 1980. "The Galilean Moons of Jupiter." *Sc. Am., 242* No. 1, Jan., p. 88.

Stone, E. C., and E. D. Miner. 1989. "The Voyager 2 Encounter with the Neptunian System (1989)." *Science 246:* 1417.

Trafton, L. 1977. "Uranus' Rotational Period." *Icarus 32:* 402.

Trafton, L., T. Parkinson, and W. Macy. 1974. "The Spatial Extent of Sodium Emission around Io." *Astrophys. J. 190:* L85.

Walker, M. F., and R. Hardie. 1955. "A Photometric Determination of the Rotational Period of Pluto." *Publ. Astron. Soc. Pacific 67:* 224.

Witteborn, F., J. Bregman, and J. Pollack. 1979. "Io: An Intense Bightening near 5μm." *Science 203:* 643.

Young, A. T. 1973. "Are the Clouds of Venus Sulfuric Acid?" *Icarus 18:* 564.

Young, L. D. G., and A. T. Young. 1973. "Comment on 'The Composition of the Venus Cloud Tops in Light of Recent Spectroscopic Data.' " *Astrophys. J. 179:* L39.

CHAPTER THREE

Aggarwal, H. R., and V. Oberbeck. 1974. "Roche Limit of a Solid Body." *Astrophys. J. 191:* 577.

Biermann, L. 1951. "Kometenschweife und solare Korpuskularstrahlung." *Z. Astrophys. 29:* 274.

Bruman, J. R. 1969. "A Lunar Libration Point Experiment." *Icarus 10:* 197.

Burns, J. A. 1981. "Planetary Rings." In *The New Solar System,* ed. J. K. Beatty and others. Cambridge, Mass.: Sky Publishing.

Burns, J. A., P. Lamy, and S. Soter. 1979. "Radiation Forces on Small Particles in the Solar System." *Icarus 40:* 1.

Columbo, G., P. Goldreich, and A. Harris. 1976. "Spiral Structure as an Explanation for the Asymmetric Brightness of Saturn's A Ring." *Nature 264:* 344.

Cuzzi, J. 1978. "The Rings of Saturn." In *The Saturn System,* ed. D. Hunten and D. Morrison. Washington, D.C.: NASA CP 2068.

Darwin, G. H. 1962. *The Tides and Kindred Phenomena in the Solar System.* San Francisco: W. H. Freeman. (Originally published 1898.)

Dermott, S., and C. Murray. 1981. "The Dynamics of Tadpole and Horsehoe Orbits. I and II." *Icarus 48*: 1 and 12.

Franklin, F. A., and G. Colombo. 1978. "On the Azimuthal Brightness Variations of Saturn's Rings." *Icarus 33:* 279.

Gehrels, T. 1977. "Some Interrelations of Asteroids, Trojans, and Satellites." In *Comets, Asteroids, Meteorites,* ed. A. Delsemme. Toledo: University of Toledo Press.

Goldreich, P., and S. Tremaine. 1978. "The Velocity Dispersion in Saturn's Rings." *Icarus 34:* 227.

———. 1979. "Towards a Theory for the Uranian Rings." *Nature 277:* 97.

Greenberg, R. 1978. "Orbital Resonance in a Dissipative Medium." *Icarus 33:* 62.

Greenberg, R., and D. R. Davis. 1978. "Stability at Potential Maxima: The L_4 and L_5 Points of the Restricted Three Body Problem." *Amer. J. Phys.*, no. 7.

Greenberg, R., D. Davis, W. Hartmann, and C. Chapman. 1977. "Size Distribution of Particles in Planetary Rings." *Icarus 30:* 796.

Harrington, R., and P. Seidelmann. 1981. "The Dynamics of the Saturnian Satellites 1980S1 and 1980S3." *Icarus 47:* 97.

Heppenheimer, T. A. 1977. *Colonies in Space.* Harrisburg, Pa.: Stackpole Books.

Johnson, R. D., and C. Holbrow. 1977. *Space Settlements—A Design Study.* Washington, D.C.: NASA SP-413.

Kaula, W. M., and A. Harris. 1975. "Dynamics of Lunar Origin and Orbital Evolution." *Rev. Geophys. Spa. Phys. 3:* 363.

Kordylewski, K. 1961. "Summary of Letter to the Editor." *Sky and Telescope 22:* 63.

Lago, B., and A. Cazanave. 1979. "Possible Dynamical Evolution of the Rotation of Venus since Formation." *Moon and Planets 21:* 127.

Melosh, H. J. 1981. "Atmospheric Breakup of Terrestrial Impactors." In *Multi-ring Basins,* ed. P. Schultz and R. Merrill. New York: Pergamon Press.

Newton, I. 1962. *Principia.* A. Motte, trans. Berkeley: University of California Press. (Originally published 1687, translated 1729.)

O'Neill, G. K. 1974. "The Colonization of Space." *Physics Today,* September, p. 32.

———. 1975. "Space Colonies and Energy Supply to the Earth." *Science 10:* 943.

———. 1977. *The High Frontier.* New York: Morrow.

Öpik, E. J. 1951. "Collision Probabilities with the Planets and the Distribution of Interplanetary Matter." *Proc. Roy. Irish Acad. 54:* 165.

Peale, S. J. 1976. "Orbital Resonance in the Solar System." *Ann. Rev. Astron. Astrophys. 14:* 215.

Peterson, C. 1976. "A Source Mechanism for Meteorites Controlled by the Yarkovsky Effect." *Icarus 29:* 91.

Poynting, J. 1903. "Radiation in the Solar System: Its Effect on Temperature and Its Pressure on Small Bodies." *Monthly Notices Roy. Astron. Soc. 64,* app. 1.

Rietsma, H. J., R. Beebe, and B. Smith. 1976. "Azimuthal Brightness Variations in Saturn's Rings." *Astron. J. 81:* 209.

Robertson, H. P. 1937. "Dynamical Effects of Radiation in the Solar System." *Monthly Notices Roy. Astron. Soc. 97:* 423.

Roche, E. 1850. "La Figure d'une Masse Fluide soumise à l'attraction d'un pointe eloigne." *Mem. Acad. Montpellier 1 (Sciences):* 1847.

Roosen, R. G. 1968. "A Photographic Investigation of the Gegenschien and the Earth-Moon Libration Point L_5." *Icarus 9:* 429.

Weisel, W. E. 1981. "The Origin and Evolution of the Great Resonance in the Jovian Satellite System." *Astron. J. 86:* 611.

Williams, J. G. 1969. Ph.D. dissertation, University of California, Los Angeles.

Wyatt, S. P., and F. L. Whipple. 1950. "The Poynting-Robertson Effect on Meteor Orbits." *Astrophys. J. 111:* 134.

CHAPTER FOUR

Aanestad, P. A., and S. Kenyon. 1978. "Temperature Fluctuations in Interstellar Dust Grains." In *Thermodynamics and Kinetics of Dust Formation in the Space Medium.* Houston: Lunar and Planetary Institute.

Abt, H. A. 1978. "The Binary Frequency Along the Main Sequence." In *Protostars and Planets,* ed. T. Gehrels and M. Matthews. Tucson: University of Arizona Press.

Allen, C. W. 1973. *Astrophysical Quantities.* London: Athlone Press.

Arny, T., and P. Weissman. 1973. "Interaction of Protostars in a Collapsing Cluster." *Astron. J. 78:* 309.

Batten, A. H. 1973. *Binary and Multiple Systems of Stars.* Oxford: Pergamon Press.

Bodenheimer, P. 1976. "Contraction Models for the Evolution of Jupiter." *Icarus 29:* 165.

Bodenheimer, P., and D. C. Black. 1978. "Numerical Calculations of Protostellar Hydrodynamic Collapse." In *Protostars and Planets,* ed. T. Gehrels and M. Matthews. Tucson: University of Arizona Press.

Böhm, K. H. 1978. "Herbig-Haro Objects and Their Interpretation." In *Protostars and Planets,* ed. T. Gehrels and M. Matthews. Tucson: University of Arizona Press.

Bok, B. J., C. Cordwell, and R. Cromwell. 1971. "Globules." In *Dark Nebulae, Globules, and Protostars,* ed. B. Lynds. Tucson: University of Arizona Press.

Bonsack, W. K., and J. L. Greenstein. 1960. "The Abundance of Lithium in T Tauri Stars and Related Objects." *Astrophys. J. 131:* 83.

Cameron, A. G. W. 1975. "The Origin and Evolution of the Solar System." *Sci. Am.* September, p. 32.

Campbell, B. 1989. "Search for Planetary-Mass Companions to Nearby Stars." *Highlights of Astronomy 8:* 109.

Davidson, K., and M. Harwit. 1967. "Infrared and Radio Appearance of Cocoon Stars." *Astrophys. J. 133:* 443.

Donn, B. 1978. "Condensation Processes and the Formation of Cosmic Grains." In *Protostars and Planets,* ed. T. Gehrels and M. Matthews. Tucson: University of Arizona Press.

Fertel, J. F. "Silicon Monoxide Bands in Some Low-Temperature Stars." *Astrophys. J. 159:* L7.

Field, G. B. 1975. "The Composition of Interstellar Dust." In *The Dusty Universe,* ed. G. Field and A. Cameron. New York: Neale Watson Academic Publications.

———. 1978. "Conditions in Collapsing Clouds." In *Protostars and Planets,* ed. T. Gehrels and M. Matthews. Tucson: University of Arizona Press.

Fleck, R. C. 1978. "On the Origin of Close Binary and Planetary Systems." *Astrophys. J. 225:* 198.

Gatewood, G. 1976. "On the Astrometric Detection of Neighboring Planetary Systems." *Icarus 27:* 1.

Gatewood, G., and H. Eichhorn. 1973. "An Unsuccessful Search for a Planetary Companion of Barnard's Star." *Astron. J. 78:* 769.

Gibson, E. G. 1973. *The Quiet Sun.* Washington, D.C.: NASA SP-303.

Greenberg, R., W. Hartmann, C. Chapman, and J. Wacker. 1978. "The Accretion of Planets from Planetesimals." In *Protostars and Planets,* ed. T. Gehrels and M. Matthews. Tucson: University of Arizona Press.

Handbury, M. J., and I. P. Williams. 1976. "The Peculiar Binary System Epsilon Aurigae." *Astrophys. Spa. Sci. 45:* 439.

Hansen, R., B. Jones, and D. Lin. 1983. "The Astrometric Position of T Tauri and the Nature of Its Companion." *Astrophys. Journ. Letters 270:* L27.

Harrington, R. 1977. "Planetary Orbits in Binary Stars." *Astron. J. 82:* 753.

Harrington, R. S., and A. L. Behall. 1973. "The Mass Ratio of L726-8." *Astron. J. 78:* 1096.

Hartmann, W. K. 1970. "Growth of Planetesimals in Nebulae Surrounding Young Stars." In *Evolution stellaire avant la sequence principale.* Memoires de la Société Royale des Sciences de Liège. Collection in −8°, 5th sér, *19:* 215.

———. 1978. "Planet Formation: Mechanism of Early Growth." *Icarus 33:* 50.

Hayashi, C. 1961. "Stellar Evolution in Early Phases of Gravitational Contraction." *Publ. Astron. Soc. Japan 13:* 450.

Henyey, L., R. LeLevier, and R. Levee. 1955. "The Early Phases of Stellar Evolution." *Publ. Astron. Soc. Pacific 67:* 154.

Herbig, G. H. 1968. "The Structure and Spectrum of R Monocerotis." *Astrophys. J. 152:* 439.

———. 1970. "Introductory Remarks." In *Evolution stellaire avant la sequence principale.* Memoires de la Société Royale des Sciences de Liège. Collection in −8°, 5th sér, *19:* 13.

———. 1977. "Eruptive Phenomenon in Early Stellar Evolution." *Astrophys. J. 217:* 693.

Herbst, W., and G. E. Assousa. 1978. "The Role of Supernovae in Star Formation and Spiral Structure." In *Protostars and Planets,* ed. T. Gehrels and M. Matthews. Tucson: University of Arizona Press.

Hills, J. G. 1977. "Exchange Collisions between Binary and Single Stars." *Astron. J. 82:* 606.

Hoxie, D. 1969. "The Structure and Evolution of Stars of Very Low Mass." Ph.D. dissertation, University of Arizona.

Hoyle, F., and N. Wickramasinghe. 1962. "Graphite Particles as Interstellar Grains." *Monthly Notices Roy. Astron. Soc. 124:* 417.

Hyland, A., E. Becklin, G. Neugebauer, and G. Wallerstein. 1969. "Observations of the Infrared Object VY Canis Majoris." *Astrophys. J. 158:* 618.

Imhoff, C. L. 1978. "T Tauri Star Evolution and Evidence for Planetary Formation." In *Protostars and Planets,* ed. T. Gehrels and M. Matthews. Tucson: University of Arizona Press.

Joy, A. H. 1945. "T Tauri Variable Stars." *Astrophys. J. 102:* 168.

Kamijo, F. 1963. "A Theoretical Study on the Long Period Variable Stars. III. Formation of Solid and Liquid Particles in the Circumstellar Envelope." *Publ. Astron. Soc. Japan 15:* 440.

Knacke, R. 1978. "Mineralogical Similarities between Interstellar Dust and Primitive Solar System Material." In *Protostars and Planets,* ed. T. Gehrels and M. Matthews. Tucson: University of Arizona Press.

Knacke, R., and J. E. Gaustad. 1969. "Possible Identification of Interstellar Silicate Absorption in the Infrared Spectrum of 119 Tauri." *Astrophys. J. 155:* L189.

Kuhi, L. V. 1966. "T Tauri Stars: A Short Review." *J. Roy. Astron. Soc. Can. 60:* 1.

———. 1978. "Spectral Characteristics of T Tauri Stars." In *Protostars and Planets,* ed. T. Gehrels and M. Matthews. Tucson: University of Arizona Press.

Kuiper, G. P. 1955. "On the Origin of Binary Stars." *Publ. Astron. Soc. Pacific 67:* 387.

Kumar, S. S. 1964. "On the Nature of Planetary Companions of Stars." *Z. Astrophys. 58:* 248.

Larimer, J. W. 1967. "Chemical Fractionations in Meteorites. I. Condensation of the Elements." *Geochim. Cosmochim. Acta 31:* 1215.

Larimer, J. W., and E. Anders. 1967. "Chemical Fractionations in Meteorites. II. Abundance Patterns and Their Interpretation." *Geochim. Cosmochim. Acta 31:* 1239.

Larson, R. B. 1968. "Numerical Calculations of the

Dynamics of a Collapsing Protostar." *Monthly Notices Roy. Astron. Soc. 145:* 271.

———. 1972. "The Evolution of Spherical Protostars with Masses of 0.25 to 10 $M_{\odot}$." *Monthly Notices Roy. Astron. Soc. 157:* 121.

———. 1978. "The Stellar State: Formation of Solar-Type Stars." In *Protostars and Planets*, ed. T. Gehrels and M. Matthews. Tucson: University of Arizona Press.

Low, F., and B. Smith. 1966. "Infrared Observations of a Preplanetary System." *Nature 212:* 675.

Lucy, L. B. 1977. "A Numerical Approach to the Testing of the Fission Hypothesis." *Astron. J. 82:* 1013.

Mendoza, E. E. V. 1966. "Infrared Photometry of T Tauri Stars and Related Objects." *Astrophys. J. 143:* 1010.

Ney, E. P. 1972. "Infrared Excesses in Supergiant Stars: Evidence for Silicates." *Publ. Astr. Soc. Pacific 84:* 613.

Oort, J. H., and H. C. van de Hulst. 1946. "Gas and Smoke in Interstellar Space." *Bull. Astron. Inst. Neth. 10:* 187.

Poveda, A. 1965. "The H-R Diagram of Young Clusters and the Formation of Planetary Systems." *Bol. Observ. Tonantz. Tacubaya 4:* 15.

Probst, R. G. 1977. "Parallax, Orbit, and Mass of Ross 614." *Astron. J. 82:* 656.

Reeves, H., and others. 1972. "On the Origin of Light Elements." Cal Tech Preprint OAP-296.

Rydgren, A. E. 1978. "Interpreting Infrared Observations of T Tauri Stars." In *Protostars and Planets*, ed. T. Gehrels and M. Matthews. Tucson: University of Arizona Press.

Serkowski, K. 1976. "Feasibility of a Search for Planets around Solar-Type Stars with a Polarimetric Velocity Meter." *Icarus 27:* 13.

Simon, T., and H. M. Dyck. 1975. "Silicate Absorption at 18 μm in Two Peculiar Infrared Sources." *Nature 253:* 101.

Srinivasan, B., and E. Anders. 1978. "Noble Gases in the Murchison Meteorite: Possible Relics of s-Process Nucleosynthesis." *Science 201:* 51.

Strom, S. E. 1972. "Optical and Infrared Observations of Young Stellar Objects—an Informal Review." *Publ. Astron. Soc. Pacific 84:* 745.

Ulrich, R. K. 1978. "The Status of T Tauri Models." In *Protostars and Planets*, ed. T. Gehrels and M. Matthews. Tucson: University of Arizona Press.

van de Kamp, P. 1963. "Astrometric Study of Barnard's Star with Plates Taken with the Sproul 24-inch Refractor." *Astron. J. 68:* 515.

Ward, W. R. 1981. "Solar Nebula Dispersal and the Stability of the Planetary System, 1." *Icarus 47:* 234.

Welin, G. 1978. "The FU Orionis Phenomenon." In *Protostars and Planets*, ed. T. Gehrels and M. Matthews. Tucson: University of Arizona Press.

Willner, S. P., R. Puettner, and R. Russell. 1978. "Unidentified Infrared Spectral Features." In *Thermodynamics and Kinetics of Dust Formation in the Space Medium.* Houston: Lunar and Planetary Institute.

Wright, A. 1970. "Results of a Computer Program for Gravito-Gas Dynamic Collapse." In *Evolution stellaire avant la sequence principale.* Memoires de la Société Royale des Sciences de Liège. Collection in −8°, 5th sér, *19:* 75.

CHAPTER FIVE

Aggarwal, H. R., and V. Oberbeck. 1974. "Roche Limit of a Solid Body." *Astrophys. J. 191:* 577.

Aitekeeva, Z. Z. 1968. "Anomalous Satellites of Planets." *Solar System Res. 2:* 19.

Albee, A. L., A. J. Gancarz, and A. A. Chodos. 1973. "Metamorphism of Apollo 16 and 17 and Luna 20 Metaclastic Rocks at about 3.95 AE: Samples 61156, 64423, 14-2, 65015, 67483, 15-2, 76055, 22006, and 22007." *Proc. Lun. Sci. Conf. 3:* 569–595.

Alfvén, H. 1954. *The Origin of the Solar System.* London: Oxford Press.

Alfvén, H., and G. Arrhenius. 1976. *Evolution of the Solar System.* Washington, D.C.: NASA SP-345.

Arnold, J. R. 1965. "The Origin of Meteorites as Small Bodies II." *Astrophys. J. 141:* 1536.

Bailey, J. M. 1971a. "Jupiter: Its Captured Satellites." *Science 173:* 812.

———. 1971b. "Origin of the Outer Satellites of Jupiter." *J. Geophys. Res. 76:* 7827.

Binder, A. B. 1974. "On the Origin of the Moon by Rotational Fission." *Moon 11:* 53.

———. 1975. "On the Petrology and Structure of a Gravitationally Differentiated Moon of Fission Origin." *Moon 13:* 431.

Blander, M., and J. Katz. 1967. "Condensation of Primordial Dust." *Geochim. Cosmochim. Acta* 31: 1025.

Boynton, W. V. 1985. "Meteoritic Evidence Concerning Conditions in the Solar Nebula." In *Protostars and Planets II*, ed. D. Black and M. Matthews. Tucson: University of Arizona Press.

Bronshten, V. A. 1968. "Origin of Irregular Satellites of Jupiter." *Solar System Res. 2:* 23.

Burns, J. A. 1973. "Where Are the Satellites of the Inner Planets?" *Nature 242:* 23.

———. 1978. "The Dynamical Evolution and Origin of the Martian Moons." *Vistas in Astronomy 22:* 193.

———. 1980. "A Model of the Jovian Ring." Paper presented at International Astronomical Union Colloquium 57, May 1980, Kona, Hawaii.

———. 1990. "Planetary Rings." In *The New Solar System*, ed. J. Beatty and A. Chaikin. Cambridge, Mass.: Sky Publishing Corp.

Carlson, R. W., and G. Lugmair. 1988. "The Age of Ferroan Anorthosite 60025: Oldest Crust on a Young Moon." *Earth Planet. Sci. Lett. 90:* 119.

Cameron, A. G. W. 1962. "The Formation of the Sun and the Planets." *Icarus 1:* 13.

———. 1975. "The Origin and Evolution of the Solar System." *Sci. Am.*, September, p. 32.

———. A. G. W. 1985. "Formation and Evolution of the Primitive Solar Nebula." In *Protostars and Planets II*, ed. D. Black and M. Matthews. Tucson: University of Arizona Press.

Cameron, A. G. W., and M. Pine. 1973. "Numerical Models of the Primitive Solar Nebula." *Icarus 18:* 377.

Cameron, A. G. W., and J. Truran. 1977. "The Supernova Trigger for Formation of the Solar System." *Icarus 30:* 447.

Cameron, A. G. W., and W. Ward. 1976. "The Origin of the Moon [abstract]." In *Lunar Science VII*. Houston: Lunar Science Institute.

Carusi, A., and G. Valsecchi. 1980. "Comets and Temporary Satellites of Jupiter." Paper presented at International Astronomical Union Colloquium 57, May 1980, Kona, Hawaii.

Chandrasekhar, S. 1946. "On a New Theory of Weizsäcker on the Origin of the Solar System." *Rev. Mod. Phys. 18:* 94.

Christy, J., and R. Harrington. 1978. "The Satellite of Pluto." *Astron. J. 83:* 1005.

Chyba, C., and Nicolson. 1987. Abstract, Division of Planetary Sciences of Am. Astronom. Soc., p. 821.

Clayton, R., L. Grossman, and T. Mayeda. 1973. "A Component of Primitive Nuclear Composition in Carbonaceous Meteorites." *Science 182:* 485.

Clayton, R., T. Mayeda, and A. Rubin. 1984. "Oxygen Isotopic Compositions of Enstatite Chondrites and Aubrites." *J. Geophys. Res. Suppl. 89:* C245.

Consolmagno, G. J. 1981. "Io: Thermal Models and Chemical Evolution." *Icarus 47:* 36.

Consolmagno, G., and J. Jokopii. 1978. "^{26}Al and the Partial Ionization of the Solar Nebula." *Moon and Planets 19:* 253.

Consolmagno, G. J., and J. Lewis. 1977. "Preliminary Thermal History Models of Icy Satellites." In *Planetary Satellites*, ed. J. Burns, p. 492. Tucson: University of Arizona Press.

Cruikshank, D. P., and D. Morrison. 1976. "The Galilean Satellites of Jupiter." *Sci. Am.*, May, p. 108.

Davis, D. R. 1974. "Secular Changes in Jovian Eccentricity: Effect on the Size of Capture Orbits." *J. Geophys. Res. 79:* 4442.

Day, K. L., and B. Donn. 1978. "Condensation of Nonequilibrium Phases of Refractory Silicates from the Vapor." *Science 202:* 307.

Dermott, S. F. 1968. "On the Origin of Commensurabilities in the Solar System I." *Monthly Notices Roy. Astron. Soc. 141:* 349.

Elliot, J. L., and others. 1978. "The Radii of Uranian Rings α, β, γ, δ, ϵ, η, 4, 5, and 6 from Their Occultations by SAO 158687." *Astron. J. 83:* 980.

Gancarz, A. J., and G. Wasserburg. 1977. "Initial Pb of the Amîtsoq Gneiss, West Greenland, and Implications for the Age of the Earth." *Geochim. Cosmochim. Acta 41:* 1283.

Gehrels, T., and others. 1980. "Imaging Photopolarimeter on Pioneer Saturn." *Science 207:* 436.

Goettel, K. A., and S. S. Barshay. 1978. "The Chemical Equilibrium Model for Condensation in the Solar Nebula: Assumptions, Implications, and Limitations." In *The Origin of the Solar System*, ed. S. F. Dermot. New York: Wiley.

Goldreich, P. 1965. "An Explanation of the Frequent Occurrence of Commensurable Mean Motions in the Solar System." *Monthly Notices Roy. Astron. Soc. 130:* 159.

Goldreich, P., and S. Tremaine. 1978a. "The Formation of the Cassini Division in Saturn's Rings." *Icarus 34:* 240.

———. 1978b. "The Velocity Dispersion in Saturn's Rings." *Icarus 34:* 227.

Goldreich, P., and W. Ward. 1973. "The Formation of Planetesimals." *Astrophys. J. 183:* 1051.

Greenberg, R. J. 1977. "Orbit-Orbit Resonances in the Solar System: Varieties and Similarities." *Vistas in Astronomy 21:* 209.

Greenberg, R., W. Hartmann, C. Chapman, J. Wacker. 1978a. "The Accretion of Planets from Planetesimals." In *Protostars and Planets*, ed. T. Gehrels and M. Matthews. Tucson: University of Arizona Press.

Greenberg, R., J. Wacker, W. Hartmann, and C. Chapman. 1978b. "Planetesimals to Planets: Numerical Simulation of Collisional Evolution." *Icarus 35:* 1.

Grossman, L. 1973. "Refractory Trace Elements in Ca-Al-rich Inclusions in the Allende Chondrite." *Geochim. Cosmochim. Acta 37:* 1119.

———. 1975. "The Most Primitive Objects in the Solar System: Carbonaceous Chondrites." *Sci. Am.*, February, p. 30.

———. 1977. "Chemical Fractionation in the Solar Nebula." In *The Soviet-American Conferences on Cosmochemistry of the Moon and Planets*, ed. J. Pomeroy and N. Hubbard. Washington, D.C.: NASA SP-370, pt. 2.

Gurevich, L. E., and A. Lebedinskii. 1950. "Formation of the Planets." *Izv. AN SSSR, seriya fizich., 14:* 765.

Haggarty, S. E. 1978. "The Allende Meteorite: Solid Solution Characteristics." *Proc. Lun. Planet. Sci. Conf. 9:* 1331.

Harris, A. W. 1977. "An Analytical Theory of Planetary Rotation Rates." *Icarus 31:* 168.

———. 1978. "Satellite Formation II." *Icarus 34:* 128.

Hartmann, W. K. 1969. "Terrestrial, Lunar, and Interplanetary Rock Fragmentation." *Icarus 10:* 201.

———. 1970. "Growth of Planetesimals in Nebulae Surrounding Young Stars." In *Evolution stellaire avant la sequence principale.* Mémoires de la Société Royale des Sciences de Liège. Collection in −8°, 5th Ser., *19:* 215.

———. 1975. "Lunar 'Cataclysm': A Misconception?" *Icarus 24:* 181.

———. 1976. "Planet Formation: Composition Mixing and Lunar Compositional Anomalies." *Icarus 27:* 553.

———. 1977. "Large Planetesimals in the Early Solar System." In *Comets, Asteroids, Meteorites,* ed. A. Delsemme. Toledo, Ohio: University of Toledo Press.

———. 1978. "Planet Formation: Mechanism of Early Growth." *Icarus 33:* 50.

———. 1987. "A Satellite-Asteroid Mystery and a Possible Early Flux of Scattered C-class Asteroids." *Icarus 71:* 57.

———. 1990. "Additional Evidence about an Early Intense Flux of C Asteroids and the Origin of Phobos." *Icarus 87:* 236.

Hartmann, W. K., and D. R. Davis. 1975. "Satellite-sized Planetesimals and Lunar Origin." *Icarus 24:* 504.

Hartmann, W. K., R. Strom, S. Weidenschilling, K. Blasius, A. Woronow, M. Dence, R. Grieve, J. Diaz, C. R. Chapman, E. M. Shoemaker, and K. Jones. 1981. "Cratering Chronology of the Planets." In *Basaltic Volcanism of the Planets,* ed. R. Merrill and others. New York: Pergamon Press.

Hartmann, W. K., R. Phillips, and G. Taylor. 1986. *Origin of the Moon.* Houston: Lunar and Planetary Institute.

Herbig, H. G. 1970. "VY Canis Majoris II. Interpretation of the Energy Distribution." *Contrib. Lick Observ. 317.*

Hoyle, F. 1963. "Formation of the Planets." In *Origin of the Solar System,* ed. R. Jastrow and A. Cameron. New York: Academic Press.

Hunten, D. M. 1979. "Capture of Phobos and Deimos by Protoatmospheric Drag." *Icarus 37:* 113.

Isaacman, R., and C. Sagan. 1977. "Computer Simulations of Planetary Accretion Dynamics: Sensitivity to Initial Conditions." *Icarus 31:* 510.

Jewitt, D. C., and G. E. Danielson. 1981. "The Jovian Ring." *J. Geophys. Res. 86:* 8691.

Jewitt, D., G. E. Danielson, and S. Synnott. 1979. "Discovery of a New Jupiter Satellite." *Science 206:* 951.

Jewitt, D., and P. Goldreich. 1980. "The Ring of Jupiter." Paper presented at the International Astronomical Union Colloquium 57, May 1980, Kona, Hawaii.

Kaula, W. M. 1968. *An Introduction to Planetary Physics.* New York: Wiley.

Kaula, W. M., and P. Bigeleisen. 1975. "Early Scattering by Jupiter and Its Collision Effects in the Terrestrial Zone." *Icarus 25:* 18.

Kerridge, J. F., and J. Vedder. 1972. "Accretionary Processes in the Early Solar System: An Experimental Approach." *Science 177:* 161.

Kuhi, L. V. 1966. "T Tauri Stars: A Short Review." *J. Roy. Astron. Soc. Can. 60:* 1.

———. 1978. "Rotation of Pre-main-sequence Stars in NGC2264 and the Solar Angular Momentum Problem." *Moon and Planets 19:* 199.

Kuiper, G. P. 1951. "On the Origin of the Solar System." In *Astrophysics,* ed. J. A. Hynek. New York: McGraw-Hill.

———. 1956a. "The Formation of the Planets, I, II, III." *J. Roy. Astron. Soc. Can. 50:* 57, 105, 158.

———. 1956b. "On the Origin of the Satellites and the Trojans." *Vistas in Astronomy 2:* 1631.

Kuiper, G. P., D. P. Cruikshank, and U. Fink. 1970. Letter in *Sky and Telescope 39:* 80.

Lange, D. E., and J. Larimer. 1973. "Chondrules: An Origin by Impacts between Dust Grains." *Science 182:* 920.

Larimer, J. W. 1967. "Chemical Fractionations in Meteorites I. Condensation of the Elements." *Geochim. Cosmochim. Acta 31:* 1215.

Larimer, J. W., and E. Anders. 1967. "Chemical Fractionations in Meteorites II. Abundance Patterns and Their Interpretations." *Geochim. Cosmochim. Acta 31:* 1239.

Larson, R. B. 1972. "The Evolution of Spherical Protostars with Masses 0.25 to 10 $M_\odot$." *Monthly Notices Roy. Astron. Soc. 157:* 121.

Lebofsky, L. A., T. V. Johnson, and T. B. McCord. 1970. "Saturn's Rings: Spectral Reflectivity and Compositional Implications." *Icarus 13:* 226.

Lewis, J. S. 1972a. "Low Temperature Condensation from the Solar Nebula." *Icarus 16:* 241.

———. 1972b. "Metal/silicate fractionation in the solar system." *Earth Planet. Sci. Lett. 15:* 286.

———. 1974. "The Temperature Gradient in the Solar System." *Science 186:* 440.

Lord, H. C. 1965. "Molecular Equilibria and Condensation in a Solar Nebula and Cool Stellar Atmospheres." *Icarus 4*: 279.

Mason, B. 1962. *Meteorites*. New York: Wiley.

McCord, T. B. 1966. "The Dynamical Evolution of the Neptunian System." *Astron. J. 71:* 585.

———. 1968. "The Loss of Retrograde Satellites in the Solar System." *J. Geophys. Res. 73:* 1497.

McSween, H. Y., Jr. 1989. "Chondritic Meteorites and the Formation of Planets." In *American Scientist 77:* 146.

Millis, R., and L. Wasserman. 1978. "The Occultation of BD $-15°$ 3969 by the Rings of Uranus." *Astron. J. 83:* 993.

Mizuno, H. 1980. "Formation of Giant Planets." *Progress of Theor. Physics, 64:* 544.

Neugebauer, G., and others. 1981. "Spectra of the Jovian Ring and Amalthea." *Astron. J. 86:* 607.

Nicholson, P. D., and others. 1978. "The Rings of Uranus: Results of the 10 April 1978 Occultation." *Astron. J. 83:* 1240.

Öpik, E. J. 1963. "Survival of Comet Nuclei and the Asteroids." *Adv. Astron. Astrophys. 2:* 219.

———. 1966. "The Stray Bodies in the Solar System II. The Cometary Origin of Meteorites." *Adv. Astron. Astrophys. 4:* 301.

Papanastassiou, D. A., and G. Wasserburg. 1976. "Rb-Sr Age of Troctolite 76535." *Proc. Lun. Sci. Conf. 7:* 2035.

Peale, S. J. 1978. "An Observational Test for the Origin of the Titan-Hyperion Orbital Resonance." *Icarus 36:* 240.

Pepin, R. O. 1991. "On the Origin and Early Evolution of Terrestrial Planet Atmospheres and Meteoritic Volatiles." *Icarus 92:* 2.

Pepin, R. O., and D. Phinney. 1975. "The Formation Interval of the Earth." *Lun. Sci. Conf. Abstracts 6:* 682.

Piironen, J. O. 1978. "The Profile of Saturn's Ring A as Derived from Stellar Occultations." *Moon and Planets 19:* 61.

Pilcher, C., C. Chapman, L. Lebofsky, and H. Kieffer. 1970. "Saturn's Rings: Identification of Water Frost." *Science 167:* 1372.

Podosek, F. A. 1970. "Dating of Meteorites by the High-Temperature Release of Iodine-Correlated Xe^{129}." *Geochim. Cosmochim. Acta 34:* 341.

Pollack, J. B. 1975. "The Rings of Saturn." *Space Sci. Rev. 18:* 3.

———. 1978. "The Rings of Saturn." *Am. Scientist 66:* 30.

Pollack, J. B., J. Burns, M. Tauber. 1979. "Gas Drag in Primordial Circumplanetary Envelopes: A Mechanism for Satellite Capture." *Icarus 37:* 587.

Poveda, A. 1965. "H-R Diagram of Young Clusters and the Formation of Planetary Systems." *Bol. Observ. Tonantz. Tacubaya 4:* 15.

Reynolds, J. G. 1960. "Determination of the Age of the Elements." *Phys. Rev. Lett. 4:* 8.

Ringwood, A. E. 1970. "Origin of the Moon: The Precipitation Hypothesis." *Earth Planet. Sci. Lett. 8:* 131.

Safronov, V. S. 1966. "Sizes of the Largest Bodies Falling onto the Planets during Their Formation." *Sov. Astron. -AJ 9:* 987.

———. 1972. *Evolution of the Protoplanetary Cloud and Formation of the Earth and the Planets*. Jerusalem: Israel Program for Scientific Translations. (Originally published 1969 in Russian.)

Schonfeld, E. 1976. "Chronology of the Lunar Crust." *Proc. Lun. Sci. Conf. 7:* 2093.

Schwartz, K., and G. Schubart. 1969. "The Early De-spinning of the Sun." *Astrophys. Space Sci. 5:* 444.

Slattery, W. L. 1978. "Protoplanetary Core Formation by Rain-out of Iron Drops." *Moon and Planets 19:* 443.

Smith, P. K., and P. Buseck. 1981. "Graphitic Carbon in the Allende Meteorite: A Microstructural Study." *Science 212:* 322.

Smoluchowski, R. 1978. "Width of a Planetary Ring System and C-ring of Saturn." *Nature 274:* 669.

———. 1979. "Planetary Rings." In *Comments on Astrophys. 8:* 69.

Synnott, S. P. 1980. "Discovery of a 15th Satellite of Jupiter." Paper presented at the International Astronomical Union Colloquium 57, May 1980, Kona, Hawaii.

Taylor, S. R. 1975. *Lunar Science: A Post-Apollo View*. New York: Pergamon Press.

Tscharnuter, W. M. 1978. "Collapse of the Pre-Solar Nebula." *Moon and Planets 19:* 229.

Urey, H. C. 1952. *The Planets: Their Origin and Development*. New Haven: Yale University Press.

Van Allen, J., and others. "Saturn's Magnetosphere, Rings, and Inner Satellites." *Science 207:* 415.

Veverka, J., P. Thomas, and T. Duxbury. 1978. "The Puzzling Moons of Mars." *Sky and Telescope 56:* 186.

Wanke, H., and G. Dreibus. 1986. "Geochemical Evidence for the Formation of the Moon by Impact-induced Fission of the Proto-Earth." In *Origin of the Moon*, ed. W. Hartmann, R. Phillips, and G. Taylor. Houston: Lunar and Planetary Institute.

Ward, W. R., and M. Reid. 1973. "Solar Tidal Friction and Satellite Loss." *Monthly Notices Roy. Astron. Soc. 164:* 21.

Wasserburg, G. 1985. "Short-Lived Nuclei in the Early Solar System." In *Protostars and Planets II,* ed. D. Black and M. Matthews. Tucson: University of Arizona Press.

Wasson, J. T. 1974. *Meteorites.* New York: Springer-Verlag.

Watson, K., B. C. Murray, and H. Brown. 1963. "The Stability of Volatiles in the Solar System." *Icarus 1:* 317.

Weidenschilling, S. J. 1980. "Dust to Planetesimals: Settling and Coagulation in the Solar Nebula." *Icarus 44:* 172.

Wetherill, G. W. 1975. "Late Heavy Bombardment of the Moon and Terrestrial Planets." *Proc. Lun. Sci. Conf. 6:* 1539.

———. 1976. "The Role of Large Bodies in the Formation of the Earth and Moon." *Proc. Lun. Sci. Conf. 7:* 3245.

———. 1977. "Evolution of the Earth's Planetesimal Swarm Subsequent to the Formation of the Earth and Moon." *Proc. Lun. Sci. Conf. 8:* 1.

———. 1981. "The Formation of the Earth from Planetesimals." *Sci. Am.,* June, p. 163.

———. 1990. "Formation of the Earth." *Ann. Rev. Earth Planet. Sci. 18:* 205.

Whipple, F. L. 1964. "History of the Solar System." *Proc. Natl. Acad. Sci. 52:* 517.

Wise, D. U. 1966. "Origin of the Moon by Fission." In *The Earth-Moon System,* ed. B. Marsden and A. Cameron. New York: Plenum.

Wood, J. A. 1963. "On the Origin of Chondrules and Chondrites." *Icarus* 2: 152.

———. 1985. "Meteoritic Constraints on Processes in the Solar Nebula." In *Protostars and Planets II,* ed. D. Black and M. Matthews. Tucson: University of Arizona Press.

CHAPTER SIX

Anders, E. 1963. "On the Origin of Carbonaceous Chondrites." *Ann. N.Y. Acad. Sci. 108:* 514.

———. 1978. "Most Stony Meteorites Come from the Asteroid Belt." In *Asteroids: An Exploration Assessment,* ed. D. Morrison and W. Wells. Washington, D.C.: NASA CP-2053.

Arnold, J. R. 1965. "The Origin of Meteorites as Small Bodies II." *Astrophys. J. 141:* 1536.

Baldwin, R. B. 1949. *The Face of the Moon.* Chicago: University of Chicago Press.

———. 1963. *The Measure of the Moon.* Chicago: University of Chicago Press.

Ballabh, G., A. Bhatnagar, and N. Bhandari. 1978. "The Orbit of the Dhajala Meteorite." *Icarus 33:* 361.

Binder, A., and others. 1977. "The Geology of the Viking I Lander Site." *J. Geophys. Res. 82:* 4439.

Blander, M., and M. Abdel-Gawad. 1969. "The Origin of Meteorites and the Constrained Equilibrium Condensation Theory." *Geochim. Cosmochim. Acta 33:* 710.

Bogard, D. D. 1976. "Fragment from a Young Asteroid?" *Astronomy 4:* 14.

Bogard, D. D., and L. Husain. 1977. "Ar-40/Ar-39 Dating of Plainview Brecciated Chondrite: Evidence of Regolith Formation since 3.7 Aeons B.P." Paper presented at the 8th annual meeting of the Division of Planetary Sciences, American Astronomical Society, 19–22 January, Honolulu.

Bogard, D., and P. Johnson. 1983. "Martian Gases in an Antarctic Meteorite?" *Science 221:* 651.

Chapman, C. R., and D. R. Davis. 1975. "Asteroid Collisional Evolution: Evidence for a Much Larger Early Population." *Science 190:* 553.

Chapman, C. R., T. McCord, and C. Pieters. 1973. "Minor Planets and Related Objects X. Spectrophotometric Study of the Composition of 1685 Toro." *Astron. J. 78:* 502.

Chapman, C. R., J. Williams, and W. Hartmann. 1978. "The Asteroids." *Ann. Rev. Astron. Astrophys. 16:* 33.

Chapman, D., and H. Larson. 1963. "On the Lunar Origin of Tektites." *J. Geophys. Res. 68:* 4305.

Clepecha, Z. 1981. "Possible Meteorite Fall, Alps Region." *Scientific Event Alert Network Bulletin 6,* no. 10, p. 12.

Crabb, J., and L. Schultz. 1981. "Cosmic Ray Exposure Ages of Ordinary Chondrites and Their Significance for Parent Body Stratiography." *Geochim. Cosmochim. Acta. 45*: 215.

Cruikshank, D., D. Tholen, W. Hartmann, J. Bell, and R. Brown. 1991. "Three Basaltic Earth-Approaching Asteroids and the Source of the Basaltic Meteorites." *Icarus 89:* 1.

Davis, D. R., C. Chapman, R. Greenberg, S. Weidenschilling, and A. Harris. 1979. "Collisional Evolution of Asteroids: Populations, Rotations, and Velocities." In *Asteroids,* ed. T. Gehrels. Tucson: University of Arizona Press.

Dodd, R. T. 1965. "Preferred Orientation of Chondrules in Chondrites." *Icarus 4:* 308.

———. 1971. "The Petrology of Chondrules in the Sharps Meteorite." *Contrib. Mineral. Petrol. 31:* 201.

———. 1981. *Meteorites.* Cambridge: Cambridge University Press.

Dohnanyi, J. S. 1970. "On the Origin and Distribution of Meteoroids." *J. Geophys. Res. 75:* 3468.

———. 1972. "Interplanetary Objects in Review: Statistics of Their Masses and Dynamics." *Icarus 17:* 1.

DuFresne, E., and E. Anders. 1963. "Chemical Evolution of the Carbonaceous Chondrites." In *The Moon, Meteorites and Comets,* ed. B. M. Middlehurst and G. P. Kuiper. Chicago: University of Chicago Press.

Eberhardt, P., and D. Hess. 1960. "Helium in Stone Meteorites." *Astrophys. J. 131:* 38.

Faul, H. 1966. "Tektites Are Terrestrial." *Science 152:* 1341.

Feierberg, M., and M. Drake. 1980. "The Meteorite-Asteroid Connection: The Infrared Spectra of Eucrites, Shergottites, and Vesta." *Science 209:* 805.

Feierberg, M., H. Larson, and C. Chapman. 1981. "Spectroscopic Evidence for Undifferentiated S-type Asteroids." Submitted to *Astrophys. J.*

Fireman, E. L., L. Rancitelli, and T. Kirsten. 1979. "Terrestrial Ages of Four Allan Hills Meteorites: Consequences for Antarctic Ice." *Science 203:* 453.

Fredriksson, K. 1963. "Chondrules and the Meteorite Parent Bodies." *Trans. N.Y. Acad. Sci. 25:* 756.

Ganapathy, R., D. Brownlee, and P. Hodge. 1978. "Silicate Spherules from Deepsea Sediments: Confirmation of Extraterrestrial Origin." *Science 201:* 1119.

Ganapathy, R., R. Keays, J. Laul, and E. Anders. 1970. "Trace Elements in Apollo 11 Lunar Rocks: Implications of Meteorite Influx and Origin of Moon." *Proc. Apollo 11 Lunar Sci. Conf. 2:* 1117.

Goldstein, J. I., and J. M. Short. 1967. "The Iron Meteorites, Their Thermal History and Parent Bodies." *Geochim. Cosmochim. Acta 31:* 1733.

Goplan, K., and G. Wetherill. 1971. "Rubidium-Strontium Studies on Black Hypersthene Chondrites." *J. Geophys. Res. 76:* 8484.

Hartmann, W. K. 1970. "Lunar Cratering Chronology." *Icarus 13:* 299.

———. 1975. "Lunar 'Cataclysm': A Misconception?" *Icarus 24:* 181.

———. 1979. "A Special Class of Planetesimal Collisions and a Possible Confirmation." *Proc. Lun. Planet. Sci. Conf. 10:* 1897.

———. 1980. "Dropping Stones in Magma Oceans: Effects of Early Lunar Cratering." In *Proceedings of Conference on the Lunar Highlands Crust,* ed. J. Papike and R. Merrill. New York: Pergamon Press.

Hartmann, W. K., and G. P. Kuiper. 1962. "Concentric Systems Surrounding Lunar Basins." *Comm. Lunar Planet. Lab. 1:* 51.

Hartmann, W. K., R. Strom, S. Weidenschilling, K. Blasius, A. Woronow, M. Dence, R. Grieve, J. Diaz, C. R. Chapman, E. M. Shoemaker, and K. Jones. 1981. "Cratering Chronology of the Planets." In *Basaltic Volcanism of the Planets,* ed. R. Merrill and others. New York: Pergamon Press.

Hartmann, W. K., and L. Wilkening. 1981. "Chondrule-sized Spherules from an Explosion Crater." *Meteoritics 15:* 299 (abstract).

Hartmann, W. K., and C. A. Wood. 1971. "Moon: Origin and Evolution of Multi-ring Basins." *Moon 3:* 3.

Hartmann, W. K., and D. J. Tholen. 1990. "Comet Nuclei and Trojan Asteroids, A New Link and a Possible Mechanism for Comet Splitting." *Icarus 86:* 448.

Heineman, R., and L. Brady. 1929. "The Winona Meteorite." *Amer. J. Sci. 18* (5th ser.): 477.

Herndon, J. M., and L. Wilkening. 1978. "Conclusions Derived from the Evidence on Accretion in Meteorites." In *Protostars and Planets,* ed. T. Gehrels and M. Matthews. Tucson: University of Arizona Press.

Heymann, D. 1967. "The Origin of Hypersthene Chondrites: Ages and Shock Effects of Black Chondrites." *Icarus 6:* 189.

———. 1978. "Solar Gases in Meteorites: The Origin of Chondrites and Cl Carbonaceous Chondrites." *Meteoritics 13:* 291.

Housen, K. R., L. Wilkening, C. R. Chapman, and R. Greenberg. 1979. "Regolith Development and Evolution on Asteroids and the Moon." In *Asteroids,* ed. T. Gehrels. Tucson: University of Arizona Press.

Hunten, D. M. 1979. "Capture of Phobos and Deimos by Protoatmospheric Drag." *Icarus 37:* 113.

Ip, W. 1978. "Model Consideration of the Bombardment Event of the Asteroid Belt by the Planetesimals Scattered from the Jupiter Zone." *Icarus 34:* 117.

Jacchia, L. G. 1974. "A Meteorite That Missed the Earth." *Sky and Telescope 48:* 4.

Kaula, W., and P. Bigeleisen. 1975. "Early Scattering by Jupiter and Its Collision Effects in the Terrestrial Zone." *Icarus 25:* 18.

Keay, C. S. 1980. "Anomalous Sounds from the Entry of Meteor Fireballs." *Science 210:* 11.

Kelly, W. R., and J. Larimer. 1977. "Chemical Fractionation in Meteorites VIII." *Geochim. Cosmochim. Acta 41:* 93.

Kerridge, F., and M. Matthews, eds. 1988. *Meteorites and the Early Solar System.* Tucson: University of Arizona Press.

Kieffer, S. W. 1975. "Droplet Chondrules." *Science 189:* 333.

Krinov, E. L. 1960. *Principles of Meteorites.* Elmsford, N.Y.: Pergamon Press.

———. 1966. *Giant Meteorites.* New York: Pergamon Press.

Levin, B. J., A. Simonenko, and E. Anders. 1976. "Farmington Meteorite: A Fragment of an Apollo Asteroid?" *Icarus 28:* 307.

Lewis, R. S., and others. 1979. "Stellar Condensates in Meteorites: Isotopic Evidence from Noble Gases." *Astrophys. J. 234:* L165.

Lorin, J., and P. Pellas. 1979. "Preirradiation History of Djermaia (H) Chondritic Breccia." *Icarus 40:* 502.

Mason, B. 1962. *Meteorites.* New York: Wiley.

Masursky, H., G. Colton, and F. El-Baz, eds. 1978. *Apollo over the Moon.* Washington, D.C.: NASA SP-362.

McSween, H. Y. Jr. 1987. *Meteorites and Their Parent Planets.* Cambridge University Press.

Nininger, H. H. 1952. *Out of the Sky.* Denver: University of Denver Press.

O'Keefe, J., ed. 1963. *Tektites.* Chicago: University of Chicago Press.

———. 1978. "The Tektite Problem." *Sci. Am.,* February, p. 116.

Öpik, E. J. 1963. "Survival of Comet Nuclei and the Asteroids." *Adv. Astron. Astrophys. 2:* 219.

———. 1966. "The Stray Bodies in the Solar System II. The Cometary Origin of Meteorites." *Adv. Astron. Astrophys. 4:* 301.

Pellegrino, C. R., and J. Stoff. 1979. "Organic Clues in Carbonaceous Meteorites." *Sky and Telescope 57:* 330.

Pepin, R. O., and M. Carr. 1991. In *Mars,* ed. H. Kieffer and B. Jakosky. Tucson: University of Arizona Press.

Prior, G. T. 1920. "The Classification of Meteorites." *Mineral. Mag. 19:* 51.

Richardson, S. M. 1978. "Vein Formation in the C1 Carbonaceous Chondrites." *Meteoritics 13:* 141.

Ringwood, A. E. 1966. "Genesis of Chondritic Meteorites." *Rev. Geophys. Spa. Phys. 4:* 113.

Sagan, C. 1975. "Kalliope and the Kaába: The Origin of Meteorites." *Natural History 84:* 8.

Scott, E. R. D. 1977. "Parent Bodies of Iron Meteorites." In *Comets, Asteroids, and Meteorites,* ed. A. Delsemme, p. 439. Toledo: University of Toledo Press.

———. 1978. "Tabulation of Meteorites." *Proc. Lun. Planet. Sci. Conf. 9:* pt. 1, endpaper.

Smith, P. P. K., and P. Buseck. 1981. "Graphitic Carbon in the Allende Meteorite: A Microstructural Study." *Science 212:* 322.

Suess, H. E. 1949. "Chemistry of the Formation of Planets." *Z. Electrochem. 53:* 237.

Taylor, S. R. 1975. *Lunar Science: A Post-Apollo View.* New York: Pergamon Press.

Taylor, S. R., and D. Heymann. 1969. "Shock Reheating and the Gas Retention Ages of Chondrites." *Earth Planet. Sci. Lett. 7:* 151.

Taylor, S. R., and M. Kaye. 1969. "Genetic Significance of the Chemical Composition of Tektites: A Review." *Geochim. Cosmochim. Acta 7:* 34.

Tera, F., D. Papanastassiou, and G. Wasserburg. 1974. "Isotopic Evidence for a Terminal Lunar Cataclysm." *Earth Planet. Sci. Lett. 22:* 1.

Uhlig, H. H. 1955. "Contribution of Metallurgy to the Study of Meteorites I. Structure of Metallic Meteorites, Their Composition and the Effect of Pressure." *Geochim. Cosmochim. Acta 6:* 282.

Urey, H. C., and H. Craig. 1953. "The Composition of the Stone Meteorites and the Origin of the Meteorites." *Geochim. Cosmochim. Acta 4:* 36.

Van Schmus, W. R., and J. A. Wood. 1967. "A Chemical-Petrologic Classification for the Chondritic Meteorites." *Geochim. Cosmochim. Acta 31:* 747.

Vortman, L. J. 1977. "Craters from Surface Explosions and Energy Dependence—a Retrospective." In *Impact and Explosion Cratering,* ed. D. Roddy, R. Pepin, and R. Merrill. New York: Pergamon Press.

Wahl, W. 1952. "The Brecciated Stony Meteorites and Meteorites Containing Foreign Fragments." *Geochim. Cosmochim. Acta 2:* 91.

Wasson, J. T. 1974. *Meteorites.* New York: Springer-Verlag.

———. 1985. *Meteorites: Their Record of Early Solar-System History.* New York: Freeman.

Wetherill, G. W. 1975. "Late Heavy Bombardment of the Moon and Terrestrial Planets." *Proc. Lun. Sci. Conf. 6:* 1539.

———. 1976. "Where Do the Meteorites Come From? A Reevaluation of the Earth-crossing Apollo Objects as Sources of Chondritic Meteorites." *Geochim. Cosmochim. Acta 40:* 1297.

———. 1977a. "Evolution of the Earth's Planetesimal Swarm Subsequent to the Formation of the Earth and Moon." *Proc. Lun. Sci. Conf. 8:* 1.

———. 1977b. "Pre-mare Cratering and Early Solar System History." In *The Soviet-American Conference on Cosmochemistry of the Moon and Planets,* ed. J. Pomeroy and N. Hubbard. Washington: NASA SP-378.

Whipple, F. L. 1966. "Chondrules: Suggestion Concerning the Origin." *Science 153:* 54.

Whitaker, E. 1978. "Galileo's Lunar Observations and the Dating of the Composition of Sidereus Nuncius." *J. Hist. Astron. 9:* 155.

Wiik, H. B. 1956. "The Chemical Composition of Some Stony Meteorites." *Geochim. Cosmochim. Acta 9:* 279.

Wilhelms, D. E., C. Hodges, and R. Pike. 1977. "Nested-Crater Model of Lunar Ringed Basins." In *Impact and Explosion Cratering,* ed. D. Roddy, and others. New York: Pergamon Press.

Wilkening, L. 1977. "Meteorites in Meteorites: Evidence for Mixing among the Asteroids." In *Comets, Asteroids, Meteorites,* ed. A. H. Delsemme. Toledo, Ohio: University of Toledo Press.

——— . 1978. "Carbonaceous Chondritic Material in the Solar System." *Naturwissenschaften 65:* 73.

Williams, J. G. 1973. "Meteorites from the Asteroid Belt?" *EOS, Trans. Am. Geophys. Union, 54:* 233 (abstract).

Wood, C. A. 1963. "Physics and Chemistry of Meteorites." In *The Moon, Meteorites, and Comets,* ed. B. M. Middlehurst and G. P. Kuiper. Chicago: University of Chicago Press.

Wood, J. A. 1968. *Meteorites and the Origin of Planets.* New York: McGraw-Hill.

Wood, J. A., and H. McSween. 1977. "Chondrules as Condensation Products." In *Comets, Meteorites, Asteroids,* ed. A. H. Delsemme. Toledo, Ohio: University of Toledo Press.

CHAPTER SEVEN

Anders, E. 1964. "Origin, Age, and Composition of Meteorites." *Space Sci. Rev. 4:* 583.

——— . 1965. "Fragmentation History of Asteroids." *Icarus 4:* 399.

Arnold, J. R. 1965. "The Origin of Meteorites as Small Bodies." *Astrophys. J. 141:* 1536.

——— . 1970. "Asteroid Families and Jet Streams." *Astron. J. 14:* 1235.

Binzel, Richard. In press. "Trojan, Hilda, and Cybele Asteroids: New Light Curve Observations and a Bias-Corrected Analysis." *Icarus.*

Binzel, R. P., and L. M. Sauter. 1992. "Trojan, Hilda, and Cybele Asteroids: New Lightcurve Observation and Analysis." *Icarus 95:* 222.

Binzel, R. P., and T. van Flandern. 1979. "Minor Planets: The Discovery of Minor Satellites." *Science 203:* 903.

Bowell, E., C. Chapman, J. Gradie, D. Morrison, and B. Zellner. 1978. "Taxonomy of Asteroids." *Icarus 35:* 313.

Brouwer, D. 1951. "Secular Variations of the Orbital Elements of Minor Planets." *Astron. J. 56:* 9.

Chapman, C. R. 1975. "The Nature of Asteroids." *Sci. Am.,* January, p. 24.

——— . 1979. "The Asteroids: Nature, Interrelations, Origin, and Evolution." In *Asteroids,* ed. T. Gehrels. Tucson: University of Arizona Press.

Chapman, C. R., and D. Davis. 1975. "Asteroid Collisional Evolution: Evidence for a Much Larger Early Population." *Science 190:* 553.

Chapman, C. R., T. McCord, and C. Pieters. 1973. "Minor Planets and Related Objects X. Spectrophotometric Study of the Composition of 1685 Toro." *Astron. J. 78:* 502.

Chapman, C. R., J. Williams, and W. Hartmann. 1978. "The Asteroids." *Ann. Rev. Astron. Astrophys. 16:* 33.

Clark, R. N. 1980. "The Spectral Reflectance of Water-Mineral Mixtures at Low Temperature." *J. Geophys. Res. 86:* 3074.

Criswell, D. R., ed. 1978. *New Moons.* Houston: Lunar and Planetary Institute.

Cruikshank, D., D. Tholen, W. Hartmann, J. Bell, and R. Brown. 1991. "Three Basaltic Earth-Approaching Asteroids and the Source of the Basaltic Meteorites." *Icarus 89:* 1.

Davis, D. R., C. Chapman, R. Greenberg, S. Weidenschilling, and A. Harris. 1979. "Collisional Evolution of Asteroids: Populations, Rotations, and Velocities." In *Asteroids,* ed. T. Gehrels. Tucson: University of Arizona Press.

Degewij, J. 1977. "Lightcurve Analyses for 170 Small Asteroids." *Proc. Lun. Sci. Conf. 8:* 145.

Degewij, J., J. Gradie, T. Lebertre, W. Wisniewski, and B. Zellner. 1977. Abstract from 150th meeting, American Astronomy Society.

Dunlap, J. L., and T. Gehrels. 1969. "Minor Planets III: Lightcurves of a Trojan Asteroid." *Astron. J. 74:* 797.

Everhart, E. 1979. "Chaotic Orbits in the Solar System." In *Asteroids,* ed. T. Gehrels. Tucson: University of Arizona Press.

Feierberg, M., H. Larson, and C. Chapman. 1981. "Spectroscopic Evidence for Undifferentiated S-type Asteroids." Submitted to *Astrophys. J.*

Gaffey, M. J., and T. McCord. 1977. "Mining the Asteroids." *Mercury 6,* no. 6, p. 1.

——— . 1978. "Asteroid Surface Materials: Mineralogical Characterizations from Reflectance Spectra." *Space Sci. Rev. 21:* 555.

Gault, D. E., E. M. Shoemaker, and H. J. Moore. 1963. "Spray Ejected from the Lunar Surface by Meteoroid Impact." NASA Technical Note Dd 767.

Gradie, J., and E. Tedesco. 1982. "Compositional Structure of the Asteroid Belt." *Science 216:* 1405.

Gradie, J., and J. Veverka. 1980. "The Composition of the Trojan Asteroids." *Nature 283:* 840.

Harris, A. W. 1977. "An Analytical Theory of Planetary Rotation Rates." *Icarus 31:* 168.

Harris, A. W., and J. A. Burns. 1979. "Asteroid Rotation I. Tabulation and Analysis of Data." *Icarus 40:* 115.

Hartmann, W. K. 1969a. "Angular Momentum of Icarus." *Icarus 10:* 445.

———. 1969b. "Terrestrial, Lunar, and Interplanetary Rock Fragmentation." *Icarus 10:* 201.

Hartmann, W. K. 1979a. "A Special Class of Planetary Collisions: Theory and Evidence." *Proc. Lun. Planet. Sci. Conf. 10:* 1897.

———. 1979b. "Diverse Puzzling Asteroids and a Possible Unified Explanation." In *Asteroids*, ed. T. Gehrels. Tucson: University of Arizona Press.

Hartmann, W. K., and D. P. Cruikshank. 1978. "The Nature of Trojan Asteroid 624 Hektor." *Icarus 36*: 353.

———. 1980. "Hektor: The Largest Highly Elongated Asteroid." *Science 207:* 976.

Hartmann, W. K., and A. C. Hartmann. 1968. "Asteroid Collisions and Evolution of Asteroidal Mass Distribution and Meteoritic Flux." *Icarus 8:* 361.

Hartmann, W. K., R. Miller, and Pamela Lee. 1984. *Out of the Cradle: Exploring the Frontiers beyond Earth.* New York: Workman Publishing Co.

Hartmann, W., D. Tholen, and D. Cruikshank. 1987. "The Relationship of Active Comets, 'Extinct' Comets, and Dark Asteroids." *Icarus 69:* 33.

Hartmann, W., D. Tholen, J. Goguen, R. Binzel, and D. Cruikshank. 1988. "Trojan and Hilda Asteroid Lightcurves 1." *Icarus 73:* 487.

Hartmann, W., D. Tholen, Karen Meech, and D. Cruikshank. 1990. "2060 Chiron: Colorimetry and Cometary Behavior." *Icarus 83:* 1.

Hartmann, W. K., and D. J. Tholen. 1990. "Comet Nuclei and Trojan Asteroids, A New Link and a Possible Mechanism for Comet Splitting." *Icarus 86:* 448.

Helin, E. F., and E. Shoemaker. 1979. "The Palomar Planet-Crossing Asteroid Survey, 1973–1978." *Icarus 40:* 321.

Housen, K. R., L. L. Wilkening, C. R. Chapman, and R. Greenberg. 1979. "Asteroid Regoliths." *Icarus 39:* 317.

Kuiper, G. P., Y. Fujita, T. Gehrels, I. Groeneveld, J. Kent, G. van Biesbroeck, and C. van Houten. 1958. "Survey of Asteroids." *Astrophys. J. Suppl. 3:* 289.

Lebofsky, L. A. 1977. "Asteroid 1 Ceres: Evidence for Water of Hydration." *Monthly Notices Roy. Astron. Soc. 82:* 17.

Lebofsky, L. A., and others. 1982. "The 1.7 to 4.2-μm Spectrum of Asteroid 1 Ceres: Evidence for Structural Water in Clay Minerals." *Icarus 41:* in press.

Morrison, D., and W. C. Wells. 1978. *Asteroids: An Exploration Assessment.* Washington, D.C.: NASA CP-2053.

Mueller, Beatrice, D. Tholen, W. Hartmann, and D. Cruikshank. 1992. "Extraordinary Colors of Asteroidal Object (5145) 1992 AD." Submitted to *Icarus.*

Öpik, E. J. 1963. "The Stray Bodies in the Solar System I. Survival of Cometary Nuclei and the Asteroids." *Adv. Astron. Astrophys. 2:* 219.

Ostro, S. J., and 5 others. 1990. "Radar Images of Asteroid 1989 PB." *Science 248:* 1523.

Piotrowski, S. 1953. "The Collisions of Asteroids." *Acta Astron. A5:* 115.

Ryan, E., W. Hartmann, D. Davis. 1991. "Impact Experiments II. Catastrophic Fragmentation of Aggregate Targets and Relation to Asteroids." *Icarus 94:* 283–298.

Scholl, H. 1979. "History and Evolution of Chiron's Orbit." *Icarus 40:* 345.

Shoemaker, E., J. Williams, E. Helin, and R. Wolfe. 1979. "Earth-Crossing Asteroids: Orbital Classes, Collision Rates with Earth, and Origin." In *Asteroids*, ed. T. Gehrels. Tucson: University of Arizona Press.

Steir, M. T., W. Traub, G. Fazio, E. Wright, and F. Low. 1978. "Far-IR Observations of Uranus, Neptune, and Ceres." *Astrophys. J. 226:* 347.

Taylor, R. C. 1979. "Pole Orientations of Asteroids." In *Asteroids*, ed. T. Gehrels. Tucson: University of Arizona Press.

Tedesco, E. 1979a. "Binary Asteroids: Evidence for Their Existence from Lightcurves." *Science 203:* 905.

———. 1979b. "Lightcurve Parameters of Asteroids." In *Asteroids*, ed. T. Gehrels. Tucson: University of Arizona Press.

———. 1979c. "The Origin of the Flora Family." *Icarus 40:* 375.

Tedesco, E., and J. Gradie, 1981. "Composition of Outer Belt Asteroids: Implications for Comets." Submitted to *Icarus.*

Tedesco, E., and others. 1978. "1580 Betulia: An Unusual Asteroid with an Extraordinary Lightcurve." *Icarus 35:* 344.

van Houten, C. J., I. van Houten-Groeneveld, and T. Gehrels. 1970. "Minor Planets and Related Objects V. The Density of Trojans near the Preceding Lagrangian Point." *Astron. J. 75:* 659.

Veverka, J., P. Thomas, and T. Duxbury. 1978. "The Puzzling Moons of Mars." *Sky and Telescope 56:* 186.

Wasson, J. T. 1974. *Meteorites.* New York: Springer-Verlag.

Weidenschilling, S. J. 1980. "Hektor: Nature and Origin of a Binary Asteroid." *Icarus 44*: 807.

Weidenschilling, S. J. 1981. "How Fast Can an Asteroid Spin?" *Icarus 46:* 124.

Wetherill, G. 1977. "Evolution of the Earth's Planetesimal Swarm Subsequent to the Formation of the Earth and Moon." *Proc. Lun. Sci. Conf. 8:* 1.

———. 1979. "Apollo Objects." *Sci. Am. 240:* March, p. 54.

Wiesel, W. 1978. "Fragmentation of Asteroids and Artificial Satellites in Orbit." *Icarus 34:* 99.

Zellner, B. 1979. "Asteroid Taxonomy and the Distribution of the Compositional Types." In *Asteroids*, ed. T. Gehrels. Tucson: University of Arizona Press.

Zellner, B., M. Leake, D. Morrison, and J. Williams. 1977. "The E Asteroids and the Origin of the Enstatite Achondrites." *Geochim. Cosmochim. Acta 41:* 1759.

CHAPTER EIGHT

Ballabh, G., A. Bhatnagar, and N. Bhandari. 1978. "The Orbit of the Dhajala Meteorite." *Icarus 33:* 361.

Biermann, L. 1951. "Komentenschweife und solare Korpuskularstrahlung." *Z. Astrophys. 29:* 274.

Brandt, John C. 1990. "Comets." In *The New Solar System.* Cambridge, Mass: Sky Publishing Corp.

Brandt, J. C., and R. D. Chapman. 1981. *Introduction to Comets.* Cambridge: Cambridge University Press.

Brower, K. 1978. *The Starship and the Canoe.* New York: Holt, Rinehart & Winston.

Brownlee, D., R. Rajan, and D. Tomandl. 1977. "A Chemical and Textural Comparison between Carbonaceous Chondrites and Interplanetary Dust." In *Comets, Asteroids, Meteorites*, ed. A. Delsemme. Toledo, Ohio: University of Toledo Press.

Cameron, A. G. W. 1973. "Accumulation Processes in the Primitive Solar Nebula." *Icarus 18*: 407.

Ceplecha, A., R. Rajche, and L. Sehnal. 1959. "New Czechoslovak Meteorite, Luhy." *Bull. Astron. Inst. Czech. 10:* 147.

Criswell, D. R., ed. 1977. *New Moons.* Houston: Lunar and Planetary Institute.

Davies, J. G., and A. Lovell. 1955. "The Giacobinid Meteor Stream." *Monthly Notices Roy. Astron. Soc. 15:* 23.

Davies, J. G., and W. Turski. 1962. "The Formation of the Giacobinid Meteor Stream." *Monthly Notices of the Roy. Astron. Soc. 123:* 459.

Delsemme, A. H. 1977. "The Pristine Nature of Comets." In *Comets, Asteroids, Meteorites*, ed. A. Delsemme. Toledo, Ohio: University of Toledo Press.

Everhart, E. 1977. "The Evolution of Comet Orbits as Perturbed by Uranus and Neptune." In *Comets, Asteroids, Meteorites*, ed. A. Delsemme. Toledo, Ohio: University of Toledo Press.

Fay, T., and W. Wisniewski. 1978. "The Light Curve of the Nucleus of Comet d'Arrest." *Icarus 34:* 1.

Fayet, G. 1910. "Recherches concernant les excentricités des comètes." *Paris Mém. 26:* A.1 a A.134.

Fernández, J. A. 1978. "Mass Removed by the Outer Planets in the Early Solar System." *Icarus 34:* 173.

Hartmann, W., D. Cruikshank, and J. Degewij. 1982. "Remote Comets and Related Bodies: VJHK Colorimetry and Surface Materials." *Icarus*, in press.

Hartmann, W. K., and D. J. Tholen. 1990. "Comet Nuclei and Trojan Asteroids, A New Link and a Possible Mechanism for Comet Splitting." *Icarus 86:* 448.

Humes, D. H., J. Alvarez, R. O'Neal, and W. Kinard. 1974. "The Interplanetary and Near-Jupiter Meteoroid Environments." *J. Geophys. Res. 79:* 3677.

Jacchia, L. G., F. Verniani, and R. E. Briggs. 1967. "An Analysis of 413 Precisely Reduced Photographic Meteors." *Smithsonian Contrib. Astrophys. 10*, no. 1.

Kresak, L. 1977. "Asteroid vs. Comet Discrimination from Orbital Data." In *Comets, Asteroids, Meteorites*, ed. A. Delsemme. Toledo, Ohio: University of Toledo Press.

Krinov, E. L. 1963. "The Tunguska and Sikhote-Alin Meteorites." In *The Moon, Meteorites, and Comets*, eds. B. M. Middlehurst and G. P. Kuiper. Chicago: University of Chicago Press.

Lebofsky, L. A. 1975. "Stability of Frosts in the Inner Solar System." *Icarus 25:* 205.

Levin, B. J., A. Simonenko, and E. Anders. 1976. "Farmington Meteorite: A Fragment of an Apollo Asteroid?" *Icarus 28:* 307.

McCrosky, R. E. 1967. "Orbits of Photographic Meteors." *Smithsonian Astrophys. Observ. Spec. Rept. 252.*

McLean, D. R. 1967. "The Leonid Meteor Shower of November 17, 1966." *Comm. Lunar Planet. Lab. 6:* 43.

Millman, P. M. 1979. "Interplanetary Dust." *Naturwissenschaften 66:* 134.

Niven, L., and J. Pournelle. 1977. *Lucifer's Hammer.* New York: Fawcett Crest.

Oberg, J. 1977. "Tunguska: Collision with a Comet." *Astronomy 5:* 18.

Oort, J. H. 1950. "The Structure of the Cloud of Comets Surrounding the Solar System and a Hypothesis Concerning its Origin." *Bull. Astron. Inst. Netherlands 11:* 91.

———. 1963. "Empirical Data on the Origin of Comets." In *The Moon, Meteorites, and Comets*, ed. B. Middlehurst and G. Kuiper. Chicago: University of Chicago Press.

Öpik, E. J. 1963. "The Stray Bodies in the Solar System I. Survival of Cometary Nuclei and the Asteroids." *Adv. Astron. Astrophys. 2:* 219.

———. 1966. "The Stray Bodies in the Solar System II. The Cometary Origin of Meteorites." *Adv. Astron. Astrophys. 4:* 301.

Oppenheimer, M. 1978. "What Are Comets Made Of?" *Natural History*, March, p. 42.

Pohn, H. A. 1966. "Observations of a Double Nucleus in Comet Ikeya-Seki." *Sky and Telescope 31:* 376.

Richter, N. B. 1978. "Periodic Comet Schwassmann-Wachmann I: One of the Most Interesting and Puzzling Objects of Our Solar System." International Astronomical Union Commission 15, *Circular Letter* no. 1.

Ridpath, I. 1977. "The Tunguska Mystery—Solved?" *Astronomy 5:* 22 (Dec.).

Russell, H. N. 1920. "On the Origin of Periodic Comets." *Astron. J. 33:* 49.

Sagdeev, R. V., and others. 1986. "Encounters with Comet Halley—The First Results." *Nature 321:* 259.

Schiaparelli, G. B. 1866. *Bull. Meteorol. Observ. Coll. Romano 5:* 10.

Sekanina, Z. 1977. "Differential Nongravitational Forces in the Motions of the Split Comets." In *Comets, Meteorites, and Asteroids*, ed. A. Delsemme. Toledo, Ohio: University of Toledo Press.

Soberman, R., S. Neste, and K. Lichtenfeld. 1974. "Optical Measurement of Interplanetary Particulates from Pioneer 10." *J. Geophys. Res. 79:* 3685.

Stromgren, E. 1914. "Über den Ursprung der Kometen." *Publ. Copenhagen Observ. 19:* 62.

Van Flandern, T. 1977. "A Former Major Planet of the Solar System." In *Comets, Asteroids, Meteorites*, ed. A. Delsemme. Toledo, Ohio: University of Toledo Press.

Van Woerkom, A. J. J. 1948. "On the Origin of Comets." *Bull. Astron. Inst. Neth. 10:* 445.

Verniani, F. 1969. "Structure and Fragmentation of Meteoroids." *Space Sci. Rev. 10:* 230.

Vsekhsvyatsky, S. K. 1977. "Comets and the Cosmogony of the Solar System." In *Comets, Asteroids, Meteorites*, ed. A. Delsemme. Toledo, Ohio: University of Toledo Press.

Watson, K., B. C. Murray, and H. Brown. 1963. "The Stability of Volatiles in the Solar System." *Icarus 1:* 317.

Weissman, Paul R. 1985. "The Origin of Comets: Implications for Planetary Formation." In *Protostars and Planets II*, ed. D. Black and M. Matthews. Tucson: University of Arizona Press.

Whipple, F. 1950. "A Comet Model I. The Acceleration of Comet Encke." *Astrophys. J. 111:* 375.

———. 1963. "On the Structure of the Cometary Nucleus." In *The Moon, Meteorites, and Comets*, ed. B. M. Middlehurst and G. P. Kuiper. Chicago: University of Chicago Press.

———. 1967. "On Maintaining the Meteoritic Complex." In *Zodiacal Light and the Interplanetary Medium*, ed. J. Weinberg. Washington, D.C.: NASA SP-150.

———. 1975. "Do Comets Play a Role in Galactic Chemistry and γ-ray Bursts?" *Astron. J. 80:* 525.

———. 1980. "The Spin of Comets." *Sci. Am.*, March, p. 124. (Reprinted in *Comets*, ed. J. Brandt. San Francisco: W. H. Freeman.)

CHAPTER NINE

Anderson, D. L., and others. 1977. "Seismology on Mars." *J. Geophys. Res. 82:* 4524.

Bacon, F. 1620. *Novum Organum*. London.

Basu, A. R., and others. 1981. "Eastern Indian 3800-My-old Crust and Early Mantle Differentiation." *Science 212:* 1502.

Bazilevskiy, A. T. 1989. "The Planet Next Door." *Sky and Telescope*, April, p. 360.

Beatty, J. K. 1982. "Venus: The Mystery Continues." *Sky and Telescope 63:* 134.

Ben-Avraham, Z. 1981. "The Movement of Continents." *Am. Scientist 69:* 291.

Carey, W. 1962. "The Scale of Geotectonic Phenomena." *J. Geol. Soc. India 3:* 97.

Cole, G. H. A. 1978. *The Structure of Planets.* New York: Crane, Russak.

Consolmagno, G., and J. Lewis. 1977. "Preliminary Thermal History Models of Icy Satellites." In *Planetary Satellites*, ed. J. Burns. Tucson: University of Arizona Press.

———. 1978. "The Evolution of Icy Satellite Interiors and Surfaces." *Icarus 34:* 280.

Cox, K. G. 1978. "Kimberlite Pipes." *Sci. Am.*, April, p. 120.

Cruikshank, D. P. 1979. "The Surfaces and Interiors of Saturn's Satellites." *Rev. Geophys. Spa. Phys. 17:* 165.

DeMarcus, W. C. 1958. "The Constitution of Jupiter and Saturn." *Astron. J. 63:* 2.

Dewey, J. F., and J. Bird. 1970. "Mountain Belts and the New Global Tectonics." *J. Geophys. Res. 75:* 2625.

Erickson, E., D. Goorvitch, J. Simpson, and D. Strecker. 1978. "Far Infrared Spectrophotometry of Jupiter and Saturn." *Icarus 35:* 61.

Fairbridge, R. W. 1972. *The Encyclopedia of Geochemistry and Environmental Sciences.* New York: Van Nostrand Reinhold.

Flint, R. F., and B. J. Skinner. 1977. *Physical Geology*, 2nd ed. New York: Wiley.

Frey, H. 1977. "Origin of the Earth's Ocean Basins." *Icarus 32:* 235.

Fricker, P., R. Reynolds, and A. Summers. 1974. "On the Thermal Evolution of the Terrestrial Planet." *Moon 9:* 211.

Goodwin, A. M. 1976. "Giant Impacting and the Development of Continental Crust." In *The Early History of the Earth*, ed. B. F. Windley. New York: Wiley.

Grinspoon, D. H. 1987. "Was Venus Wet? Deuterium Reconsidered." *Science 238:* 1702.

Gutenberg, B. 1926. "Untersuchungen zur Frage, bis zu welcher tiefe die erde Kristallin ist." *Z. Geophysik 2:* 24.

Hanel, R., and others. 1979. "Infrared Observations of the Jovian System from Voyager 1." *Science 204:* 972.

Herbert, F., M. Drake, and C. Sonett. 1978. "Geophysical and Geochemical Evolution of the Lunar Magma Ocean." *Proc. Lun. Sci. Conf. 9:* 249.

Herbert, F., and C. Sonett. 1979. "Electromagnetic Heating of Minor Planets in the Early Solar System." In *Asteroids*, ed. T. Gehrels. Tucson: University of Arizona Press.

Hooke, R. 1705. "Lectures and Discourses on Earthquakes." In *Posthumous Works*. London.

Hsui, A., and M. Toksöz. 1977. "Thermal Evolution of Planetary Size Bodies." *Proc. Lun. Sci. Conf. 8:* 447.

Hubbard, W. 1970. "Structure of Jupiter: Chemical Composition, Contraction and Rotation." *Astrophys. J. 162:* 687.

———. 1977. "The Jovian Surface Condition and Cooling Rate." *Icarus 30:* 305.

———. 1981. "Interiors of the Giant Planets." *Science 214:* 145.

———. 1990. "Interiors of the Giant Planets." In *The New Solar System*, ed. J. Beatty and A. Chaikin. Cambridge, Mass.: Sky Publishing Corp.

Hurley, P. M. 1968. "The Confirmation of Continental Drift." *Sci. Am.*, April, p. 52.

Jacobsen, S. B., and G. Wasserburg. 1979. "The Mean Age of the Mantle and Crustal Reservoirs." *J. Geophys. Res. 84:* 7411.

Jeffreys, H. 1924. "On the Internal Constitution of the Earth." *Monthly Notices Roy. Astron. Soc. 84:* 534.

———. 1962. *The Earth.* New York: Cambridge University Press.

Kaula, W. M. 1968. *An Introduction to Planetary Physics.* New York: Wiley.

———. 1990. "Venus: A Contrast in Evolution to Earth." *Science 247:* 1191.

Kuiper, G. P. 1952. "Planetary Atmospheres and Their Origin." In *The Atmospheres of the Earth and Planets*, ed. G. Kuiper. Chicago: University of Chicago Press.

Latham, G. V. 1971. "Lunar Seismology." *EOS, Trans. Am. Geophys. Union 52:* 162.

Lewis, J. S. 1971a. "Satellites of the Outer Planets: Their Physical and Chemical Nature." *Icarus 15:* 174.

———. 1971b. "Satellites of the Outer Planets: Thermal Models." *Science 172:* 1127.

Lupo, M. J., and J. Lewis. 1979. "Mass-Radius Relationships in Icy Satellites." *Icarus 40:* 157.

Marsh, B. D. 1979. "Island-Arc Volcanism." *Am. Scientist 67:* 161.

Masursky, H., E. Eliason, R. Jordan, G. Pettingill, P. Ford, and G. McGill. 1979. "Pioneer Venus Radar Observations—The First Year." *Bull. Amer. Astron. Soc. 11:* 549 (abstract.)

Masursky, H., and others. 1980. "Pioneer Venus Radar Results: Altimetry and Surface Properties." *J. Geophys. Res. 85:* 8232.

Menard, H. W. 1969. "The Deep-Ocean Floor." *Sci. Am.*, September, p. 127.

Mohorovičić, A. 1909. *Jb. Met. Observ. Agram. 9:* 1. (Zagreb, Croatia).

Morrison, D. A., and J. Warner. 1978. "Planetary Tectonics II. A Gravitational Effect." *Lun. Planet. Sci. Abstracts 9:* 769.

Muller, P. M., and W. L. Sjogren. 1968. "Mascons: Lunar Mass Concentrations." *Science 161:* 680.

Murthy, V. R., and H. T. Hall. 1972. "The Origin and Chemical Composition of the Earth's Core." *Phys. Earth Planet. Interiors 6:* 123.

Nierenberg, W. A. 1978. "The Deep Sea Drilling Project After Ten Years." *Am. Scientist 66:* 20.

Oldham, R. D. 1900. *Phil. Trans. Roy. Soc. London A194:* 135.

———. 1906. "The Constitution of the Earth." *Quart. J. Geol. Soc. 62:* 456.

Öpik, E. J. 1962. "Jupiter: Chemical Composition, Structure, and Origin of a Giant Planet." *Icarus 1:* 200.

Owen, N. B., and J. E. Martin. 1966. "The Selection, Orientation, and Mounting of Diamonds for Use as Bridgman Anvils." *J. Sci. Instr. 43:* 197.

Parker, E. N. 1970. "The Origin of Magnetic Fields." *Astrophys. J. 160:* 383.

Peale, S. J., P. Cassen, and R. Reynolds. 1979. "Melting of Io by Tidal Dissipation." *Science 203:* 892.

———. 1980. "Tidal Dissipation, Orbital Evolution, and the Nature of Saturn's Inner Moons." *Icarus 43:* 65.

Phillips, R., W. Kaula, G. McGill, and M. Malin. 1981. "Tectonics and Evolution of Venus." *Science 212:* 879.

Phillips, R., R. Grimm, and M. Malin. 1991. "Hot-Spot Evolution and the Global Tectonics of Venus." *Science 252:* 651.

Podolak, M. 1978. "Models of Saturn's Interior: Evidence for Phase Separation." *Icarus 33:* 342.

Podolak, M., and A. Cameron. 1974. "Possible Formation of Meteoritic Chondrules and Inclusions in the Precollapse Jovian Protoplanetary Atmosphere." *Icarus 22:* 123.

Podolak, M., and R. Reynolds. 1981. "On the Structure and Composition of Uranus and Neptune." *Icarus 46:* 40.

Poisson, S. D. 1829. *Mem. Acad. Sci. Paris 8:* 623.

Pollack, James B. 1985. "Formation of the Giant Planets and Their Satellite-Ring Systems: An Overview." In *Protostars and Planets II,* ed. D. Black and M. Matthews. Tucson: University of Arizona Press.

Ramsey, W. H. 1967. "On the Constitutions of Uranus and Neptune." *Planet. Space Sci. 15:* 1609.

Rayleigh, Lord (J. W. Strutt). 1887. "On Waves Propagated along the Plane Surface of an Elastic Solid." *Proc. London Math. Soc. (1), 17:* 4.

Reasenberg, R. D. 1977. "The Moment of Inertia and Isostasy of Mars." *J. Geophys. Res. 82:* 369.

Ringwood, A. E. 1956. "The Olivine-Spinel Transition in the Earth's Mantle." *Nature 178:* 1303.

Ringwood, A. E., and A. Major. 1966. "High-Pressure Transformations in Pyroxenes." *Earth Planet. Sci. Lett. 1:* 351.

Ritsema, A. R. 1954. "A Statistical Study of the Seismicity of the Earth." *Meterol. Geophys. Serv., Indonesia, Verhandl. 46.*

Saunders, R. S., and M. Malin. 1977. "Geologic Interpretation of New Observations of the Surface of Venus." *Geophys. Res. Lett. 4:* 547.

Scarf, F. L., and others. 1981. "Jupiter Tail Phenomena Upstream from Saturn." *Nature 292:* 585.

Schubert, G., D. Stevenson, and K. Ellsworth. 1981. "Internal Structures of the Galilean Satellites." *Icarus 47:* 46.

Sharp, R. P. 1971. "The Surface of Mars 2. Uncratered Terrains." *J. Geophys. Res. 76:* 331.

Sinton, W. M. 1979. "Io: Confirmation of 5-μm Outbursts and a Possible Explanation." *Bull. Amer. Astron. Soc. 11:* 585 (abstract).

Slattery, W., W. DeCampli, and A. Cameron. 1980. "Protoplanetary Core Formation by Rainout of Minerals." *Moon and Planets 23:* 381.

Smith, B. A. 1979. "Crustal Structure and Volcanic Mechanisms." *Bull. Amer. Astron. Soc. 11:* 585 (abstract).

Smith, B. A., and others (Voyager Imaging Team). 1979. "The Jupiter System through the Eyes of Voyager I." *Science 204:* 951.

Smith, B. A. and others. 1982. "A New Look at the Saturn System: The Voyage 2 Images." *Science* 215: 504.

Solomon, S. 1976. "Some Aspects of Core Formation in Mercury." *Icarus 28:* 509.

———. 1979. "Formation, History, and Energetics of Cores in Terrestrial Planets." *Phys. Earth Planet. Interiors 19:* 168.

Solomon, S., and J. Head. 1979. "Vertical Movement in Mare Basins." *J. Geophys. Res. 84:* 1667.

Sonett, C. P., D. Colburn, K. Schwartz, and K. Keil. 1970. "The Melting of Asteroidal-sized Bodies by Unipolar Dynamo Induction from a Primordial T Tauri Sun." *Astrophys. Spa. Sci. 7:* 446.

Sonett, C., B. Smith, D. Colburn, G. Schubert, and K. Schwartz. 1971. "Preliminary Assessment of the Lunar Lithospheric Thermal Gradient, Heat Flux, Deep Temperature and Compositional Gradation." Paper presented at the Lunar Science Conference, Houston.

Stevenson, D. J. 1981. "Models of the Earth's Core." *Science 214:* 611.

Stone, E. C., and E. Miner. 1989. "The Voyager 2 Encounter with the Neptunian System." *Science 246:* 1417.

Strom, R., N. Trask, and J. Guest. 1975. "Tectonism and Volcanism on Mercury." *J. Geophys. Res. 80:* 2478.

Taylor, D. J. 1965. "Spectrophotometry of Jupiter's 3400–10,000 Å Spectrum and a Bolometric Albedo for Jupiter." *Icarus 4:* 362.

Taylor, S. R. 1982. *Planetary Science: A Lunar Perspective.* Houston: Lunar and Planetary Institute.

Tazieff, H. 1964. *When the Earth Trembles.* London: Hart-Davis.

Toksöz, M., and A. Hsui. 1978. "Thermal History and Evolution of Mars." *Icarus 34:* 537.

Urey, H. C. 1952. *The Planets: Their Origin and Development.* New Haven: Yale University Press.

Van Allen, J. A. 1961. "The Geomagnetically Trapped Corpuscular Radiation." In *Science in Space,* eds. L. V. Berkner and H. Odishaw. New York: McGraw-Hill.

———. 1990. "Magnetospheres, Cosmic Rays, and the Interplanetary Medium." In *The New Solar System,* ed. J. Beatty and A. Chaikin. Cambridge, MA: Sky Publishing.

Vine, F. J., and D. Matthews. 1963. "Magnetic Anomalies over Ocean Ridges." *Nature 199:* 947.

Warner, J. L., and D. A. Morrison. 1978. "Planetary Tectonics I: The Role of Water." *Lun. Planet. Sci. Abstracts 9:* 1217.

Warren, P. H., and J. Wasson. 1979. "The Compositional-Petrographic Search for Pristine Nonmare Rocks, 3rd Foray." *Proc. Lunar Planet. Sci. Conf. 10:* 583.

Wasson, J. T. 1974. *Meteorites.* New York: Springer-Verlag.

Wildt, R. 1938. "On the State of Matter in the Interior of the Planets." *Astrophys. J. 85:* 508.

———. 1972. "Hydrogen Planets and High-Pressure Physics." *Phys. Earth Planet. Interiors 6:* 1.

Windley, B. F. 1976. "New Tectonic Models for the Evolution of Archaean Continents and Oceans." In *The Early History of the Earth,* ed. B. F. Windley. New York: Wiley.

Wood, J. A. 1977. "A Survey of Lunar Rock Types and Comparison of the Crusts of Earth and Moon." In *The*

Soviet-American Conference on Geochemistry of the Moon and Planets, ed. J. Pomeroy and N. Hubbard. Washington, D.C.: NASA SP-370.

Wu, S. C. 1978. "Mars Synthetic Topographic Mapping." *Icarus 33:* 417.

Wyllie, P. J. 1971. *The Dynamic Earth: Textbook in Geosciences.* New York: Wiley.

Yoder, C. F. 1979. "How Tidal Heating in Io Drives the Galilean Orbital Resonance Locks." *Nature 279:* 767.

CHAPTER TEN

Arnold, J. 1977. "Lunar versus Asteroidal Resources." In *New Moons*, ed. D. Criswell. Houston: Lunar and Planetary Institute.

Baldwin, R. B. 1949. *The Face of the Moon.* Chicago: University of Chicago Press.

———. 1963. *The Measure of the Moon.* Chicago: University of Chicago Press.

Bell, J. F., D. Cruikshank, and M. Gaffey. 1985. *Icarus 61:* 192.

Binder, A. B. 1977. "The Geology of the Viking 1 Lander Site." *J. Geophys. Res. 82:* 4439.

Bowen, N. L. 1928. "The Evolution of the Igneous Rocks." Princeton: Princeton University Press. Reprint. New York: Dover, 1956.

Burns, B. 1982. "Cratering Analysis of the Surface of Venus as Mapped by 12.6-cm Radar." Thesis (NAIC 167). Arecibo, P.R.: Natl. Astron. and Ionosph. Center.

Carr, M. H., and others. 1977. "Martian Impact Craters and Emplacement of Ejecta by Surface Flow." *J. Geophys. Res. 82:* 4055.

Chapman, C. R., and K. Jones. 1977. "Cratering and Obliteration History of Mars." *Ann. Rev. Earth Planet. Sci. 5:* 515.

Clark, B. C., and others. 1977. "The Viking X-Ray Fluorescence Experiment: Analytic Methods and Early Results." *J. Geophys. Res. 82:* 4577.

Clark, R. N. 1980. "Ganymede, Europa, Callisto, and Saturn's Rings: Compositional Analysis from Reflectance Spectroscopy." *Icarus 44:* 388.

Cruikshank, D. P. 1979. "The Surfaces and Interiors of Saturn's Satellites." *Rev. Geophys. Spa. Phys. 17:* 165.

Cruikshank, D. P., and P. M. Silvaggio. 1979. "Triton: A Satellite with an Atmosphere." *Astrophys. J. 233:* 1016.

———. 1980. "The Surface and Atmosphere of Pluto." *Icarus 41:* 96.

Cruikshank, D. P., and 6 others. 1983. "The Dark Side of Iapetus." *Icarus 53:* 90.

Cruikshank, D. P., and D. Morrison. 1990. "Icy Bodies of the Outer Solar System." In *The New Solar System*, ed. J. Beatty and A. Chaikin. Cambridge, Mass: Sky Publishing Corp.

Cruikshank, D. P., and 6 others. 1991. "Solid C≡N Bearing Material on Outer Solar System Body." *Icarus 94:* 345.

Delsemme, A. H. 1977. "The Pristine Nature of Comets." In *Comets, Asteroids, Meteorites*, ed. A. Delsemme. Toledo, Ohio: University of Toledo Press.

Fink, U., and others. 1980. "Detection of a CH_4 Atmosphere on Pluto." *Icarus 44:* 62.

Fink, U., and 4 others. 1992. "The Steep Red Spectrum of 1992AD: An Asteroid Covered with Organic Material?" Submitted to *Icarus.*

Ganapathy, R., and others. 1970. "Trace Elements in Apollo 11 Lunar Rocks: Implications for Meteorite Influx and Origin of the Moon." *Proc. Apollo 11 Lunar Sci. Conf. 1:* 1117.

Gault, D. E., E. D. Heitowit, and H. J. Moore. 1964. "Some Observations of Hypervelocity Impacts with Porous Media." In *The Lunar Surface Layer*, ed. J. W. Salisbury and P. E. Glaser. New York: Academic Press.

Geikie, A. 1905. *The Founders of Geology*, 2nd ed. New York: Macmillan. Reprint. New York: Dover.

Gibson, E. G. 1973. *The Quiet Sun.* Washington, D.C.: NASA SP-303.

Hapke, B. 1965. "Effects of a Simulated Solar Wind on the Photometric Properties of Rocks and Powders." *Ann. N.Y. Acad. Sci. 123:* 711.

Hargraves, R., and others. 1977. "The Viking Magnetic Properties Experiment: Primary Mission Results." *J. Geophys. Res. 82:* 4547.

Hartmann, W. K. 1966. "Martian Cratering." *Icarus 5:* 565.

———. 1973. "Ancient Lunar Mega-Regolith and Subsurface Structure." *Icarus 18:* 634.

Hartmann, W. K. 1982. *Astronomy: The Cosmic Journey*, 2nd ed. Belmont, CA.: Wadsworth.

———. 1980a. "Dropping Stones in Magma Oceans: Effects of Early Lunar Cratering." In *Proc. Conf. Lunar Highlands Crust*, ed. J. Papike and R. Merrill. New York: Pergamon Press.

———. 1980b. "Surface Evolution of Two-Component Stone/Ice Bodies in the Jupiter Region." *Icarus 44:* 441.

Hartmann, W. K., and G. P. Kuiper. 1962. "Concentric Structures Surrounding Lunar Basins." *Comm. Lunar Planet. Lab. 1:* 51.

Hartmann, W. K., and C. A. Wood. 1971. "Moon: Origin and Evolution of Multi-Ring Basins." *Moon 3:* 3.

Hartmann, W. K. and 10 others. 1981 "Chronology of Planetary Volcanism by Comparative Studies of Planetary Cratering." In *Basaltic Volcanism on the Terrestrial Planets*, ed. Basaltic Volcanism Study Project. New York: Pergamon Press.

Herbert, F., and others. 1977. "Some Constraints on the Thermal History of the Lunar Magma Ocean." *Proc. Lun. Sci. Conf. 8:* 573.

KenKnight, C., D. Rosenberg, and G. Wehner. 1967. "Parameters of the Optical Properties of the Lunar Surface Powder in Relation to Solar Wind Bombardment." *J. Geophys. Res. 72:* 3105.

Lofgren, G. E., and others. 1981. "Petrology and Chemistry of Terrestrial, Lunar, and Meteoritic Basalts." In *Basaltic Volcanism on the Terrestrial Planets*, ed. Basaltic Volcanism Study Project. New York: Pergamon Press.

Mason, B. 1958. *Principles of Geochemistry.* New York: Wiley.

———. 1962. *Meteorites.* New York: Wiley.

McCracken, C., and M. Dubin. 1964. "Dust Bombardment on the Lunar Surface." In *The Lunar Surface Layer*, ed. J. Salisbury and P. Glaser. New York: Academic Press.

Meuller, B., D. Tholen, W. Hartmann, D. Cruikshank. 1992. "Extraordinary Colors of Asteroid 5145." Submitted to *Icarus.*

Morrison, D. A., and J. L. Warner. 1978. "Planetary Tectonics II. A Gravitational Effect." *Lun. Planet. Sci. Abstracts 9.*

Mutch, T. A. 1970. *Geology of the Moon.* Princeton: Princeton University Press.

Mutch, T. A., R. Arvidson, A. Binder, E. Guinness, and E. Morris. 1977. "The Geology of the Viking Lander 2 Site." *J. Geophys. Res. 82:* 4452.

Nakamura, N., and M. Tatsumoto. 1977. "The History of the Apollo 17 Station 7 Boulder." *Proc. Lun. Sci. Conf. 8:* 2301.

Neukum, G., and others. 1975. "A Study of Lunar Impact Crater Size Distributions." *Moon 12:* 201.

Parmentier, E., and J. Head. 1981. "Viscous Relaxation of Impact Craters on Icy Planetary Surfaces: Determination of Viscosity Variation with Depth." *Icarus 47:* 100.

Pieters, Carle. 1982. "Copernicus Crater Central Peak: Lunar Mountain of Unique Composition." *Science 215:* 59.

Rittmann, A. 1962. *Volcanoes and Their Activity.* New York: Wiley.

Salisbury, J., and V. Smalley. 1964. "The Lunar Surface Layer." In *The Lunar Surface Layer*, ed. J. Salisbury and P. Glaser. New York: Academic Press, Inc.

Shoemaker, E. M. 1965. "Preliminary Analysis of the Fine Structure of the Lunar Surface." In *Ranger VII, part II, Experimenters' Analyses and Interpretations.* Pasadena, Calif.: Jet Propulsion Laboratory, JPL TR 32-700.

Shoemaker, E. M., and others. 1970. "Lunar Regolith at Tranquillity Base." *Science 167:* 452.

Shoemaker, E. M., and R. Hackman. 1962. "Stratigraphic Basis for a Lunar Timescale." In *The Moon*, ed. Z. Kopal and Z. Mikhailov. London: Academic Press.

Short, N. M. 1975. *Planetary Geology.* Englewood Cliffs, N.J.: Prentice-Hall.

Short, N. M., and M. Forman. 1972. "Thickness of Crater Impact Ejecta on the Lunar Surface." *Mod. Geol. 3:* 69.

Smith, B. A., and others (Voyager Imaging Team). 1979. "The Galilean Satellites of Jupiter: Voyager 2 Imaging Science Results." *Science 206:* 927.

———. 1982. "A New Look at the Saturn System: The Voyager 2 Images." *Science 215:* 504.

Smith, B. A., and 39 others. 1986. "Voyager 2 in the Uranian System: Imaging Science Results." *Science 233:* 43.

Soderblom, L., and L. Lebofsky. 1972. "Technique for Rapid Determination of Relative Ages of Lunar Areas from Orbital Photography." *J. Geophys. Res. 77:* 279.

Solomon, S. C., and J. Longhi. 1977. "Magma Oceanography 1. Thermal Evolution." *Proc. Lun. Sci. Conf. 8:* 583.

Soter, S. 1974. Paper presented at Satellite Colloquium, Cornell University, August.

Stone, E. C., and E. D. Miner. 1989. "The Voyager 2 Encounter with the Neptunian System." *Science 246:* 1417.

Strazzulla, G. 1986. "Organic Material from Phoebe to Iapetus." *Icarus 66:* 397.

Strom, R. B. 1987. "The Solar System Cratering Record: Voyager 2 Results at Uranus and Implications for the Origin of Impacting Objects." *Icarus 70:* 517.

Taylor, S. R. 1982. *Planetary Science: A Lunar Perspective.* Lunar and Planetary Science Institute.

Toulmin, T., and others. 1977. "Geochemical and Mineralogical Interpretation of the Viking Inorganic Chemical Results." *J. Geophys. Res. 82:* 4625.

Vortman, L. J. 1977. "Craters from Surface Explosions and Energy Dependence—A Retrospective View." In *Impact and Explosion Cratering*, ed. D. Roddy, R. Pepin, and R. Merrill. New York: Pergamon Press.

Warren, P. H., and Wasson, J. T. 1979. "Effects of Pressure on the Crystallization of a 'Chondritic' Magma Ocean and Implications for the Bulk Composition of the Moon." *Proc. Lun. Planet. Sci. Conf. 10:* 2051.

Wasson, J. T. 1974. *Meteorites.* New York: Springer-Verlag.

Whitaker, E. A. 1978. "Galileo's Lunar Observations and the Dating of the Composition of the Sidereus Nuncius." *J. Hist. Astron. 9:* 155.

Wilhelms, D. E., C. Hodges, and R. Pike. 1977. "Nested-Crater Model of Lunar Ringed Basins." In *Impact and Explosion Cratering*, ed. D. Roddy, R. Pepin, and R. Merrill. New York: Pergamon Press.

Williams, R. J. 1976. "Hydrogen Resources for the Moon." In *Lunar Utilization*, ed. D. Criswell. Houston: Lunar Science Institute.

Wood, J. A. 1977. "A Survey of Lunar Rock Types and Comparison of the Crusts of Earth and Moon." In *The Soviet-American Conference on Cosmochemistry of the Moon and Planets*, ed. J. Pomeroy and N. Hubbard. Washington, D.C.: NASA SP-370, p. 35.

Woronow, A. 1978. "A General Cratering History Model and Its Implications for the Lunar Highlands." *Icarus 34:* 76.

Wyllie, P. J. 1971. *The Dynamic Earth.* New York: Wiley.

CHAPTER ELEVEN

Bagnold, R. A. 1941. *The Physics of Blown Sand and Desert Dunes.* London: Chapman and Hall.

Baldwin, R. B. 1949. *The Face of the Moon.* Chicago: University of Chicago Press.

Binder, A. B. 1966. "Mariner IV: Analysis of Preliminary Photographs." *Science 152:* 1053.

Binzel, R. P. 1990. "Pluto." *Scientific Am. 262:* 50.

Breed, C. S. 1978. "Terrestrial Analogs of the Hellespontus Dunes, Mars." *Icarus 30:* 326.

Burns, J. A., ed. 1977. *Planetary Satellites.* Tucson: University of Arizona Press.

Carr, M., and G. Schaber. 1977. "Martian Permafrost Features." *J. Geophys. Res. 82:* 4039.

Carr, M, H. Masursky, R. Strom, and R. Terrile. 1979. "Volcanic Features of Io." *Nature 280:* 729.

Clark, B. C., and others. 1977. "The Viking X-ray Fluorescence Experiment: Analytical Methods and Early Results." *J. Geophys. Res. 82:* 4577.

Connes, P., J. Connes, W. Benedict, and L. Kaplan. 1967. "Traces of HCl and HF in the Atmosphere of Venus." *Astrophys. J. 147:* 1230.

Cruikshank, D. P. 1979. "The Surfaces and Interiors of Saturn's Satellites." *Rev. Geophys. Space Phys. 17:* 165.

Cruikshank, D., and P. Silvaggio. 1979. "Triton: A Satellite with an Atmosphere." *Astrophys. J. 233:* 1016.

———. 1980. "The Surface and Atmosphere of Pluto." *Icarus 41:* 2367.

Cruikshank, D. P., R. Brown, L. Giver, and A. Tokunaga. 1989. "Triton: Do We See to the Surface?" *Science 245:* 283.

Cutts, J. A., and others. 1976. "North Polar Region of Mars: Imaging Results from Viking 2." *Science 194:* 1329–1337.

Cutts, J. A., K. Blasius, and W. J. Roberts. 1979. "Evolution of Martian Polar Landscapes: Interplay of Long-Term Variations in Perennial Ice Cover and Dust Storm Intensity." *J. Geophys. Res. 84:* 2975.

Dermott, S., R. Malhotra, and C. Murray. 1988. "Dynamics of the Uranian and Saturnian Satellite Systems: A Chaotic Route for Melting Miranda?" *Icarus 76:* 295.

Eaton, J., and J. Murata. 1960. "How Volcanoes Grow." *Science 132:* 925.

Epstein, S., and H. P. Taylor, Jr. 1970. "$^{18}O/^{16}O$, $^{30}Si/^{28}Si$, D/H, and $^{13}C/^{12}C$ Studies of Lunar Rocks and Minerals." *Science 167:* 533.

Esposito, L. W. 1984. "Sulfur Dioxide: Episodic Injection Shows Evidence for Active Venus Volcanism." *Science 223:* 1072.

Fanale, F., T. Johnson, and D. Matson. 1974. "Io: A Surface Evaporite Deposit?" *Science 186:* 922.

Fanale, F., and J. Salvail. 1990. "Evolution of the Water Regime of Phobos." *Icarus 88:* 383.

Farmer, C. B., and P. Doms. 1979. "Global Seasonal Variation of Water Vapor on Mars and the Implications for Permafrost." *J. Geophys. Res. 84:* 2881.

Fielder, G. 1961. *Structure of the Moon's Surface.* Elmsford, N.Y.: Pergamon Press.

Goldreich, P., and 4 others. 1989. "Neptune's Story." *Science 245:* 500.

Gradie, J., P. Thomas, and J. Veverka. 1980. "The Surface Composition of Amalthea." *Icarus 44:* 373.

Greeley, R., and others. 1974. "Wind Tunnel Simulations of Light and Dark Streaks on Mars." *Science 183:* 847.

———. 1991. "Wind Streaks on Venus: Magellan Results." Abstracts, 23rd Meeting, Div. of Plan. Sci. of Am. Astr. Soc.: 122.

Greeley, R., R. Papson, and J. Veverka. 1978. "Crater Streaks in the Chryse Planetia Region of Mars: Early Viking Results." *Icarus 34:* 556.

Greenacre, J. A. 1963. "A Recent Observation of Lunar Color Phenomena." *Sky and Telescope 26:* 316. [See also: *Sky and Telescope 27:* 3.]

Greenberg, R., and R. Marcialis. 1987. Div. Planet. Sci., Am. Astron. Soc. Abstracts, p. 820.

Hanel, R., and others. 1979a. "Infrared Observations of the Jovian System from Voyager 2." *Science 206:* 952–956.

———. 1979b. "Infrared Observations of the Jovian System from Voyager 1." *Science 204:* 972–976.

———. 1982. "Infrared Observations of the Saturnian System from Voyager 2." *Science 215:* 544.

Hartmann, W. K. 1963. "Radial Structures Surrounding Lunar Basins I." *Comm. Lunar Planet. Lab. 2:* 1.

———. 1964. "Radial Structures Surrounding Lunar Basins II." *Comm. Lunar Planet. Lab. 2:* 175.

———. 1978. "Martian Cratering V. Toward an Empirical Martian Chronology, and Its Implications." *Geophys. Res. Letters 5:* 450.

———. 1980. "Surface Evolution of Two-Component Stone/Ice Bodies in the Jupiter Region." *Icarus 44:* 441.

———. 1990. "Additional Evidence about an Early Intense Flux of C Asteroids and the Origin of Phobos." *Icarus 87:* 236.

Hawke, B. R., P. Spudis, and P. Clark. 1985. "The Origin of Selected Lunar Geochemical Anomalies." *Earth, Moon, and Planets,* pp. 257–273.

Hess, S. L., and others. 1977. "Meteorological Results from the Surface of Mars: Viking 1 and 2." *J. Geophys. Res. 82:* 4559–4574.

Hunten, D. 1973. "The Escape of Light Gases from Planetary Atmospheres." *J. Atmosph. Sci. 30:* 1481.

Johnson, T. V., A. Cook, C. Sagan, and L. Soderblom. 1979. "Volcanic Resurfacing Rates and Implications for Volatiles on Io." *Nature 280:* 746.

Johnson, T. V., and 5 others. 1988. "Io: Evidence for Silicate Volcanism in 1986." *Science 242:* 1280.

Keldysh, M. V. 1977. "Venus Exploration with the Venera 9 and Venera 10 Spacecraft." *Icarus 30:* 605.

Khodakovsky, I., and others. 1979. "Venus: Preliminary Prediction of the Mineral Composition of Surface Rocks." *Icarus 39:* 352.

Kieffer, H. H. 1977. "Thermal and Albedo Mapping of Mars during the Viking Primary Mission." *J. Geophys. Res. 82:* 4249.

King, T., and 4 others. 1992. "Evidence for Ammonium-bearing Minerals on Ceres." *Science 255:* 1551.

Komar, P. D. 1979. "Comparisons of the Hydraulics of Water Flows in Martian Outflow Channels with Flows of Similar Scale on Earth." *Icarus 37:* 156.

Kozyrev, N. 1959. "Observation of a Volcanic Process on the Moon." *Sky and Telescope 18:* 184.

Lewis, J. S. 1968. "An Estimate of the Surface Conditions of Venus." *Icarus 8:* 434.

MacDonald, G. A. 1972. *Volcanoes.* Englewood Cliffs, N.J.: Prentice-Hall.

Marcialis, R., and R. Greenberg. 1987. "Warming of Miranda during Chaotic Rotation." *Nature 328:* 227.

Masursky, H., J. Boyce, A. Dial, G. Schaber, and M. Strobell. 1977. "Classification and Time of Formation of Martian Channels Based on Viking Data." *J. Geophys. Res. 82:* 4016.

Masursky, H., and others. 1979. "Pioneer Venus Radar Observations—The First Year." *Bull. Amer. Astron. Soc. 11:* 549.

———. 1980. "Pioneer Venus Radar Results: Geology from Images and Altimetry." *J. Geophys. Res. 85:* 8232.

McCauley, J., B. Smith, and L. Soderblom. 1979. "Erosional Scarps on Io." *Nature 280:* 736.

Middlehurst, B., and W. Chapman. 1968. "Tidal Cycles and Lunar Event Mechanisms." *Astron. J. 73:* 192.

Milton, D. J. 1967. "Slopes on the Moon." *Science 156:* 1135.

Morrison, D. 1977. "Radiometry of Satellites and of the Rings of Saturn." In *Planetary Satellites,* ed. J. A. Burns. Tucson: University of Arizona Press.

Morrison, D., D. Cruikshank, and J. Burns. 1977. "Introducing the Satellites." In *Planetary Satellites,* ed. J. A. Burns. Tucson: University of Arizona Press.

Mutch, T. A., and others. 1976. *The Geology of Mars.* Princeton: Princeton University Press.

———. 1977. "The Geology of the Viking Lander 2 Site." *J. Geophys. Res. 82:* 4452.

Orton, G. 1978. "Planetary Atmospheres." *Proc. Lunar Planet. Sci. Conf. 9:* endpaper.

Pang, K., J. Rhoads, A. Lane, and J. Ajello. 1980. "Spectral Evidence for a Carbonaceous Chondrite Surface Composition on Deimos." *Nature 283:* 277.

Peale, S. J., P. Cassen, and R. Reynolds. 1979. "Melting of Io by Tidal Dissipation." *Science 203:* 892.

Pettengill, G. H., and others. 1979. "Venus: Preliminary Topographic and Surface Imaging Results from the Pioneer Orbiter." *Science 205:* 90–93.

Phillips, R. J., and 6 others. 1991. "Impact Craters on Venus: Initial Analysis from Magellan." *Science 252:* 288.

Pieri, D. 1976. "Distribution of Small Channels on the Martian Surface." *Icarus 37:* 351.

Pollack, J. 1977. "Phobos and Deimos." In *Planetary Satellites,* ed. J. Burns. Tucson: University of Arizona Press.

Rittmann, A. 1962. *Volcanoes and Their Activity.* New York: Wiley.

Sagan, C. 1979. "Sulfur Flows on Io." *Nature 280:* 750.

Sagan, C., and J. B. Pollack. 1969. "Windblown Dust on Mars." *Nature 223:* 791.

Saunders, R. S., and M. Malin. 1977. "Geologic Interpretation of New Observations of the Surface of Venus." *Geophys. Res. Lett. 4:* 547.

Saunders, R. S., and 5 others. 1991. "An Overview of Venus Geology." *Science 252:* 249.

Schultz, P. H., and P. Spudis. 1979. "Evidence for Ancient Mare Volcanism." *Proc. Lun. Planet. Sci. Conf. 10:* 2899.

Seiff, A., and others. 1979. "Structure of the Atmosphere of Venus up to 110 Kilometers: Preliminary Results from the Four Pioneer Venus Entry Probes." *Science 203:* 787–790.

Sill, G. T. 1973. "Sulfuric Acid in the Venus Clouds." *Comm. Lunar Planet. Lab. 9:* 191.

Sinton, W. M. 1979. "Io: Confirmation of 5 μm Outbursts and a Possible Explanation." *Bull. Amer. Astron. Soc. 11:* 598 (abstract).

———. 1981. "The Thermal Emission Spectrum of Io and a Determination of the Heat Flux from Its Hot Spots." *J. Geophys. Res. 86:* 3122.

Smith, B. A., and others (Voyager Imaging Team). 1979a. "The Galilean Satellites and Jupiter: Voyager 2 Imaging Science Results." *Science 206:* 927.

———. 1979b. "The Role of SO_2 in Volcanism on Io." *Nature 280:* 738–743.

———. 1981. "Encounter with Saturn: Voyager 1 Imaging Science Results." *Science 212:* 163.

———. 1982. "A New Look at the Saturn System: The Voyager 2 Images." *Science 215:* 504.

Smith, B. A., and 39 others. 1986. "Voyager 2 in the Uranian System: Imaging Science Results." *Science, 233:* 43.

Smith, B. A., and 64 others. 1989. "Voyager 2 at Neptune: Imaging Science Results." *Science 246:* 1422.

Soter, S. 1971. "The Dust Belts of Mars." Ph.D. dissertation, Cornell University.

Squyres, S. W. 1979. "The Evolution of Dust Deposits in the Martian North Polar Region." *Icarus 40:* 244.

Strom, R. G. 1964. "Tectonic Maps of the Moon I." *Comm. Lunar Planet. Lab. 2:* 205.

———. 1979. "Mercury: A Post-Mariner 10 Assessment." *Space Sci. Rev. 24:* 3.

Taylor, F. W. 1979. "Infrared Remote Sounding of the Middle Atmosphere of Venus from the Pioneer Orbiter." *Science 203:* 779.

Taylor, S. R. 1975. *Lunar Science: A Post-Apollo View.* New York: Pergamon Press.

Urey, H. C. 1952. *The Planets: Their Origin and Development.* New Haven: Yale University Press.

Van Houten, F. B. 1973. "Origin of Red Beds: A Review—1961–1972." *Ann. Rev. Earth Planet. Sci. 1:* 39.

Veverka, J., P. Thomas, and T. Duxbury. 1978. "The Puzzling Moons of Mars." *Sky and Telescope 56:* 186.

Veverka, J., P. Thomas, and R. Greeley. 1977. "A Study of the Variable Features on Mars during the Viking Primary Mission." *J. Geophys. Res. 82:* 4167.

Wakita, H., Y. Nakamura, K. Notsu, M. Noguchi, and T. Asada. 1980. "Radon Anomaly: A Possible Precursor of the 1978 Izu-Oshima-Kinkai Earthquake." *Science 207:* 882.

Wallace, D., and C. Sagan. 1979. "Evaporation of Ice in Planetary Atmospheres: Ice-Covered Rivers on Mars." *Icarus 39:* 385.

White, B. R. 1981. "Venusian Saltation." *Icarus 46:* 226.

Wilson, L., and J. Head. 1981. "Ascent and Emplacement of Basaltic Magma on the Earth and Moon." *J. Geophys. Res. 86:* 2971.

Wise, D. U., M. Golombek, and G. McGill. 1979. "Tharsis Province of Mars: Geologic Sequence Geometry, and a Deformation Mechanism." *Icarus 38:* 456.

Witteborn, F., J. Bregman, and J. Pollack. 1979. "Io: An Intense Brightening Near 5 μm." *Science 203:* 643.

Young, A. T. 1973. "Are the Clouds of Venus Sulfuric Acid?" *Icarus 18:* 564.

CHAPTER TWELVE

Anders, E., and T. Owen. 1977. "Mars and Earth: Origin and Abundance of Volatiles." *Science 198:* 453.

Atreya, S., T. Donahue, and W. Kuhn. 1978. "Evolution of a Nitrogen Atmosphere on Titan." *Science 201:* 611.

Barbato, J. P., and E. Ayer. 1981. *Atmospheres.* New York: Pergamon Press.

Bar-Nun, A. 1975. "Thunderstorms on Jupiter." *Icarus 24:* 86.

Belton, M., D. Hunten, and M. McElroy. 1967. "A Search for an Atmosphere on Mercury." *Astrophys. J. 150:* 1111.

Briggs, G., and others. 1977. "Martian Dynamical Phenomena During June–November 1976: Viking Orbiter Imaging Results." *J. Geophys. Res. 82:* 4121.

Broadfoot, A., and others. 1980. "Extreme Ultraviolet Observations from Voyager 1 Encounter with Jupiter." *Science 204:* 979.

Brown, H. 1948. "Rare Gases and the Formation of the Earth's Atmosphere." In *The Atmospheres of the Earth and Planets,* ed. G. P. Kuiper. Chicago: University of Chicago Press.

Carr, M. H. 1981. *The Surface of Mars.* New Haven: Yale University Press.

Cess, R., V. Ramanathan, and T. Owen. 1980. "The Martian Paleoclimate and Enhanced Atmospheric CO_2." *Icarus 41:* 159.

Chamberlain, J. W. 1962. "Upper Atmospheres of the Planets." *Astrophys. J. 136:* 382.

Chapman, C. R., and K. Jones. 1977. "Cratering and Obliteration History of Mars." *Ann. Rev. Earth Planet. Sci. 5:* 515.

Combes, M., and others. 1981. "Upper Limit of the Gaseous CH_4 Abundance on Triton." *Icarus, 47:* 139.

Connes, P., J. Connes, W. Benedict, and L. Kaplan. 1967. "Traces of HCl and HF in the Atmosphere of Venus." *Astrophys. J. 147:* 1230.

Cruikshank, D. P., and P. Silvaggio. 1979. "Triton: A Satellite with an Atmosphere." *Astrophys. J. 233:* 1016.

———. 1980. "The Surface and Atmosphere of Pluto." *Icarus 41:* 96.

Cutts, J., and others. 1976. "North Polar Region of Mars: Imaging Results from Viking 2." *Science 194:* 1329.

Donahue, T., and others. 1982. "Venus Was Wet: A Measurement of the Ratio of Deuterium to Hydrogen." *Science 216*: 630.

Fanale, F., and W. Cannon. 1979. "Mars: CO_2 Absorption and Capillary Condensation on Clays—Significance for Volatile Storage and Atmospheric History." *J. Geophys. Res. 84:* 8404.

Fink, U., and others. 1980. "Detection of a CH_4 Atmosphere on Pluto." *Icarus 44:* 62.

Gierasch, P., and R. Goody. 1973. "A Model of a Martian Great Dust Storm." *J. Atmos. Sci. 30:* 169.

Golitsyn, G. 1973. "On the Martian Dust Storms." *Icarus 18:* 113.

Grinspoon, D. H. 1987. "Was Venus Wet?: Deuterium Reconsidered." *Science 238:* 1702.

Hanel, R., and others. 1979. "Infrared Observations of the Jovian System from Voyager 1." *Science 204:* 972.

———. 1981. "Infrared Observations of the Saturnian System from Voyager 1." *Science 212:* 192.

Hanson, W. B., S. Sanatani, and D. Zuccaro. 1977. "The Martian Ionosphere as Observed by the Viking Retarding Potential Analyzers." *J. Geophys. Res. 82:* 4351.

Hartmann, W. 1973. "Martian Cratering 4. Mariner 9 Initial Analysis of Cratering Chronology." *J. Geophys. Res. 78:* 4096.

———. 1978. "Martian Cratering 5: Toward an Empirical Martian Chronology, and Its Implications." *Geophys. Res. Lett. 5:* 450.

Hays, J., J. Imbrie, and N. Shackleton. 1976. "Variations in the Earth's Orbit: Pacemaker of the Ice Ages." *Science 194:* 1121.

Hess, S., R. Henry, and J. Tillman. 1979. "The Seasonal Variation of Atmospheric Pressure on Mars as Affected by the South Polar Cap." *J. Geophys. Res. 84:* 2923.

Hoffert, M. I., and others. 1981. "Liquid Water on Mars: An Energy Balance Climate Model for CO_2/H_2O Atmospheres." *Icarus 47:* 112.

Holland, H. D. 1962. "Model for the Evolution of the Earth's Atmosphere." In *Petrologic Studies.* New York: Geological Society of America.

Hunten, D. M. 1977. "Titan's Atmosphere and Surface." In *Planetary Satellites*, ed. J. Burns. Tucson: University of Arizona Press.

Ingersoll, A. 1981, 1990. "Jupiter and Saturn." In *The New Solar System*, ed. J. Beatty, B. O'Leary, and A. Chaikin (3rd ed., 1990). Cambridge, Mass.: Sky Publishing Corp.

James, P. B., G. Briggs, J. Burnes, and A. Spruck. 1979. "Seasonal Recession of Mars' South Polar Cap As Seen by Viking." *J. Geophys. Res. 84:* 2889.

Jeans, J. H. 1916. *Dynamical Theory of Gases.* Cambridge: Cambridge University Press.

Khodakovsky, I. L., and others. 1979. "Venus: Preliminary Prediction of the Mineral Composition of Surface Rocks." *Icarus 39:* 352.

Kieffer, H., and others. 1976. "Martian North Polar Summer Temperatures: Dirty Water Ice." *Science 194:* 1341.

Knollenberg, R., and D. Hunten. 1979a. "Clouds of Venus: Particle Size Distribution Measurements." *Science 203:* 792.

———. 1979b. "Clouds of Venus: A Preliminary Assessment of Microstructure." *Science 205:* 70.

Ksanfomaliti, L. V. 1980. "Discovery of Frequent Lightning Discharges on Clouds of Venus." *Nature 284:* 244.

Kuiper, G. P. 1944. "Titan: A Satellite with an Atmosphere." *Astrophys. J. 100:* 378.

———. 1952. "Planetary Atmospheres and Their Origin." In *Atmospheres of the Earth and Planets*, ed. G. P. Kuiper. Chicago: University of Chicago Press.

Kumar, S. 1979. "The Stability of an SO_2 Atmosphere on Io." *Nature 280:* 758.

Levine, J. S. 1969. "The Ashen Light: An Auroral Phenomenon on Venus." *Planet. Space Sci. 17:* 1081.

Lewis, J. S. 1980. "Lightning Synthesis of Organic Compounds on Jupiter." *Icarus 43:* 85.

Lewis, J. S., and R. G. Prinn. 1970. "Jupiter's Clouds: Structure and Composition." *Science 169:* 472.

Malin, M. C. 1976. "Age of Martian Channels." *J. Geophys. Res. 81:* 4825.

Masursky, H., and others. 1977. "Classification and Time of Formation of Martian Channels Based on Viking Data." *J. Geophys. Res. 82:* 4016.

McElroy, M. B., Y. Yung, and A. Nier. 1976. "Isotopic Composition of Nitrogen: Implications for the Past History of Mars' Atmosphere." *Science 194:* 70.

Meech, K. and M. Belton. 1990. "The Atmosphere of 2060 Chiron." *Astron. Journ.* 100: 1323.

Murray, B., W. Ward, and S. Yeung. 1973. "Periodic Insolation Variations on Mars." *Science 80:* 638.

Mutch, T., R. Arvidson, J. Head, K. Jones, and R. S. Saunders. 1976. *The Geology of Mars.* Princeton: Princeton University Press.

Orton, G. 1978. "Data Tables." *Proc. Lun. Planet. Sci. Conf. 9:* endpapers.

Owen, T. 1970. "The Atmosphere of Jupiter." *Science 167:* 1675.

———. 1982. "Titan." *Sc. Amer.*, February, p. 98.

Owen, T., and others. 1977. "The Composition of the Atmosphere at the Surface of Mars." *J. Geophys. Res. 82:* 4635.

Oyama, V., and others. 1979. "Venus Lower Atmosphere Composition: Analysis by Gas Chromatography." *Science 203:* 802. (Revision supplied by NASA)

Pepin, R. O. 1991. "On the Origin and Early Evolution of Terrestrial Planet Atmospheres and Meteoritic Volatiles." *Icarus 92:* 2.

Podolak, M., N. Noy, and A. Bar-Nun. 1979. "Photochemical Aerosols in Titan's Atmospheres." *Icarus 40:* 193.

Pollack, J. B. 1979. "Climatic Change on the Terrestrial Planets." *Icarus 37:* 479.

Pollack, J. B., and D. Black. 1979. "Implications of the Gas Compositional Measurements of Pioneer Venus for the Original Planetary Atmospheres." *Science 205:* 56.

Prinn, R. G., and T. Owen. 1976. "Chemistry and Spectroscopy of the Jovian Atmosphere." In *Jupiter*, ed. T. Gehrels. Tucson: University of Arizona Press.

Rages, K., and J. Pollack. 1980. "Titan Aerosols: Optical Properties and Vertical Distribution." *Icarus 41:* 119.

Rasool, S. I., and C. de Bergh. 1970. "The Runaway Greenhouse and the Accumulation of CO_2 in the Venus Atmosphere." *Nature 226:* 1037.

Ross, F. 1928. "Photographs of Venus." *Astrophys. J. 68:* 57.

Rossabacher, L. A., and S. Judson. 1981. "Ground Ice on Mars: Inventory, Distribution, and Resulting Landforms." *Icarus 45:* 39.

Rubey, W. W. 1951. "Geologic History of Sea Water." *Bull. Geol. Soc. Am. 62:* 1111. Reprinted in *The Origin and Evolution of Atmospheres and Oceans*, ed. P. Brancazio and A. Cameron. New York: Wiley, 1964.

Sagan, C. 1962. "Structure of the Lower Atmosphere of Venus." *Icarus 1:* 151.

Seiff, A., and D. Kirk. 1977. "Structure of the Atmosphere of Mars in Summer at Mid-Latitudes." *J. Geophys. Res. 82:* 4364.

Seiff, A., and others. 1979. "Thermal Contrast in the Atmosphere of Venus: Initial Appraisal from Pioneer Venus Probe Data." *Science 205:* 46.

Sharonov, V. V. 1958. *The Nature of Planets.* Trans. Israel Program for Scientific Translations. Washington, D.C.: Department of Commerce, Office of Technical Services.

Smith, B. A., and others. 1979. "The Galilean Satellites and Jupiter: Voyager 2 Imaging Science Results." *Science 206*: 927.

Smith, B. A., and 64 others. 1989. "Voyager 2 at Neptune: Imaging Science Results." *Science 248:* 1422.

Spitzer, L. 1952. "The Terrestrial Atmosphere above 300 Kilometers." In *The Atmospheres of the Earth and Planets*, ed. G. P. Kuiper. Chicago: University of Chicago Press.

Stewart, A., and others. 1979. "Ultraviolet Spectroscopy of Venus: Initial Results from the Pioneer Venus Orbiter." *Science 203:* 777.

Stone, E. C., and E. D. Miner. 1981. "Voyager 1 Encounter with the Saturnian System." *Science 212:* 159.

———. 1986. "The Voyager 2 Encounter with the Uranian System." *Science 233:* 39.

———. 1989. "The Voyager 2 Encounter with the Neptunian System." *Science 246:* 1417.

Suess, H. E. 1949. "Die Haufigkeit der Edelgase auf der Erde und im Kosmos." *J. Geol. 57:* 600.

Toon, O. B. 1979. "A Physical-Chemical Model of the Venus Clouds." *Bull. Amer. Astron. Soc. 11:* 544 (abstract).

Toon, O. B., and others. 1980. "The Astronomical Theory of Climate Change on Mars." *Icarus 44:* 552.

Trafton, L. M. 1972. "On the Possible Detection of H_2 in Titan's Atmosphere." *Astrophys. J. 175:* 285.

Veverka, J. 1973. "Titan: Polarimetric Evidence for an Optically Thick Atmosphere?" *Icarus 18:* 657.

Veverka, J., and others. 1981. "Voyager Search for Post-eclipse Brightening on Io." *Icarus 47:* 60.

von Zahn, V., and others. 1979. "Venus Thermosphere: In Situ Composition Measurements, the Temperature Profile, and the Homopause Altitude." *Science 203:* 768.

Walker, J. C. G. 1977. *Evolution of the Atmosphere.* New York: Macmillan.

Ward, W. R. 1974. "Climatic Variations on Mars 1. Astronomical Theory of Insolation." *J. Geophys. Res. 79:* 3375.

Ward, W., J. Burns, and O. Toon. 1979. "Past Obliquity Oscillations of Mars: Role of the Tharsis Uplift." *J. Geophys. Res. 84:* 243.

Young, A. T. 1982. "Rayleigh Scattering." *Phys. Today*, January, p. 42.

Zellner, B. 1973. "The Polarization of Titan." *Icarus 18:* 661.

CHAPTER THIRTEEN

Ahrens, T. J., and J. D. O'Keefe. 1982. "Impact of an Asteroid or Comet in the Ocean and Extinction of Terrestrial Life." *Lunar Planet. Sci. Abstracts 13:* 3 (see also press abstracts, p. 1).

Allen, W. 1980. *Side Effects.* New York: Ballantine Books.

Alvarez, L. W., W. Alvarez, F. Asaro, and H. Michel. 1980. "Extraterrestrial Cause for the Cretaceous-Tertiary Extinction." *Science 208:* 1095.

Condon, E. U., and others. 1969. *Scientific Study of Unidentified Flying Objects.* New York: Dutton.

Cronin, F. R., S. Pizzarello, and C. B. Moore. 1979. "Amino Acids in an Antarctic Carbonaceous Chondrite." *Science 206:* 335.

Dole, S. H. 1964. *Habitable Planets for Man.* Waltham, Mass.: Blaisdell.

Engel, M., and B. Nagy. 1982. "Distribution and Enantiometric Composition of Amino Acids in the Murchison Meteorite." *Nature 296*: 837.

Gartner, S., and J. McGuirk. 1980. "Terminal Cretaceous Extinction Scenario for a Catastrophe." *Science 206:* 1272.

Goldsmith, D., and T. Owen. 1979. *The Search for Life in the Universe.* Menlo Park: Benjamin/Cummings.

Gould, S. J. 1974. "The Great Dying." *Natural History 83* (October): 22.

———. 1978. "An Early Start." *Natural History 87* (February): 10.

Hartmann, W. K., and Ron Miller. 1991. *The History of Earth.* New York: Workman Publishing Co.

Hildebrand, A. R., and W. Boynton. 1991. "Cretaceous Ground Zero." *Natural History,* June, p. 47.

Horowitz, N. H. 1977. "The Search for Life on Mars." *Sci. Am.,* November, p. 52.

Horowitz, P. 1978. "A Search for Ultra-Narrowband Signals of Extraterrestrial Origin." *Science 201:* 733.

Hoyle, F. 1972. *From Stonehenge to Modern Cosmology.* San Francisco: W. H. Freeman.

Hoyle, F., and N. C. Wickramasinghe. 1978. *Lifecloud.* New York: Harper & Row.

Imbrie, J., and J. Z. Imbrie. 1980. "Modeling the Climatic Response to Orbital Variations." *Science 207:* 943.

Karl, D., C. Wirsen, and H. Jannasch. 1980. "Deep-sea Primary Production at the Galápagos Hydrothermal Vents." *Science 207:* 1345.

Kerr, R. A. 1981. "Impact Looks Real, the Catastrophe Smaller." *Science 214:* 896.

Kuiper, T., and M. Morris. 1977. "Searching for Extraterrestrial Civilizations." *Science 196:* 616.

Kvenvolden, K., and others. 1970. "Evidence for Extraterrestrial Amino Acids and Hydrocarbons in the Murchison Meteorite." *Nature 228:* 923.

Kyte, F. T., N. Z. Zhou, and J. Wasson. 1980. "Siderophile-Enriched Sediments from the Cretaceous-Tertiary Boundary." *Nature 228:* 651.

Levin, G., and P.. Straat. 1981. "A Search for a Nonbiological Explanation of the Viking Labeled Release Life Detection Experiment." *Icarus 45*: 494.

Miller, S. L. 1955. "Production of Some Organic Compounds under Possible Primitive Earth Conditions." *J. Amer. Chem. Soc. 77:* 2351.

Morrison, P. 1973. "Discussion." In *Communication with Extraterrestrial Intelligence,* ed. C. Sagan. Cambridge, Mass.: MIT Press, p. 120.

Nisbet, E. G. 1980. "Achaean Stromatolites and the Search for the Earliest Life." *Nature 284:* 395.

Oparin, A. I. 1962. *Life: Its Nature, Origin, and Development.* New York: Academic Press.

Parker, P. 1977. "An Ecological Comparison of Marsupial and Placental Patterns of Reproduction." In *The Biology of Marsupials,* ed. B. Stonehouse and D. Gilmore. London: Macmillan.

Pollard, W. G. 1979. "The Prevalence of Earthlike Planets." *Am. Scientist 67:* 653.

Raup, D. M. 1979. "Size of the Permo-Triassic Bottleneck and Its Evolutionary Implications." *Science 206:* 217.

Raup, D. M., and J. Sepkoski, Jr. 1986. "Periodic Extinction of Families and Genera." *Science 231:* 833.

Russell, D. A. 1982. "The Mass Extinctions of the Late Mesozoic." *Sci. Am.,* January, p. 58.

Sagan, C. 1970. "Life." In *Encyclopaedia Brittanica.* Chicago: Encyclopaedia Britannica, Inc.

———, ed. 1973. *Communication with Extraterrestrial Intelligence.* Cambridge, Mass.: MIT Press.

Shklovskii, I. S., and C. Sagan. 1966. *Intelligent Life in the Universe.* San Francisco: Holden-Day.

Siegel, S. M. 1970. "Experimental Biology of Extreme Environments and Its Significance for Space Bioscience 2." *Spaceflight 12:* 256.

Simpson, G. G. 1964. "The Nonprevalence of Humanoids." *Science 143:* 769.

———. 1973. "Added Comments on 'The Nonprevalence of Humanoids.'" In *Communication with Extraterrestrial Intelligence,* ed. C. Sagan. Cambridge, Mass.: MIT Press.

Story, R. 1976. *The Space-Gods Revealed.* New York: Harper & Row.

Turco, R., and others. 1981. "Tunguska Meteor Fall of 1908: Effects on Stratospheric Ozone." *Science 212:* 19.

Underwood, J. 1975. *Biocultural Interactions and Human Variations.* Dubuque, Iowa: W. C. Brown.

Walter, M. R., R. Buick, and J. Dunlop. 1980. "Stromatolites 3400–3500 Myr Old from the North Pole Area, Western Australia." *Nature 284:* 443.

Wright, I. P., M. Grady, and C. Pillinger. 1989. "Organic Materials in a Martian Meteorite." *Nature 340:* 220.

INDEX